BASIC TRIGONOMETRY

MILWAUKEE AREA TECHNICAL COLLEGE MATHEMATICS SERIES

BASIC TRIGONOMETRY

THOMAS J. McHALE
PAUL T. WITZKE
Milwaukee Area Technical College, Milwaukee, Wisconsin

ADDISON-WESLEY PUBLISHING COMPANY
Reading, Massachusetts
Menlo Park, California • London • Amsterdam • Don Mills, Ontario • Sydney

Milwaukee Area Technical College Mathematics Series

BASIC ALGEBRA

CALCULATION AND SLIDE RULE

BASIC TRIGONOMETRY

ADVANCED ALGEBRA

Reproduced by Addison-Wesley from camera-ready copy prepared by the authors.

ISBN 0-201-04631-8
JKLMNOPQRS-VB-798

FOREWORD

The Milwaukee Area Technical College Mathematics Series is the product of a five-year project whose goal has been the development of a method of communicating the mathematics skills needed in basic science and technology to a wide range of students, including average and below-average students. This Series is not just a set of textbooks. It is a highly-organized, highly-assessed, and highly-successful system of instruction which has been designed to cope with the learning process of individual students by a combination of programmed instruction, continual diagnostic assessment, and tutoring. Though this system of instruction is different than conventional mathematics instruction, a deliberate effort has been made to keep it realistic in the following two ways:

1. It can be used in a regular classroom by regular teachers without any special training.
2. Its cost has been minimized by avoiding the use of educational hardware.

History. The five-year project was originally funded by the Carnegie Corporation of New York and subsequently funded by the Wisconsin State Board of Vocational, Technical, and Adult Education and the Milwaukee Area Technical College (MATC). The system of instruction offered in this Series was specifically developed for a two-semester Technical Mathematics course at MATC. It has been used in that capacity for the past five years with the following general results:

1. The dropout rate in the course has been reduced by 50%.
2. Average scores on equivalent final exams have increased from 55% to 85%.
3. The rate of absenteeism has decreased.
4. Student motivation and attitudes have been very favorable.

Besides its use in the Technical Mathematics course at MATC, parts or all of the system have been field tested in various other courses at MATC, in other technical institutes in Wisconsin and neighboring states, and in secondary schools in the Milwaukee area. These field tests with over 4,000 students have provided many constructive comments by teachers and students, plus a wealth of test data which has been item-analyzed and error-analyzed. On the basis of all this feedback, the textbooks and tests have been revised several times until a high level of learning is now achieved by a high percentage of students.

Uses of the System. Though specifically prepared for a two-semester Technical Mathematics course, the system of instruction can be used in a variety of contexts. At the college level, parts of it can be used in pre-technical, health occupations, apprentice, trade, and developmental programs, or in an intermediate algebra course. At the secondary school level, the whole system can be used in a two-year Technical Mathematics course in Grades 11 and 12; parts of it can be used in other courses or as supplementary materials in other courses. The system is highly flexible because it can be used either in a conventional classroom or in a learning center, with either a paced or self-paced schedule.

Written for students who have completed one year of algebra and one year of geometry in high school, the system has been used successfully with students without these prerequisites. Though not designed for students with serious deficiencies in arithmetic, it has been successfully used with students who were below average in arithmetic skills.

Three Main Elements in the System. The three main elements in the instructional system are programmed textbooks, diagnostic assessment, and a teacher. Though discussed in greater detail in the Teacher's Manual - MATC Mathematics Series, each main element is described briefly below:

1. Programmed Textbooks - The Series includes four programmed textbooks: BASIC ALGEBRA, CALCULATION AND SLIDE RULE, BASIC TRIGONOMETRY, and ADVANCED ALGEBRA. These textbooks include frequent short self-tests (with answers) which are an integral part of the instruction.

2. Diagnostic Assessment - Each textbook is accompanied by a test book which includes a diagnostic test for each assignment, chapter tests (three parallel forms), and a comprehensive final examination. Pre-tests in arithmetic and algebra are included in two of the test books, TESTS FOR BASIC ALGEBRA and TESTS FOR CALCULATION AND SLIDE RULE. A complete set of keys is provided in each test book for all tests in that book.

   Note: Test books are provided only to teachers, not to individual students. Copies of the tests for student use must be made by the Xerox process or some similar process.

3. A Teacher - A teacher is essential to the success of this system. Its success depends on the teacher's ability to diagnose learning difficulties and to remedy them by means of tutoring. Its success also depends on the teacher's skill in maintaining a suitable effort by each student. This new role for the teacher can be best described by calling him a "manager of the learning process". Teachers who use the system correctly find that this new role is a highly professional and satisfying one because it permits them to deal successfully with the wide range of individual differences which exist in any group of students.

Special Features. Some of the major features of the system of instruction are listed below:

Individual Attention - The system is designed so that the teacher can deal with the learning problems of individual students.

Learnability - When the system is used correctly, teachers can expect (and will obtain) a high level of learning in a high percentage of students.

Relevant Content - All of the content has been chosen on the basis of its relevance for basic science and technology. The content also provides a basis for the study of more advanced topics in mathematics.

Assessment - The many tests which are provided enable the teacher to maintain a constant assessment of each student's progress and the overall success of the system.

Student Motivation - Because students receive individual attention and have a high probability of success, motivation problems are minimized and student attitudes are generally quite positive.

Besides these main features, the system also provides a mechanism for dealing with absenteeism, and it lends itself quite easily to the use of paraprofessionals.

HOW TO USE THE SYSTEM

Though discussed in greater detail in the Teacher's Manual - MATC Mathematics Series, the procedure for using the system is briefly outlined below. All of the tests mentioned are available in the book TESTS FOR BASIC TRIGONOMETRY. Both books are available from the Addison-Wesley Publishing Company, Reading, Massachusetts 01867.

1. Each chapter is covered in a number of assignments. Each assignment can be assessed by the diagnostic test which is provided. The assignments for BASIC TRIGONOMETRY are listed at the bottom of this page. Though the best results have been obtained with a paced schedule, whether daily or otherwise, the assignments can also be covered on a self-paced schedule. Since the diagnostic tests are designed to take only 15 to 25 minutes, ample time is left for correction and tutoring within a normal class period. The diagnostic tests need not be graded since they are simply a teaching tool.

2. After all assignments for a chapter are completed, one of the three equivalent forms of the chapter test can be administered. Ordinarily, these tests should be graded. If the diagnostic tests are used in conjunction with the assignments, high scores should be obtained on the chapter tests, and grades can be assigned on a percentage basis.

3. When all of the chapters in a book are completed in the manner above, the comprehensive test for that book can be given. Or, if the teacher prefers, the items in that comprehensive test can be included in a final exam at the end of the semester.

ASSIGNMENTS FOR BASIC TRIGONOMETRY

Ch. 1: #1 (pp. 1-14)
#2 (pp. 14-31)
#3 (pp. 32-47)
#4 (pp. 47-63)

Ch. 2: #5 (pp. 64-80)
#6 (pp. 80-93)
#7 (pp. 94-111)
#8 (pp. 112-127)

Ch. 3: #9 (pp. 128-146)
#10 (pp. 146-165)
#11 (pp. 166-188)
#12 (pp. 188-206)
*#13 (pp. 207-219)

Ch. 4: #14 (pp. 220-242)
#15 (pp. 243-260)
#16 (pp. 261-273)
#17 (pp. 273-283)
#18 (pp. 283-298)

Ch. 5: #19 (pp. 299-315)
#20 (pp. 316-330)
#21 (pp. 331-348)
#22 (pp. 348-361)
#23 (pp. 362-380)

Ch. 6: #24 (pp. 381-404)
#25 (pp. 405-419)
#26 (pp. 420-440)
#27 (pp. 440-461)
#28 (pp. 462-478)

Ch. 7: #29 (pp. 479-505)
#30 (pp. 505-526)
#31 (pp. 527-547)

Ch. 8: #32 (pp. 548-566)
#33 (pp. 566-582)
#34 (pp. 583-600)
#35 (pp. 600-614)
#36 (pp. 614-620)

*Optional assignment. Though an assignment test is provided, no items from this assignment appear in the chapter test.

BASIC TRIGONOMETRY

PREREQUISITES: Since signed numbers, basic equations, the coordinate system, and other algebraic principles are used in BASIC TRIGONOMETRY, a knowledge of the content of Chapters 1 through 7 of BASIC ALGEBRA is essential.

The concepts in BASIC TRIGONOMETRY can be learned without performing all of the calculations. However, since the use of these concepts in applied situations always requires calculations, the full benefit of this book cannot be gained if the calculations are avoided. To obtain this full benefit, slide rule skills (similar to those presented in Chapters 1 through 7 of CALCULATION AND SLIDE RULE) are almost essential. Though the calculations could conceivably be performed by long arithmetic methods, progress through the book would be seriously hampered by those methods.

SEQUENCING: Chapters 1, 2, and 3 should be studied in the order listed.

Chapters 1 through 3 are prerequisites for Chapters 4 through 8.

Chapters 4 through 8 can be studied in any order, with the single exception that Chapter 4 (Oblique Triangles) is a prerequisite for Chapter 5 (Applied Geometric Problems).

FEATURES: The general approach to trigonometry is numerical rather than analytical. Basic trigonometric concepts, however, are emphasized.

Only the sine, cosine, and tangent of angles are initially defined and used. The definitions of cosecant, secant, and cotangent are not only delayed but deemphasized.

The treatment of interpolation in trigonometric tables is delayed to avoid unnecessary complexity.

An extensive treatment of vectors is given. Furthermore, a vector-approach is used both in the definitions of the trigonometric ratios of standard-position angles and in the treatment of complex numbers.

Obtuse oblique triangles are not introduced until the Law of Sines and the Law of Cosines are fully explored in the context of acute oblique triangles.

The basic properties of circles and half-tangents are reviewed in the context of applied problems, many of which require the use of trigonometry.

Though identities and trigonometric equations are discussed, the treatment of them is not exhaustive.

An extensive treatment of the sine wave and its properties is given. However, except for a brief introduction to the cosine wave, the graphs of all other trigonometric functions are omitted.

ACKNOWLEDGMENTS

The Carnegie Corporation of New York made this Series possible by originally funding the project in 1965.

Dr. George A. Parkinson, now Director Emeritus of MATC, was instrumental in obtaining the original grant and totally supported the efforts of the project staff. Dr. William L. Ramsey, present Director of MATC, has continued this support.

Dr. Lawrence M. Stolurow, Professor and Director of Graduate Studies, Division of Educational Research and Development, State University of New York at Stony Brook, New York, served as the original chairman of the project.

Mr. Gail W. Davis of the project staff was largely responsible for the development of an efficient classroom procedure and the organization of a learning center. He also contributed many suggestions which helped to improve the instructional materials.

Mr. Keith J. Roberts, Mr. Allan A. Christenson, and Mr. Joseph A. Colla, members of the project staff, offered many constructive criticisms and helpful suggestions.

Many other teachers and administrators contributed to the implementation, field testing, and improvement of the instructional system.

Mrs. Arleen A. D'Amore typed the camera-ready copy of the textbooks and tests.

ABOUT THE AUTHORS

THOMAS J. McHALE

He has been director of the MATC Mathematics Project since its beginning in June, 1965. He received his doctorate in experimental psychology at the University of Illinois, with major emphasis on the psychology of learning. He is currently a part-time member of the Psychology Department at Marquette University. He has taught mathematics at the secondary-school level.

PAUL T. WITZKE

He has been a member of the MATC Mathematics Project staff since its beginning. He has taught mathematics, physics, electronics, and other technical courses at the Milwaukee Area Technical College for 25 years. He has written numerous instructional manuals and has developed several televised courses in mathematics.

CONTENTS

Chapter 1 RIGHT TRIANGLES AND TRIGONOMETRIC RATIOS

Trigonometry is the branch of mathematics which deals with the solution of triangles and related topics. By the "solution of triangles", we mean finding the lengths of sides and the sizes of angles in triangles. Though we will eventually solve triangles of all types, this chapter is devoted to the solution of right triangles. Right triangles can be solved by using the Pythagorean Theorem, the angle-sum principle, or the three basic trigonometric ratios: sine, cosine, and tangent. After reviewing the Pythagorean Theorem and the angle-sum principle and introducing the three basic trigonometric ratios, we will discuss the strategies for solving right triangles.

1-1 RIGHT TRIANGLES AND THE ANGLE-SUM PRINCIPLE

In this section, we will study the angle-sum principle for triangles. Then, we will define what is meant by a right triangle. Finally, we will use the angle-sum principle to find the size of unknown angles in right triangles.

---

1. There are three angles in any triangle. The angles are usually labeled with capital letters.

   Any angle between 0° and 90° is called an acute angle.
   Any angle with exactly 90° is called a right angle.
   Any angle between 90° and 180° is called an obtuse angle.

   In triangle ABC:

   Angle A is an acute angle.

   (a) Angle B is an acute angle.

   (b) Angle C is an obtust angle.

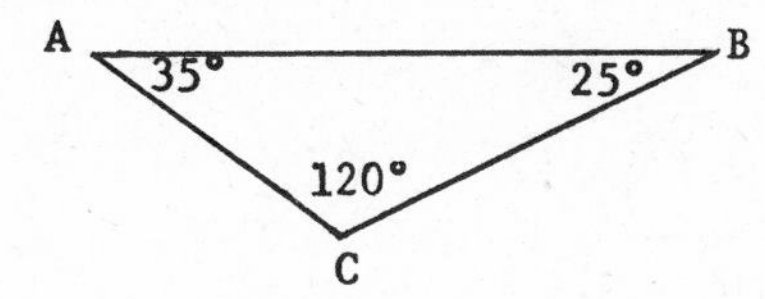

Answers to frame 1:

a) acute

b) obtuse

2. In triangle CDE:

   (a) There are two acute angles, angle C and angle E.

   (b) The obtuse angle is angle D.

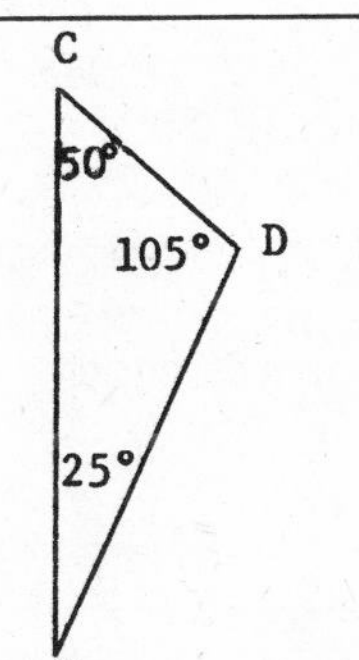

Answers to frame 2:

a) Angles C and E

b) Angle D

3. The angle-sum principle for triangles is this:

> THE SUM OF THE THREE ANGLES OF ANY TRIANGLE IS 180°.

Using this fact, we can compute the number of degrees in the third angle of a triangle if the size of each of the other two angles is known. To do so, we simply subtract the total number of degrees in the two known angles from 180°.

One angle is unknown in each triangle below.

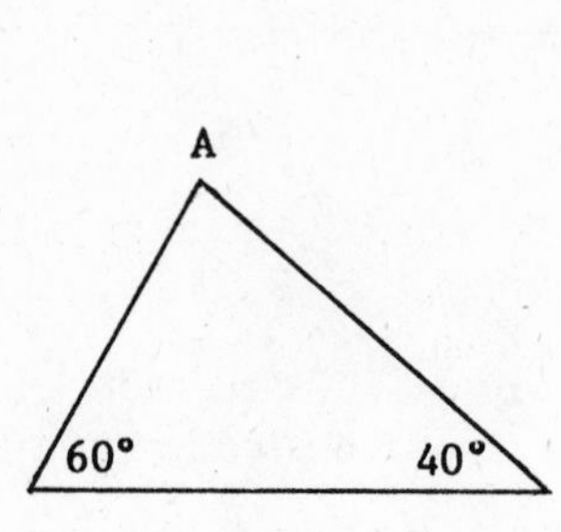

Figure 1

Figure 2

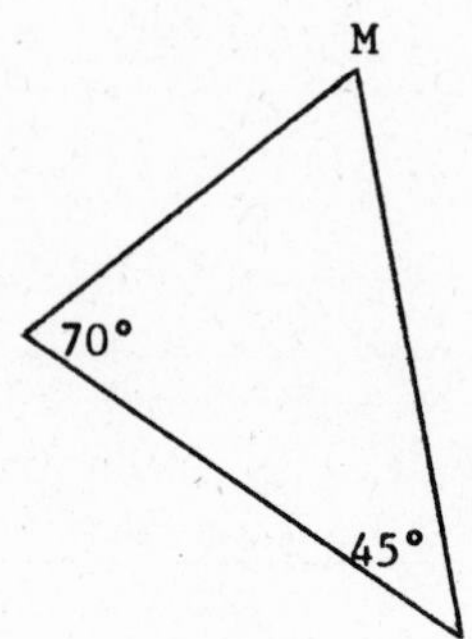

Figure 3

(a) In Figure 1, angle A contains 80 °.

(b) In Figure 2, angle T contains 100 °.

(c) In Figure 3, angle M contains 65 °.

---

4. Use the angle-sum principle to answer these:

(a) Could a triangle contain these three angles: 40°, 60°, 70°? NO

(b) Could a triangle contain these three angles: 100°, 50°, 40°? NO

(c) Could a triangle contain two obtuse angles, like 91° and 92°? NO

a) 80°
b) 100°
c) 65°

---

5. A triangle is called a right triangle if it contains a right angle (90°).

Triangle MPR is a right triangle since angle P is a right angle (or 90° angle). (Notice that the right angle is identified with a small square.)

The "angle-sum principle" also applies to right triangles. Therefore, in the triangle at the right:

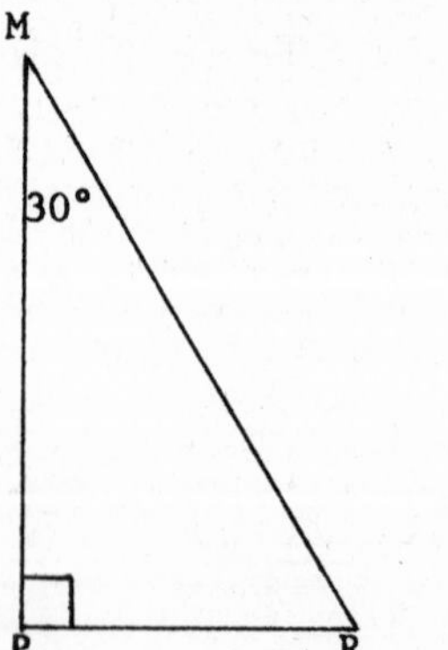

(a) The sum of the three angles is 180 °.

(b) Angle R contains 60 °.

a) No, because their sum is 170°.
b) No, because their sum is 190°.
c) No, because their sum is greater than 180°.

---

a) 180°
b) 60°

6. Triangle AFD is a right triangle.

(a) The right angle is angle __F__.

(b) Angle D contains __40__°.

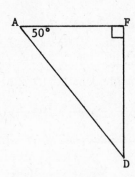

7. Could a triangle contain a right angle and an obtuse angle? __NO__

a) angle F b) 40°

---

8. In a right triangle, the right angle contains 90°. Since the sum of the other two angles must be 90°, the other two angles must be acute angles.

In right triangle TBC:

(a) Angles T and B are both __acute__ angles.

(b) The sum of angles T and B must be __90__°.

No, because their sum is greater than 180°.

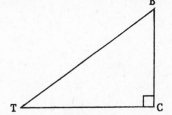

9. The sum of the two acute angles in a right triangle is 90°. Therefore, it is easy to compute the number of degrees in one acute angle if we know the number of degrees in the other acute angle. To do so, we simply subtract the known angle from 90°.

In right triangle ABC:

(a) If angle B contains 55°, angle C contains 90° - 55° = __35__°

(b) If angle C contains 37°, angle B contains 90° - 37° = __53__°

a) acute

b) 90°

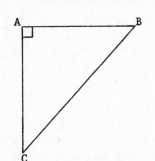

10. In triangle BDS:

(a) Angle D contains __90__°.

(b) Angle S contains __45__°.

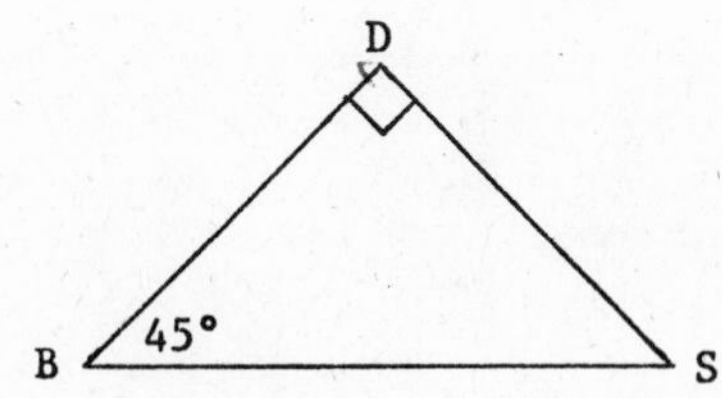

a) 35°

b) 53°

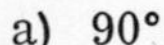

a) 90°

b) 45°

11. In a right triangle:

(a) If one acute angle contains 40°, the other acute angle contains 50 °.

(b) If one acute angle contains 67°, the other acute angle contains 23 °.

Answer to Frame 11: a) 50° b) 23°

SELF-TEST 1 (Frames 1-11)

1. Find the size of angle R:

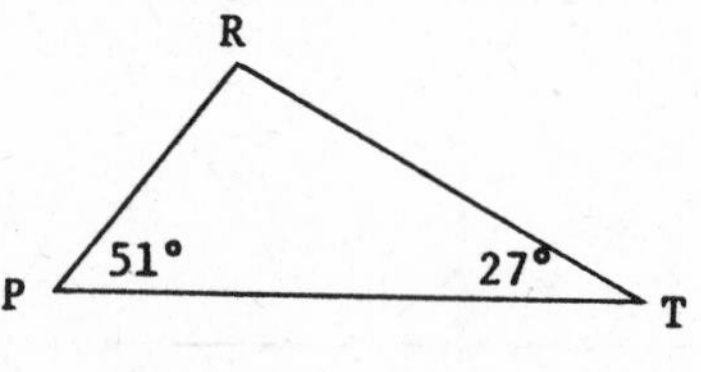

Angle R = 102°

2. Find the size of angle H:

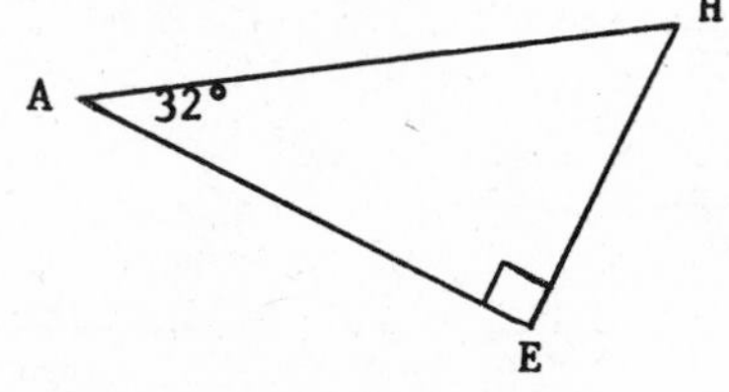

Angle H = 58°

ANSWERS: 1. Angle R = 102° 2. Angle H = 58°

1-2 RIGHT TRIANGLES AND THE PYTHAGOREAN THEOREM

In this section, we will review some terminology about the sides of right triangles. Then we will use the Pythagorean Theorem to find the lengths of unknown sides of right triangles.

12. There is a conventional way of labeling the angles and sides in any triangle. This convention has been followed in labeling the triangle on the right. Notice these points:

(1) Each <u>angle</u> is labeled with a <u>capital letter</u>.

(2) Each <u>side</u> is labeled with the <u>small letter</u> corresponding to the capital letter of the angle opposite it. That is:

The side opposite angle R is labeled "r".
The side opposite angle B is labeled "b".
The side opposite angle D is labeled "d".

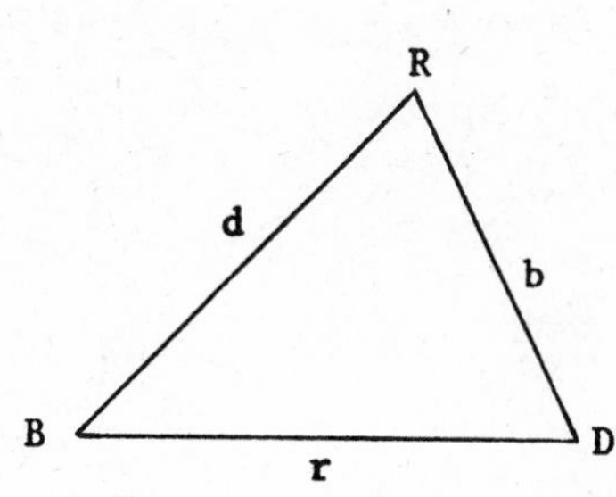

Following the convention above, label the sides in the triangle on the right.

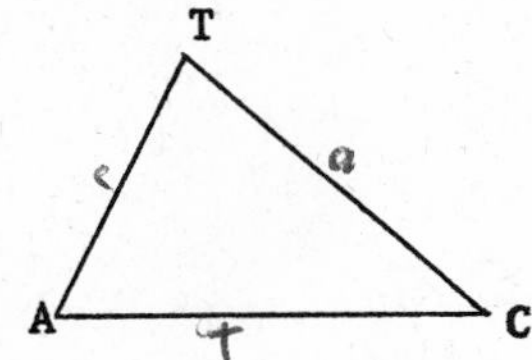

---

13. The triangle on the right is labeled according to the accepted convention.

Though the sides of a triangle are ordinarily represented by small letters, we can use the two capital letters on each end of a side to represent that side. For example:

Instead of "q", we could use "DM" or "MD".

Instead of "d", we could use MQ or QM.

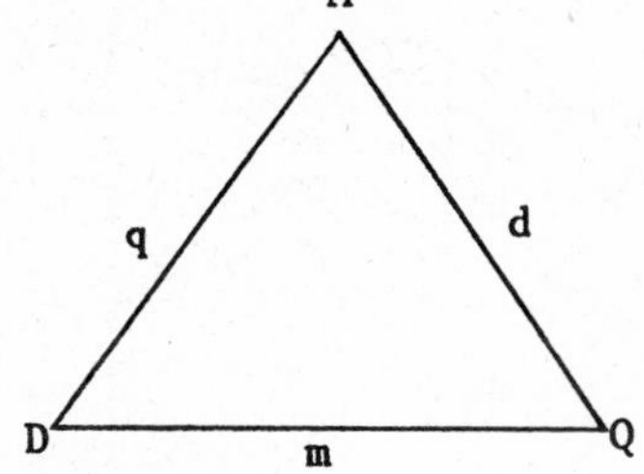

Your triangle should look like this:

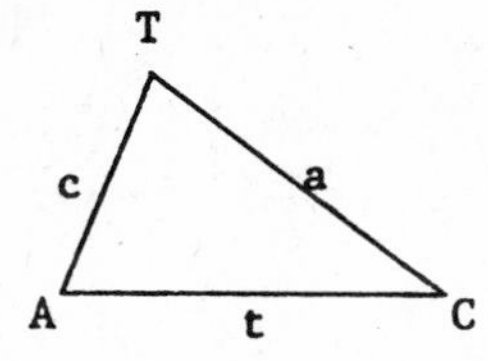

---

14. Occasionally the sides of a triangle are not labeled with small letters. In such cases, we must use two capital letters to represent the sides.

In the triangle on the right:

(a) The side opposite angle P is SV.

(b) The side opposite angle S is PV.

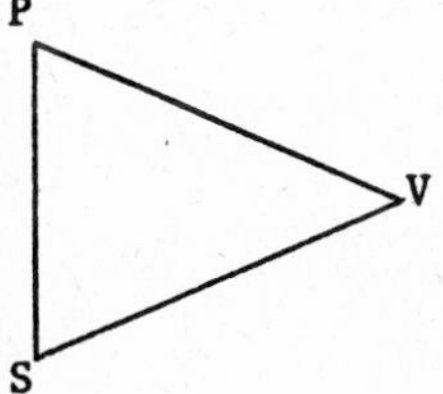

MQ or QM

---

a) SV (or VS)

b) PV (or VP)

15. In any triangle:

THE LONGEST SIDE IS OPPOSITE THE LARGEST ANGLE.

THE SHORTEST SIDE IS OPPOSITE THE SMALLEST ANGLE.

In triangle DFH:

(a) Angle F is the largest angle. Therefore, DH is the longest side.

(b) Angle H is the smallest angle. Therefore, DF is the shortest side.

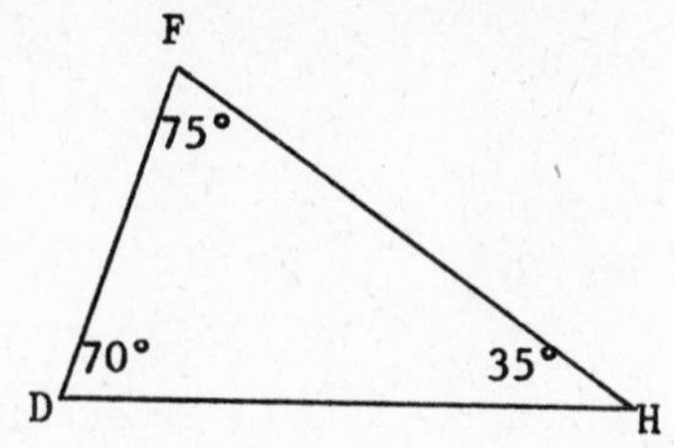

---

16. In the triangle on the right:

(a) The longest side is m.

(b) The shortest side is q.

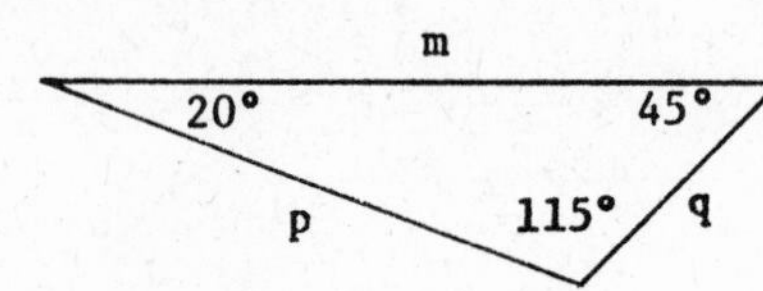

a) DH

b) DF

---

17. In a right triangle, the side opposite the right angle is called the "hypotenuse". The sides opposite the two acute angles are simply called "legs". For example:

In right triangle DFH:

"f" is the hypotenuse, and "d" and "h" are the legs.

(a) The hypotenuse "f" can be labeled in two other ways: DH or HD

(b) Leg "d" can be labeled in two other ways: FH or HF

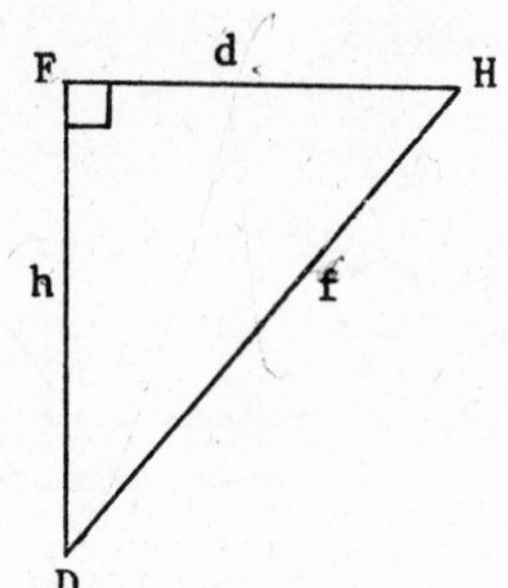

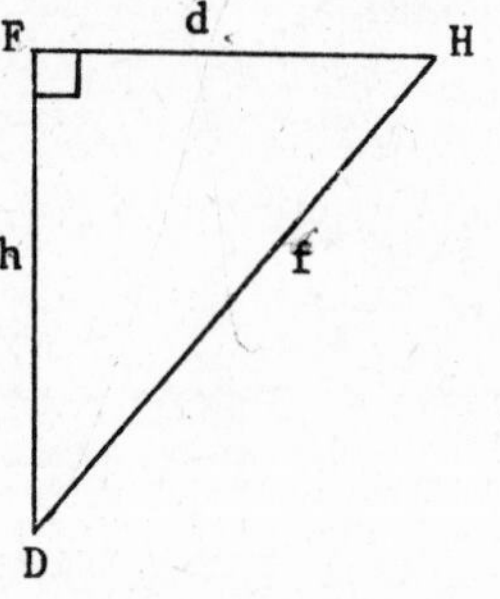

a) m

b) q

---

18. Ordinarily we use small letters to represent the hypotenuse and legs of a right triangle. Use small letters to complete the questions below.

In right triangle FET:

(a) The hypotenuse is e.

(b) The legs are f and t.

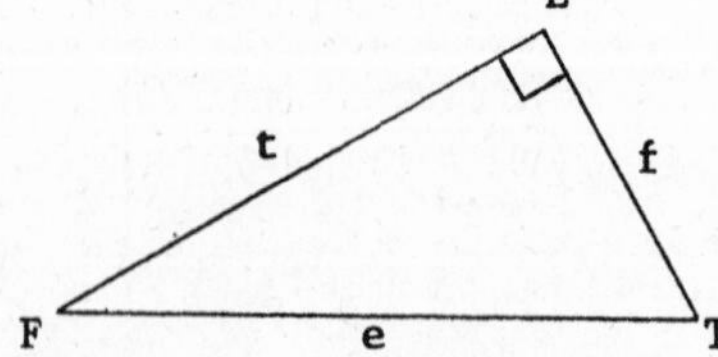

a) DH or HD

b) FH or HF

---

a) e

b) t and f

19. Since the right angle is the largest angle in any right triangle, the hypotenuse is the longest side of any right triangle.

In right triangle DMF:

(a) The longest side is ___.

(b) The shortest side is ___.

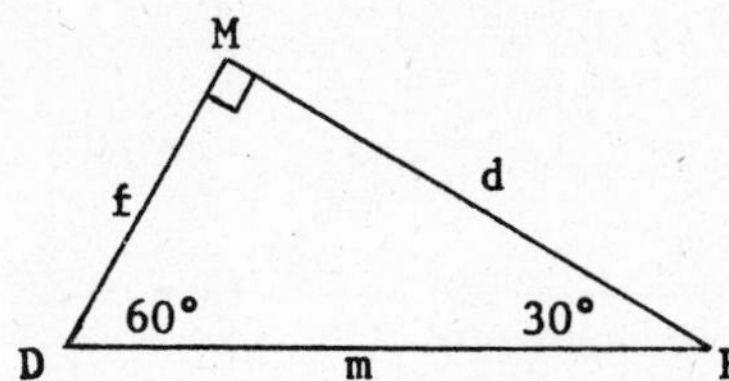

a) m (the hypotenuse)

b) f

20. The Pythagorean Theorem states a useful relationship among the three sides of a right triangle. The Theorem states:

IN ANY RIGHT TRIANGLE, THE "SQUARE" OF THE LENGTH OF THE HYPOTENUSE IS EQUAL TO THE SUM OF THE "SQUARES" OF THE LENGTHS OF THE TWO LEGS.

For right triangle ABC, the Pythagorean Theorem says:

$c^2 = a^2 + b^2$

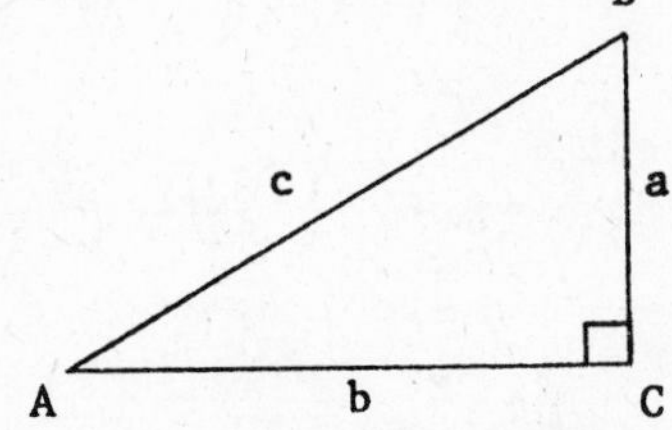

Below each of these right triangles, write the Pythagorean Theorem in symbol form:

(a)

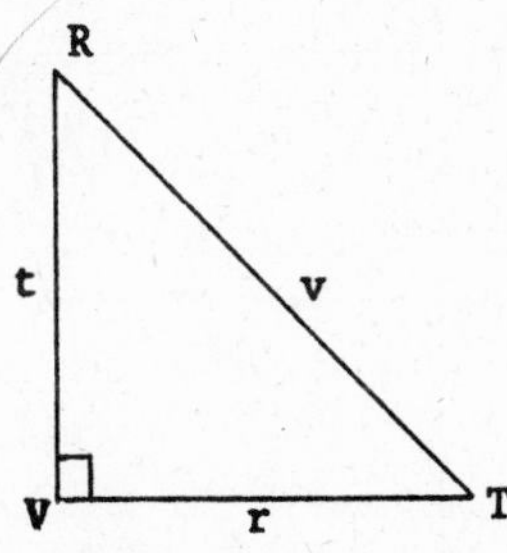

(b)

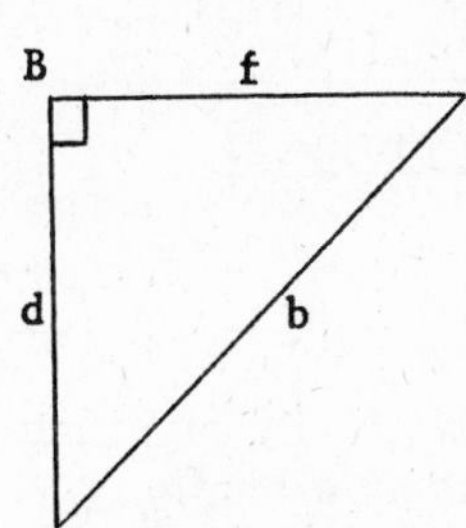

(c)

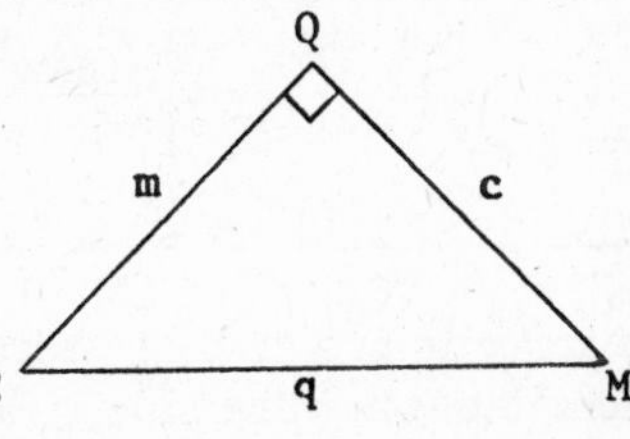

___ = ___ + ___   ___ = ___ + ___   ___ = ___ + ___

a) $v^2 = t^2 + r^2$

b) $b^2 = d^2 + f^2$

c) $q^2 = m^2 + c^2$

21. The lengths of the three sides are given for right triangle CDE. (Note: The symbol " means inches. That is, 3.0" means 3.0 inches.) For this triangle, the Pythagorean Theorem says:

$$d^2 = e^2 + c^2$$

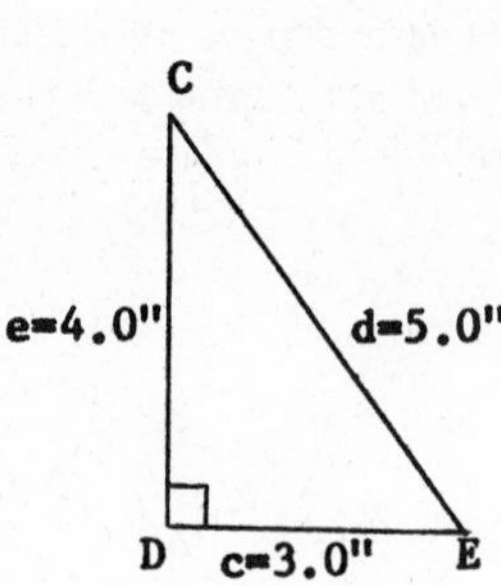

To show that the Pythagorean Theorem makes sense, we can plug in the known lengths for the letters. We get:

$$(5.0'')^2 = (4.0'')^2 + (3.0'')^2$$
$$25 \text{ sq. in.} = 16 \text{ sq. in.} + 9.0 \text{ sq. in.}$$
$$25 \text{ sq. in.} = 25 \text{ sq. in.}$$

The lengths of the three sides are given for right triangle RST. The Pythagorean Theorem says:

$$s^2 = t^2 + r^2$$

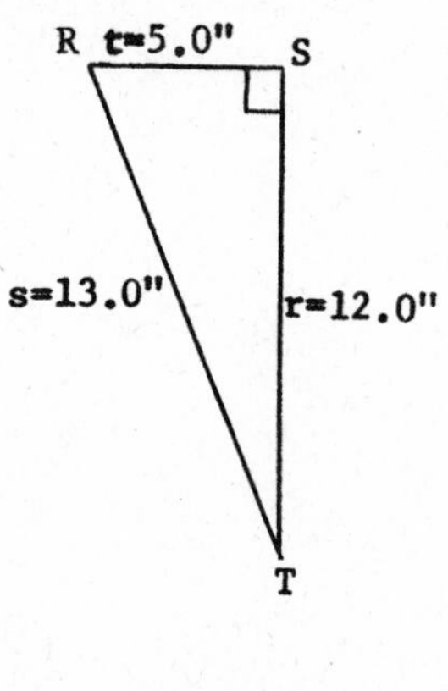

Plug in the known values to show that this statement makes sense:

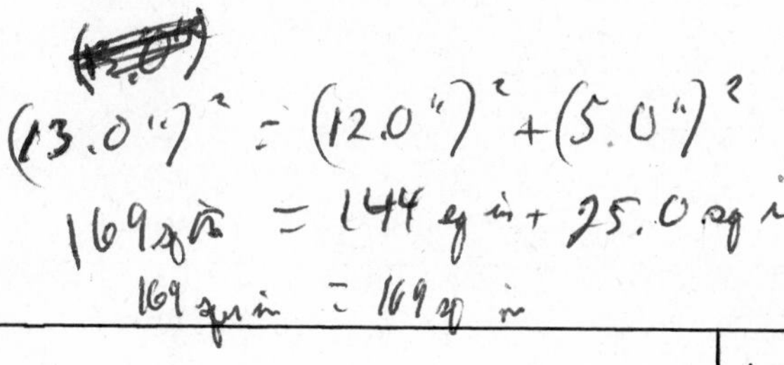

Answer to Frame 21:

$$(13.0'')^2 = (5.0'')^2 + (12.0'')^2$$
$$169 \text{ sq. in.} = 25 \text{ sq. in.} + 144 \text{ sq. in.}$$
$$169 \text{ sq. in.} = 169 \text{ sq. in.}$$

22. If two sides of a right triangle are known, we can use the Pythagorean Theorem to find the length of the third side. Here is an example:

In right triangle BDF, the lengths of the two legs are known. We can use the Pythagorean Theorem to find the length of the hypotenuse "b".

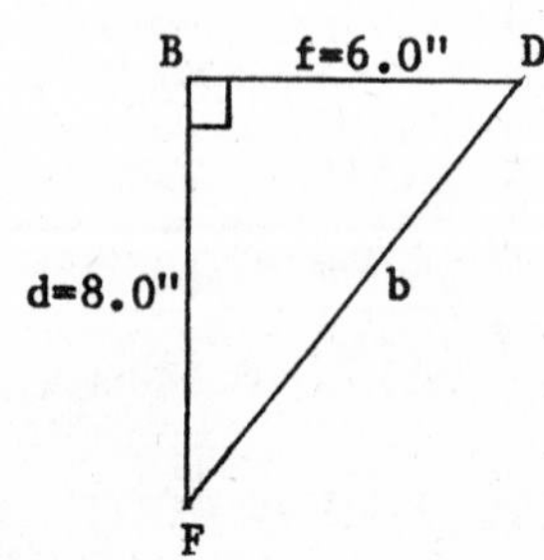

The Pythagorean Theorem says:

$$b^2 = f^2 + d^2$$

Plugging in the known values and simplifying, we get:

$$b^2 = (6.0'')^2 + (8.0'')^2$$
$$b^2 = 36 \text{ sq. in.} + 64 \text{ sq. in.}$$
$$b^2 = 100 \text{ sq. in.}$$

Taking the square root of both sides of the last equation, we can find the value of "b". b = 10"

10.0"

23. The reported lengths of the sides of right triangles are usually measurements. Therefore, the computed length of an unknown side should contain as many significant digits as there are in the lengths of the known sides. For example, if the two known sides each contain three significant digits, then the computed length of the third side should contain three significant digits.

Let's use the Pythagorean Theorem to to find the length of the hypotenuse of right triangle MBT. (Note: The symbol ' means feet. That is, 5.68' means 5.68 feet.)

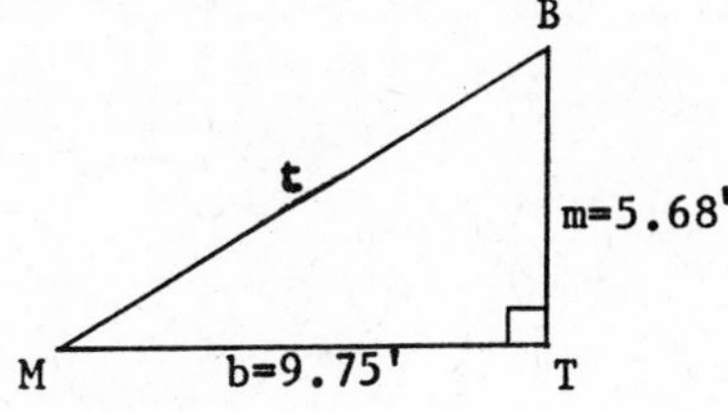

(a) Write the Pythagorean Theorem for right triangle MBT in letter-form:

$T^2 = m^2 + b^2$

(b) Plug in the known values and solve for "t".

$T^2 = (9.75')^2 + (5.68')^2$

$T^2 = 127.3$

t = 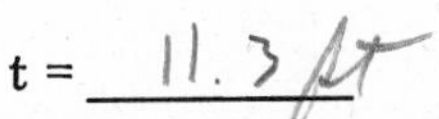 11.3 ft

---

a) $t^2 = b^2 + m^2$

b) t = 11.3 ft.
(from $t^2 = 127.3$ sq. ft.)

24. In right triangle DMR, the lengths of the hypotenuse and one leg are given. We can use the Pythagorean Theorem to find the length of leg "d":

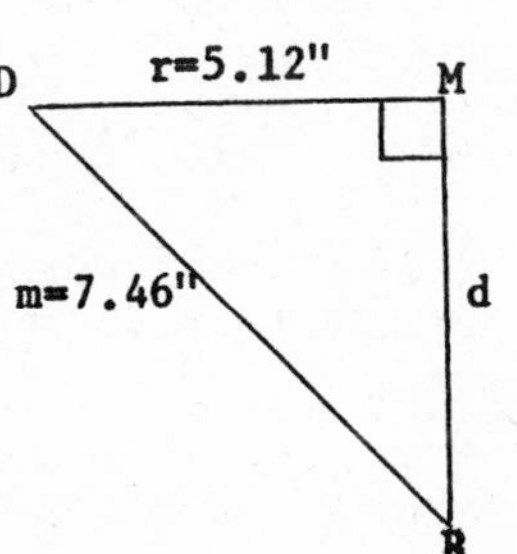

The Pythagorean Theorem says:

$$m^2 = r^2 + d^2$$

Plugging in the known values, we get:

$$(7.46")^2 = (5.12")^2 + d^2$$

Solving for "d^2" and simplifying, we get:

$$d^2 = (7.46")^2 - (5.12")^2$$
$$d^2 = 55.7 \text{ sq. in.} - 26.2 \text{ sq. in.}$$
$$d^2 = 29.5 \text{ sq. in.}$$

Taking the square root of both sides, we can find the length of "d".
d = 5.43

---

d = 5.43 in.

25. Let's use the Pythagorean Theorem to find the length of leg "a" in right triangle ARM:

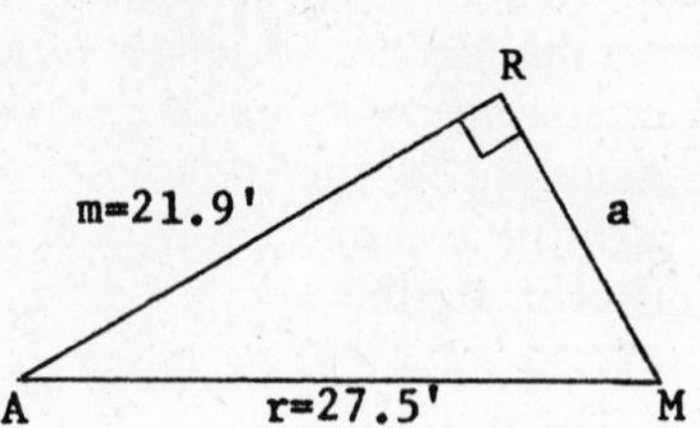

(a) Write the Pythagorean Theorem in letter-form:

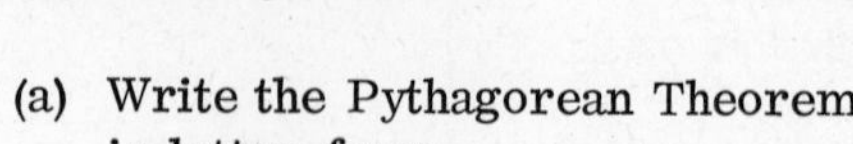

r^2 = m^2 + a^2

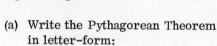

(b) Plug in the known values and solve for "a":

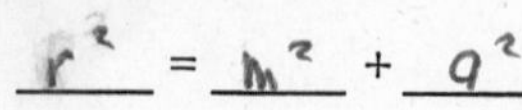

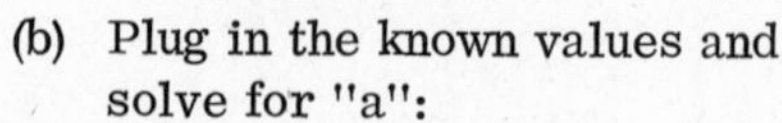

$(27.5)^2 = (21.9)^2 + (a)^2$

$a^2 = (27.5)^2 - (21.9)$

$a^2 =$

a = 16.6 ft

	a) $r^2 = m^2 + a^2$ b) a = 16.6 ft. (from a^2 = 276 sq. ft.)

26. In right triangle TMB, the values of two sides and two angles are given.

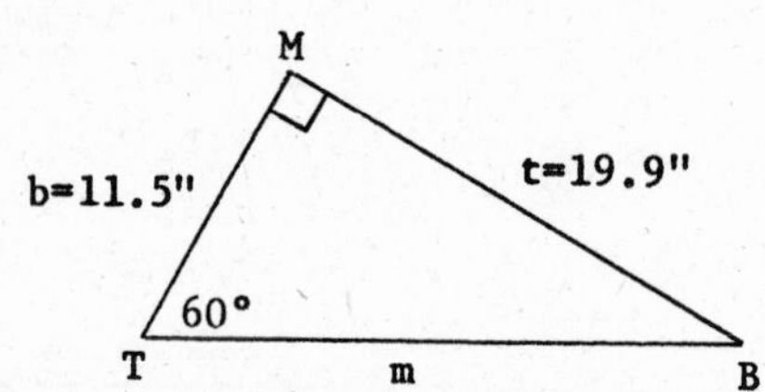

(a) Using the angle-sum principle, find the size of angle B. 30

(b) Using the Pythagorean Theorem, find the length of side "m".

$(m^2) = (b^2) + (t^2)$

$m^2 = 11.5^2 + 19.9^2$

$m^2 = 528$ sq ft

m = 23.0

a) Angle B = 30°

b) m = 23.0"
(from m^2 = 528 sq. in.)

27. The values of two sides and two angles are given for right triangle RQS.

(a) Using the angle-sum principle, find the size of angle S. ______

(b) Using the Pythagorean Theorem, find the length of side "s".

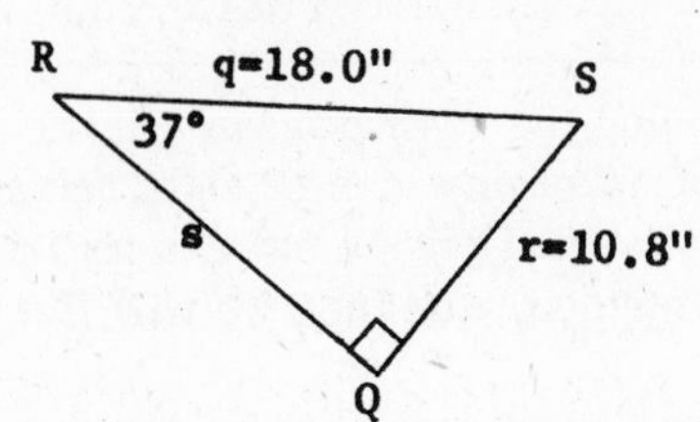

s = ______

Answer to Frame 27: a) Angle S = 53° b) s = 14.4"

SELF-TEST 2 (Frames 12-27)

1. Find the length of side w.

w
1.34"
2.18"

w = ______

2. Find the length of side d.

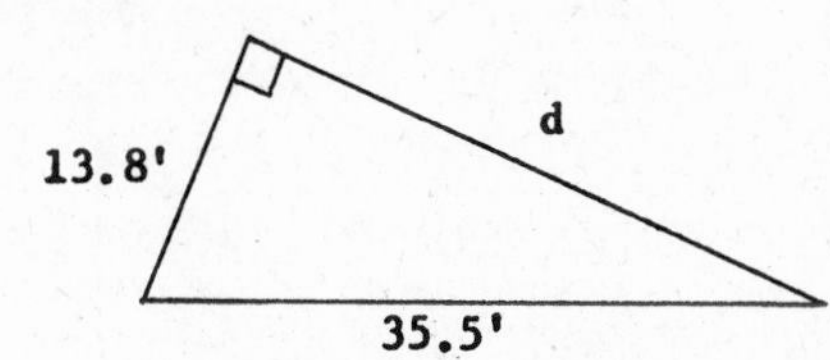

d = ______

ANSWERS: 1. w = 2.56" 2. d = 32.7 ft.

1-3 THE SCALING METHOD AND RIGHT TRIANGLES

Though the angle-sum principle and the Pythagorean Theorem can frequently be used to find the sizes of unknown angles or the lengths of unknown sides of right triangles, they cannot be used in all cases. When they cannot be used, the scaling method can be used to find unknown angles and sides. In this section, we will briefly discuss the scaling method, and then we will list the reasons why this method is not commonly used.

28. The angle-sum principle can be used to find the size of a third angle of a triangle only if the size of each of the other two angles is known.

In which triangle or triangles below can we find the size of angle A by means of the angle-sum principle? b

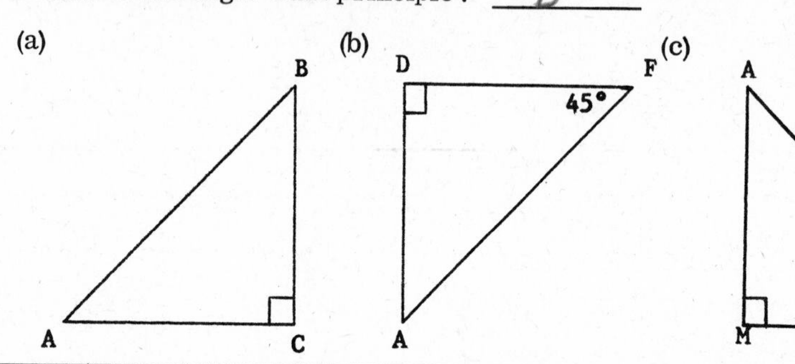

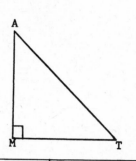

Answer to Frame 28: Only in (b)

29. In right triangle DFM in Figure 1 below, only the size of right angle F is known. Therefore, the angle-sum principle cannot be used to find the size of angle D (or angle M).

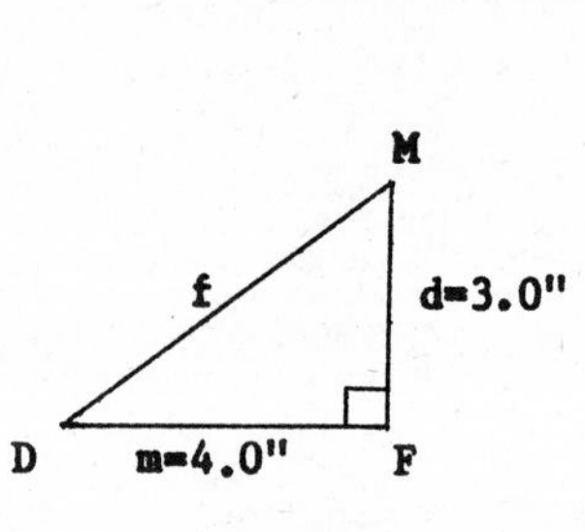

Figure 1

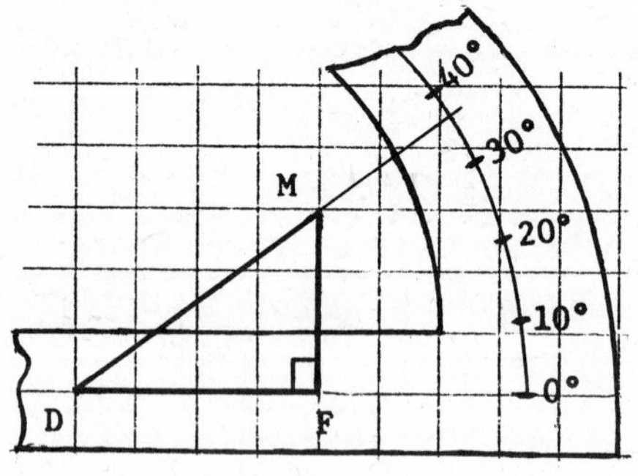

Figure 2

One way to find the size of angle D is to construct the triangle to scale, and then measure the size of angle D with a protractor. Triangle DFM has been constructed to scale in Figure 2 above. A sketch of a protractor is provided.

By reading the protractor scale, determine the size of angle D. Angle D = 37°

Angle D = 37°

30. The Pythagorean Theorem can be used to find the length of a side of a right triangle only if the lengths of the other two sides are known.

In which triangle or triangles below can we use the Pythagorean Theorem to find the length of "t"? a, c

(a) (b) (c)

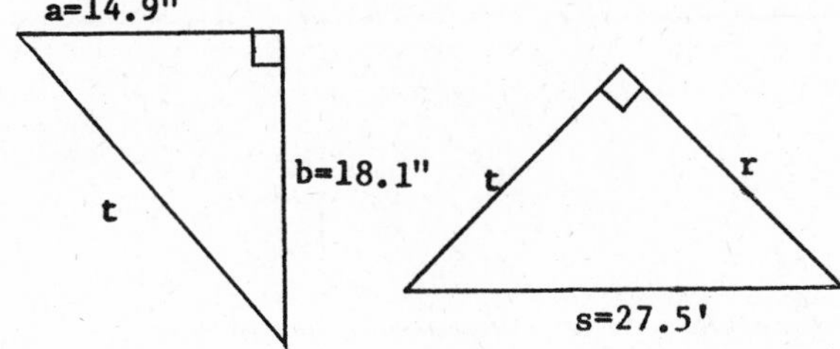

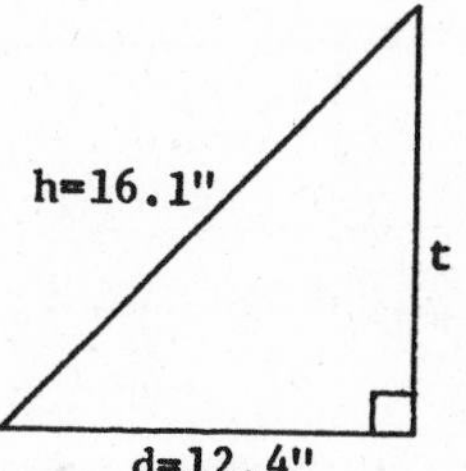

Answer to Frame 30: Only in (a) and (c)

31. In right triangle PST in Figure 1 below, only the length of side "t" is given. Therefore, we cannot use the Pythagorean Theorem to find the length of side "p".

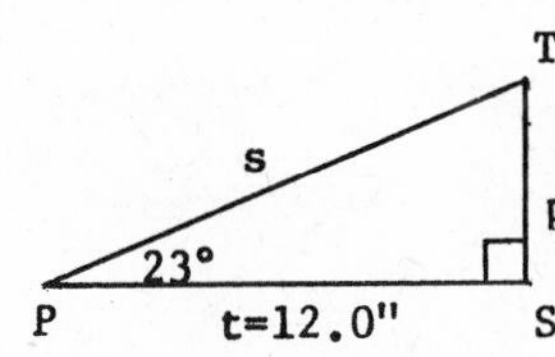

Figure 1

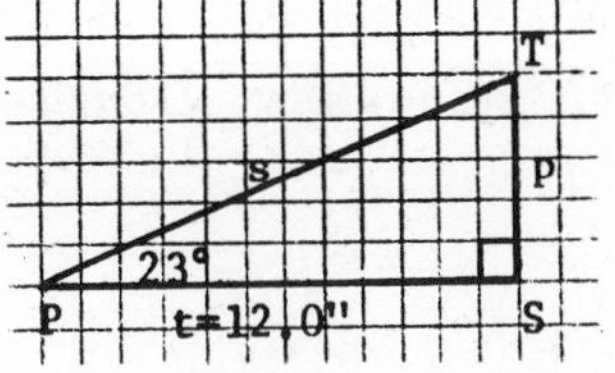

Figure 2

One way to find the length of "p" is to construct the triangle to scale, and then measure the length of "p" from the scaled diagram. Triangle PST has been constructed to scale on square-ruled paper in Figure 2 above.

If the side of each square represents 1 inch, how long is side "p"? p = 5 in

Answer to Frame 31: p = 5 inches

32. Since only one side of right triangle ABC (in Figure 1) is known, we cannot use the Pythagorean Theorem to find the length of side "a". However, we can find the length of "a" by the scaling method. The triangle has been drawn to scale in Figure 2.

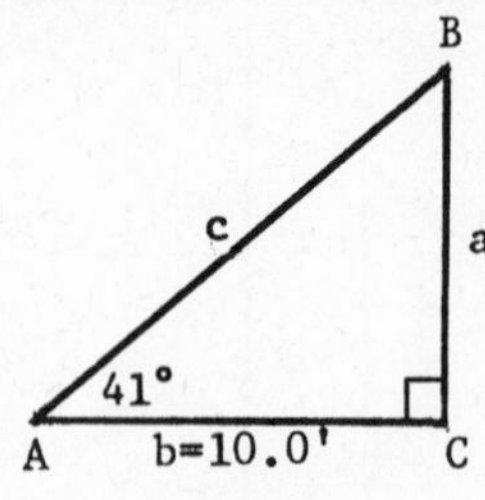

Figure 1

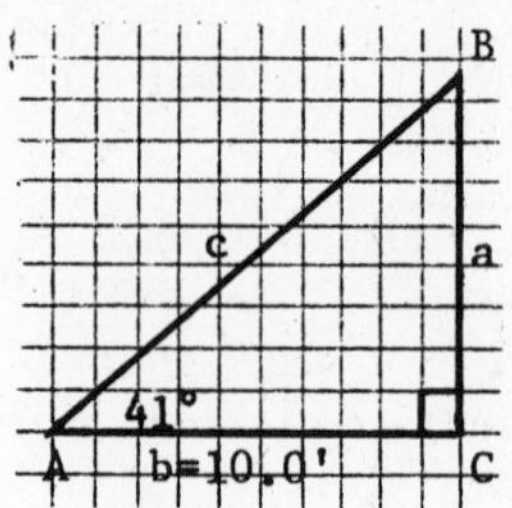

Figure 2

If the side of each square represents 1 foot, how long is side "a"? a = 8.7

Answer to Frame 32: Approximately 8.7 feet

33. When the angle-sum principle or the Pythagorean Theorem cannot be used, we can find the size of an angle or the length of a side by the "scaling method". However, the scaling method is not ordinarily used for these reasons:

(1) It is laborious and time-consuming.

(2) It is difficult to draw a triangle precisely to scale.

(3) "Scaling" solutions are only approximate since there is a need to estimate when reading lengths or angles on scaled triangles.

(4) A simpler and more accurate algebraic method can be used.

For these reasons, from here on, we will not use the scaling method for finding unknown sides and angles of right triangles.

The algebraic method mentioned above in (4) involves the use of trigonometric ratios. The rest of this chapter will be devoted to a discussion of these ratios and their use in solving right triangles.

1-4 THE DEFINITION OF THE TANGENT RATIO

The trigonometric ratios are comparisons of the lengths of two sides of a right triangle. These comparisons (or ratios) are identified in terms of a specific angle in the triangle. In this section, we will define the "tangent ratio", which is one of the three basic trigonometric ratios. Before doing so, however, we must briefly discuss the meaning of the "side opposite" and "side adjacent" for angles in right triangles.

34. In right triangle ABC, "c" is the hypotenuse, and "a" and "b" are the two legs.

"a" is the "side opposite" angle A.

"b" is the "side adjacent" to angle A.

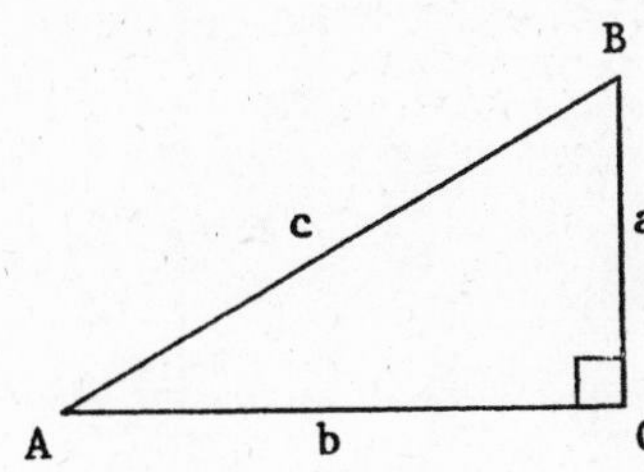

Note: The word "adjacent" means "next to". Though both "c" and "b" are "next to" angle A, only "b" is called the "side adjacent" since "c" is the hypotenuse.

In right triangle MFD:

(a) The hypotenuse is _____.

(b) The side opposite angle M is _____.

(c) The side adjacent to angle M is _____.

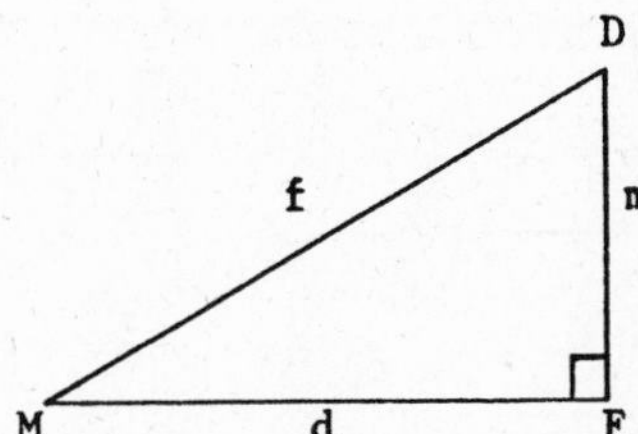

---

a) f

b) m

c) d

35. In right triangle ABC:

"b" is the "side opposite" angle B.

"a" is the "side adjacent" to angle B.

Note: Both "a" and "c" are "next to" angle B, but "c" is the hypotenuse.

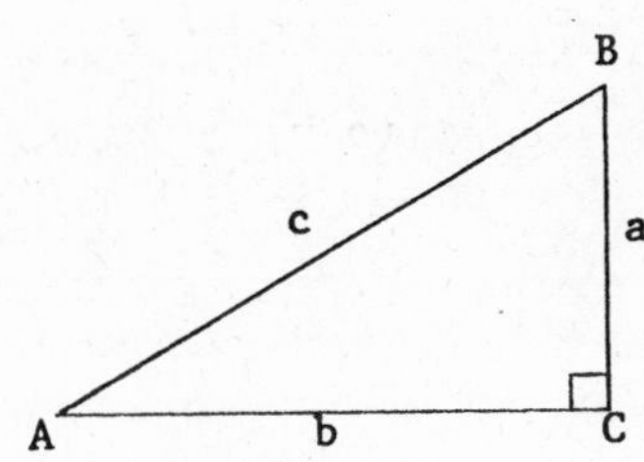

In right triangle DTR:

(a) The hypotenuse is _____.

(b) The side opposite angle R is _____.

(c) The side adjacent to angle R is _____.

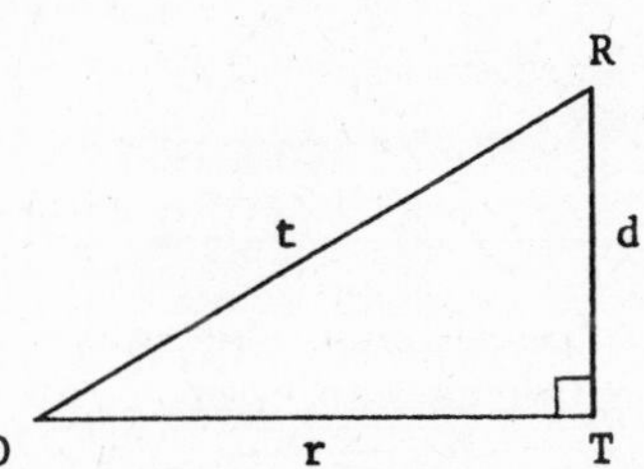

---

a) t

b) r

c) d

36. In right triangle BQM:

(a) The hypotenuse is ____.

(b) The side opposite angle B is ____.

(c) The side opposite angle M is ____.

(d) The side adjacent to angle B is ____.

(e) The side adjacent to angle M is ____.

---

37. In right triangle CMT:

(a) The hypotenuse is ____.

(b) The side opposite angle C is ____.

(c) The side adjacent to angle C is ____.

(d) The side opposite angle T is ____.

(e) The side adjacent to angle T is ____.

a) q
b) b
c) m
d) m
e) b

---

38. In right triangle STY:

(a) The hypotenuse is ____.

(b) The side opposite angle S is ____.

(c) The side adjacent to angle S is ____.

(d) The side opposite angle Y is ____.

(e) The side adjacent to angle Y is ____.

a) m
b) c
c) t
d) t
e) c

---

39. A ratio is a comparison of two quantities by means of a division. Like any division, a ratio can be written as a fraction. For example:

The ratio of "a" to "b" is written $\frac{a}{b}$.

The ratio of "m" to "t" is written ____.

a) t
b) s
c) y
d) y
e) s

$\frac{m}{t}$

40. The <u>tangent</u> <u>ratio</u> is one of the three basic trigonometric ratios. In a right triangle, the "<u>tangent of an angle</u>" is a comparison of the "<u>length of the side opposite</u>" the angle to the "<u>length of the side adjacent</u>" to the angle. That is:

$$\text{THE TANGENT OF AN ANGLE} = \frac{\text{LENGTH OF THE SIDE OPPOSITE}}{\text{LENGTH OF THE SIDE ADJACENT}}$$

or

$$\text{THE TANGENT OF AN ANGLE} = \frac{\text{SIDE OPPOSITE}}{\text{SIDE ADJACENT}}$$

In right triangle MDT:

The side opposite angle M is "m".
The side adjacent to angle M is "t".
Therefore, <u>the tangent of angle M</u> is $\frac{m}{t}$.

The side opposite angle T is "t".
The side adjacent to angle T is "m".
Therefore, <u>the tangent of angle T</u> is $\frac{t}{m}$.

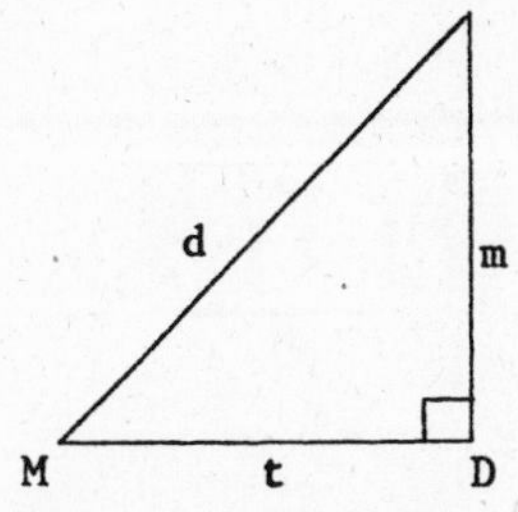

Answer: $\frac{t}{m}$

---

41. In right triangle RBF:

(a) The tangent of angle F is $\frac{f}{r}$.

(b) The tangent of angle R is $\frac{r}{f}$.

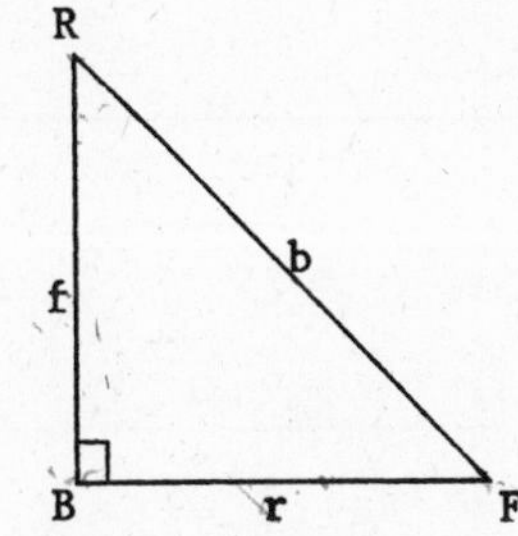

Answers: a) $\frac{f}{r}$ b) $\frac{r}{f}$

---

42. In right triangle CAM:

(a) The tangent of angle C is $\frac{c}{m}$.

(b) The tangent of angle M is $\frac{m}{c}$.

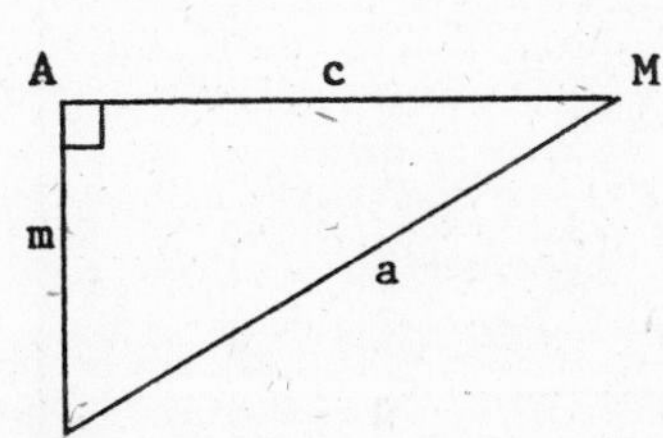

Answers: a) $\frac{c}{m}$ b) $\frac{m}{c}$

---

43. In right triangle HFM, the sides are not labeled with small letters. However, we can use the two capital letters at each end of a side to label the side. For example:

The hypotenuse is HM.
The side opposite angle H is FM.
The side adjacent to angle H is HF.

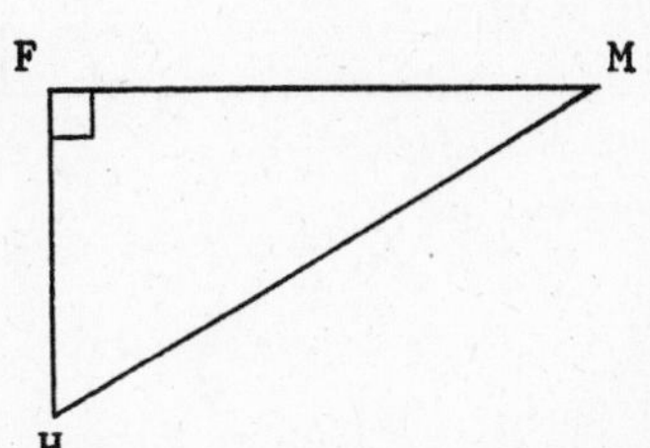

Answer: FH

44. In right triangle PST:

(a) The tangent of angle T is ______.

(b) The tangent of angle P is ______.

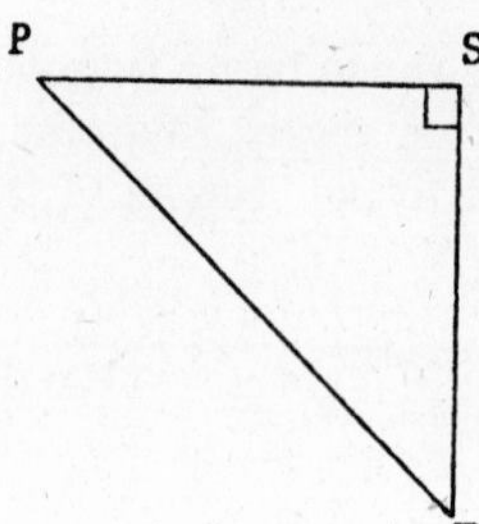

a) $\frac{PS}{ST}$

b) $\frac{ST}{PS}$

---

45. In right triangle ADQ:

(a) The tangent of angle A is ______.

(b) The tangent of angle Q is ______.

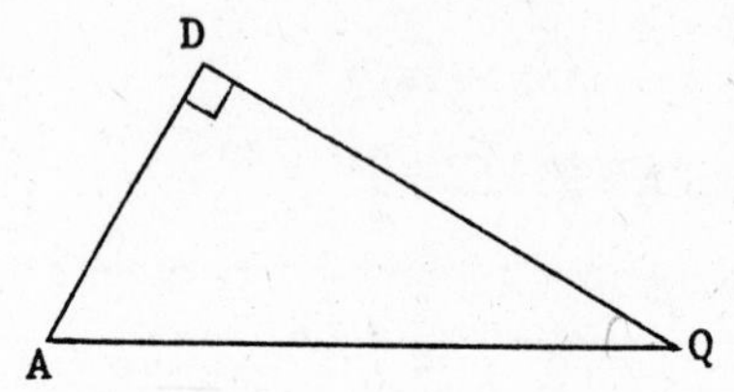

<u>Answer to Frame 45:</u> a) $\frac{DQ}{AD}$ b) $\frac{AD}{DQ}$

<u>SELF-TEST 3 (Frames 34-45)</u>

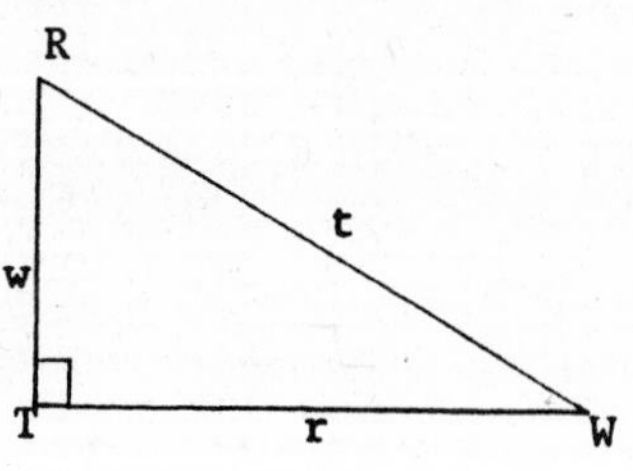

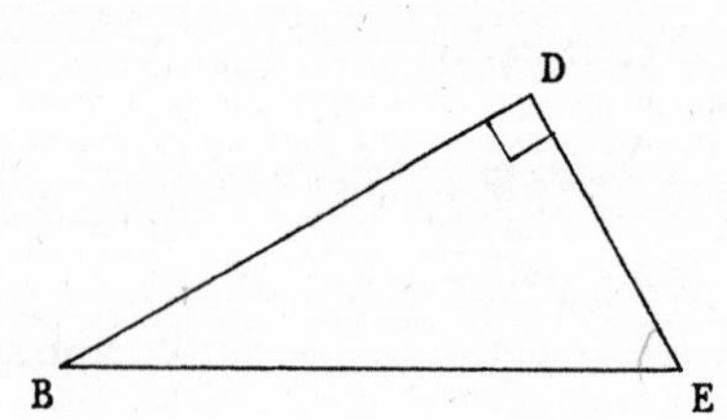

1. The tangent of angle W is ______.

2. The tangent of angle R is ______.

3. The tangent of angle B is ______.

4. The tangent of angle E is ______.

<u>ANSWERS:</u> 1. $\frac{w}{r}$ 2. $\frac{r}{w}$ 3. $\frac{DE}{BD}$ 4. $\frac{BD}{DE}$

1-5 THE TANGENT RATIO OF THREE SPECIFIC ANGLES

In the last section, we defined what is meant by the "tangent of an angle". In this section, we will discuss the tangents of three specific angles: 18°, 37°, and 68°. The tangent ratios of these angles will be used to find the lengths of sides of right triangles which contain an angle of either 18°, 37°, or 68°.

46. Let's compute the tangent of angle A in right triangle ABC.

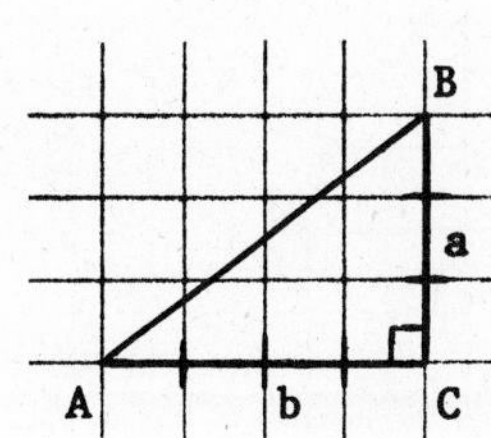

(a) The side opposite angle A is "a". By counting units, the length of side "a" is _____ units.

(b) The side adjacent to angle A is "b". The length of side "b" is _____ units.

(c) The tangent of angle A $= \frac{a}{b} =$ _____.

Answer to Frame 46: a) 3 b) 4 c) $\frac{3}{4}$

47. In the diagram on the right, there are four right triangles: ABI, ACH, ADG, and AEF. Each of these right triangles contains the common angle A, which is a 37° angle.

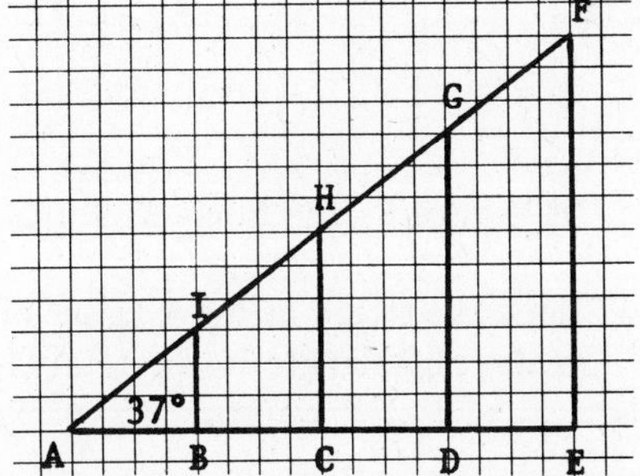

In the four figures below, we have separated the four triangles above. We will compute the tangent of angle A in each triangle. The purpose of doing so is to show that the value of the tangent of a 37° angle is the same in right triangles of any size.

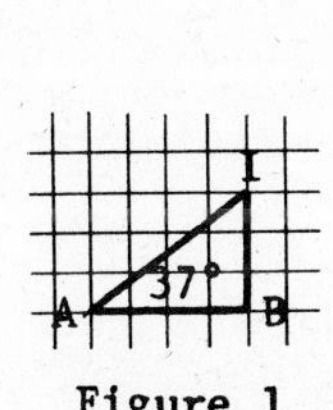

Figure 1

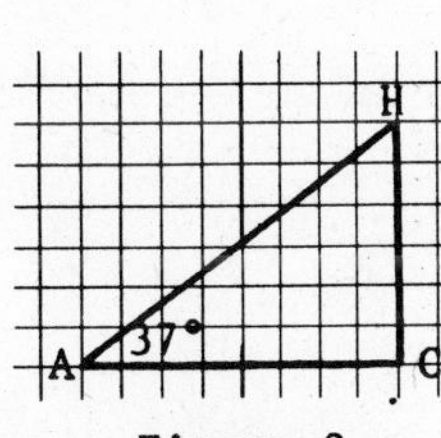

Figure 2

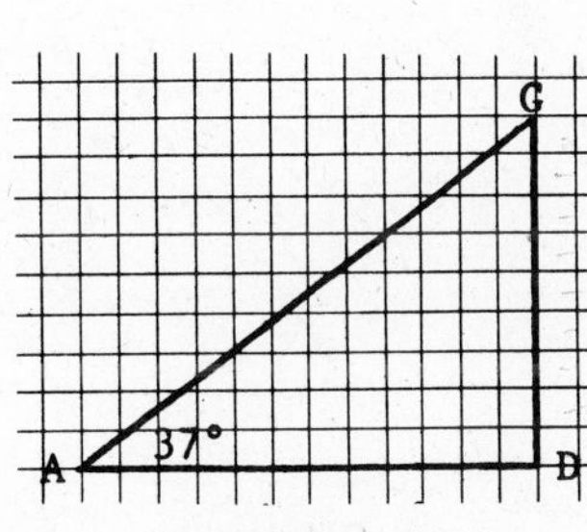

Figure 3

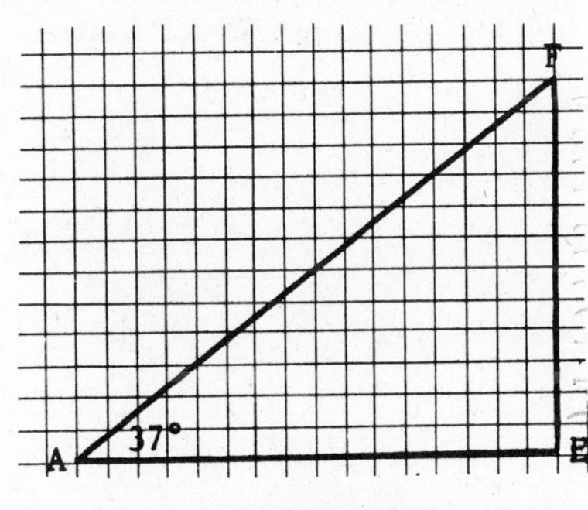

Figure 4

The tangent of angle A in each figure can be computed by counting units.

(a) In Figure 1, $\frac{BI}{AB} =$ 3/4

(b) In Figure 2, $\frac{CH}{AC} =$ 6/8

(c) In Figure 3, $\frac{DG}{AD} =$ 9/12

(d) In Figure 4, $\frac{EF}{AE} =$ 12/16

(e) Even though the four triangles are of different sizes, is the computed value of the tangent of angle A the same for each triangle? yes

48. The value of the tangent of a 37° angle is $\frac{3}{4}$ in right triangles of all sizes. This value of $\frac{3}{4}$ is also independent of the position of the right angle in the triangle. Here are some examples:

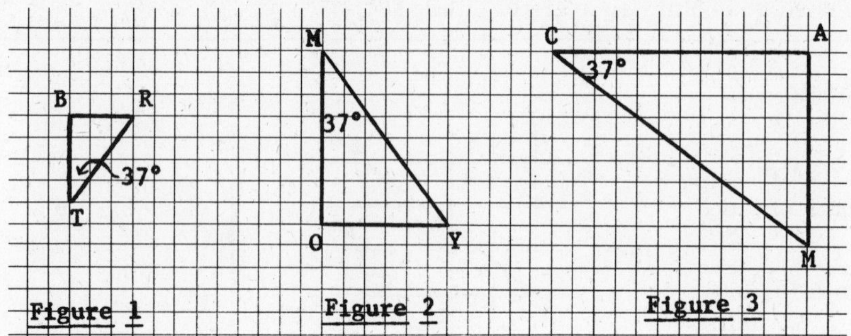

The numerical value of the tangent of the 37° angle in each triangle above can be obtained by counting units.

(a) In Figure 1, the tangent of angle T = $\frac{BR}{BT}$ = ______.

(b) In Figure 2, the tangent of angle M = $\frac{OY}{MO}$ = ______.

(c) In Figure 3, the tangent of angle C = $\frac{AM}{CA}$ = ______.

(d) Does the value of the tangent of a 37° angle depend upon the position of the right angle in the triangle? ______

a) $\frac{3}{4}$

b) $\frac{6}{8} = \frac{3}{4}$

c) $\frac{9}{12} = \frac{3}{4}$

d) $\frac{12}{16} = \frac{3}{4}$

e) Yes. It is $\frac{3}{4}$ for each.

---

49. We have seen that the tangent of a 37° angle is $\frac{3}{4}$. This value does not depend on the size of the right triangle or the position of the right angle. <u>It also does not depend on the position of the 37° angle.</u> Some scaled examples are given below.

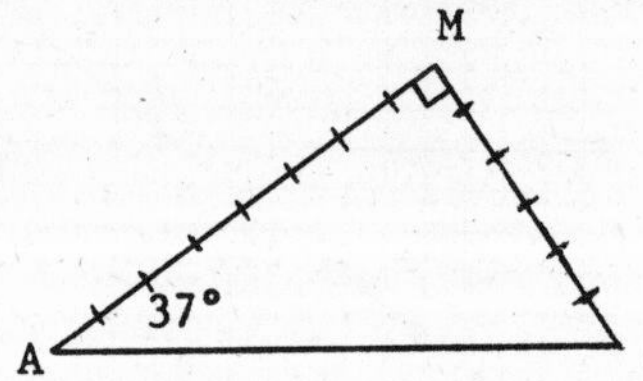

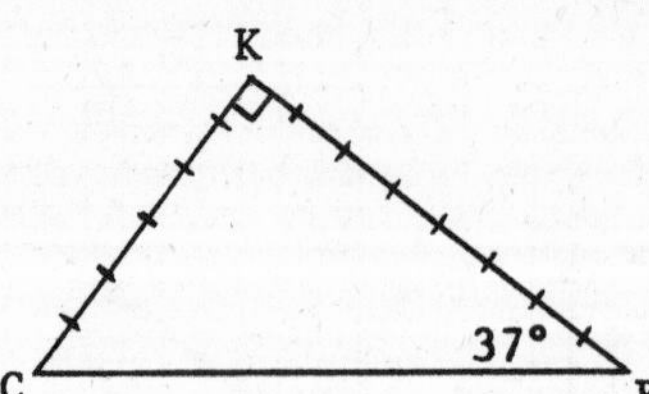

Each triangle above has a right angle at the top. We can compute the tangent of the 37° angle in each case by counting units.

(a) In triangle AMT, the tangent of angle A = $\frac{MT}{AM}$ = ______.

(b) In triangle CKP, the tangent of angle P = $\frac{CK}{KP}$ = ______.

(c) Does the value of the tangent of a 37° angle change because of the position of the 37° angle in the right triangle? ______

a) $\frac{3}{4}$

b) $\frac{6}{8} = \frac{3}{4}$

c) $\frac{9}{12} = \frac{3}{4}$

d) No. It is $\frac{3}{4}$ in all cases.

50. The point we are making is this:

> The tangent of a 37° angle is always $\frac{3}{4}$, no matter how large the triangle is, where the right angle is, or where the 37° angle is.

Here is a right triangle without scaled sides. Angle J is a 37° angle.

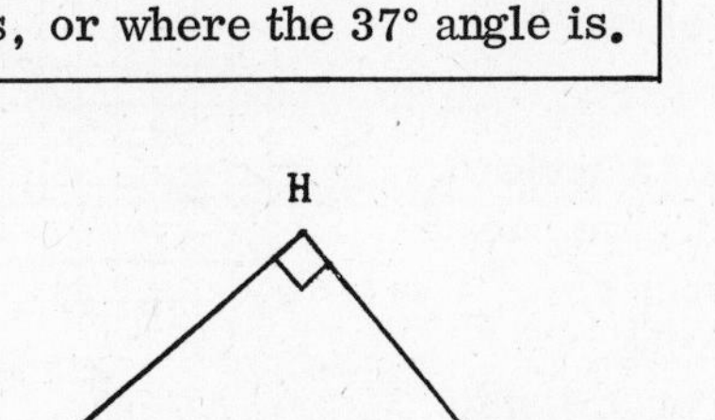
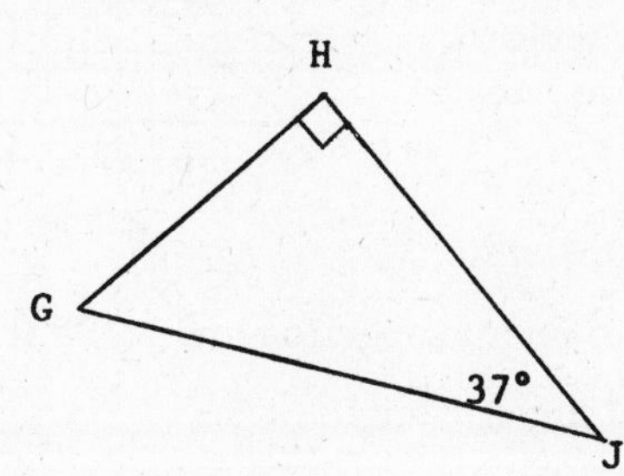

(a) The side opposite angle J is ______.

(b) The side adjacent to angle J is ______.

(c) The tangent of angle J is $\frac{GH}{HJ}$.

Even though we do not know the lengths of these two sides, we know that the numerical value of this ratio is ______

a) $\frac{6}{8} = \frac{3}{4}$

b) $\frac{6}{8} = \frac{3}{4}$

c) No

---

51. The right triangle in the scaled diagram contains a 37° angle. The length of side MQ is 20 units. We want to find the length of side PQ. We can do so by either of two methods, the <u>scaling method</u> or an <u>algebraic method</u>:

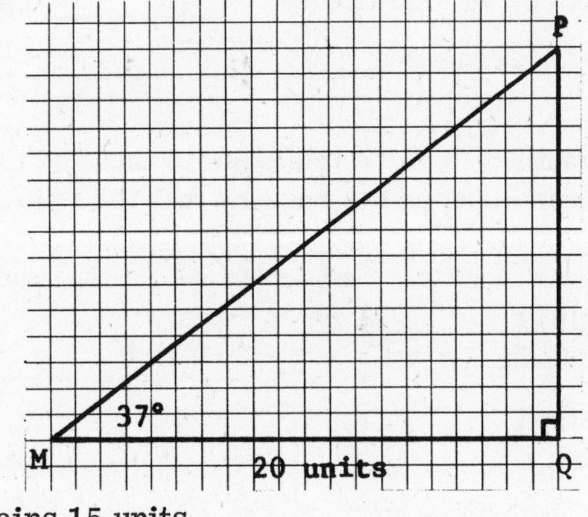

<u>Scaling Method.</u> To find the number of units in side PQ by this method, we simply count them. PQ contains 15 units.

<u>Algebraic Method.</u> In this method, we make use of the fact that the tangent of a 37° angle is <u>always</u> $\frac{3}{4}$ in order to set up a proportion. The length of PQ is calculated as follows:

Since: (1) The tangent of a 37° angle is always $\frac{3}{4}$,

and (2) the tangent of angle M $= \frac{\text{side opposite}}{\text{side adjacent}} = \frac{PQ}{MQ} = \frac{PQ}{20}$,

we can set up the following proportion:

$$\frac{PQ}{20} = \frac{3}{4}$$

(a) Solving the proportion, we find that PQ = ______ units.

(b) Is this the same answer we obtained by the scaling method? ______

a) GH

b) HJ

c) $\frac{3}{4}$

---

a) 15 units

b) Yes

52. Triangle PQR is a right triangle. Angle P contains 37°. The length of side PR is 13 units. We want to find the length of side QR. We can again do so by either of two methods.

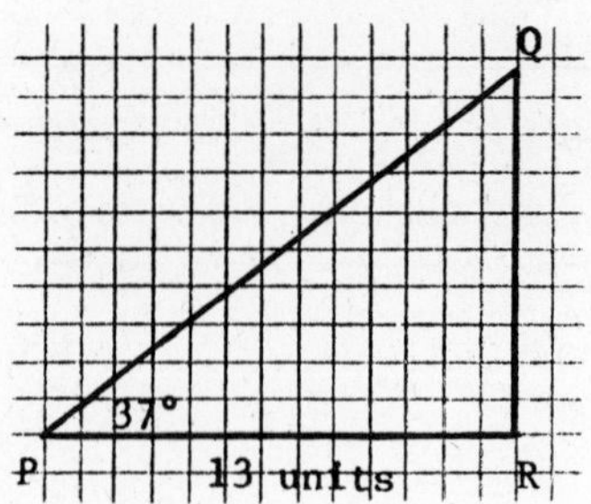

Scaling Method. Since the triangle is constructed to scale, we can simply count the number of units in QR. By counting, QR is approximately 9.7 or 9.8 units.

Algebraic Method. We can set up a proportion and solve it, as follows:

Since: (1) The tangent of any 37° angle is $\frac{3}{4}$,

and (2) the tangent of angle P $= \frac{\text{side opposite}}{\text{side adjacent}} = \frac{QR}{PR} = \frac{QR}{13}$,

we can set up the proportion:  $\frac{QR}{13} = \frac{3}{4}$.

(a) Solving the proportion, we find that QR = 9.75 units.

(b) Which method is more exact? algebraic

a) 9.75 units

b) The algebraic method

---

53. We will use the algebraic method to solve triangles in the rest of this chapter. It is more exact, and it does not require that the triangle be drawn to scale.

In right triangle DMK, angle D contains 37°. Side "k" is 118 ft.

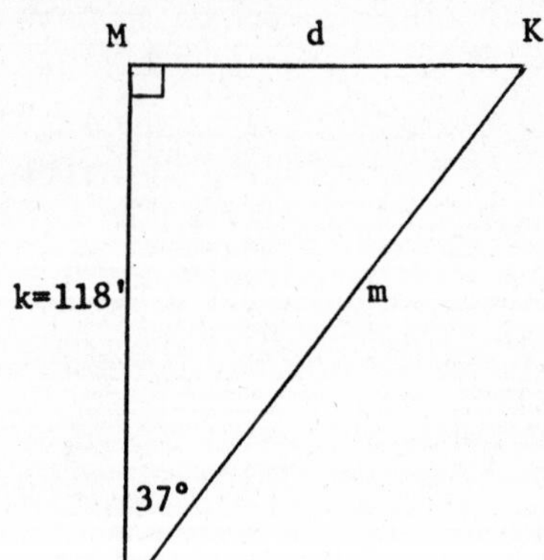

Let's use the algebraic method to find the length of "d", the side opposite angle D:

The tangent of angle D $= \frac{d}{k} = \frac{d}{118 \text{ ft.}}$

The tangent of a 37° angle is $\frac{3}{4}$.

Therefore: $\frac{d}{118 \text{ ft.}} = \frac{3}{4}$

Solving the proportion, d = 88.5 feet.

88.5 ft.

54. In right triangle ABC, angle A contains 37°. Side "a" is 69.0 inches.

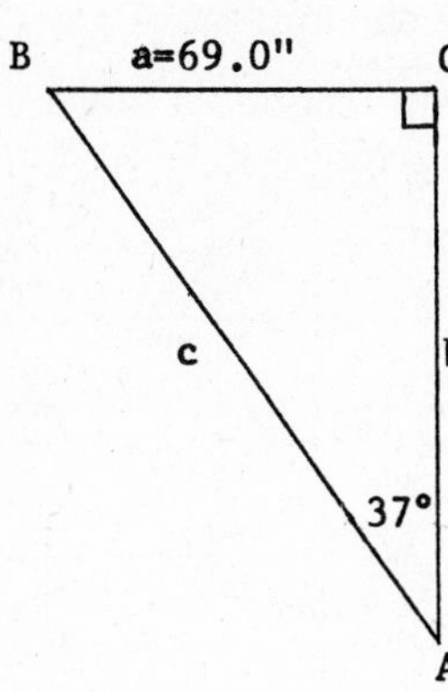

Let's use the algebraic method to find the length of "b", the side adjacent to angle A:

The tangent of angle A $= \frac{a}{b} = \frac{69.0''}{b}$.

The tangent of any 37° angle is $\frac{3}{4}$.

Therefore: $\frac{69.0''}{b} = \frac{3}{4}$

Solving the proportion, b = 92 inches.

---

55. We have seen that the tangent of any 37° angle is $\frac{3}{4}$. Let's examine the tangent ratio for any 68° angle. In the figure below, there are four right triangles: ABJ, ACK, ADL, and AEM. Each triangle contains the common angle A, which contains 68°.

92.0 inches

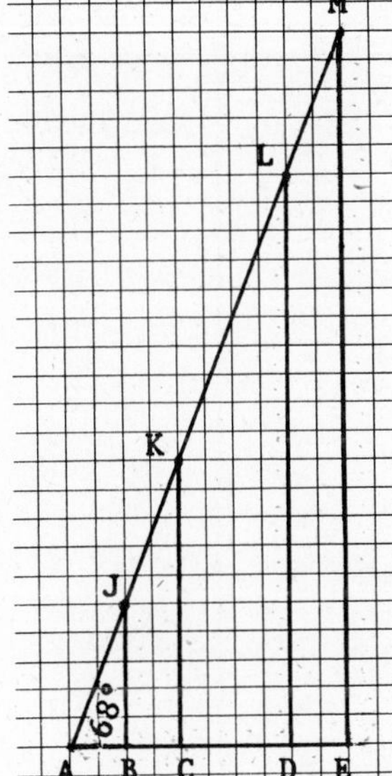

We can compute the tangent of angle A in each right triangle by counting units.

(a) In triangle ABJ, $\frac{BJ}{AB} = \frac{5}{2}$

(b) In triangle ACK, $\frac{CK}{AC} = \frac{10}{4}$

(c) In triangle ADL, $\frac{DL}{AD} = \frac{20}{8}$

(d) In triangle AEM, $\frac{EM}{AE} = \frac{25}{10}$

(e) The tangent of any 68° angle is $\frac{5}{2}$.

---

a) $\frac{5}{2}$

b) $\frac{10}{4} = \frac{5}{2}$

c) $\frac{20}{8} = \frac{5}{2}$

d) $\frac{25}{10} = \frac{5}{2}$

e) $\frac{5}{2}$

56. In the last frame, we saw that the tangent of any 68° angle is $\frac{5}{2}$. We can use this fact to find the unknown lengths of sides of right triangles which contain a 68° angle. For example:

In right triangle CDM, angle C contains 68°. Side "m" is 11.8 inches. Find side "c".

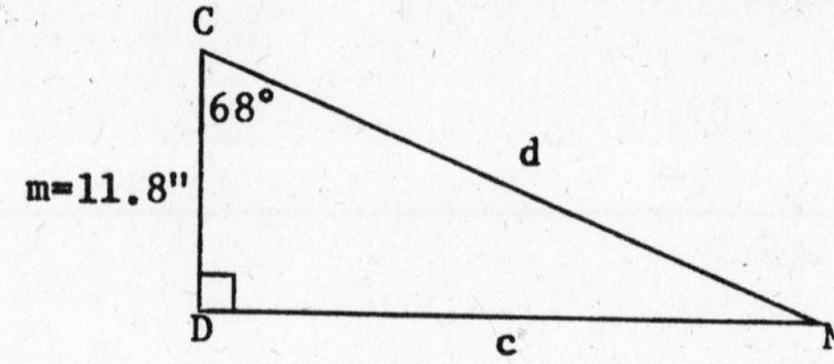

We can find side "c" by the algebraic method:

The tangent of angle C is $\frac{c}{m}$ or $\frac{c}{11.8"}$.

The tangent of any 68° angle is $\frac{5}{2}$.

Therefore: $\frac{c}{11.8"} = \frac{5}{2}$

Solving the proportion, "c" is 29.5 inches.

	29.5 inches

57. In right triangle RTV, angle V contains 68°. Side RT is 59.0 feet.

R 59.0' T

68°

V

We can find side TV by the algebraic method:

The tangent of angle V is $\frac{RT}{TV} = \frac{59.0'}{TV}$.

The tangent of any 68° angle is $\frac{5}{2}$.

Therefore: $\frac{59.0'}{TV} = \frac{5}{2}$

Solving the proportion, TV = 23.6 feet.

23.6 feet

58. In the figure on the right, there are four right triangles: ABP, ACQ, ADR, and AES. Each triangle contains a common angle of 18°.

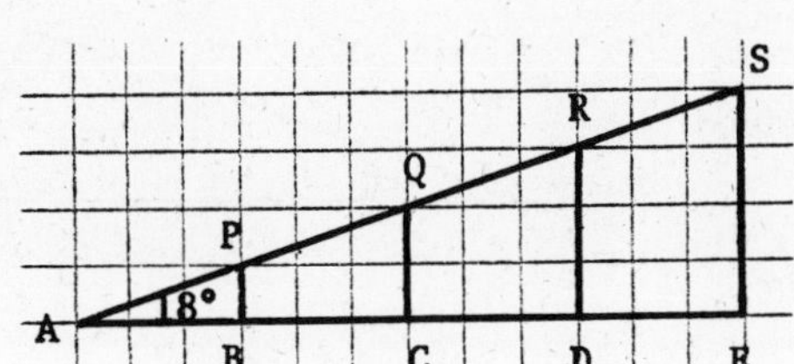

We can compute the tangent ratio for the 18° angle in each right triangle by counting units.

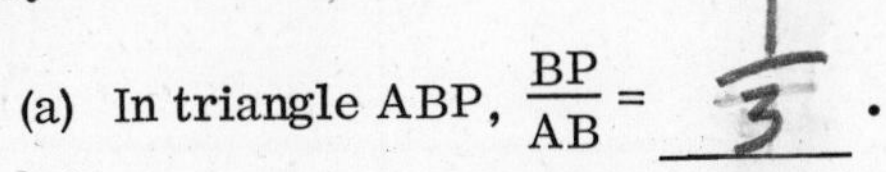

(a) In triangle ABP, $\frac{BP}{AB} =$ $\frac{1}{3}$.

(b) In triangle ACQ, $\frac{CQ}{AC} =$ $\frac{2}{6}$.

(c) In triangle ADR, $\frac{DR}{AD} =$ $\frac{3}{9}$.

(d) In triangle AES, $\frac{ES}{AE} =$ $\frac{4}{12}$.

(e) The tangent of any 18° angle is $\frac{1}{3}$.

---

59. The tangent of any 18° angle is $\frac{1}{3}$. We can use this fact to find the unknown lengths of sides of right triangles which contain an 18° angle.

In right triangle DBK, angle K contains 18°. BK is 40.8".

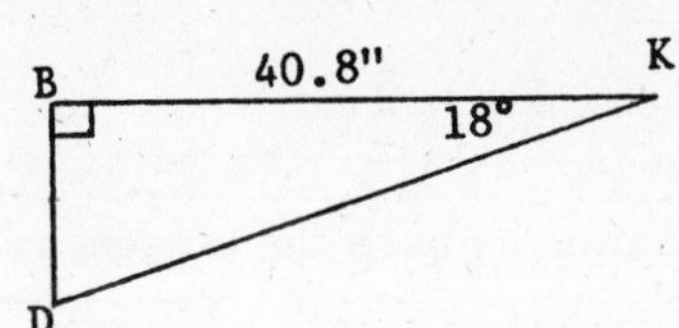

We can find BD by the algebraic method:

The tangent of angle K is $\frac{BD}{BK} = \frac{BD}{40.8''}$.

The tangent of any 18° angle is $\frac{1}{3}$.

Therefore: $\frac{BD}{40.8''} = \frac{1}{3}$

Solving the proportion, BD is 13.6 inches.

Answers to 58:

a) $\frac{1}{3}$

b) $\frac{2}{6} = \frac{1}{3}$

c) $\frac{3}{9} = \frac{1}{3}$

d) $\frac{4}{12} = \frac{1}{3}$

e) $\frac{1}{3}$

Answer to 59: 13.6 inches

60. In right triangle DMY, angle Y contains 18°. MD is 27.8 feet.

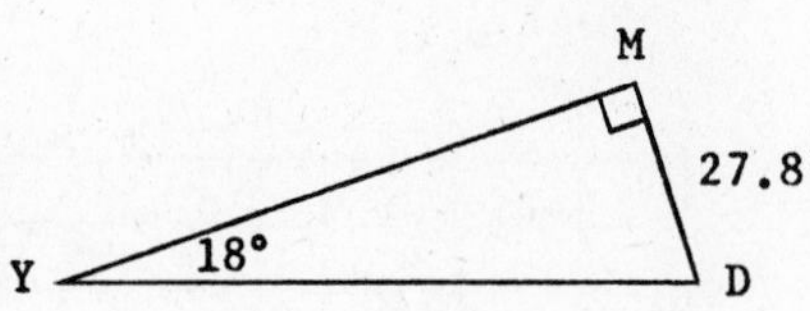

We can use the algebraic method to find MY:

The tangent of angle Y is $\frac{MD}{MY} = \frac{27.8'}{MY}$.

The tangent of any 18° angle is $\frac{1}{3}$.

Therefore: $\frac{27.8'}{MY} = \frac{1}{3}$

Solving the proportion, MY = ________ feet.

Answer to Frame 60: 83.4 ft.

In this section, we have examined and used the following three facts:

The tangent of any 18° angle is $\frac{1}{3}$.

The tangent of any 37° angle is $\frac{3}{4}$.

The tangent of any 68° angle is $\frac{5}{2}$.

In later sections, we will examine and use the tangent ratios of angles of other sizes.

SELF-TEST 4 (Frames 46-60)

1. Find the length of side "h".

h
37°
2,640 ft.

h = ________

2. Find the length of side "d".

68°
d
7.15"

d = ________

ANSWERS: 1. h = 1,980 ft. 2. d = 2.86"

1-6 A PARTIAL TABLE OF TANGENT RATIOS

In the last section, we discussed the tangents of three specific angles. Their tangents were written as ratios in fraction form. In this section, we will introduce a table which contains the numerical values of the tangents of some other angles of different sizes. Though still ratios, the tangents in the table are reported as regular decimal numbers. The values in the table will be used to find the lengths of sides and the sizes of angles in right triangles.

---

61. Instead of the phrase "the tangent of an 18° angle", mathematicians write: "tan 18°"

Therefore: "tan 37°" means: "The tangent of a 37° angle".
"tan 68°" means: ______________________

The tangent of a 68° angle.

62. In the last section, we discussed the tangents of three specific angles. In that section, the tangents were written as fractions. Ordinarily, however, tangents are written as decimal numbers. For example:

$$\tan 18° = \frac{1}{3} = 0.33$$

$$\tan 37° = \frac{3}{4} = 0.75$$

$$\tan 68° = \frac{5}{2} = 2.5$$

The decimal numbers are more convenient to use when solving right triangles.

Go to next frame.

63. Mathematicians have prepared a table which includes the numerical values of tangents of angles of any size. Part of that table is given on the right. The Greek letter "θ" (pronounced "thay-ta") is used as the general symbol for any angle.

Angle θ	tan θ
.	.
.	.
.	.
25°	0.4663
30°	0.5774
35°	0.7002
40°	0.8391
45°	1.000
50°	1.192
55°	1.428
.	.
.	.
.	.

Note: (1) The decimal value of each tangent is given.
(2) The decimal value is reported to four significant digits.

Using the table, complete these:

(a) The tangent of any 35° angle is .7002 .

(b) 1.192 is the tangent of any 50° angle.

---

a) 0.7002

b) 50°

64. By using the table in the last frame, we can solve many problems. Here is an example:

In right triangle PMR, PR is 53.0", and angle P is 25°. Find MR.

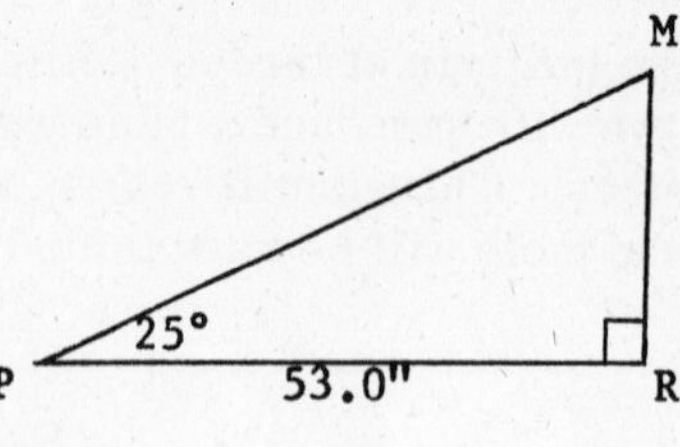

The steps of the solution are:

(1) From the table, tan 25° = 0.4663.

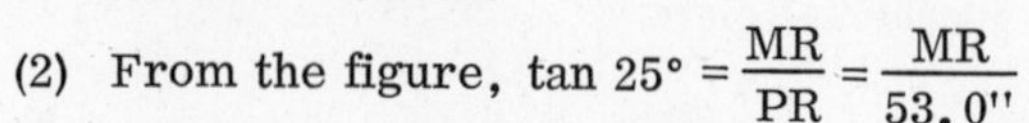

(2) From the figure, $\tan 25° = \frac{MR}{PR} = \frac{MR}{53.0''}$.

(3) Therefore: $0.4663 = \frac{MR}{53.0''}$

MR = 24.7 inches

MR = 24.7 inches

Solution:

$$0.4663 = \frac{MR}{53.0''}$$

$$MR = (0.4663)(53.0'')$$

$$MR = 24.7 \text{ inches}$$

---

65. Here is another problem we can solve by using the table:

If PQ is 83.0 feet and angle R is 50°, find QR.

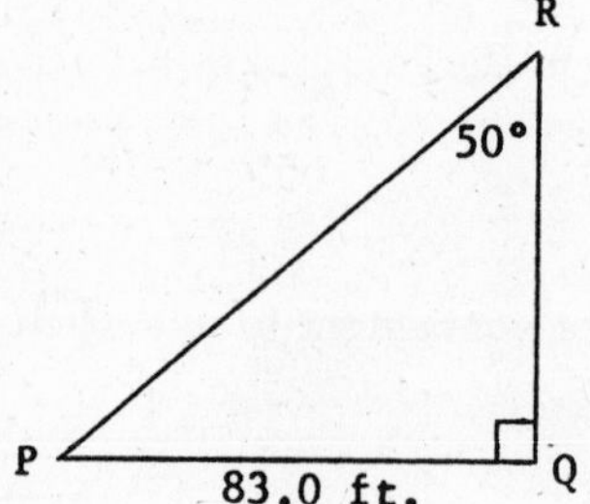

Steps: (1) From the table, tan 50° = 1.192.

(2) From the figure, $\tan 50° = \frac{PQ}{QR} = \frac{83.0'}{QR}$.

(3) Therefore: $1.192 = \frac{83.0'}{QR}$

QR = 69.6 feet

QR = 69.6 feet

Solution:

$$1.192 = \frac{83.0}{QR}$$

$$(1.192)(QR) = 83.0$$

$$QR = \frac{83.0}{1.192}$$

$$QR = 69.6$$

---

66. We can also use the table to find the size of an unknown angle. Here is an example:

If BC is 63.8 inches and AC is 76.0 inches, how large is angle A?

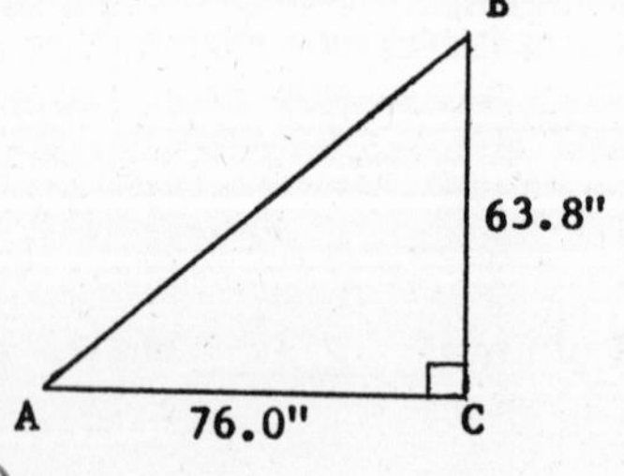

Steps: (1) From the figure, $\tan A = \frac{BC}{AC} = \frac{63.8}{76.0} \doteq 0.8391$.

(2) From the table, if tan A = 0.8391, angle A is 40 degrees.

---

40 degrees (or 40°)

67. The table says this: tan 25° = 0.4663

Though the tangent is reported as a decimal number, we must remember that this decimal number stands for the following ratio:

$$\tan 25° = \frac{\text{side opposite the 25° angle}}{\text{side adjacent to the 25° angle}}$$

To emphasize the fact that a tangent is a ratio, we can convert the decimal to a fraction. We get:

$$0.4663 = \frac{4,663}{10,000}$$

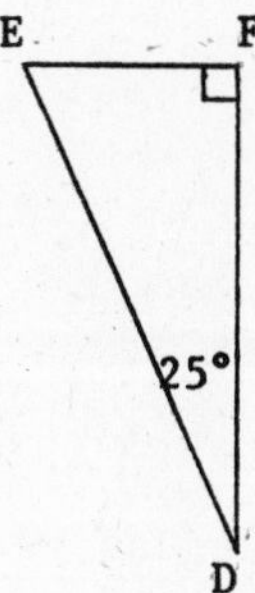

Using this fact:

If EF = 4,663 feet,

then DF = 10,000 feet.

Answer: 10,000 feet

68. Referring to the table, we find that: tan 55° = 1.428

Converting this decimal to a fraction, we get: $1.428 = \frac{1,428}{1,000}$

Therefore, in right triangle MNL:

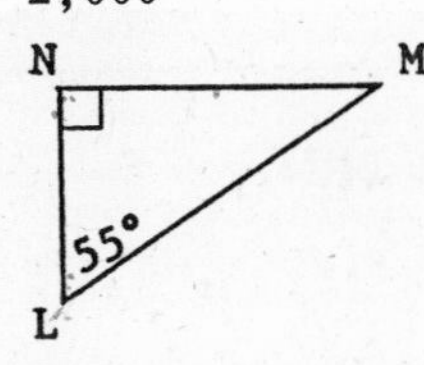

If LN = 1,000 inches,

then MN = 1,428 inches.

Answer: 1,428 inches

69. Here is the same incomplete tangent table.

Angle θ	$\tan \theta$
.	.
.	.
.	.
25°	0.4663
30°	0.5774
35°	0.7002
40°	0.8391
45°	1.000
50°	1.192
55°	1.428
.	.
.	.
.	.

Problem: In right triangle PQR, angle R is 35° and side QR is 417 feet. How long is side PQ?

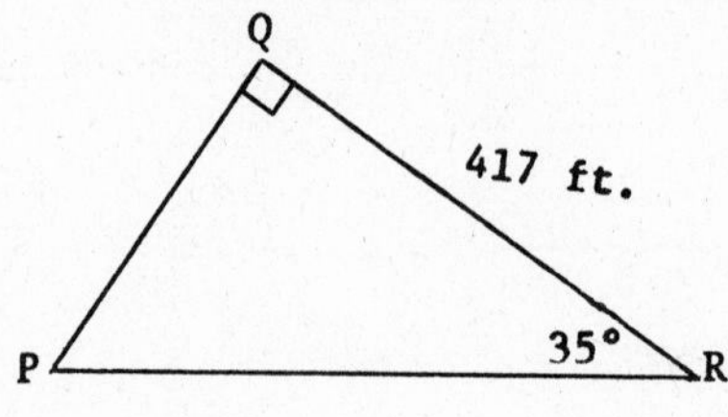

Steps: (1) From the table, tan 35° = 0.7002.

(2) From the figure, $\tan 35° = \frac{PQ}{QR} = \frac{PQ}{417'}$

(3) Therefore: $0.7002 = \frac{PQ}{417'}$

PQ = 292 feet

292 feet

PQ = (0.7002)(417)

= 292 ft.

70. Problem: In right triangle LNM, angle L = 55° and MN = 16.0".

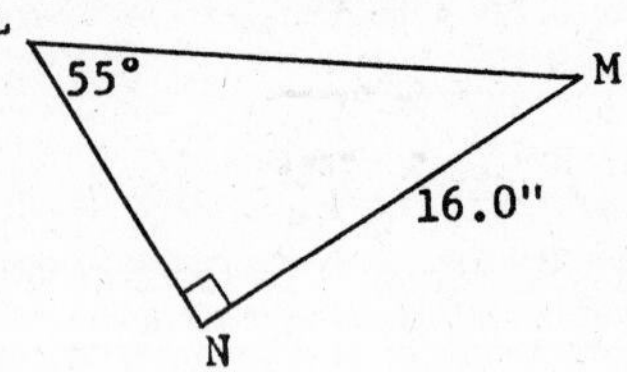

(a) Set up the equation needed to solve for LN.

$\frac{mn}{Ln} = \frac{16.0}{Ln}$

(b) Solve for LN.

$1.428 = \frac{16.0}{LN}$; $LN = \frac{16.0}{1.428}$

LN = 11.2"

(a) $\frac{16.0''}{LN} = 1.428$

(b) LN = 11.2 inches

71. If RS = 34.3 feet and ST = 19.8 feet, how large is angle R? ________

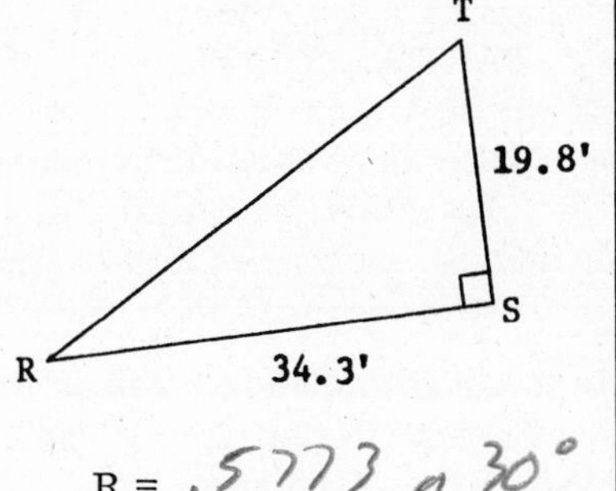

R = ________

Answer to Frame 71: R = 30° (From: $\tan R = \frac{19.8}{34.3} \doteq 0.5773$)

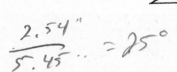

SELF-TEST 5 (Frames 61–71)

1. Find side a.

50°
34.8 ft.
a

a = ________

2. Find angle θ.

5.45"
2.54"
θ

θ = ________

ANSWERS: 1. a = 29.2 ft. 2. θ = 25°

1-7 THE SINE RATIO - A PARTIAL TABLE

The tangent of an angle is one of the three basic trigonometric ratios. The "sine" of an angle is a second basic trigonometric ratio. In this section, we will define what is meant by the sine of an angle. We will also introduce a partial sine table and use the values in it to find the unknown sides and angles in right triangles.

72. The tangent of an angle in a right triangle involves the "side opposite" the angle and the "side adjacent" to the angle. The other two basic trigonometric ratios involve the "hypotenuse".

The "sine" of an angle is a comparison of the "side opposite" the angle to the "hypotenuse". That is:

$$\text{THE SINE OF AN ANGLE} = \frac{\text{LENGTH OF THE SIDE OPPOSITE}}{\text{LENGTH OF THE HYPOTENUSE}}$$

or

$$\text{THE SINE OF AN ANGLE} = \frac{\text{SIDE OPPOSITE}}{\text{HYPOTENUSE}}$$

In right triangle ABC:

The side opposite angle A is "a".

The hypotenuse is "c".

Therefore, the sine of angle A is $\frac{a}{c}$.

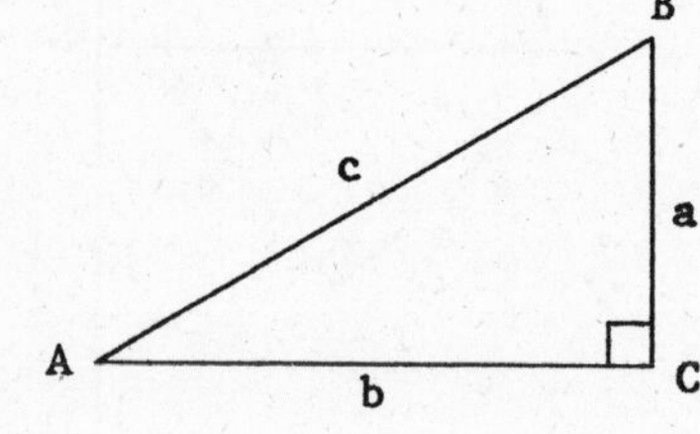

What ratio is the sine of angle B? $\frac{b}{c}$

73. The word "sine" is pronounced "sign". It is not pronounced "sin".

Using the expressions "side opposite" and "hypotenuse" of a right triangle, write the ratio for the sine of an angle:

The sine of an angle = $\frac{\text{side opp.}}{\text{hypotenuse}}$

Answer to 72: $\frac{b}{c}$

74. In right triangle TDM:

(a) The sine of angle T is $\frac{t}{d}$.

(b) The sine of angle M is $\frac{m}{d}$.

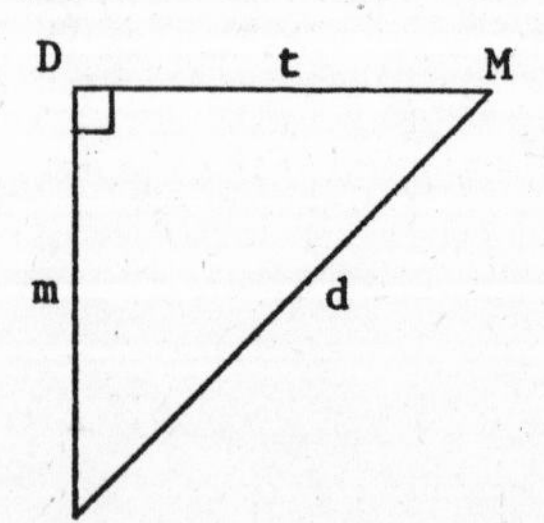

Answer to 73: $\frac{\text{side opposite}}{\text{hypotenuse}}$

Answer to 74:

a) $\frac{t}{d}$

b) $\frac{m}{d}$

75. In right triangle CFH:

(a) The sine of angle C is ______.

(b) The sine of angle H is ______.

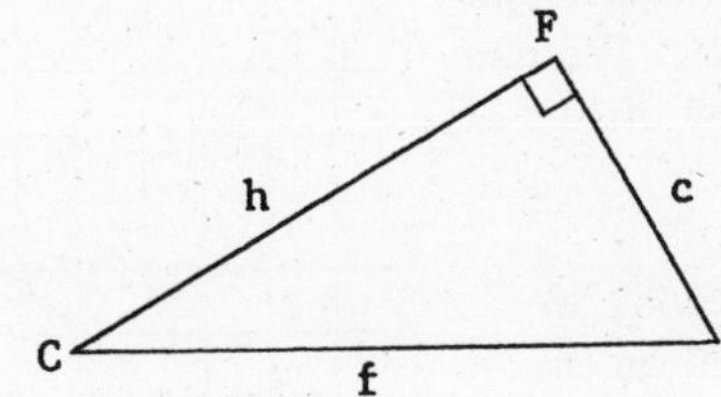

---

a) $\frac{c}{f}$

b) $\frac{h}{f}$

76. In the figure on the right, there are three right triangles: ABG, ACF, and ADE. Angle A, which is common to each triangle, contains 37°.

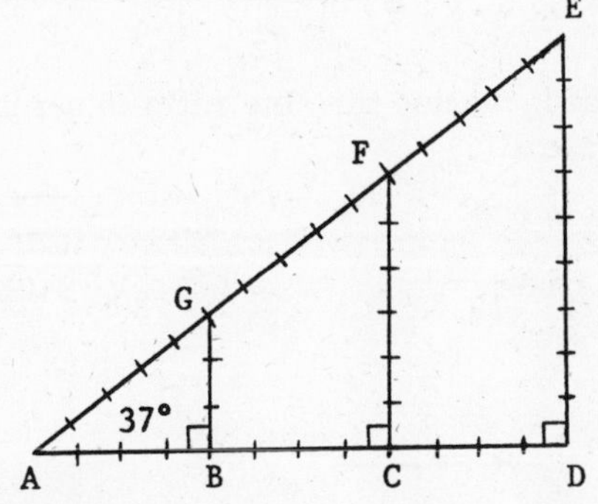

To show that the sine of a 37° angle has the same numerical value in right triangles of any size, we will compute numerical values of the following ratios by counting units:

(a) In triangle ABG, the sine of angle A = $\frac{BG}{AG}$ = ______.

(b) In triangle ACF, the sine of angle A = $\frac{CF}{AF}$ = ______.

(c) In triangle ADE, the sine of angle A = $\frac{DE}{AE}$ = ______.

(d) Even though the three right triangles have different sizes, is the sine of 37° the same in each? ______

---

a) $\frac{3}{5} = 0.6$

b) $\frac{6}{10} = \frac{3}{5} = 0.6$

c) $\frac{9}{15} = \frac{3}{5} = 0.6$

d) Yes

77. Just like the tangent of an angle, the sine of an angle has the same numerical value no matter what size the right triangle is, where the right angle is, or where the angle is. For example, the sine of a 37° angle is 0.6 in all cases.

(a) The sine of an angle is the ratio of the side opposite the angle to the ______.

(b) The tangent of an angle is the ratio of the side opposite the angle to the ______.

---

a) hypotenuse

b) side adjacent to the angle

78. Instead of writing "the sine of a 37° angle", mathematicians simply write "sin 37° ".

Note: Though the abbreviation of "sine" is "sin", the word is still pronounced "sign".

Just as "sin 37° "means "the sine of a 37° angle",

"sin 42° "means ______.

---

the sine of a 42° angle

79. In right triangle MFD, the hypotenuse "f" is 12.0". Angle M is 37°. We can find the length of side "m" by either of two methods:

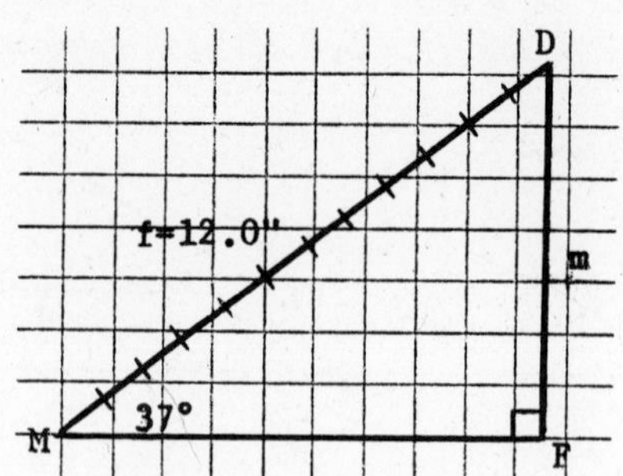

Scaling Method. In this method, we simply count the units in "m" on the scaled diagram. Side "m" is approximately 7.1 or 7.2 or 7.3 inches.

Algebraic Method. In this method, we use the sine ratio to set up an equation which is solved algebraically.

The sine of a 37° angle is 0.6.

The sine of angle M is $\frac{m}{f} = \frac{m}{12.0''}$.

Therefore: $\frac{m}{12.0''} = 0.6$

(a) Solving the equation, m = 7.2 inches.

(b) Which method gives a more exact answer? algebraic method

a) 7.2 inches

b) The algebraic method.

---

80. Mathematicians have also prepared a table of numerical values for the sines of angles. Here is part of that table. Note that each sine contains four significant digits.

Angle θ	sin θ
.	.
.	.
.	.
25°	0.4226
30°	0.5000
35°	0.5736
40°	0.6428
45°	0.7071
50°	0.7660
55°	0.8192
.	.
.	.
.	.

(a) If the sine of an angle is 0.5736, how large is the angle? 35°

(b) If $\sin \theta = 0.7660$, $\theta =$ 50°

(c) If $\sin \theta = 0.4226$, $\theta =$ 25°

(d) If $\theta = 45°$, $\sin \theta =$ 7071

a) 35°

b) 50°

c) 25°

d) 0.7071

81. In the table in the last frame, the sines are given as decimal numbers. Remember that these decimal numbers stand for ratios or comparisons.

For example, $\sin 55° = 0.8192 = \frac{8,192}{10,000}$

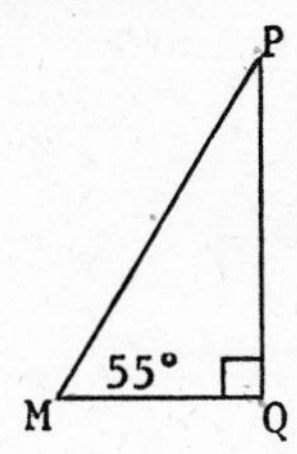

In right triangle MPQ:

If PQ = 8,192 miles,

then MP = 10,000 miles.

---

82. Using the partial table of sines, solve this one:

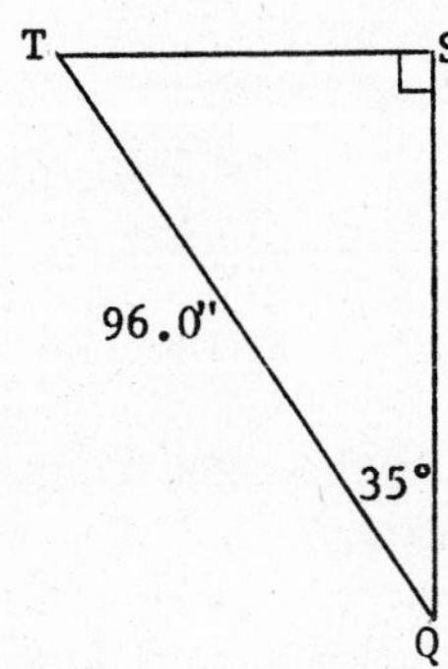

If angle Q is 35° and QT is 96.0", find TS.

(a) Set up the sine equation.

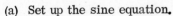

TS / 96.0 = .5736

(b) Solving the equation, TS = 55.1

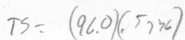

TS = (96.0)(.5736)

Answer to 81: 10,000 miles

---

83. Using the table, complete this one:

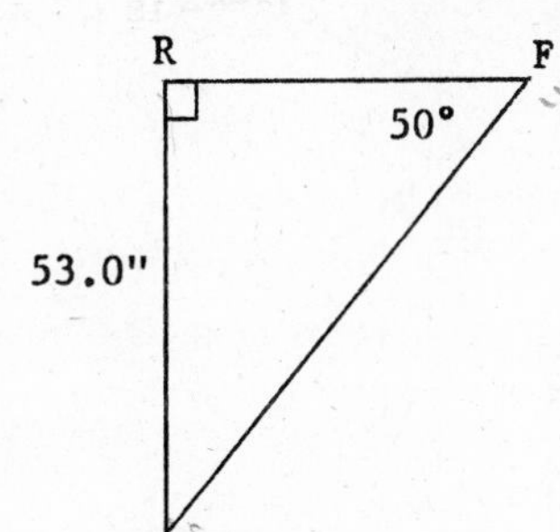

If angle F is 50° and RC is 53.0", find the length of FC.

(a) Set up the sine equation.

53.0 / FC = .7660

(b) Solving the equation, FC = 69.2

FC = (53.0) / .7660

Answer to 82:

a) $\frac{TS}{96.0''} = 0.5736$

b) TS = 55.1 inches

Answer to 83:

a) $\frac{53.0''}{FC} = 0.7660$

b) FC = 69.2 inches

84. We can also use the table for this one:

If PR is 70.7" and DR is 100", find angle D.

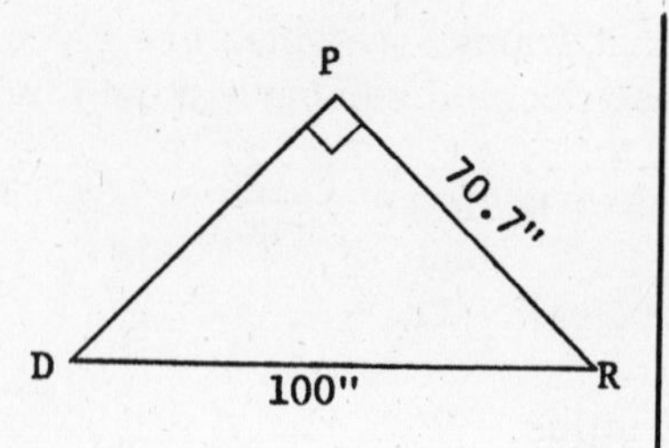

(a) The sine of angle D is $\frac{PR}{DR} = \frac{70.7''}{100''} =$ .707 .

(b) Therefore, angle D is 45° .

Answer to Frame 84: a) 0.707, which is approximately 0.7071 b) 45°

SELF-TEST 6 (Frames 72-84)

Using sides r, s, and t of the right triangle shown, define each of the following:

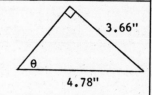

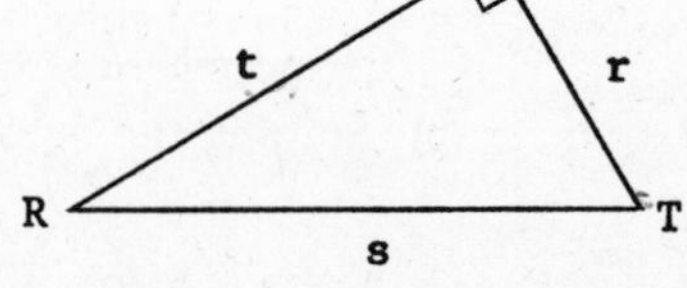

1. sin R = r/s 2. tan R = r/t 3. sin T = t/s 4. tan T = t/R

5. Find side w.

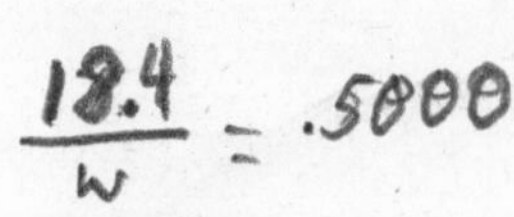

w = ______

6. Find angle θ.

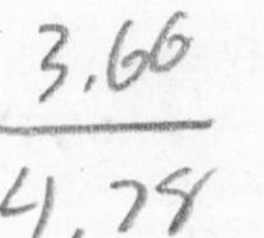

θ = ______

ANSWERS:

1. $\sin R = \frac{r}{s}$
2. $\tan R = \frac{r}{t}$
3. $\sin T = \frac{t}{s}$
4. $\tan T = \frac{t}{r}$
5. w = 36.8"
6. θ = 50°

1-8 THE COSINE RATIO - A PARTIAL TABLE

The cosine ratio is the third basic trigonometric ratio. We will define what is meant by the cosine of an angle in this section. A partial table of cosines will be introduced. The values in this table will be used to find the unknown sides and angles in right triangles.

---

85. The cosine of an angle is a comparison of the "side adjacent" to the angle to the "hypotenuse". That is:

$$\text{THE }\underline{\text{COSINE}}\text{ OF AN ANGLE} = \frac{\text{LENGTH OF THE SIDE ADJACENT}}{\text{LENGTH OF THE HYPOTENUSE}}$$

or

$$\text{THE }\underline{\text{COSINE}}\text{ OF AN ANGLE} = \frac{\text{SIDE ADJACENT}}{\text{HYPOTENUSE}}$$

In right triangle PRQ:

The side adjacent to angle P is "q".

The hypotenuse is "r".

Therefore, the cosine of angle P is $\frac{q}{r}$.

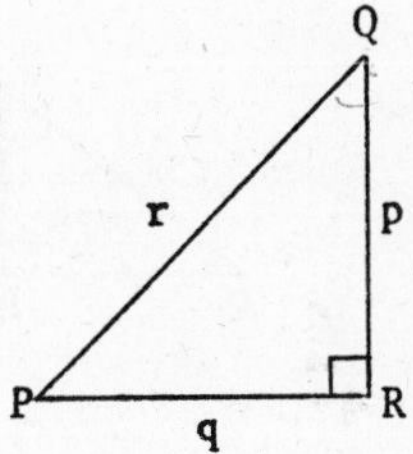

What ratio is the cosine of angle Q? ______

---

86. The word "cosine" is pronounced "co-sign". Its abbreviation is "cos". That is:

Just as "cos 37°" means "the cosine of a 37° angle",

"cos 77°" means ______________________.

Answer to 85: $\frac{p}{r}$

---

87. The sine of an angle is the ratio of the side opposite to the hypotenuse.

The cosine of an angle is the ratio of the ______________ to the hypotenuse.

Answer to 86: the cosine of a 77° angle

---

88. In right triangle CDB:

(a) cos B = ______

(b) cos C = ______

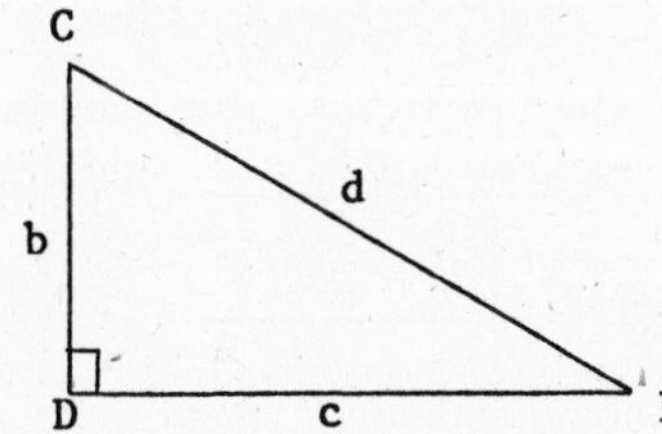

Answer to 87: side adjacent

Answer to 88:

a) $\frac{c}{d}$

b) $\frac{b}{d}$

89. In right triangle TAM:

(a) cos T = m/a

(b) cos M = _____

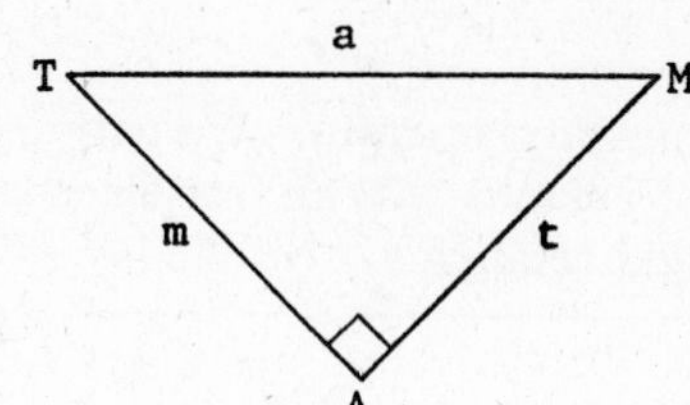

90. In the figure on the right, there are three right triangles: ABG, ACF, and ADE. Angle A, which is common to each triangle, contains 37°.

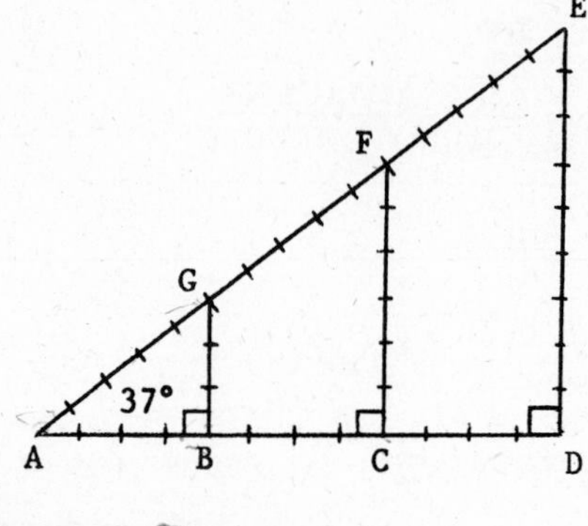

We can compute the cosine of angle A in each triangle by counting units.

(a) In triangle ABG, $\cos A = \frac{AB}{AG} =$ _____

(b) In triangle ACF, $\cos A = \frac{AC}{AF} =$ _____

(c) In triangle ADE, $\cos A = \frac{AD}{AE} =$ _____

(d) Even though the triangles have different sizes, is cos A the same in each case? _____

Answers to 89:

a) $\frac{m}{a}$

b) $\frac{t}{a}$

91. Just like the tangent and the sine of an angle, the numerical value of the cosine of an angle is the same no matter how large the triangle is, where the right angle is, or where the angle is. For example, the cosine of a 37° angle is 0.8 in all cases.

Mathematicians have prepared a table of values for the cosines of angles. Here is part of that table.

Angle θ	Cos θ
•	•
•	•
•	•
25°	0.9063
30°	0.8660
35°	0.8192
40°	0.7660
45°	0.7071
50°	0.6428
55°	0.5736
•	•
•	•
•	•

(a) All cosines in the table are written with an accuracy of _____ significant digits.

(b) $\cos 40° =$ _____

(c) If $\cos \theta = 0.7071$, $\theta =$ _____.

(d) If $\cos \theta = 0.9063$, $\theta =$ _____.

Answers to 90:

a) $\frac{4}{5} = 0.8$

b) $\frac{8}{10} = \frac{4}{5} = 0.8$

c) $\frac{12}{15} = \frac{4}{5} = 0.8$

d) Yes

92. The values in the partial table can be used to solve various problems. Here is an example:

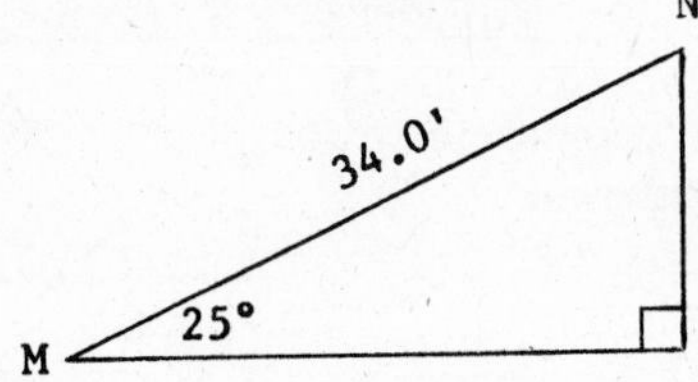

If angle M is 25° and MN is 34.0', find side MO.

(a) Set up the cosine equation.

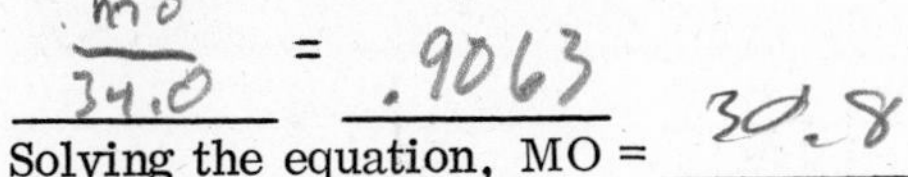

$\overline{\qquad} = \overline{\qquad}$

(b) Solving the equation, MO = ______

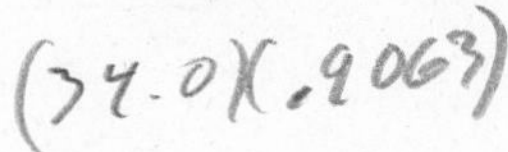

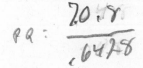

a) four

b) 0.7660

c) 45°

d) 25°

---

93. The table can be used to solve this one:

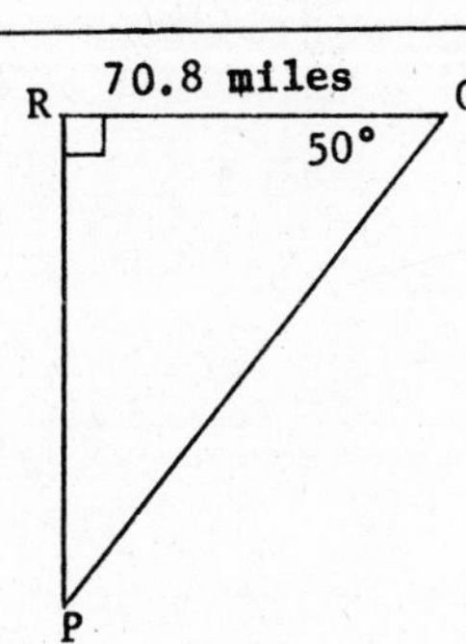

If angle Q is 50° and RQ is 70.8 miles, find PQ.

(a) Set up the cosine equation.

$\overline{\qquad} = \overline{\qquad}$

(b) Solving this equation, PQ = ______

a) $\dfrac{\text{MO}}{34.0'} = 0.9063$

b) MO = 30.8 feet

---

94. The table can also be used to solve this one:

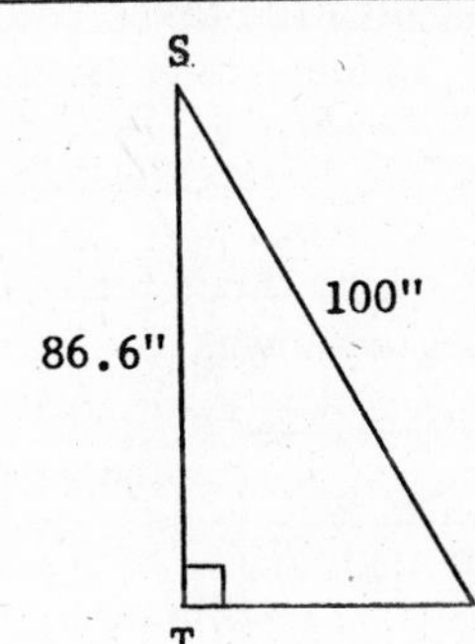

If ST is 86.6 inches and SM is 100 inches, find angle S.

(a) The cosine of angle S is $\dfrac{\text{ST}}{\text{SM}} = \dfrac{86.6''}{100''} =$ ______.

(b) Therefore, angle S is ______.

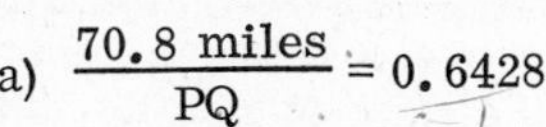

a) $\dfrac{70.8 \text{ miles}}{\text{PQ}} = 0.6428$

b) PQ = 110 miles

---

a) 0.8660

b) 30°

SELF-TEST 7 (Frames 85-94)

Define each of these trigonometric ratios, using the right triangle shown:

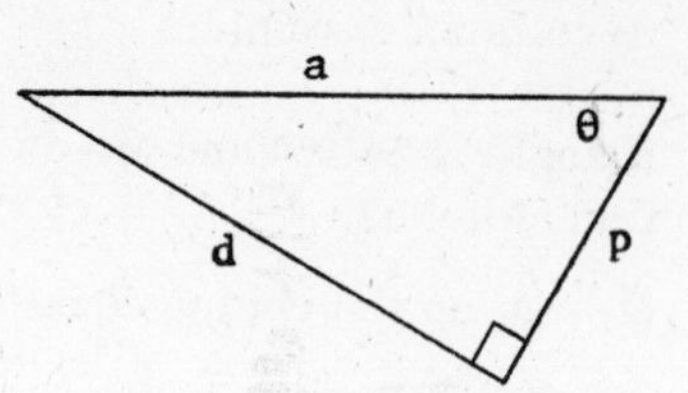

1. cos θ = ______ 2. sin θ = ______ 3. tan θ = ______

4. Find side b.

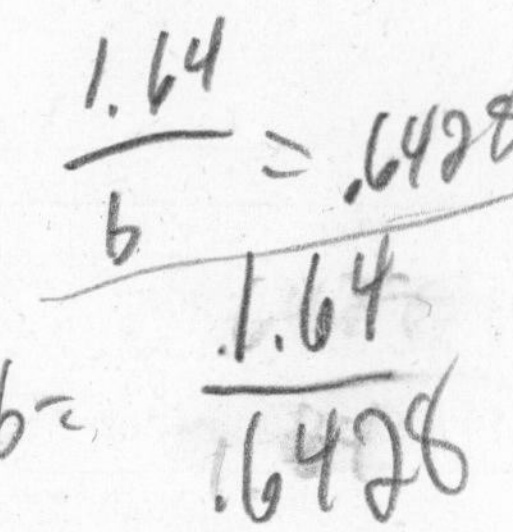

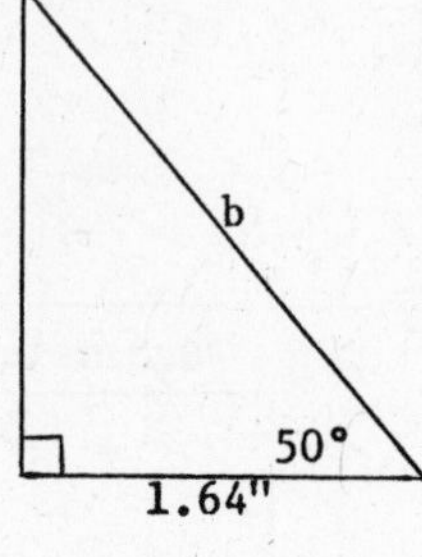

b = ______

5. Find angle G.

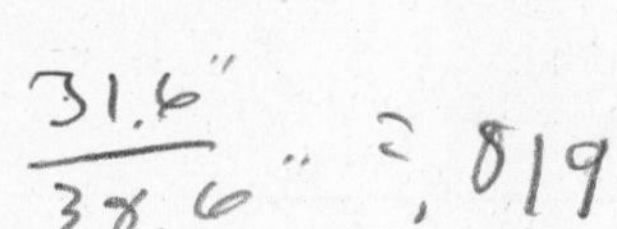

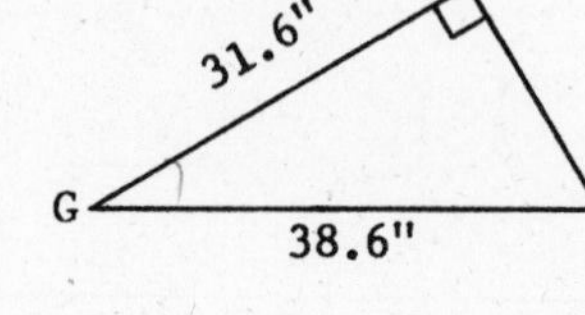

G = ______

ANSWERS: 1. $\cos\theta = \frac{p}{a}$ 2. $\sin\theta = \frac{d}{a}$ 3. $\tan\theta = \frac{d}{p}$ 4. b = 2.55" 5. G = 35°

1-9 CONTRASTING THE BASIC TRIGONOMETRIC RATIOS

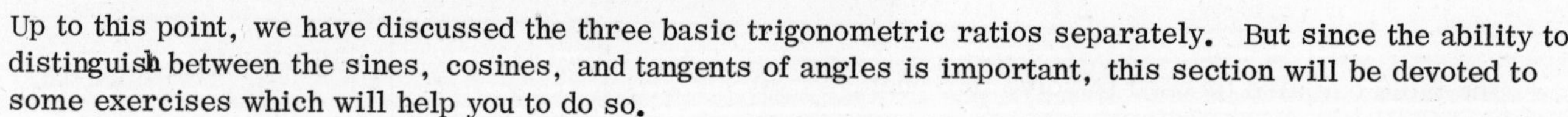

Up to this point, we have discussed the three basic trigonometric ratios separately. But since the ability to distinguish between the sines, cosines, and tangents of angles is important, this section will be devoted to some exercises which will help you to do so.

95. Here are the definitions of the three basic trigonometric ratios for angle θ in the right triangle shown:

$$\tan\theta = \frac{\text{side opposite angle }\theta}{\text{side adjacent to angle }\theta}$$

$$\sin\theta = \frac{\text{side opposite angle }\theta}{\text{hypotenuse}}$$

$$\cos\theta = \frac{\text{side adjacent to angle }\theta}{\text{hypotenuse}}$$

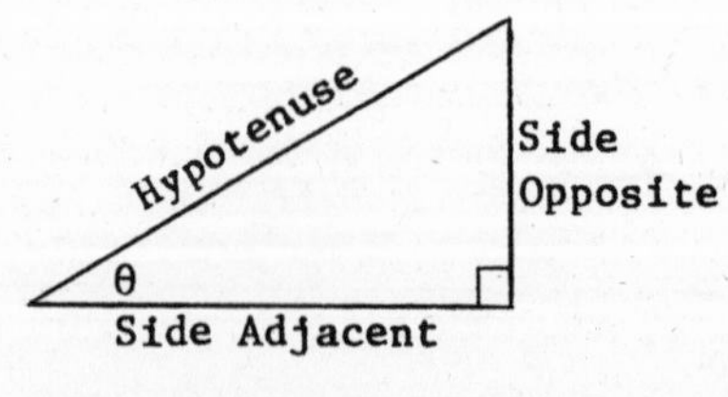

Go to next frame.

96. When writing the trig ratios for angles in a right triangle, it is helpful to locate the hypotenuse first. For example, in triangle CTR:

(a) The hypotenuse is ______.

(b) $\sin C =$ ______

(c) $\tan C =$ ______

(d) $\cos R =$ ______

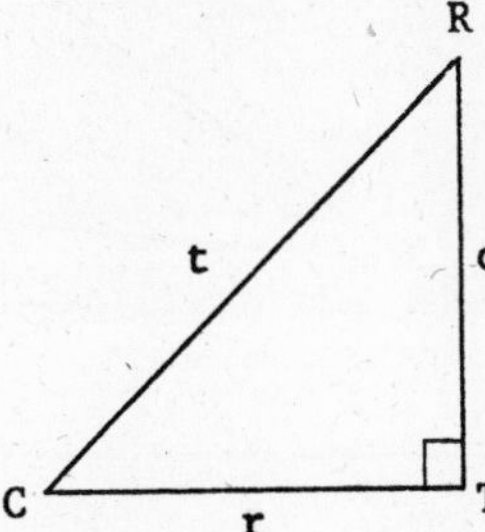

---

a) t

b) $\frac{c}{t}$

c) $\frac{c}{r}$

d) $\frac{c}{t}$

97. Use capital letters for the sides in defining these trig ratios:

(a) $\cos P =$ ______

(b) $\tan S =$ ______

(c) $\sin P =$ ______

(d) $\cos S =$ ______

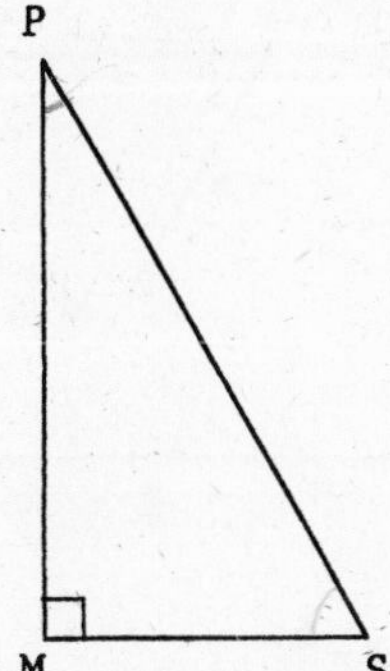

---

a) $\frac{PM}{PS}$

b) $\frac{PM}{MS}$

c) $\frac{MS}{PS}$

d) $\frac{MS}{PS}$

98. In mathematics, we frequently label angles with Greek letters. Some of the common letters used are:

α (alpha) - pronounced "al-fa"
β (beta) - pronounced "bay-ta"
γ (gamma) - pronounced "gam-ma"
θ (theta) - pronounced "thay-ta"
φ (phi) - pronounced "fee"

These Greek letters are general symbols which can stand for angles of any size.

Using capital letters for the sides, complete these:

(a) $\tan \alpha =$ ______

(b) $\tan \theta =$ ______

(c) $\sin \alpha =$ ______

(d) $\sin \theta =$ ______

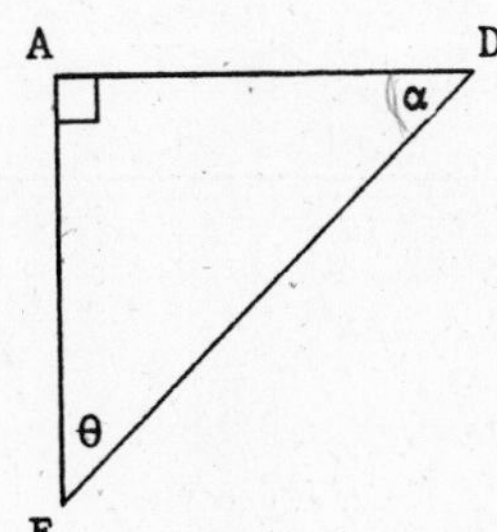

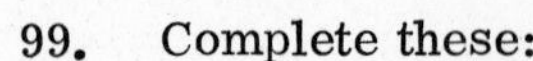

99. Complete these:

(a) $\tan \beta =$ ______

(b) $\tan \gamma =$ ______

(c) $\cos \beta =$ ______

(d) $\cos \gamma =$ ______

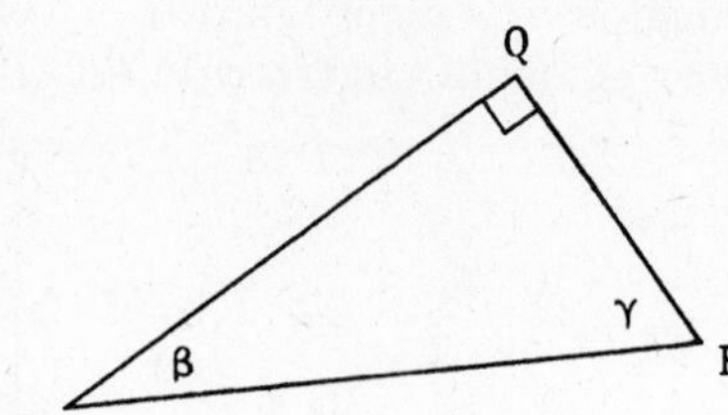

a) $\frac{AF}{AD}$

b) $\frac{AD}{AF}$

c) $\frac{AF}{DF}$

d) $\frac{AD}{DF}$

---

100. Use small letters to write these ratios:

(a) $\sin \theta =$ ______

(b) $\sin \beta =$ ______

(c) $\tan \theta =$ ______

(d) $\cos \beta =$ ______

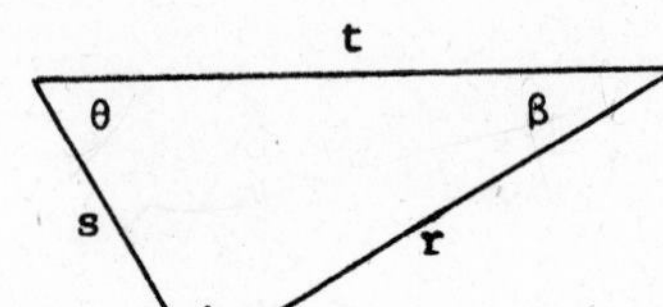

a) $\frac{QR}{PQ}$

b) $\frac{PQ}{QR}$

c) $\frac{PQ}{PR}$

d) $\frac{QR}{PR}$

---

101. Complete these:

(a) $\sin \beta =$ ______

(b) $\sin \gamma =$ ______

(c) $\tan \beta =$ ______

(d) $\cos \gamma =$ ______

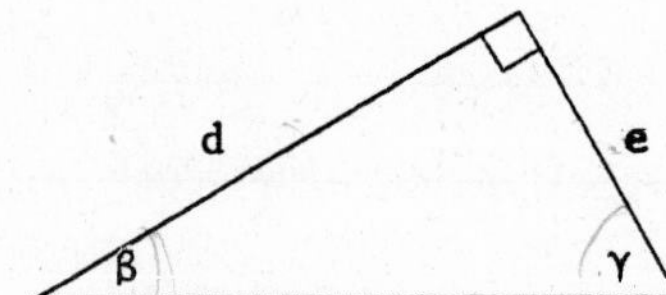

a) $\frac{r}{t}$

b) $\frac{s}{t}$

c) $\frac{r}{s}$

d) $\frac{r}{t}$

---

102. If a ratio does not involve the hypotenuse, it is a "tangent" ratio. For example, in right triangle FGH:

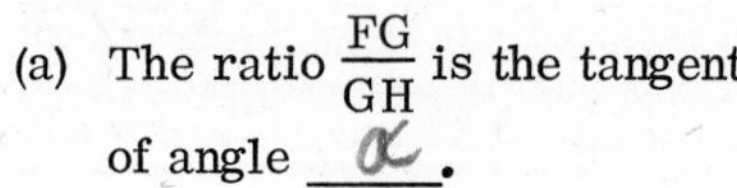

(a) The ratio $\frac{FG}{GH}$ is the tangent of angle ______.

(b) The ratio $\frac{GH}{FG}$ is the tangent of angle ______.

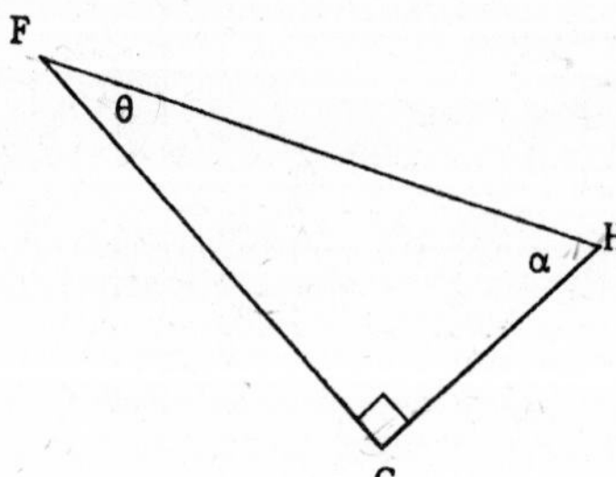

a) $\frac{e}{m}$

b) $\frac{d}{m}$

c) $\frac{e}{d}$

d) $\frac{e}{m}$

---

a) α

b) θ

103. If a ratio involves the hypotenuse, it is the sine of one angle and the cosine of the other. For example, in the triangle on the right:

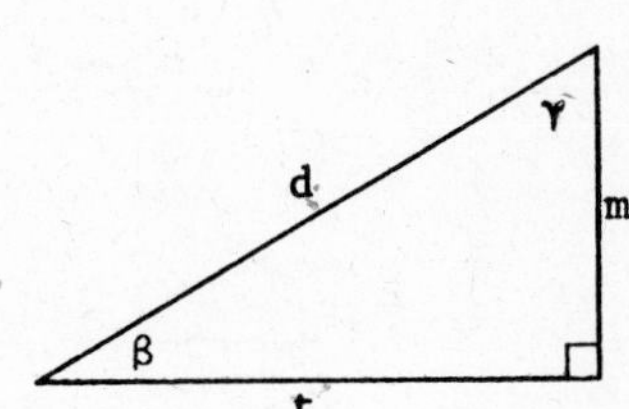

$\frac{m}{d}$ is the <u>sine</u> of angle β and the <u>cosine</u> of angle γ.

$\frac{t}{d}$ is the <u>sine</u> of angle __γ__ and the <u>cosine</u> of angle __β__.

---

sine of angle γ,
cosine of angle β

104. In this right triangle:

$\frac{MN}{NP}$ is $\tan \theta$.

(a) $\frac{NP}{MN}$ is __Tan β__.

(b) $\frac{MN}{MP}$ is either ______ or ______.

---

a) $\tan \beta$

b) $\sin \theta$ or $\cos \beta$

105. In this right triangle:

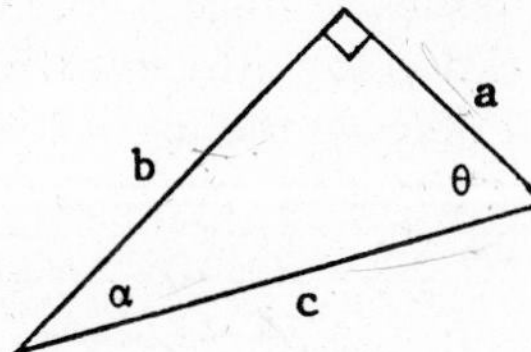

(a) $\frac{a}{b}$ = __tan α__

(b) $\frac{b}{a}$ = __tan θ__

(c) $\frac{a}{c}$ = ______ or ______.

---

a) $\tan \alpha$

b) $\tan \theta$

c) $\sin \alpha$ or $\cos \theta$

106. In this right triangle:

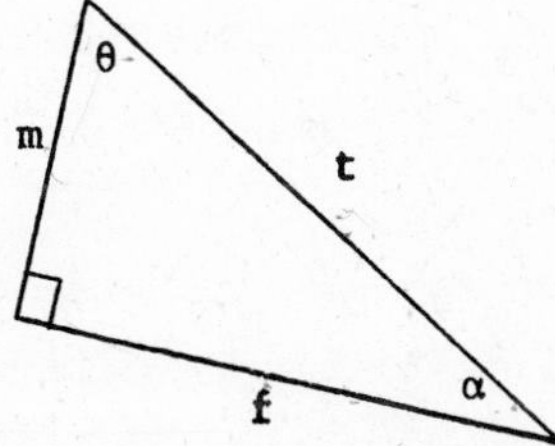

(a) $\frac{m}{t}$ = __cos θ__ or __sin α__

(b) $\frac{f}{t}$ = ______ or ______

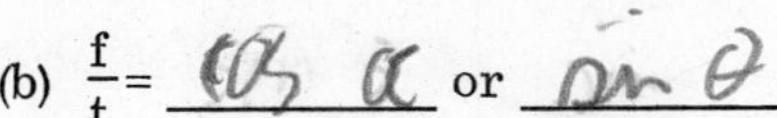

---

a) $\sin \alpha$ or $\cos \theta$

b) $\sin \theta$ or $\cos \alpha$

SELF-TEST 8 (Frames 95-106)

In the right triangle shown:

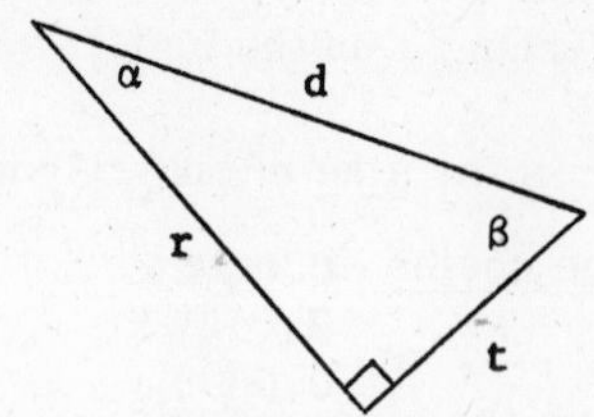

1. $\sin \alpha =$ $\frac{t}{d}$
2. $\cos \beta =$ $\frac{t}{d}$
3. $\sin \beta =$ $\frac{r}{d}$
4. $\frac{r}{t} =$ tan β
5. $\frac{t}{r} =$ tan α

ANSWERS: 1. $\frac{t}{d}$ 2. $\frac{t}{d}$ 3. $\frac{r}{d}$ 4. $\tan \beta$ 5. $\tan \alpha$

1-10 THE COMPLETE TABLE OF TRIG RATIOS

We have seen and used some partial tables of sines, cosines, and tangents. Mathematicians have computed the numerical values of these three ratios for all angles from 0° to 90°. These values are presented in a single table. In this section, we will discuss the use of this complete table. The values in the table will be used to make some general comments about the numerical values of the three basic ratios.

---

107. The complete table of trig ratios for angles from 0° to 90° is given on the next page. Refer to this table and answer the following:

(a) $\sin 11° =$ .1908
(b) $\cos 82° =$ .1392
(c) $\tan 65° =$ 2.145
(d) If $\sin \theta = 0.9986$, $\theta =$ 87
(e) If $\cos \theta = 0.8829$, $\theta =$ 28
(f) If $\tan \theta = 0.7536$, $\theta =$ 37

---

108. Use the table to complete these:

(a) $\sin 0° =$ 0.000 $\sin 90° =$ 1.000

As the angle increases from 0° to 90°, does the sine increase or decrease? increase

(b) $\cos 0° =$ 1.000 $\cos 90° =$ 0.000

As the angle increases from 0° to 90°, does the cosine increase or decrease? decreases

(c) $\tan 0° =$ 0.000 $\tan 89° =$ ______

As the angle increases from 0° to 89°, does the tangent increase or decrease? increases

a) 0.1908 d) 87°
b) 0.1392 e) 28°
c) 2.145 f) 37°

TRIGONOMETRIC RATIOS

Angle θ	Sin θ	Cos θ	Tan θ	Angle θ	Sin θ	Cos θ	Tan θ
0°	0.0000	1.0000	0.0000	45°	0.7071	0.7071	1.0000
1°	0.0175	0.9998	0.0175	46°	0.7193	0.6947	1.036
2°	0.0349	0.9994	0.0349	47°	0.7314	0.6820	1.072
3°	0.0523	0.9986	0.0524	48°	0.7431	0.6691	1.111
4°	0.0698	0.9976	0.0699	49°	0.7547	0.6561	1.150
5°	0.0872	0.9962	0.0875	50°	0.7660	0.6428	1.192
6°	0.1045	0.9945	0.1051	51°	0.7771	0.6293	1.235
7°	0.1219	0.9925	0.1228	52°	0.7880	0.6157	1.280
8°	0.1392	0.9903	0.1405	53°	0.7986	0.6018	1.327
9°	0.1564	0.9877	0.1584	54°	0.8090	0.5878	1.376
10°	0.1736	0.9848	0.1763	55°	0.8192	0.5736	1.428
11°	0.1908	0.9816	0.1944	56°	0.8290	0.5592	1.483
12°	0.2079	0.9781	0.2126	57°	0.8387	0.5446	1.540
13°	0.2250	0.9744	0.2309	58°	0.8480	0.5299	1.600
14°	0.2419	0.9703	0.2493	59°	0.8572	0.5150	1.664
15°	0.2588	0.9659	0.2679	60°	0.8660	0.5000	1.732
16°	0.2756	0.9613	0.2867	61°	0.8746	0.4848	1.804
17°	0.2924	0.9563	0.3057	62°	0.8829	0.4695	1.881
18°	0.3090	0.9511	0.3249	63°	0.8910	0.4540	1.963
19°	0.3256	0.9455	0.3443	64°	0.8988	0.4384	2.050
20°	0.3420	0.9397	0.3640	65°	0.9063	0.4226	2.145
21°	0.3584	0.9336	0.3839	66°	0.9135	0.4067	2.246
22°	0.3746	0.9272	0.4040	67°	0.9205	0.3907	2.356
23°	0.3907	0.9205	0.4245	68°	0.9272	0.3746	2.475
24°	0.4067	0.9135	0.4452	69°	0.9336	0.3584	2.605
25°	0.4226	0.9063	0.4663	70°	0.9397	0.3420	2.747
26°	0.4384	0.8988	0.4877	71°	0.9455	0.3256	2.904
27°	0.4540	0.8910	0.5095	72°	0.9511	0.3090	3.078
28°	0.4695	0.8829	0.5317	73°	0.9563	0.2924	3.271
29°	0.4848	0.8746	0.5543	74°	0.9613	0.2756	3.487
30°	0.5000	0.8660	0.5774	75°	0.9659	0.2588	3.732
31°	0.5150	0.8572	0.6009	76°	0.9703	0.2419	4.011
32°	0.5299	0.8480	0.6249	77°	0.9744	0.2250	4.331
33°	0.5446	0.8387	0.6494	78°	0.9781	0.2079	4.705
34°	0.5592	0.8290	0.6745	79°	0.9816	0.1908	5.145
35°	0.5736	0.8192	0.7002	80°	0.9848	0.1736	5.671
36°	0.5878	0.8090	0.7265	81°	0.9877	0.1564	6.314
37°	0.6018	0.7986	0.7536	82°	0.9903	0.1392	7.115
38°	0.6157	0.7880	0.7813	83°	0.9925	0.1219	8.144
39°	0.6293	0.7771	0.8098	84°	0.9945	0.1045	9.514
40°	0.6428	0.7660	0.8391	85°	0.9962	0.0872	11.43
41°	0.6561	0.7547	0.8693	86°	0.9976	0.0698	14.30
42°	0.6691	0.7431	0.9004	87°	0.9986	0.0523	19.08
43°	0.6820	0.7314	0.9325	88°	0.9994	0.0349	28.64
44°	0.6947	0.7193	0.9657	89°	0.9998	0.0175	57.29
45°	0.7071	0.7071	1.0000	90°	1.0000	0.0000	-----

109. Answer the following by referring to the table:

(a) For two ratios of angles between 0° and 90°, the decimal value is some number between 0.0000 and 1.0000. The names of these two ratios are ________ and ________.

(b) For one ratio, the decimal value can be a number greater than 1. The name of this ratio is ________.

a) sin 0° = 0.0000
sin 90° = 1.0000
Increases

b) cos 0° = 1.0000
cos 90° = 0.0000
Decreases

c) tan 0° = 0.0000
tan 89° = 57.29
Increases

---

110. $\frac{x}{10}$ is a ratio.

(a) The value of this ratio is 1 when x is ________.

(b) When is the value of this ratio greater than 1? ________

(c) When is the value of this ratio less than 1? ________

a) sine and cosine

b) tangent

---

111. (a) In a right triangle, can a leg ever be longer than the hypotenuse? ________

(b) In right triangle ABC, hypotenuse AB = 100".

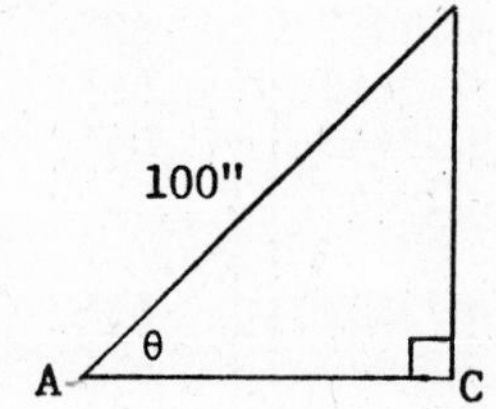

$\sin \theta = \frac{BC}{100''}$

$\cos \theta = \frac{AC}{100''}$

Since legs BC and AC cannot be greater than 100", the sine and cosine ratios cannot be greater than ________.

a) 10

b) When x is greater than 10.

c) When x is less than 10.

---

112. Here are two right triangles:

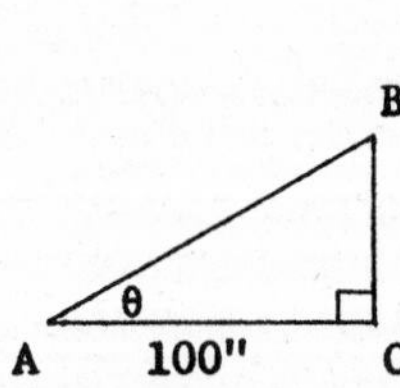

Triangle #1

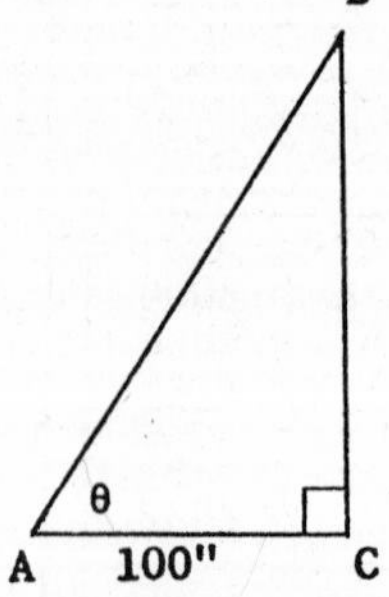

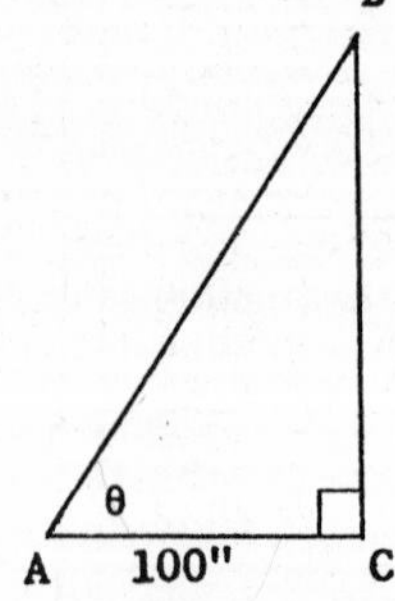

Triangle #2

In each triangle: $\tan \theta = \frac{BC}{AC} = \frac{BC}{100''}$

(a) The ratio $\frac{BC}{100''}$ will be greater than 1 when BC is greater than ________.

(b) In which triangle is BC obviously greater than AC? ________

(c) Therefore, in which triangle is tan θ greater than 1? ________

a) No

b) 1

113. In right triangle PQR, angle P is 45°.

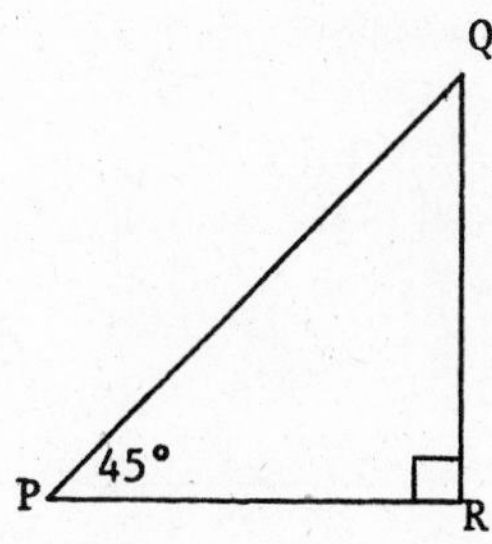

(a) From the table, $\tan 45° =$ ________.

(b) $\tan P = \frac{QR}{PR}$. This ratio equals 1.0000 only under what condition? ________________

(c) Since QR = PR in the triangle above, triangle PQR is an ____________ triangle.

a) 100"
b) #2
c) #2

---

114. (a) The sine and cosine of an angle cannot be greater than 1 because the side opposite and the side adjacent cannot be greater than the ____________.

(b) The tangent of an angle can be greater than 1 because the side opposite can be greater than the ________________.

a) 1.0000
b) When QR = PR.
c) isosceles

Answer to Frame 114: a) hypotenuse b) side adjacent

SELF-TEST 9 (Frames 107-114)

Answer the following by referring to the trig table:

1. cos 59° = ________	4. If tan θ = 1.000, θ = ______.
2. tan 87° = ________	5. If sin θ = 0.9511, θ = ______.
3. sin 12° = ________	6. If cos θ = 0.9511, θ = ______.

ANSWERS: 1. 0.5150 2. 19.08 3. 0.2079 4. 45° 5. 72° 6. 18°

1-11 FINDING UNKNOWN SIDES IN RIGHT TRIANGLES

The lengths of unknown sides of right triangles can be found either by using one of the trig ratios or by using the Pythagorean Theorem. In this section, we will begin with the cases in which a trig ratio must be used. Then we will review the cases in which the Pythagorean Theorem should be used.

115. In a right triangle, the hypotenuse is always the longest side. Therefore, in the right triangle shown, we know that both "m" and "d" are less than 11.5.

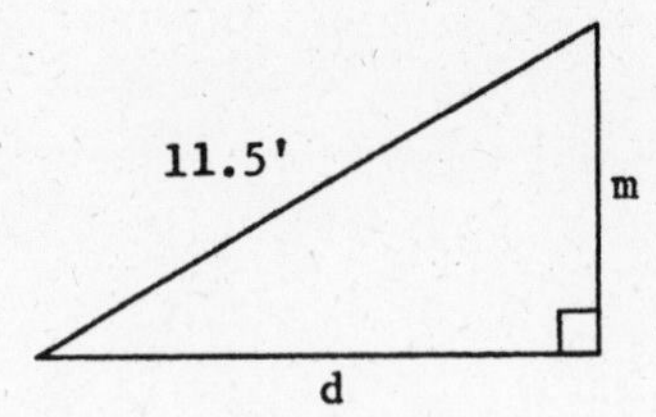

---

116. Suppose someone computes the length of "h" in the triangle on the right and gets 1.72". What is wrong with this answer?
It is too small
it must be greater than 1.98

a=1.57"
c=1.98"
h

11.5'

---

117. In a right triangle, the leg opposite the larger acute angle is the longer leg.

In the right triangle shown, which leg must be longer, "a" or "b"? b

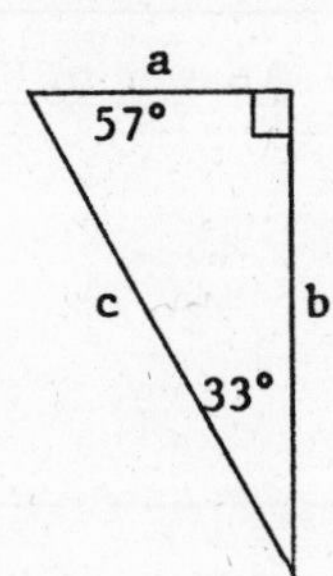

"h" must be greater than 1.98" since "h" is the longest side.

---

118. In right triangle BDF, DF is 27.9'.

(a) Could BF be 23.6'? NO

(b) Could BD be 29.1'? NO

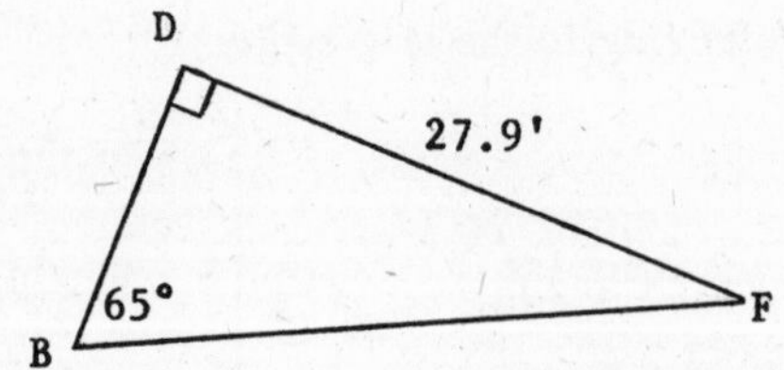

Leg "b", since it is opposite the 57° angle.

---

119. When computing the sides of triangles, always check to see whether your computed length makes sense.

In right triangle ACB:

(a) Does a computed length of 6.5" make sense for "b"? No

(b) Does a computed length of 12.3" make sense for "c"? Yes

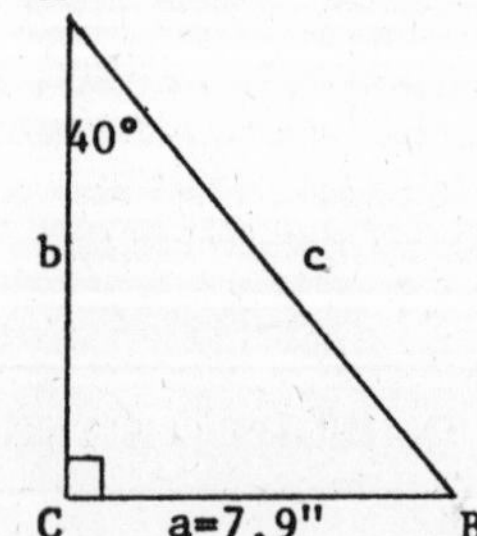

a) No. BF is the hypotenuse. It must be longer than 27.9'.

b) No. BD must be shorter than DF since it is opposite a smaller angle.

120. When dealing with triangles, the legs and hypotenuse are lengths. These lengths are measurements. When solving for unknown lengths, we must report the computed length with the correct number of significant digits.

Here is an example of a case in which we can use the sine ratio to find BC.

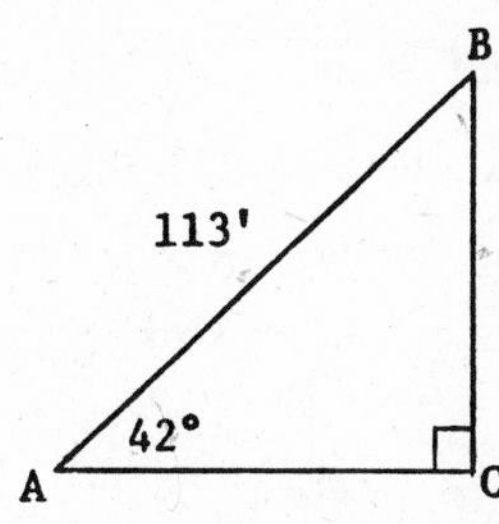

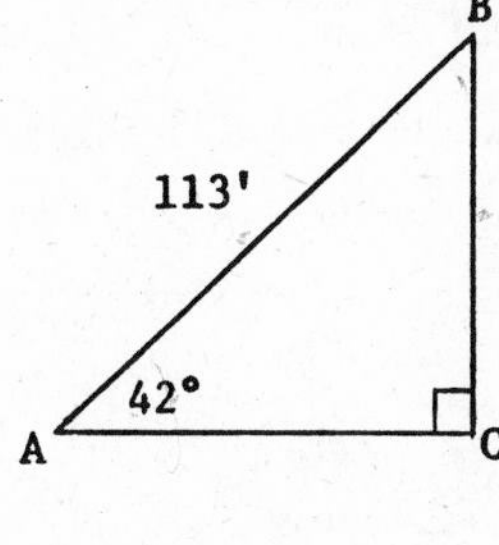

(1) From the table, sin 42° = 0.6691.

The value "0.6691" is not a measurement. It is a pure number. We don't have to worry about the number of significant digits in it.

(2) AB is 113 feet. 113 feet is a measurement with three significant digits. Therefore, when we compute the length of BC, we should report three significant digits.

$$\sin A = \frac{BC}{AB}$$

$$\sin 42° = \frac{BC}{113'}$$

$$0.6691 = \frac{BC}{113'}$$

$$BC = (113')(0.6691)$$

Use your slide rule. To do so, round 0.6691 to 0.669.

BC = 75.6 feet

a) No. "b" must be larger than "a" since "b" is opposite a larger angle.

b) Yes. "c" is the hypotenuse and therefore it is larger than either leg.

---

121. To find PS in this right triangle, we can use the cosine ratio. The solution is:

$$\cos P = \frac{PR}{PS}$$

$$\cos 70° = \frac{12.3''}{PS}$$

$$0.3420 = \frac{12.3''}{PS}$$

$$PS = \frac{12.3''}{0.3420}$$

Use your slide rule and report the answer to three significant digits.
PS = 36

75.6 feet

122. Before using a trig ratio to find an unknown side of a right triangle, we must decide which of the three ratios (sine, cosine, or tangent) to use.

36.0 inches

We will discuss the strategy for making this decision by working an actual problem. The problem is: Find "p" in right triangle PRQ.

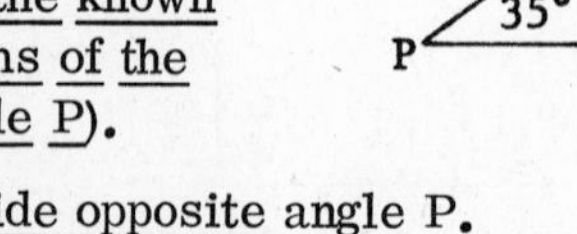

Step 1: Identify the side you want to find ("p") and the known side ("q") in terms of the known angle (angle P).

"p" is the side opposite angle P.
"q" is the side adjacent to angle P.

Step 2: Determine which ratio of angle P includes both "side opposite" and "side adjacent".

The ratio is the "tangent" of angle P.

Step 3: Use this ratio to set up the trig-ratio equation.

$$\tan P = \frac{p}{q} \quad \text{or} \quad \tan 35° = \frac{p}{19.0''}$$

Using the trig table to find tan 35°, complete the solution.

$.7002 = \frac{p}{19.0} \rightarrow p = (.7002)(19.0)$

p = 13.3"

---

123. Let's follow the same strategy to determine which ratio to use to find TS in right triangle TVS.

13.3 inches

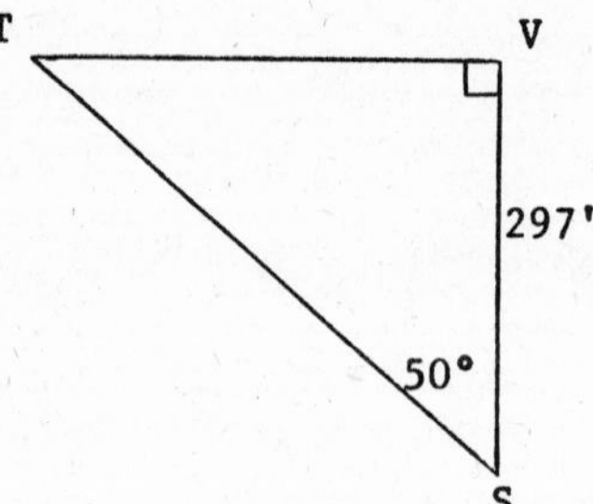

Step 1: Identify the side you want to find (TS) and the known side (VS) in terms of the known angle (angle S).

TS is the hypotenuse.
VS is the side adjacent to angle S.

Step 2: Determine which ratio of angle S includes both "side adjacent" and "hypotenuse".

The ratio is the "cosine" of angle S.

Step 3: Using this ratio, set up the trig equation.

$$\cos S = \frac{VS}{TS} \quad \text{or} \quad \cos 50° = \frac{297'}{TS}$$

Using the trig table to find cos 50°, complete the solution.

$.6428 = \frac{297'}{TS} \rightarrow TS = \frac{297}{.6428}$

TS = 462'

462 feet

(from $\frac{297'}{0.6428}$)

---

124. We want to find DF in right triangle DFE. The known angle is D.

"DF" is the <u>side</u> <u>adjacent</u> to angle D.
"FE" is the <u>side</u> <u>opposite</u> angle D.

Which trig ratio of angle D includes both "side adjacent" and "side opposite"?

Tan D

D
27°
F 11.7 miles E

tan D

---

125. We want to find side "b" in right triangle ABC. The known angle is A.

"b" is the <u>hypotenuse.</u>
"a" is the <u>side</u> <u>opposite</u> angle A.

Which trig ratio of angle A includes both "hypotenuse" and "side opposite"?

sin A

B
c
a=9.77"
25°
A b C

sin A

---

126. To solve for "t", should you use sin 35°, cos 35°, or tan 35°?

Tan 35

q
r=7.69'
35°
t

tan 35°

---

127. To solve for "h", should you use sin 67°, cos 67°, or tan 67°?

cos 67°

h
67°
7.9 miles

cos 67°

---

128. To solve for "n", should you use sin 27°, cos 27°, or tan 27°?

sin 27°

m
n
27°
d=10.7"

sin 27°

129. Set up the trig equation needed to solve for "m" in this right triangle. Do not solve.

sin 35 = $\frac{m}{12.7}$

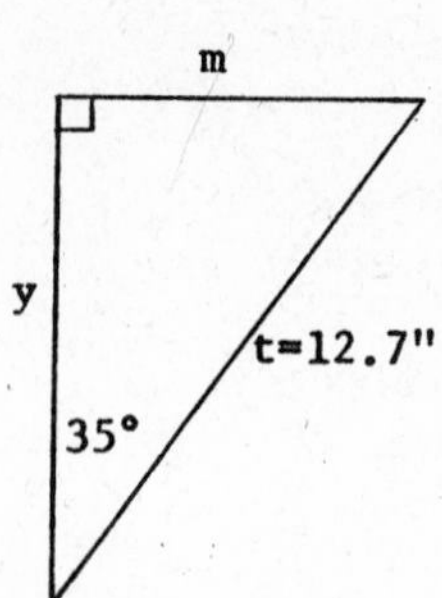

---

130. Set up the trig equation needed to solve for "g" in this right triangle. Do not solve.

tan 38° = $\frac{g}{5.9}$

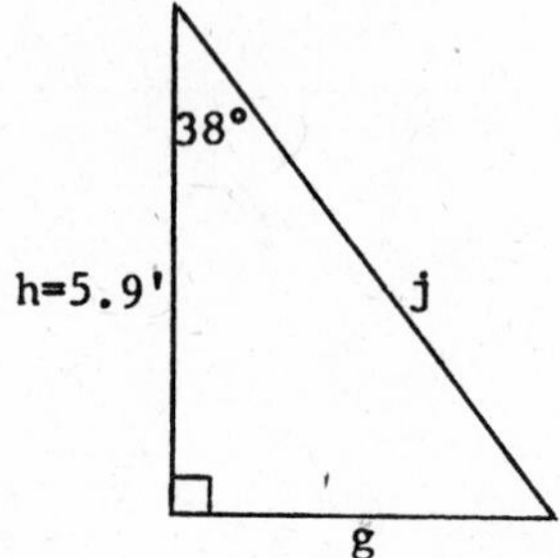

$\sin 35° = \frac{m}{12.7''}$

---

131. Set up the trig equation needed to solve for "c" in this right triangle. Do not solve.

cos 52 = $\frac{12.4}{c}$

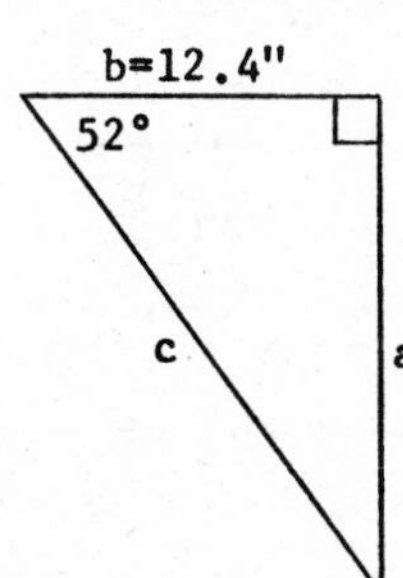

$\tan 38° = \frac{g}{5.9'}$

---

132. (a) Set up the trig equation needed to solve for "p" in this right triangle.

sin 75° = $\frac{p}{5.6}$

(b) Solve for "p".

$.9659 = \frac{p}{5.6} \Rightarrow p = (.9659)(5.6)$

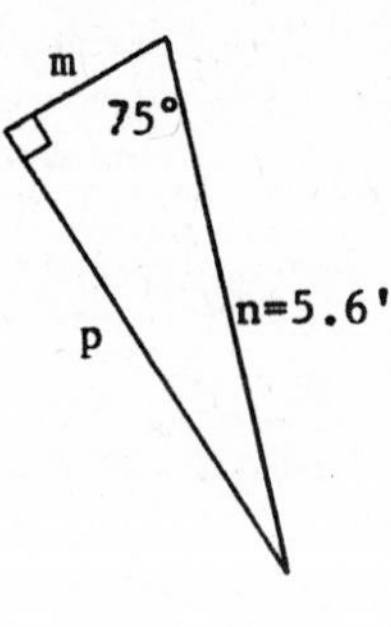

p = 5.4

$\cos 52° = \frac{12.4''}{c}$

---

a) $\sin 75° = \frac{p}{5.6'}$

b) p = 5.4 feet (two significant digits)

133. If DH is 58.9":

(a) Set up the trig equation needed to solve for HM.

_______ = _______

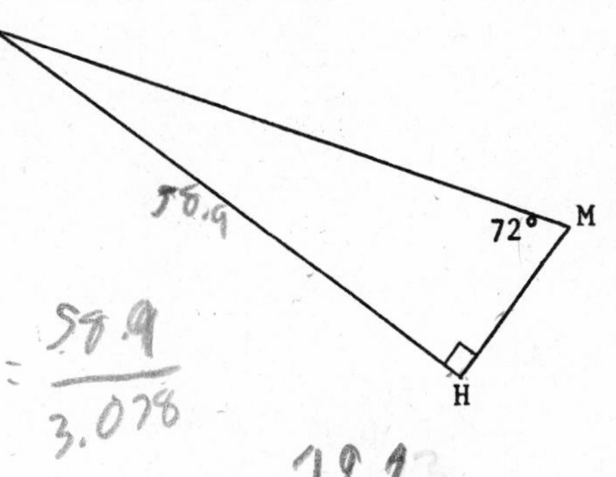

(b) Solve for HM.

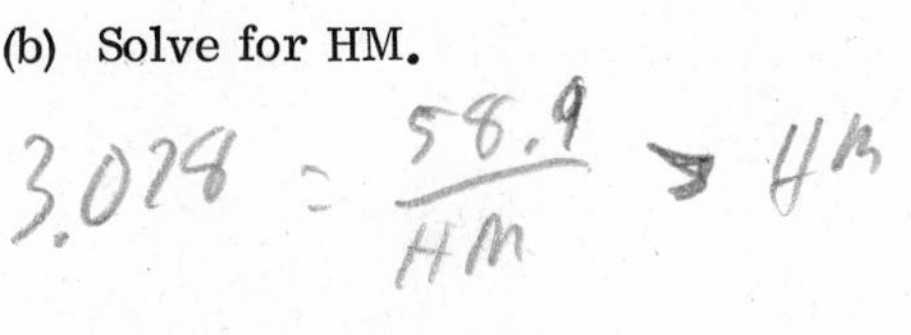

HM = _______

---

134. If LM is 546 miles:

(a) Set up the trig equation needed to solve for LN.

_______ = _______

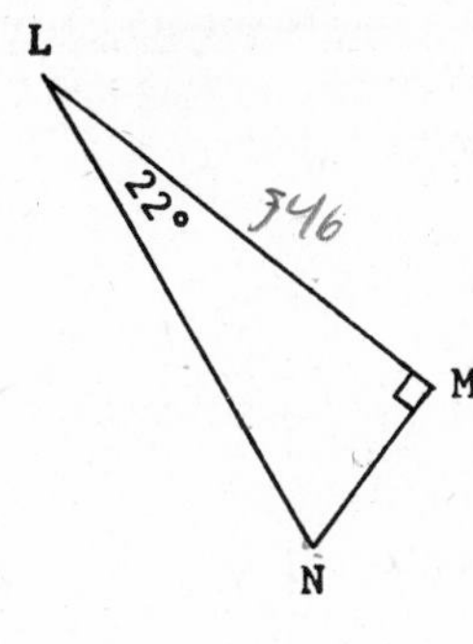

(b) Solve for LN.

LN = _______

a) $\tan 72° = \frac{58.9''}{HM}$

b) HM = 19.1 inches (from $\frac{58.9''}{3.078}$)

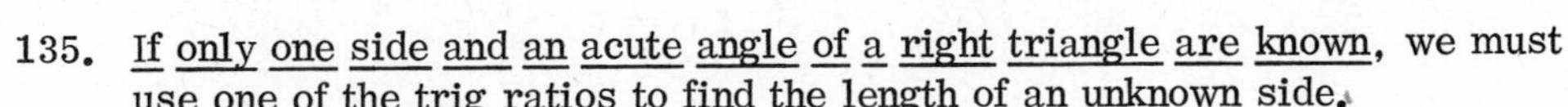

135. If only one side and an acute angle of a right triangle are known, we must use one of the trig ratios to find the length of an unknown side.

If only two sides of a right triangle are known, we should use the Pythagorean Theorem to find the third side.

In which triangles below would we use the Pythagorean Theorem in order to find the length of MP? _______

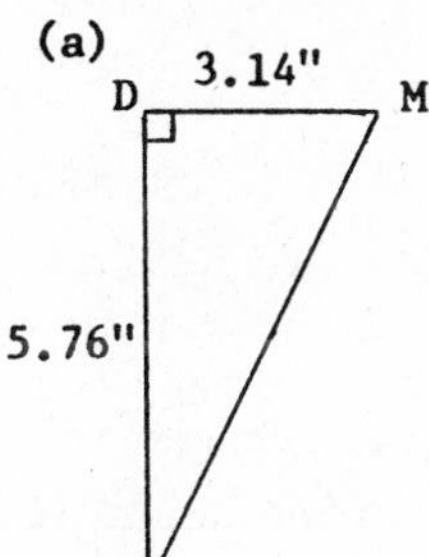

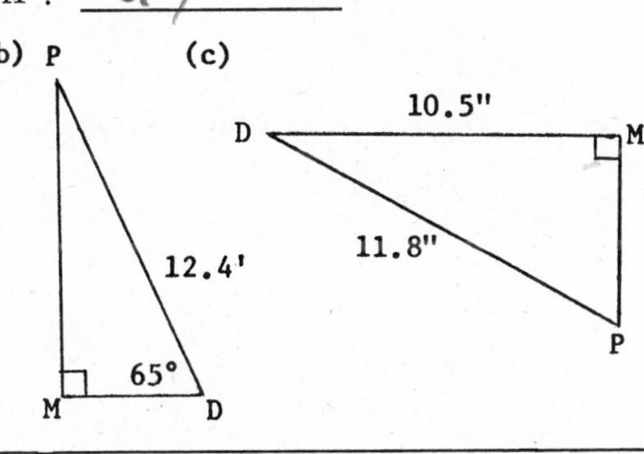

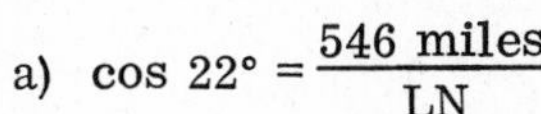

a) $\cos 22° = \frac{546 \text{ miles}}{LN}$

b) LN = 589 miles

In (a) and (c)

136. In which triangles below would we have to use a trigonometric ratio to find the length of CD? A, B

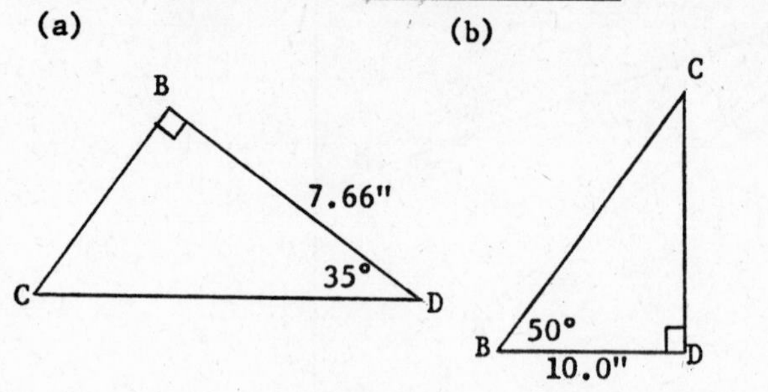

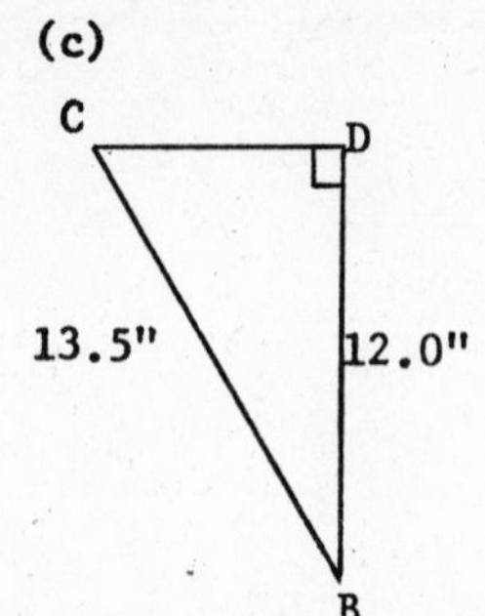

---

137. If we know two sides plus one acute angle of a right triangle, we can use either of two trig ratios or the Pythagorean Theorem to find the third side. Here is an example:

(a) Use cos 33° to find the length of "t".

p=21.7"　m=11.8"　33°　t

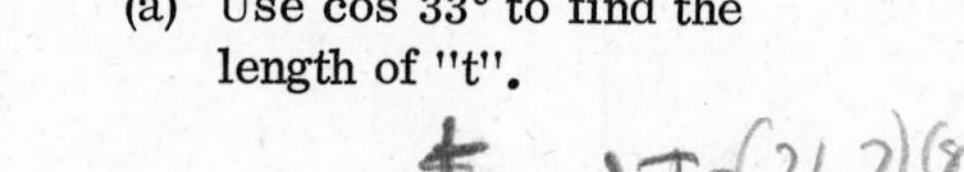

t = 18.2

(b) Use tan 33° to find the length of "t".

t = 18.2

(c) Use the Pythagorean Theorem to find the length of "t".

t = 11.2

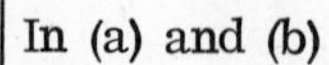
In (a) and (b)

---

a) $t = 18.2''$ (from $\cos 33° = \frac{t}{21.7''}$)

b) $t = 18.2''$ (from $\tan 33° = \frac{11.8''}{t}$)

c) $t = 18.2''$ [from $t^2 = (21.7'')^2 - (11.8'')^2$]

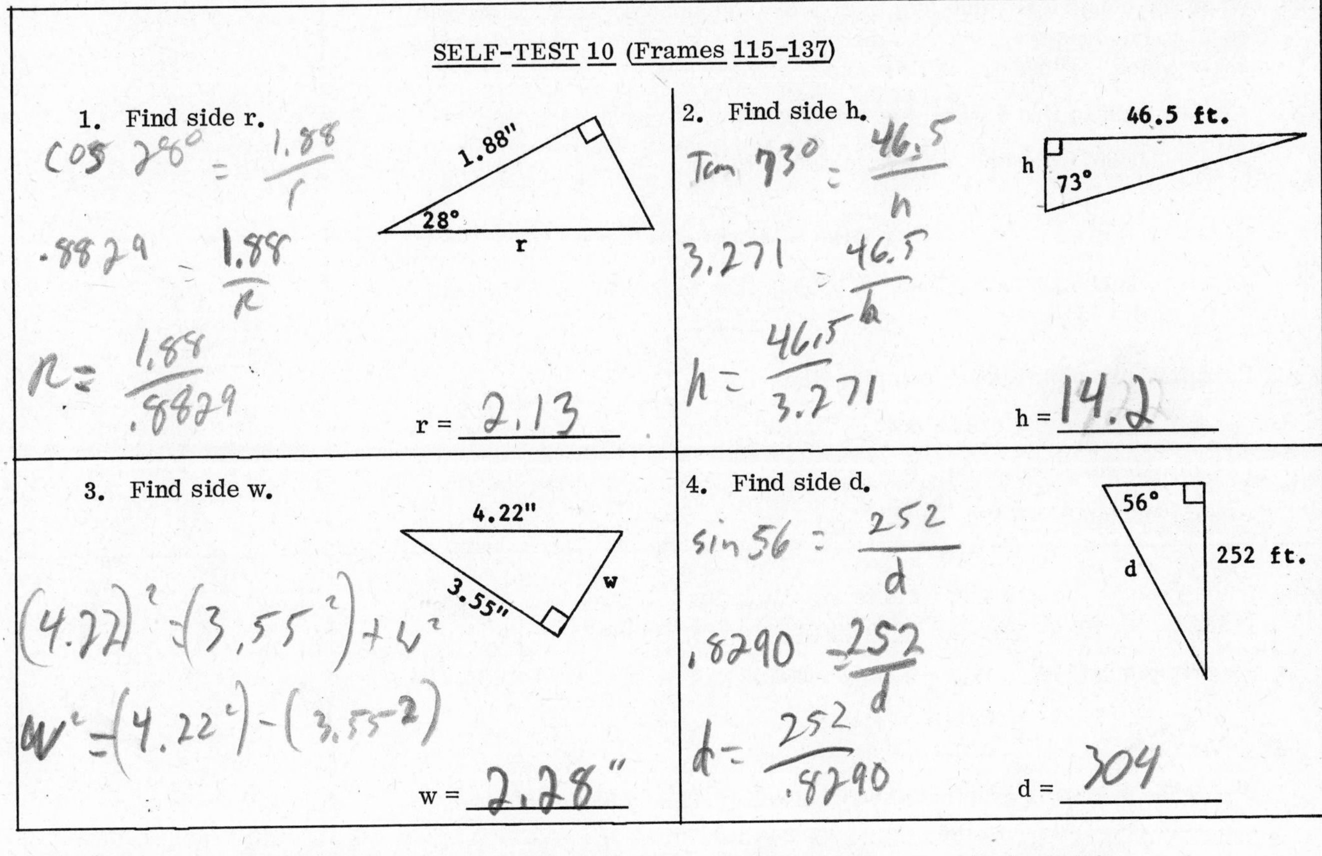

SELF-TEST 10 (Frames 115-137)
1. Find side r.
1.88"
28°
r
r = ____
2. Find side h.
46.5 ft.
h
73°
h = ____
3. Find side w.
4.22"
3.55"
w
w = ____
4. Find side d.
56°
252 ft.
d
d = ____

ANSWERS: 1. r = 2.13" 2. h = 14.2 ft. 3. w = 2.28" 4. d = 304 ft.

1-12 FINDING UNKNOWN ANGLES IN RIGHT TRIANGLES

We can find an unknown acute angle in a right triangle in either of two ways. If the other acute angle is known, we can use the angle-sum principle. If only two sides are known, we can use these two known sides to compute the numerical value of one of the three trig ratios of the unknown angle, and then use the trig table to find the angle. The latter method will be emphasized in this section.

---

138. If we know the numerical value of one of the trig ratios for a given angle, we can find the size of the angle by using the trig table. For example:

(a) If $\sin \theta = 0.9135$, $\theta =$ ____.

(b) If $\cos \alpha = 0.9397$, $\alpha =$ ____.

(c) If $\tan \beta = 6.314$, $\beta =$ ____.

a) 66°

b) 20°

c) 81°

139. When the numerical value of a trig ratio is computed, sometimes that exact value does not appear in the table. In such cases, we report the angle which is closest. For example:

We want to find θ when $\sin \theta = 0.9071$.

0.9071 does not appear in the table.

From the table: $0.9063 = \sin 65°$
$0.9135 = \sin 66°$

Since 0.9071 is closer to 0.9063 than to 0.9135, we say that $\theta = 65°$.

Using the principle above, complete these:

(a) If $\sin \theta = 0.6680$, $\theta =$ 42°.

(b) If $\cos \theta = 0.9700$, $\theta =$ 14°.

(c) If $\tan \theta = 4.420$, $\theta =$ 77°.

---

140. If the given value of a trig ratio is exactly halfway between two values in the table, we choose the "even" degree. For example:

$\sin \theta = 0.9170$ In the table we find:

$0.9135 = \sin 66°$
$0.9205 = \sin 67°$

Since 0.9170 is exactly halfway between 0.9135 and 0.9205, we report 66° which is the "even" degree.

Using the principle above, complete these:

(a) If $\sin \theta = 0.9935$, $\theta =$ 84°.

(b) If $\cos \theta = 0.9170$, $\theta =$ 24°.

(c) If $\tan \theta = 1.570$, $\theta =$ 58°.

Answers to 139: a) 42° b) 14° c) 77°

---

141. If one acute angle in a right triangle is known, we can easily find the other acute angle by means of the angle-sum principle. We simply subtract the known angle from 90°.

When neither acute angle in a right triangle is known, we can find an unknown angle by computing one of its trig ratios if two sides of the triangle are known. Here is an example:

To find angle θ in this right triangle, we can use the sine ratio. There are two steps.

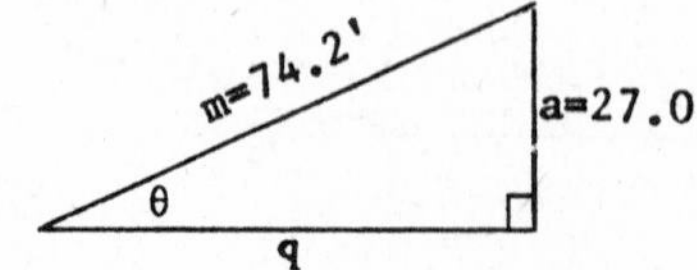

(Continued on following page.)

Answers to 140: a) 84° b) 24° c) 58°

141. (Continued)

<u>Step 1</u>: <u>We compute the sine of angle θ.</u>

$$\sin\theta = \frac{a}{m} = \frac{27.0'}{74.2'} = 0.364$$

<u>Note</u>: (1) The computed sine contains only <u>three</u> significant digits since the sides are reported to three significant digits.

(2) Since the ratios in the trig table contain <u>four</u> significant digits, we add a "0" after the "4" and get "0.3640", to make it easier to use the table.

<u>Step 2</u>: <u>We use the table to find the angle with that particular sine.</u>

If $\sin\theta = 0.3640$, θ = 21°.

---

142. To find angle Q in this right triangle, we can compute its cosine and then use the trig table.

21°

<u>Step 1</u>: $\cos Q = \frac{7.3'}{9.7'} = 0.75$

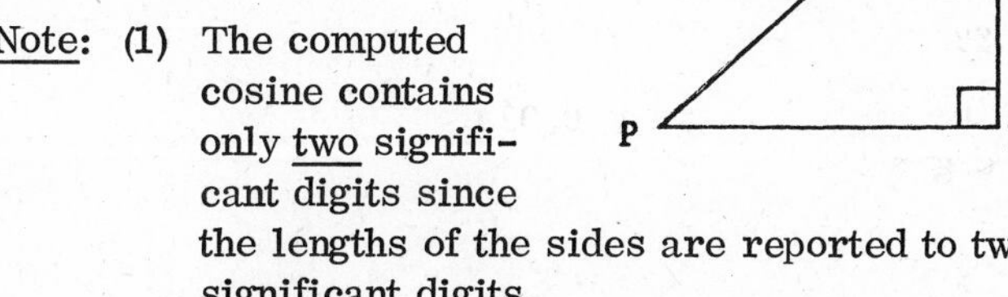

<u>Note</u>: (1) The computed cosine contains only <u>two</u> significant digits since the lengths of the sides are reported to two significant digits.

(2) Before using the table, we add two 0's after the "5" to get "0.7500".

<u>Step 2</u>: If $\cos Q = 0.7500$, Q = 41°.

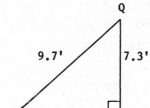

---

143. When using two sides of a triangle to compute one of the ratios of a given angle, it is easy to decide which ratio can be computed. To do so, we merely identify the known sides in terms of that angle. Here is an example:

41°

To find angle β in this triangle, we identify the two known sides in terms of angle β.

"t" is the <u>side opposite</u> β.
"x" is the <u>side adjacent</u> to β.

Therefore, we compute <u>tan</u> β.

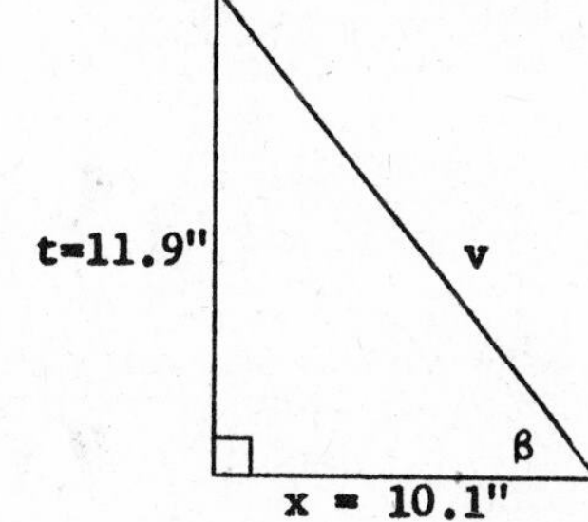

(Continued on following page.)

143. (Continued)

In the blank under each right triangle below, write the ratio of angle α which can be computed for that triangle:

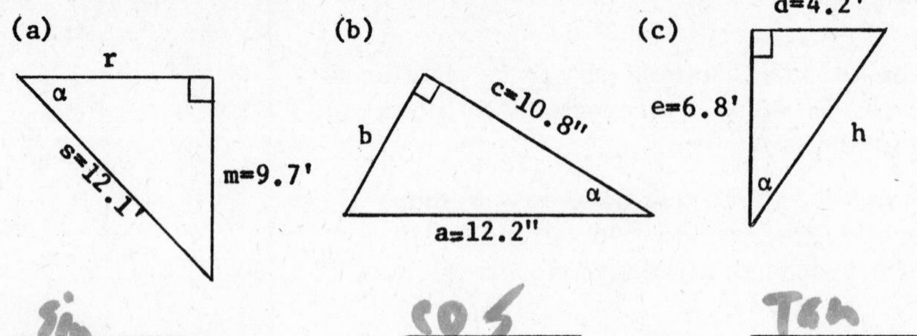

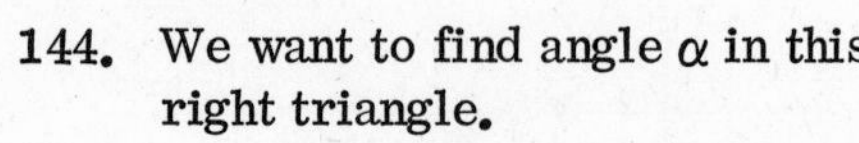

---

144. We want to find angle α in this right triangle.

(a) Which ratio of angle α can be computed? ________

(b) Do so. Angle α = ________

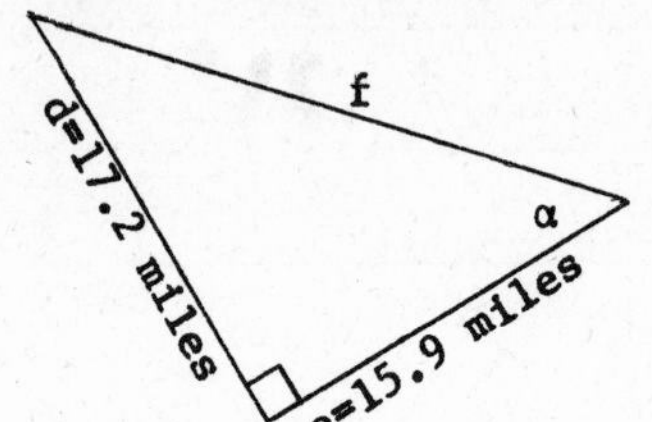

a) $\sin\alpha$

b) $\cos\alpha$

c) $\tan\alpha$

---

145. In right triangle MNT:

MT is 783 miles.
MN is 699 miles.

(a) Which ratio of angle β can be computed? ________

(b) Do so. Angle β = ________

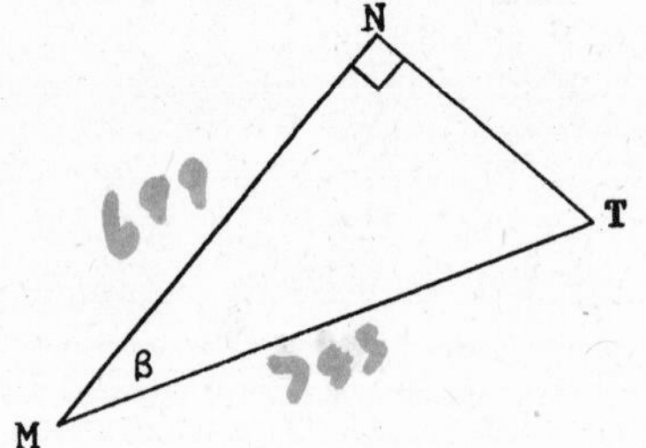

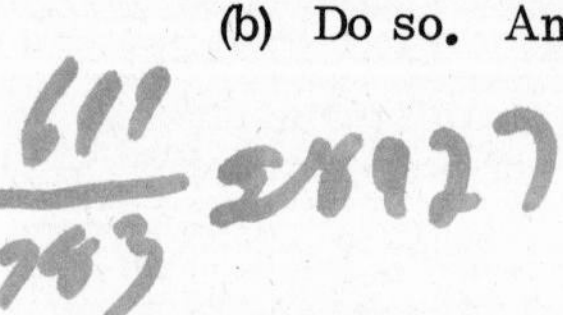

a) $\tan\alpha$

b) $\alpha = 47°$, since: $\tan\alpha = 1.08$

---

146. In right triangle AGH:

AG is 7.9'.
AH is 9.8'.

(a) Which ratio of angle γ can be computed? ________

(b) Do so. Angle γ = ________

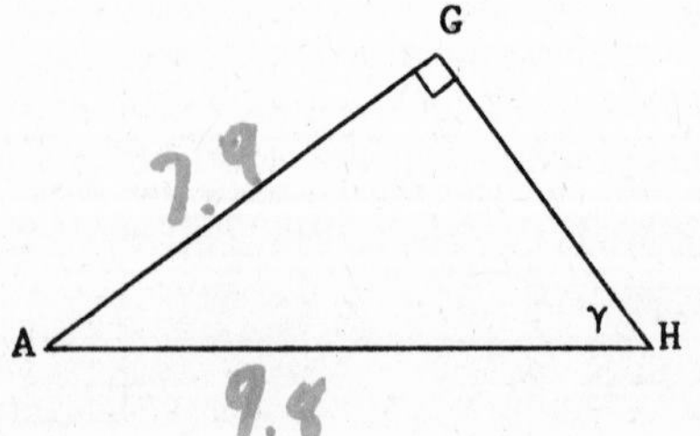

a) $\cos\beta$

b) $\beta = 27°$, since: $\cos\beta = 0.893$

147. After computing the size of one acute angle in a right triangle, it is easy to find the size of the other acute angle, since the sum of the two acute angles in any right triangle is 90°.

In right triangle PDQ:

DQ is 14.3 miles.
PQ is 21.9 miles.

(a) Angle P = ________.

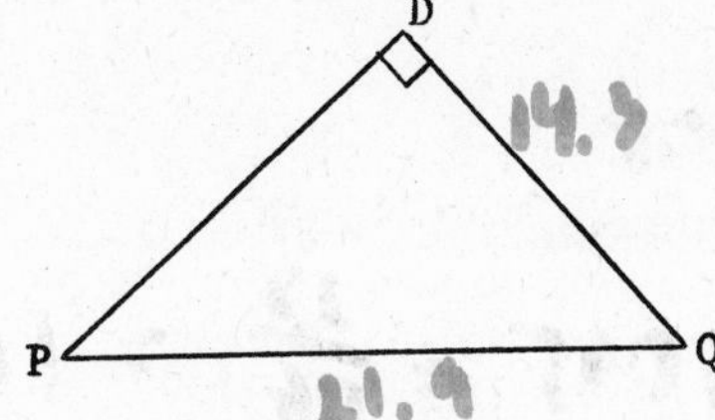

(b) Angle Q = ________.

a) $\sin \gamma$

b) $\gamma = 54°$, since:
$\sin \gamma = 0.81$
(Note: 0.81 contains two significant digits.)

---

148. In right triangle RST:

Angle T is 53°.
RS is 39.9 feet.

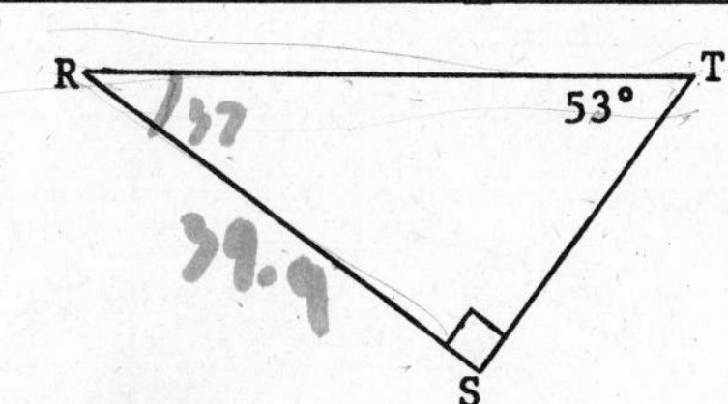

(a) Angle R = ________.

(b) RT = ________ feet.

a) 41°

b) 49°

---

149. In right triangle LDS:

Angle L is 71°.
LS is 68 inches.

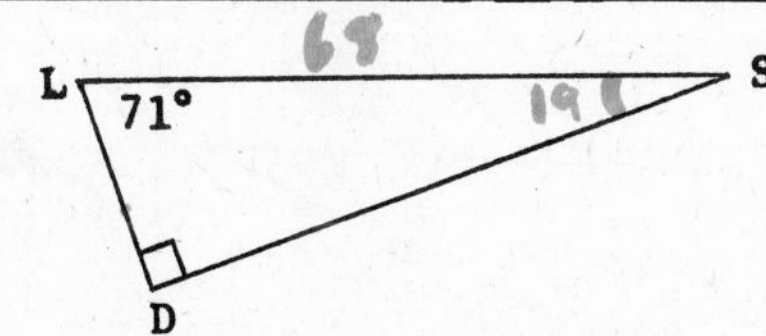

(a) Angle S = ________.

(b) LD = ________ inches.

a) 37°

b) 50.0 feet

---

a) 19°

b) 22 inches

150. In right triangle FGH:

Angle H = 63°.
GH = 2.8 centimeters.

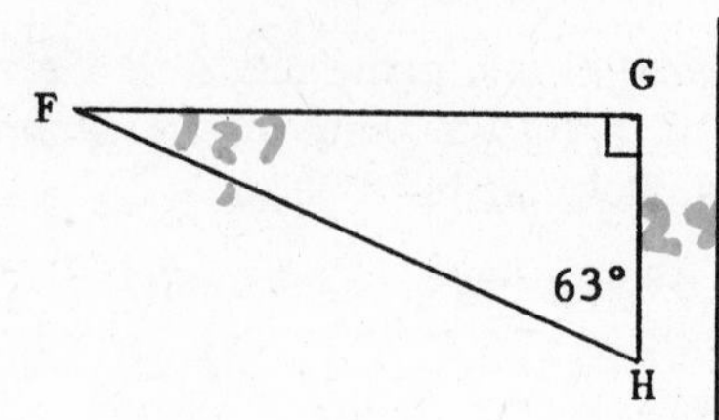

(a) Angle F = ______.

(b) FG = ______ centimeters.

Answer to Frame 150: a) 27° b) 5.5 centimeters

SELF-TEST 11 (Frames 138-150)

1. Find angle θ.

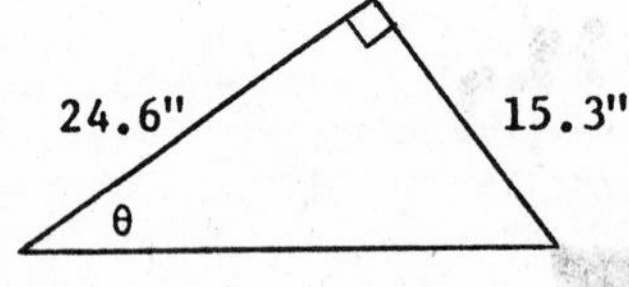

θ = ______

2. Find angle α.

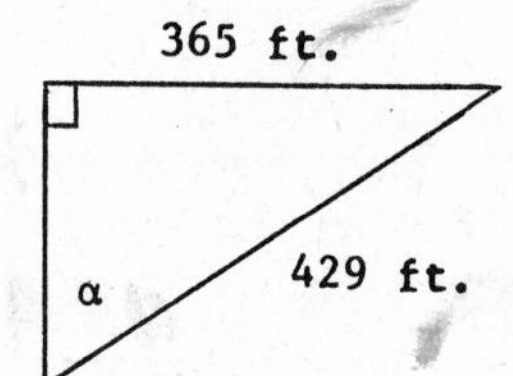

α = ______

3. Find angle β.

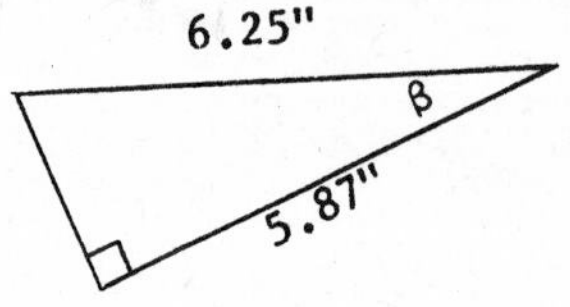

β = ______

4. (a) Find angle θ.
   (b) Find side h.
   (c) Find side w.

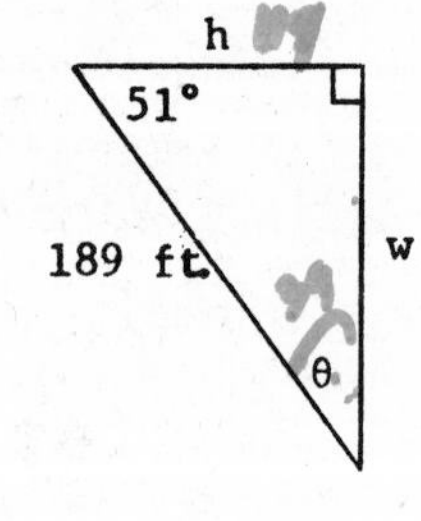

(a) θ = ______ (b) h = ______ (c) w = ______

ANSWERS: 1. θ = 32° 2. α = 58° 3. β = 20° 4. (a) θ = 39°
(b) h = 119 ft.
(c) w = 147 ft.

1-13 SOME APPLIED PROBLEMS

Many applied problems involve right triangles. The trigonometric ratios are frequently used to solve these problems. We will give a sample of some simple problems of this type in this section.

151. The end-view of the roof of a building is an isosceles triangle, as shown at the right.

Find angle θ, the angle of slope of the roof.

θ = ________

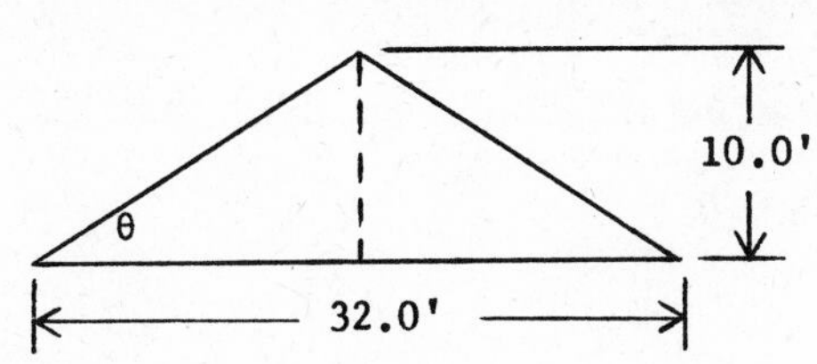

---

152. In this metal bracket, find D, the distance between the two holes.

D = ________

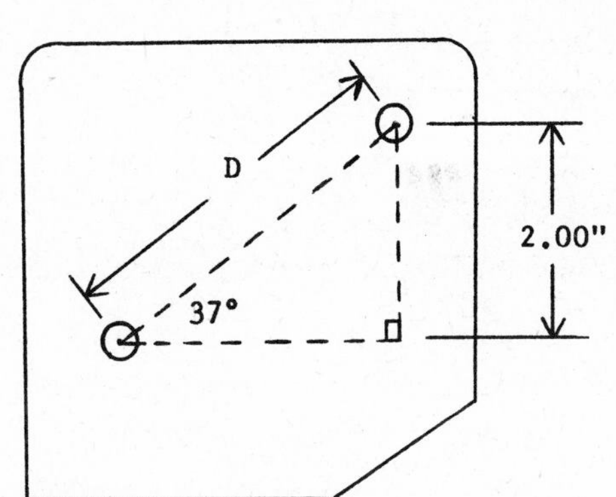

θ = 32°

Solution:

$$\tan\theta = \frac{10.0}{16.0} = 0.625$$

$$\theta = 32°$$

---

153. The metal shape at the right is called a "template". Figure EFHG is a rectangle.

Given: Angle D = 68°
Side DG = 4.8"
Side FH = 2.9"

Find: Side DF. DF = ________

(Note: DE = 4.8" - 2.9")

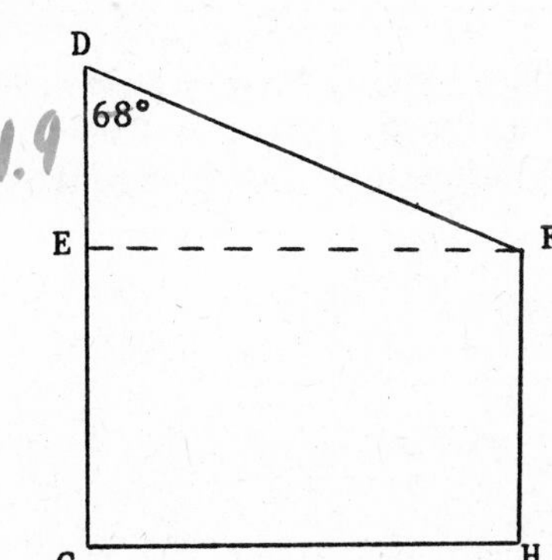

D = 3.32"

Solution:

$$\sin 37° = \frac{2.00}{D}$$

$$D = \frac{2.00}{\sin 37°} = \frac{2.00}{0.6018}$$

D = 3.32"

154. A right triangle called an "impedance" triangle is used in analyzing alternating current circuits. The angle θ, shown in the diagram, is the "phase angle" of the circuit.

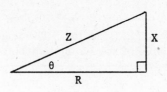

In a particular circuit, $X = 2{,}630$ and $Z = 6{,}820$. Find angle θ.

θ = ________

DF = 5.1"

Solution:

In right triangle DEF:

$$\cos 68° = \frac{DE}{DF} = \frac{1.9''}{DF}$$

$$0.3746 = \frac{1.9''}{DF}$$

$$DF = \frac{1.9''}{0.3746} = 5.1''$$

---

155. Here is a cross-sectional view of a metal shaft with a tapered end.

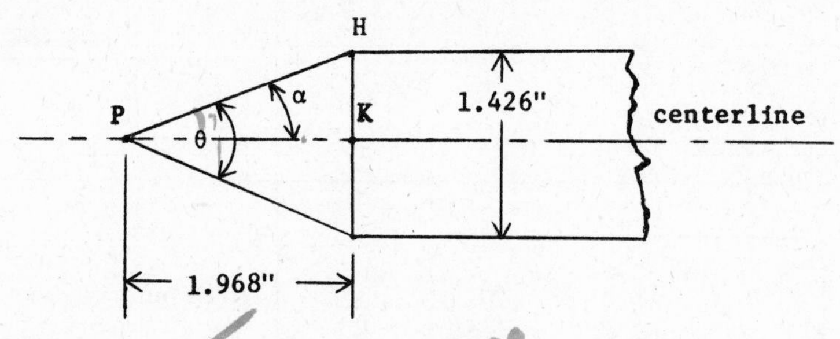

Find: θ, the taper angle. θ = ________

(Hint: In right triangle PKH, find angle α, which is half of angle θ.)

$\theta = 23°$

Solution:

$$\sin \theta = \frac{X}{Z} = \frac{2{,}630}{6{,}820} = 0.386$$

$$\theta = 23°$$

---

156. To measure the height of a tall building, a surveyor set his transit (angle-measuring instrument) 100 feet horizontally from the base of the building, and measured the angle of elevation to the top of the building and found it to be 72°.

Problem: Find "h", the height of the building.

h = ________

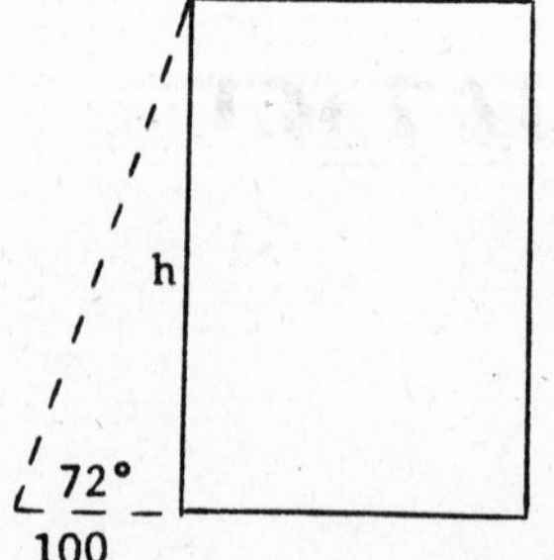

$\theta = 40°$

Solution:

$$\tan \alpha = \frac{HK}{PK} = \frac{\frac{1.426}{2}}{1.968}$$

$$\tan \alpha = \frac{0.713}{1.968} = 0.362$$

$$\alpha = 20°$$

$$\theta = 2\alpha = 40°$$

Answer to Frame 156: h = 308 ft. Solution: $\frac{h}{100} = \tan 72°$

$$h = 100 \tan 72° = 308 \text{ ft.}$$

SELF-TEST 12 (Frames 151-156)

1. In the isosceles triangle below, find altitude "a".

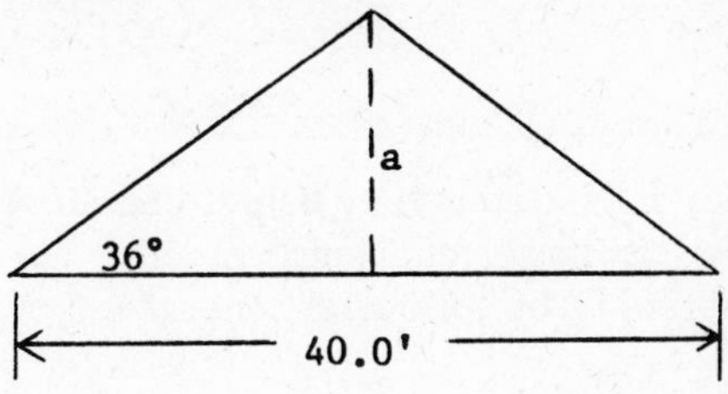

a = 14.5

2. Two holes are drilled in a metal plate, as shown below. Find horizontal distance "h" and vertical distance "v".

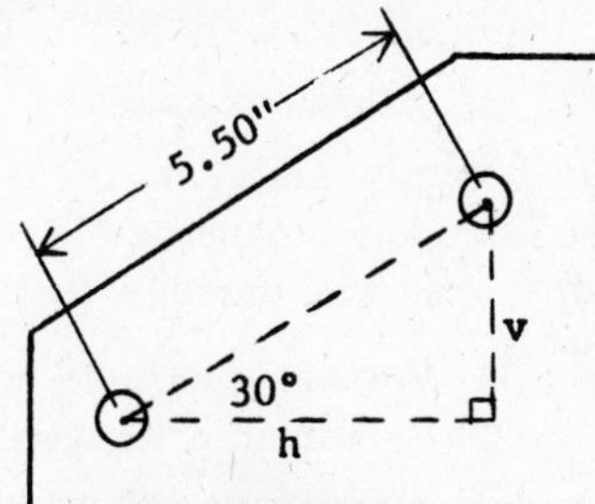

h = 4.76 v = 2.75

ANSWERS: 1. a = 14.5' 2. h = 4.76" v = 2.75"

Chapter 2 VECTORS

Though vector mathematics is a pure mathematical topic, it is also a very useful one since vectors are used in science and technology to represent concepts like force, voltage, current, velocity, acceleration, and others. We will discuss vector mathematics in this chapter. The following general topics are emphasized:

(1) Vectors and their components on the coordinate system.
(2) Vector addition by the parallelogram method and by the component method.
(3) Vectors in a state of equilibrium.

In this chapter, the direction of vectors is specified in terms of reference angles only. The use of standard-position angles to specify the direction of vectors is delayed until the following chapter.

2-1 VECTORS ON THE COORDINATE SYSTEM

Horizontal, vertical, and slanted vectors can be drawn on the coordinate system. All three types of vectors have both length and direction. All three types can be represented by signed numbers. In this section, we will discuss vectors on the coordinate system and show how these vectors are named.

---

1. The rectangular coordinate system contains both a horizontal and a vertical axis. Both axes are number lines. Any "move" or "change" on these axes can be represented by an arrow which is called a "vector".

Two vectors (0A and BC) are drawn on the horizontal axis below:

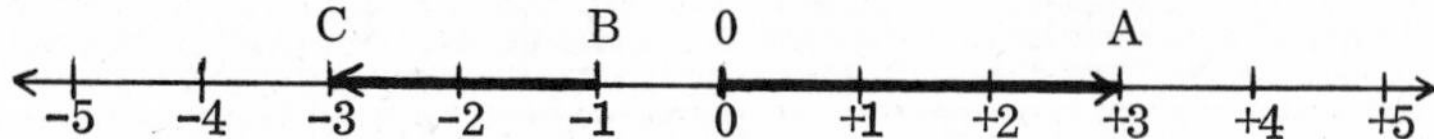

Any vector has both a length and a direction.

Vector 0A is a 3-unit move to the right.
Vector BC is a 2-unit move to the left.

Any vector can be represented by a signed number.

On the horizontal axis: (1) All vectors directed to the right are positive.
(2) All vectors directed to the left are negative.

Therefore: Vector 0A is represented by "+3".
Vector BC is represented by "-2".

When a horizontal vector is represented by a signed number:

(1) The absolute value of the signed number tells you the length of the vector.
(2) The sign tells you the direction of the vector.

Go to next frame.

2. All vectors have two properties, length (or distance) and direction. Therefore, vectors are directed lengths or directed distances.

Two vectors are drawn on the horizontal axis below.

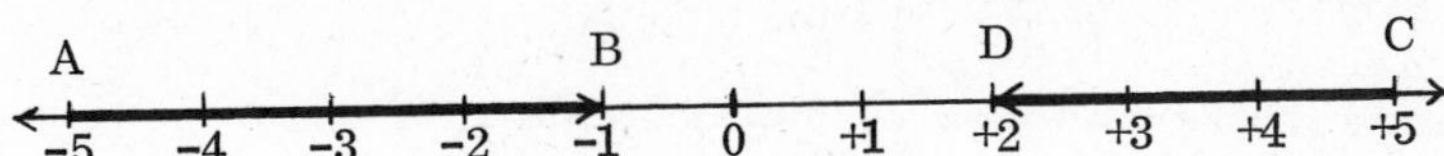

(a) Vector AB is a 4-unit vector to the right. Therefore, it is represented by what signed number? +4

(b) Vector CD is a 3-unit vector to the left. Therefore, it is represented by what signed number? -3

---

a) +4

b) -3

3. All three vectors on the horizontal axis below are 2-unit vectors to the right.

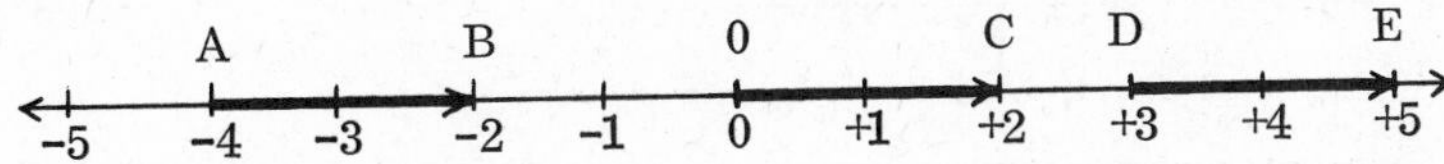

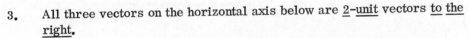

What signed number would be used to represent each of the three vectors? +2

+2

4. Vectors can also be drawn on the vertical axis of the coordinate system. These vectors are also represented by signed numbers:

Upward vectors are positive.
Downward vectors are negative.

On the vertical axis shown at the right:

(a) Vector AB is represented by what signed number? -2

(b) Vector CD is represented by what signed number? +3

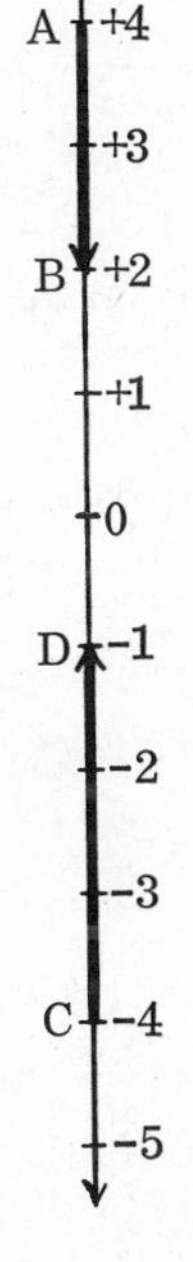

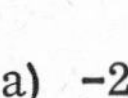

---

a) -2

b) +3

5. The horizontal and vertical axes divide the coordinate plane into four parts called "quadrants". The quadrants are labeled I, II, III, and IV, as shown at the right.

II | I
III | IV

Horizontal and vertical vectors can be drawn anywhere on the coordinate system. They need not be drawn on the axes. However, the directions for determining their signs are still the same.

(a) The sign of any vector to the right or upwards is  ________.

(b) The sign of any vector to the left or downwards is ________.

a) positive (or +)

b) negative (or -)

---

6. Some horizontal and vertical vectors are drawn on the coordinate system at the right.

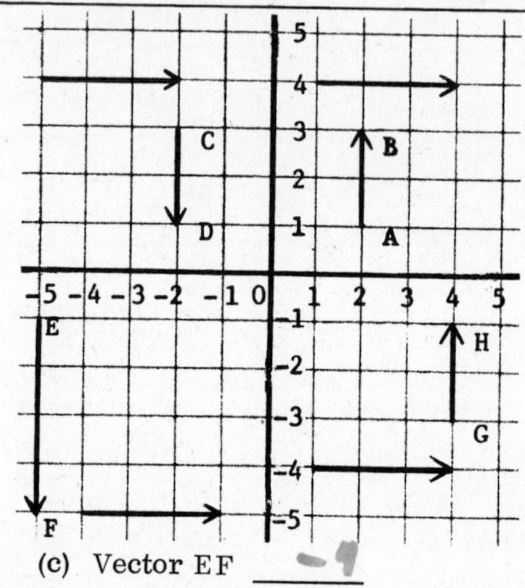

There is one horizontal vector in each quadrant. Since each of these vectors is 3 units to the right, each is represented by "+3".

What signed number would be used to represent each of the vertical vectors?

(a) Vector AB ________

(b) Vector CD ________

(c) Vector EF ________

(d) Vector GH ________

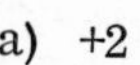

a) +2

b) -2

c) -4

d) +2

---

7. Slanted vectors can also be drawn on the coordinate system. Though slanted vectors can begin at places other than the origin, we will limit our discussion to those vectors which begin at the origin since they are more common and more important.

The slanted vector 0A, shown at the right, is drawn on the coordinate system in the first quadrant.

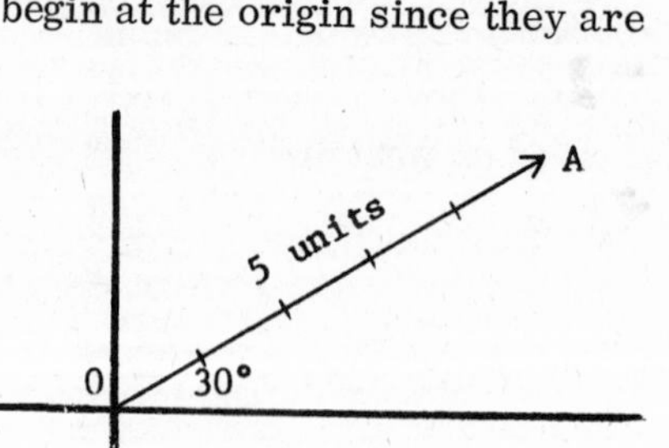

The direction of slanted vectors which begin at the origin is stated in terms of:

(1) The angle formed by the vector and the horizontal axis, and

(2) The quadrant in which that angle is formed.

Therefore: (a) The length of vector 0A is ______ units.

(b) The direction of vector 0A is a ______° angle in the ________ quadrant.

8. Vector 0B on the right is a slanted vector in the second quadrant.

(a) The length of vector 0B is ______ units.

(b) The direction of vector 0B is a ______° angle in the ______ quadrant.

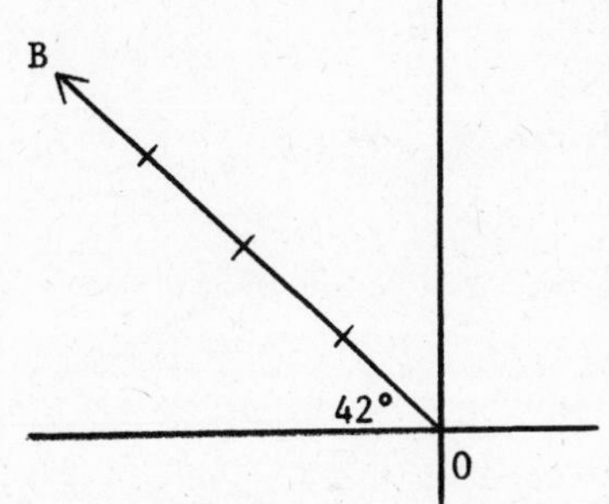

a) 5

b) 30° ... first

---

9. Here are two more slanted vectors which begin at the origin:

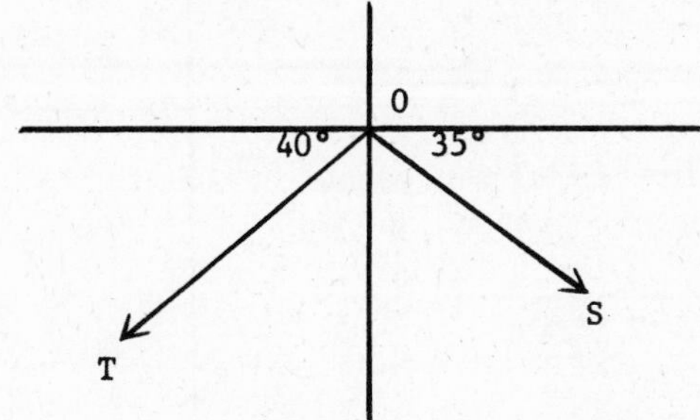

(a) The direction of vector 0T is a ______° angle in the ______ quadrant.

(b) The direction of vector 0S is a ______° angle in the ______ quadrant.

a) 4

b) 42° ... second

---

10. The angle which specifies the direction of a slanted vector is called its reference angle.

A reference angle is always the angle formed by the vector and the ______ axis.

a) 40° ... third

b) 35° ... fourth

---

11. To describe the reference angle of a slanted vector, you must know:

(1) Its size in degrees.

(2) The quadrant in which it appears.

If you are told that a slanted vector is 3 units long and has a 50° reference angle, could you draw the vector on the coordinate system? ______

horizontal

No, because you must know in what quadrant the vector lies.

12. All slanted vectors have positive lengths. Therefore, in the diagram at the right, both vector 0F and vector 0K are +5 vectors.

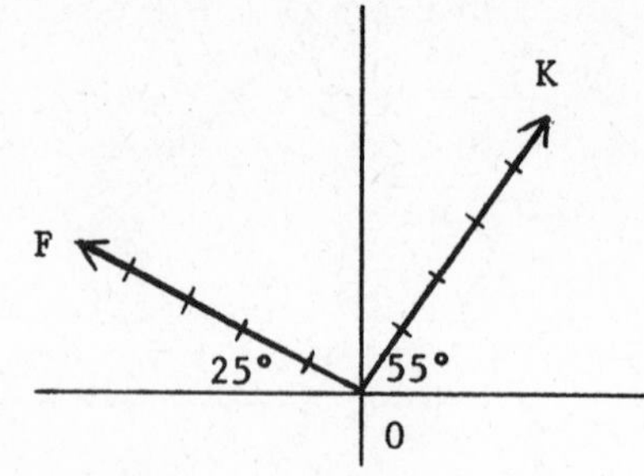

Though both vectors are +5 vectors, their reference angles are different.

(a) Specify the reference angle of vector 0F.

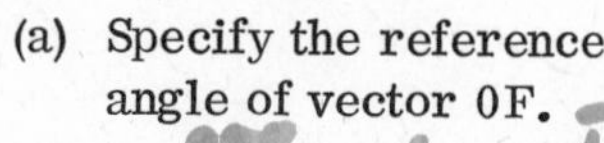

____

(b) Specify the reference angle of vector 0K.

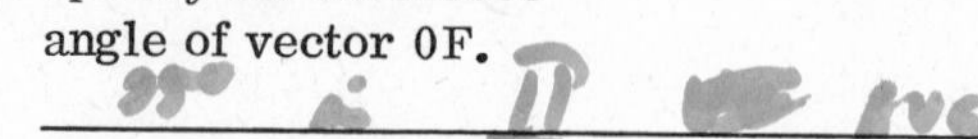

____

---

13. When naming a vector, we use two letters. The first letter represents the point where the vector begins. The second letter represents the tip of the vector. For example:

The slanted vector in the second quadrant is called "vector 0D" (not "vector D0") because it begins at the origin "0".

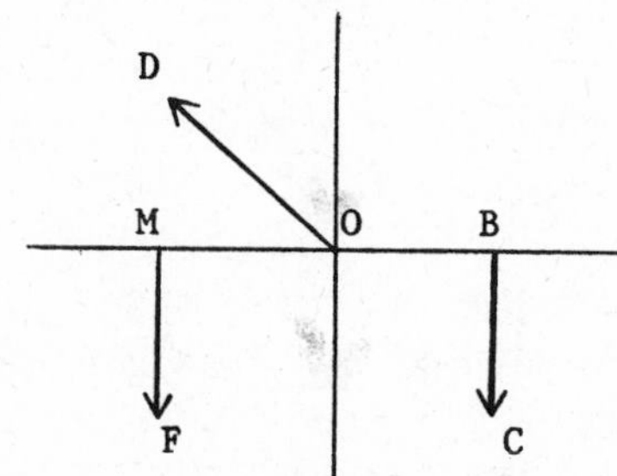

(a) The name of the vector in the third quadrant is vector ______.

(b) The name of the vector in the fourth quadrant is vector ______.

a) A 25° angle in the second quadrant.

b) A 55° angle in the first quadrant.

---

14. Instead of writing "vector 0D", we can simply write $\overrightarrow{0D}$.

The arrow above "0D" signifies that "0D" is a vector. This arrow is always drawn horizontally, with the tip of the arrow to the right. It has nothing to do with the actual direction of the vector represented.

$\overrightarrow{DT}$ means "vector DT". Does the arrow above "DT" mean that the vector is a horizontal one to the right? ______

a) vector MF (not vector FM)

b) vector BC (not vector CB)

---

15. To name the vectors on the right:

We can write $\overrightarrow{0D}$ instead of "vector 0D".

(a) We can write ______ instead of "vector BD".

(b) We can write ______ instead of "vector EC".

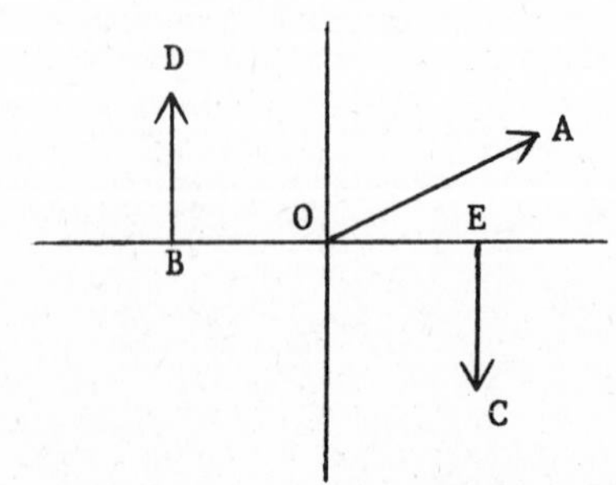

No. It could have any direction.

---

a) $\overrightarrow{BD}$

b) $\overrightarrow{EC}$

Self-Test 1 (Frames 1-15)

In the diagram at the right, each vector is 3 units long and is either a horizontal or a vertical vector. Write the signed number which represents each of the following vectors:

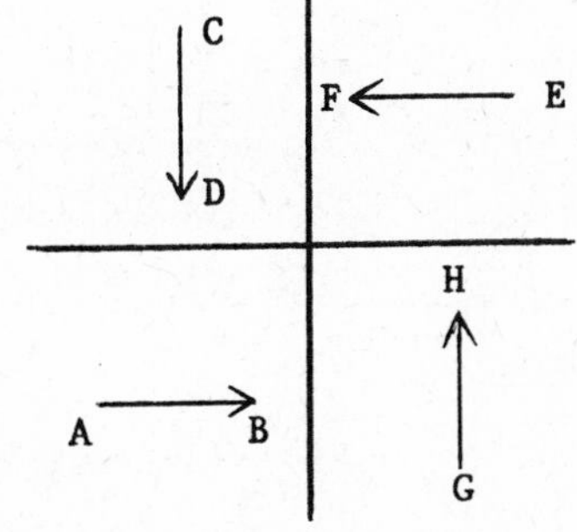

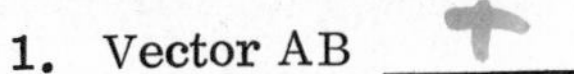

1. Vector AB ________ 3. Vector EF ________
2. Vector CD ________ 4. Vector GH ________

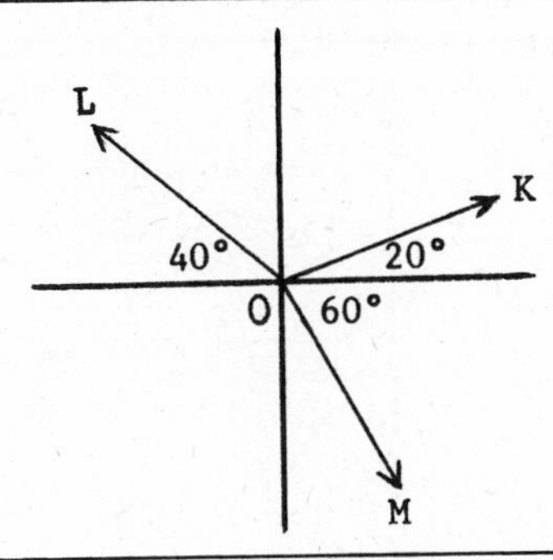

In the diagram at the left, each slanted vector is 4 units long.

5. What signed number represents the length of each vector? ________
6. The reference angle of vector 0L is a ________ angle in the ____________ quadrant.
7. The reference angle of vector 0M is a ________ angle in the ____________ quadrant.

8. "Vector HK" can be written in shorter form as ________.

ANSWERS:

1. +3	3. -3	5. +4	6. 40° ... second	8. $\overrightarrow{HK}$
2. -3	4. +3		7. 60° ... fourth	

2-2 THE COMPONENTS OF SLANTED VECTORS

Any slanted vector on the coordinate system has both a horizontal and a vertical component. Slanted vectors and their components form right triangles. The coordinates of the tip of any slanted vector are the signed numbers which represent each component vector. We will discuss the components of slanted vectors in this section.

16.

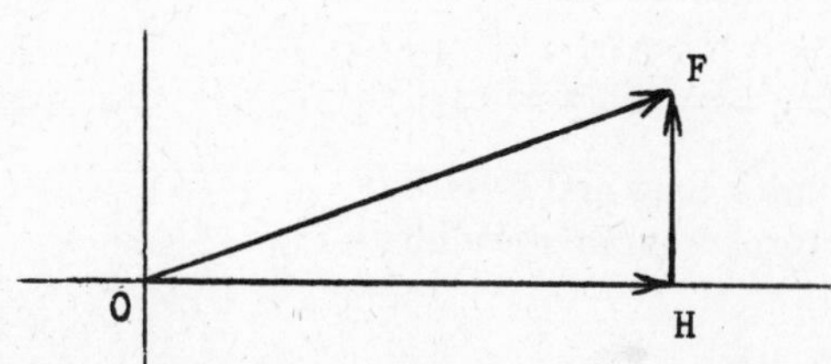

On the coordinate system above, we have drawn $\overrightarrow{0F}$, a slanted vector in the first quadrant. We have added two additional vectors, called the components of $\overrightarrow{0F}$.

(1) $\overrightarrow{0H}$ is the horizontal component of $\overrightarrow{0F}$.

(2) $\overrightarrow{HF}$ is the vertical component of $\overrightarrow{0F}$.

The three vectors, $\overrightarrow{0F}$, $\overrightarrow{0H}$, and $\overrightarrow{HF}$, form what type of a triangle? Right

A right triangle, since angle 0HF = 90°.

---

17. On the coordinate system at the right, we have drawn $\overrightarrow{0P}$, a slanted vector in the second quadrant.

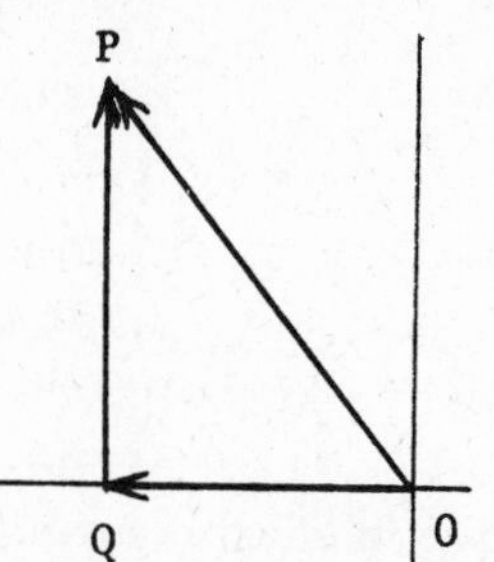

(a) $\overrightarrow{QP}$ is the vertical component of $\overrightarrow{0P}$.

(b) $\overrightarrow{0Q}$ is the Horizontal component of $\overrightarrow{0P}$.

(c) Angle 0QP is a Right angle.

a) vertical

b) horizontal

c) right or 90°

---

18. $\overrightarrow{0V}$ is a slanted vector in the third quadrant.

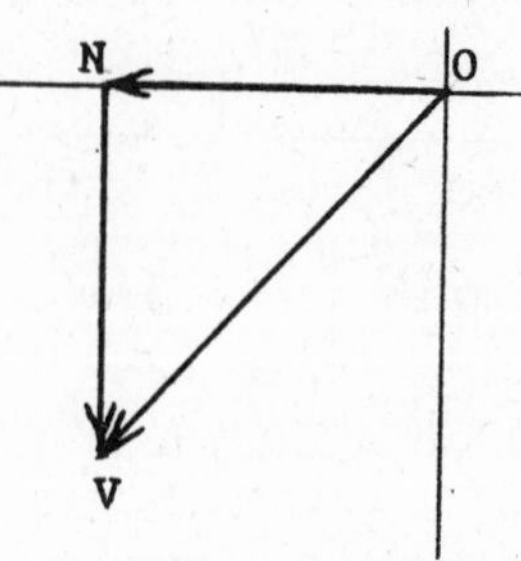

(a) Is the horizontal component of $\overrightarrow{0V}$ written as $\overrightarrow{N0}$ or $\overrightarrow{0N}$? $\overrightarrow{0N}$

(b) Is the vertical component of $\overrightarrow{0V}$ written as $\overrightarrow{NV}$ or $\overrightarrow{VN}$? $\overrightarrow{NV}$

(c) Name the right angle which is formed by two of the vectors above. Angle ONV

a) $\overrightarrow{0N}$, since it begins at point 0.

b) $\overrightarrow{NV}$, since it begins at point N.

c) Angle 0NV

---

19. $\overrightarrow{0G}$ is a slanted vector in the fourth quadrant.

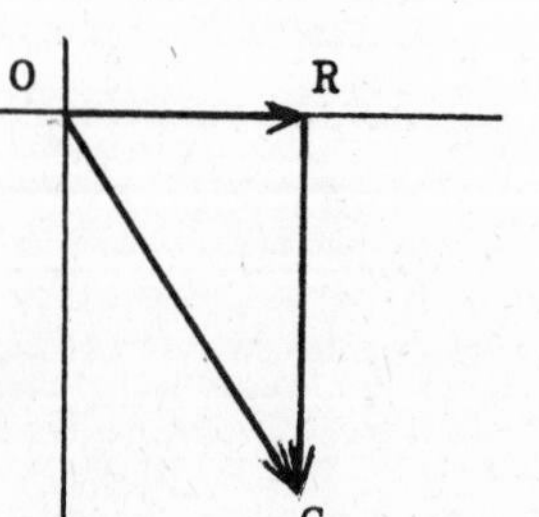

(a) Write the horizontal component of $\overrightarrow{0G}$. 0R

(b) Write the vertical component of $\overrightarrow{0G}$. RG

(c) The reference angle for $\overrightarrow{0G}$ is angle ROG in the IV quadrant.

(d) $\overrightarrow{0G}$ is a positive (positive/negative) vector.

20. Slanted vectors and their components form right triangles. Since all three sides of these right triangles are vectors on the coordinate system, the sides can be represented by signed numbers.

Here is a vector in the second quadrant with its components drawn:

(a) Is $\overrightarrow{0Q}$ positive or negative? positive

(b) Is $\overrightarrow{RQ}$ positive or negative? positive

(c) Is $\overrightarrow{0R}$ positive or negative? Neg.

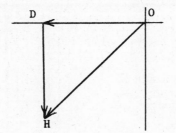

a) $\overrightarrow{0R}$ (not $\overrightarrow{R0}$)

b) $\overrightarrow{RG}$ (not $\overrightarrow{GR}$)

c) angle R0G ... fourth

d) positive
(All slanted vectors are positive.)

---

21. Which vectors on the right are negative? $\overrightarrow{0D}$, $\overrightarrow{DH}$

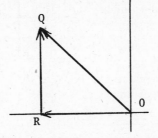

a) Positive, since all slanted vectors are positive.

b) Positive (upward)

c) Negative (to the left)

---

22. Which vectors are negative? $\overrightarrow{EV}$

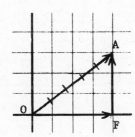

Both $\overrightarrow{0D}$ (to the left) and $\overrightarrow{DH}$ (downward)

($\overrightarrow{0H}$ is positive, because a slanted vector is always positive, by definition.)

---

23. On the graph on the right:

The slanted vector $\overrightarrow{0A}$ is +5 units.

The component $\overrightarrow{0F}$ is +4 units.

The component $\overrightarrow{FA}$ is +3 units.

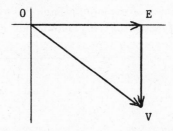

We want you to see that the coordinates of point A, the tip of the slanted vector, give you the signed-value of the two components.

(a) What are the coordinates of point A? (4 , 3)

(b) The horizontal-coordinate of point A, which is +4, has the same signed-value as $\overrightarrow{0F}$, the Horizontal (horizontal/vertical) component of $\overrightarrow{0A}$.

(c) The vertical-coordinate of point A, which is +3, has the same signed-value as $\overrightarrow{FA}$, the vertical (horizontal/vertical) component of $\overrightarrow{0A}$.

Only $\overrightarrow{EV}$ (downward)

24. On the graph on the right:

(a) The coordinates of point C are (−5 , +4).

(b) The horizontal component, $\overrightarrow{0H}$, has the same signed-value as the Horizontal-coordinate of point C.

(c) The vertical component, $\overrightarrow{HC}$, has the same signed-value as the Vertical-coordinate of point C.

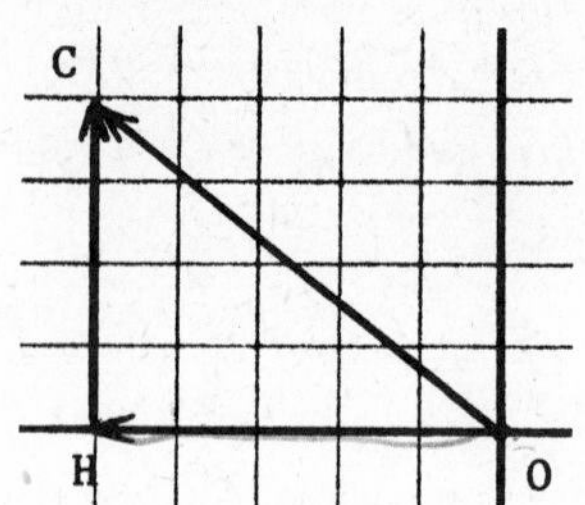

a) (+4, +3)

b) horizontal

c) vertical

---

25. (a) The coordinates of point V are (−4 , −7).

(b) The coordinates of point S are (+8 , −5).

(c) Does the horizontal coordinate of each point have the same signed-value as the horizontal component of the slanted vector drawn to that point? yes

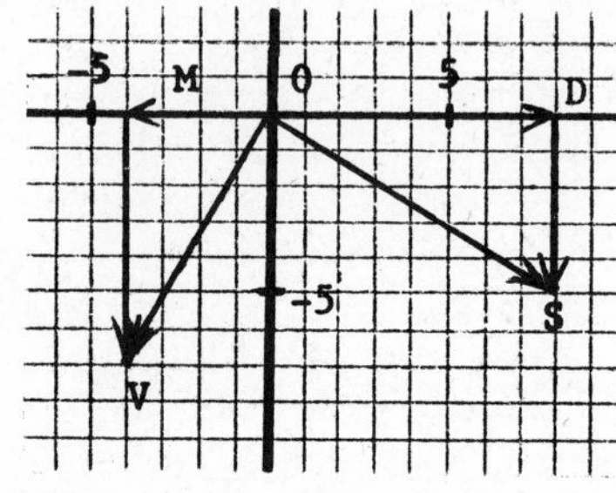

(d) Does the vertical coordinate of each point have the same signed-value as the vertical component of the slanted vector drawn to that point? yes

a) (−5, 4)

b) horizontal

c) vertical

---

26. If the coordinates of point G are (4.7, 3.1):

(a) What signed number represents $\overrightarrow{0X}$? 4.7

(b) What signed number represents $\overrightarrow{XG}$? 3.1

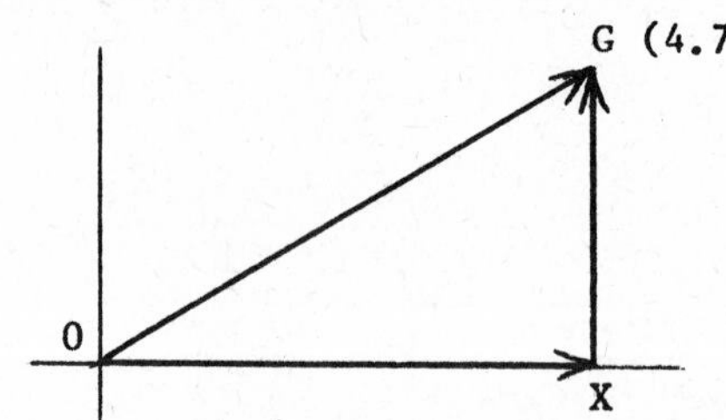

a) (−4, −7)

b) (+8, −5)

c) Yes

d) Yes

---

27. If the coordinates of point D are (−3.8, −3.4):

(a) What signed number represents $\overrightarrow{AD}$? −3.4

(b) What signed number represents $\overrightarrow{0A}$? −3.4

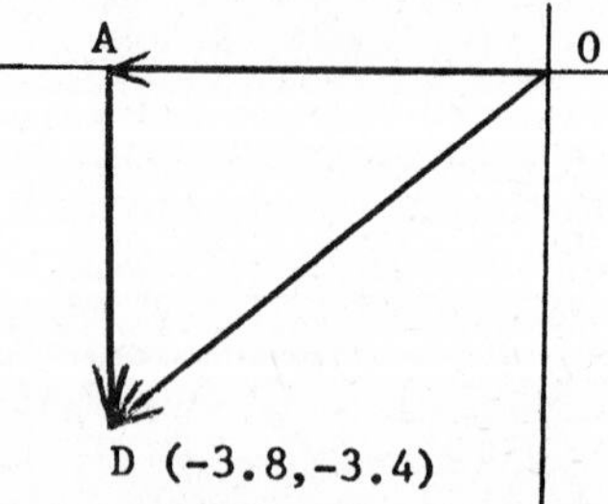

a) +4.7

b) +3.1

---

a) −3.4

b) −3.8

28. If $\overrightarrow{0B}$ is a -5 vector and $\overrightarrow{BF}$ is a +3 vector, what are the coordinates of point F? (-5 , +3)

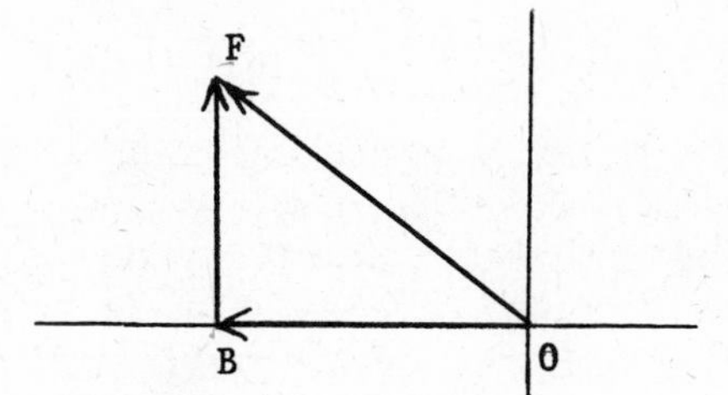

29. If $\overrightarrow{MP}$ is a -6.3 vector and $\overrightarrow{0M}$ is a +5.4 vector, the coordinates of point P are (+5.4, -6.3).

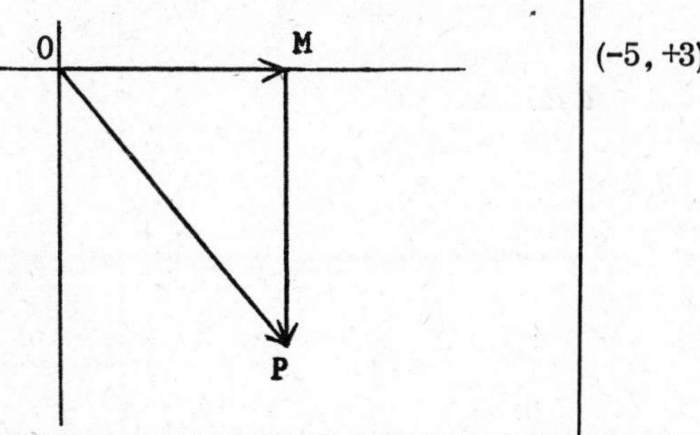

(-5, +3)

Answer to Frame 29: (5.4, -6.3)

SELF-TEST 2 (Frames 16-29)

Refer to the vector diagram at the right:

1. The horizontal component of $\overrightarrow{0P}$ is $\overrightarrow{0R}$.
2. The vertical component of $\overrightarrow{0P}$ is $\overrightarrow{RP}$.

The coordinates of point P are (4, -2).

3. The numerical value of the vertical component of $\overrightarrow{0P}$ is -2.
4. The numerical value of the horizontal component of $\overrightarrow{0P}$ is +4.

O R P

ANSWERS: 1. $\overrightarrow{0R}$ 2. $\overrightarrow{RP}$ 3. -2 4. +4

2-3 THE TRIG RATIOS OF REFERENCE ANGLES

The reference angle of a slanted vector is an acute angle in a right triangle. Like any acute angle in a right triangle, a reference angle has a sine, cosine, and tangent. However, since the sides of the right triangles in which reference angles appear are vectors, the trig ratios of reference angles can be either positive or negative. We will discuss the trig ratios of reference angles in this section.

30. In the figure on the right, we have drawn $\overrightarrow{0L}$ with its two components. Angle L0N is the reference angle for $\overrightarrow{0L}$. <u>We will use the general symbol "α" for all reference angles.</u>

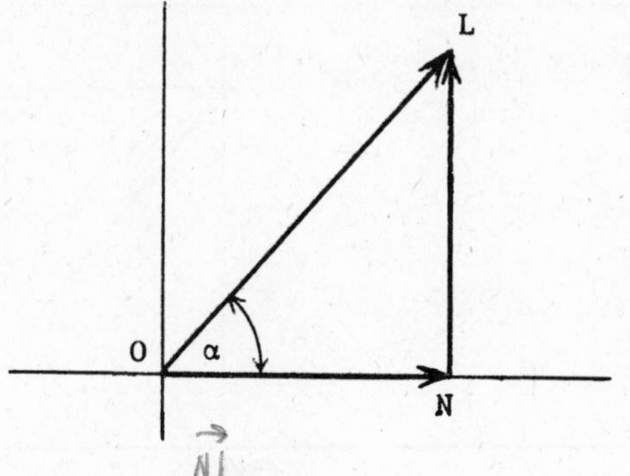

Since the three vectors form a right triangle:

$$\sin\alpha = \frac{\overrightarrow{NL}}{\overrightarrow{0L}}$$

(a) $\cos\alpha =$ ______ (b) $\tan\alpha =$ ______

31. In the triangle on the right:

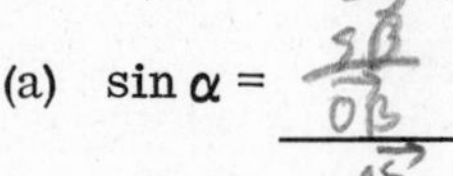

(a) $\sin\alpha =$ ______

(b) $\cos\alpha =$ ______

(c) $\tan\alpha =$ ______

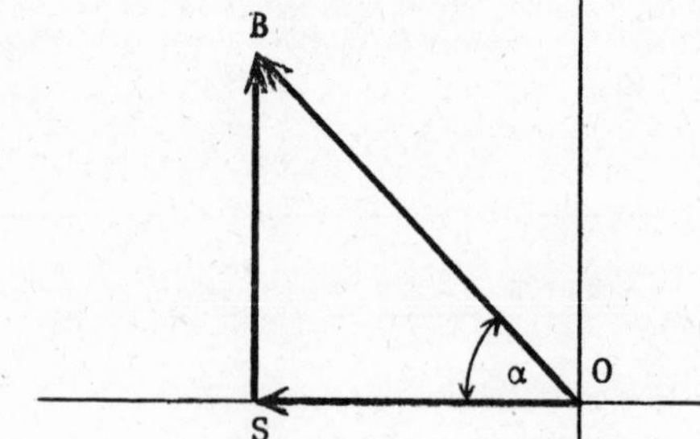

Answers to 30: a) $\frac{\overrightarrow{0N}}{\overrightarrow{0L}}$ b) $\frac{\overrightarrow{NL}}{\overrightarrow{0N}}$

32. In the triangle on the right:

(a) $\tan\alpha =$ ______

(b) $\sin\alpha =$ ______

(c) $\cos\alpha =$ ______

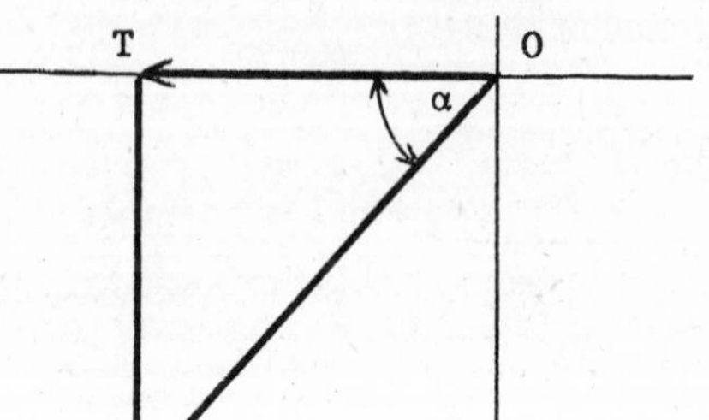

Answers to 31: a) $\frac{\overrightarrow{SB}}{\overrightarrow{0B}}$ b) $\frac{\overrightarrow{0S}}{\overrightarrow{0B}}$ c) $\frac{\overrightarrow{SB}}{\overrightarrow{0S}}$

33. In the triangle on the right:

(a) $\cos\alpha =$ ______

(b) $\tan\alpha =$ ______

(c) $\sin\alpha =$ ______

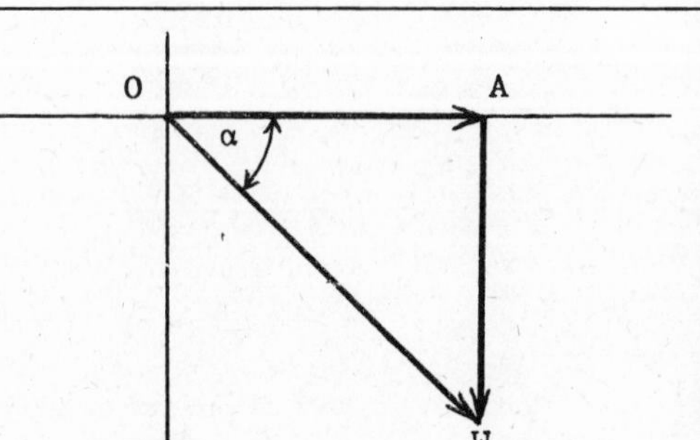

Answers to 32: a) $\frac{\overrightarrow{TJ}}{\overrightarrow{0T}}$ b) $\frac{\overrightarrow{TJ}}{\overrightarrow{0J}}$ c) $\frac{\overrightarrow{0T}}{\overrightarrow{0J}}$

34. To answer the following, look back at the last four frames, if you have to.

For any reference angle (which we call "α"),

$$\sin \alpha = \frac{\text{vertical component}}{\text{slanted vector}}$$

(a) $\cos \alpha = \frac{\text{Horizontal component}}{\text{slanted vector}}$

(b) $\tan \alpha = \frac{\text{vertical component}}{\text{Horizontal component}}$

a) $\frac{\overrightarrow{0A}}{\overrightarrow{0W}}$

b) $\frac{\overrightarrow{AW}}{\overrightarrow{0A}}$

c) $\frac{\overrightarrow{AW}}{\overrightarrow{0W}}$

---

35. A reference angle is an acute angle in a right triangle whose sides are vectors. Since vectors can be either positive or negative, <u>the trig ratios of a reference angle can be either positive or negative.</u>

What sign will a ratio have:

(a) If both vectors are positive? positive

(b) If both vectors are negative? Positive

(c) If one vector is positive and one is negative? Neg

a) $\frac{\text{horizontal component}}{\text{slanted vector}}$

b) $\frac{\text{vertical component}}{\text{horizontal component}}$

---

36. Here is a <u>second-quadrant</u> vector with its components.

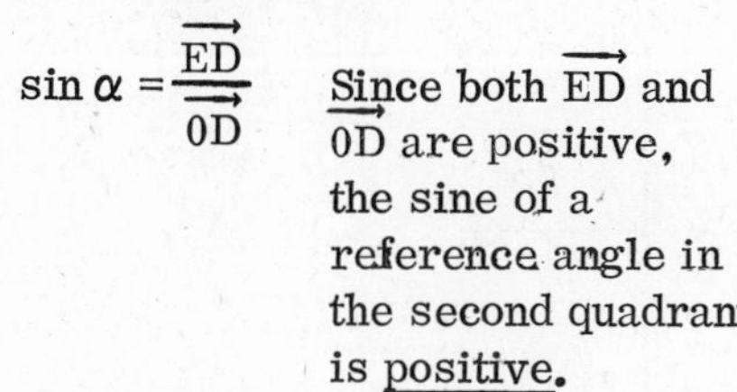

$\sin \alpha = \frac{\overrightarrow{ED}}{\overrightarrow{0D}}$ Since both $\overrightarrow{ED}$ and $\overrightarrow{0D}$ are positive, the sine of a reference angle in the second quadrant is <u>positive</u>.

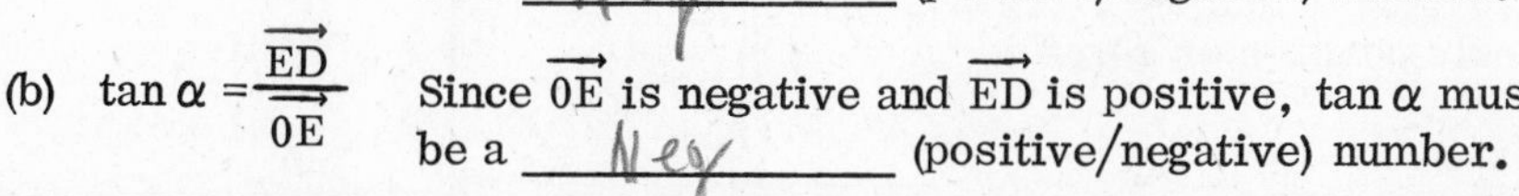

(a) $\cos \alpha = \frac{\overrightarrow{0E}}{\overrightarrow{0D}}$ Since $\overrightarrow{0E}$ is negative and $\overrightarrow{0D}$ is positive, $\cos \alpha$ must be a Neg (positive/negative) number.

(b) $\tan \alpha = \frac{\overrightarrow{ED}}{\overrightarrow{0E}}$ Since $\overrightarrow{0E}$ is negative and $\overrightarrow{ED}$ is positive, $\tan \alpha$ must be a Neg (positive/negative) number.

a) Positive

b) Positive

c) Negative

---

37. Here is a <u>third-quadrant</u> vector with its reference angle.

$\overrightarrow{0F}$ is positive.

$\overrightarrow{0B}$ and $\overrightarrow{BF}$ are negative.

In the third quadrant:

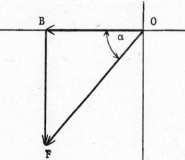

(a) Which of the ratios ($\sin \alpha$, $\cos \alpha$, $\tan \alpha$) for the reference angle are positive? tan

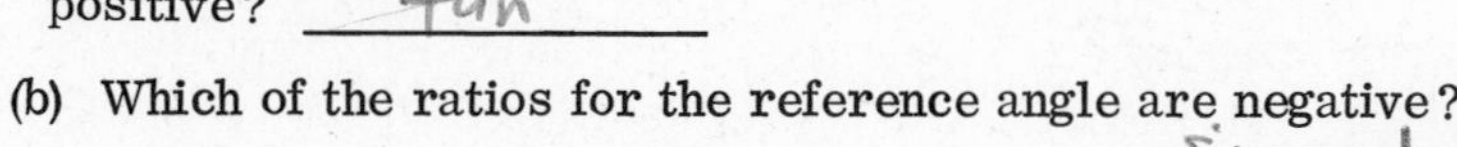

(b) Which of the ratios for the reference angle are negative? sin and cos

a) negative

b) negative

38. Here is a vector and its components in the fourth quadrant.

$\overrightarrow{0G}$ and $\overrightarrow{0J}$ are positive.

$\overrightarrow{GJ}$ is negative.

Therefore, in the fourth quadrant:

(a) sin α is ________ (positive/negative).

(b) cos α is ________ (positive/negative).

(c) tan α is ________ (positive/negative).

a) tan α

b) sin α and cos α

---

39. Here is a vector in the first quadrant.

In the first quadrant:

(a) All three vectors are ________ (positive/negative).

(b) All three ratios are ________ (positive/negative).

a) negative

b) positive

c) negative

---

40. For the next three frames, make your own sketches of vectors and their components in the four quadrants, if you need them.

Is the sine of a reference angle positive or negative:

(a) In Quadrant I? ________

(b) In Quadrant II? ________

(c) In Quadrant III? ________

(d) In Quadrant IV? ________

a) positive

b) positive

---

41. Is the cosine of a reference angle positive or negative:

(a) In Quadrant I? ________

(b) In Quadrant II? ________

(c) In Quadrant III? ________

(d) In Quadrant IV? ________

a) Positive

b) Positive

c) Negative

d) Negative

---

42. Is the tangent of a reference angle positive or negative:

(a) In Quadrant I? ________

(b) In Quadrant II? ________

(c) In Quadrant III? ________

(d) In Quadrant IV? ________

a) Positive

b) Negative

c) Negative

d) Positive

43. Here is a 46° reference angle in the first quadrant.

Since all three vectors are positive vectors, all three trigonometric ratios for this reference angle are positive numbers.

Therefore, from the trig table:

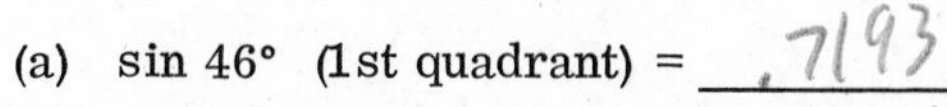

(a) sin 46° (1st quadrant) = ________

(b) cos 46° (1st quadrant) = ________

(c) tan 46° (1st quadrant) = ________

a) Positive

b) Negative

c) Positive

d) Negative

---

44. The absolute values of the three ratios (sine, cosine, and tangent) are the same for any 46° reference angle in any quadrant. However, the signs of the ratios depend on the particular quadrant in which the 46° angle lies.

Here is a 46° reference angle in the second quadrant.

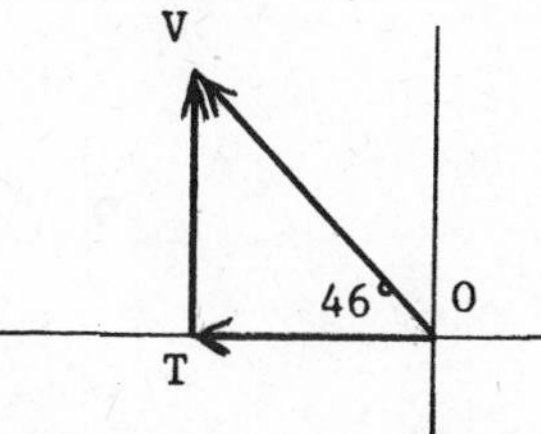

$\overrightarrow{TV}$ and $\overrightarrow{0V}$ are positive vectors.

$\overrightarrow{0T}$ is a negative vector.

(a) In the 2nd quadrant, which of the three ratios are positive? ________

(b) In the 2nd quadrant, which of the three ratios are negative? ________

(c) sin 46° (2nd quadrant) = ________

(d) cos 46° (2nd quadrant) = ________

(e) tan 46° (2nd quadrant) = ________

a) 0.7193

b) 0.6947

c) 1.036

---

45. Here is a 46° reference angle in the third quadrant.

(a) Which of the three vectors are negative? ________

(b) In the 3rd quadrant, which of the three ratios are negative? ________

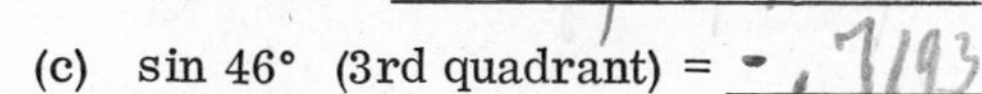

(c) sin 46° (3rd quadrant) = ________

(d) cos 46° (3rd quadrant) = ________

(e) tan 46° (3rd quadrant) = ________

a) sine

b) cosine and tangent

c) 0.7193

d) -0.6947

e) -1.036

---

a) $\overrightarrow{0W}$ and $\overrightarrow{WD}$

b) sine and cosine

c) -0.7193

d) -0.6947

e) 1.036

46. Here is a 46° reference angle in the fourth quadrant.

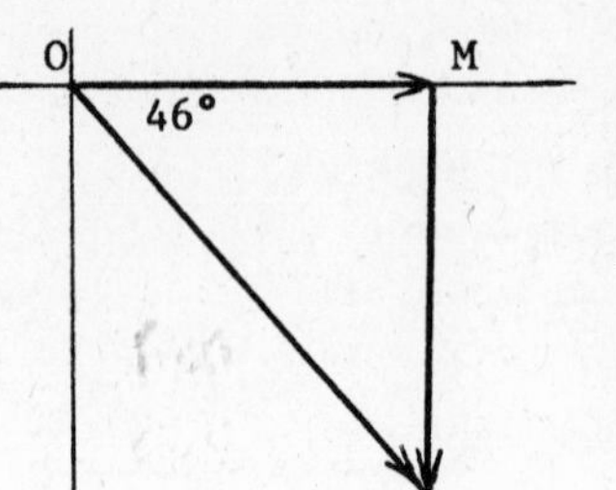

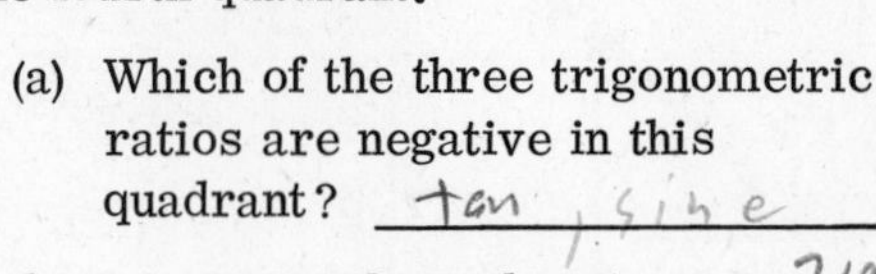

(a) Which of the three trigonometric ratios are negative in this quadrant? tan, sine

(b) sin 46° (4th quadrant) = -.7193

(c) cos 46° (4th quadrant) = .6947

(d) tan 46° (4th quadrant) = -1.036

---

47. Find the numerical value of each of the following. Using the axes at the right, make a rough sketch of the three vectors in order to determine whether each ratio is positive or negative. The absolute values are given in your trig table.

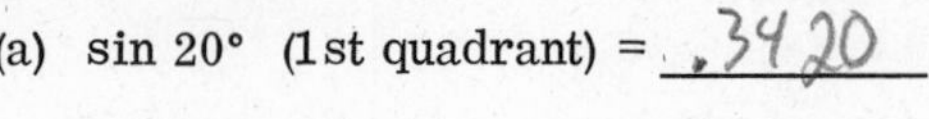

(a) sin 20° (1st quadrant) = .3420

(b) sin 20° (2nd quadrant) = .3420

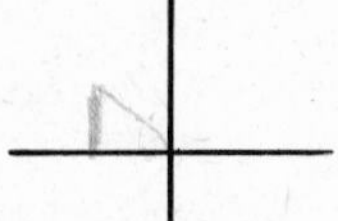

(c) sin 20° (3rd quadrant) = -.3420

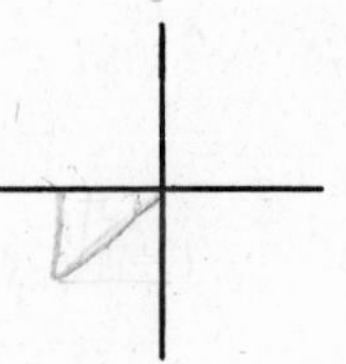

(d) sin 20° (4th quadrant) = -.3420

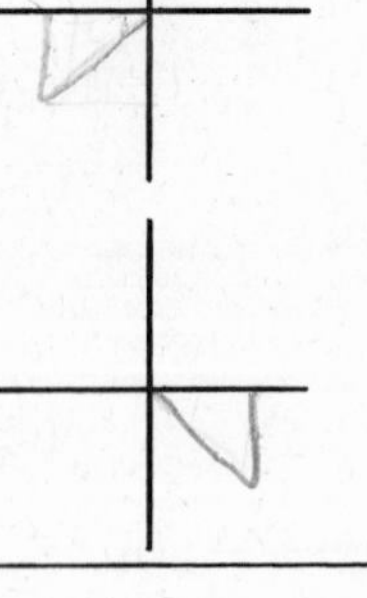

a) sine and tangent
b) -0.7193
c) 0.6947
d) -1.036

---

48. Make a rough sketch of the vectors using the axes at the bottom of this frame.

(a) cos 63° (1st quadrant) = .4540

(b) cos 63° (2nd quadrant) = -.4540

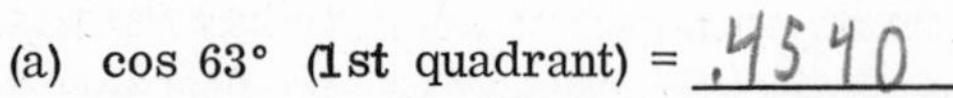

(c) cos 63° (3rd quadrant) = -.4540

(d) cos 63° (4th quadrant) = .4540

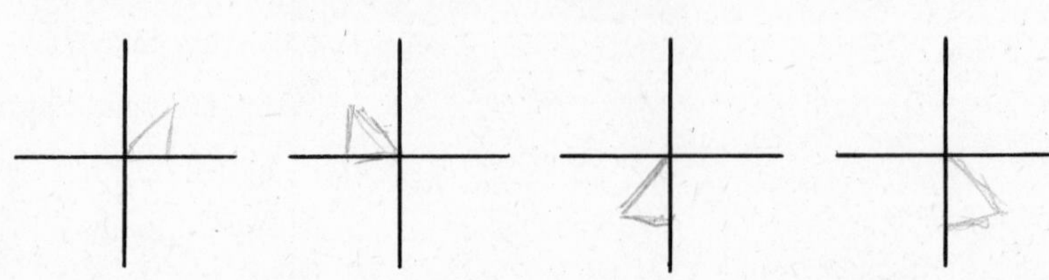

a) 0.3420
b) 0.3420
c) -0.3420
d) -0.3420

49. Since "sign errors" are very common when the ratios of reference angles are determined, <u>always make a rough sketch of the vectors.</u>

(a) tan 31° (1st quadrant) = ________

(b) tan 31° (2nd quadrant) = ________

(c) tan 31° (3rd quadrant) = ________

(d) tan 31° (4th quadrant) = ________

a) 0.4540
b) -0.4540
c) -0.4540
d) 0.4540

---

50. Instead of writing 1st quadrant, 2nd quadrant, etc., we will merely write Q1, Q2, Q3, Q4. Remember to make a rough sketch to determine the sign of each ratio.

(a) sin 41° (Q2) = ________

(b) cos 41° (Q4) = ________

(c) tan 41° (Q3) = ________

a) 0.6009
b) -0.6009
c) 0.6009
d) -0.6009

---

51. (a) sin 57° (Q3) = ________

(b) cos 57° (Q2) = ________

(c) tan 57° (Q4) = ________

a) 0.6561
b) 0.7547
c) 0.8693

---

52. (a) sin 81° (Q4) = ________

(b) cos 81° (Q3) = ________

(c) tan 81° (Q2) = ________

a) -0.8387
b) -0.5446
c) -1.540

---

53. All three ratios (sine, cosine, and tangent) are positive for reference angles in only one quadrant. Which quadrant is it? ________

a) -0.9877
b) -0.1564
c) -6.314

Q1, the 1st quadrant.

SELF-TEST 3 (Frames 30-53)

In Problems 1-6, state whether each ratio is positive or negative. To determine this, make a rough sketch of each angle.

1. sin α (Q2) Pos
2. sin α (Q3) Neg
3. cos α (Q1) Pos
4. cos α (Q4) pos
5. tan α (Q3) pos
6. tan α (Q2) Neg

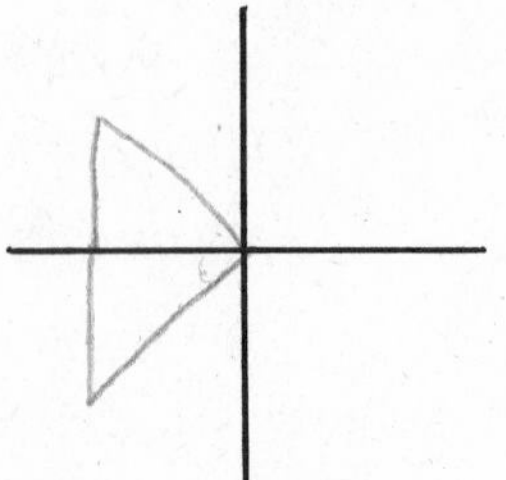
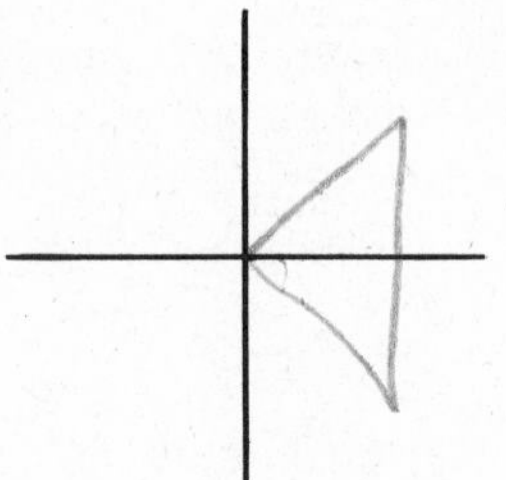
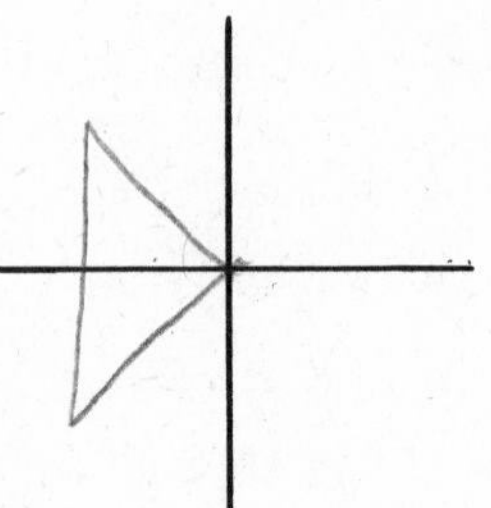

Using a table of "Trigonometric Ratios", find the numerical value of each of these ratios. Be sure to determine whether each ratio is positive or negative.

7. cos 27° (Q2) = -.8910
8. sin 59° (Q2) = .8572
9. tan 12° (Q3) = .2126
10. cos 35° (Q4) = .8192
11. sin 83° (Q3) = -.9925
12. tan 45° (Q2) = -1.0000

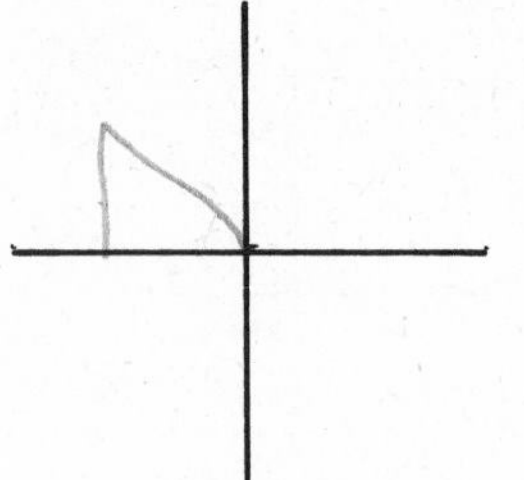
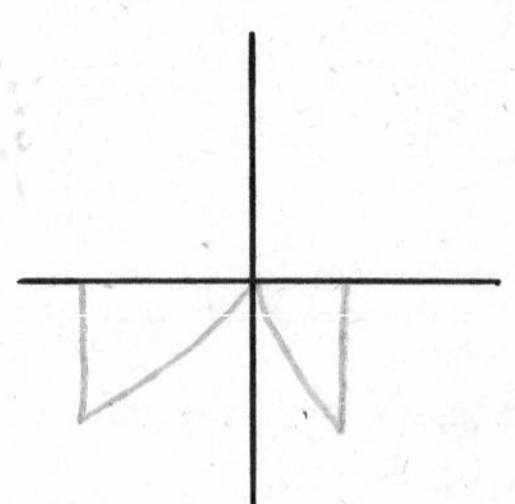
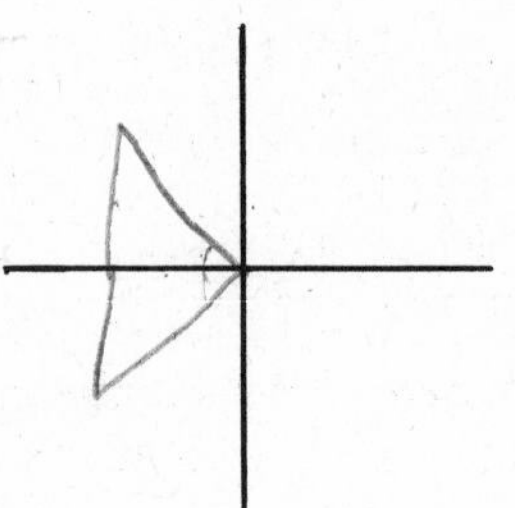

ANSWERS:

1. positive	4. positive	7. -0.8910	10. 0.8192
2. negative	5. positive	8. 0.8572	11. -0.9925
3. positive	6. negative	9. 0.2126	12. -1.0000

2-4 FINDING THE COMPONENTS OF A SLANTED VECTOR

If the length and direction of a slanted vector are known, we can find its components by using the sine and cosine ratios. We will show the method in this section. Since the components are "signed" vectors, we must be careful to avoid "sign" errors.

54. We know the length and direction of the first-quadrant vector on the right.

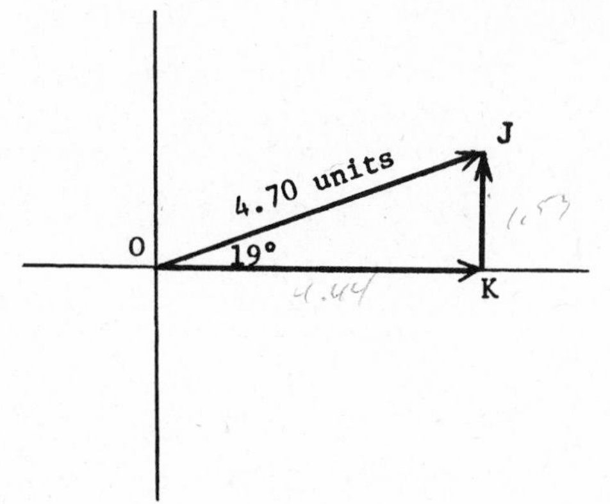

We want to find the signed-length of its two components, $\overrightarrow{KJ}$ and $\overrightarrow{0K}$.

<u>To find the length of $\overrightarrow{KJ}$, we use the sine ratio:</u>

$$\text{Since } \sin 19° = \frac{\overrightarrow{KJ}}{\overrightarrow{0J}},\ \overrightarrow{KJ} = (\overrightarrow{0J})(\sin 19°)$$
$$= (4.70)(0.3256)$$
$$= 1.53 \text{ units}$$

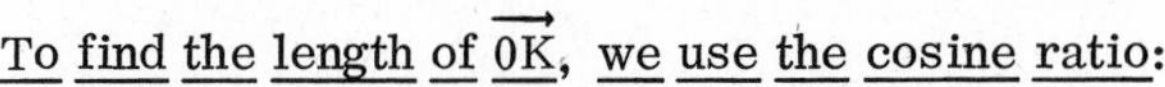

<u>To find the length of $\overrightarrow{0K}$, we use the cosine ratio:</u>

$$\text{Since } \cos 19° = \frac{\overrightarrow{0K}}{\overrightarrow{0J}},\ \overrightarrow{0K} = (\overrightarrow{0J})(\cos 19°)$$
$$= (4.70)(0.9455)$$
$$= 4.44 \text{ units}$$

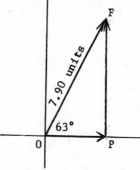

Given the lengths of the components, the coordinates of point J must be (____).

---

55. Here is a problem similar to the one in the last frame.

Here are some questions which you should always ask yourself so that you do not compute "blindly".

(a) Since angle 0PF is a 90° angle, angle 0FP is a ____° angle.

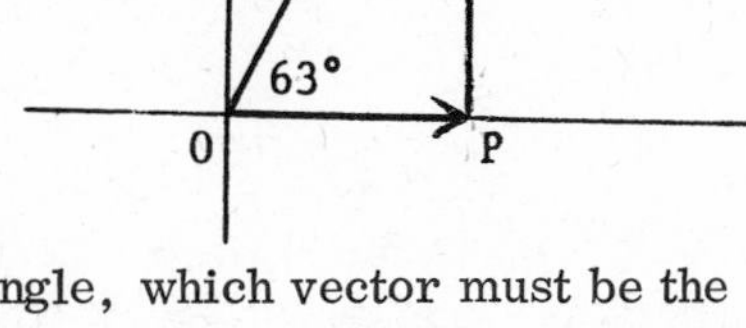

The longest side of a triangle is always opposite the largest angle. Therefore:

(b) Of the three vectors in this triangle, which vector must be the longest? ________

(c) Which vector must be longer, $\overrightarrow{0P}$ or $\overrightarrow{PF}$? ________

(d) If you compute the length of $\overrightarrow{0P}$ and obtain 8.43 units as an answer, does this answer make sense? ________

(e) If you compute the length of $\overrightarrow{0P}$ and $\overrightarrow{PF}$ and obtain 5.48 units and 3.92 units, respectively, do these answers make sense? ________

(4.44, 1.53)

---

a) 27°

b) $\overrightarrow{0F}$, the hypotenuse

c) $\overrightarrow{PF}$

d) No. A leg of a right triangle cannot be longer than the hypotenuse.

e) No. $\overrightarrow{PF}$ must be longer than 0P, since it is opposite a larger angle.

56. Here is a second-quadrant vector.

To find the lengths of its components, $\overrightarrow{0N}$ and $\overrightarrow{NA}$, we use the following two ratios:

sin 49° (Q2) = +0.7547

cos 49° (Q2) = -0.6561

(Notice the signs of these two ratios.)

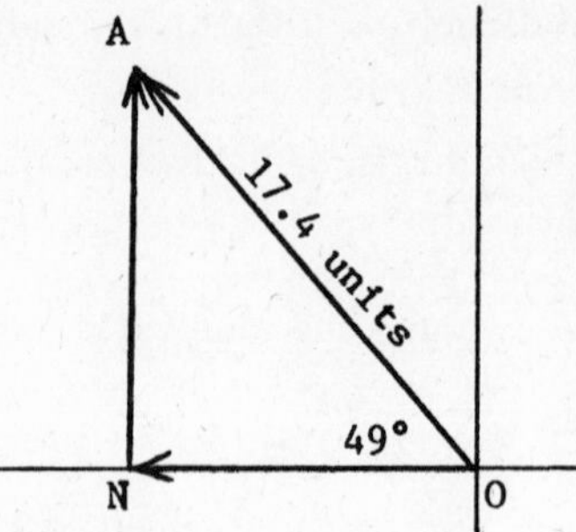

(a) $\overrightarrow{0N}$ is a __Neg__ (positive/negative) vector.

(b) $\overrightarrow{NA}$ is a __pos__ (positive/negative) vector.

(c) To find $\overrightarrow{0N}$, the horizontal component, we use the cosine ratio.

$$\text{Since } \cos 49° \text{ (Q2)} = \frac{\overrightarrow{0N}}{\overrightarrow{0A}},\ \overrightarrow{0N} = (\overrightarrow{0A})[\cos 49° \text{ (Q2)}]$$

$$= (17.4)(-0.6561)$$

= __-11.4__

(d) To find $\overrightarrow{NA}$, the vertical component, we use the sine ratio.

$$\text{Since } \sin 49° \text{ (Q2)} = \frac{\overrightarrow{NA}}{\overrightarrow{0A}},\ \overrightarrow{NA} = (\overrightarrow{0A})[\sin 49° \text{ (Q2)}]$$

$$= (17.4)(+0.7547)$$

= __13.1__

---

a) negative

b) positive

c) -11.4 units

d) +13.1 units

57. Here is a 3rd-quadrant vector. We want to find the signed-lengths of its components.

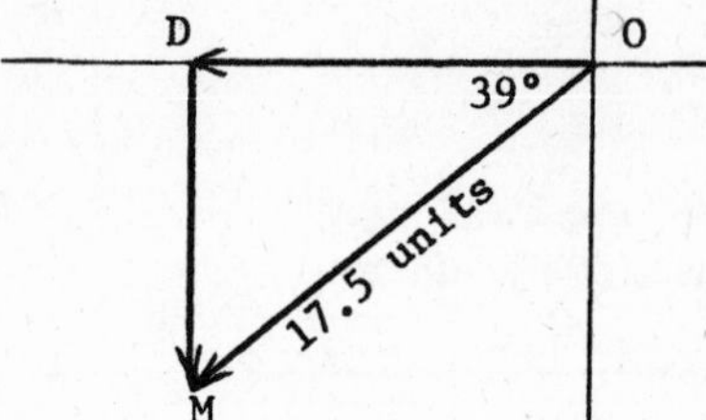

(a) The length of $\overrightarrow{0D}$ is __Neg__ (positive/negative).

(b) The length of $\overrightarrow{DM}$ is __Neg__ (positive/negative).

(c) sin 39° (Q3) = __-.6293__ (d) cos 39° (Q3) = __-.7771__

(e) Finding $\overrightarrow{0D}$, the horizontal component:

$$\text{Since } \cos 39° \text{ (Q3)} = \frac{\overrightarrow{0D}}{\overrightarrow{0M}},\ \overrightarrow{0D} = (\overrightarrow{0M})[\cos 39° \text{ (Q3)}]$$

$$= (17.5)(-0.7771)$$

= __13.6__

(f) Finding $\overrightarrow{DM}$, the vertical component:

$$\text{Since } \sin 39° \text{ (Q3)} = \frac{\overrightarrow{DM}}{\overrightarrow{0M}},\ \overrightarrow{DM} = (\overrightarrow{0M})[\sin 39° \text{ (Q3)}]$$

$$= (17.5)(-0.6293)$$

= __11.0__

58. Here is a 4th-quadrant vector. Again, we want to find the signed-lengths of its components.

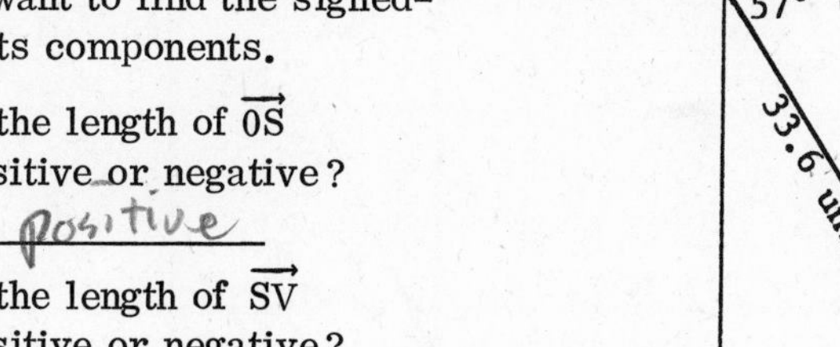

(a) Is the length of $\overrightarrow{0S}$ positive or negative? ______

(b) Is the length of $\overrightarrow{SV}$ positive or negative? ______

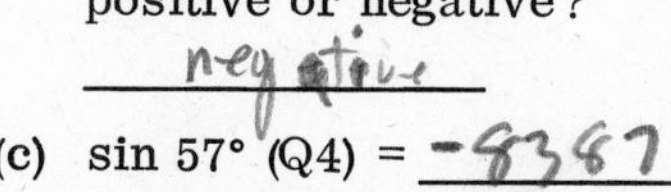

(c) sin 57° (Q4) = ______ (d) cos 57° (Q4) = ______

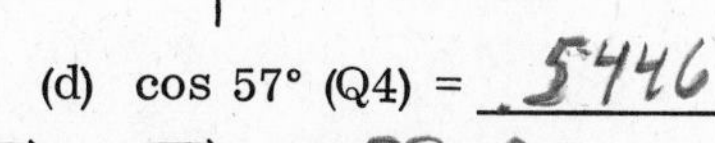

(e) Calculate the signed-length of $\overrightarrow{SV}$. $\overrightarrow{SV}$ = ______

(f) Calculate the signed-length of $\overrightarrow{0S}$. $\overrightarrow{0S}$ = ______

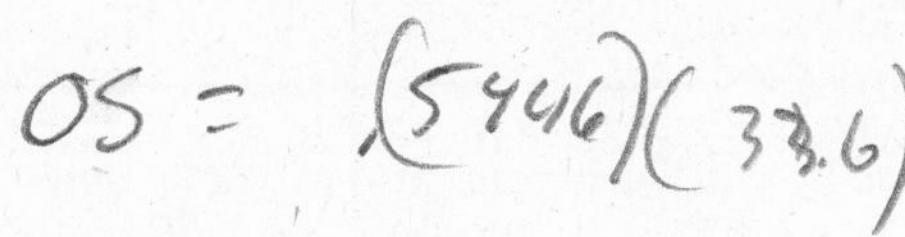

a) negative
b) negative
c) -0.6293
d) -0.7771
e) -13.6 units
f) -11.0 units

---

59. After computing the signed-lengths of the components of a slanted vector, always check your sketch to make sure that you have not made a "sign" error.

Here is a vector in the 3rd quadrant. By examining the sketch, you can check the "signs" of the computed lengths for the two components.

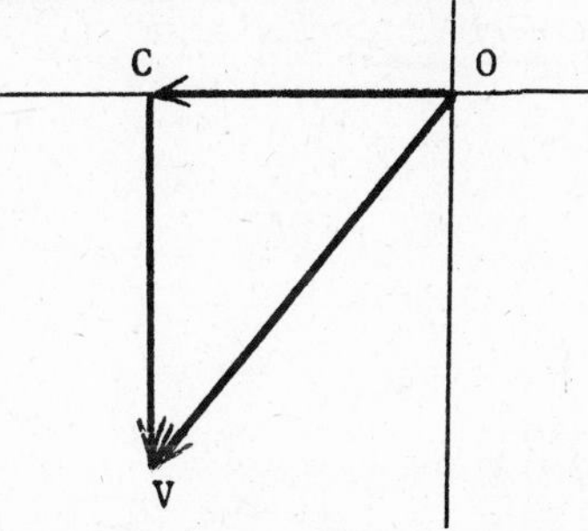

(a) $\overrightarrow{0C}$ must be a ______ (positive/negative) length.

(b) $\overrightarrow{CV}$ must be a ______ (positive/negative) length.

a) positive
b) negative
c) -0.8387
d) +0.5446
e) -28.2 units
f) +18.3 units

---

60. Find the horizontal and vertical components of $\overrightarrow{0H}$.

(a) Horizontal component $\overrightarrow{0G}$ = ______

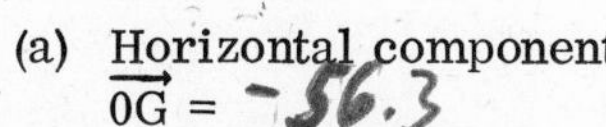

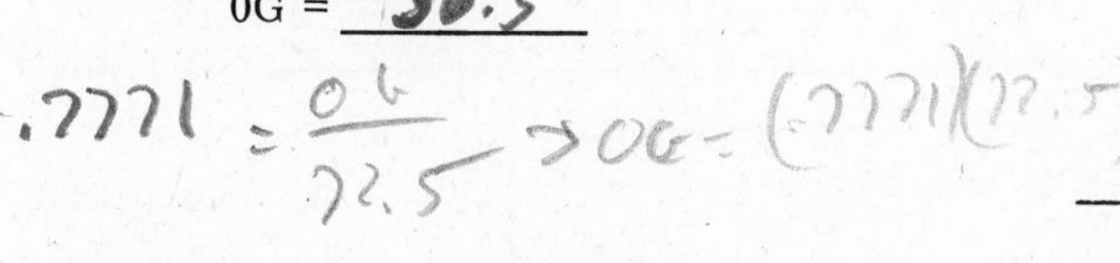

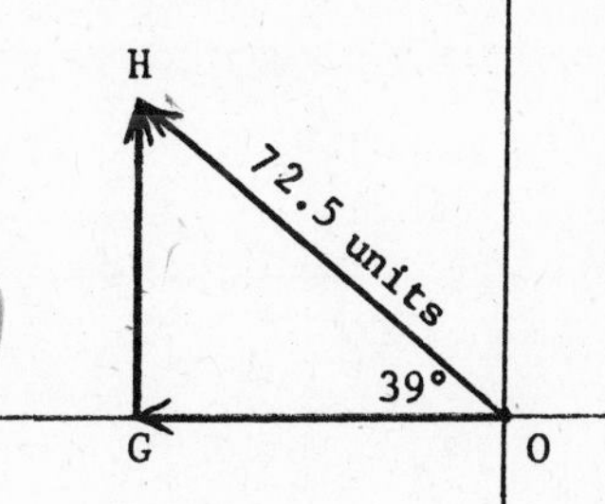

(b) Vertical component $\overrightarrow{GH}$ = ______

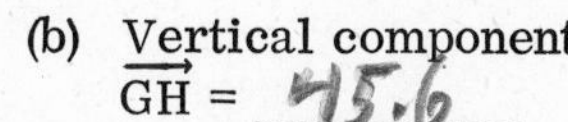

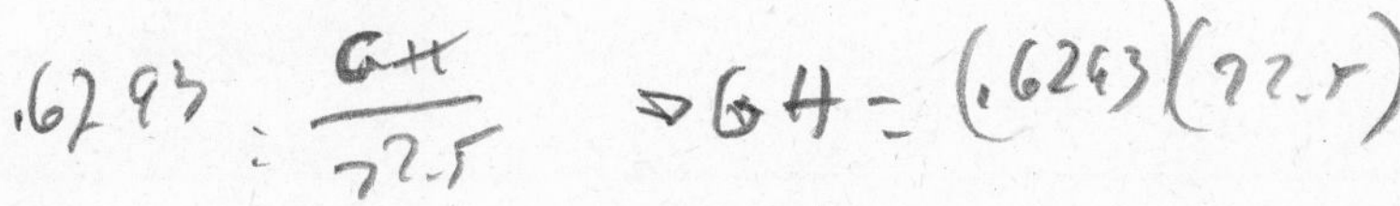

a) negative
b) negative

61. Find the horizontal and vertical components of this vector.

(a) Horizontal component = -177

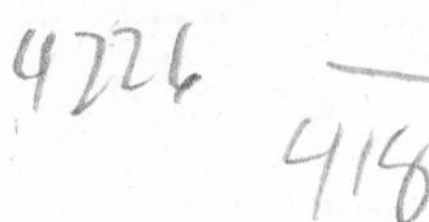

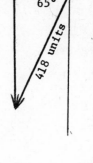

(b) Vertical component = 379

a) $\overrightarrow{0G} = -56.3$

b) $\overrightarrow{GH} = +45.6$

---

62. A vector 8.35 units long has a reference angle of 18° and lies in Quadrant IV.

(a) Sketch the vector and its components on the right.

(b) Calculate its horizontal component. 7.94

(c) Calculate its vertical component. 2.58

a) -177 units

b) -379 units

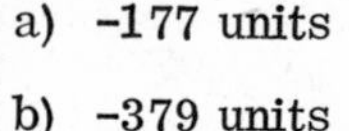

---

a)

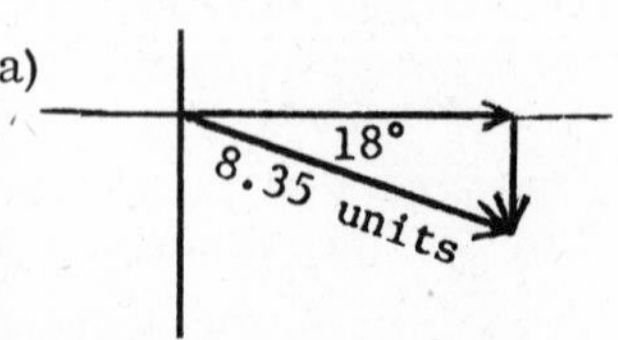

b) +7.94 units

c) -2.58 units

SELF-TEST 4 (Frames 54-62)

Find the horizontal and vertical components of $\overrightarrow{OR}$.

1. Horizontal component = ________
2. Vertical component = ________

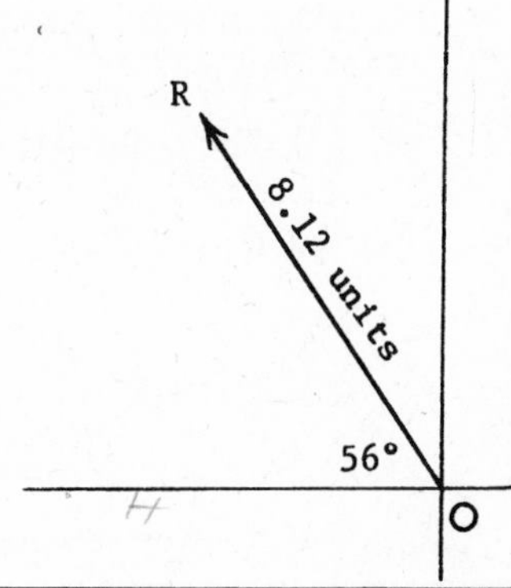

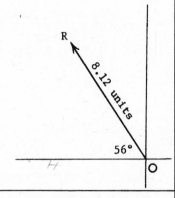

A slanted vector has a length of 600 units. Its reference angle is 16° (Q4). Find its horizontal and vertical components.

3. Horizontal component = ________
4. Vertical component = ________

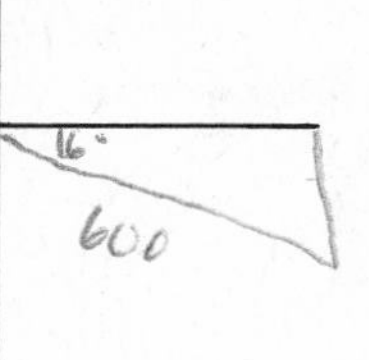

ANSWERS:

1. -4.54 units	3. 577 units
2. 6.73 units	4. -165 units

2-5 FINDING THE LENGTH AND DIRECTION OF A SLANTED VECTOR

If the components of a slanted vector are known, we can determine its length and direction (or reference angle). We will show the method in this section.

63. Here is a slanted vector in the first quadrant.

Since the coordinates of point B are (5.0, 4.0),

$\overrightarrow{0D}$, the horizontal component, must be a +5.0 vector.

$\overrightarrow{DB}$, the vertical component, must be a +4.0 vector.

We want to find: (1) The size of the reference angle α.
(2) The length of the slanted vector, $\overrightarrow{0B}$.

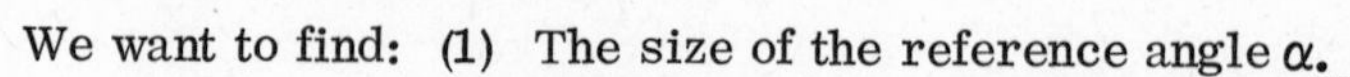

(a) To find the size of α, we use the tangent ratio.

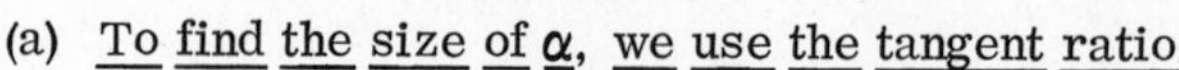

Since $\tan\alpha = \dfrac{\overrightarrow{DB}}{\overrightarrow{0D}} = \dfrac{4.0}{5.0}$, $\tan\alpha = 0.8000$

Therefore, α contains 39°.

(b) To find the length of $\overrightarrow{0B}$, we use the Pythagorean Theorem.

$$\text{Since } \left(\overrightarrow{0B}\right)^2 = \left(\overrightarrow{0D}\right)^2 + \left(\overrightarrow{DB}\right)^2$$

$$\overrightarrow{0B} = \sqrt{\left(\overrightarrow{0D}\right)^2 + \left(\overrightarrow{DB}\right)^2}$$

$$= \sqrt{(5.0)^2 + (4.0)^2}$$

$$= \sqrt{41} = 6.4$$

Answer to Frame 63: a) 39° b) 6.4 units

64. In the last frame, we solved for the length of the slanted vector, $\overrightarrow{0B}$, by means of the Pythagorean Theorem. Another method is possible. After determining the size of the reference angle α, we could then use either $\sin\alpha$ or $\cos\alpha$ to calculate the length of $\overrightarrow{0B}$.

For example, referring to the diagram in the last frame:

(1) Since $\sin\alpha = \dfrac{\overrightarrow{DB}}{\overrightarrow{0B}}$: $\overrightarrow{0B} = \dfrac{\overrightarrow{DB}}{\sin\alpha} = \dfrac{4.0}{\sin 39°} = \dfrac{4.0}{0.6293} = 6.4$ units

(2) Since $\cos\alpha = \dfrac{\overrightarrow{0D}}{\overrightarrow{0B}}$: $\overrightarrow{0B} = \dfrac{\overrightarrow{0D}}{\cos\alpha} = \dfrac{5.0}{\cos 39°} = \dfrac{5.0}{0.7771} = 6.4$ units

However, using the Pythagorean Theorem is better for the following reason:

To determine the size of angle α, we computed its tangent ratio and obtained 0.8000.

We than picked 39°, the angle in the table whose tangent ratio is closest to 0.8000. However, some rounding was involved because tan 39° (from the table) is 0.8098. Since α in our problem is not exactly 39°, $\sin\alpha$ and $\cos\alpha$ are not exactly 0.6293 and 0.7771. These are the exact values for a 39° angle.

<u>We can avoid this rounding problem by using the Pythagorean Theorem, and it is better to do so.</u>

Go to next frame.

65. In this case, we know that:

$\overrightarrow{0T}$ is 6.9 units.

$\overrightarrow{TR}$ is 4.7 units.

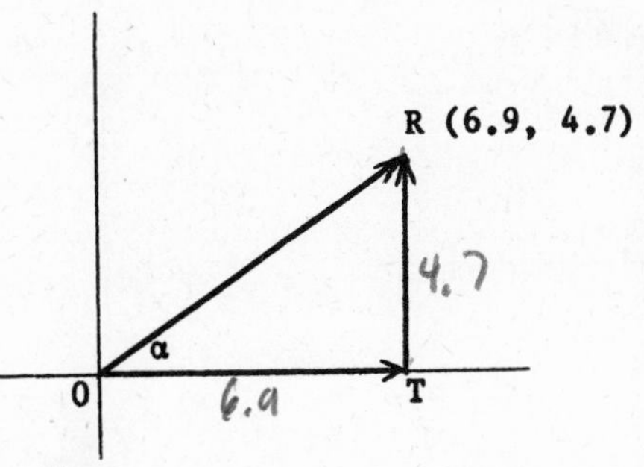

If you were to calculate $\overrightarrow{0R}$ and angle α, here are some questions you could ask yourself to see whether or not your answers make sense.

(a) Can $\overrightarrow{0R}$ be less than 6.9 units? NO

(b) Which angle must be larger, angle α or angle 0RT? Why? ORT because it is opposite a longer side

(c) Suppose you compute the size of angle α and report that it contains 49°. Does this make sense? NO

---

66. Here is a vector in the 2nd quadrant. We know the lengths of its two components.

To find the length of $\overrightarrow{0K}$, we use the Pythagorean Theorem. Notice that squaring a negative number always gives a positive number.

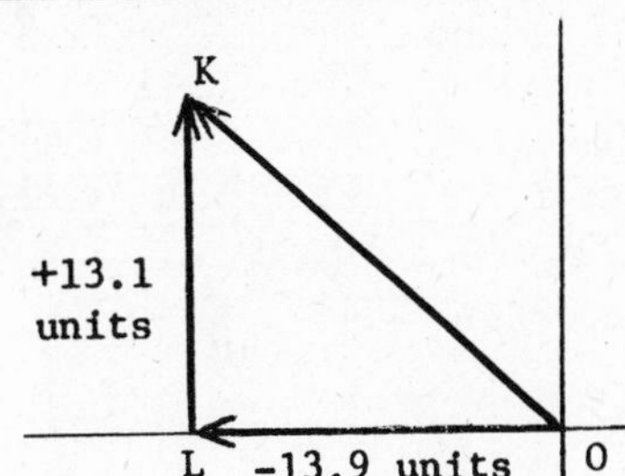

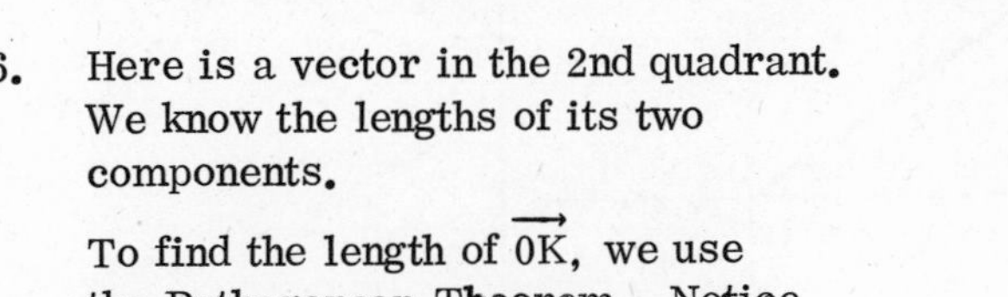

$$(\overrightarrow{0K})^2 = (\overrightarrow{0L})^2 + (\overrightarrow{LK})^2$$

$$\overrightarrow{0K} = \sqrt{(\overrightarrow{0L})^2 + (\overrightarrow{LK})^2} = \sqrt{(-13.9)^2 + (13.1)^2}$$

$$= \sqrt{(+193) + 172}$$

$$= \sqrt{365} = \underline{\quad 19.1 \quad}$$

Answers to 65:

a) No. $\overrightarrow{0R}$ is the hypotenuse and it must be longer than the longest side of the right triangle.

b) Angle 0RT, since it is opposite a longer side.

c) No. If this were true, angle 0RT would contain only 41°. But angle 0RT must be larger than angle α.

---

67. Here is a vector in the 3rd quadrant. We know the lengths of its two components.

Watch the negative signs disappear when we use the Pythagorean Theorem to find the length of $\overrightarrow{0T}$.

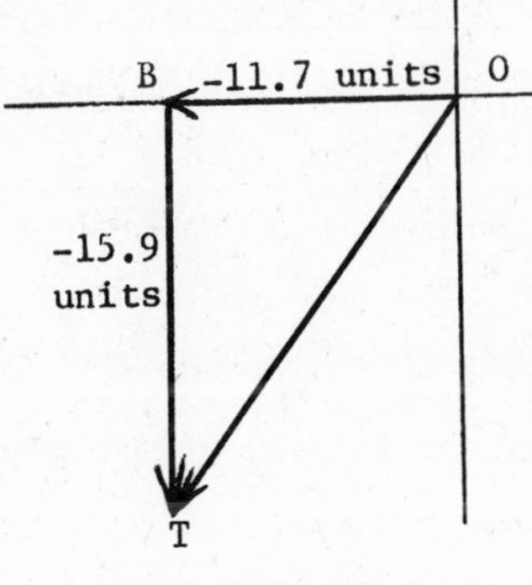

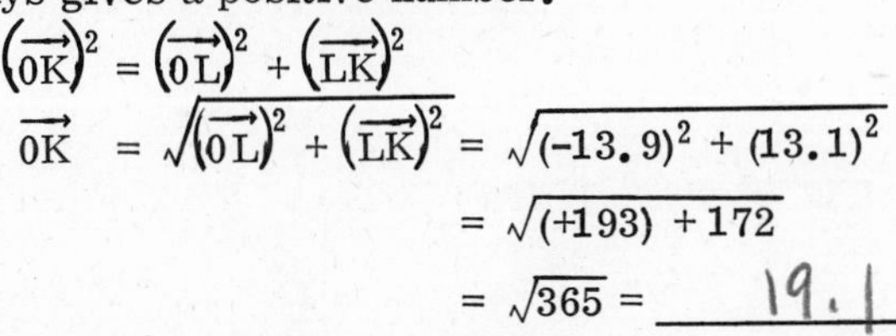

$$\overrightarrow{0T} = \sqrt{(\overrightarrow{0B})^2 + (\overrightarrow{BT})^2} = \sqrt{(-11.7)^2 + (-15.9)^2}$$

$$= \sqrt{137 + 253}$$

$$= \sqrt{390} = \underline{\quad 19.7 \quad}$$

19.1 units

19.7 units

68. Let's find the length of $\overrightarrow{0S}$.

$$\overrightarrow{0S} = \sqrt{(6.30)^2 + (-4.10)^2}$$
$$= \sqrt{39.7 + 16.8}$$
$$= \sqrt{56.5} = \_\_\_\_\_\_\_\_$$

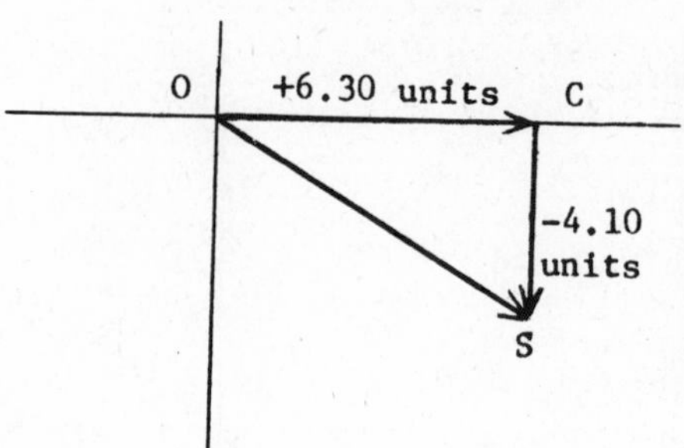

7.52 units

69. We want to find reference angle α for slanted vector $\overrightarrow{0E}$.

From the coordinates of point E, we know that:

The length of $\overrightarrow{0F}$ is -6 units.

The length of $\overrightarrow{FE}$ is +4 units.

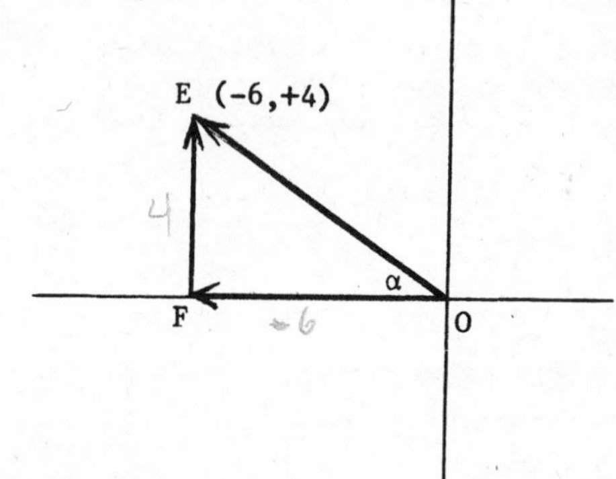

To find the size of the reference angle α, we use $\tan \alpha$.

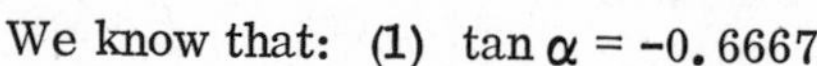

$$\tan \alpha = \frac{\overrightarrow{FE}}{\overrightarrow{0F}} = \frac{+4}{-6} = -0.6667$$

We know that: (1) $\tan \alpha = -0.6667$

(2) α is an angle in the 2nd quadrant.

To find α: (1) We ignore the negative sign and look up "0.6667" in the table. (34° is the closest angle.)

(2) We report the reference angle as: 34° (Q2)
(Be sure to list the quadrant.)

Go to next frame.

70. (a) The signed length of $\overrightarrow{0D}$ is ______ units.

(b) The signed length of $\overrightarrow{DH}$ is ______ units.

(c) Since $\tan \alpha = \frac{\overrightarrow{DH}}{\overrightarrow{0D}} = \frac{-7}{-5} = 1.4000$

$\alpha =$ ________

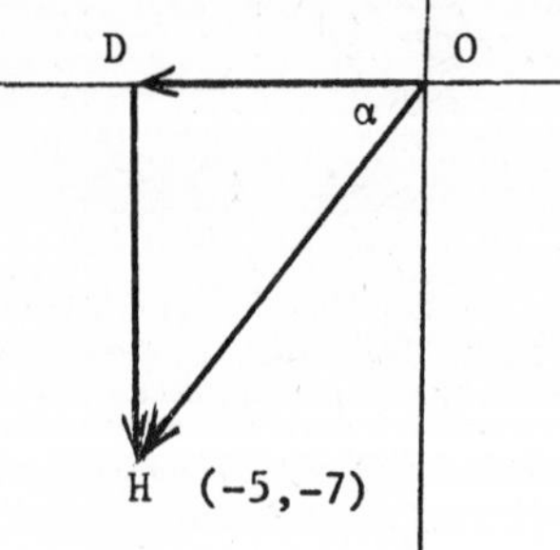

---

a) -5

b) -7

c) 54° (Q3)
(Be sure to list the quadrant.)

71. To specify a reference angle, you must state its quadrant. If you only state the size of the angle, it could be in any of the four quadrants.

Specify the reference angle for $\overrightarrow{0M}$.

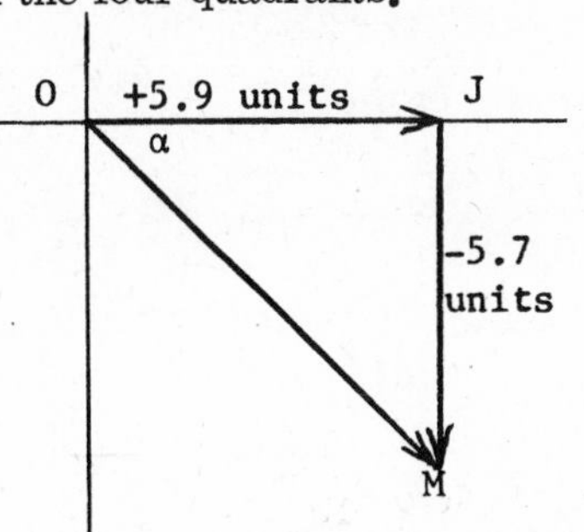

α = 44° (Q4)

72. The components of $\overrightarrow{0P}$ are given.

(a) Calculate the length of $\overrightarrow{0P}$.

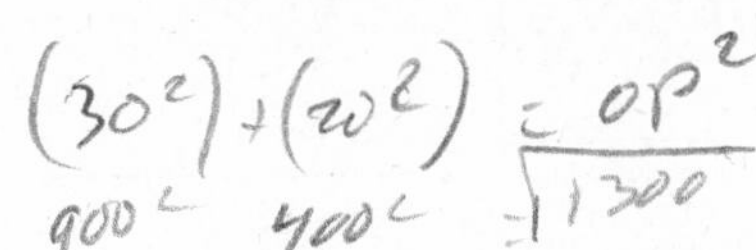

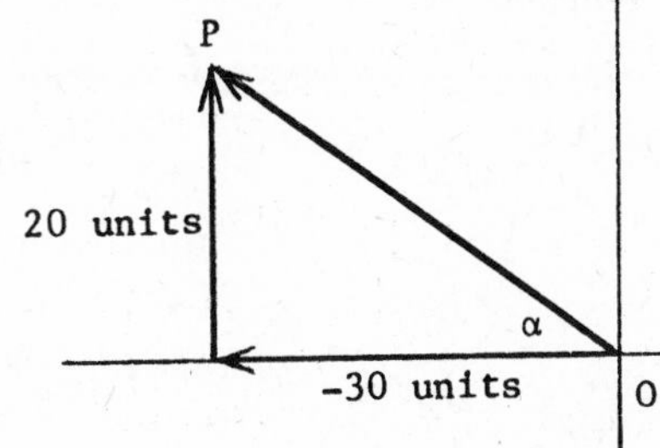

$\overrightarrow{0P}$ = 36

(b) Calculate the size of reference angle α.

α = 34° (Q2)

44° (Q4)
(The quadrant must be listed.)

73. Referring to the diagram:

(a) The horizontal component of $\overrightarrow{0H}$ is -2.00.

(b) The vertical component of $\overrightarrow{0H}$ is -4.30.

(c) The length of $\overrightarrow{0H}$ is 4.74.

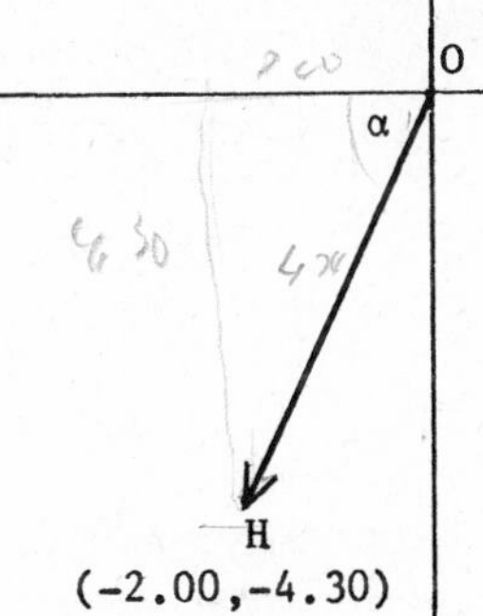

18.49 + 4.00

(d) Find reference angle α.

$\frac{4.30}{2.00}$

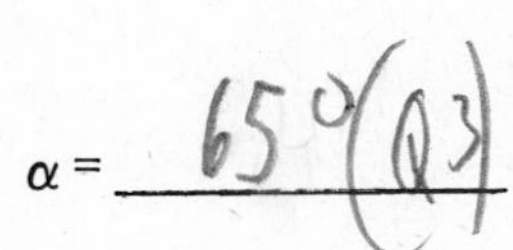

α = 65° (Q3)

a) $\overrightarrow{0P}$ = 36 units
(from $\sqrt{1300}$)

b) α = 34° (Q2)

74. A slanted vector begins at the origin and ends at the point (840, -320).

(a) Sketch the vector at the right.

(b) Calculate the length of the vector.

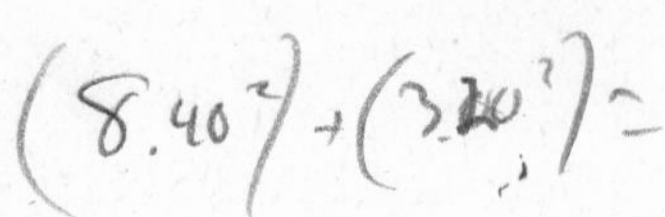

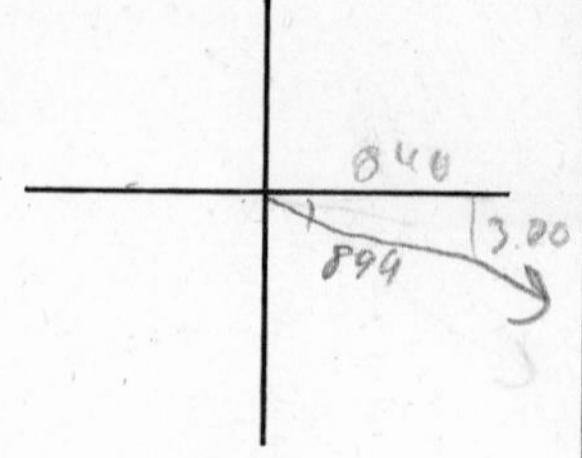

(c) Calculate the reference angle of the vector.

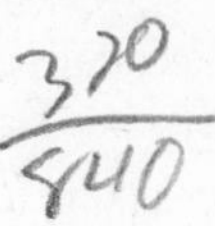

a) -2.00 units

b) -4.30 units

c) $\overrightarrow{0H} = 4.74$ units

d) $\alpha = 65°$ (Q3)

Answer to Frame 74: a)

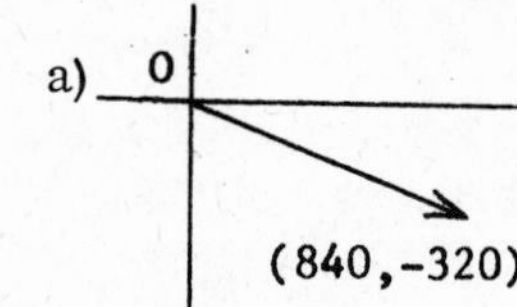

b) 899 units

c) 21° (Q4)

SELF-TEST 5 (Frames 63-74)

In the diagram at the right, $\overrightarrow{0P} = 41.6$ units, and $\overrightarrow{PM} = -26.2$ units.

1. Find the length of $\overrightarrow{0M}$.

2. Find reference angle α.

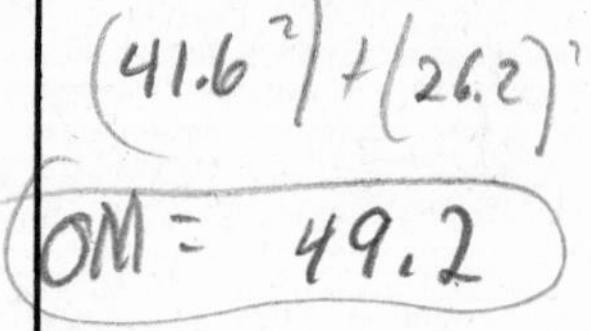

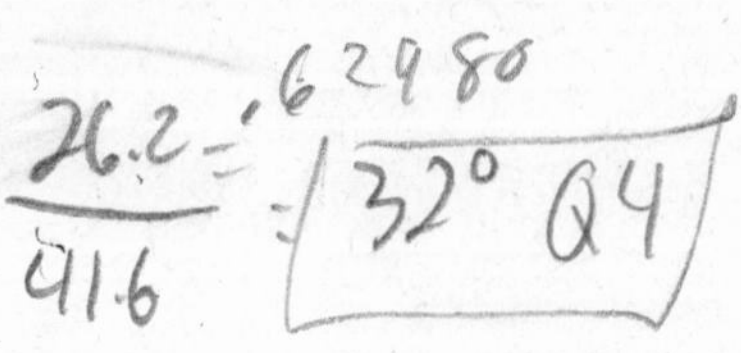

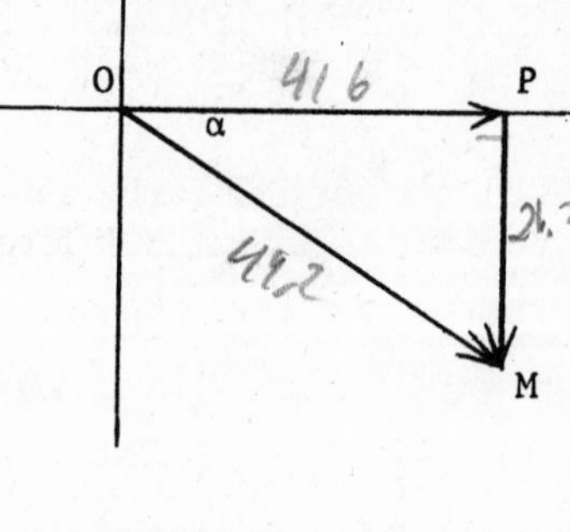

A vector begins at the origin and ends at the point (-50, -80).

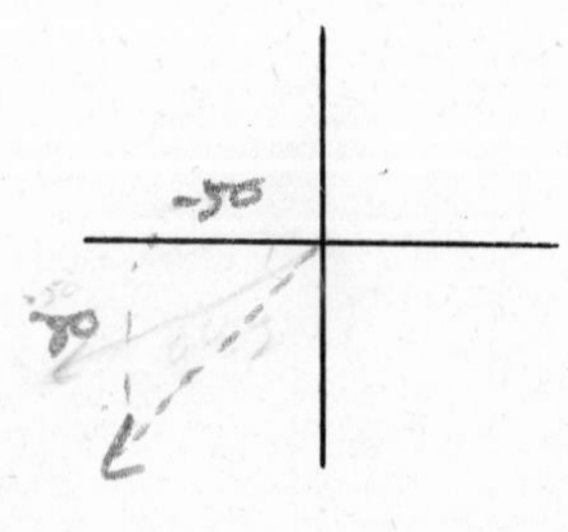

3. The length of the vector is ________.

4. The reference angle of the vector is ________.

ANSWERS:

1. 49.2 units
2. 32° (Q4)
3. 94.3 units
4. 58° (Q3)

2-6 ADDITION OF VECTORS BY THE PARALLELOGRAM METHOD

In this section, we will define what is meant by the addition of two vectors. Vectors will be added by the "parallelogram method".

75. There are two vectors, $\overrightarrow{0S}$ and $\overrightarrow{0P}$, in Figure 1 below. We will use these two vectors to show what is meant by the "addition of two vectors".

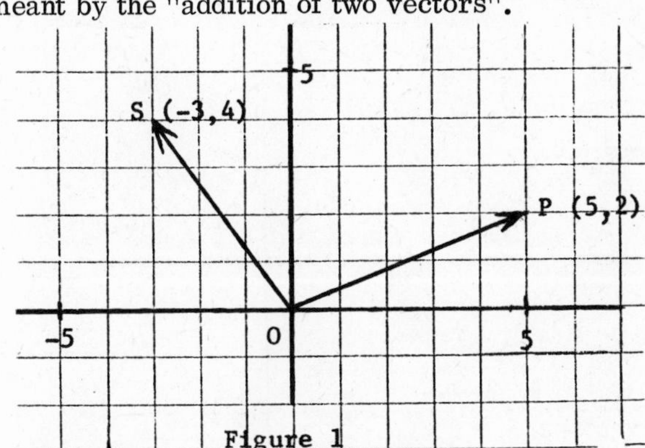

Figure 1

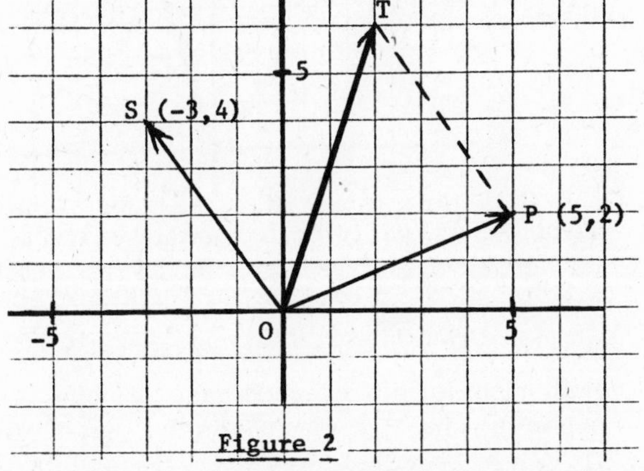

Figure 2

In Figure 2, the same two vectors have been added graphically. The sum of the two vectors is a third vector, $\overrightarrow{0T}$. To find this sum, we used two steps:

Step 1: We drew the dotted vector PT which is both parallel to $\overrightarrow{0S}$ and equal in length to $\overrightarrow{0S}$.

Step 2: We drew a vector from the origin to point T. This vector, $\overrightarrow{0T}$, is the sum of the two original vectors, $\overrightarrow{0S}$ and $\overrightarrow{0P}$.

From the graph, the coordinates of point T are (2, 6).

Answer to Frame 75: (2, 6)

76. In Figure 1 below, we have drawn the same two vectors, $\overrightarrow{0S}$ and $\overrightarrow{0P}$. In Figure 2, we have added the two vectors by a different procedure.

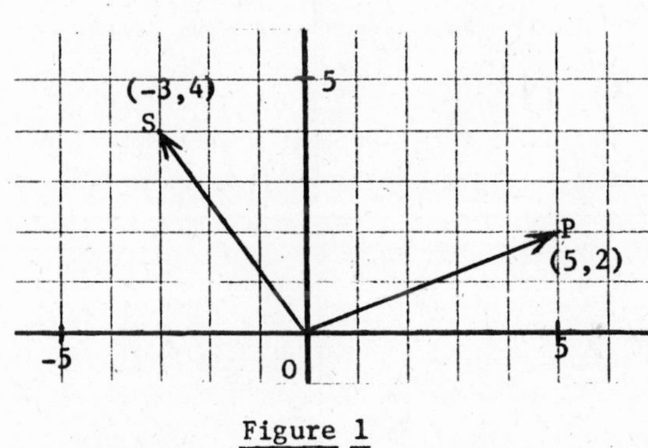

Figure 1

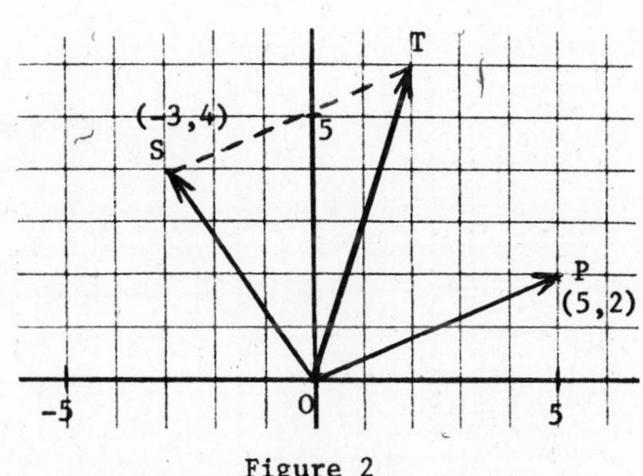

Figure 2

Two steps were also used to find the sum, vector $\overrightarrow{0T}$, by this method. They are:

Step 1: We drew the dotted vector $\overrightarrow{ST}$ which is both parallel to $\overrightarrow{0P}$ and equal in length to $\overrightarrow{0P}$.
Step 2: We drew a vector from the origin to point T.

(a) The coordinates of point T are (2 , 6).

(b) Are these the same coordinates we obtained for point T in the last frame? yes

(c) The sum of $\overrightarrow{0S}$ and $\overrightarrow{0P}$ is what vector? $\overrightarrow{0T}$

77. We can use the graph below to summarize what was done in the last two frames.

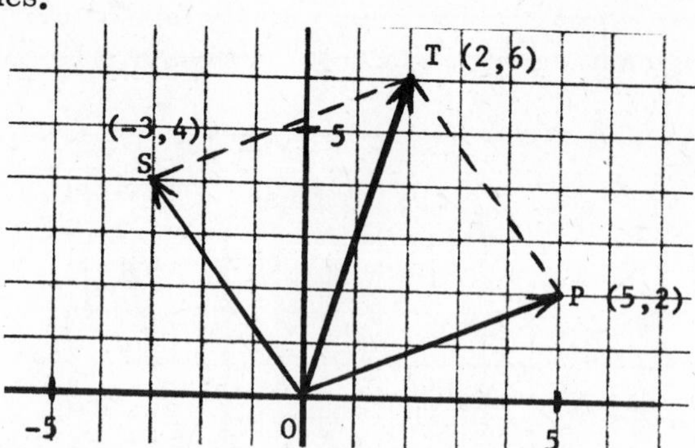

To add $\overrightarrow{0S}$ and $\overrightarrow{0P}$, we can do either of two things:

(1) Draw $\overrightarrow{PT}$ parallel to and equal to $\overrightarrow{0S}$.

(2) Draw $\overrightarrow{ST}$ parallel to and equal to $\overrightarrow{0P}$.

Both of these dotted vectors meet at point T. The vector from the origin to point T (vector $\overrightarrow{0T}$) is the sum of the two vectors.

Since $\overrightarrow{ST}$ is parallel to $\overrightarrow{0P}$ and $\overrightarrow{PT}$ is parallel to $\overrightarrow{0S}$:

(1) Figure STP0 is a parallelogram,
and (2) Vector $\overrightarrow{0T}$ is a diagonal of the parallelogram.

Therefore, we can find the sum of two vectors by:

(1) Completing the parallelogram,
and (2) Drawing its diagonal from the origin.

The sum of two vectors is itself a vector. A vector which is the sum of vectors is called the resultant vector.

In the figure above, since $\overrightarrow{0T}$ is the sum of $\overrightarrow{0S}$ and $\overrightarrow{0P}$, $\overrightarrow{0T}$ is called the resultant vector.

a) (2,6)

b) Yes

c) $\overrightarrow{0T}$

---

78. Here is a different addition problem. Add $\overrightarrow{0B}$ and $\overrightarrow{0D}$.

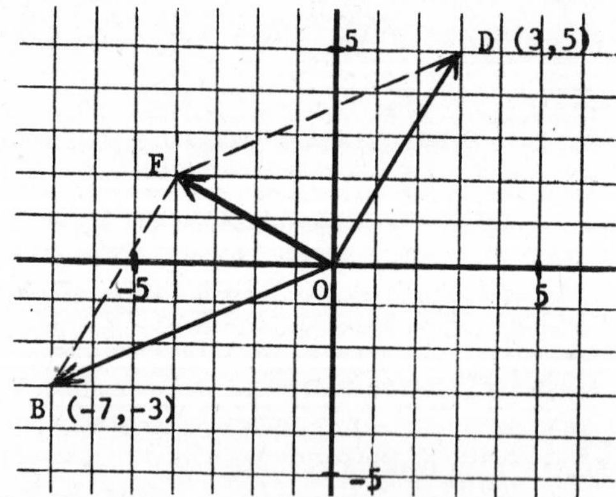

We have completed the parallelogram by:

(1) Drawing $\overrightarrow{BF}$ parallel to and equal to $\overrightarrow{0D}$.

(2) Drawing $\overrightarrow{DF}$ parallel to and equal to $\overrightarrow{0B}$.

(Continued on following page.)

resultant vector

78. (Continued)

(a) The coordinates of point F are (,).

(b) Which vector is the sum of the two vectors? ______

(c) Since $\overrightarrow{0F}$ is the sum of $\overrightarrow{0B}$ and $\overrightarrow{0D}$, it is called the ______.

---

79. Here is another vector addition problem. Add $\overrightarrow{0M}$ and $\overrightarrow{0R}$.

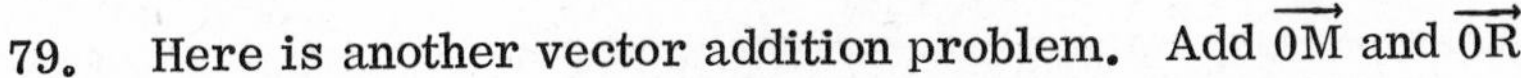

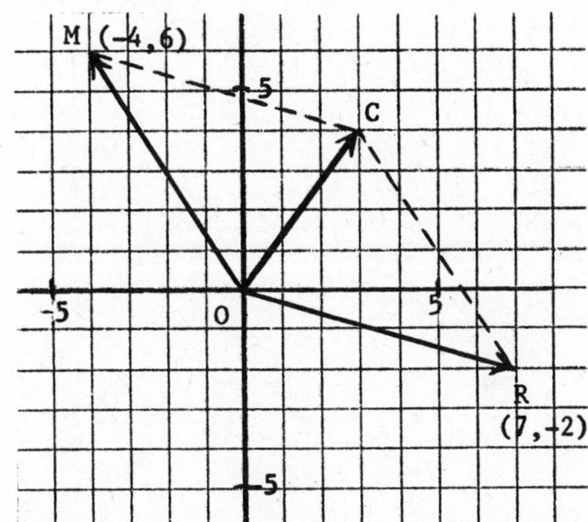

We have completed the parallelogram by drawing $\overrightarrow{MC}$ and $\overrightarrow{RC}$.

(a) The coordinates of point C are (,).

(b) The resultant vector is ______.

Answers to 78:
a) (-4, 2)
b) Vector 0F or $\overrightarrow{0F}$
c) resultant vector

---

80. Two vectors can always be added by the "parallelogram method".

However, there are two reasons why we do not ordinarily use this method:

(1) It is difficult to construct the parallelogram accurately.
(2) There is an alternate method which avoids this construction and which gives a more accurate result.

Answers to 79:
a) (3, 4)
b) $\overrightarrow{0C}$

SELF-TEST 6 (Frames 75-80)

1. Using the parallelogram method, sketch the resultant of the two vectors shown at the right.

2. In what quadrant does the resultant lie? ______

3. Is the length of the resultant greater than or less than the length of vector 0E? ______

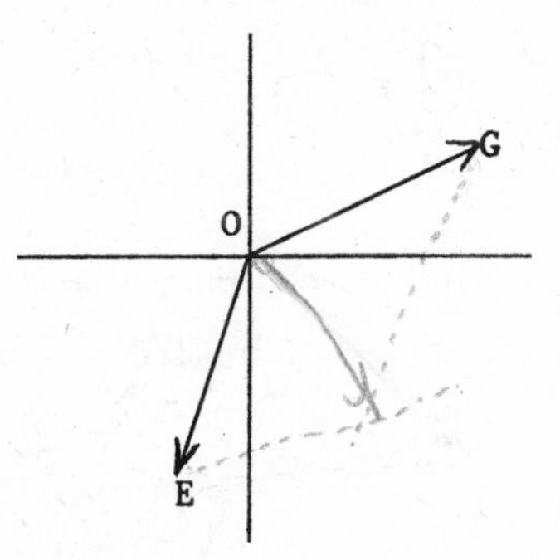

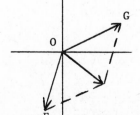

ANSWERS: 1. 2. Quadrant IV 3. Less than.

2-7 ADDITION OF VECTORS BY THE COMPONENT METHOD

Vectors are ordinarily added by the "component method". The component method provides accurate answers without the necessity of constructing parallelograms. We will discuss the component method of vector addition in this section.

---

81. Here is the first addition we did by the parallelogram method. The two vectors added are $\overrightarrow{0S}$ and $\overrightarrow{0P}$. Their resultant vector, or sum, is $\overrightarrow{0T}$.

Since the coordinates of point T are (2, 6):

"2" is the horizontal component of $\overrightarrow{0T}$.

"6" is the vertical component of $\overrightarrow{0T}$.

We can easily find these components of the resultant vector by calculation.

<u>To calculate the horizontal component of $\overrightarrow{0T}$ (which is "2"):</u>

We add the horizontal components of $\overrightarrow{0S}$ and $\overrightarrow{0P}$.

The horizontal component of $\overrightarrow{0S}$ is "-3".
The horizontal component of $\overrightarrow{0P}$ is "5".

Adding: $(-3) + 5 = \underline{2}$

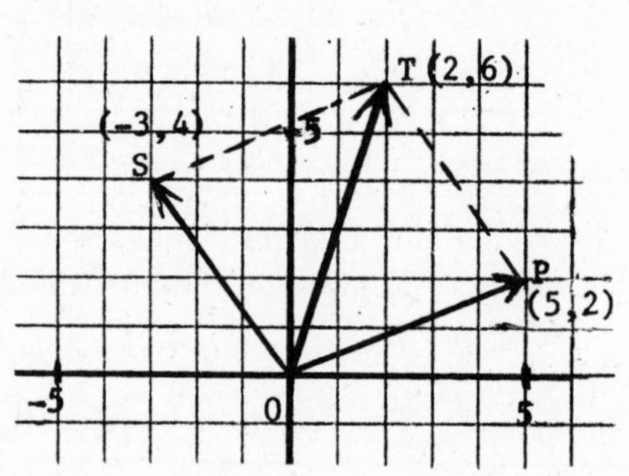

<u>To calculate the vertical component of $\overrightarrow{0T}$ (which is "6"):</u>

We add the vertical components of $\overrightarrow{0S}$ and $\overrightarrow{0P}$.

The vertical component of $\overrightarrow{0S}$ is "4".
The vertical component of $\overrightarrow{0P}$ is "2".

Adding: $4 + 2 = \underline{6}$

---

82. In the last frame, we saw that we could find the components of the resultant vector by adding the components of the two original vectors. Let's see if this method works in the second addition we did by the parallelogram method.

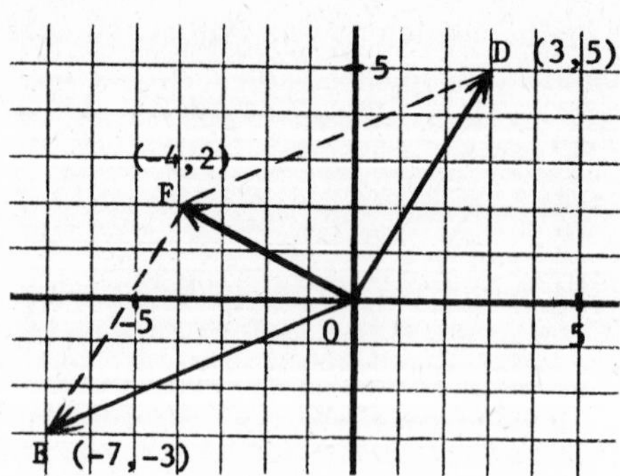

The two vectors added are $\overrightarrow{0B}$ and $\overrightarrow{0D}$, and their resultant vector is $\overrightarrow{0F}$.

(a) From the coordinates of point F, the horizontal component of $\overrightarrow{0F}$ is __-4__.

(b) If we add the horizontal components of $\overrightarrow{0B}$ and $\overrightarrow{0D}$ (-7 and 3), do we get the horizontal component of $\overrightarrow{0F}$? __-4__

(c) The vertical component of $\overrightarrow{0F}$ is __2__.

(d) If we add the vertical components of $\overrightarrow{0B}$ and $\overrightarrow{0D}$ (-3 and 5), do we get the vertical component of $\overrightarrow{0F}$? __2__

Go to next frame.

83. Finding the components of the resultant by adding the components of the two original vectors is called the "component method" of adding two vectors. Let's check the sensibleness of the "component method" with the third addition performed by the parallelogram method.

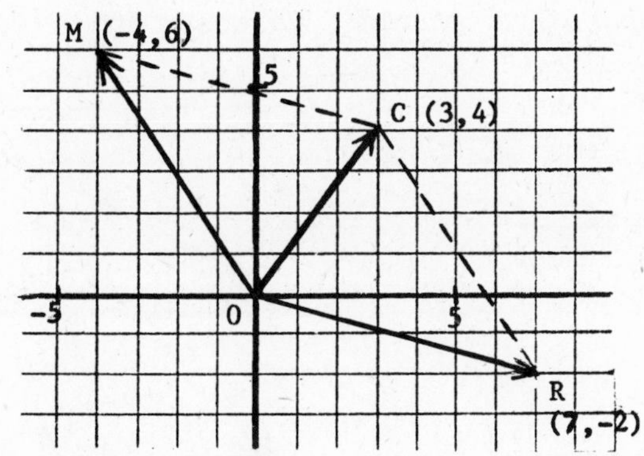

(a) The horizontal component of $\overrightarrow{0C}$, the resultant vector, is 3.

(b) If we add the horizontal components of $\overrightarrow{0M}$ and $\overrightarrow{0R}$, do we get the horizontal component of $\overrightarrow{0C}$? 3

(c) The vertical component of $\overrightarrow{0C}$, the resultant vector, is 4.

(d) If we add the vertical components of $\overrightarrow{0M}$ and $\overrightarrow{0R}$, do we get the vertical component of $\overrightarrow{0C}$? 4

a) -4

b) Yes, since (-7) + 3 = -4

c) 2

d) Yes, since (-3) + 5 = 2

---

84. We can add two vectors whose components are known without using a graph. To do so, we simply add their horizontal and vertical components. For example:

If the horizontal and vertical components of $\overrightarrow{0A}$ are -7.9 and +3.1 and the horizontal and vertical components of $\overrightarrow{0B}$ are +5.2 and -4.6:

(a) The horizontal component of their resultant is -2.7.

(b) The vertical component of their resultant is -1.5.

a) 3

b) Yes, since (-4) + 7 = 3

c) 4

d) Yes, since 6 + (-2) = 4

---

85. In the figure on the right, we have not drawn the components of vectors #1 and #2. However, the coordinates of the two points give us the lengths of their components.

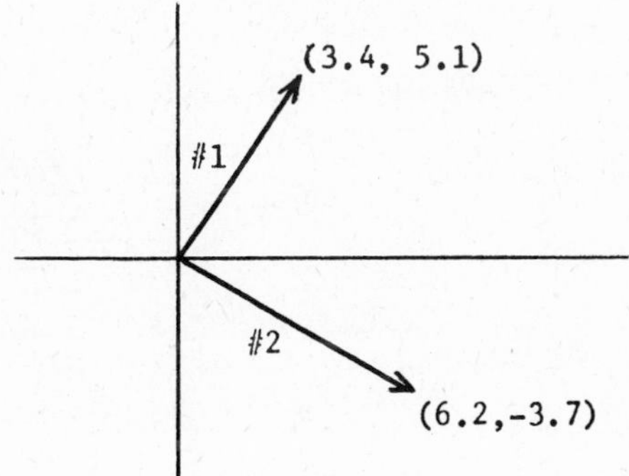

(a) How long is the horizontal component of:

Vector #1? 3.4
Vector #2? 6.2
The resultant? 9.6

(b) How long is the vertical component of:

Vector #1? 5.1
Vector #2? -3.7
The resultant? 1.4

a) -2.7, since: (-7.9) + 5.2 = -2.7

b) -1.5, since: 3.1 + (-4.6) = -1.5

86. Here are two vectors in the third quadrant. We want to find the components of their resultant.

(a) List the horizontal component of:

Vector #1 -5.6
Vector #2 -1.9
The resultant -7.5

(b) List the vertical component of:

Vector #1 -3.1
Vector #2 -7.3
The resultant 10.4

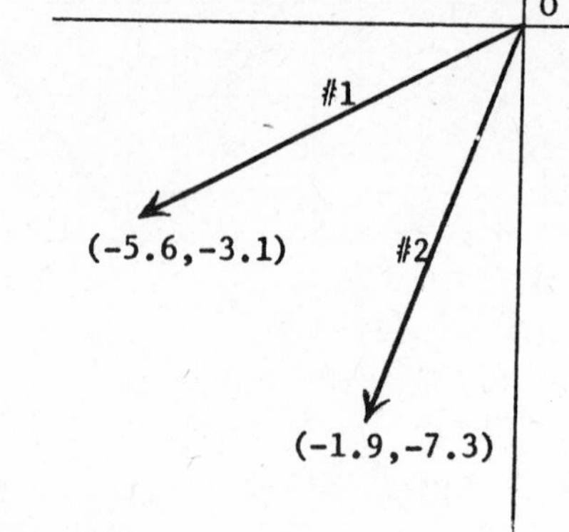

a) 3.4 units
6.2 units
9.6 units

b) 5.1 units
-3.7 units
1.4 units

Answer to Frame 86:

a)	-5.6 units	b)	-3.1 units
	-1.9 units		-7.3 units
	-7.5 units		-10.4 units

87. Ordinarily, we are not given the components of the two vectors which we want to add. We must find these components first. Here are two such vectors, $\overrightarrow{0D}$ and $\overrightarrow{0H}$. We are given their lengths and their reference angles, and will calculate their components.

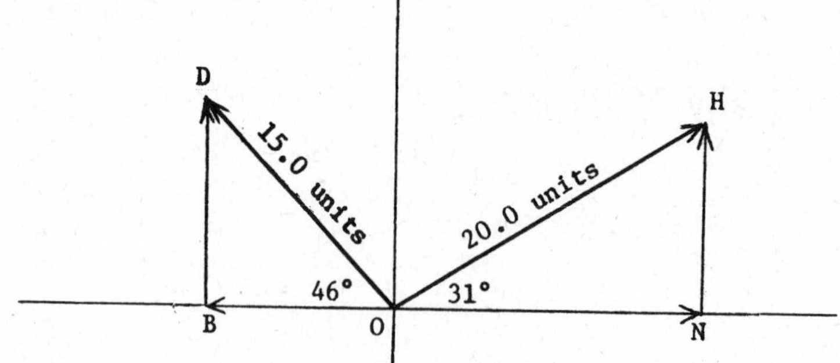

Finding the horizontal and vertical components of $\overrightarrow{0H}$:

(1) Horizontal component, $\overrightarrow{0N}$:

$$\frac{\overrightarrow{0N}}{\overrightarrow{0H}} = \cos 31° \text{ (Q1)}$$

$$\overrightarrow{0N} = (\overrightarrow{0H})[\cos 31° \text{ (Q1)}]$$

$$= (20.0)(0.8572)$$

$$= 17.1 \text{ units}$$

(2) Vertical component, $\overrightarrow{NH}$:

$$\frac{\overrightarrow{NH}}{\overrightarrow{0H}} = \sin 31° \text{ (Q1)}$$

$$\overrightarrow{NH} = (\overrightarrow{0H})[\sin 31° \text{ (Q1)}]$$

$$= (20.0)(0.5150)$$

$$= 10.3 \text{ units}$$

Finding the horizontal and vertical components of $\overrightarrow{0D}$:

(1) Horizontal component, $\overrightarrow{0B}$:

$$\frac{\overrightarrow{0B}}{\overrightarrow{0D}} = \cos 46° \text{ (Q2)}$$

$$\overrightarrow{0B} = (\overrightarrow{0D})[\cos 46° \text{ (Q2)}]$$

$$= (15.0)(-0.6947)$$

$$= -10.4 \text{ units}$$

(2) Vertical component, $\overrightarrow{BD}$:

$$\frac{\overrightarrow{BD}}{\overrightarrow{0D}} = \sin 46° \text{ (Q2)}$$

$$\overrightarrow{BD} = (\overrightarrow{0D})[\sin 46° \text{ (Q2)}]$$

$$= (15.0)(0.7193)$$

$$= 10.8 \text{ units}$$

(Continued on following page.)

87. (Continued)

Now you can easily answer these questions:

(a) The horizontal component of the resultant is 6.7 units.

(b) The vertical component of the resultant is 21.1 units.

(c) The coordinates of the tip of the resultant are (6.7, 21.1).

Answer to Frame 87:

a) 6.7 units
From:
17.1 + (-10.4) = 6.7

b) 21.1 units
From:
10.3 + 10.8 = 21.1

c) (6.7, 21.1)

---

88. In this diagram the lengths and directions of $\overrightarrow{0R}$ and $\overrightarrow{0S}$ are known. We want to find the components of their resultant.

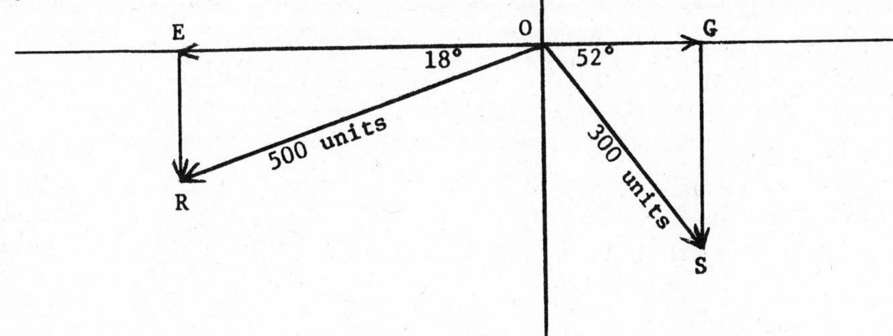

Finding the horizontal and vertical components of $\overrightarrow{0R}$:

(1) Horizontal component, $\overrightarrow{0E}$:

$$\frac{\overrightarrow{0E}}{500} = \cos 18° \text{ (Q3)}$$

$$\overrightarrow{0E} = (500)[\cos 18° \text{ (Q3)}]$$

$$= (500)(-0.9511)$$

$$= -476 \text{ units}$$

(2) Vertical Component, $\overrightarrow{ER}$:

$$\frac{\overrightarrow{ER}}{500} = \sin 18° \text{ (Q3)}$$

$$\overrightarrow{ER} = (500)[\sin 18° \text{ (Q3)}]$$

$$= (500)(-0.3090)$$

$$= -154 \text{ units}$$

Finding the horizontal and vertical components of $\overrightarrow{0S}$:

(1) Horizontal component, $\overrightarrow{0G}$:

$$\frac{\overrightarrow{0G}}{300} = \cos 52° \text{ (Q4)}$$

$$\overrightarrow{0G} = (300)[\cos 52° \text{ (Q4)}]$$

$$= (300)(0.6157)$$

$$= 185 \text{ units}$$

(2) Vertical component, $\overrightarrow{GS}$:

$$\frac{\overrightarrow{GS}}{300} = \sin 52° \text{ (Q4)}$$

$$\overrightarrow{GS} = (300)[\sin 52° \text{ (Q4)}]$$

$$= (300)(-0.7880)$$

$$= -236 \text{ units}$$

Now answer these:

(a) The horizontal component of the resultant is -291 units.

(b) The vertical component of the resultant is -390 units.

(c) The coordinates of the tip of the resultant are (-291, -390).

---

a) -291 units

b) -390 units

c) (-291, -390)

89. State whether each of the following is positive or negative:

(a) Horizontal component of $\overrightarrow{0M}$.

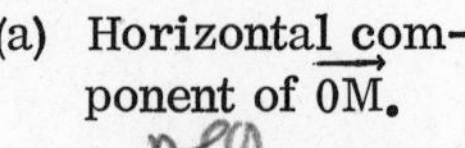

_____

(b) Vertical component of $\overrightarrow{0T}$.

_____

(c) Vertical component of $\overrightarrow{0M}$. _____

(d) Horizontal component of $\overrightarrow{0T}$. _____

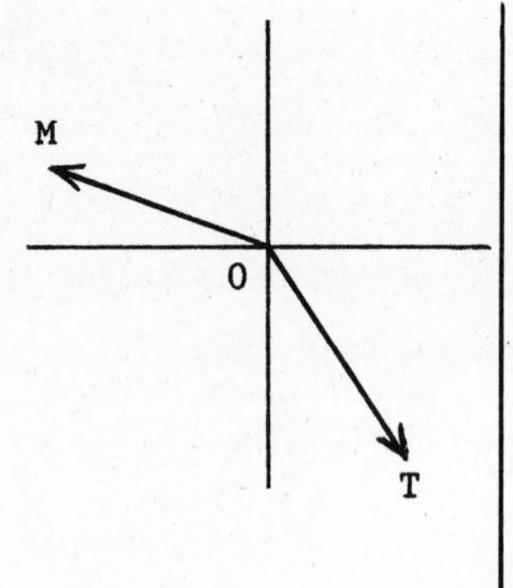

Answer to Frame 89: a) Negative b) Negative c) Positive d) Positive

90. It is extremely important that vector components be labeled with the proper sign, either positive or negative. If they are not, then the components of the resultant will be incorrect. In this addition, watch the signs carefully.

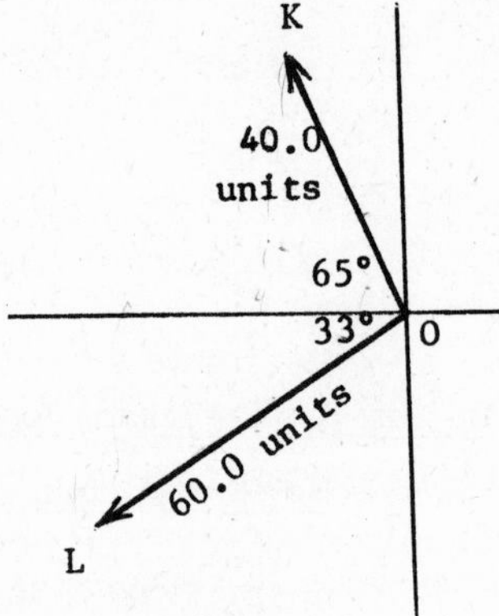

Calculate the horizontal and vertical components of $\overrightarrow{0K}$:

(a) Horizontal component = _____ (b) Vertical component = _____

Calculate the horizontal and vertical components of $\overrightarrow{0L}$:

(c) Horizontal component = _____ (d) Vertical component = _____

Calculate the horizontal and vertical components of the resultant vector:

(e) Horizontal component = _____ (f) Vertical component = _____

Answer to Frame 90:

a) -16.9 units
b) 36.2 units
c) -50.3 units
d) -32.7 units
e) -67.2 units From: -16.9 + (-50.3)
f) 3.5 units From: 36.2 + (-32.7)

---

91. For this vector system, calculate these:

(a) Horizontal component of $\overrightarrow{0F}$ = ______

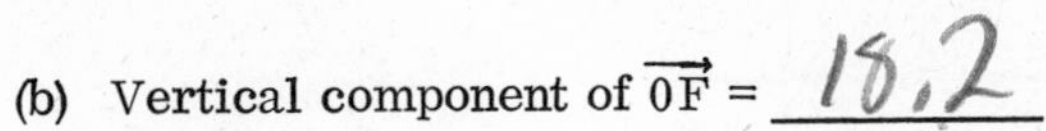

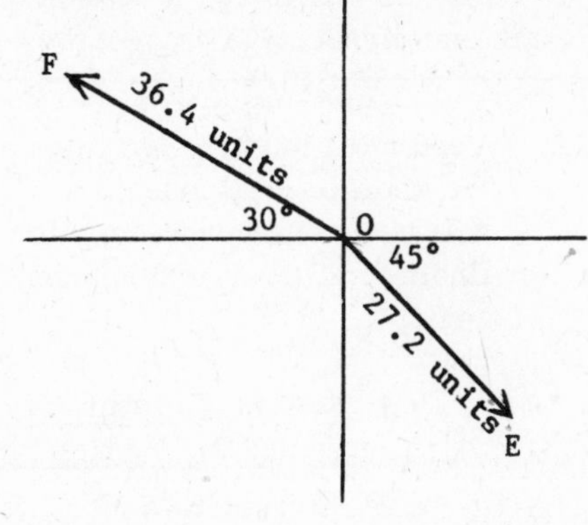

(b) Vertical component of $\overrightarrow{0F}$ = ______

(c) Horizontal component of $\overrightarrow{0E}$ = ______

(d) Vertical component of $\overrightarrow{0E}$ = ______

(e) Horizontal component of resultant = ______

(f) Vertical component of resultant = ______

Answer to Frame 91:

a) -31.5 units
b) 18.2 units
c) 19.2 units
d) -19.2 units
e) -12.3 units
f) -1.0 units

SELF-TEST 7 (Frames 81-91)

In this diagram, the length of $\overrightarrow{0R}$ is 30.0 units and its reference angle is 23° (Q1). The length of $\overrightarrow{0V}$ is 50.0 units and its reference angle is 60° (Q3).

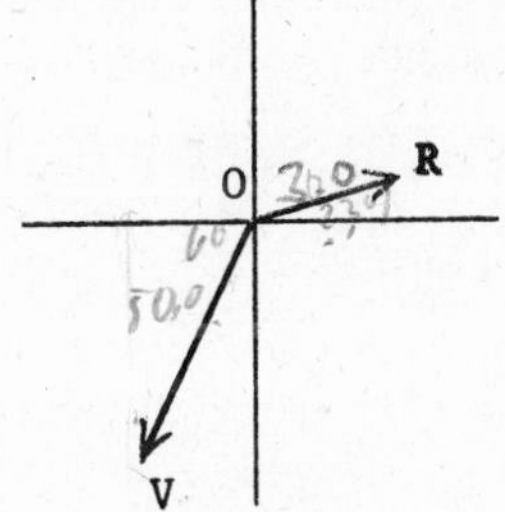

1. Find the horizontal component of the resultant vector.

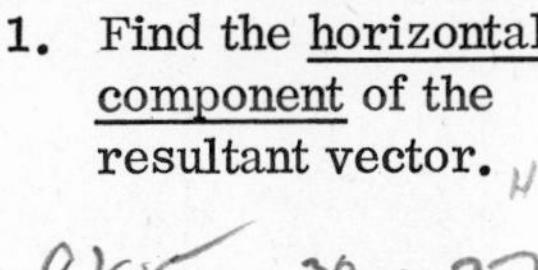

2. Find the vertical component of the resultant vector.

ANSWERS: 1. +2.6 [from 27.6 + (-25.0)] 2. -31.6 [from 11.7 + (-43.3)]

2-8 FINDING THE LENGTH AND DIRECTION OF A RESULTANT

Up to this point, we have only found the components of a resultant vector. However, we are usually interested in finding the length and direction of a resultant. In this section, we will review the methods of finding the length and direction of any vector, and then we will find the length and direction of resultant vectors.

92. Vector 0C is in the first quadrant. We know its components. We will review the procedure for finding its length and reference angle.

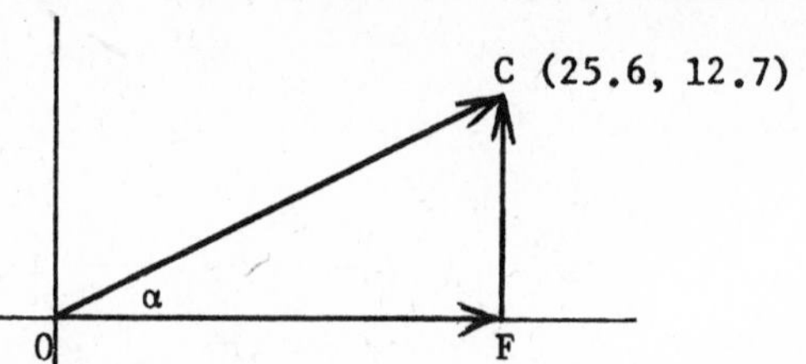

(a) To find the length of $\overrightarrow{0C}$, we use the Pythagorean Theorem:

$$\left(\overrightarrow{0C}\right)^2 = \left(\overrightarrow{0F}\right)^2 + \left(\overrightarrow{FC}\right)^2,$$

$$\overrightarrow{0C} = \sqrt{\left(\overrightarrow{0F}\right)^2 + \left(\overrightarrow{FC}\right)^2}$$

$$= \sqrt{(25.6)^2 + (12.7)^2}$$

$$= \sqrt{655 + 161}$$

$$= \sqrt{816} =$$ 28.6

(b) To find the reference angle of $\overrightarrow{0C}$, we use the tangent ratio:

$$\tan \alpha = \frac{\overrightarrow{FC}}{\overrightarrow{0F}} = \frac{12.7}{25.6} = 0.496 \text{ (slide-rule accuracy)}$$

Specify the reference angle. 

Answers: a) 28.6 units b) 26° (Q1) (You must identify the quadrant.)

93. The phrase "the direction of a vector" means the same thing as "specifying the reference angle of a vector".

If the reference angle of a vector is 35° (Q3), we say that the direction of the vector is 35° Q3.

Answer: 35° (Q3)

94. To find reference angle α for $\overrightarrow{0F}$, we use the tangent ratio:

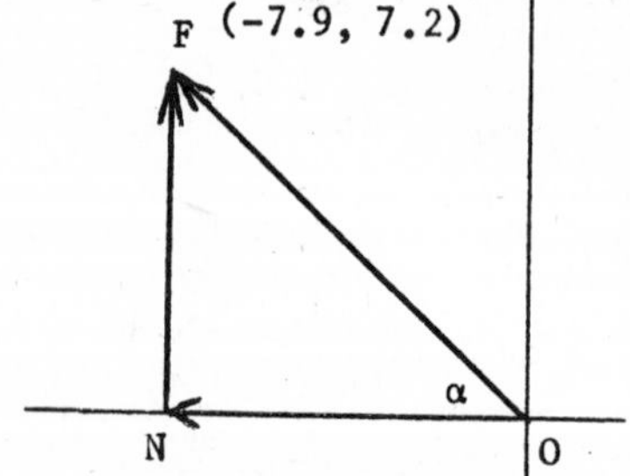

$$\tan \alpha = \frac{\overrightarrow{NF}}{\overrightarrow{0N}} = \frac{7.2}{-7.9} = -0.911$$

$$\alpha = 42° \text{ (Q2)}$$

In finding any reference angle, the following formula is used:

$$\boxed{\tan \alpha = \frac{\text{vertical component}}{\text{horizontal component}}}$$

Since these components are the same as the coordinates of the tip of the vector, we do not have to draw the components. We can simply plug the coordinates into the formula above. There is an example on the next page.

(Continued on following page.)

94. (Continued)

In the diagram at the right, without drawing the components of $\overrightarrow{0H}$, we can find the tangent of reference angle α.

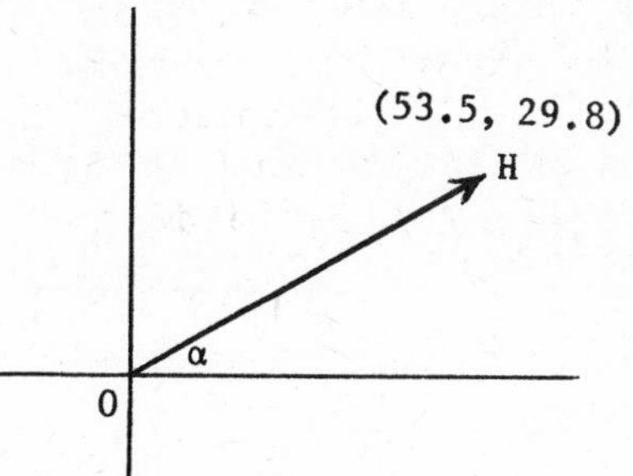

$$\tan\alpha = \frac{29.8}{53.5} = 0.557$$

The direction of $\overrightarrow{0H}$ is 29° Q1

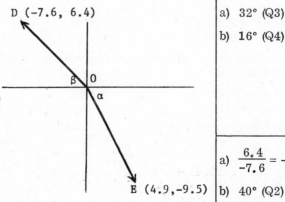

29° (Q1)

95. On the graph below, we have drawn $\overrightarrow{0F}$ and $\overrightarrow{0M}$ without showing their components.

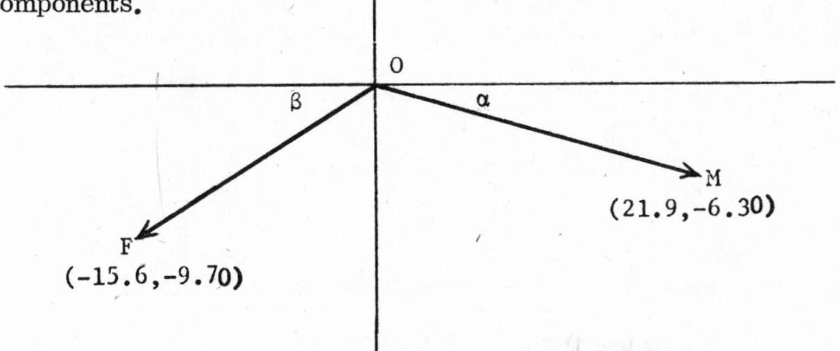

The reference angle for $\overrightarrow{0F}$ is β. The reference angle for $\overrightarrow{0M}$ is α.

$$\tan\beta = \frac{\text{vertical component}}{\text{horizontal component}} = \frac{-9.70}{-15.6} = +0.622$$

$$\tan\alpha = \frac{\text{vertical component}}{\text{horizontal component}} = \frac{-6.30}{21.9} = -0.288$$

Complete: (a) The direction of $\overrightarrow{0F}$ is 32° Q3

(b) The direction of $\overrightarrow{0M}$ is 16° Q4

---

a) 32° (Q3)

b) 16° (Q4)

96. (a) $\tan\beta =$ 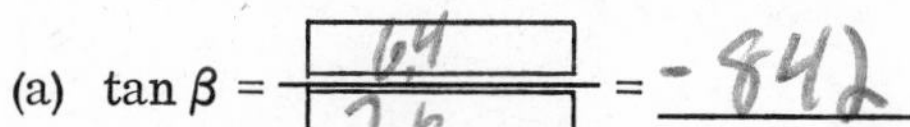 $\frac{6.4}{7.6}$ = -.842

(b) The direction of $\overrightarrow{0D}$ is 40° Q2

(c) $\tan\alpha =$ 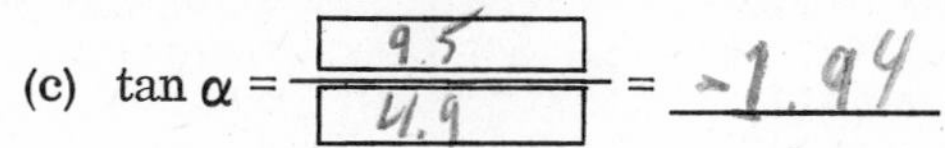 $\frac{9.5}{4.9}$ = -1.94

(d) The direction of $\overrightarrow{0E}$ is 63° Q4

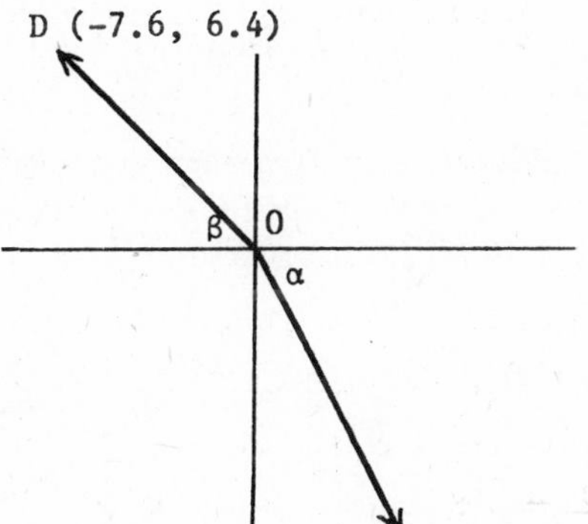

---

a) $\frac{6.4}{-7.6} = -0.842$

b) 40° (Q2)

c) $\frac{-9.5}{4.9} = -1.94$

d) 63° (Q4)

97. If we know the components of a vector, we can find its direction without drawing its components. We can also find its length without drawing its components. To do so, we simply plug the values of the components into the Pythagorean Theorem. Here is an example:

Though the components of $\overrightarrow{0T}$ are not drawn:

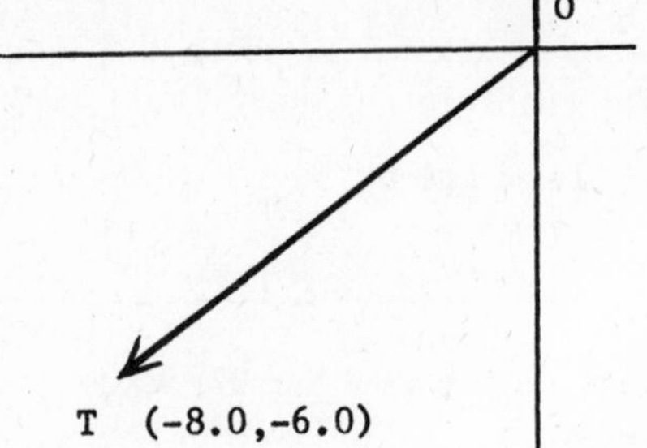

$$\overrightarrow{0T} = \sqrt{(-8.0)^2 + (-6.0)^2}$$
$$= \sqrt{64 + 36}$$
$$= 10 \text{ units}$$

Note: Any step in a vector addition can be performed without drawing the components of the vectors on the diagram. Therefore, from this point on, we will draw vectors without drawing their components. Feel free, however, to draw in the components if seeing them helps you.

---

98. In the steps below, we will add the two vectors at the right and then find the length and direction of their resultant.

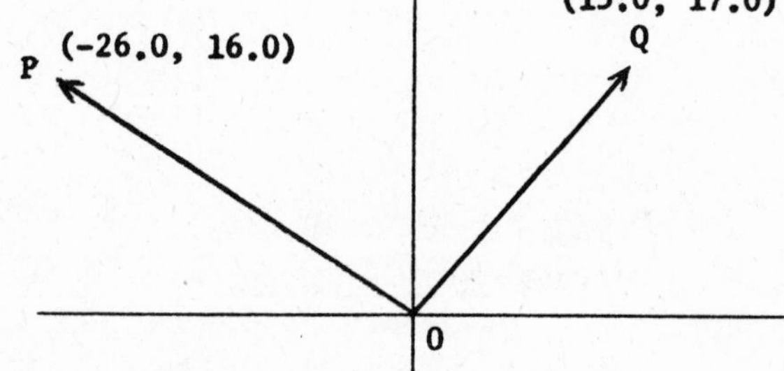

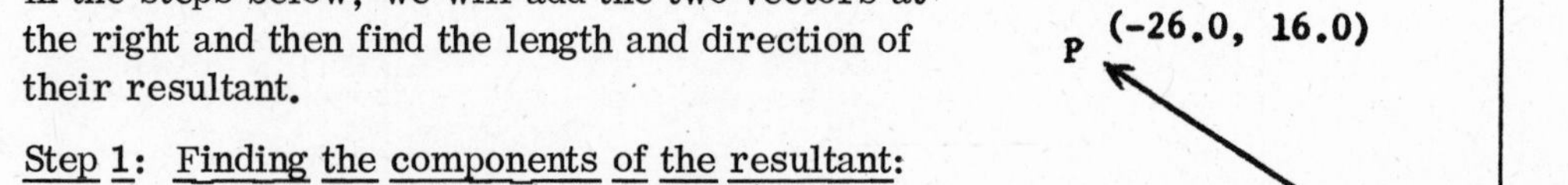

Step 1: Finding the components of the resultant:

The horizontal component is (-26.0) + 15.0 or -11.0 units.
The vertical component is 16.0 + 17.0 or 33.0 units.

Step 2: Roughly completing the parallelogram and drawing the resultant:

We do so to check the sensibleness of the computed components and to determine the quadrant of the resultant.

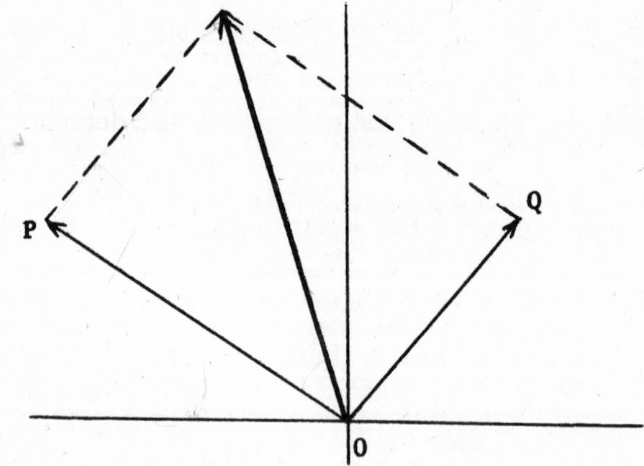

Step 3: Determining the length and direction of the resultant:

Since a resultant is simply a vector like any other vector, we use the same steps to find its length and direction.

(a) How long is the resultant? 34.8

(b) The resultant is in what quadrant? 2

(c) What is the tangent of the reference angle of the resultant? tan α = -3.000

(d) The direction of the resultant is 72° Q2.

Answer to Frame 98: a) $\sqrt{1,210} = 34.8$ units b) Quadrant II c) $\tan\alpha = \frac{33}{-11} = -3.00$ d) 72° (Q2)

---

99. When performing a vector-addition, you should always draw a sketch of the vectors. Furthermore, you should roughly complete the parallelogram and sketch the resultant, for two reasons:

(1) To check the sensibleness of the computed components.
(2) To determine the quadrant of the resultant so that you can determine its direction.

In each case below, roughly complete the parallelogram and then identify the quadrant of the resultant.

(a)

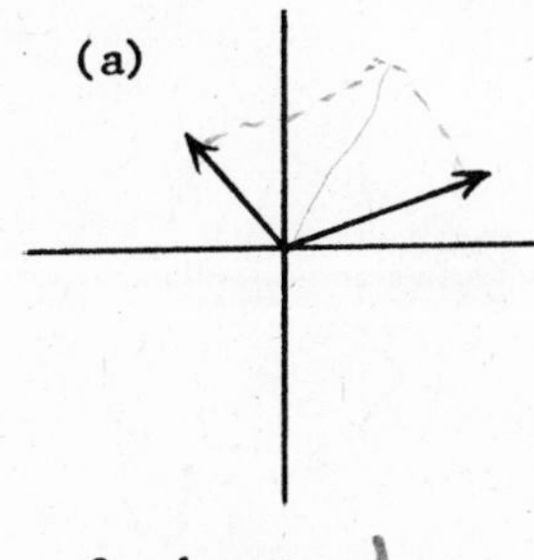

Quadrant 1

(b)

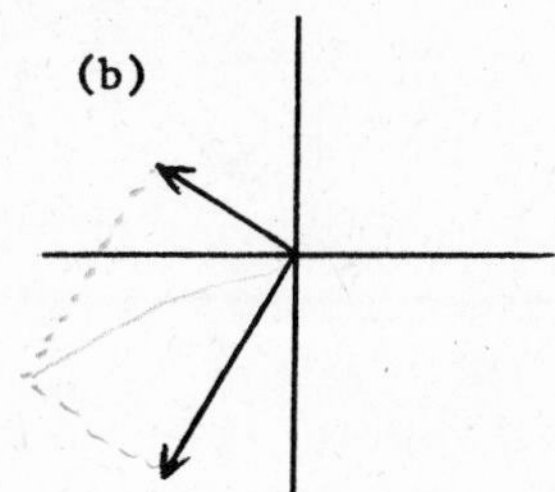

Quadrant 3

(c)

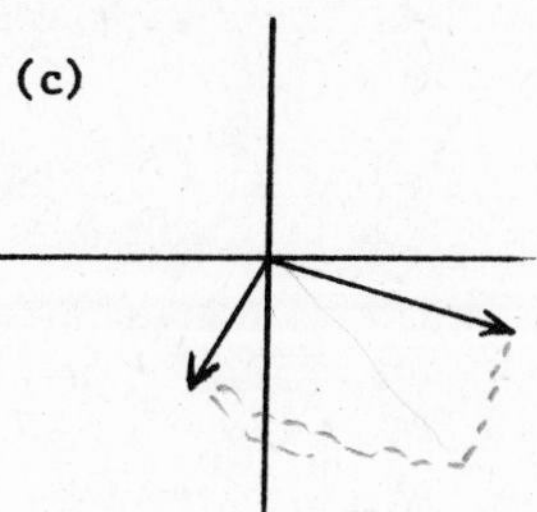

Quadrant 4

---

100. In the steps below, we will add $\overrightarrow{0T}$ and $\overrightarrow{0V}$ and find the length and direction of their resultant.

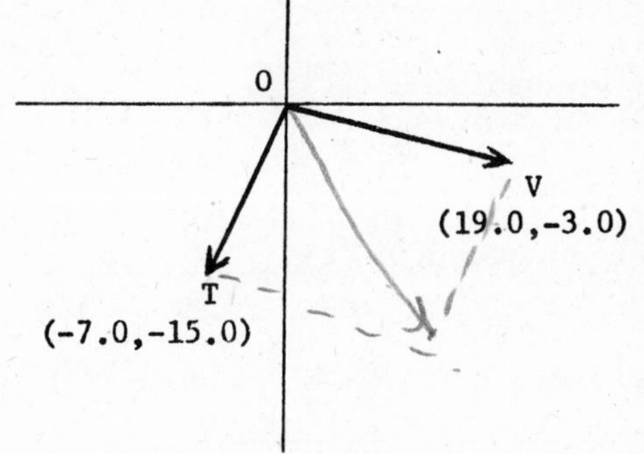

(a) The horizontal component of the resultant is 12.

(b) The vertical component of the resultant is -14.

(c) Roughly complete the parallelogram and draw the resultant to check the computed components and determine the quadrant of the resultant. The resultant is in what quadrant? 4

(d) How long is the resultant? 21.6 units

(e) Find the tangent of the reference angle of the resultant.

$\tan\alpha =$ 1.5

(f) The direction of the resultant is 56.3.

a)

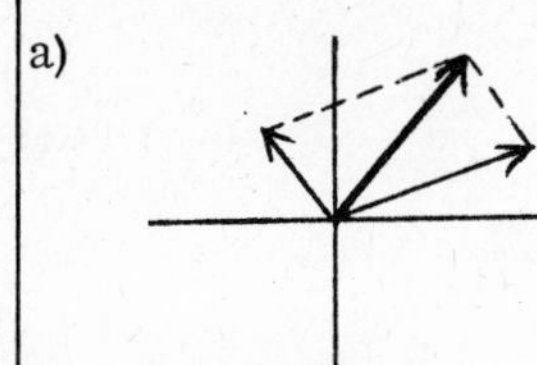

Quadrant I

b)

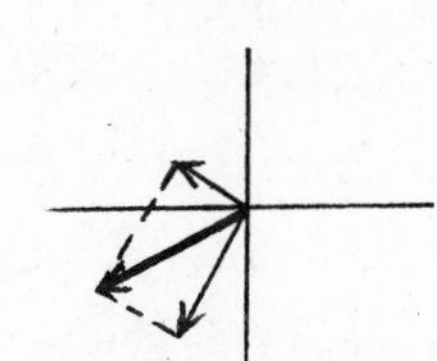

Quadrant III

c)

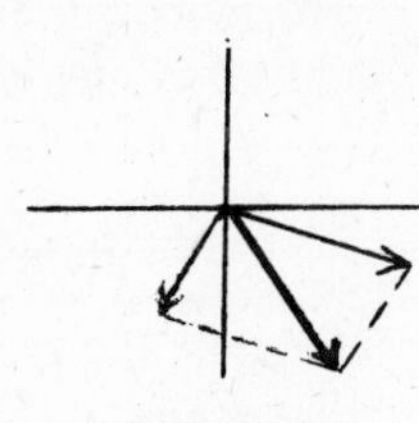

Quadrant IV

101. In a typical vector-addition problem,

(1) You are given the lengths and directions (reference angles) of two vectors.

(2) You must find the length and direction of the resultant.

The information given at the right is typical. Only the length and direction of $\overrightarrow{0B}$ and $\overrightarrow{0F}$ are given. The steps for finding the length and direction of their resultant are outlined below.

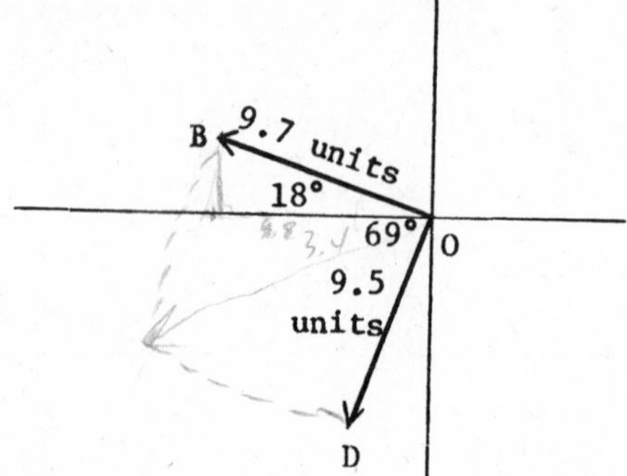

Step 1: Finding the components of the two vectors:

(a) The horizontal component of $\overrightarrow{0B}$ is -9.2.

(b) The vertical component of $\overrightarrow{0B}$ is 3.0.

(c) The horizontal component of $\overrightarrow{0D}$ is -3.4.

(d) The vertical component of $\overrightarrow{0D}$ is -8.9.

Step 2: Finding the components of the resultant:

(a) The horizontal component of the resultant is 12.6.

(b) The vertical component of the resultant is -5.9.

Step 3: Finding the length and direction of the resultant:

(a) Complete the parallelogram and roughly sketch the resultant on the axes at the right. The resultant lies in Quadrant 3.

(b) The length of the resultant is 13.9.

(c) The direction of the resultant is 25° Q3

a) 12.0 units

b) -18.0 units

c) Quadrant IV

d) $\sqrt{468} = 21.6$ units

e) $\tan \alpha = \frac{-18.0}{12.0} = -1.50$

f) 56° (Q4)

102. To find the length and direction of the resultant of these two vectors, follow the steps below.

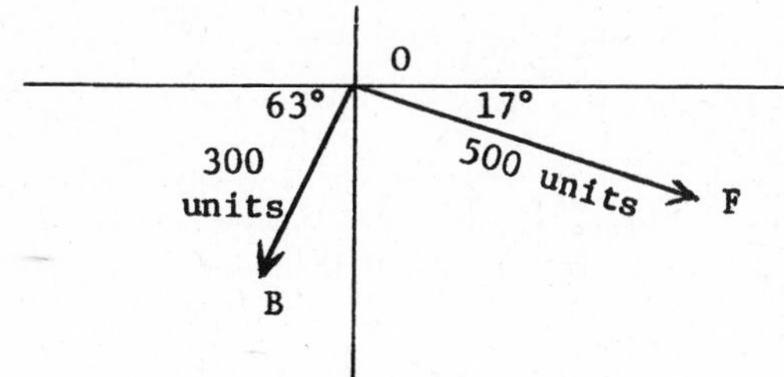

Finding the components of the known vectors:

(a) Horizontal component of $\overrightarrow{0B}$ = ___________

(b) Vertical component of $\overrightarrow{0B}$ = ___________

(c) Horizontal component of $\overrightarrow{0F}$ = ___________

(d) Vertical component of $\overrightarrow{0F}$ = ___________

Finding the components of the resultant:

(e) Horizontal component of resultant = ___________

(f) Vertical component of resultant = ___________

(g) The coordinates of the end-point of the resultant are (,   ).

Finding length and direction of resultant:

(h) Complete the parallelogram and roughly sketch the resultant on the axes at the right. The resultant lies in Quadrant ______.

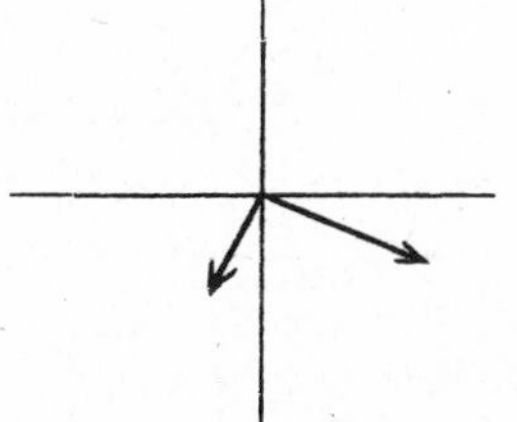

(i) The length of the resultant is _____________.

(j) The direction of the resultant is ___________.

Step 1

a) -9.2 units
b) +3.0 units
c) -3.4 units
d) -8.9 units

Step 2

a) -12.6 units
b) -5.9 units

Step 3

a) Quadrant III
b) $\sqrt{193.8} = 13.9$ units
c) 25° (Q3) Since:

$$\tan \alpha = \frac{-5.9}{-12.6} = +0.468$$

Answer to Frame 102:

a) -136 units
b) -267 units
c) 478 units
d) -146 units
e) 342 units
f) -413 units
g) (342, -413)
h) Quadrant IV
i) 536 units (or 537 units)
j) 50° (Q4)

SELF-TEST 8 (Frames 92-102)

1. Sketch the resultant of vectors $\overrightarrow{0R}$ and $\overrightarrow{0T}$ on the diagram at the right. Then calculate the following:

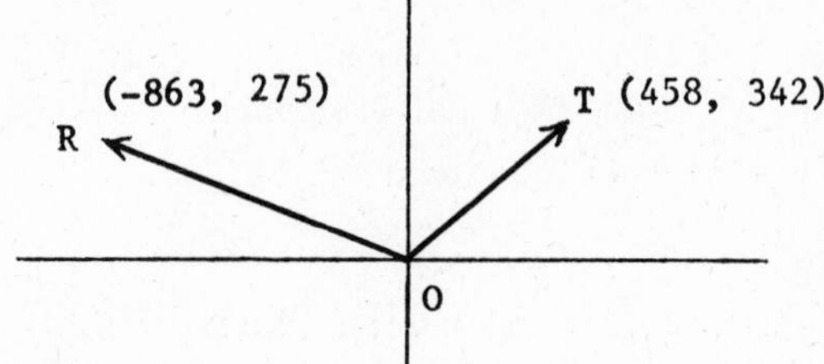

(a) Coordinates of the end-point of the resultant: ____________

(b) Length of the resultant: ________ units

(c) Direction of the resultant: ____________

2. On the diagram at the right, sketch the resultant of the two vectors. Then calculate the following.

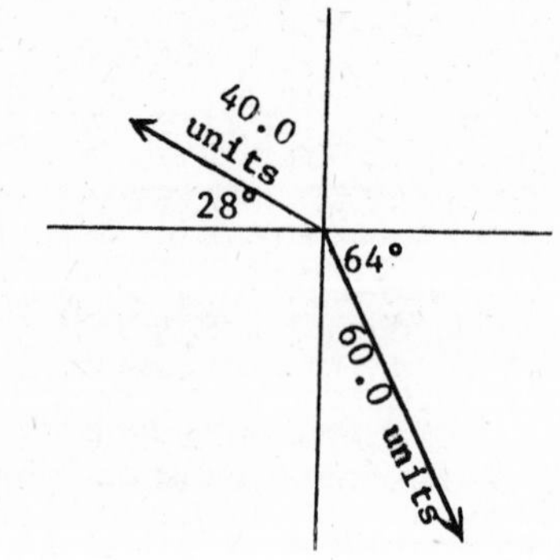

(a) Horizontal and vertical components of the resultant: ________ and ________

(b) Length of the resultant: ________ units

(c) Direction of the resultant: ________

ANSWERS:

1. (a) (-405, 617)
   (b) 738 units
   (c) 57° (Q2)

2. (a) -9.0 units and -35.1 units
   (b) 36.2 units
   (c) 76° (Q3)

2-9 ADDING VECTORS WHICH LIE ON THE COORDINATE AXES

Up to this point, we have avoided vectors which lie on either of the two axes. In this section, we will discuss the length and direction of vectors of this type. We will also perform vector-additions which involve vectors of this type.

---

103. On the axes at the right, we have drawn $\overrightarrow{0F}$ and $\overrightarrow{0N}$. Both vectors are on the horizontal axis. The co-ordinates of points F and N are given.

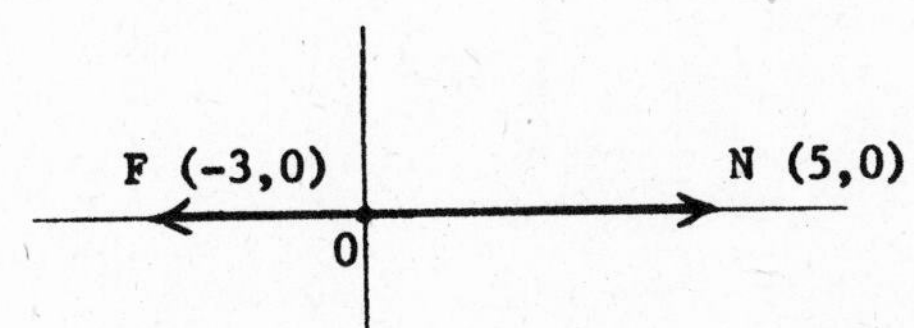

For $\overrightarrow{0N}$: The horizontal component is 5 units.
The vertical component is 0 units.

For $\overrightarrow{0F}$: The horizontal component is -3 units.
The vertical component is 0 units.

Reference angles are the angles formed by a vector and the horizontal axis. For each of these vectors, the reference angle is 0°. But we specify these 0° reference angles in the following way:

Since $\overrightarrow{0N}$ lies between quadrants I and IV, its direction is 0° (Q1 and Q4).

Since $\overrightarrow{0F}$ lies between quadrants II and III, its direction is 0° (Q2 and Q3).

If a vector lies on the horizontal axis, its vertical component must always be _______ units.

Answer: 0

---

104. At the right, we have drawn two vectors on the vertical axis.

For $\overrightarrow{0B}$: (1) The horizontal component is 0 units.
(2) The vertical component is 4 units.
(3) The reference angle is 90° (Q1 and Q2), where "Q1 and Q2" means "between quadrants I and II".

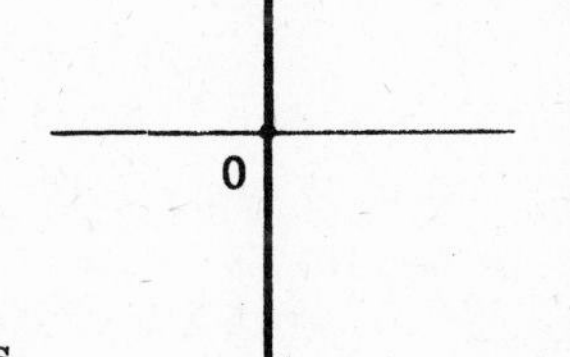

For $\overrightarrow{0C}$: (a) The horizontal component is _____ units.
(b) The vertical component is _____ units.
(c) The reference angle is ____________________.

C (0,-5)

a) 0

b) -5

c) 90° (Q3 and Q4)

105. Refer to the figure on the right for these:

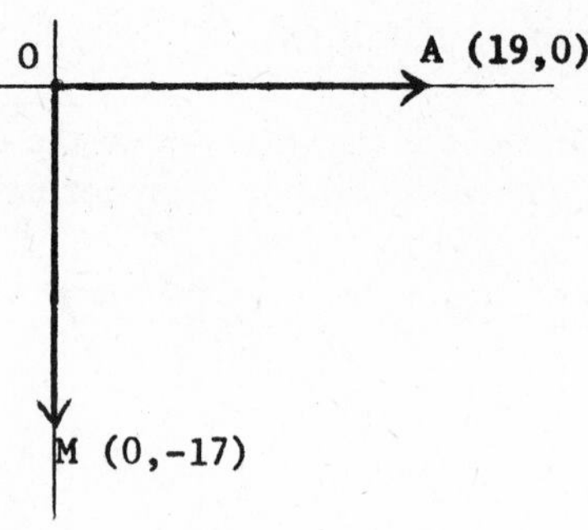

(a) The reference angle for $\overrightarrow{0A}$ is ________________.

(b) The direction of $\overrightarrow{0M}$ is ________________.

(c) The horizontal component of $\overrightarrow{0A}$ is __________.

(d) The vertical component of $\overrightarrow{0A}$ is __________.

(e) The horizontal component of $\overrightarrow{0M}$ is __________.

(f) The vertical component of $\overrightarrow{0M}$ is __________.

---

106. $\overrightarrow{0L}$ is a vector on the horizontal axis. To add $\overrightarrow{0L}$ and $\overrightarrow{0K}$, we have completed the parallelogram and drawn the resultant $\overrightarrow{0P}$.

K (3,4)
P
0
L (5,0)

Since the horizontal component of $\overrightarrow{0K}$ is 3

and the horizontal component of $\overrightarrow{0L}$ is 5,

the horizontal component of $\overrightarrow{0P}$ is 8.

Since the vertical component of $\overrightarrow{0K}$ is 4

and the vertical component of $\overrightarrow{0L}$ is 0,

the vertical component of $\overrightarrow{0P}$ is 4.

(a) The coordinates of point P are (,   ).

(b) The length of $\overrightarrow{0P}$ is __________ units.

(c) The direction of $\overrightarrow{0P}$ is ___________.

a) 0° (Q1 and Q4)

b) 90° (Q3 and Q4)

c) 19

d) 0

e) 0

f) -17

---

a) (8,4)

b) $\sqrt{80} = 8.9$ units

c) 27° (Q1), since:

$\tan \alpha = \frac{4}{8} = 0.500$

107. $\overrightarrow{0S}$ is a vector on the vertical axis. To add $\overrightarrow{0G}$ and $\overrightarrow{0S}$, we have completed the parallelogram and drawn the resultant $\overrightarrow{0T}$.

By adding components:

(a) The horizontal component of $\overrightarrow{0T}$ is ______________.

(b) The vertical component of $\overrightarrow{0T}$ is ______________.

(c) How long is $\overrightarrow{0T}$?

____

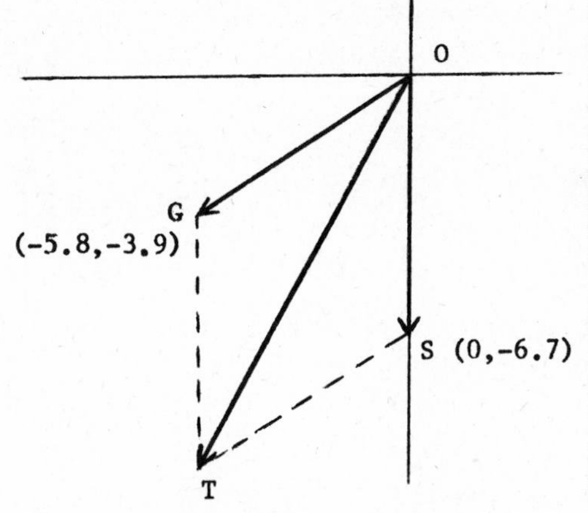

(d) The direction of $\overrightarrow{0T}$ is ______________.

---

a) -5.8 units, since: (-5.8) + 0 = -5.8

b) -10.6 units, since: (-3.9) + (-6.7) = -10.6

c) $\sqrt{146} = 12.1$ units

d) 61° (Q3), since:

$\tan \alpha = \frac{-10.6}{-5.8} = 1.83$

---

108. $\overrightarrow{0B}$ is on the horizontal axis. $\overrightarrow{0F}$ is on the vertical axis. To add $\overrightarrow{0B}$ and $\overrightarrow{0F}$, we have completed the parallelogram and drawn the resultant $\overrightarrow{0D}$.

(a) The coordinates of point D are (,   ).

(b) The length of $\overrightarrow{0D}$ is __________ units.

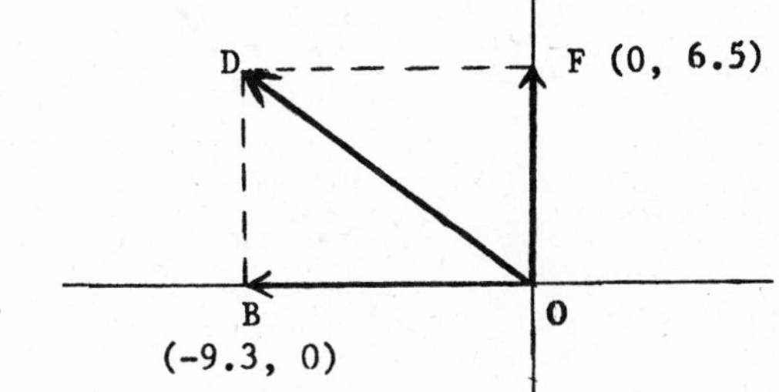

(c) The direction of $\overrightarrow{0D}$ is ______________.

---

a) (-9.3, 6.5)

b) $\sqrt{129} = 11.3$ units (or 11.4 units)

c) 35° (Q2), since:

$\tan \alpha = \frac{6.5}{-9.3} = -0.699$

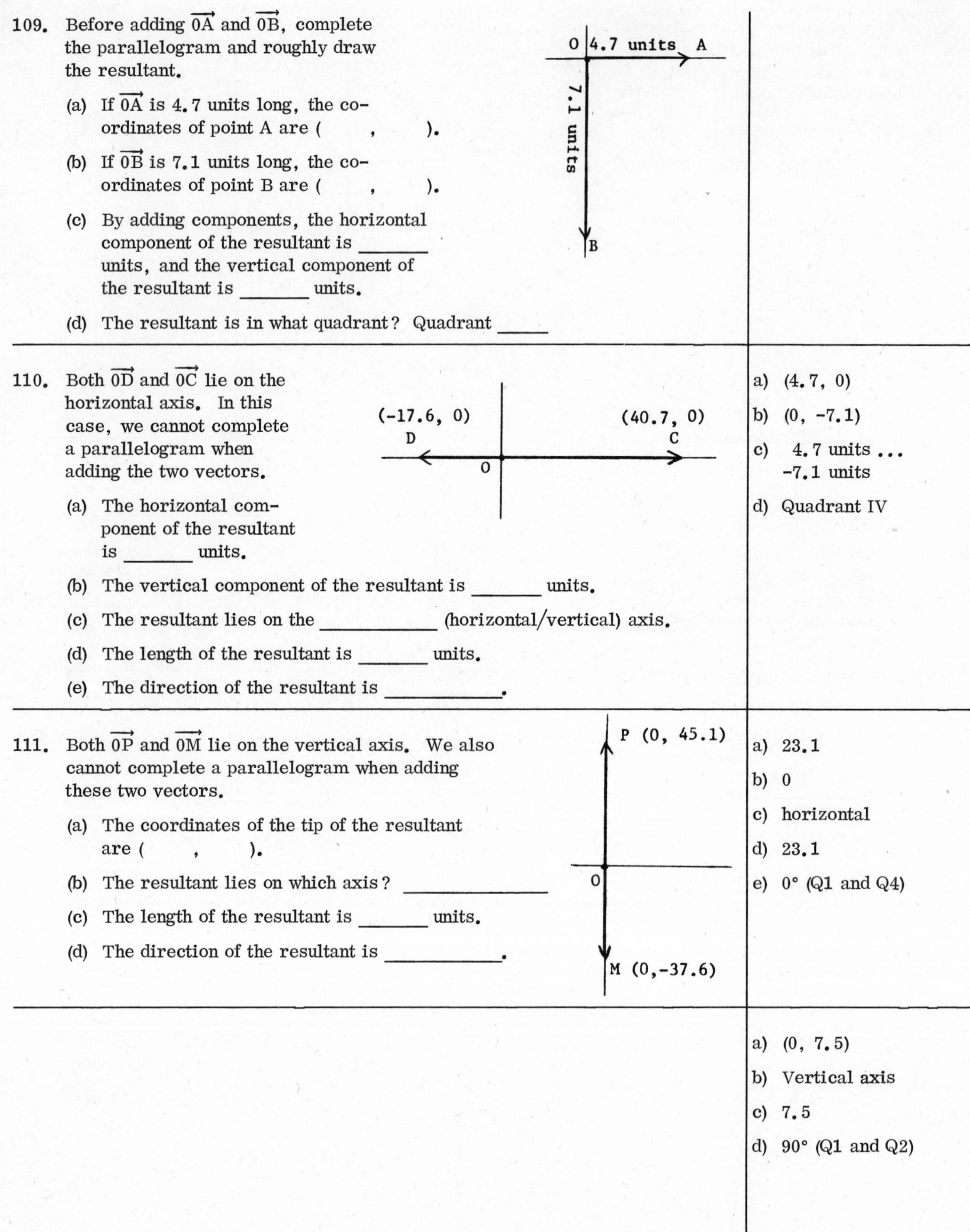

109. Before adding $\overrightarrow{0A}$ and $\overrightarrow{0B}$, complete the parallelogram and roughly draw the resultant.

(a) If $\overrightarrow{0A}$ is 4.7 units long, the coordinates of point A are (,).

(b) If $\overrightarrow{0B}$ is 7.1 units long, the coordinates of point B are (,).

(c) By adding components, the horizontal component of the resultant is _______ units, and the vertical component of the resultant is _______ units.

(d) The resultant is in what quadrant? Quadrant _____

---

110. Both $\overrightarrow{0D}$ and $\overrightarrow{0C}$ lie on the horizontal axis. In this case, we cannot complete a parallelogram when adding the two vectors.

(a) The horizontal component of the resultant is _______ units.

(b) The vertical component of the resultant is _______ units.

(c) The resultant lies on the ___________ (horizontal/vertical) axis.

(d) The length of the resultant is _______ units.

(e) The direction of the resultant is ___________.

a) (4.7, 0)
b) (0, -7.1)
c) 4.7 units ...
-7.1 units
d) Quadrant IV

---

111. Both $\overrightarrow{0P}$ and $\overrightarrow{0M}$ lie on the vertical axis. We also cannot complete a parallelogram when adding these two vectors.

(a) The coordinates of the tip of the resultant are (,).

(b) The resultant lies on which axis? ____________

(c) The length of the resultant is _______ units.

(d) The direction of the resultant is ___________.

a) 23.1
b) 0
c) horizontal
d) 23.1
e) 0° (Q1 and Q4)

---

a) (0, 7.5)
b) Vertical axis
c) 7.5
d) 90° (Q1 and Q2)

SELF-TEST 9 (Frames 103-111)

The endpoint of $\overrightarrow{0A}$ is at (-200, 0).
The endpoint of $\overrightarrow{0B}$ is at (500, 0).

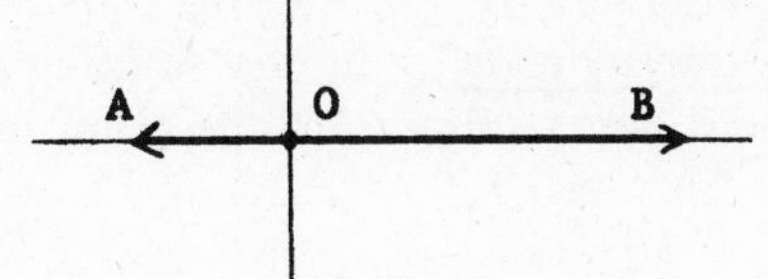

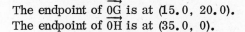

1. Find the coordinates of the endpoint of the resultant of $\overrightarrow{0A}$ and $\overrightarrow{0B}$. ____________
2. Find the length of the resultant. ____________
3. Find the direction of the resultant. ____________

The endpoint of $\overrightarrow{0R}$ is at (-6.00, 0).
The endpoint of $\overrightarrow{0S}$ is at (0, -4.00).

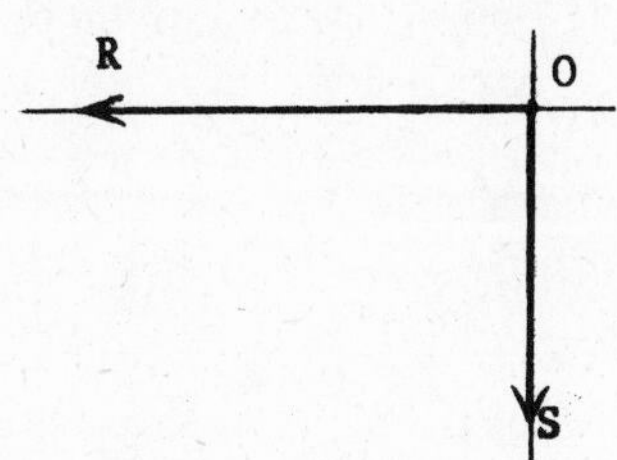

4. Find the coordinates of the endpoint of the resultant of $\overrightarrow{0R}$ and $\overrightarrow{0S}$. ____________
5. Find the length of the resultant. ____________
6. Find the direction of the resultant. ____________

The endpoint of $\overrightarrow{0G}$ is at (15.0, 20.0).
The endpoint of $\overrightarrow{0H}$ is at (35.0, 0).

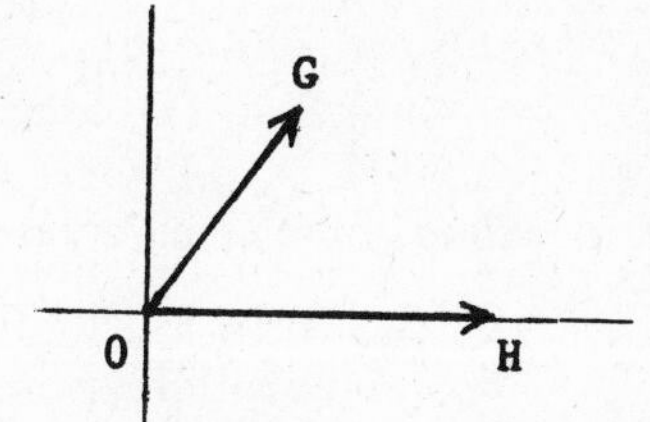

7. Find the coordinates of the endpoint of the resultant of $\overrightarrow{0G}$ and $\overrightarrow{0H}$. ____________
8. Find the length of the resultant. ____________
9. Find the direction of the resultant. ____________

ANSWERS:

1. (300, 0)
2. 300 units
3. 0° (Q1 and Q4)
4. (-6.00, -4.00)
5. 7.21 units
6. 34° (Q3)
7. (50.0, 20.0)
8. 53.9 units
9. 22° (Q1)

2-10 FINDING A VECTOR-ADDEND

The sum of a vector addition is called the "resultant". The two vectors which are added can be called "vector-addends". In this section, we will find the components of one vector-addend when the components of both the resultant and the other vector-addend are given.

112. In a vector-addition, the two vectors which are added can be called "vector-addends". In Figure 1 below, we are given one vector-addend $\overrightarrow{0A}$ and the resultant $\overrightarrow{0R}$. In Figure 2, we have completed the parallelogram to get a rough idea of the length and direction of the second vector-addend $\overrightarrow{0B}$.

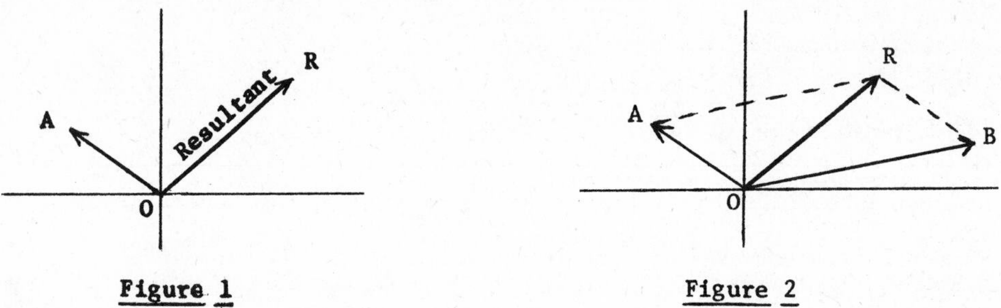

Figure 1 Figure 2

When completing a parallelogram to locate a missing vector-addend:

(1) Draw a dotted line connecting the tips of the two known vectors. This helps to see the parallelogram.

(2) Remember that the resultant is the diagonal of the parallelogram.

Complete the parallelograms below to locate the missing vector-addend $\overrightarrow{0B}$. $\overrightarrow{0R}$ in each case is the resultant.

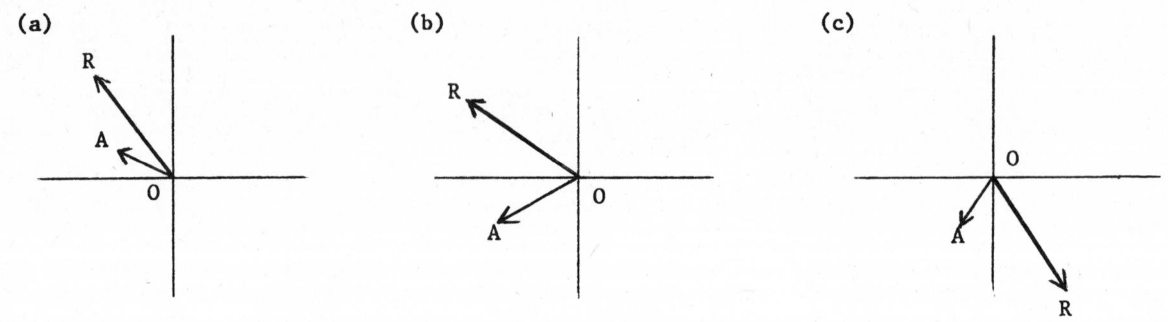

Answer to Frame 112:

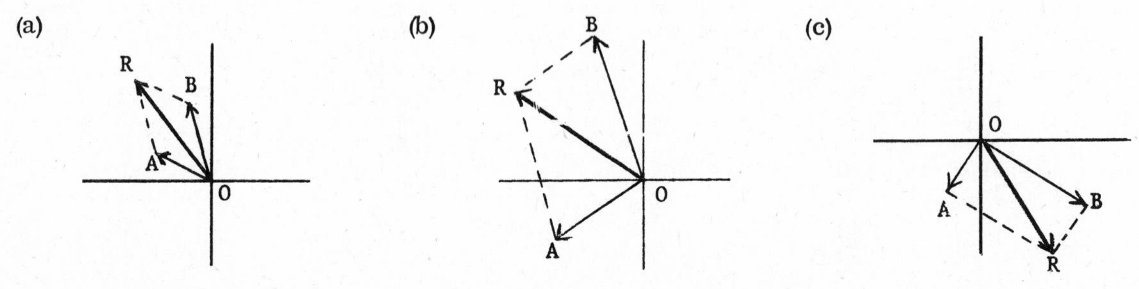

113. If the components of the resultant and one vector-addend are given, we can find the components of the second vector-addend.

In this case, we know the components of $\overrightarrow{0R}$ (the resultant) and $\overrightarrow{0T}$ (a vector-addend). The steps for finding the components of $\overrightarrow{0S}$ (the other vector-addend) are given below.

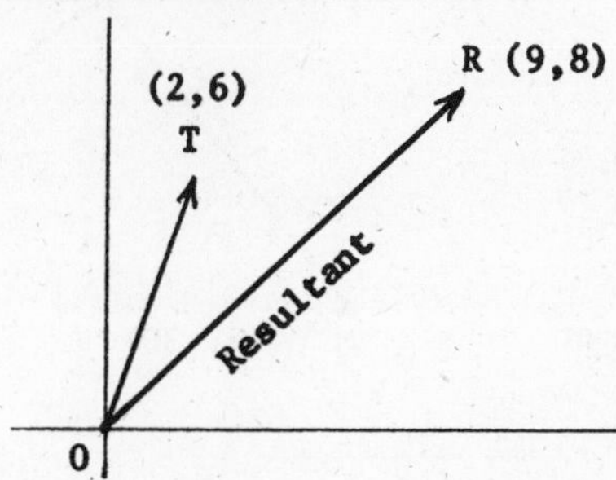

Step 1: Complete the parallelogram, to roughly locate the position of $\overrightarrow{0S}$, as we have done on the right.

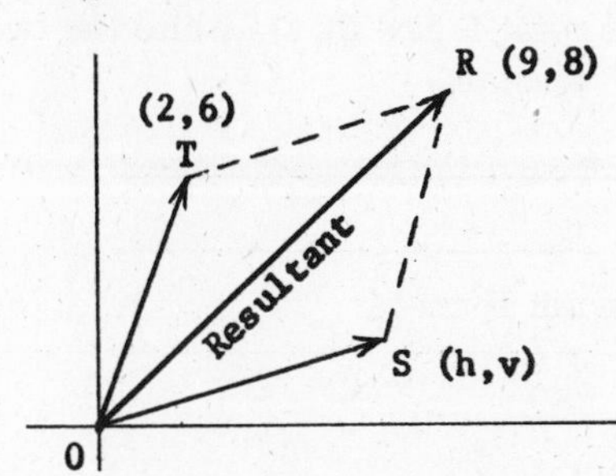

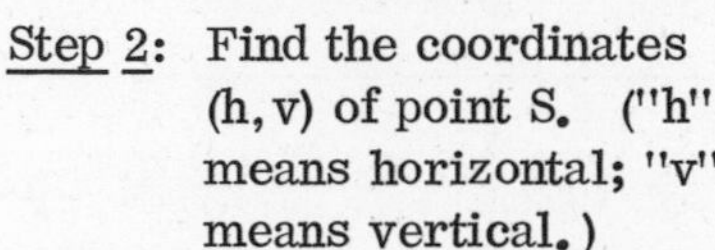

Step 2: Find the coordinates (h, v) of point S. ("h" means horizontal; "v" means vertical.)

Since the components of $\overrightarrow{0R}$ are the sums of the components of $\overrightarrow{0T}$ and $\overrightarrow{0S}$, these equations can be set up:

$2 + h = 9$ $(h = 7)$

$6 + v = 8$ $(v = 2)$

Therefore, the coordinates of point S are (,).

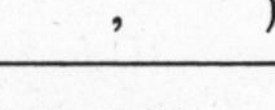

Answer to 113: (7, 2)

114. On the graph at the right:

$\overrightarrow{0B}$ is a resultant.

$\overrightarrow{0C}$ is a vector-addend.

Find the components of the vector that must be added to $\overrightarrow{0C}$ to give the resultant $\overrightarrow{0B}$. (Call this other vector $\overrightarrow{0D}$.)

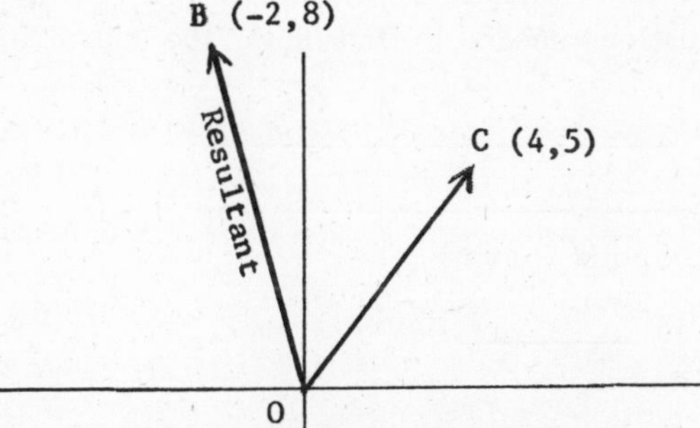

(a) On the graph above, complete the parallelogram and roughly locate $\overrightarrow{0D}$.

(b) Let (h, v) be the coordinates of point D. Write the two equations in which "h" and "v" appear.

(1) ______________________

(2) ______________________

(c) The coordinates of point D are (,).

(d) The horizontal component of $\overrightarrow{0D}$ is ______ units.

(e) The vertical component of $\overrightarrow{0D}$ is ______ units.

115. On the graph at the right:

$\overrightarrow{0M}$ is a resultant.

$\overrightarrow{0Q}$ is a vector-addend.

Find the other vector-addend. (Call it $\overrightarrow{0P}$.)

0

Resultant

Q (15,-10)

M (-25,-30)

(a) Make a rough sketch of the parallelogram on the graph above. In what quadrant is 0P? ____________

(b) If the coordinates of point P are (h, v), write the two equations in which "h" and "v" appear.

(1) ______________________

(2) ______________________

(c) The coordinates of point P are (,).

a)

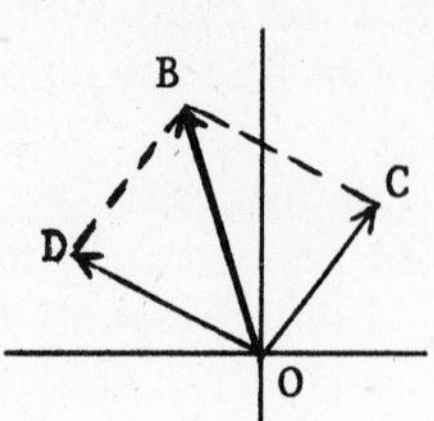

b) (1) h + 4 = -2
(2) v + 5 = 8

c) (-6, 3)

d) -6

e) 3

---

116. On the graph at the right:

$\overrightarrow{0L}$ is a resultant.

$\overrightarrow{0K}$ is a vector-addend.

Find $\overrightarrow{0G}$, the other vector-addend.

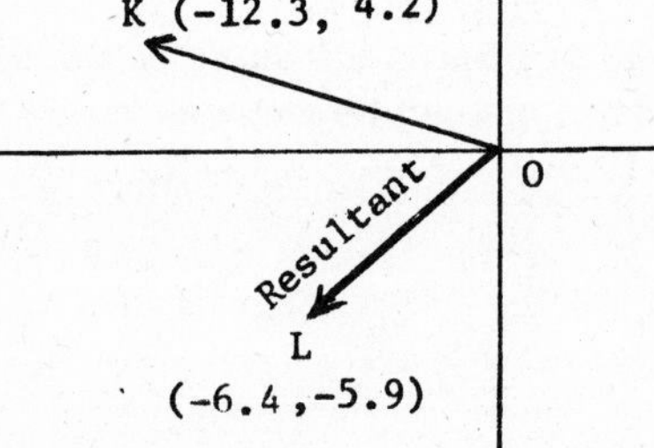

(a) Make a rough sketch of the parallelogram on the graph above. $\overrightarrow{0G}$ lies in what quadrant? __________

(b) Write the two equations needed to find (h, v), the coordinates of point G.

(1) ______________________

(2) ______________________

(c) The coordinates of point G are (,).

(d) The length of $\overrightarrow{0G}$ is __________ units.

(e) The direction of $\overrightarrow{0G}$ is ____________.

a) Quadrant III

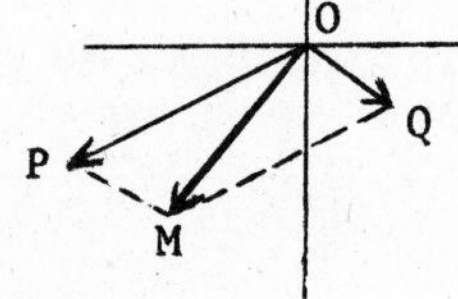

b) (1) 15 + h = -25
(2) (-10) + v = -30

c) (-40, -20)

Answer to Frame 116:

a) Quadrant IV

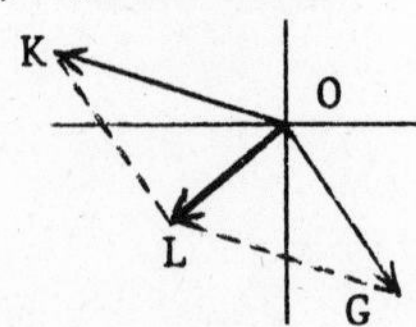

b) (1) h + (-12.3) = 6.4
(2) v + 4.2 = -5.9

c) (5.9, -10.1)

d) $\sqrt{137}$ = 11.7 units

e) 60° (Q4), since:

$$\tan \alpha = \frac{-10.1}{5.9} = -1.71$$

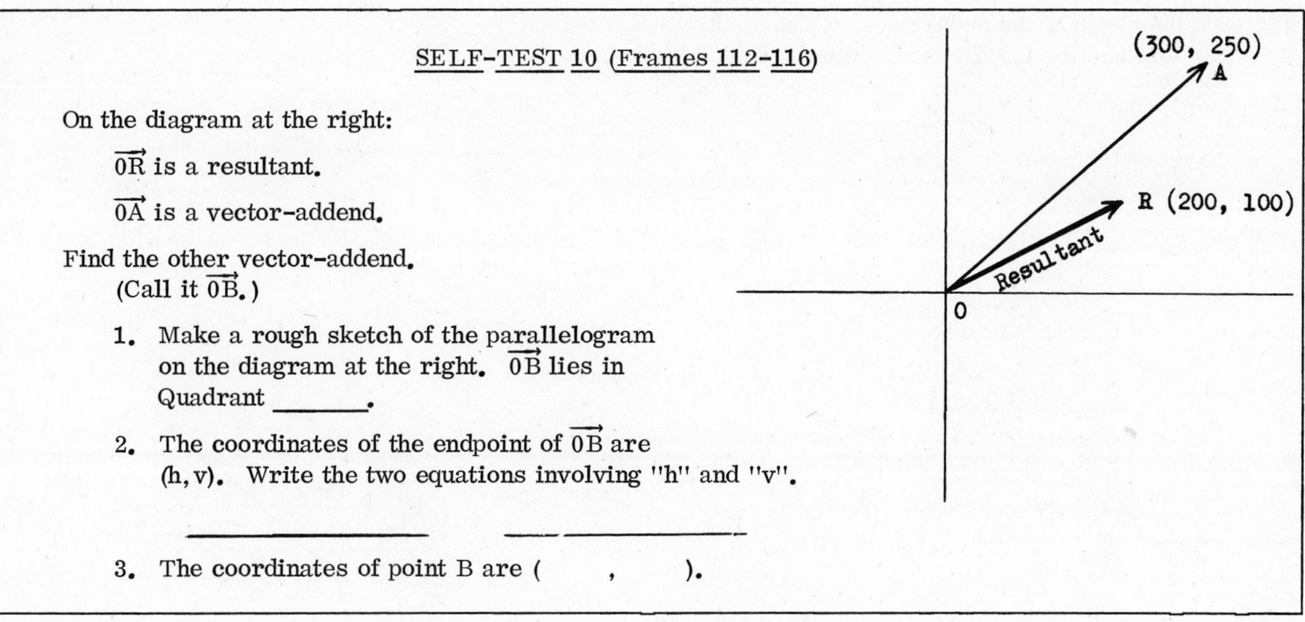

SELF-TEST 10 (Frames 112-116)

On the diagram at the right:

$\overrightarrow{0R}$ is a resultant.

$\overrightarrow{0A}$ is a vector-addend.

Find the other vector-addend.
(Call it $\overrightarrow{0B}$.)

1. Make a rough sketch of the parallelogram on the diagram at the right. $\overrightarrow{0B}$ lies in Quadrant ________.

2. The coordinates of the endpoint of $\overrightarrow{0B}$ are (h, v). Write the two equations involving "h" and "v".

   ________________ ________________

3. The coordinates of point B are (,).

ANSWERS: 

1\.

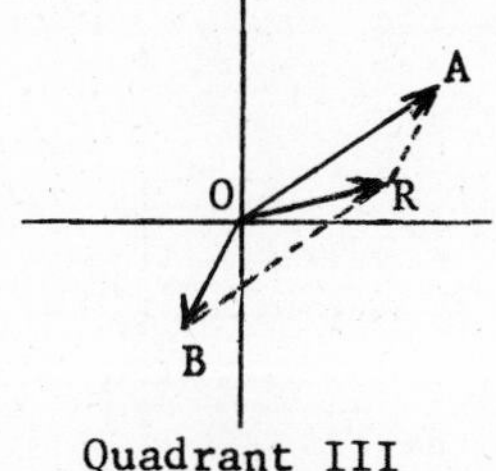

Quadrant III

2\. $h + 300 = 200$
$v + 250 = 100$

3\. (-100, -150)

2-11 ADDING MORE THAN TWO VECTORS

In this section, we will discuss the procedure for adding three or more vectors. Though three or more vectors can be added by the parallelogram method, they are usually added by the component method.

117. On the graph at the right, we have drawn three vectors. We will use the parallelogram method to add them and obtain a single resultant. The steps are described below.

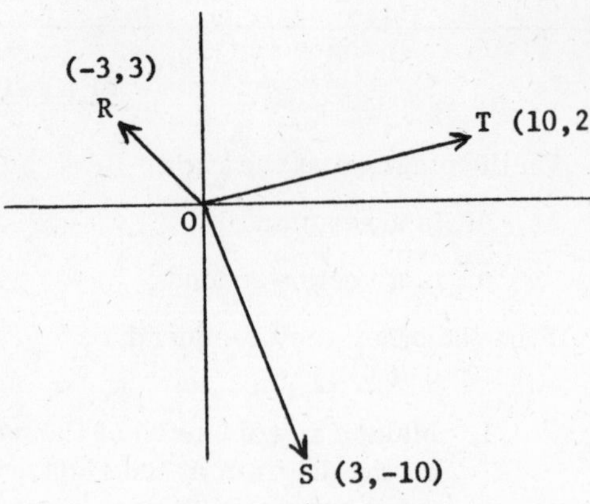

Step 1: Add any two vectors.

We will add $\overrightarrow{0R}$ and $\overrightarrow{0T}$. We have done so on the graph at the right. $\overrightarrow{0B}$ is the resultant. The coordinates of point B are (7, 5).

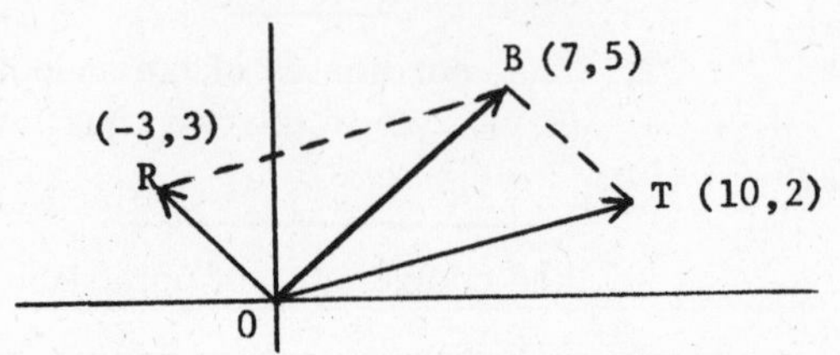

Step 2: Add $\overrightarrow{0B}$, the resultant of $\overrightarrow{0R}$ and $\overrightarrow{0T}$, to $\overrightarrow{0S}$.

We have done so on the graph at the right. $\overrightarrow{0A}$ is the resultant of the three vectors. The coordinates of point A are (10, -5).

$\overrightarrow{0A}$ is the sum of $\overrightarrow{0R}$, $\overrightarrow{0T}$, and $\overrightarrow{0S}$. $\overrightarrow{0A}$ is the single resultant vector which represents the sum of the original three vectors.

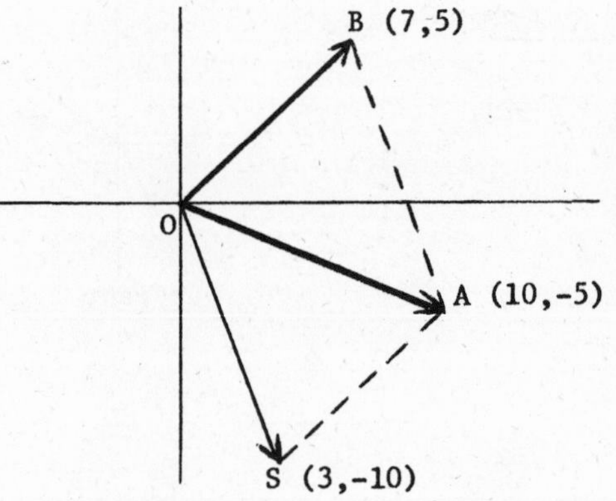

---

118. The graph at the right shows the same three vectors we added in the last frame ($\overrightarrow{0R}$, $\overrightarrow{0S}$, and $\overrightarrow{0T}$).

We found that their sum or resultant was $\overrightarrow{0A}$, which is also shown at the right. Point A had the coordinates (10, -5), as determined by the parallelogram method.

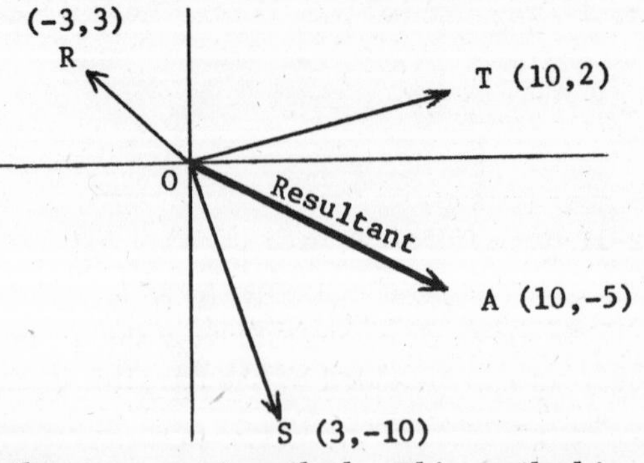

The coordinates of point A can be found much more quickly by the component method. This method is as follows:

In the table at the right, we listed the horizontal and vertical components of each original vector. Note the following:

(1) If we add the three horizontal components, we obtain the horizontal component of the resultant $\overrightarrow{0A}$.

(2) If we add the three vertical components, we obtain the vertical component of the resultant $\overrightarrow{0A}$.

From the table, the coordinates of the endpoint of resultant $\overrightarrow{0A}$ are (10, -5).

	Horizontal Component	Vertical Component
$\overrightarrow{0R}$	-3	3
$\overrightarrow{0T}$	10	2
$\overrightarrow{0S}$	3	-10
Resultant $\overrightarrow{0A}$	10	-5

119. When adding three or more vectors, we use the component method because it is simpler and more exact.

Let's add the three vectors on the left below by the component method. To do so, fill in the table on the right.

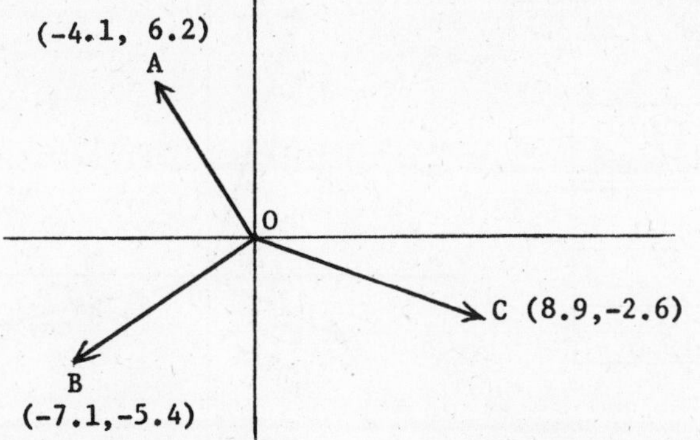

	Horizontal Component	Vertical Component
$\overrightarrow{0A}$		
$\overrightarrow{0B}$		
$\overrightarrow{0C}$		
Resultant		

(a) The horizontal component of the resultant is _________ units.

(b) The vertical component of the resultant is _________ units.

(c) The resultant lies in what quadrant? ____________

Answer to Frame 119: a) -2.3 units b) -1.8 units c) Quadrant III

120. On the graph at the left below, we have drawn four vectors. The procedure for adding four vectors is the same as that for adding three vectors. To do so, fill in the table on the right.

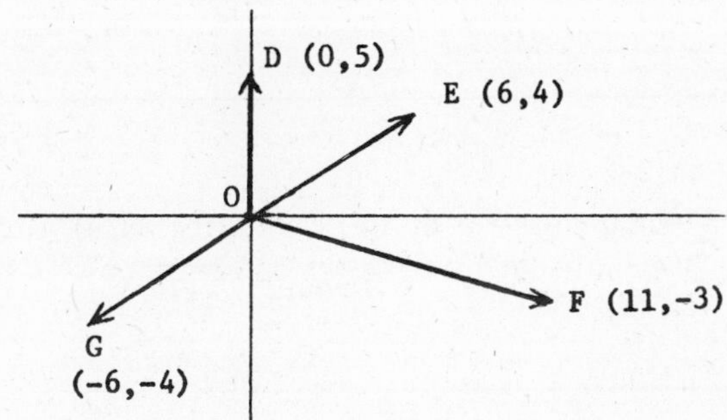

	Horizontal Component	Vertical Component
$\overrightarrow{0D}$		
$\overrightarrow{0E}$		
$\overrightarrow{0F}$		
$\overrightarrow{0G}$		
Resultant		

(a) The horizontal component of the resultant is _________ units.

(b) The vertical component of the resultant is _________ units.

(c) The resultant is a vector in Quadrant _____.

a) 11

b) 2

c) Quadrant I

121. For this system of three vectors, we will calculate the length and direction of the resultant. Note that the length and direction of each vector is given.

(a) To find the resultant, we must calculate the components of each vector first. Complete this table:

Vector	Horizontal Component	Vertical Component
$\overrightarrow{OF}$		+69.1
$\overrightarrow{OG}$	-37.3	
$\overrightarrow{OH}$	+82.4	-19.0

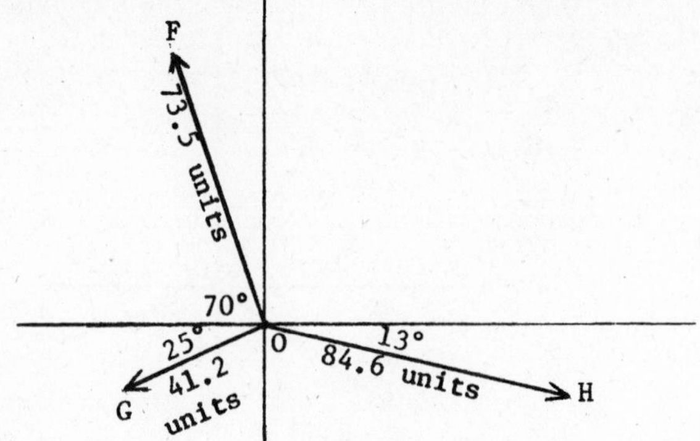

(b) From the table, determine:

Horizontal component of resultant = __________ units.

Vertical component of resultant = __________ units.

(c) The length of the resultant is __________ units.

(d) The direction of the resultant is __________.

Answer to Frame 121:

a)

$\overrightarrow{OF}$	-25.1	+69.1
$\overrightarrow{OG}$	-37.3	-17.4
$\overrightarrow{OH}$	+82.4	-19.0

b) Hor. Comp. = +20.0 units
Vert. Comp. = +32.7 units

c) 38.3 units (or 38.4)
(From: $\sqrt{1470}$)

d) 59° (Q1)

SELF-TEST 11 (Frames 117-121)

Three vectors which begin at the origin have their ends at the following points:

$\overrightarrow{OA}$: (50,10) $\overrightarrow{OB}$: (-30,0) $\overrightarrow{OC}$: (0,-20)

(a) Find the coordinates of the endpoint of their resultant. (,)

(b) The length of the resultant is __________.

(c) The direction of the resultant is __________.

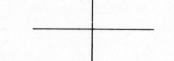

ANSWERS: a) (20,-10) b) 22.4 units (From: $\sqrt{500}$) c) 27° (Q4)

2-12 VECTOR OPPOSITES, THE ZERO-VECTOR, AND THE STATE OF EQUILIBRIUM

In this section, we will define what is meant by a pair of vector-opposites, the zero-vector, and a state of equilibrium. We will show that any system of two or more vectors is in a state of equilibrium if the resultant of the system is the zero-vector.

122. Two numbers are a pair of opposites if their sum is "0".

For example: +5 and -5 are a pair of opposites.

Two vectors are a pair of vector-opposites if both:

(1) The sum of their horizontal components is "0".
and (2) The sum of their vertical components is "0".

$\overrightarrow{0P}$ and $\overrightarrow{0D}$ are graphed on the right.

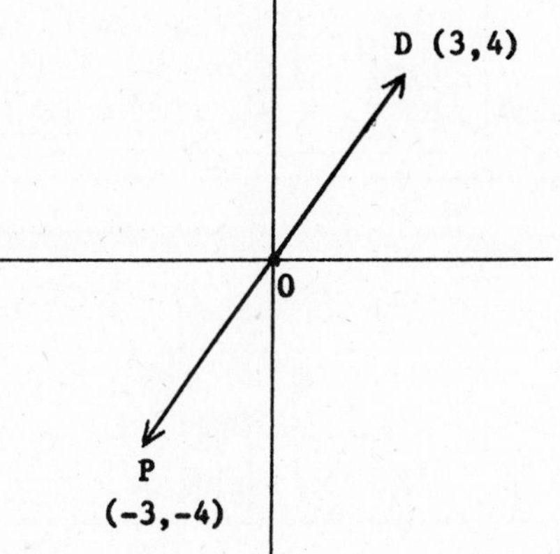

The sum of their horizontal components (+3 and -3) is "0".
The sum of their vertical components (+4 and -4) is "0".

Therefore, $\overrightarrow{0P}$ and $\overrightarrow{0D}$ are a pair of vector-opposites.

When two vector-opposites are graphed:

(1) They have the same length.
(2) They are in opposite directions.

---

123. The horizontal and vertical components of $\overrightarrow{0C}$ are 15 and -10.
The horizontal and vertical components of $\overrightarrow{0D}$ are -15 and 9.

Are $\overrightarrow{0C}$ and $\overrightarrow{0D}$ a pair of vector-opposites? ________

---

124. The vector-opposite of:

(a) $\overrightarrow{0S}$ is ______
(b) $\overrightarrow{0Q}$ is ______
(c) $\overrightarrow{0T}$ is ______
(d) $\overrightarrow{0R}$ is ______

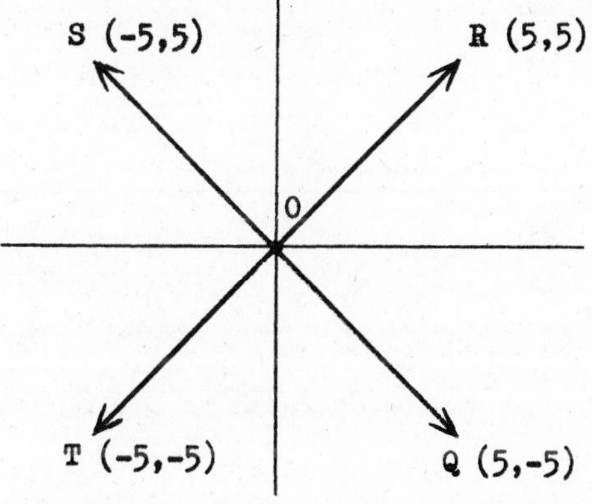

Answer to 123: No. The sum of their vertical components is not "0".

---

125. The coordinates of point C for $\overrightarrow{0C}$ are (-5, 7).

If $\overrightarrow{0F}$ is the vector-opposite of $\overrightarrow{0C}$, the coordinates of point F must be (,).

Answer to 124:
a) $\overrightarrow{0Q}$
b) $\overrightarrow{0S}$
c) $\overrightarrow{0R}$
d) $\overrightarrow{0T}$

Answer to 125: (5, -7)

126. $\overrightarrow{0G}$ and $\overrightarrow{0H}$ are a pair of vector-opposites.

If we add these two vectors,

(1) The horizontal component of the resultant is "0".

(2) The vertical component of the resultant is "0".

This particular resultant is a vector with no length. However, it is a vector, and it is called the zero-vector.

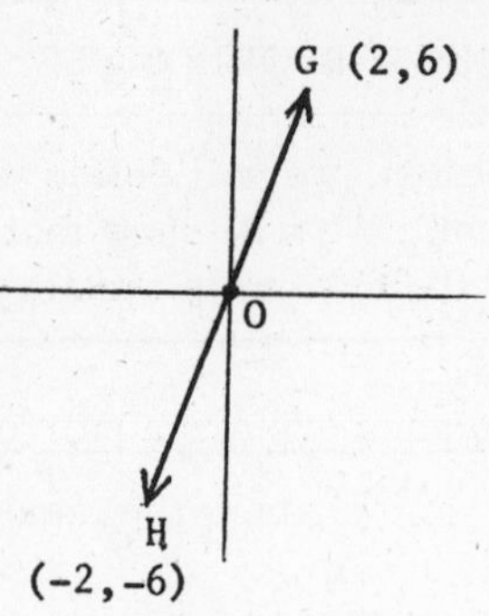

WHEN THE RESULTANT OF TWO VECTORS IS THE ZERO-VECTOR, WE SAY THAT THE TWO VECTORS ARE IN A STATE OF EQUILIBRIUM.

---

127. (a) The horizontal component of a zero-vector is _____ units.

(b) The vertical component of a zero-vector is _____ units.

(c) The length of a zero-vector is _____ units.

(d) If the resultant of two vectors is the zero-vector, we say that the two vectors are in a state of ________________.

---

a) 0
b) 0
c) 0
d) equilibrium

128. We want to add a vector (call it $\overrightarrow{0F}$) to $\overrightarrow{0R}$ so that we produce a state of equilibrium.

R (2,4)

0

(a) For two vectors to be in equilibrium, their resultant must be the ______________.

(b) To produce a state of equilibrium, we add $\overrightarrow{0F}$ to $\overrightarrow{0R}$. The coordinates of point F must be (,   ).

(c) $\overrightarrow{0F}$ must be the ____________________ of $\overrightarrow{0R}$.

---

a) zero-vector.
b) (-2, -4)
c) vector-opposite

129. If two vectors are in a state of equilibrium,

(a) Their resultant is the ____________________.

(b) The two-vectors must be a pair of ____________________.

---

a) zero-vector
b) vector-opposites

130. We have added the components of these three vectors in the table on the right.

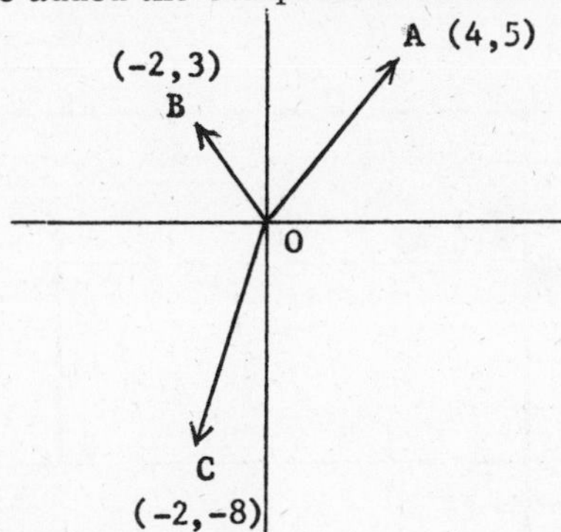

	Horizontal Component	Vertical Component
$\overrightarrow{0A}$	4	5
$\overrightarrow{0B}$	-2	3
$\overrightarrow{0C}$	-2	-8
Resultant	0	0

Since the resultant of these three vectors is the zero-vector, these three vectors are in a state of ____________________.

Answer to Frame 130: equilibrium

---

131. Here are the two key points to notice about vectors in a state of equilibrium:

(1) Among the horizontal components, those "to-the-left" cancel out those "to-the-right".

(2) Among the vertical components, those "upward" cancel out those "downward".

We have drawn two vectors in equilibrium, $\overrightarrow{0A}$ and $\overrightarrow{0B}$, with their components on the graph below.

In Projection #1, we have projected the two vertical components to the right. Notice how $\overrightarrow{CA}$ and $\overrightarrow{DB}$ cancel each other out because they are a pair of opposites.

In Projection #2, we have projected the two horizontal components downward. Notice how $\overrightarrow{0D}$ and $\overrightarrow{0C}$ cancel each other out because they are a pair of opposites.

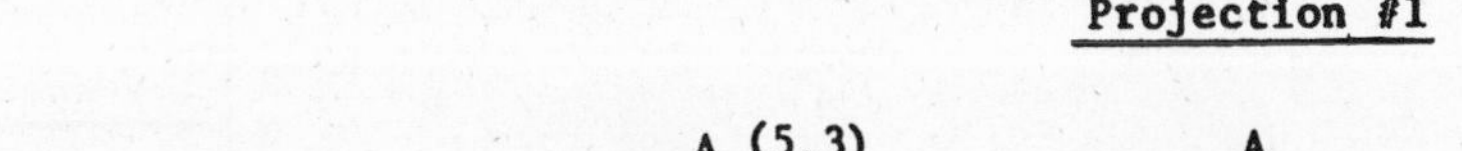

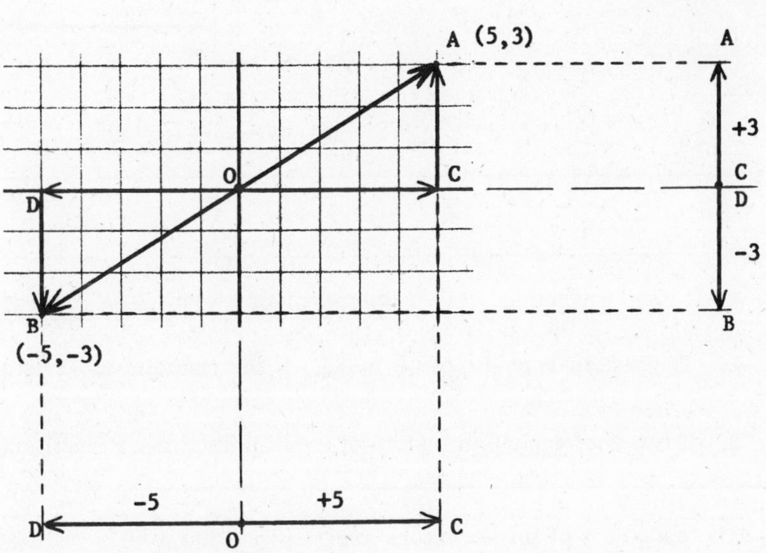

Projection #2

132. The three vectors ($\overrightarrow{0M}$, $\overrightarrow{0P}$, $\overrightarrow{0R}$) shown below are in a state of equilibrium.

Their components have also been drawn, with a projection of the vertical components to the right, and a projection of the horizontal components downward.

In Projection #1, we have projected the vertical components to the right. Notice how $\overrightarrow{BP}$ and $\overrightarrow{AM}$ cancel out $\overrightarrow{CR}$, since the sum of $\overrightarrow{BP}$ and $\overrightarrow{AM}$ (4 + 7) is the opposite of $\overrightarrow{CR}$ (-11).

In Projection #2, we have projected the horizontal components downward. Notice how $\overrightarrow{0C}$ and $\overrightarrow{0A}$ cancel out $\overrightarrow{0B}$, since the sum of $\overrightarrow{0C}$ and $\overrightarrow{0A}$ [(-2) + (-4)] is the opposite of $\overrightarrow{0B}$ (+6).

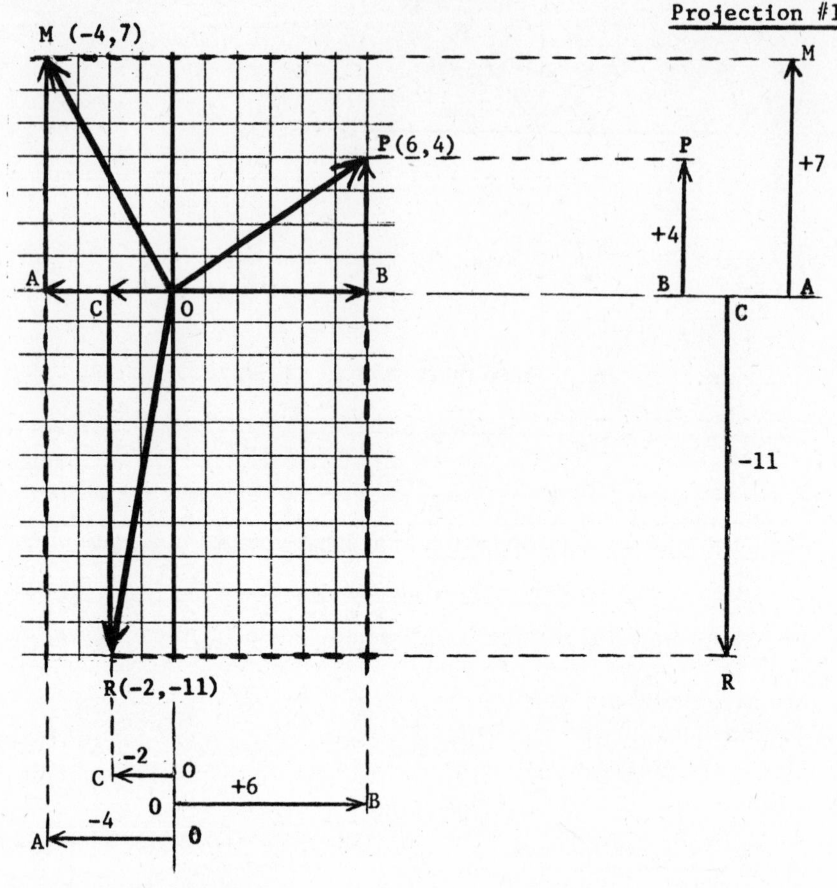

Projection #2

SELF-TEST 12 (Frames 122-132)

1. The vector-opposite of $\overrightarrow{0P}$ is $\overrightarrow{0Q}$. If the endpoint of $\overrightarrow{0P}$ is at (19,-27), the endpoint of $\overrightarrow{0Q}$ is at (,).
2. If two vectors are in a state of equilibrium, their resultant is called the ________________.

The endpoints of three vectors are: $\overrightarrow{0F}$: (-200,100) $\overrightarrow{0G}$: (500,-100) $\overrightarrow{0H}$: (-300,-200)

3. The sum of their horizontal components is ________.
4. The sum of their vertical components is ________.
5. Are the three vectors in a state of equilibrium? ________

ANSWERS: 1. (-19,27) 2. zero-vector 3. 0 4. -200 5. No

2-13 EQUILIBRANTS AND THE STATE OF EQUILIBRIUM

If a system of vectors is not in a state of equilibrium, we can always add one more vector which produces a state of equilibrium. This "added" vector is called an "equilibrant". We will briefly discuss equilibrants in this section.

133. If three vectors are in a state of equilibrium, each vector is the vector-opposite of the resultant of the other two vectors.

This principle is illustrated in the figures below. In each figure, the three vectors in equilibrium are $\overrightarrow{0A}$, $\overrightarrow{0B}$, and $\overrightarrow{0C}$.

Figure 1

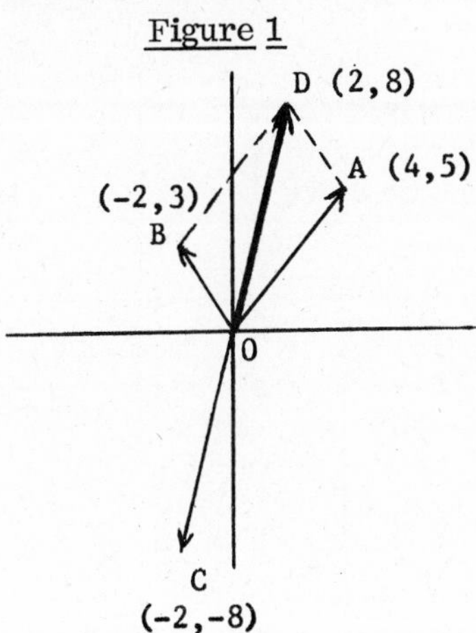

In Figure 1, the resultant of $\overrightarrow{0A}$ and $\overrightarrow{0B}$ is $\overrightarrow{0D}$.

Compare $\overrightarrow{0C}$ with $\overrightarrow{0D}$. Note that $\overrightarrow{0C}$ is the vector-opposite of $\overrightarrow{0D}$.

Figure 2

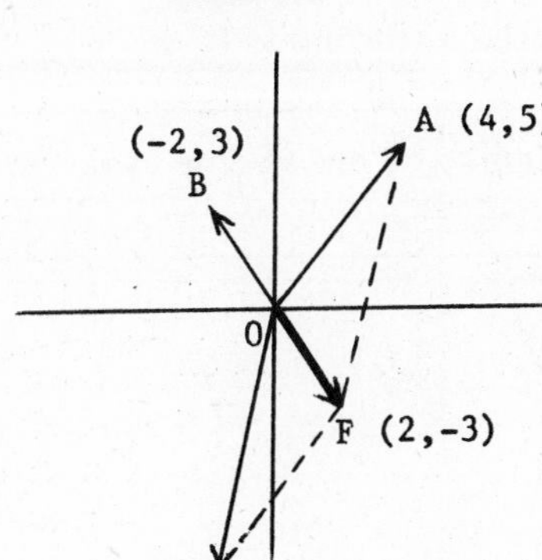

In Figure 2, the resultant of $\overrightarrow{0A}$ and $\overrightarrow{0C}$ is $\overrightarrow{0F}$.

Compare $\overrightarrow{0B}$ with $\overrightarrow{0F}$. Note that $\overrightarrow{0B}$ is the vector-opposite of $\overrightarrow{0F}$.

Figure 3

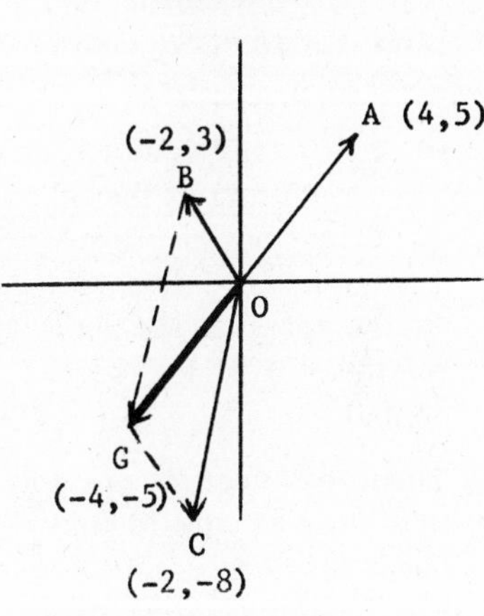

In Figure 3, the resultant of $\overrightarrow{0B}$ and $\overrightarrow{0C}$ is $\overrightarrow{0G}$.

Compare $\overrightarrow{0A}$ with $\overrightarrow{0G}$. Note that $\overrightarrow{0A}$ is the vector-opposite of $\overrightarrow{0G}$.

134. The four vectors on the left below were added by the component method. The results are shown in the table on the right below.

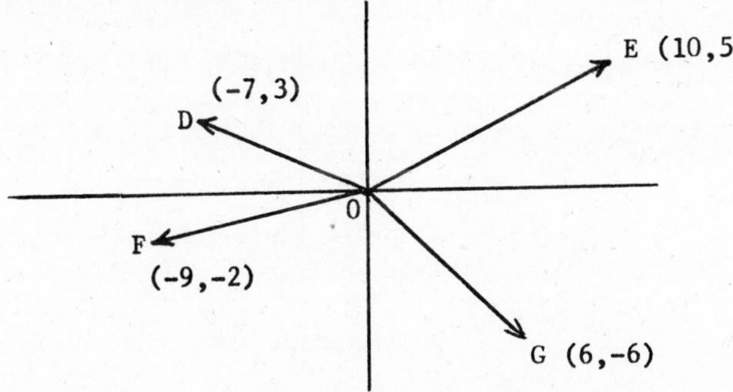

	Horizontal Component	Vertical Component
$\overrightarrow{0D}$	-7	3
$\overrightarrow{0E}$	10	5
$\overrightarrow{0F}$	-9	-2
$\overrightarrow{0G}$	6	-6
Resultant	0	0

Since the resultant of these four vectors is the zero-vector, the vectors are in a state of equilibrium.

(Continued on following page.)

134. (Continued)

Let's look at this system of four vectors more closely. We will compare $\overrightarrow{0D}$ with the resultant of the other three vectors ($\overrightarrow{0E}$, $\overrightarrow{0F}$, and $\overrightarrow{0G}$). (This resultant is computed in the table on the right below.)

Notice these points:

(1) The components of $\overrightarrow{0D}$ are -7 and 3.

(2) The components of the resultant of the other three vectors are 7 and -3.

(3) Therefore, $\overrightarrow{0D}$ and this resultant are a pair of vector-opposites.

	Horizontal Component	Vertical Component
$\overrightarrow{0E}$	10	5
$\overrightarrow{0F}$	-9	-2
$\overrightarrow{0G}$	6	-6
Resultant	7	-3

If we made a similar comparison for each of the other three vectors, we could show that the following is true:

> For four vectors in equilibrium, any one vector is the vector-opposite of the resultant of the other three vectors.

---

135. On the graph at the right, we have drawn two vectors, $\overrightarrow{0S}$ and $\overrightarrow{0T}$.

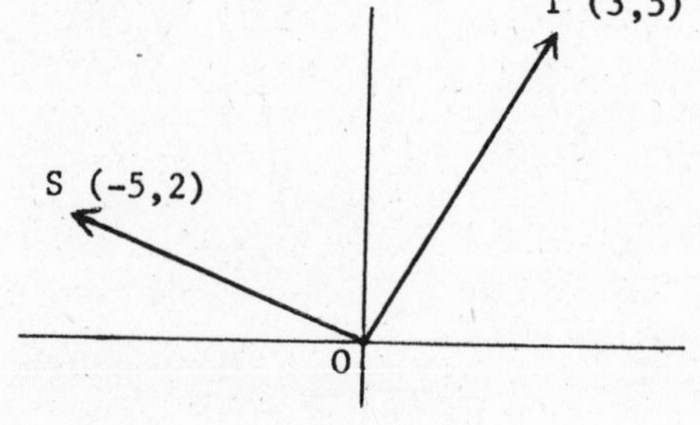

These two vectors are not in a state of equilibrium because:

(1) The sum of their horizontal components is "-2".

(2) The sum of their vertical components is "7".

We want to add a third vector (call it $\overrightarrow{0D}$) to $\overrightarrow{0S}$ and $\overrightarrow{0T}$ to produce a state of equilibrium.

(a) To do so, the third vector, $\overrightarrow{0D}$, must be the ________________ of the resultant of $\overrightarrow{0S}$ and $\overrightarrow{0T}$.

(b) The horizontal component of $\overrightarrow{0D}$ must be ______ units.

(c) The vertical component of $\overrightarrow{0D}$ must be ______ units.

(d) The coordinates of the endpoint of $\overrightarrow{0D}$ are (,).

---

a) vector-opposite

b) +2

c) -7

d) (2, -7)

136. On the graph at the right, we have drawn two vectors, $\overrightarrow{0C}$ and $\overrightarrow{0D}$.

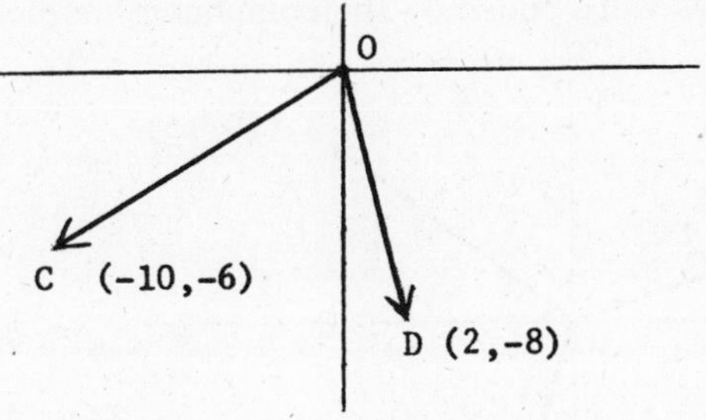

If we add $\overrightarrow{0C}$ and $\overrightarrow{0D}$,

(a) The horizontal component of the resultant is _____ units.

(b) The vertical component of the resultant is _______ units.

We want to add a third vector, $\overrightarrow{0F}$, to the two vectors above to produce a state of equilibrium.

(c) To produce a state of equilibrium, the coordinates of the endpoint of $\overrightarrow{0F}$ must be (,).

---

a) -8

b) -14

c) (8,14)

137. When we add another vector to produce a state of equilibrium, the added vector is called an equilibrant.

Here are two vectors:

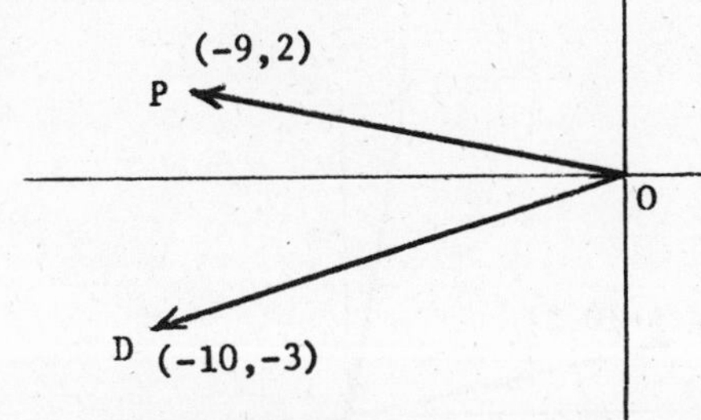

The coordinates of the endpoint of their equilibrant are (,).

(19,1)

138. Here are two more vectors:

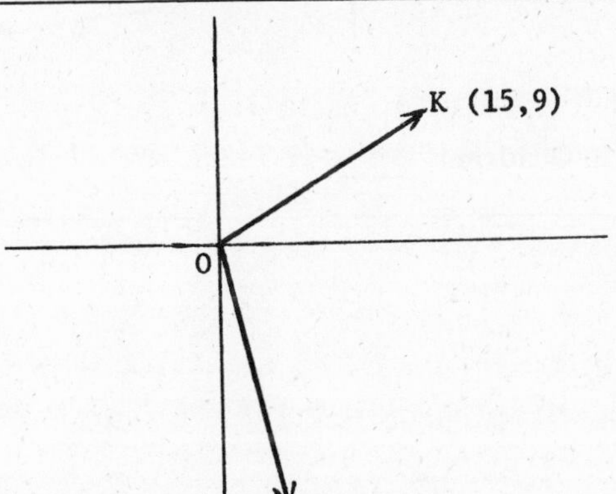

(a) The horizontal component of their equilibrant is _____ units.

(b) The vertical component of their equilibrant is _____ units.

(c) Their equilibrant lies in Quadrant _____.

---

a) -20

b) 9

c) II

139. The three vectors below were added by the component method in the table on the right.

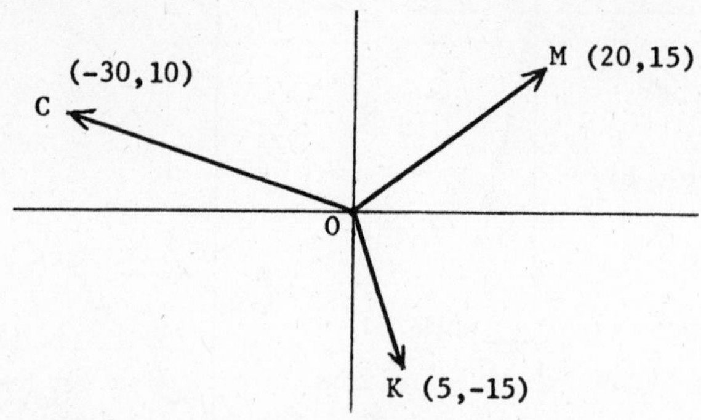

	Horizontal Component	Vertical Component
$\overrightarrow{0C}$	-30	10
$\overrightarrow{0M}$	20	15
$\overrightarrow{0K}$	5	-15
Resultant	-5	10

To produce a state of equilibrium, we want to add a fourth vector called the <u>equilibrant.</u>

(a) The horizontal component of the equilibrant must be _____ units.

(b) The vertical component of the equilibrant must be _____ units.

(c) The equilibrant lies in Quadrant _____.

---

a) +5

b) -10

c) IV

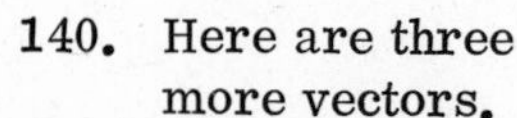

140. Here are three more vectors.

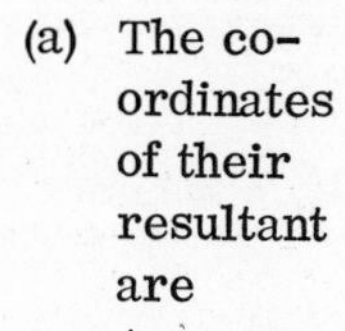

(a) The co-ordinates of their resultant are (,).

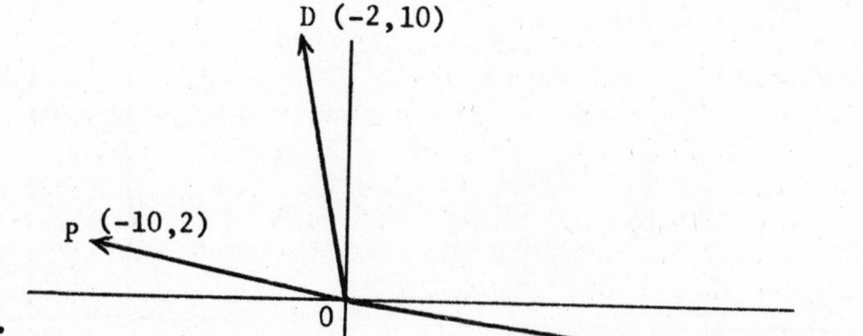

(b) The co-ordinates of their equilibrant are (,).

(c) The equilibrant lies in Quadrant _____.

---

a) (2,10)

b) (-2,-10)

c) III

SELF-TEST 13 (Frames 133-140)

Refer to the two-vector system shown at the right.

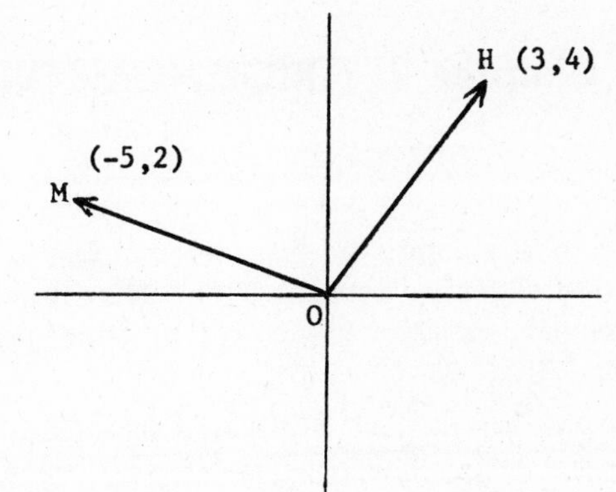

For the resultant, find its:

1. Coordinates ___________
2. Length ___________
3. Direction ___________

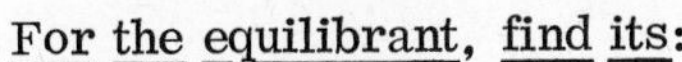

For the equilibrant, find its:

4. Coordinates ___________
5. Length ___________
6. Direction ___________

Refer to the four-vector system shown at the right.

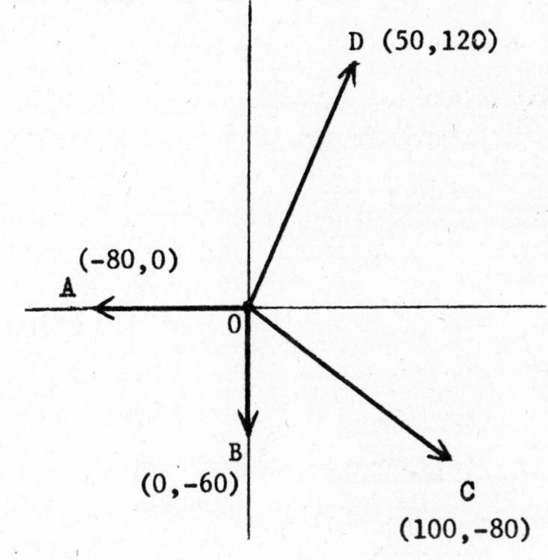

For the resultant, find its:

7. Coordinates ___________
8. Length ___________
9. Direction ___________

For the equilibrant, find its:

10. Coordinates ___________
11. Length ___________
12. Direction ___________

ANSWERS:

1. (-2, 6)	4. (2, -6)	7. (70, -20)	10. (-70, 20)
2. 6.32 units	5. 6.32 units	8. 72.8 units	11. 72.8 units
3. 72° (Q2)	6. 72° (Q4)	9. 16° (Q4)	12. 16° (Q2)

Chapter 3 TRIGONOMETRIC RATIOS OF STANDARD-POSITION ANGLES

Up to this point, the discussion of angles on the coordinate system has been limited to reference angles, all of which lie between 0° and 90°. Reference angles, however, are related to standard-position angles, and standard-position angles can be either positive angles of any size (like 120° or 452°) or negative angles of any size (like -78° or -246°).

In this section, we will define what is meant by a standard-position angle and examine the trigonometric ratios of standard-position angles. We will briefly review vector addition, showing that the direction of a vector is ordinarily specified by its standard-position angle. We will introduce the basic sine-wave graph, and show the procedure for finding trigonometric ratios on the slide rule.

3-1 STANDARD POSITION ANGLES BETWEEN 0° AND 360°

Standard-position angles are angles which are generated on the coordinate system. We will show what is meant by a standard-position angle in this section. Though such angles can be of any size, we will limit our discussion to standard-position angles between 0° and 360°.

1. In this frame, we will introduce the concept of "the standard position of an angle on the coordinate system". It is a fairly simple concept.

In Figure 1, $\overrightarrow{0A}$ is a vector on the positive side of the horizontal axis. An angle is generated by rotating a vector in a counterclockwise direction from this position.

In Figure 2, a 60° angle has been generated by rotating $\overrightarrow{0A}$ 60° in a counterclockwise direction. The curved arrow shows the path of the rotation.

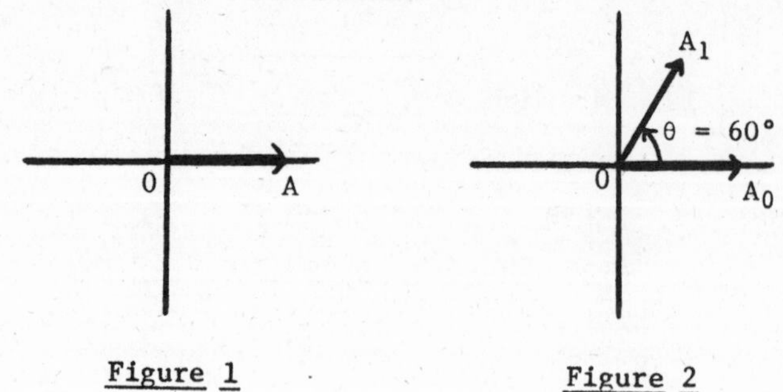

Figure 1 Figure 2

To distinguish the initial position of $\overrightarrow{0A}$ and the terminal position of $\overrightarrow{0A}$, we use the subscripts "0" and "1".

Note: (1) We will use the symbol "θ" for the rotated angle. "θ" is the Greek letter theta, and is pronounced "thay-ta".

(2) $\overrightarrow{0A_0}$ is called the initial side of angle θ. It lies on the positive or right-hand part of the horizontal axis.

(3) $\overrightarrow{0A_1}$ is called the terminal side of angle θ. After the rotation of 60°, the terminal side is in the first quadrant.

2. Here is another angle in standard position. The angle of rotation is 135°. The curved arrow shows the path of rotation.

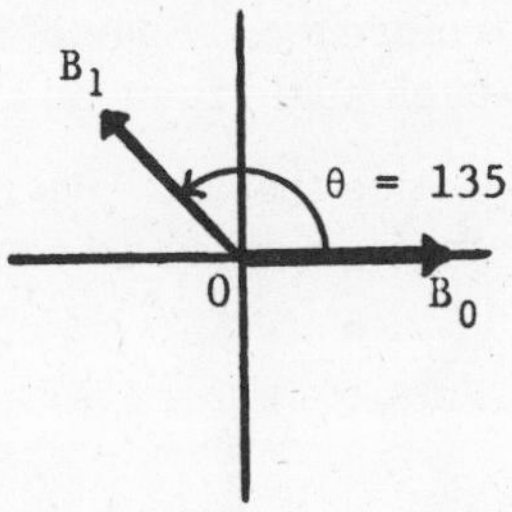

(a) $\overrightarrow{0B_0}$ is called the __________ (initial/terminal) side of angle θ.

(b) $\overrightarrow{0B_1}$ is called the __________ (initial/terminal) side of angle θ.

(c) The terminal side of a 135° angle lies in what quadrant? ______

(d) When this vector was rotated to generate the 135° angle, the direction of rotation was ________________ (clockwise/counterclockwise).

B_1 $\theta = 135°$ 0 B_0

---

a) initial

b) terminal

c) Quadrant II

d) counterclockwise

---

3. Here is another angle in standard position. The angle of rotation is 240°.

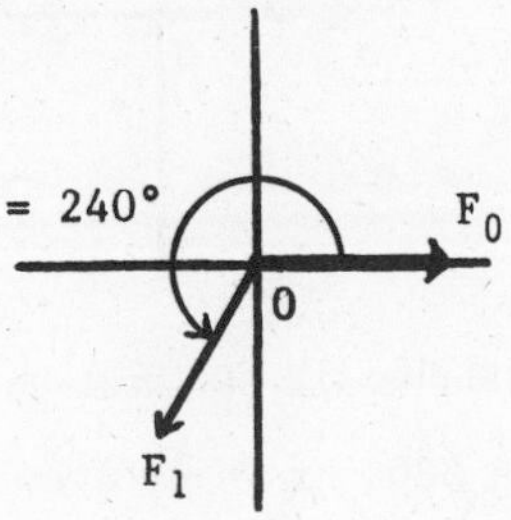

(a) $\overrightarrow{0F_0}$ is the __________ side.

(b) $\overrightarrow{0F_1}$ is the __________ side.

(c) The angle θ was generated by a 240° rotation in a ________________ direction.

(d) The terminal side lies in what quadrant? __________

$\theta = 240°$ F_0 0 F_1

---

a) initial

b) terminal

c) counterclockwise

d) Quadrant III

---

4. The following angle is also in standard position. The angle of rotation is 300°.

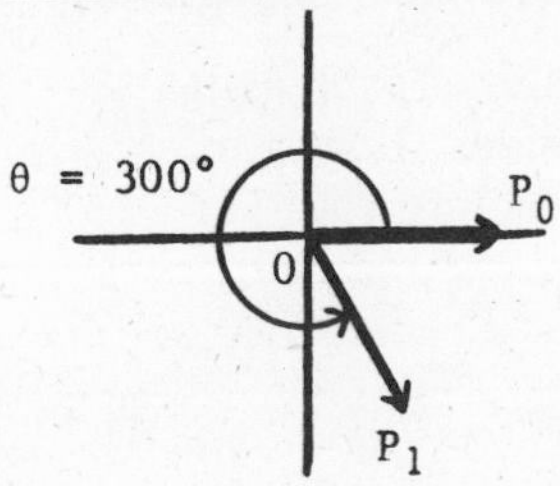

(a) Which vector is the terminal side? ______

(b) Which vector is the initial side? ______

(c) Can a terminal side lie in the 4th quadrant? ________

$\theta = 300°$ P_0 0 P_1

---

a) $\overrightarrow{0P_1}$

b) $\overrightarrow{0P_0}$

c) Yes, $\overrightarrow{0P_1}$ obviously does.

---

5. For an angle in standard position:

(a) The initial side always lies on the ________________ (right-hand, left-hand) side of the horizontal axis.

(b) Can the terminal side lie in any quadrant? ________

(c) The direction of rotation is always ________________ (clockwise/counterclockwise).

---

a) right-hand

b) Yes

c) counterclockwise

6. When discussing standard-position angles, there are some basic angles which can be used as points of reference. Four of these basic angles (90°, 180°, 270°, 360°) are drawn below. The terminal side of each angle is on an axis.

Notice that we have not drawn the initial sides of the angles. Since the initial side is the same for every angle in standard position, we will not draw initial sides from now on.

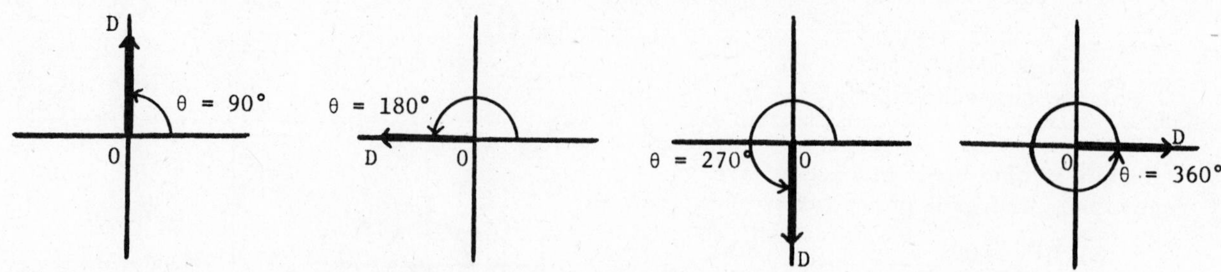

Notice these points: (1) There are 90° in each of the four quadrants.

(2) A 360° angle involves one complete rotation through all four quadrants. For a 360° angle, the terminal side is identical to the initial side.

---

7. In the diagram below at the right, an angle of 0° and an angle of 360° are shown in standard position.

Notice that the terminal side of each angle coincides with the initial side. However, a 360° angle involves a complete rotation through all four quadrants. A 0° angle involves no rotation at all.

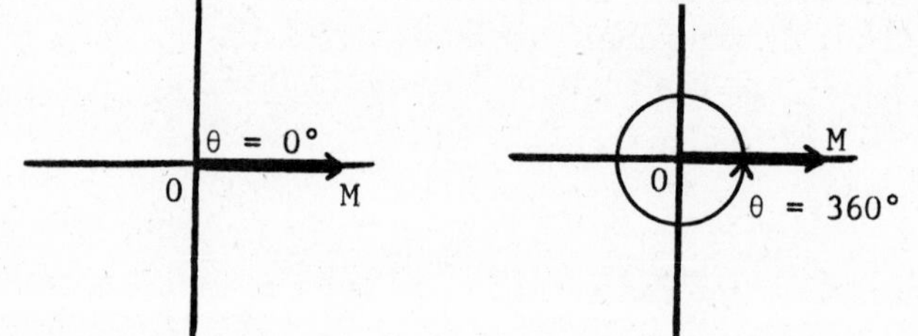

---

8. In both figures below, a standard-position angle divides a quadrant into two equal parts.

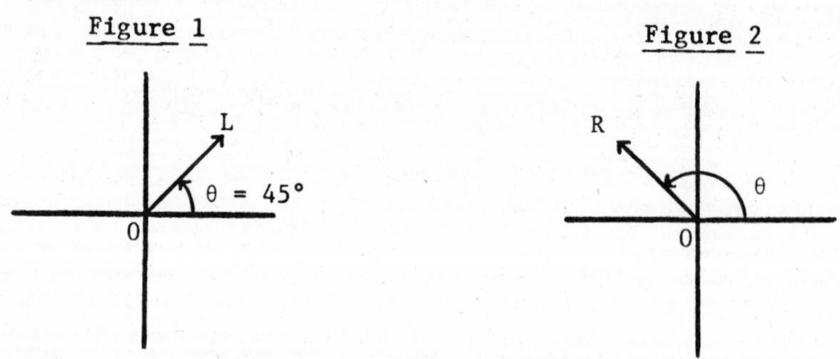

In Figure 1, $\overrightarrow{0L}$ divides the first quadrant into two equal parts. Therefore:

(1) The angle of rotation is halfway between 0° and 90°.
(2) Angle θ contains 45°.

In Figure 2, $\overrightarrow{0R}$ divides the second quadrant into two equal parts. Therefore:

(a) The angle of rotation is halfway between _____° and _____°.

(b) Angle θ contains _____°.

---

a) 90° and 180°

b) 135°

9. In each figure below, a standard-position angle divides a quadrant in two equal parts.

Figure 1   Figure 2

In Figure 1, $\overrightarrow{0N}$ divides Quadrant III into two equal parts.

(a) Therefore, angle θ is halfway between _______° and _______°.

(b) Angle θ is a _______° angle.

In Figure 2, $\overrightarrow{0T}$ divides Quadrant IV into two equal parts.

(c) Therefore, angle θ is halfway between _______° and _______°.

(d) Angle θ is a _______° angle.

---

10. Identify the standard-position angles which divide each quadrant in half.

(a) Quadrant I: _______   (c) Quadrant III: _______

(b) Quadrant II: _______   (d) Quadrant IV: _______

a) 180° and 270°
b) 225°
c) 270° and 360°
d) 315°

---

11. You can use the angles which divide the quadrants in half as reference lines when estimating the size of angles. Do so in the following.

Figure 1   Figure 2

(a) In Figure 1, $\overrightarrow{0D}$ is clearly not halfway through quadrant I. Therefore, angle θ is clearly greater than 0° and less than _______°.

(b) In Figure 2, $\overrightarrow{0E}$ is clearly not halfway through quadrant II. Therefore, angle θ is greater than 90° and less than _______°.

a) 45°   c) 225°
b) 135°   d) 315°

---

a) 45°
b) 135°

12.

Figure 1 Figure 2

(a) In Figure 1, $\overrightarrow{0S}$ is not halfway through quadrant III. Therefore, angle θ is greater than 180° but less than ______°.

(b) In Figure 2, $\overrightarrow{0M}$ is not halfway through quadrant IV. Therefore, angle θ is greater than 270° but less than ______°.

Answer to Frame 12: a) 225° b) 315°

13. In each figure below, we have drawn a terminal side which is more than halfway through one of the quadrants.

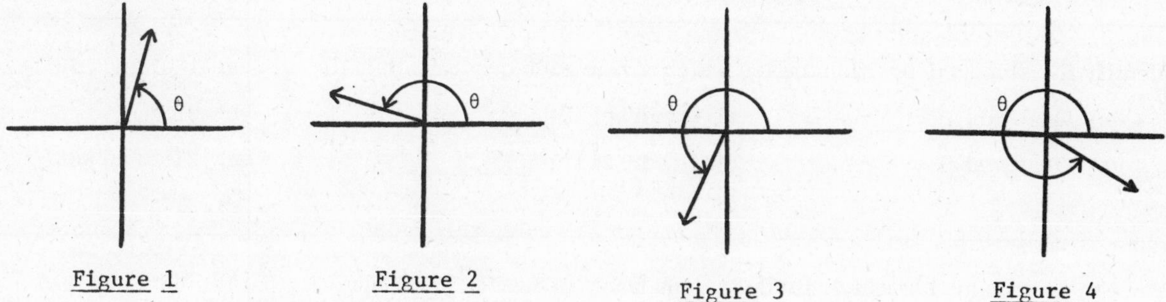

Figure 1 Figure 2 Figure 3 Figure 4

In Figure 1, angle θ must be between 45° and 90°.

(a) In Figure 2, angle θ must be between ______° and ______°.

(b) In Figure 3, angle θ must be between ______° and ______°.

(c) In Figure 4, angle θ must be between ______° and ______°.

Answer to Frame 13: a) 135° and 180° b) 225° and 270° c) 315° and 360°

14. Keep the halfway angles in mind as reference points for these.

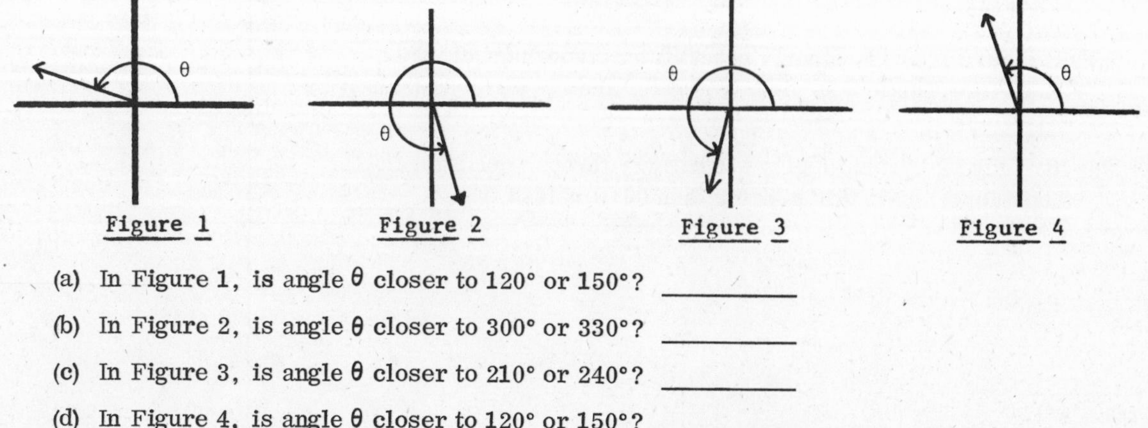

(a) In Figure 1, is angle θ closer to 120° or 150°? ______

(b) In Figure 2, is angle θ closer to 300° or 330°? ______

(c) In Figure 3, is angle θ closer to 210° or 240°? ______

(d) In Figure 4, is angle θ closer to 120° or 150°? ______

15. Use the halfway angles as reference points for these.

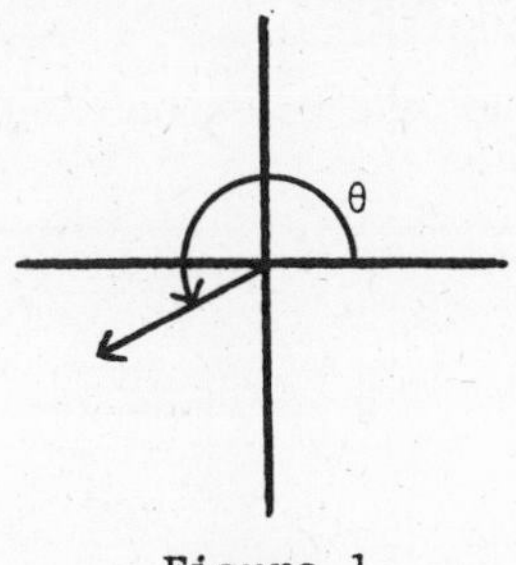

Figure 1

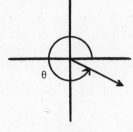

Figure 2

(a) In Figure 1, is angle θ closer to 210° or 240°? ________

(b) In Figure 2, is angle θ closer to 300° or 330°? ________

a) 150°
b) 300°
c) 240°
d) 120°

---

16. In which quadrant (I, II, III or IV) does the terminal side of each of the following standard-position angles lie?

(a) 91° ____ (b) 179° ____ (c) 271° ____ (d) 181° ____

a) 210°
b) 330°

---

17. In which quadrant does the terminal side of each of these angles lie?

(a) 88° ____ (b) 302° ____ (c) 269° ____ (d) 178° ____

a) II c) IV
b) II d) III

Answer to Frame 17: a) I b) IV c) III d) II

SELF-TEST 1 (Frames 1-17)

Sketch and label these angles:

1. 205°

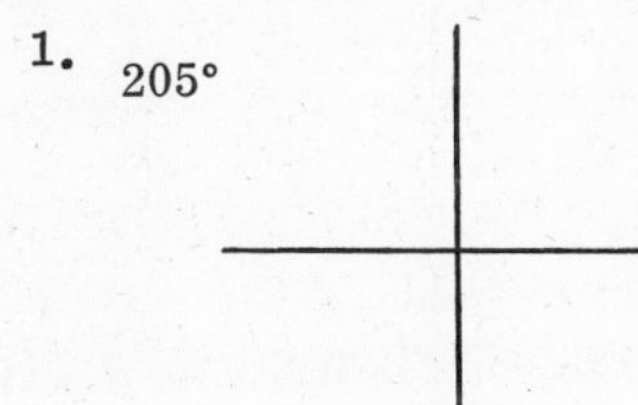

2. 302°

3. 124°

State the quadrant number (I, II, III, or IV) in which the terminal side of each angle is located.

4. 348° ______ 6. 259° ______ 8. 94° ______

5. 167° ______ 7. 87° ______ 9. 270° ______

ANSWERS:

1.

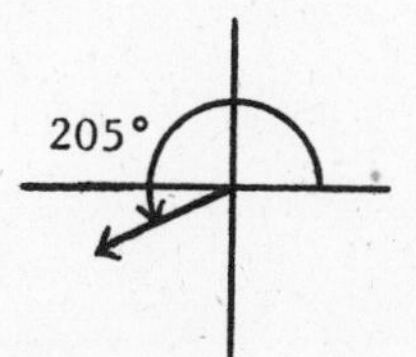

2.

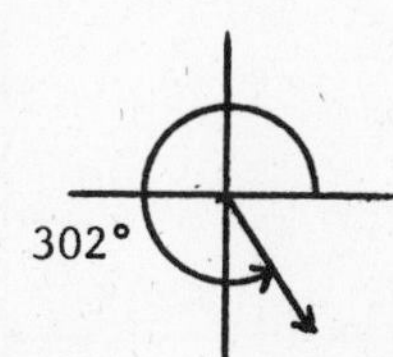

3.

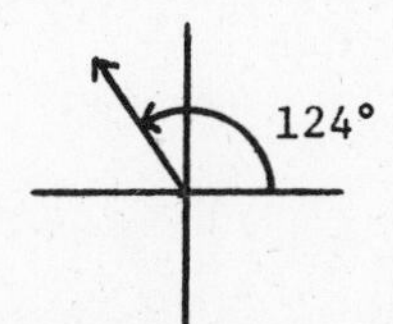

4. IV
5. II
6. III
7. I
8. II
9. Between III and IV

3-2 REFERENCE ANGLES OF STANDARD-POSITION ANGLES BETWEEN 0° AND 360°

Any standard-position angle has a reference angle. In this section, we will discuss the reference angles for standard-position angles between 0° and 360°.

18. The following three figures are discussed below.

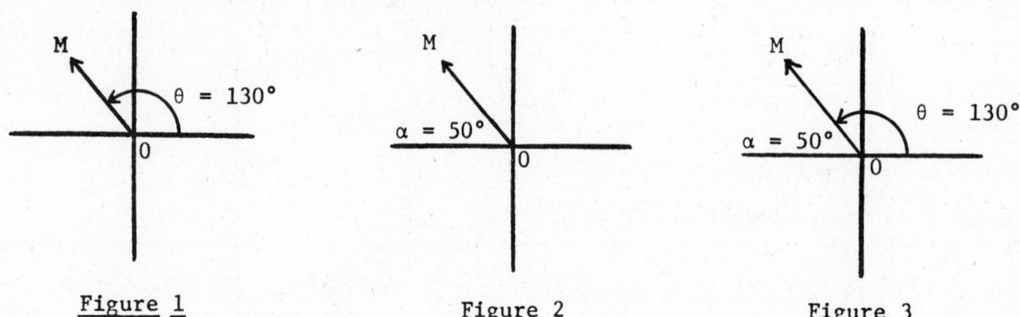

Figure 1 Figure 2 Figure 3

In Figure 1, angle θ is a standard-position angle of 130°. That is, $\overrightarrow{0M}$ has been rotated 130° to a terminal position in the second quadrant.

In Figure 2, the reference angle (α) for $\overrightarrow{0M}$ is 50° (Q2).

In Figure 3, we have combined θ and α. α is the reference angle for the standard-position angle θ.

The sum of θ and α is ________.

19. Both of the angles below are in standard position, with their terminal sides in the second quadrant. Under each figure, specify the reference angle (α) of each standard-position angle. (Note: A reference angle is not completely specified unless its quadrant is listed.) (a) 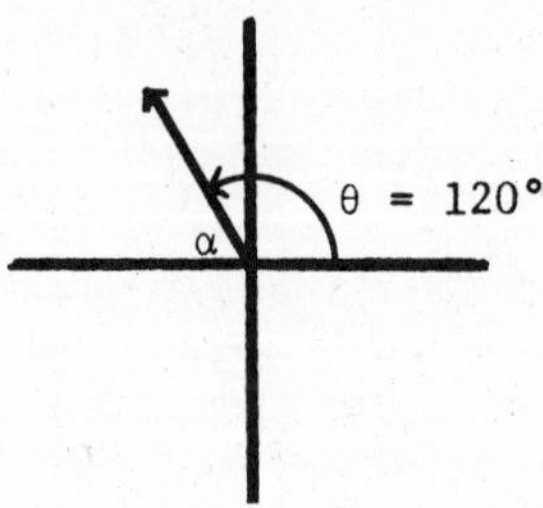 α = ________ (b) 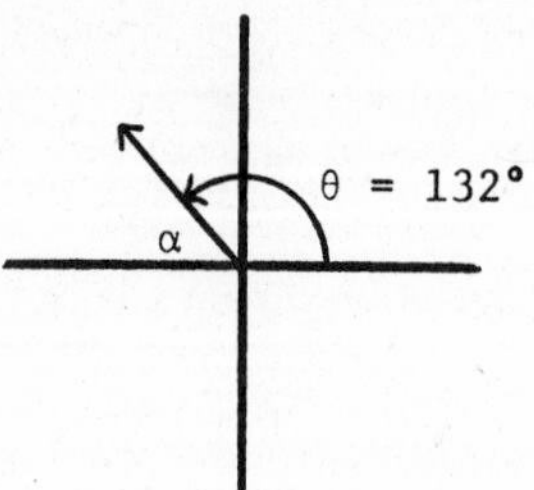 α = ________	180°
20. When the terminal side of a standard-position angle (θ) lies in the second quadrant: (a) θ is an angle between 90° and ______°. (b) The sum of θ and its reference angle (α) is ______°.	a) α = 60° (Q2) b) α = 48° (Q2)
	a) 180° b) 180°.

21. Specify the reference angle for the following standard-position angles.

(a) If $\theta = 110°$, $\alpha =$ ______
(b) If $\theta = 172°$, $\alpha =$ ______
(c) If $\theta = 131°$, $\alpha =$ ______

---

22. Name the standard-position angle for which each of the following is the reference angle.

(a) If $\alpha = 69°$ (Q2), $\theta =$ ______
(b) If $\alpha = 13°$ (Q2), $\theta =$ ______
(c) If $\alpha = 28°$ (Q2), $\theta =$ ______

Answers to 21: a) 70° (Q2) b) 8° (Q2) c) 49° (Q2)

---

23. On the right, the terminal side of θ is in the first quadrant.

$\theta = 50°$

When an angle is in standard position in the first quadrant, it is its own reference angle.

For example, in this case: $\theta = 50°$ and $\alpha = 50°$ (Q1)

Specify the reference angle for a standard-position angle of 37°. ______

Answers to 22: a) 111° b) 167° c) 152°

---

24. Specify the reference angle for each of the following standard-position angles.

(a) If $\theta = 127°$, $\alpha =$ ______
(b) If $\theta = 79°$, $\alpha =$ ______
(c) If $\theta = 93°$, $\alpha =$ ______

Answer to 23: 37° (Q1)

---

25. Name the standard-position angle for which each of the following is the reference angle.

(a) If $\alpha = 49°$ (Q1), $\theta =$ ______
(b) If $\alpha = 49°$ (Q2), $\theta =$ ______
(c) If $\alpha = 1°$ (Q2) $\theta =$ ______

Answers to 24: a) 53° (Q2) b) 79° (Q1) c) 87° (Q2)

Answers to 25: a) 49° b) 131° c) 179°

26. The following three figures are discussed below.

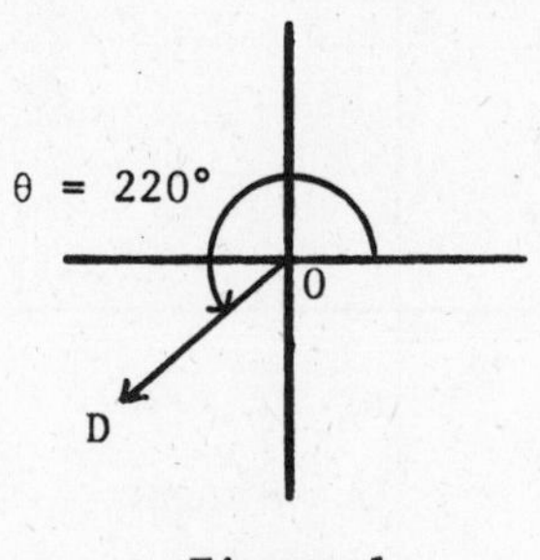

Figure 1

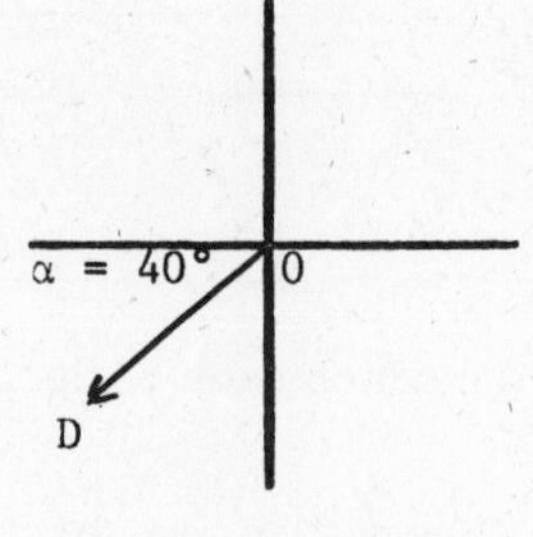

Figure 2

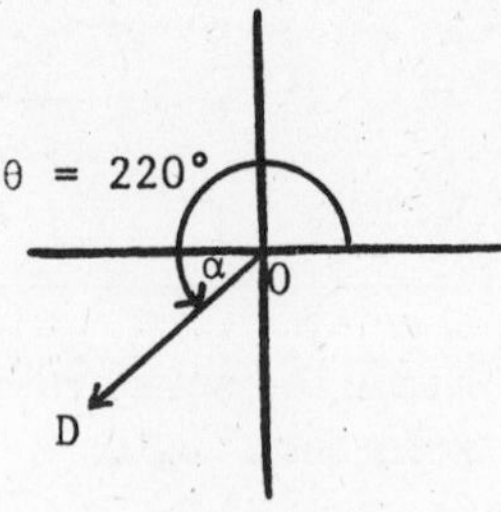

Figure 3

In Figure 1, θ is a standard-position angle of 220°. Its terminal side $\overrightarrow{0D}$ lies in the third quadrant.

In Figure 2, the reference angle (α) for $\overrightarrow{0D}$ is 40° (Q3).

In Figure 3, we have combined θ and α. α is the reference angle for θ.

Since $\theta = 220°$ and $\alpha = 40°$,
it should be obvious (from the graph) that $\theta = \alpha +$ ______°

27. Here's another standard-position angle whose terminal side lies in the third quadrant. We know these two facts: (1) $\theta = 250°$ (2) $\theta = \alpha + 180°$ Specify the reference angle α. ________	180°

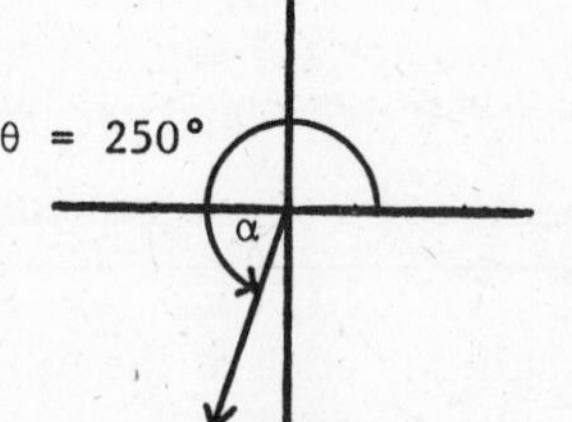

28. Any standard-position angle (θ) in the third quadrant must lie between 180° and ______°.	70° (Q3)

Answer to Frame 28: 270°

29. For the following and any similar problems, make a rough sketch of each angle. This will make the problem easier to solve.

Specify the reference angles for the following standard-position angles.

(a) If θ is 190°, α is ________

(b) If θ is 207°, α is ________

(c) If θ is 259°, α is ________

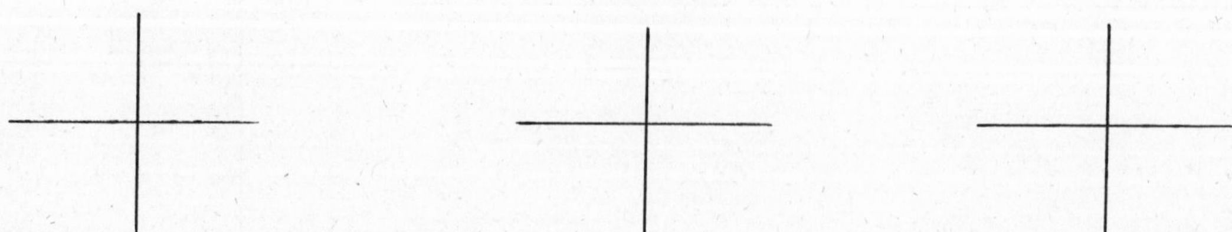

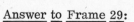

Answer to Frame 29: a) 10° (Q3) b) 27° (Q3) c) 79° (Q3)

30. The following angles are the reference angles for what standard-position angles? Make a rough sketch of each.

(a) If $\alpha = 29°$ (Q3) 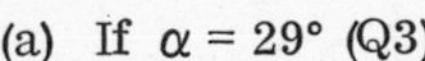

$\theta =$ ________

(b) If $\alpha = 35°$ (Q3)

$\theta =$ ________

(c) If $\alpha = 84°$ (Q3)

$\theta =$ ________

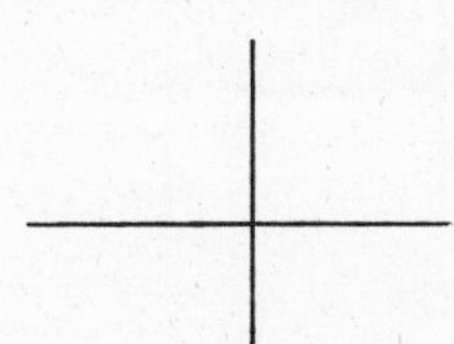

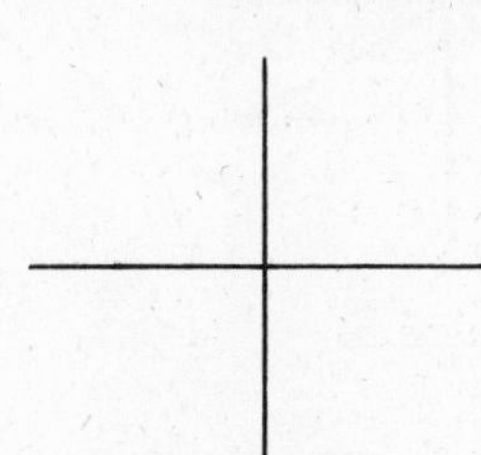

---

31. Specify the reference angle for each of the following standard-position angles. Be sure to include the quadrant number.

(a) If $\theta = 173°$, $\alpha =$ ________

(b) If $\theta = 242°$, $\alpha =$ ________

(c) If $\theta = 269°$, $\alpha =$ ________

(d) If $\theta = 93°$, $\alpha =$ ________

(e) If $\theta = 69°$, $\alpha =$ ________

a) 209°

b) 215°

c) 264°

---

32. The following reference angles are the reference angles for what standard-position angles?

(a) If $\alpha = 33°$ (Q2), $\theta =$ ______

(b) If $\alpha = 33°$ (Q3), $\theta =$ ______

(c) If $\alpha = 17°$ (Q1), $\theta =$ ______

(d) If $\alpha = 17°$ (Q2), $\theta =$ ______

(e) If $\alpha = 17°$ (Q3), $\theta =$ ______

a) 7° (Q2)

b) 62° (Q3)

c) 89° (Q3)

d) 87° (Q2)

e) 69° (Q1)

---

a) 147°

b) 213°

c) 17°

d) 163°

e) 197°

33. The following three figures are discussed below.

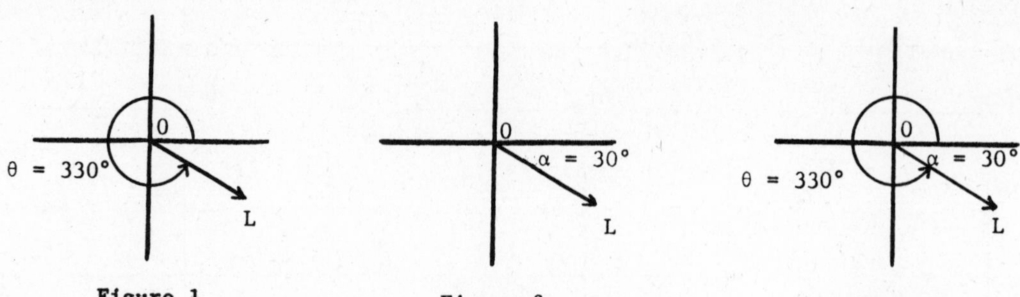

Figure 1 Figure 2 Figure 3

In Figure 1, θ is a standard-position angle of 330°. Its terminal side $\overrightarrow{0L}$ lies in the fourth quadrant.

In Figure 2, the reference angle (α) for $\overrightarrow{0L}$ is 30° (Q4).

In Figure 3, we have combined θ and α. α is the reference angle for θ.

For a standard-position angle in the fourth quadrant, the sum of θ and α is ______°. (Look at Figure 3.)

---

34. Here is another fourth-quadrant angle.

We know these two facts:

(1) $\theta = 290°$
(2) $\theta + \alpha = 360°$

Specify the reference angle (α). ________

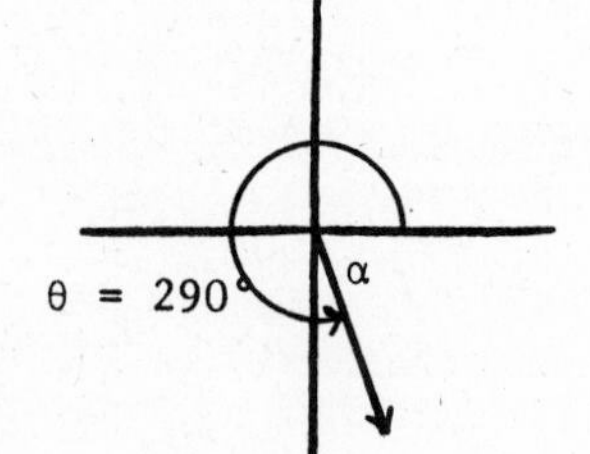

Answer: 360°

---

35. If a standard-position angle lies in the fourth quadrant, it must be an angle between 270° and ______°.

Answer: 70° (Q4)

---

36. Specify the reference angles for the following fourth-quadrant angles. Make a rough sketch of each angle.

(a) If $\theta = 341°$ $\alpha =$ ________

(b) If $\theta = 272°$ $\alpha =$ ________

(c) If $\theta = 301°$ $\alpha =$ ________

Answer: 360°

Answers:

a) 19° (Q4)
b) 88° (Q4)
c) 59° (Q4)

37. Name the corresponding standard-position angle for each of the following reference angles. As your first step, sketch each angle.

(a) If $\alpha = 29°$ (Q4), $\theta =$ ________
(b) If $\alpha = 46°$ (Q4), $\theta =$ ________
(c) If $\alpha = 79°$ (Q4), $\theta =$ ________

---

38. Specify the reference angles for each of the following standard-position angles. (Watch the quadrants.) Make a rough sketch of each angle.

(a) If $\theta = 102°$, $\alpha =$ ________
(b) If $\theta = 182°$, $\alpha =$ ________
(c) If $\theta = 358°$, $\alpha =$ ________

a) 331°
b) 314°
c) 281°

---

39. Specify the reference angles for each of the following standard-position angles.

(a) If $\theta = 281°$, $\alpha =$ ________
(b) If $\theta = 167°$, $\alpha =$ ________
(c) If $\theta = 265°$, $\alpha =$ ________

a) 78° (Q2)
b) 2° (Q3)
c) 2° (Q4)

---

40. Ordinarily, the direction of a vector is specified by its standard-position angle instead of its reference angle. The reference angles for various vectors are given below. Specify the direction of each vector by giving its standard-position angle. Make a sketch of each vector.

(a) If $\alpha = 71°$ (Q2), $\theta =$ ________
(b) If $\alpha = 71°$ (Q3), $\theta =$ ________
(c) If $\alpha = 71°$ (Q4), $\theta =$ ________

a) 79° (Q4)
b) 13° (Q2)
c) 85° (Q3)

41. Find the standard-position angle for vectors having the following reference angles.

(a) If $\alpha = 48°$ (Q3), $\theta =$ ________

(b) If $\alpha = 33°$ (Q2), $\theta =$ ________

(c) If $\alpha = 86°$ (Q4), $\theta =$ ________

a) 109°

b) 251°

c) 289°

Answer to Frame 41: a) 228° b) 147° c) 274°

SELF-TEST 2 (Frames 18-41)

Specify the reference angle for each of these standard-position angles. First sketch each angle.

1. $\theta = 120°$ $\alpha =$ ________

2. $\theta = 346°$ $\alpha =$ ________

3. $\theta = 219°$ $\alpha =$ ________

4. $\theta = 74°$ $\alpha =$ ________

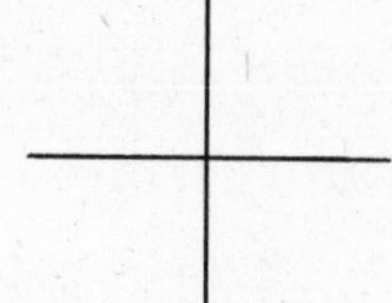

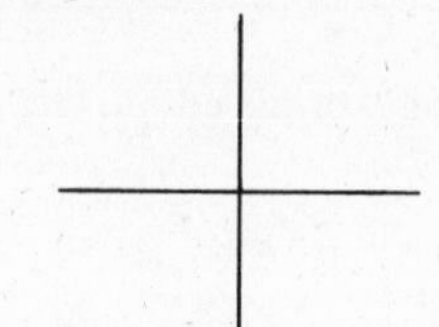

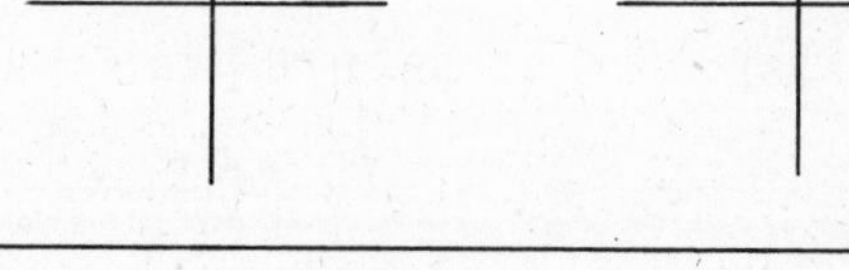

For these reference angles, specify the corresponding standard-position angle. First sketch each angle.

5. $\alpha = 30°$ (Q4) $\theta =$ ________

6. $\alpha = 17°$ (Q1) $\theta =$ ________

7. $\alpha = 85°$ (Q2) $\theta =$ ________

8. $\alpha = 70°$ (Q3) $\theta =$ ________

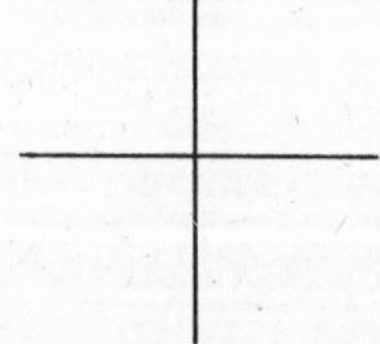

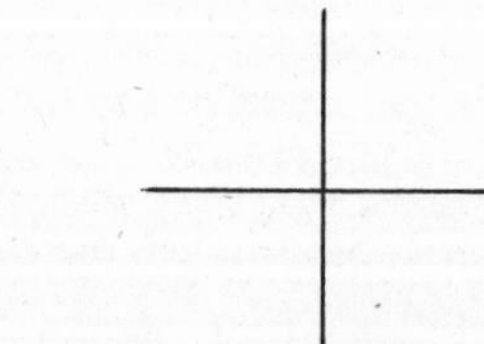

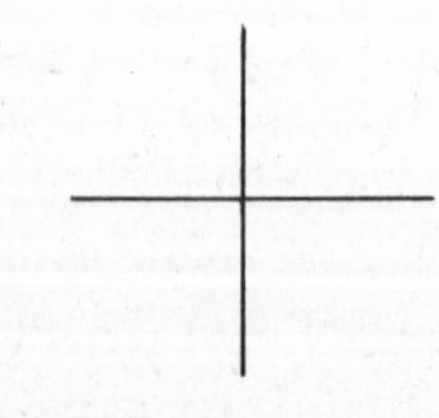

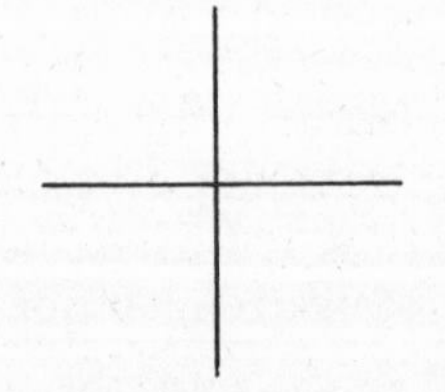

ANSWERS:

1. 60° (Q2)
2. 14° (Q4)
3. 39° (Q3)
4. 74° (Q1)
5. 330°
6. 17°
7. 95°
8. 250°

3-3 TRIGONOMETRIC RATIOS OF ANGLES BETWEEN 90° AND 360°

Up to this point, we have only considered the sines, cosines, and tangents of angles between 0° and 90°. In this section, we will discuss the same three ratios for angles between 90° and 360°. Since the trig ratios of these angles are identical to the trig ratios of their reference angles, we can use our ordinary trig table to determine their sines, cosines, and tangents.

42. Let's briefly review the sine, cosine, and tangent of reference angles in each quadrant.

Here is a reference angle of 43° in the first quadrant.

We have drawn $\overrightarrow{0P}$ with its components, $\overrightarrow{0D}$ and $\overrightarrow{DP}$.

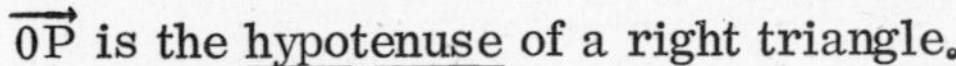

$\overrightarrow{0P}$ is the hypotenuse of a right triangle.

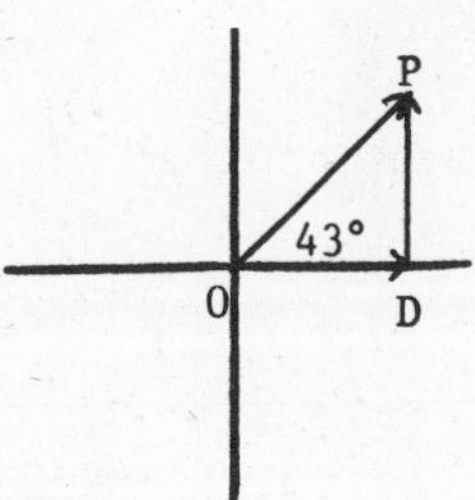

$\overrightarrow{0P}$ is positive, since it is a slanted vector.

$\overrightarrow{0D}$ and $\overrightarrow{DP}$ are both positive since:

$\overrightarrow{0D}$ is "to the right".

$\overrightarrow{DP}$ is "upward".

α is 43° (Q1) in this case. Use your table to write the trig ratios below.

(a) $\sin\alpha = \dfrac{\overrightarrow{DP}}{\overrightarrow{0P}} = \dfrac{+}{+} =$ "+". Therefore, sin 43° (Q1) = ________

(b) $\cos\alpha = \dfrac{\overrightarrow{0D}}{\overrightarrow{0P}} = \dfrac{+}{+} =$ "+". Therefore, cos 43° (Q1) = ________

(c) $\tan\alpha = \dfrac{\overrightarrow{DP}}{\overrightarrow{0D}} = \dfrac{+}{+} =$ "+". Therefore, tan 43° (Q1) = ________

Answer to Frame 42: a) 0.6820 b) 0.7314 c) 0.9325

43. Here is a reference angle of 43° in the second quadrant. We have also drawn its components.

$\overrightarrow{0K}$ is positive (a slanted vector).

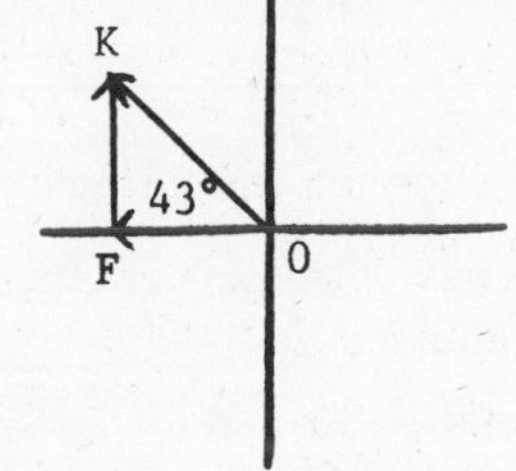

$\overrightarrow{0F}$ is negative ("to the left").

$\overrightarrow{FK}$ is positive ("upward").

α is 43° (Q2) in this case. Use your table to write its trig ratios.

(a) $\sin\alpha = \dfrac{\overrightarrow{FK}}{\overrightarrow{0K}} = \dfrac{+}{+} =$ "+". Therefore, sin 43° (Q2) = ________

(b) $\cos\alpha = \dfrac{\overrightarrow{0F}}{\overrightarrow{0K}} = \dfrac{-}{+} =$ "−". Therefore, cos 43° (Q2) = ________

(c) $\tan\alpha = \dfrac{\overrightarrow{FK}}{\overrightarrow{0F}} = \dfrac{+}{-} =$ "−". Therefore, tan 43° (Q2) = ________

a) +0.6820

b) −0.7314

c) −0.9325

44. Here is a reference angle of 43° in the <u>third</u> quadrant. The components are also drawn.

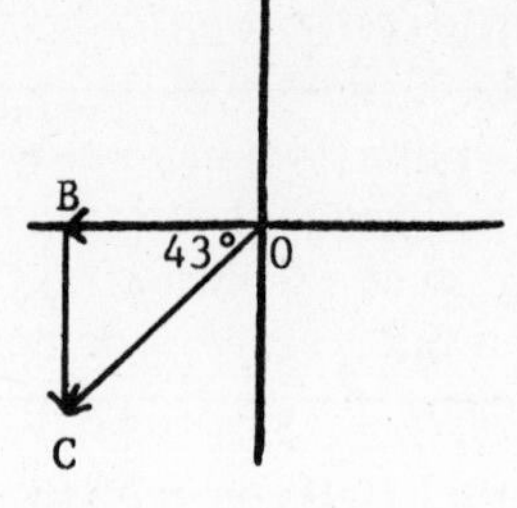

$\overrightarrow{0C}$ is <u>positive</u> (a slanted vector).

$\overrightarrow{0B}$ is <u>negative</u> ("to the left").

$\overrightarrow{BC}$ is <u>negative</u> ("downward").

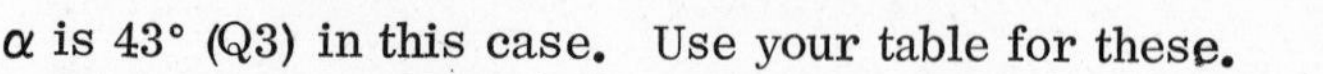

α is 43° (Q3) in this case. Use your table for these.

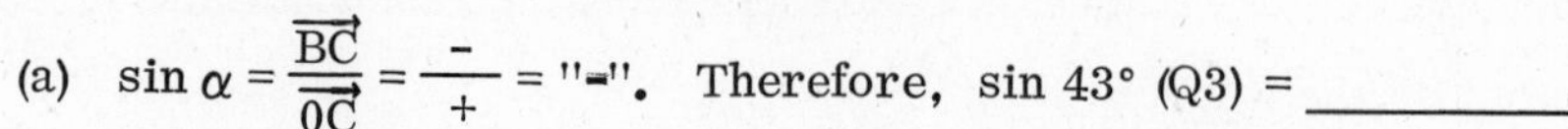

(a) $\sin \alpha = \frac{\overrightarrow{BC}}{\overrightarrow{0C}} = \frac{-}{+} =$ "−". Therefore, $\sin 43°$ (Q3) = ________

(b) $\cos \alpha = \frac{\overrightarrow{0B}}{\overrightarrow{0C}} = \frac{-}{+} =$ "−". Therefore, $\cos 43°$ (Q3) = ________

(c) $\tan \alpha = \frac{\overrightarrow{BC}}{\overrightarrow{0B}} = \frac{-}{-} =$ "+". Therefore, $\tan 43°$ (Q3) = ________

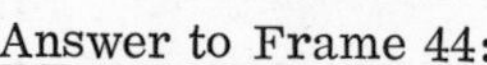

Answer to Frame 44: a) −0.6820 b) −0.7314 c) +0.9325

45. Here is a reference angle of 43° in the <u>fourth</u> quadrant.

Only $\overrightarrow{ST}$ is negative.

α is 43° (Q4) in this case. Use your table for these.

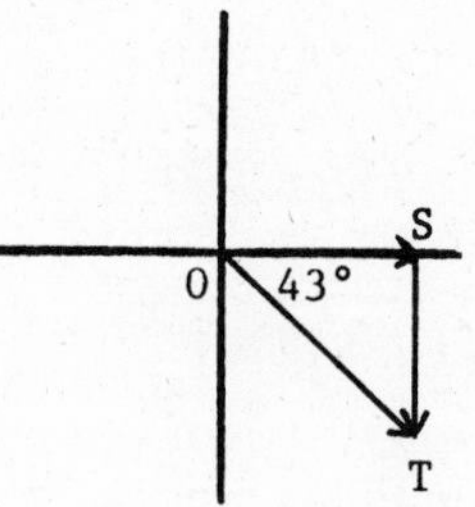

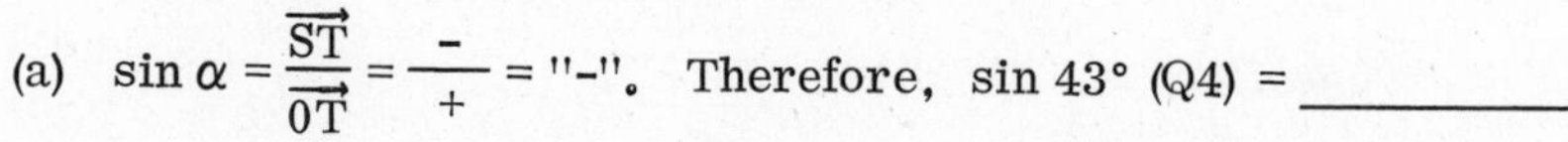

(a) $\sin \alpha = \frac{\overrightarrow{ST}}{\overrightarrow{0T}} = \frac{-}{+} =$ "−". Therefore, $\sin 43°$ (Q4) = ________

(b) $\cos \alpha = \frac{\overrightarrow{0S}}{\overrightarrow{0T}} = \frac{+}{+} =$ "+". Therefore, $\cos 43°$ (Q4) = ________

(c) $\tan \alpha = \frac{\overrightarrow{ST}}{\overrightarrow{0S}} = \frac{-}{+} =$ "−". Therefore, $\tan 43°$ (Q4) = ________

Answer to Frame 45: a) −0.6820 b) +0.7314 c) −0.9325

46. Use rough sketches for the following:

(a) The sines of reference angles are "negative" in what two quadrants? ________

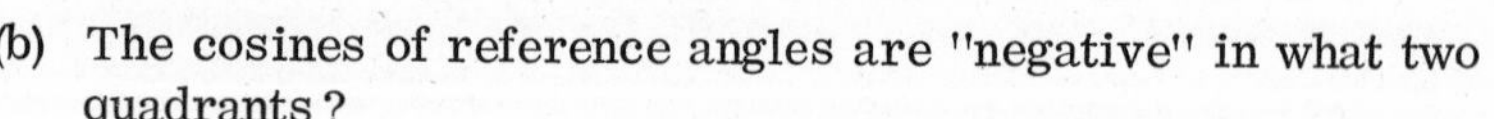

(b) The cosines of reference angles are "negative" in what two quadrants? ________

(c) The tangents of reference angles are "negative" in what two quadrants? ________

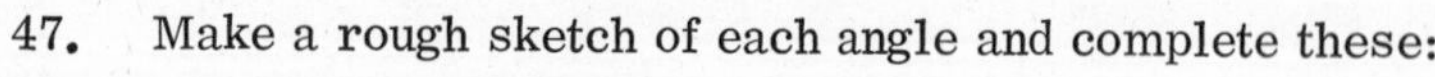

47. Make a rough sketch of each angle and complete these:

(a) $\sin 27°$ (Q2) = ________

(b) $\cos 27°$ (Q3) = ________

(c) $\tan 27°$ (Q4) = ________

a) Quadrants III and IV

b) Quadrants II and III

c) Quadrants II and IV

48. Make a rough sketch of each angle and complete these:

(a) sin 47° (Q3) = ________

(b) cos 47° (Q4) = ________

(c) tan 47° (Q2) = ________

a) +0.4540

b) -0.8910

c) -0.5095

---

49. Make a rough sketch of each angle and complete these:

(a) sin 67° (Q4) = ________

(b) cos 67° (Q2) = ________

(c) tan 67° (Q3) = ________

a) -0.7314

b) +0.6820

c) -1.072

Answer to Frame 49: a) -0.9205 b) -0.3907 c) +2.356

---

50. Here is a very important definition:

THE SINE, COSINE, AND TANGENT OF ANY STANDARD-POSITION ANGLE IS THE SAME AS THE SINE, COSINE, AND TANGENT OF ITS REFERENCE ANGLE.

For example: If $\theta = 130°$, $\alpha = 50°$ (Q2).

Since: sin 50° (Q2) = +0.7660
cos 50° (Q2) = -0.6428
tan 50° (Q2) = -1.192,

Therefore: sin 130° = +0.7660
cos 130° = -0.6428
tan 130° = -1.192.

Since a standard-position angle has the same trig ratios as its reference angle, we can find its sine, cosine, and tangent as follows:

(1) We sketch the standard-position angle.
(2) We identify the value of its reference angle.
(3) We find the sine, cosine, and tangent of its reference angle.

Sketch a 160° angle below and complete these:

(a) If $\theta = 160°$, α = ________

(b) sin 160° = ________

(c) cos 160° = ________

(d) tan 160° = ________

---

a) 20° (Q2)

b) +0.3420

c) -0.9397

d) -0.3640

51. Sketch a 220° angle below and complete these:

(a) If $\theta = 220°$, its reference angle is ________.

(b) sin 220° = ________

(c) cos 220° = ________

(d) tan 220° = ________

---

52. Sketch a 330° angle below and complete these:

(a) If $\theta = 330°$, α = ________.

(b) sin 330° = ________

(c) cos 330° = ________

(d) tan 330° = ________

a) 40° (Q3)
b) -0.6428
c) -0.7660
d) +0.8391

---

53. A standard-position angle in the first quadrant is identical to its reference angle. Therefore, the trig ratios of first-quadrant angles can be obtained immediately from the trig table. For standard-position angles in other quadrants, however, a sketch must be made to determine the reference angle.

Sketch a 258° angle below and complete these:

(a) If $\theta = 258°$, α = ________.

(b) sin 258° = ________

(c) cos 258° = ________

(d) tan 258° = ________

a) 30° (Q4)
b) -0.5000
c) +0.8660
d) -0.5774

---

54. Sketch a 304° angle below and complete these:

(a) If $\theta = 304°$, α = ________.

(b) sin 304° = ________

(c) cos 304° = ________

(d) tan 304° = ________

a) 78° (Q3)
b) -0.9781
c) -0.2079
d) +4.705

---

55. In this and the following frames, sketch each angle on the axes provided. Then find the numerical value of each trig ratio.

(a) sin 97° = ______ (b) sin 286° = ______ (c) sin 198° = ______

a) 56° (Q4)
b) -0.8290
c) +0.5592
d) -1.483

56.	(a) cos 245° = ______	(b) cos 171° = ______	(c) cos 301° = ______	a) +0.9925 b) -0.9613 c) -0.3090
57.	(a) tan 299° = ______	(b) tan 109° = ______	(c) tan 269° = ______	a) -0.4226 b) -0.9877 c) +0.5150
58.	(a) cos 129° = ______	(b) tan 358° = ______	(c) sin 239° = ______	a) -1.804 b) -2.904 c) +57.29
59.	(a) tan 175° = ______	(b) sin 287° = ______	(c) cos 261° = ______	a) -0.6293 b) -0.0349 c) -0.8572
60.	(a) sin 147° = ______	(b) cos 311° = ______	(c) tan 189° = ______	a) -0.0875 b) -0.9563 c) -0.1564
				a) +0.5446 b) +0.6561 c) +0.1584

SELF-TEST 3 (Frames 42-60)

State whether the sign of each of the following is "+" or "-":

1. sin 160° ______ 2. cos 300° ______ 3. tan 225° ______ 4. cos 95° ______

Find the numerical value of each of these trig ratios:

5. cos 158° = ________	7. tan 97° = ________	9. cos 290° = ________
6. sin 115° = ________	8. sin 348° = ________	10. sin 203° = ________

ANSWERS:

1. +	3. +	5. -0.9272	7. -8.144	9. +0.3420
2. +	4. -	6. +0.9063	8. -0.2079	10. -0.3907

3-4 FINDING STANDARD-POSITION ANGLES CORRESPONDING TO KNOWN RATIOS

There are usually four standard-position angles (one in each quadrant) which have reference angles containing the same number of degrees. Therefore, there is generally more than one angle between 0° and 360° corresponding to any numerical trig ratio (like $\sin \theta = +0.5000$). In this section, we will discuss the method of finding angles corresponding to known numerical ratios.

61. For a given reference angle in the second quadrant, the size of the corresponding standard-position angle can be found by subtracting the reference angle from 180°. For example:

For 30° (Q2), the standard-position angle is: $\theta = 180° - 30° = 150°$

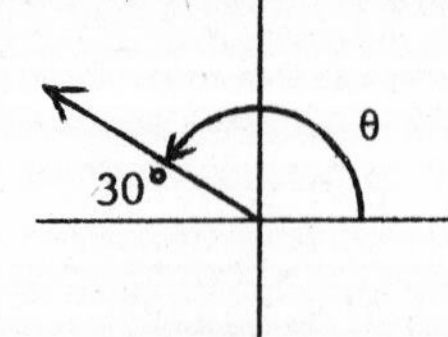

Find the standard-position angle corresponding to each reference angle:

(a) 10° (Q2) ______ (b) 45° (Q2) ______ (c) 87° (Q2) ______

a) 170°
b) 135°
c) 93°

62. For a given reference angle in the third quadrant, the size of the corresponding standard-position angle can be found by adding the reference angle to 180°. For example:

For 30° (Q3), the standard-position angle is: $\theta = 180° + 30° = 210°$

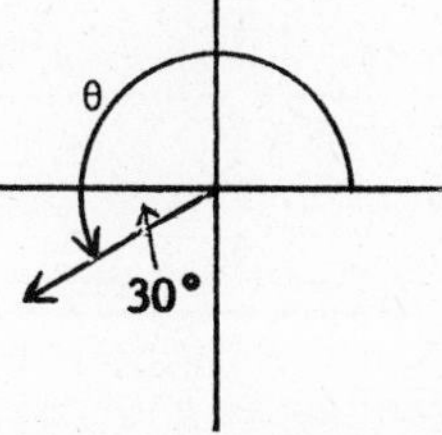

Find the standard-position angle corresponding to each reference angle:

(a) 5° (Q3) _______ (b) 60° (Q3) _______ (c) 77° (Q3) _______

---

a) 185°
b) 240°
c) 257°

63. For a given reference angle in the fourth quadrant, the size of the corresponding standard-position angle can be found by subtracting the reference angle from 360°. For example:

For 40° (Q4), the standard-position angle is: $\theta = 360° - 40° = 320°$

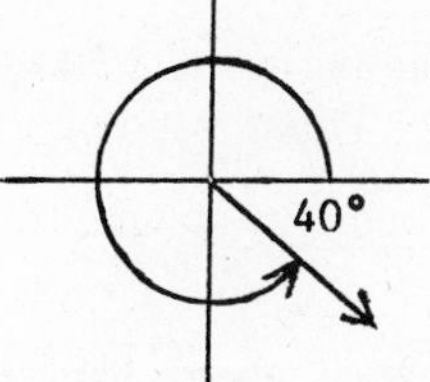

Find the standard-position angle corresponding to each reference angle:

(a) 25° (Q4) _______ (b) 52° (Q4) _______ (c) 70° (Q4) _______

---

a) 335°
b) 308°
c) 290°

64. For any standard-position angle in the first quadrant, there is a standard-position angle in each of the other three quadrants with a reference angle of the same size. For example:

For a 40° angle, the other three angles are:

Quadrant II: 180° - 40° = 140°
Quadrant III: 180° + 40° = 220°
Quadrant IV: 360° - 40° = 320°

Name the standard-position angles in Quadrants II, III, and IV which have a reference angle of the same size as an 80° angle.

(a) Quadrant II ______ (b) Quadrant III ______ (c) Quadrant IV ______

---

a) 100°
b) 260°
c) 280°

65. List the standard-position angles in each of the other three quadrants which have reference angles of the same size as these first-quadrant angles.

(a) 50°, ________, ________, ________

(b) 8°, ________, ________, ________

(c) 84°, ________, ________, ________

---

a) 130°, 230°, 310°

b) 172°, 188°, 352°

c) 96°, 264°, 276°

66. There are four standard-position angles (30°, 150°, 210°, and 330°) whose reference angles are 30°. The sine of a 30° angle is 0.5000.

Since the sines of angles are positive in Quadrants I and II and negative in Quadrants III and IV:

(a) Which two of the four angles above have +0.5000 as their sine? ________ and ________

(b) Which two of the four angles above have -0.5000 as their sine? ________ and ________

---

a) 30° and 150°

b) 210° and 330°

67. There are four standard-position angles (40°, 140°, 220°, and 320°) whose reference angles are 40°. The cosine of a 40° angle is 0.7660.

Since the cosines of angles are positive in Quadrants I and IV and negative in Quadrants II and III:

(a) Which two of the four angles above have +0.7660 as their cosine? ________ and ________

(b) Which two of the four angles above have -0.7660 as their cosine? ________ and ________

---

a) 40° and 320°

b) 140° and 220°

68. There are four standard-position angles (50°, 130°, 230°, and 310°) whose reference angles are 50°. The tangent of a 50° angle is 1.192.

Since the tangents of angles are positive in Quadrants I and III and negative in Quadrants II and IV:

(a) Which two of the four angles above have +1.192 as their tangent? ________ and ________

(b) Which two of the four angles above have -1.192 as their tangent? ________ and ________

---

a) 50° and 230°

b) 130° and 310°

69. All three trig ratios are positive for any standard-position angle in Quadrant I. At the right, complete the "table of signs" of the trig ratios for angles in the other three quadrants. Sketches of an angle in each quadrant are shown below.

	I	II	III	IV
sin θ	+			
cos θ	+			
tan θ	+			

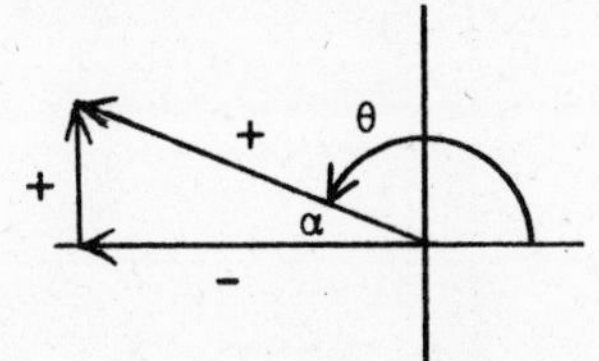

Quadrant II

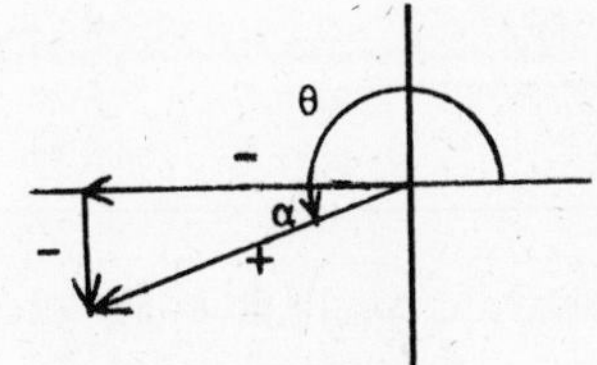

Quadrant III

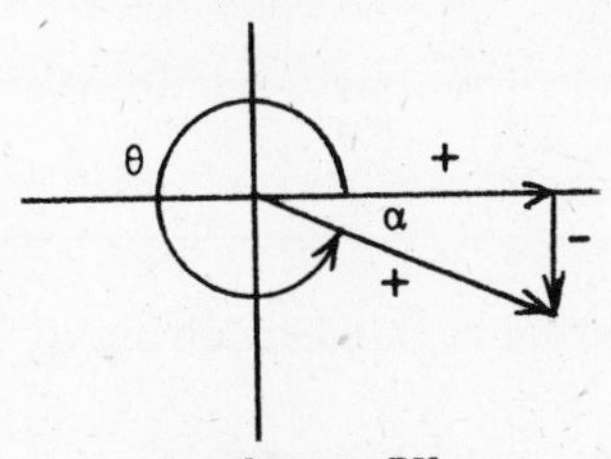

Quadrant IV

---

70. Here is the completed table of "signs" of the trig ratios in the four quadrants.

	I	II	III	IV
sin θ	+	+	-	-
cos θ	+	-	-	+
tan θ	+	-	+	-

(a) The sines of angles are negative in Quadrants ____ and ____.

(b) The cosines of angles are negative in Quadrants ____ and ____.

(c) The tangents of angles are negative in Quadrants ____ and ____.

See the next frame.

---

71. Since the sines of angles are positive in Quadrants I and II, there is an angle in Quadrant II with the same sine as a 45° angle in Quadrant I. The two angles must have reference angles of the same size.

(a) The reference angle of the second-quadrant angle is ________.

(b) The second-quadrant angle is ________.

(c) Therefore: sin 45° = sin ________

a) III and IV

b) II and III

c) II and IV

---

72. Use either a first-quadrant or second-quadrant angle to complete these:

(a) sin 55° = sin ________

(b) sin 77° = sin ________

(c) sin 100° = sin ________

(d) sin 146° = sin ________

a) 45° (Q2)

b) 135°

c) 135°
(Both sin 45° and sin 135° have the same numerical value, namely, +0.7071)

---

a) sin 125°

b) sin 103°

c) sin 80°

d) sin 34°

73. Since the sines of angles are negative in Quadrants III and IV, there must be an angle in the fourth quadrant with the same sine as a 200° angle in the third quadrant. The two angles must have reference angles of the same size.

(a) The reference angle of a 200° angle is ________.

(b) The reference angle of the fourth-quadrant angle is ________.

(c) The fourth-quadrant angle is ________.

(d) Therefore: sin 200° = sin ________

---

a) 20° (Q3)

b) 20° (Q4)

c) 340°

d) 340°
(Both sin 200° and sin 340° have the same numerical value, namely, -0.3420)

74. Use either a third-quadrant or fourth-quadrant angle to complete these:

(a) sin 190° = sin ________

(b) sin 237° = sin ________

(c) sin 325° = sin ________

(d) sin 287° = sin ________

---

a) sin 350°

b) sin 303°

c) sin 215°

d) sin 253°

75. Based on the fact that the cosines of angles are positive in Quadrants I and IV and negative in Quadrants II and III, use an angle between 0° and 360° to complete these. (If necessary, sketch the angle.)

(a) cos 60° = cos ______ (c) cos 320° = cos ______

(b) cos 110° = cos ______ (d) cos 215° = cos ______

---

a) 300° c) 40°

b) 250° d) 145°

76. Based on the fact that the tangents of angles are positive in Quadrants I and III and negative in Quadrants II and IV, use an angle between 0° and 360° to complete these. (If necessary, sketch the angle.)

(a) tan 80° = tan ______ (c) tan 230° = tan ______

(b) tan 140° = tan ______ (d) tan 355° = tan ______

---

a) 260° c) 50°

b) 320° d) 175°

77. Use an angle between 0° and 360° to complete these:

(a) sin 22° = sin _______ (d) tan 333° = tan _______

(b) cos 105° = cos _______ (e) cos 297° = cos _______

(c) tan 43° = tan _______ (f) sin 309° = sin _______

---

78. (a) There are two angles between 0° and 360° whose sines are 0.9397. The two angles are 70° and _______°.

(b) There are two angles between 0° and 360° whose sines are -0.3420. The two angles are 200° and _______°.

a) sin 158° d) tan 153°
b) cos 255° e) cos 63°
c) tan 223° f) sin 231°

---

79. (a) There are two angles between 0° and 360° whose cosines are 0.5000. They are 60° and _______°.

(b) There are two angles between 0° and 360° whose cosines are -0.9063. They are 155° and _______°.

a) 110°
b) 340°

---

80. (a) There are two angles between 0° and 360° whose tangents are 2.747. They are 70° and _______°.

(b) There are two angles between 0° and 360° whose tangents are -1.0000. They are 135° and _______°.

a) 300°
b) 205°

81. Identify the second angle in each of these:

(a) If $\sin\theta = +0.7071$, $\theta = 45°$ and $\theta =$ ________

(b) If $\cos\theta = +0.4226$, $\theta = 65°$ and $\theta =$ ________

(c) If $\tan\theta = +5.671$, $\theta = 80°$ and $\theta =$ ________

a) 250°

b) 315°

---

82. Identify the second angle in each of these:

(a) If $\sin\theta = -0.8192$, $\theta = 235°$ and $\theta =$ ________

(b) If $\cos\theta = -0.2588$, $\theta = 105°$ and $\theta =$ ________

(c) If $\tan\theta = -0.1763$, $\theta = 170°$ and $\theta =$ ________

a) 135°

b) 295°

c) 260°

---

83. Identify both angles in these:

(a) If $\sin\theta = 0.9063$, $\theta =$ ________ and ________

(b) If $\cos\theta = 0.8192$, $\theta =$ ________ and ________

(c) If $\tan\theta = 11.43$, $\theta =$ ________ and ________

a) 305°

b) 255°

c) 350°

---

84. Identify both angles in these:

(a) If $\sin\theta = -0.7660$, $\theta =$ ________ and ________

(b) If $\cos\theta = -0.8660$, $\theta =$ ________ and ________

(c) If $\tan\theta = -2.747$, $\theta =$ ________ and ________

a) 65° and 115°

b) 35° and 325°

c) 85° and 265°

85. Identify both angles: (a) If $\cos\theta = -0.6820$, $\theta =$ ________ and ________ (b) If $\sin\theta = +0.1392$, $\theta =$ ________ and ________ (c) If $\tan\theta = -0.3443$, $\theta =$ ________ and ________	a) 230° and 310° b) 150° and 210° c) 110° and 290°
86. Identify both angles: (a) If $\tan\theta = +2.246$, $\theta =$ ________ and ________ (b) If $\sin\theta = -0.0175$, $\theta =$ ________ and ________ (c) If $\cos\theta = +0.0349$, $\theta =$ ________ and ________	a) 133° and 227° b) 8° and 172° c) 161° and 341°

Answer to Frame 86: a) 66° and 246° b) 181° and 359° c) 88° and 272°

SELF-TEST 4 (Frames 61-86)

1. $\sin\theta$ is negative in Quadrants ____ and ____.
2. $\cos\theta$ is negative in Quadrants ____ and ____.
3. $\tan\theta$ is negative in Quadrants ____ and ____.

Fill in each blank with an angle lying between 0° and 360°:

4. sin 135° = sin ______ 5. tan 27° = tan ______ 6. cos 210° = cos ______

Answer each problem by writing in two angles between 0° and 360°:

7. If $\sin\theta = 0.9945$, $\theta =$ ____ and ____.
8. If $\sin\theta = -0.3256$, $\theta =$ ____ and ____.
9. If $\tan\theta = 4.331$, $\theta =$ ____ and ____.
10. If $\cos\theta = -0.7431$, $\theta =$ ____ and ____.

ANSWERS:

1. III and IV	4. sin 45°	7. 84° and 96°	9. 77° and 257°
2. II and III	5. tan 207°	8. 199° and 341°	10. 138° and 222°
3. II and IV	6. cos 150°		

3-5 TRIGONOMETRIC RATIOS OF 0°, 90°, 180°, 270°, 360°

Up to this point, we have not discussed the sines, cosines, and tangents of these specific angles: 0°, 90°, 180°, 270°, 360°. Since the terminal side of each of these angles lies on a coordinate axis, special attention must be given to determining the numerical values of their trig ratios. We will discuss their trig ratios in this section.

---

87. The trig table gives this fact: $\boxed{\sin 90° = 1.0000}$

We will use the series of four figures below to show that this fact makes sense. In each one, terminal side $\overrightarrow{0R}$ is 10 units.

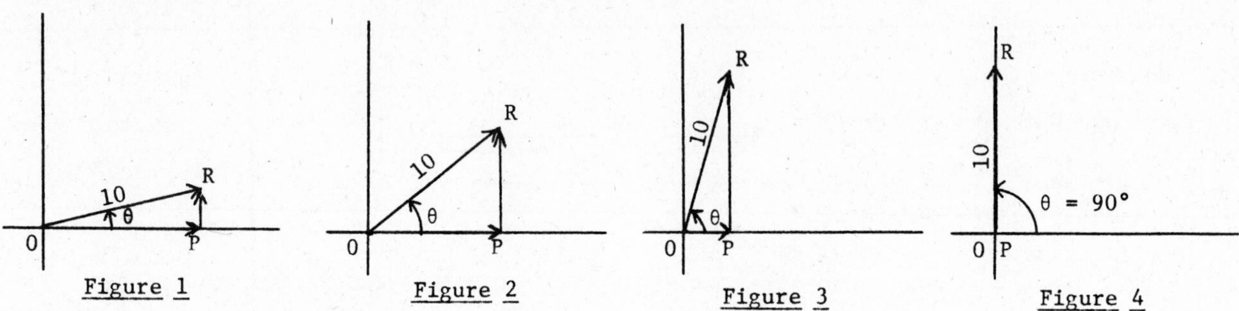

From Figure 1 to Figure 4, angle θ increases in size, and becomes 90° in Figure 4.

In each figure, $\sin\theta = \dfrac{\overrightarrow{PR}}{\overrightarrow{0R}} = \dfrac{\overrightarrow{PR}}{10}$

(a) From Figure 1 to Figure 4, does the length of $\overrightarrow{PR}$ increase or decrease? ______________

(b) From Figure 1 to Figure 4, does the numerical value of sin θ increase or decrease? ______________

(c) In Figure 4, $\overrightarrow{PR}$ is identical to $\overrightarrow{0R}$. Since $\overrightarrow{0R}$ is 10 units, $\overrightarrow{PR}$ is _______ units in that figure.

(d) In Figure 4, θ = 90°. Therefore:

$$\sin 90° = \frac{\overrightarrow{PR}}{\overrightarrow{0R}} = \frac{10}{10} = \_\_\_\_\_\_\_\_$$

---

a) Increase

b) Increase

c) 10

d) 1 or 1.0000

---

88. The trig table gives this fact: $\boxed{\cos 90° = 0.0000}$ We can use the series of figures in the last frame to show that this fact makes sense.

In each figure, $\cos\theta = \dfrac{\overrightarrow{0P}}{\overrightarrow{0R}} = \dfrac{\overrightarrow{0P}}{10}$

(a) From Figure 1 to Figure 4, does the length of $\overrightarrow{0P}$ increase or decrease? ______________

(b) From Figure 1 to Figure 4, does the numerical value of cos θ increase or decrease? ______________

(c) The length of $\overrightarrow{0R}$ is 10 units in each figure. In Figure 4, where θ = 90°, what is the length of $\overrightarrow{0P}$? _______ units

(d) When θ = 90°, $\overrightarrow{0P}$ = 0 units and $\overrightarrow{0R}$ = 10 units.

Therefore, $\cos 90° = \dfrac{\overrightarrow{0P}}{\overrightarrow{0R}} = \dfrac{0}{10} = \_\_\_\_\_\_\_\_$

89. Before discussing the tangent of a 90° angle, we must review a fact about division. The fact is this: division by "0" is impossible and meaningless because "0" has no reciprocal. For example:

$\frac{5}{0} = 5 \times$ (the reciprocal of 0), and "0" has no reciprocal.

Which of the following divisions are impossible? ______________

(a) $\frac{10}{0}$ (b) $\frac{0}{10}$ (c) $\frac{3}{0}$ (d) $\frac{0}{3}$

a) Decrease

b) Decrease

c) 0 units

d) 0 or 0.0000

---

90. The trig table gives this fact: | tan 90° = ----- | This entry means that there is no numerical value for tan 90°. (Note: Some trig tables show the word "infinite" for tan 90°, but "infinite" is not a number.)

We can use the same series of four figures (Refer to Frame 87) to show that the table entry for tan 90° makes sense.

In each figure, $\tan \theta = \frac{\overrightarrow{PR}}{\overrightarrow{0P}}$.

(a) In Figure 4, where $\theta = 90°$, $\overrightarrow{PR}$ = _____ units and $\overrightarrow{0P}$ = _____ units.

(b) Therefore: $\tan 90° = \frac{\overrightarrow{PR}}{\overrightarrow{0P}} = \frac{10}{0} =$ __________

Both (a) and (c) are impossible.

Both (b) and (d) equal "0".

---

91. In the trig table, the following values are given for the trig ratios of 0°:

| sin 0° = 0.0000 | | cos 0° = 1.0000 | | tan 0° = 0.0000 |

We can use the series of three figures below to show that these values make sense.

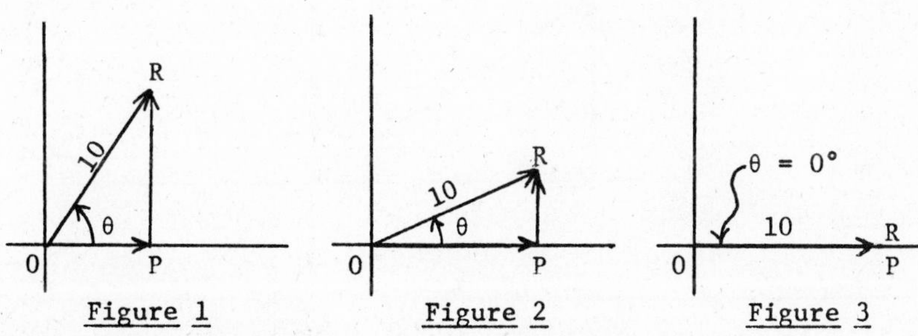

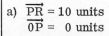

Figure 1 Figure 2 Figure 3

From Figure 1 to Figure 3, the size of θ decreases to 0° in Figure 3. In Figure 3 where $\theta = 0°$:

$\overrightarrow{0R}$ = 10 units, $\overrightarrow{0P}$ = 10 units, $\overrightarrow{PR}$ = 0 units.

Using these values, complete the following:

(a) $\sin 0° = \frac{\overrightarrow{PR}}{\overrightarrow{0R}} = \frac{0}{10} =$ __________

(b) $\cos 0° = \frac{\overrightarrow{0P}}{\overrightarrow{0R}} = \frac{10}{10} =$ __________

(c) $\tan 0° = \frac{\overrightarrow{PR}}{\overrightarrow{0P}} = \frac{0}{10} =$ __________

a) $\overrightarrow{PR}$ = 10 units
$\overrightarrow{0P}$ = 0 units

b) Division by 0 is impossible and meaningless.

92. Since either 0° or 90° is the reference angle for 180°, 270°, or 360°, it is important that you know the trig ratios for 0° and 90°. If you think of an angle decreasing to 0° or increasing to 90° (as we used in the series of figures), it will help you to recall the numerical values of the ratios for 0° and 90°.

Complete these:

sin 0° = _______ cos 0° = _______ tan 0° = _______

sin 90° = _______ cos 90° = _______ tan 90° = _______

a) 0 (or 0.0000)

b) 1 (or 1.0000)

c) 0 (or 0.0000)

---

93. Here is a summary of the trig ratios for 0° and 90°.

θ	$\sin\theta$	$\cos\theta$	$\tan\theta$
0°	0	1	0
90°	1	0	-----

As angle θ increases from 0° to 90°:

(a) $\sin\theta$ increases from _____ to _____.

(b) $\cos\theta$ decreases from _____ to _____.

See next frame.

Answer to Frame 93: a) 0 to 1 b) 1 to 0

---

94. Let's determine the numerical value of sin 180°. To do so, first examine the diagram at the right. Then complete the pattern of sines on the left below.

From the diagram:

$\boxed{\sin\ \theta = +\sin\alpha}$

sin 170° = +sin 10° = +0.1736
sin 172° = +sin 8° = +0.1392
sin 174° = +sin 6° = +0.1045
sin 176° = +sin 4° = +0.0698
sin 178° = +sin 2° = +0.0349

(a) sin 179° = ________ = ________

(b) sin 180° = ________ = ________

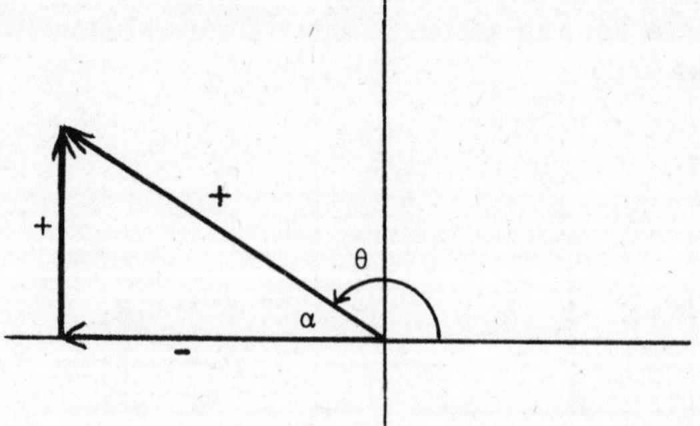

---

a) sin 179° = +sin 1° = +0.0175

b) sin 180° = +sin 0° = +0.0000

95. Let's determine the numerical values of cos 180° and tan 180°.

To do so, first examine the diagram at the right. Then complete the patterns of cosines and tangents below.

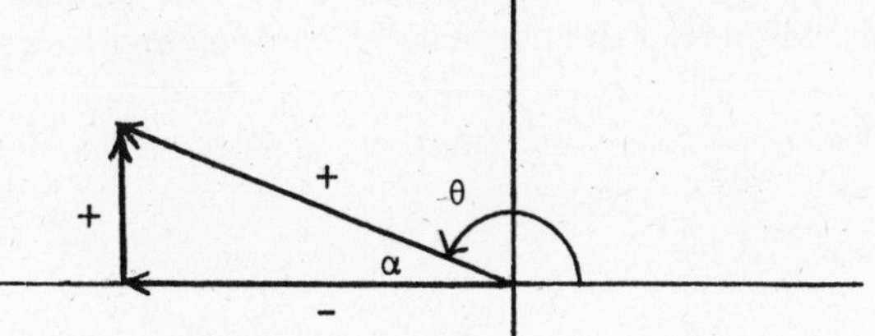

cos θ = -cos α

cos 170° = -cos 10° = -0.9848
cos 175° = -cos 5° = -0.9962

(a) cos 179° = ________ = ________

(b) cos 180° = ________ = ________

tan θ = -tan α

tan 170° = -tan 10° = -0.1763
tan 175° = -tan 5° = -0.0875

(c) tan 179° = ________ = ________

(d) tan 180° = ________ = ________

---

96. To determine the trig ratios of a standard-position angle of 180°, draw an angle in the second quadrant and think of it as increasing to 180°. The reference angle for 180° is a 0° angle. Therefore:

(1) The absolute values of the trig ratios of 180° are the same as those of 0°.

(2) The signs of the trig ratios of 180° can differ from those of 0°. You can determine these signs by means of the standard-position angle in your sketch.

Using the procedure above, complete these:

(a) sin 180° = ______

(b) cos 180° = ______

(c) tan 180° = ______

a) cos 179° = -cos 1°
= -0.9998

b) cos 180° = -cos 0°
= -1.0000
(or -1)

c) tan 179° = -tan 1°
= -0.0175

d) tan 180° = -tan 0°
= -0.0000
(or 0)

Since zero is neither positive or negative, -0.0000 = +0.0000 = 0

---

97. To find sin 270°, we will use the diagram and pattern of sines below.

From the diagram:

sin θ = -sin α

sin 250° = -sin 70° = -0.9397
sin 260° = -sin 80° = -0.9848
sin 268° = -sin 88° = -0.9994

(a) sin 269° = ________ = ________

(b) sin 270° = ________ = ________

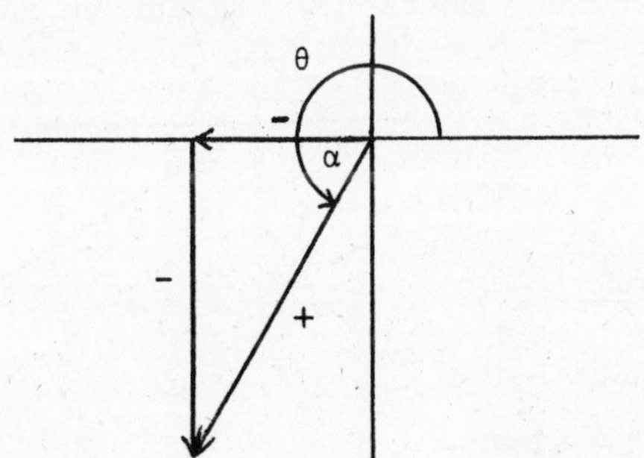

a) 0

b) -1

c) 0

Answer to Frame 97: sin 269° = -sin 89° = -0.9998

sin 270° = -sin 90° = -1.0000 (or -1)

98. To find cos 270° and tan 270°, we will use the diagram on the right and the patterns of cosines and tangents below.

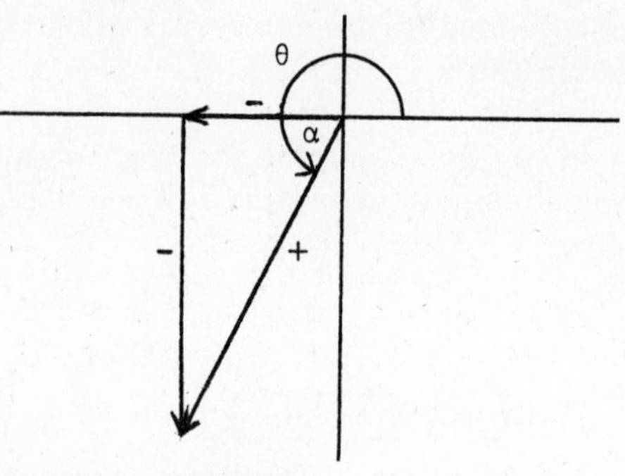

cos θ = -cos α

cos 260° = -cos 80° = -0.1736
cos 265° = -cos 85° = -0.0872

(a) cos 269° = ________ = ________

(b) cos 270° = ________ = ________

tan θ = +tan α

tan 260° = +tan 80° = +5.67
tan 265° = +tan 85° = +11.43

(c) tan 269° = ________ = ________

(d) tan 270° = ________ = ________

---

99. To determine the trig ratios of a standard-position angle of 270°, draw an angle <u>in the third quadrant</u> and think of it as increasing to 270°. The reference angle for 270° is a 90° angle. Therefore:

(1) The absolute values of the trig ratios of 270° are the same as those of 90°.

(2) The signs of the trig ratios of 270° can be determined from your sketch.

Using the procedure above, complete these:

(a) sin 270° = ______

(b) cos 270° = ______

(c) tan 270° = ______

Answers to 98:

a) cos 269° = <u>-cos 89°</u>
= <u>-0.0175</u>

b) cos 270° = <u>-cos 90°</u>
= <u>-0.0000</u> (or 0)

c) tan 269° = <u>+tan 89°</u>
= <u>+57.29</u>

d) tan 270° = <u>+tan 90°</u>
= -------
(Remember, tan 90° has no numerical value.)

---

100. To find the trig ratios of 360°, we can use the diagram and pattern of angles below.

If θ = 340°, then α = 20°.
If θ = 350°, then α = 10°.
If θ = 358°, then α = 2°.

If θ = 360°, then α = 0°.

(a) From the diagram, sin θ = -sin α.

Therefore: sin 360° = -sin 0° = ________

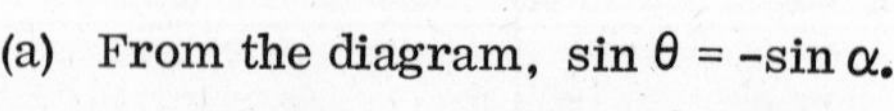
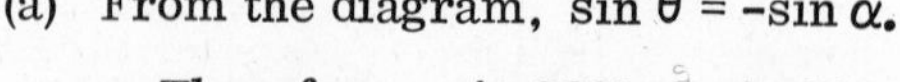

(b) From the diagram, cos θ = +cos α.

Therefore: cos 360° = +cos 0° = ________

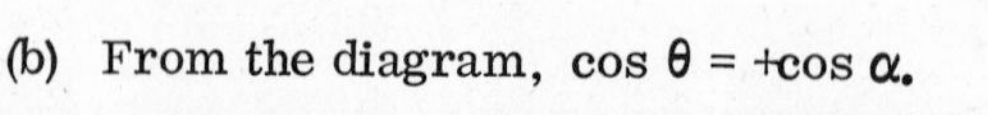

(c) From the diagram, tan θ = -tan α.

Therefore: tan 360° = -tan 0° = ________

Answers to 99:

a) sin 270° = -1

b) cos 270° = 0

c) tan 270° = -----
(There is no numerical value for tan 270°, just as there is no numerical value for tan 90°.)

101. To determine the trig ratios of a 360° angle, draw an angle in the fourth quadrant and think of it as increasing to 360°. The reference angle for 360° is a 0° angle. Therefore:

(1) The absolute values of the trig ratios of 360° are the same as those of 0°.

(2) The signs of the trig ratios of 360° can be determined from your sketch.

Using the procedure above, complete these:

(a) sin 360° = ______

(b) cos 360° = ______

(c) tan 360° = ______

a) -0.0000 (or 0)
b) +1.0000 (or 1)
c) -0.0000 (or 0)

---

102. Here is a summary of the trig ratios of the five special angles studied in this section:

	0°	90°	180°	270°	360°
sin θ	0	+1	0	-1	0
cos θ	+1	0	-1	0	+1
tan θ	0	---	0	---	0

Since two of the angles have their terminal sides in the same position, they have identical trig ratios. Which two angles are they?

______ and ______

a) sin 360° = 0
b) cos 360° = 1
c) tan 360° = 0

---

103. In which quadrant (I, II, III, or IV) should we sketch a standard-position angle to determine the trig ratios of each of the following angles:

(a) 180° ______ (b) 270° ______ (c) 360° ______

0° and 360°

---

104. Of the three angles: 180°, 270°, and 360°:

(a) 0° is the reference angle for ________ and ________.

(b) 90° is the reference angle for ________.

a) II
b) III
c) IV

---

105. Write the numerical value of each ratio.

(a) tan 0° = _____ (b) sin 90° = _____ (c) cos 180° = _____

a) 180° and 360°
b) 270°

---

106. Write the numerical value of each:

(a) sin 0° = _____ (b) sin 270° = _____ (c) cos 90° = _____

a) 0
b) +1
c) -1

107. Write the numerical value of each: (a) $\cos 360° =$ ______ (b) $\tan 90° =$ ______ (c) $\tan 180° =$ ______	a) 0 b) -1 c) 0
108. The "special angles" we have examined are: 0°, 90°, 180°, 270°, 360°. (a) There is one special angle whose sine is +1. It is ______. (b) There is one special angle whose sine is -1. It is ______. (c) There are three special angles whose sines are 0. They are ______, ______, and ______.	a) +1 b) ----- c) 0
109. Use the five special angles to complete these: (a) If $\cos\theta = +1$, $\theta =$ ______ and ______. (b) If $\cos\theta = 0$, $\theta =$ ______ and ______. (c) If $\cos\theta = -1$, $\theta =$ ______	a) 90° b) 270° c) 0°, 180°, 360°
110. Use the five special angles to complete these: (a) If $\tan\theta = 0$, $\theta =$ ______, ______, and ______. (b) If $\tan\theta =$ ---, $\theta =$ ______ and ______.	a) 0° and 360° b) 90° and 270° c) 180°

Answer to Frame 110: a) 0°, 180°, 360° b) 90° and 270°

SELF-TEST 5 (Frames 87-110)

In answering the following, use these angles: 0°, 90°, 180°, 270°, 360°

1. Name three angles whose sine ratio is 0: ______, ______, ______
2. Name one angle whose sine ratio is +1: ______
3. Name one angle whose cosine ratio is -1: ______
4. Name two angles whose tangent ratio does not exist: ______, ______
5. If $\sin\theta = -1$, $\theta =$ ______.
6. If $\cos\theta = +1$, $\theta =$ ______ and ______.

Complete:

7. $\sin 0° =$ ______
8. $\sin 90° =$ ______
9. $\sin 180° =$ ______
10. $\sin 270° =$ ______
11. $\sin 360° =$ ______

ANSWERS:

1. 0°, 180°, 360°
2. 90°
3. 180°
4. 90°, 270°
5. 270°
6. 0°, 360°
7. 0
8. +1
9. 0
10. -1
11. 0

3-6 THE BASIC CYCLE OF THE SINE-WAVE GRAPH

In this section, we will graph the sine ratio from 0° to 360°. The graph of the sine ratio is usually called a "sine wave". The graph of the sine ratio from 0° to 360° is the "basic cycle" of the sine wave.

Sine waves are an important graphical model in basic science and technology. For example, the graphs of such diverse phenomena as alternating currents, radio and television waves, sound waves, and vibration of a spring are related to the sine wave in some way or other. When the graph of a phenomenon is a sine wave, we say that the phenomenon is "sinusoidal".

111. Before graphing the sine wave for angles between 0° and 360°, we can obtain a general overview of the graph by examining the table below.

	0°	90°	180°	270°	360°
sin θ	0	+1	0	-1	0

(a) As θ increases from 0° to 90°, sin θ increases from _____ to _____.

(b) As θ increases from 90° to 180°, sin θ decreases from _____ to _____.

(c) As θ increases from 180° to 270°, sin θ decreases from _____ to _____.

(d) As θ increases from 270° to 360°, sin θ increases from _____ to _____.

112. To graph the sine ratio of angles, a table of values of θ and sin θ is needed. Such a table is given on the right. Note that various entries are included from each quadrant.

Complete the last four entries, rounding each value to the nearest hundredth.

	θ	sin θ
Angles in Quadrant I	0°	0.00
	30°	+0.50
	45°	+0.71
	60°	+0.87
	90°	+1.00
Angles in Quadrant II	120°	+0.87
	135°	+0.71
	150°	+0.50
	180°	0.00
Angles in Quadrant III	210°	-0.50
	225°	-0.71
	240°	-0.87
	270°	-1.00
Angles in Quadrant IV	300°	_____
	315°	_____
	330°	_____
	360°	_____

a) 0 to +1

b) +1 to 0

c) 0 to -1

d) -1 to 0

See next frame.

113. Here is the completed table from the last frame.

Angles in Quadrant I		Angles in Quadrant II		Angles in Quadrant III		Angles in Quadrant IV	
θ	sin θ	θ	sin θ	θ	sin θ	θ	sin θ
0°	0.00	90°	+1.00	180°	0.00	270°	-1.00
30°	+0.50	120°	+0.87	210°	-0.50	300°	-0.87
45°	+0.71	135°	+0.71	225°	-0.71	315°	-0.71
60°	+0.87	150°	+0.50	240°	-0.87	330°	-0.50
90°	+1.00	180°	0.00	270°	-1.00	360°	0.00

Some of the points in the table have been plotted on the graph below. Notice that:

The angle (θ) is plotted on the horizontal axis.
The sine of the angle (sin θ) is plotted on the vertical axis.

Plot the remaining points. Then connect the points with a smooth curve. The resulting graph is called the "basic cycle of the sine-wave graph".

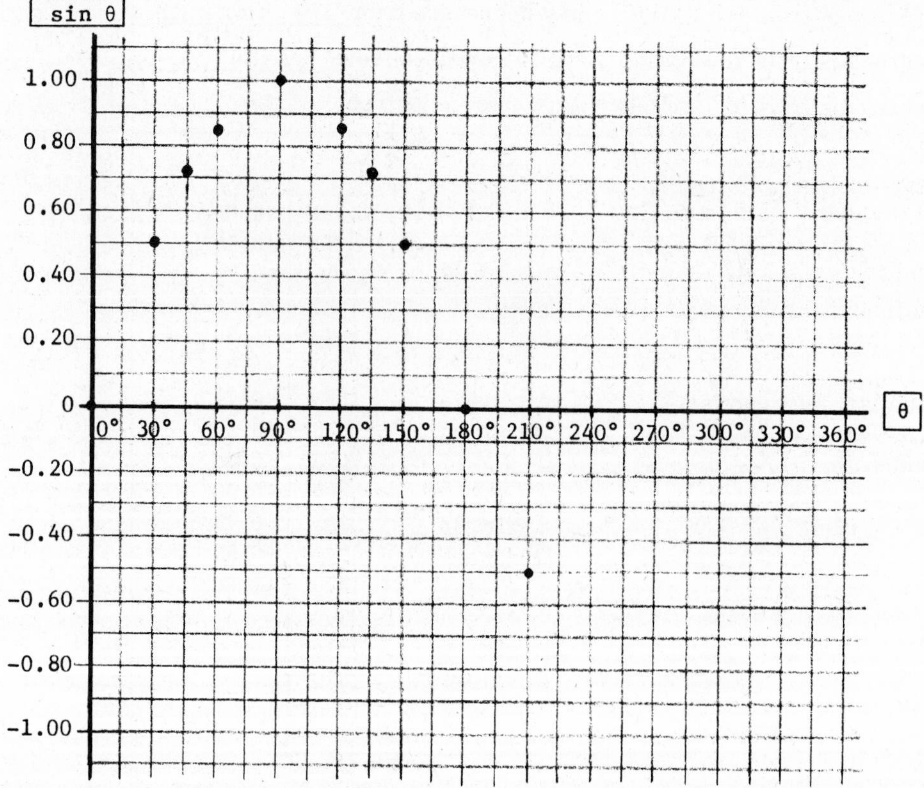

See next frame.

114. Your graph should look like this:

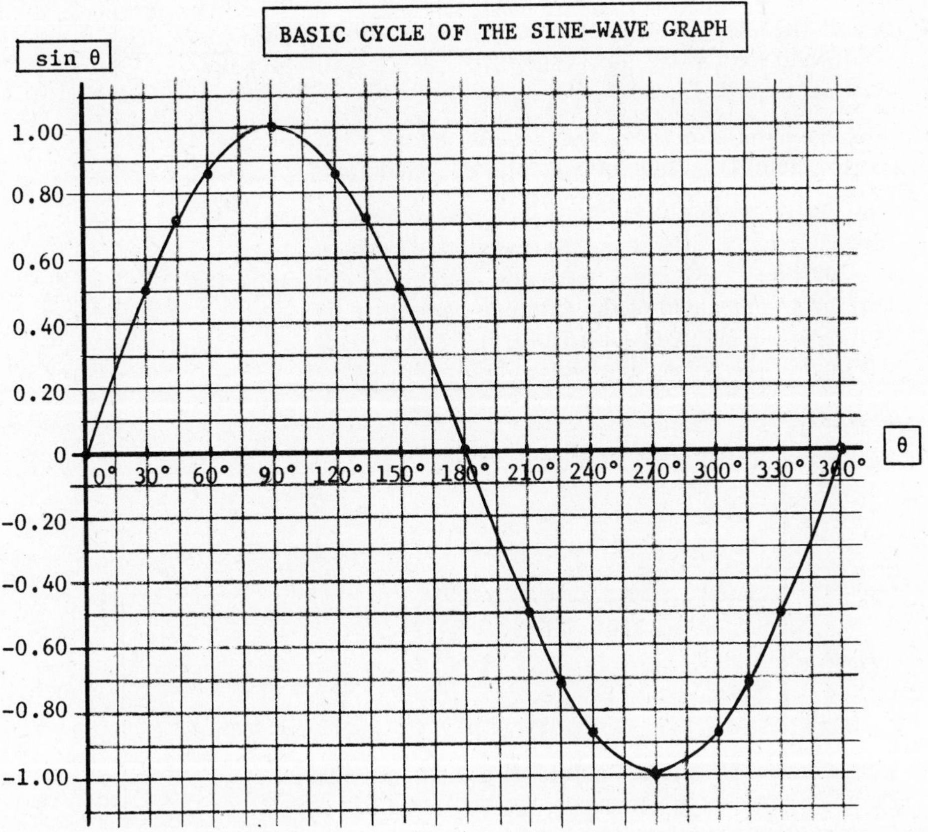

Refer to the graph and answer these:

(a) For angles between 0° and 180°, is sin θ positive or negative? ______________

(b) For angles between 180° and 360°, is sin θ positive or negative? ______________

(c) For what angle θ does sin θ = +1? __________

(d) For what angle θ does sin θ = -1? __________

(e) Is the value of sin θ ever greater than +1? __________

(f) Is the value of sin θ ever less than -1? __________

---

a) Positive

b) Negative

c) 90°

d) 270°

e) No

f) No

115. It is useful to be able to draw rough sketches of the sine-wave graph. To do so, follow these steps:

Step 1: Sketch the axes as we have done below. Calibrate the horizontal axis at 0°, 90°, 180°, 270°, 360°. Calibrate the vertical axis at +1, +0.5, 0, -0.5, -1.

Step 2: Plot the points for these five special angles.
(0°, 0), (90°, 1), (180°, 0), (270°, -1), (360°, 0)

Plot the points for these four angles whose reference angles are 30°. These points are plotted because sin 30° = 0.5. To do so, estimate distances, remembering that 30° is one-third of the distance from 0° to 90°.
(30°, 0.5), (150°, 0.5), (210°, -0.5), (330°, -0.5)
All nine of the points are plotted below.

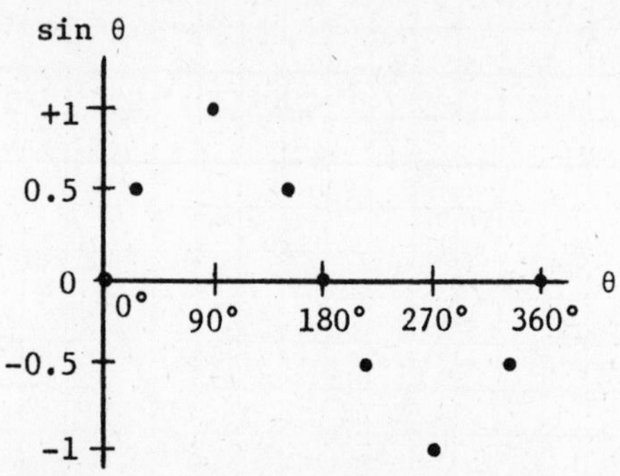

Step 3: Then connect the plotted points with a smooth curve. Do so on the sketch above.

Answer to Frame 115: See next frame.

116. Your sketch should look like the figure on the left. Make another sketch of the basic cycle of the sine-wave graph on the larger axes at the right. The two sketches should be similar.

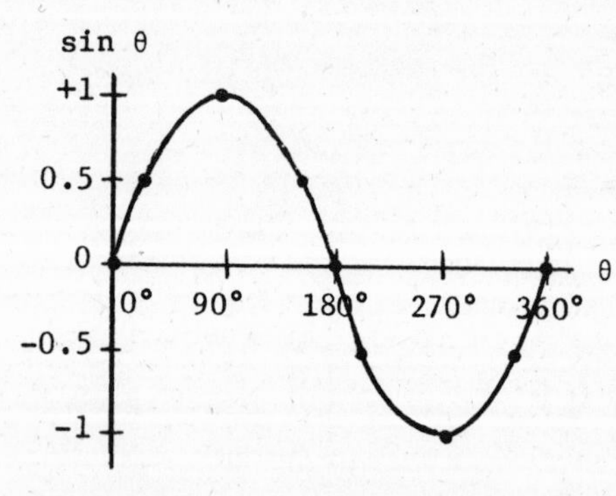

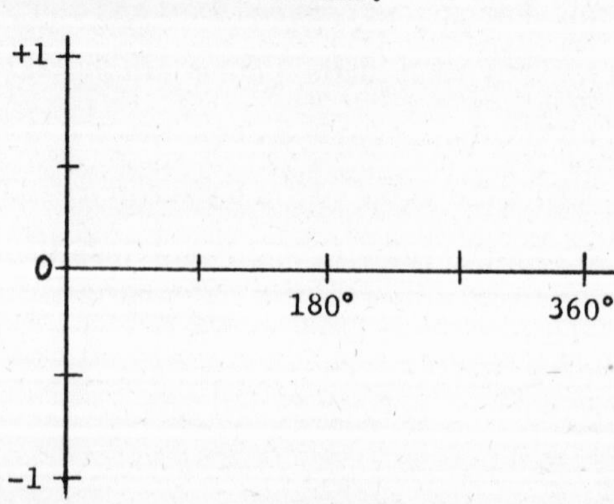

Go to next frame.

117. The data for plotting a sine-wave graph can be obtained from this equation:

$$\boxed{y = \sin\theta}$$

For example: If $\theta = 30°$, $y = 0.5000$ (from $y = \sin 30°$)

If $\theta = 90°$, $y = 1.0000$ (from $y = \sin 90°$)

If $\theta = 240°$, $y = -0.8660$ (from $y = \sin 240°$)

(a) Which variable, "y" or "θ", is the angle? _______

(b) Which variable, "y" or "θ", is the sine of the angle? _______

---

118. The equation $\boxed{y = \sin\theta}$ is called the equation of the sine wave. Its graph is a sine wave, as sketched on the right.

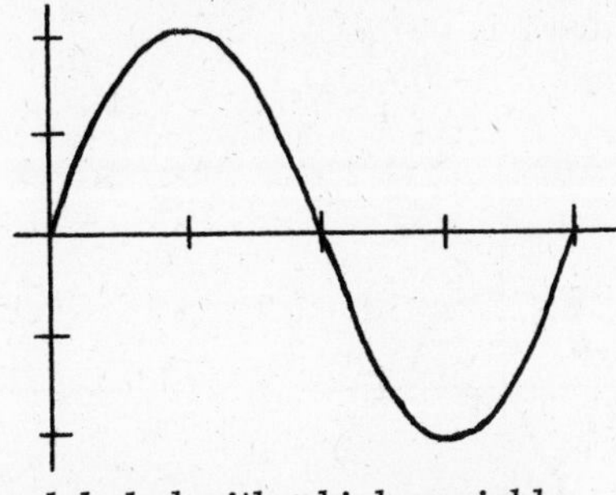

(a) The horizontal axis should be labeled with which variable, "y" or "θ"? __________

(b) The vertical axis should be labeled with which variable, "y" or "θ"? __________

Answers to Frame 117: a) θ b) y

Answer to Frame 118: a) θ Since: θ is an angle, and the angle is plotted on the horizontal axis. b) y Since: y is a sine, and sines are plotted on the vertical axis.

SELF-TEST 6 (Frames 111-118)

1. Complete the entries in the right-hand table below.

θ	sin θ
0°	0
30°	0.5
90°	1
150°	0.5
180°	0

θ	sin θ
210°	
270°	
330°	
360°	

2. Sketch the graph of $y = \sin\theta$ on the axes below.

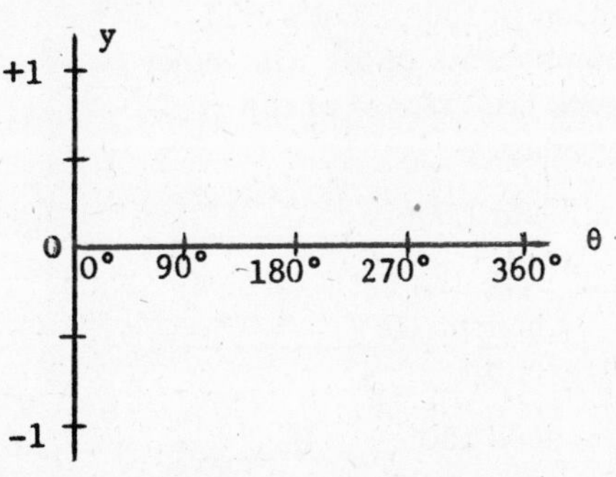

ANSWERS:

1.

θ	sin θ
210°	-0.5
270°	-1
330°	-0.5
360°	0

2. Your sketch should look like the figure in Frame 116.

3-7 VECTOR-ADDITION AND STANDARD-POSITION ANGLES

In the vector-additions performed up to this point, the directions of the vectors have been stated in terms of reference angles. Ordinarily, however, the directions of vectors are stated in terms of standard-position angles. In this section, we will use the component method to add vectors whose directions are stated in terms of standard-position angles.

---

119. In each of the three figures below, the reference angle (α) of $\overrightarrow{0L}$ is given. In terms of reference angles:

The direction of $\overrightarrow{0L}$ in Figure 1 is 40° (Q2).
The direction of $\overrightarrow{0L}$ in Figure 2 is 60° (Q3).
The direction of $\overrightarrow{0L}$ in Figure 3 is 30° (Q4).

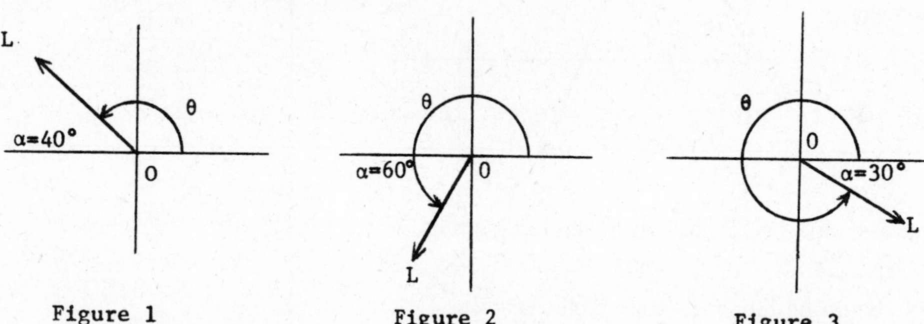

Figure 1 Figure 2 Figure 3

Ordinarily, the direction of a vector is stated in terms of its standard-position angle (θ) rather than its reference angle. Using standard-position angles:

(a) The direction of $\overrightarrow{0L}$ in Figure 1 is ________.

(b) The direction of $\overrightarrow{0L}$ in Figure 2 is ________.

(c) The direction of $\overrightarrow{0L}$ in Figure 3 is ________.

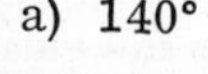

a) 140°

b) 240°

c) 330°

120. In the figure on the right, the length of $\overrightarrow{0B}$ is 10.0 units. Its direction is 130°. Let's find the components of $\overrightarrow{0B}$. Be sure to check the "signs" of the components.

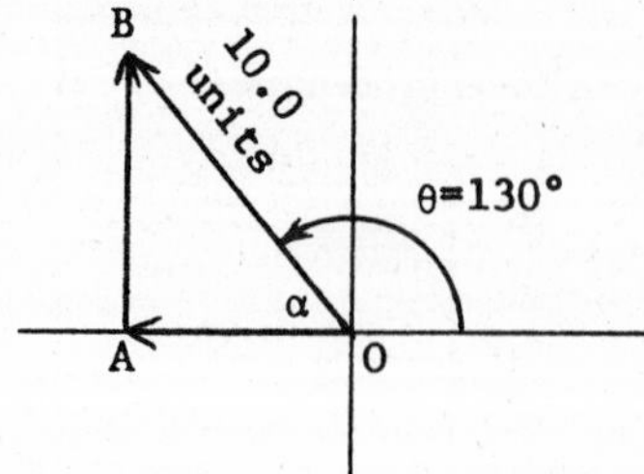

(a) sin 130° = ________

(b) $\overrightarrow{AB}$, the vertical component, is ________ units.

(c) cos 130° = ________

(d) $\overrightarrow{0A}$, the horizontal component, is ________ units.

---

a) +0.7660
b) $\overrightarrow{AB}$ = +7.66
c) -0.6428
d) $\overrightarrow{0A}$ = -6.43
(Note: Did you check the sign?)

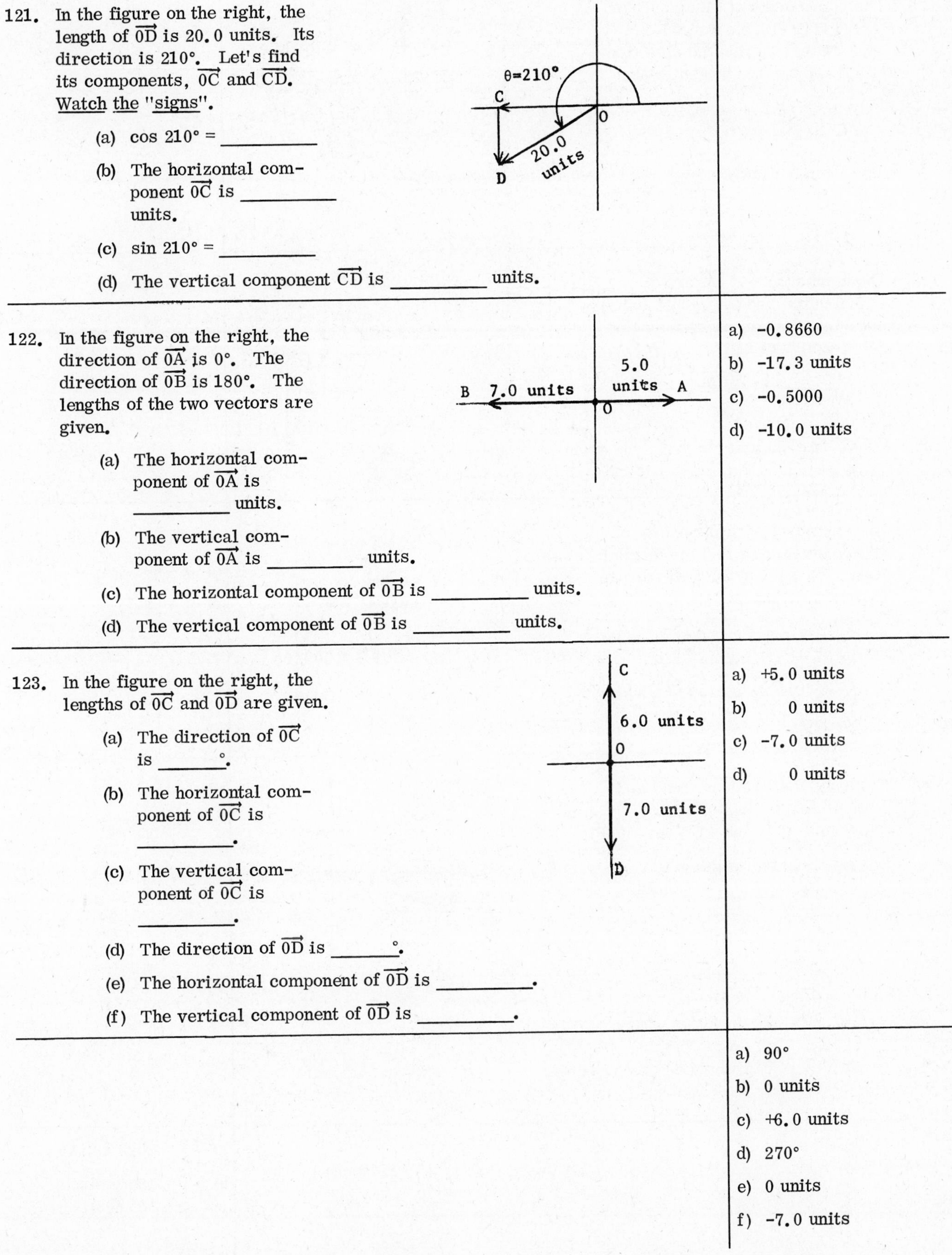

121. In the figure on the right, the length of $\overrightarrow{0D}$ is 20.0 units. Its direction is 210°. Let's find its components, $\overrightarrow{0C}$ and $\overrightarrow{CD}$. Watch the "signs".

(a) cos 210° = __________

(b) The horizontal component $\overrightarrow{0C}$ is __________ units.

(c) sin 210° = __________

(d) The vertical component $\overrightarrow{CD}$ is __________ units.

---

a) -0.8660

b) -17.3 units

c) -0.5000

d) -10.0 units

122. In the figure on the right, the direction of $\overrightarrow{0A}$ is 0°. The direction of $\overrightarrow{0B}$ is 180°. The lengths of the two vectors are given.

(a) The horizontal component of $\overrightarrow{0A}$ is __________ units.

(b) The vertical component of $\overrightarrow{0A}$ is __________ units.

(c) The horizontal component of $\overrightarrow{0B}$ is __________ units.

(d) The vertical component of $\overrightarrow{0B}$ is __________ units.

---

a) +5.0 units

b) 0 units

c) -7.0 units

d) 0 units

123. In the figure on the right, the lengths of $\overrightarrow{0C}$ and $\overrightarrow{0D}$ are given.

(a) The direction of $\overrightarrow{0C}$ is ______°.

(b) The horizontal component of $\overrightarrow{0C}$ is __________.

(c) The vertical component of $\overrightarrow{0C}$ is __________.

(d) The direction of $\overrightarrow{0D}$ is ______°.

(e) The horizontal component of $\overrightarrow{0D}$ is __________.

(f) The vertical component of $\overrightarrow{0D}$ is __________.

---

a) 90°

b) 0 units

c) +6.0 units

d) 270°

e) 0 units

f) -7.0 units

124. The components of $\overrightarrow{0F}$ are $\overrightarrow{0D}$ and $\overrightarrow{DF}$. $\overrightarrow{0D}$ is -10.0 units; $\overrightarrow{DF}$ is +6.5 units. To find the direction of $\overrightarrow{0F}$, we use the tangent ratio.

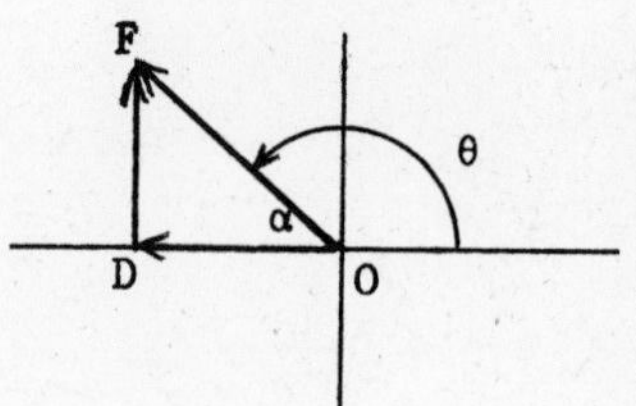

Since $\tan\theta = \tan\alpha = \dfrac{\overrightarrow{DF}}{\overrightarrow{0D}} = \dfrac{6.5}{-10.0} = -0.65$ or -0.6500:

(a) α = ________ (b) θ = ________

---

a) $\alpha = 33°$ (Q2)

b) $\theta = 147°$

125. The components of $\overrightarrow{0M}$ are $\overrightarrow{0C}$ (+10.0 units) and $\overrightarrow{CM}$ (-25.0 units). Let's use the tangent ratio to find the direction of $\overrightarrow{0M}$.

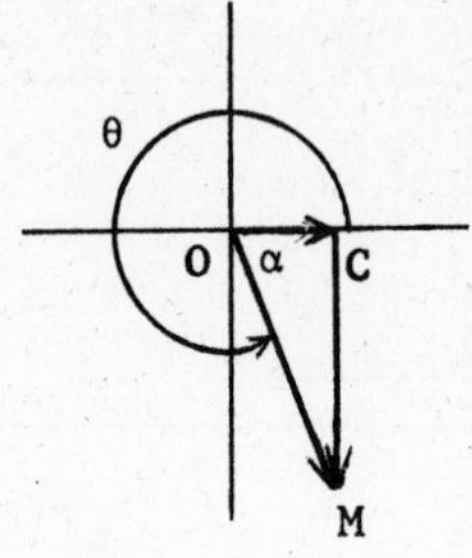

(a) $\tan\theta = \tan\alpha = \dfrac{\overrightarrow{CM}}{\overrightarrow{0C}} =$ ________

(b) α = ________

(c) θ = ________

---

a) -2.5 or -2.500

b) $\alpha = 68°$ (Q4)

c) $\theta = 292°$

126. The lengths of the components of $\overrightarrow{0T}$ are given. To find the length of $\overrightarrow{0T}$, we can use the Pythagorean Theorem. That is:

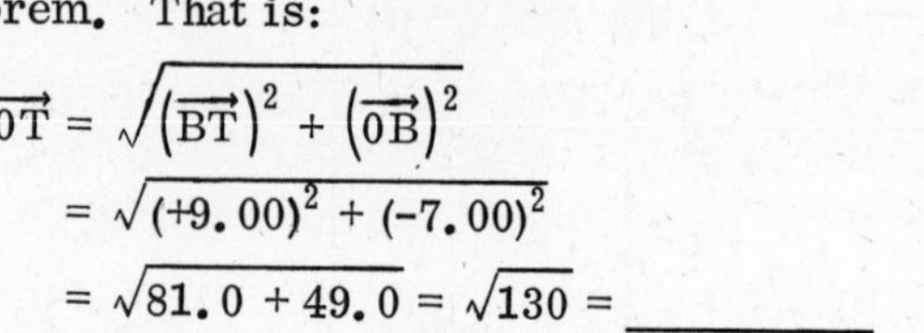

$$\overrightarrow{0T} = \sqrt{\left(\overrightarrow{BT}\right)^2 + \left(\overrightarrow{0B}\right)^2}$$
$$= \sqrt{(+9.00)^2 + (-7.00)^2}$$
$$= \sqrt{81.0 + 49.0} = \sqrt{130} = \_\_\_\_\_\_\_\_$$

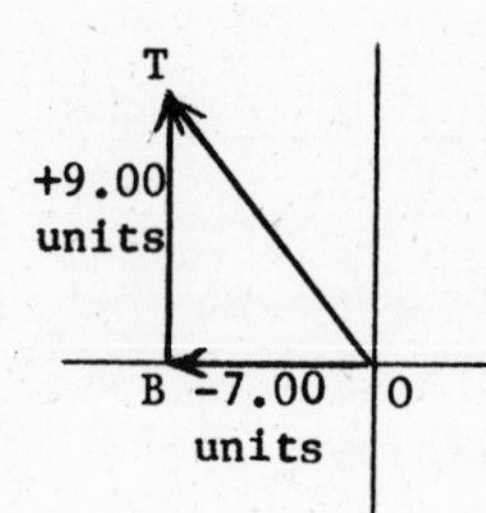

11.4 units

127. The lengths of the horizontal and vertical components of $\overrightarrow{0F}$ are -28.5 units ($\overrightarrow{0A}$) and -9.5 units ($\overrightarrow{AF}$).

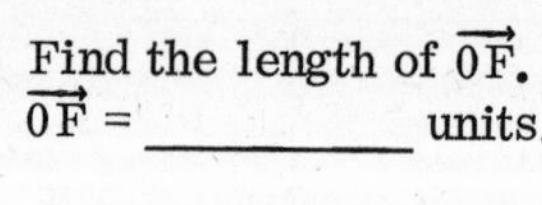

(a) Find the length of $\overrightarrow{0F}$.
$\overrightarrow{0F}$ = ________ units.

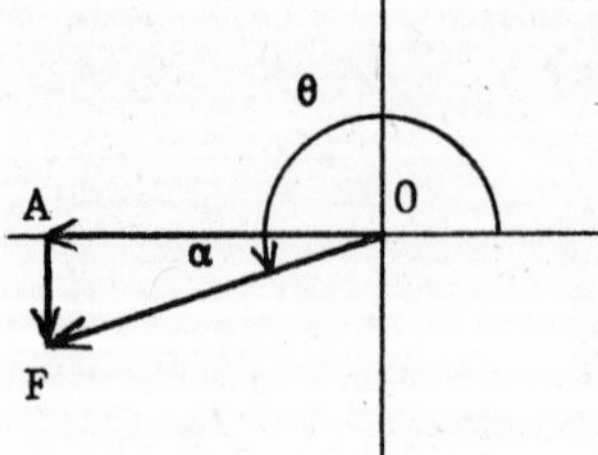

(b) Find the direction of $\overrightarrow{0F}$. θ = ________

---

a) $\overrightarrow{0F} = 30.0$ units

b) $\theta = 198°$, since: $\alpha = 18°$ (Q3)

128. In the figure on the right, the length and direction of $\overrightarrow{0A}$ are 43.0 units and 50°. The length and direction of $\overrightarrow{0B}$ are 52.0 units and 160°. Let's add the two vectors by the component method and find the length and direction of their resultant $\overrightarrow{0L}$.

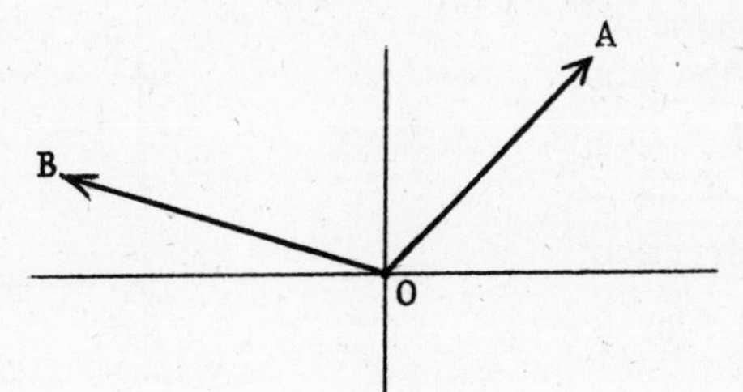

(a) Use the following table to find the components of their resultant $\overrightarrow{0L}$.

	Horizontal Component	Vertical Component
$\overrightarrow{0A}$		
$\overrightarrow{0B}$		
$\overrightarrow{0L}$		

(b) The length of $\overrightarrow{0L}$ is ______________.

(c) The direction of $\overrightarrow{0L}$ is ______________.
(Note: Roughly complete the parallelogram to check that this direction makes sense.)

---

a)

	Hor. Comp.	Vert. Comp.
$\overrightarrow{0A}$	+27.6	+32.9
$\overrightarrow{0B}$	-48.9	+17.8
$\overrightarrow{0L}$	-21.3	+50.7

b) $\overrightarrow{0L}$ = 55.0 units

c) θ = 113°, since α = 67° (Q2)

129. The lengths and directions of $\overrightarrow{0C}$, $\overrightarrow{0D}$, and $\overrightarrow{0F}$ are given in the table below.

	Length	Direction
$\overrightarrow{0C}$	10.0 units	0°
$\overrightarrow{0D}$	12.0 units	90°
$\overrightarrow{0F}$	14.0 units	260°

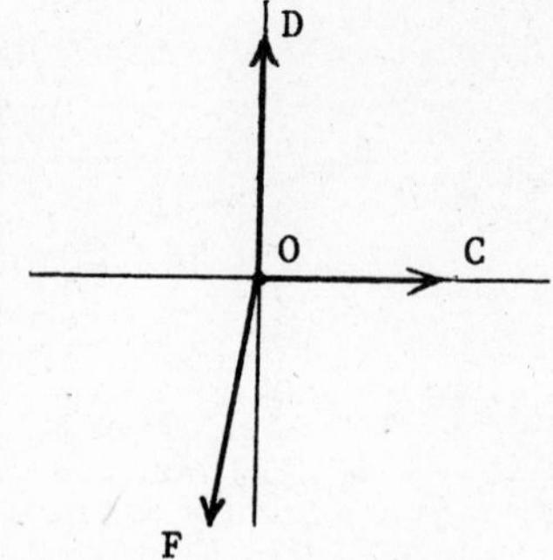

Let's find the resultant $\overrightarrow{0R}$ of these three vectors by the component method.

(a) To find the components of the resultant $\overrightarrow{0R}$, complete this table:

	Horizontal Component	Vertical Component
$\overrightarrow{0C}$		
$\overrightarrow{0D}$		
$\overrightarrow{0F}$		
$\overrightarrow{0R}$		

(b) The length of $\overrightarrow{0R}$ is ______________.

(c) The direction of $\overrightarrow{0R}$ is ______________.

---

a)

	Horiz. Comp.	Vert. Comp.
$\overrightarrow{0C}$	+10.0	0.0
$\overrightarrow{0D}$	0.0	+12.0
$\overrightarrow{0F}$	-2.43	-13.8
$\overrightarrow{0R}$	+7.6	-1.8

b) 7.8 units

c) 347°, since $\alpha = 13°$ (Q4)
Note: We can tell that it is a fourth quadrant angle because its horizontal component is positive and its vertical component is negative.

SELF-TEST 7 (Frames 119-129)

In the diagram at the right, $\overrightarrow{0A}$ is 300 units long and its direction is 180°. $\overrightarrow{0B}$ is 600 units long and its direction is 300°.

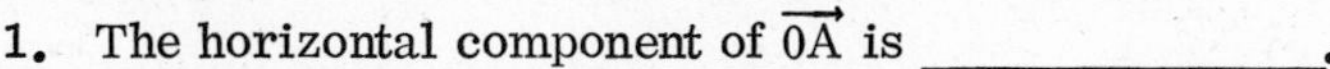

1. The horizontal component of $\overrightarrow{0A}$ is ______________.
2. The vertical component of $\overrightarrow{0A}$ is ______________.
3. The horizontal component of $\overrightarrow{0B}$ is ______________.
4. The vertical component of $\overrightarrow{0B}$ is ______________.
5. The horizontal component of the resultant is ______________.
6. The vertical component of the resultant is ______________.
7. The length of the resultant is ______________.
8. The direction of the resultant is ______________.

ANSWERS:

1. -300 units	3. +300 units	5. 0 units	7. 520 units
2. 0 units	4. -520 units	6. -520 units	8. 270°

3-8 TRIG RATIOS OF ANGLES GREATER THAN 360°

Up to this point, we have only discussed angles between 0° and 360°. In this section, we will discuss the meaning and trig ratios of angles greater than 360°.

130. On the axes at the left, we have sketched a 315° angle in standard position. Sketch a 360° angle in standard position on the axes at the right.

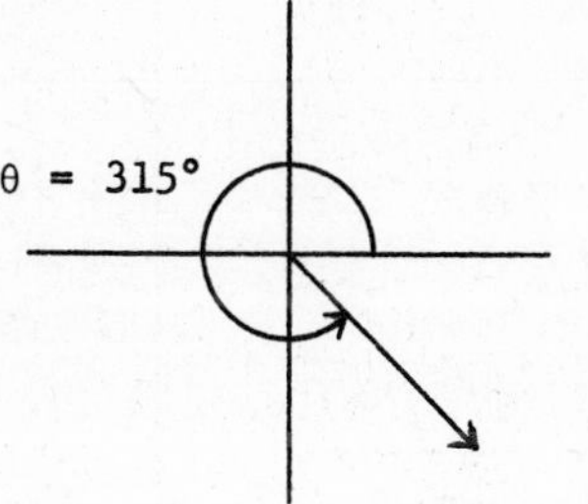

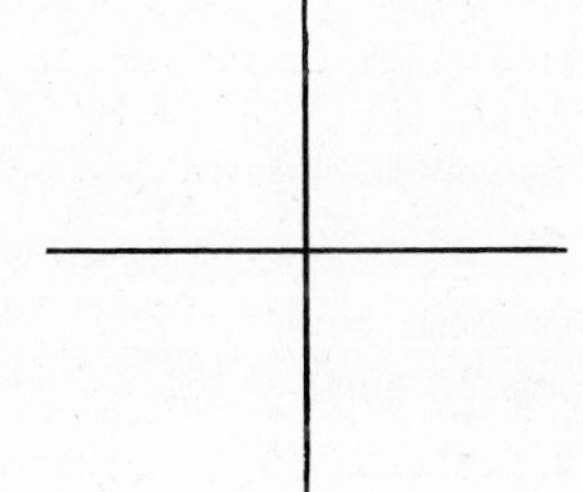

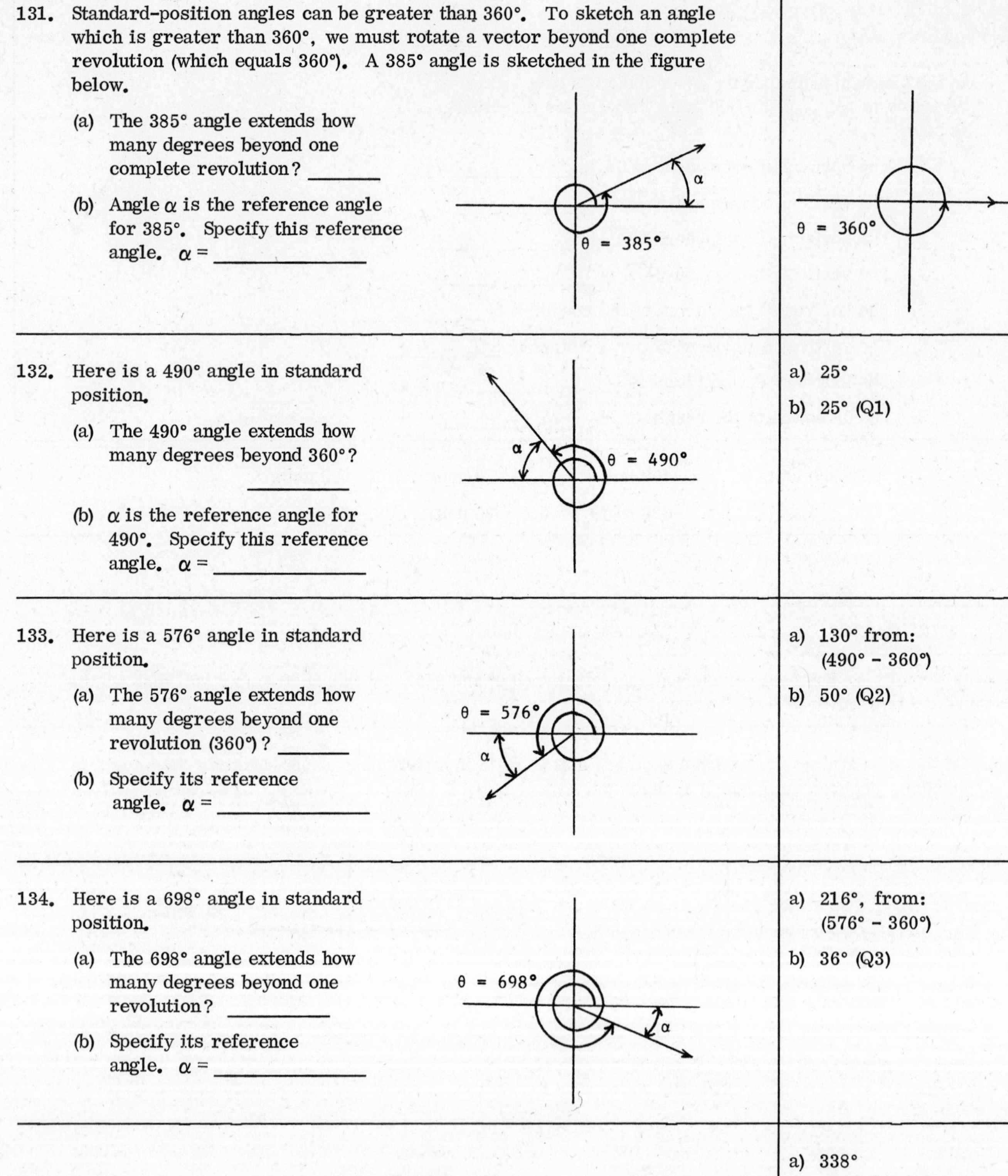

131. Standard-position angles can be greater than 360°. To sketch an angle which is greater than 360°, we must rotate a vector beyond one complete revolution (which equals 360°). A 385° angle is sketched in the figure below.

(a) The 385° angle extends how many degrees beyond one complete revolution? ______

(b) Angle α is the reference angle for 385°. Specify this reference angle. α = ______

---

132. Here is a 490° angle in standard position.

(a) The 490° angle extends how many degrees beyond 360°? ______

(b) α is the reference angle for 490°. Specify this reference angle. α = ______

a) 25°

b) 25° (Q1)

---

133. Here is a 576° angle in standard position.

(a) The 576° angle extends how many degrees beyond one revolution (360°)? ______

(b) Specify its reference angle. α = ______

a) 130° from: (490° − 360°)

b) 50° (Q2)

---

134. Here is a 698° angle in standard position.

(a) The 698° angle extends how many degrees beyond one revolution? ______

(b) Specify its reference angle. α = ______

a) 216°, from: (576° − 360°)

b) 36° (Q3)

---

a) 338°

b) 22° (Q4)

135. (a) How many revolutions in angle θ? ________

(b) How many degrees in angle θ? ________

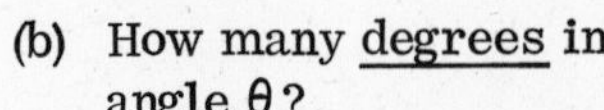

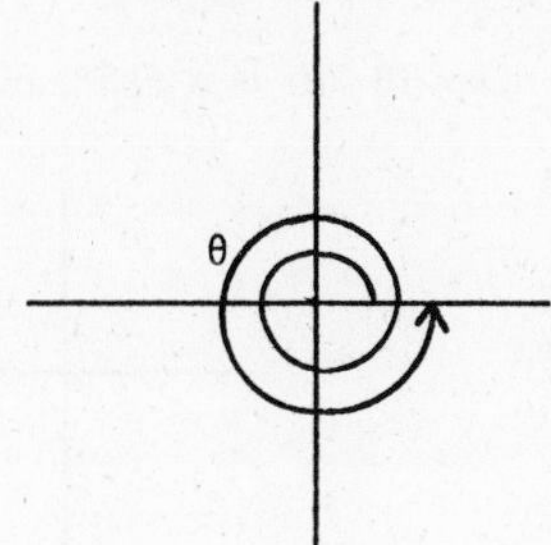

---

a) Two

b) 720°
(2 x 360° = 720°)

136. Here is a 865° angle in standard position.

(a) The 865° angle extends how many degrees beyond 720° (two revolutions)? ________

(b) Specify its reference angle.
α = ________

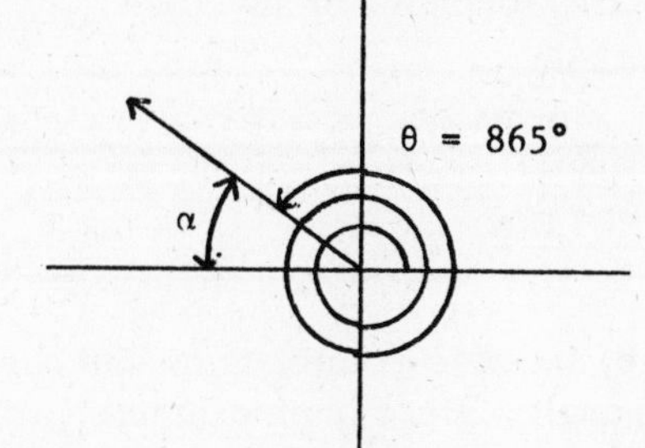

---

a) 145°

b) 35° (Q2)

137. Here is a 1100° angle.

(a) How many degrees are there in three revolutions?
3 x 360° = ________

(b) The 1100° angle extends how many degrees beyond three revolutions (or 1080°)? ________

(c) Specify its reference angle.
α = ________

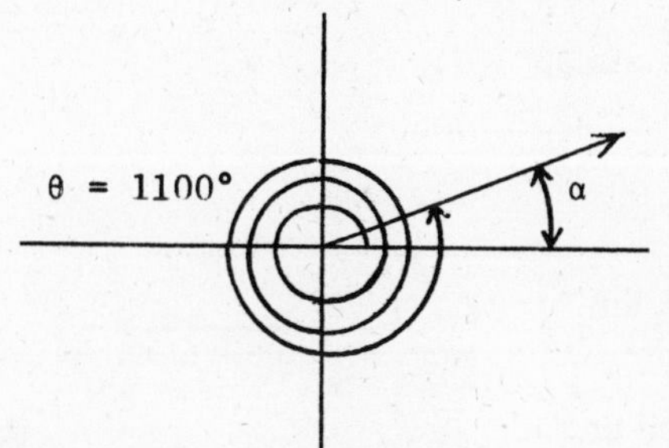

---

a) 1080°

b) 20°

c) 20° (Q1)

138. As you have just learned, angles can be of any size. In science and technology, however, angles are usually less than 1000° in size.

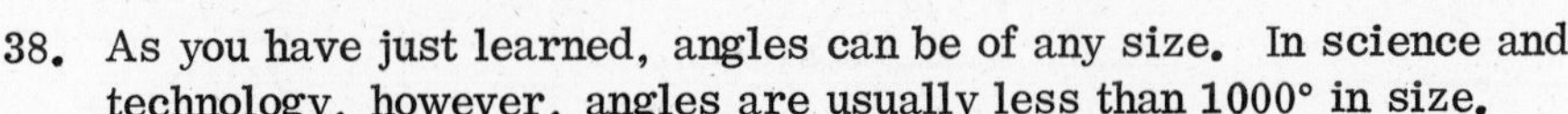

Every standard-position angle has a reference angle. You must be able to determine the size and quadrant of the reference angle. To do this, a sketch of the given angle should always be made.

Sketch each angle θ; then specify its reference angle α.

(a) If $\theta = 570°$, α = ________ (b) If $\theta = 852°$, α = ________

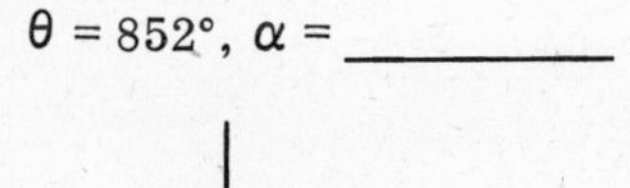

---

a) $\alpha = 30°$ (Q3)

b) $\alpha = 48°$ (Q2)

139. Determine the reference angle α for these angles.

(a) If $\theta = 652°$, α = ________ (b) If $\theta = 417°$, α = ________

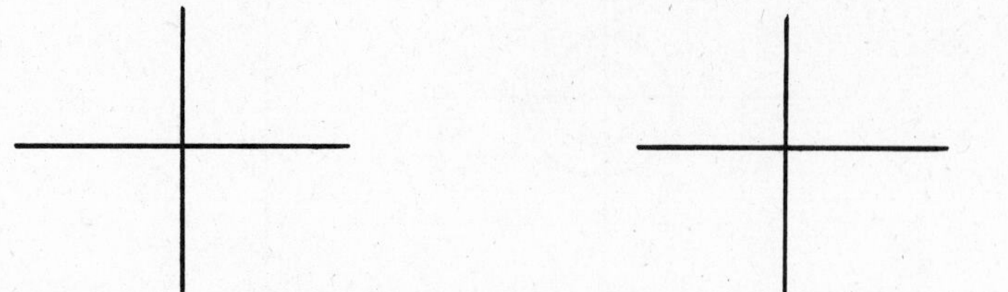

---

140. The following definition applies to angles of any size:

THE SINE, COSINE, OR TANGENT OF ANY ANGLE θ IS EQUAL TO THE SINE, COSINE, OR TANGENT OF ITS REFERENCE ANGLE α.

Therefore, the trig ratios of an angle greater than 360° are equal to the trig ratios of its reference angle. Here is an example:

We have sketched a 390° angle on the right. Its reference angle is 30° (Q1). Therefore:

(a) sin 390° = sin 30° (Q1)

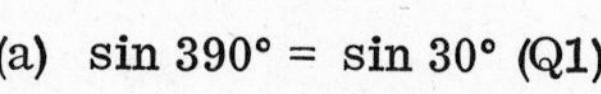

(b) cos 390° = cos 30° (Q1)

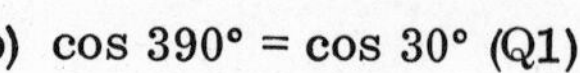

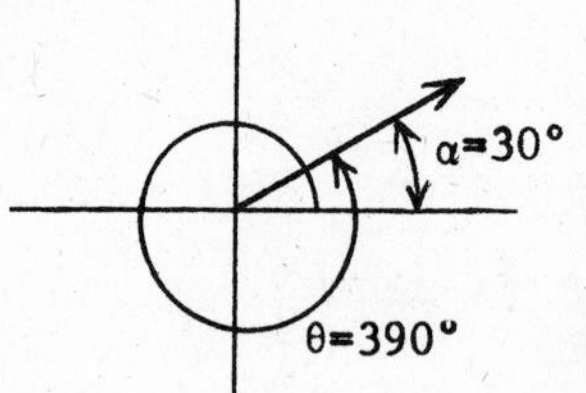

(c) tan 390° = tan 30° (Q1) = ________

Answers:

a) $\alpha = 68°$ (Q4)

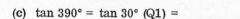

b) $\alpha = 57°$ (Q1)

---

141. A 480° angle is sketched on the right. Its reference angle is 60° (Q2). Therefore:

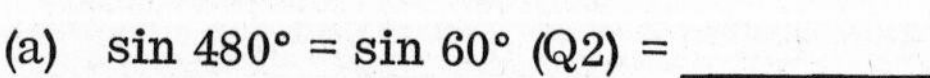

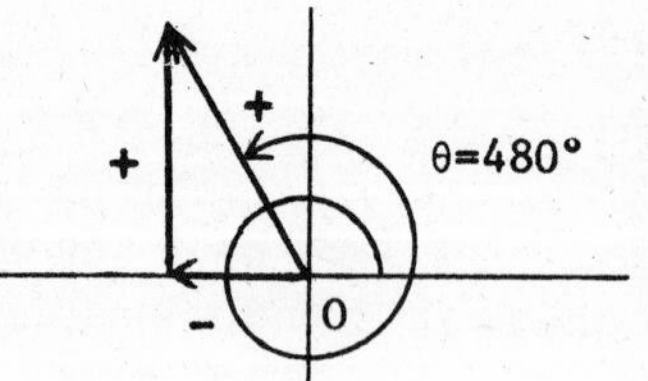

(a) sin 480° = sin 60° (Q2) = ________

(b) cos 480° = cos 60° (Q2) = ________

(c) tan 480° = tan 60° (Q2) = ________

Answers:

a) 0.5000

b) 0.8660

c) 0.5774

---

142. A 570° angle is sketched at the right. Its reference angle is 30° (Q3). Therefore:

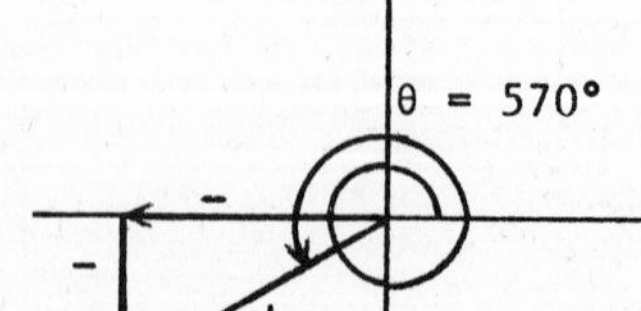

(a) sin 570° = sin 30° (Q3) = ________

(b) cos 570° = cos 30° (Q3) = ________

(c) tan 570° = tan 30° (Q3) = ________

Answers:

a) +0.8660

b) −0.5000

c) −1.732

143. A 660° angle is sketched at the right.
Its reference angle is 60° (Q4).
Therefore:

(a) $\sin 660° = \sin 60°$ (Q4) = ________

(b) $\cos 660° = \cos 60°$ (Q4) = ________

(c) $\tan 660° = \tan 60°$ (Q4) = ________

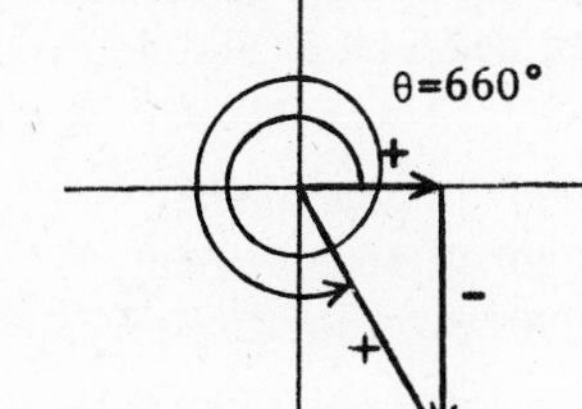

a) -0.5000
b) -0.8660
c) +0.5774

---

144. In order to find the trig ratios of any angle (θ) greater than 360°, we must find its reference angle (α). To do so, a sketch of the angle is needed.

Sketch each angle below on the axes provided. Then determine its reference angle and the appropriate trig ratio. (Be careful of the signs.)

(a) If $\theta = 405°$, $\alpha =$ ________ $\sin\theta =$ ________

(b) If $\theta = 520°$, $\alpha =$ ________ $\cos\theta =$ ________

(c) If $\theta = 580°$, $\alpha =$ ________ $\tan\theta =$ ________

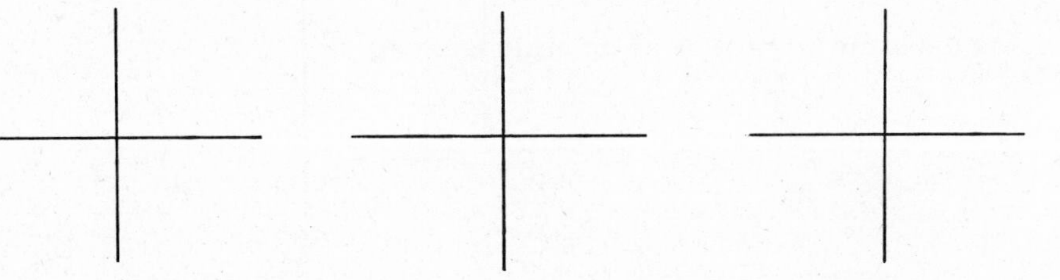

a) -0.8660
b) +0.5000
c) -1.732

---

145. To complete the following, sketch each angle below.

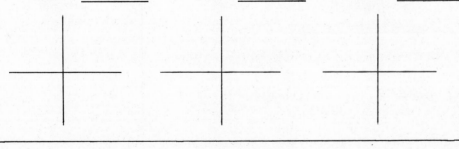

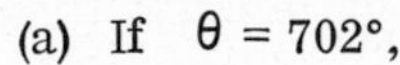

(a) If $\theta = 702°$, $\alpha =$ ________ $\sin\theta =$ ________

(b) If $\theta = 465°$, $\alpha =$ ________ $\tan\theta =$ ________

(c) If $\theta = 565°$, $\alpha =$ ________ $\cos\theta =$ ________

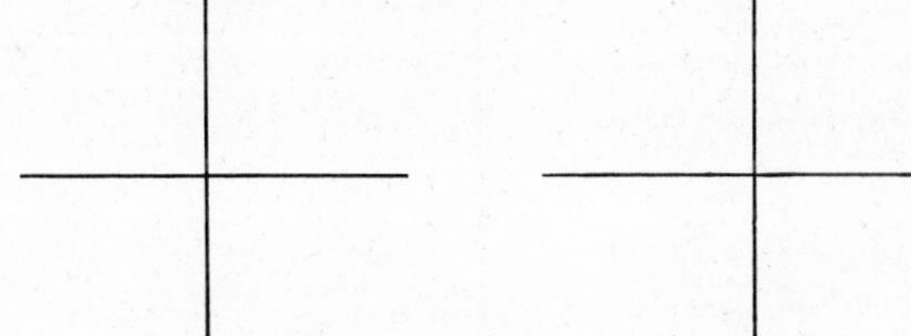

a) $\alpha = 45°$ (Q1)
$\sin\theta = +0.7071$

b) $\alpha = 20°$ (Q2)
$\cos\theta = -0.9397$

c) $\alpha = 40°$ (Q3)
$\tan\theta = +0.8391$

---

a) $\alpha = 18°$ (Q4)
$\sin\theta = -0.3090$

b) $\alpha = 75°$ (Q2)
$\tan\theta = -3.732$

c) $\alpha = 25°$ (Q3)
$\cos\theta = -0.9063$

146. It should be obvious that many angles greater than 360° have the same terminal side as any angle between 0° and 360°. For example, all of the following angles have the same terminal side as a 100° angle.

$$460° = 100° + 360° \text{ (1 revolution)}$$
$$820° = 100° + 720° \text{ (2 revolutions)}$$
$$1180° = 100° + 1080° \text{ (3 revolutions)}$$

Name three angles greater than 360° which have the same terminal side as a 200° angle.

__________, __________, and __________

560° (200° + 360°)

920° (200° + 720°)

1280° (200° + 1080°)

147. Since angles with the same terminal side have the same reference angle, they have the same trig ratios.

Name three angles greater than 360° which have the same terminal side, reference angle, and trig ratios as a 90° angle.

__________, __________, and __________

450°, 810°, 1170°

148. Since 430° - 360° = 70°, a 430° angle has the same trig ratios as a 70° angle.

Since 850° - 720° = 130°, a 850° angle has the same trig ratios as a 130° angle.

Each of the following angles has the same trig ratios as an angle between 0° and 360°. Name the angle between 0° and 360°:

(a) 520° ______ (b) 900° ______ (c) 450° ______ (d) 720° ______

---

a) 160°

b) 180°

c) 90°

d) 0° or 360°

149. Since 450° - 360° = 90°, a 450° angle has the same trig ratios as a 90° angle. Therefore:

(a) sin 450° = _____ (b) cos 450° = _____ (c) tan 450° = _____

---

a) +1

b) 0

c) -----

150. Since 540° - 360° = 180°, a 540° angle has the same trig ratios as a 180° angle. Therefore:

(a) sin 540° = _____ (b) cos 540° = _____ (c) tan 540° = _____

---

a) 0

b) -1

c) 0

151. Since 630° - 360° = 270°, a 630° angle has the same trig ratios as a 270° angle. Therefore:

(a) sin 630° = _____ (b) cos 630° = _____ (c) tan 630° = _____

---

a) -1

b) 0

c) -----

152. Since $720° - 720° = 0°$ or $720° - 360° = 360°$, a 720° angle has the same trig ratios as a 0° or 360° angle. Therefore:

(a) sin 720° = _____ (b) cos 720° = _____ (c) tan 720° = _____

---

153. Complete these:

(a) sin 450° = _______ (c) tan 720° = _______

(b) cos 630° = _______ (d) sin 540° = _______

a) 0
b) +1
c) 0

Answer to Frame 153: a) +1 b) 0 c) 0 d) 0

SELF-TEST 8 (Frames 130-153)

1. Name a standard-position angle between 0° and 360° which has the same terminal side as 1,000°. _________

2. Name the reference angle for 1,000°. __________

Find the numerical value of each of the following:

3. sin 517° = _________ 4. cos 965° = _________ 5. tan 670° = _________

ANSWERS: 1. 280° 2. 80° (Q4) 3. +0.3907 4. -0.4226 5. -1.192

3-9 TRIG RATIOS OF NEGATIVE ANGLES

All angles discussed up to this point have been positive angles. In this section, we will discuss the meaning and trig ratios of negative angles.

154. All angles discussed up to this point have been positive angles. Positive angles are generated by a counter-clockwise rotation from the initial side on the horizontal axis. For example, $\theta = +120°$ (or $120°$) is sketched in Figure 1 below.

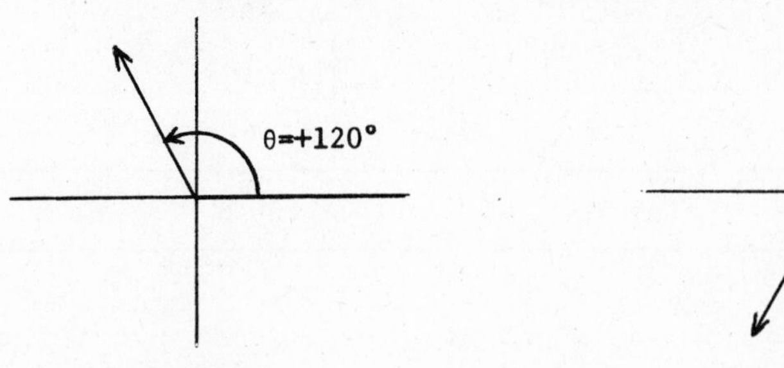

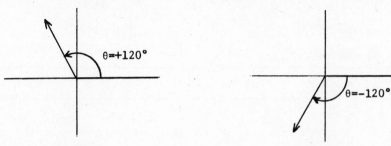

Figure 1 Figure 2

It is also possible to generate angles by a clockwise rotation from the initial side. Angles generated by a clockwise rotation are called "negative" angles. For example, $\theta = -120°$ is sketched in Figure 2 above.

Sketch and label the following negative angles:

(a) $\theta = -38°$ (b) $\theta = -243°$ (c) $\theta = -500°$

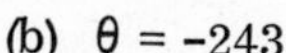

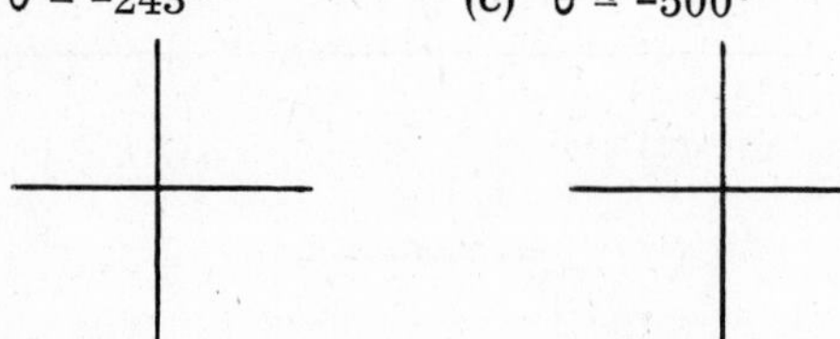

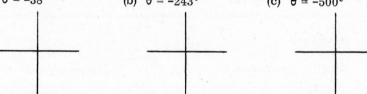

155. Just like a positive angle, the reference angle of any negative angle is the angle between its terminal side and the horizontal axis.

$\theta = -150°$ is sketched on the right. As you can see from the sketch, its reference angle (α) is $30°$ (Q3).

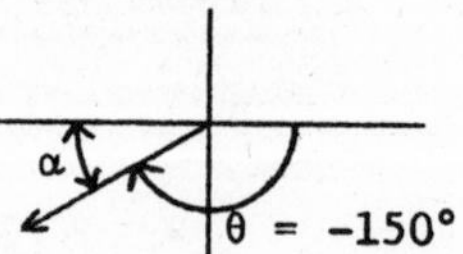

Use the sketches to determine the reference angles of each of these:

(a) $\theta = -240°$

$\alpha =$ __________

θ = -240°, α (sketch)

(b) $\theta = -317°$

$\alpha =$ __________

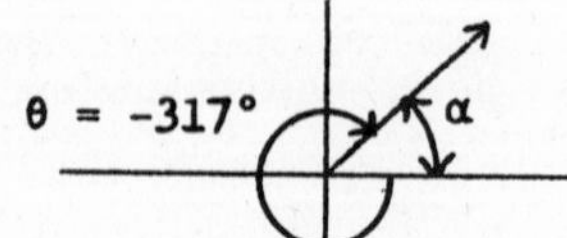

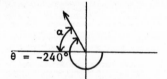

(c) $\theta = -30°$

$\alpha =$ __________

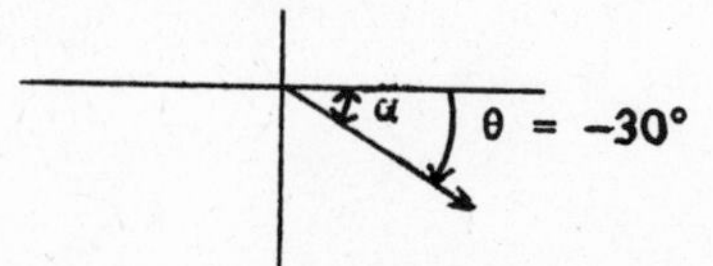

a)

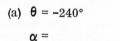

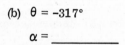

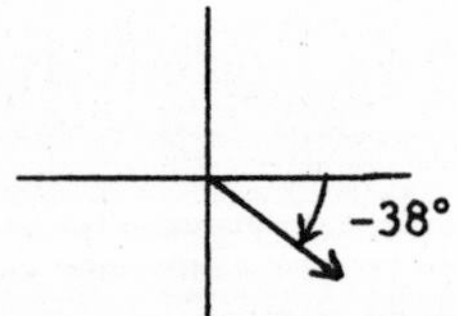

b)

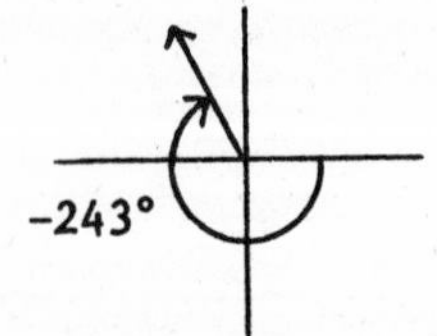

c)

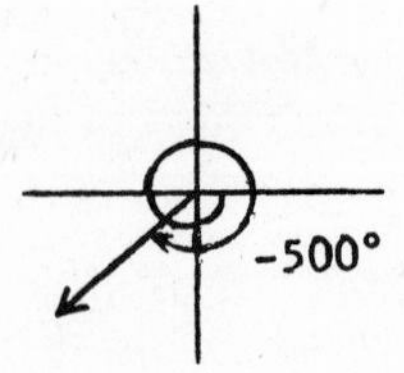

156. Just like a positive angle, <u>THE TRIG RATIOS OF A NEGATIVE ANGLE ARE THE SAME AS THE TRIG RATIOS OF ITS REFERENCE ANGLE.</u>

$\theta = -40°$ is sketched at the right.
Its reference angle is 40° (Q4).
Therefore:

(a) $\sin -40° = \sin 40°$(Q4)
$= \_\_\_\_\_\_\_\_$

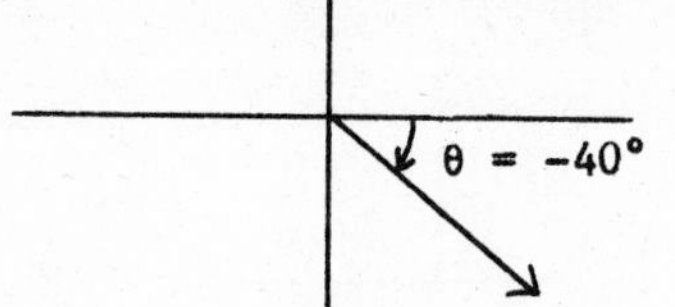

(b) $\cos -40° = \cos 40°$ (Q4)
$= \_\_\_\_\_\_\_\_$

(c) $\tan -40° = \tan 40°$ (Q4)
$= \_\_\_\_\_\_\_\_$

a) $\alpha = 60°$ (Q2)
b) $\alpha = 43°$ (Q1)
c) $\alpha = 30°$ (Q4)

---

157. $\theta = -150°$ is sketched at the right.
Its reference angle is 30° (Q3).
Therefore:

(a) $\sin -150° = \sin 30°$ (Q3)
$= \_\_\_\_\_\_\_\_$

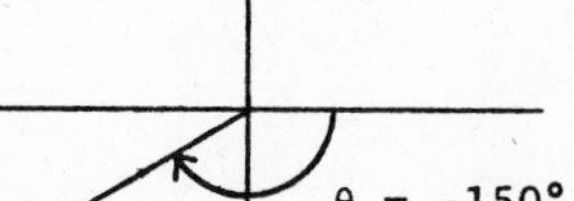
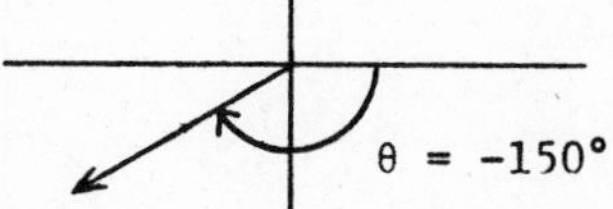

(b) $\cos -150° = \cos 30°$ (Q3)
$= \_\_\_\_\_\_\_\_$

(c) $\tan -150° = \tan 30°$ (Q3)
$= \_\_\_\_\_\_\_\_$

a) −0.6428
b) +0.7660
c) −0.8391

---

158. $\theta = -240°$ is sketched at the right.
Its reference angle is 60° (Q2).
Therefore:

(a) $\sin -240° = \sin 60°$ (Q2)
$= \_\_\_\_\_\_\_\_$

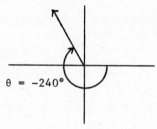

(b) $\cos -240° = \cos 60°$ (Q2)
$= \_\_\_\_\_\_\_\_$

(c) $\tan -240° = \tan 60°$ (Q2)
$= \_\_\_\_\_\_\_\_$

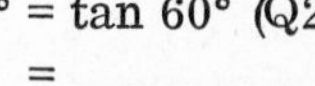

a) −0.5000
b) −0.8660
c) +0.5774

---

159. Sketch each angle on the axes below. After determining its reference angle, find the appropriate trig ratio.

(a) If $\theta = -310°$, $\alpha = \_\_\_\_\_\_\_\_$ $\sin\theta = \_\_\_\_\_\_\_\_$

(b) If $\theta = -120°$, $\alpha = \_\_\_\_\_\_\_\_$ $\cos\theta = \_\_\_\_\_\_\_\_$

(c) If $\theta = -55°$, $\alpha = \_\_\_\_\_\_\_\_$ $\tan\theta = \_\_\_\_\_\_\_\_$

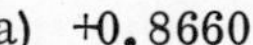

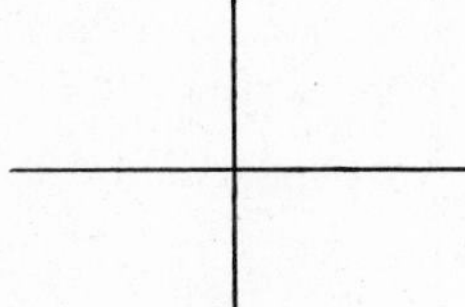
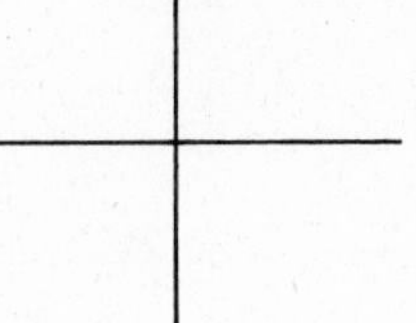
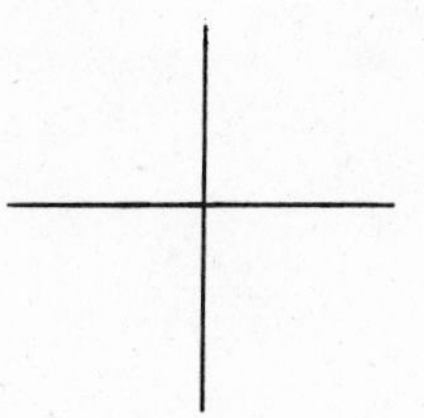

a) +0.8660
b) −0.5000
c) −1.732

160. Sketch each angle to find its reference angle. Then find the appropriate trig ratio.

(a) If $\theta = -208°$, α = ________ $\tan\theta$ = ________

(b) If $\theta = -147°$, α = ________ $\sin\theta$ = ________

(c) If $\theta = -288°$, α = ________ $\cos\theta$ = ________

a) $\alpha = 50°$ (Q1) $\sin\theta = +0.7660$

b) $\alpha = 60°$ (Q3) $\cos\theta = -0.5000$

c) $\alpha = 55°$ (Q4) $\tan\theta = -1.428$

---

161. Negative angles can also be extended beyond 360°. Of course, negative angles beyond -360° involve <u>more than one complete revolution in a clockwise direction.</u>

An angle of -390° is sketched at the right. It has the same terminal side as a -30° angle.

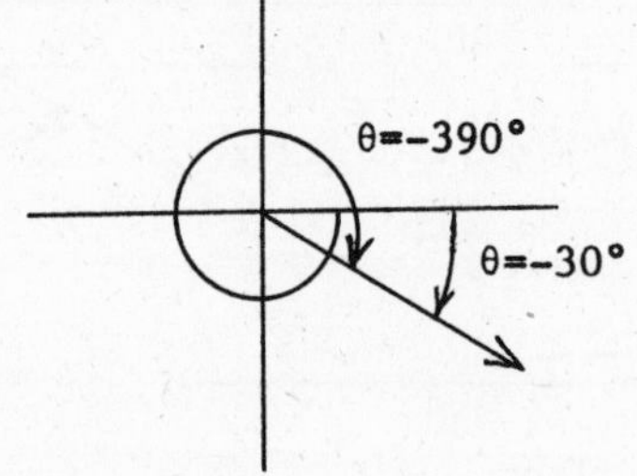

Sketch a -420° angle, a -550° angle, and a -690° angle on the axes below and then answer the following questions.

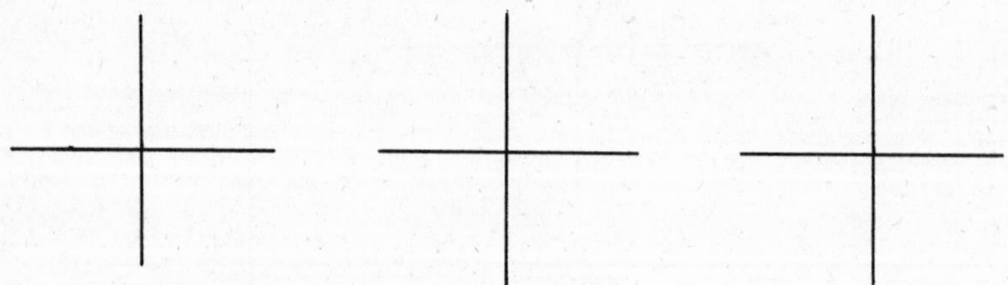

Of the <u>negative angles</u> between 0° and -360°:

(a) A -420° angle has the same terminal side as a ________ angle.

(b) A -550° angle has the same terminal side as a ________ angle.

(c) A -690° angle has the same terminal side as a ________ angle.

a) $\alpha = 28°$ (Q2) $\tan\theta = -0.5317$

b) $\alpha = 33°$ (Q3) $\sin\theta = -0.5446$

c) $\alpha = 72°$ (Q1) $\cos\theta = +0.3090$

---

a) -60°

b) -190°

c) -330°

162. There are many negative angles beyond -360° which have the same terminal side, reference angle, and trig ratios as a negative angle lying between 0° and -360°. All of the following negative angles have the same terminal side as a -120° angle:

$$-480° = (-120°) + (-360°)$$
$$-840° = (-120°) + (-720°)$$
$$-1200° = (-120°) + (-1080°)$$

Name three negative angles beyond -360° which have the same trig ratios as a -50° angle: __________, __________, and __________

Answer: -410°, -770°, -1130°

163. In each figure below, a positive angle between 0° and 360° and a negative angle between 0° and -360° have the same terminal side.

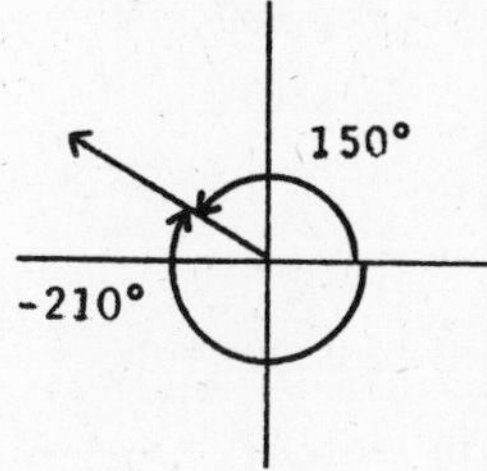

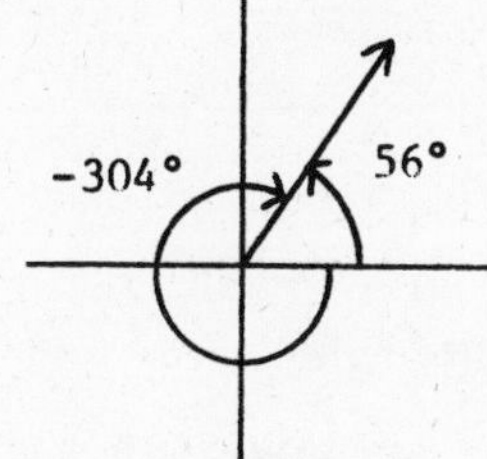

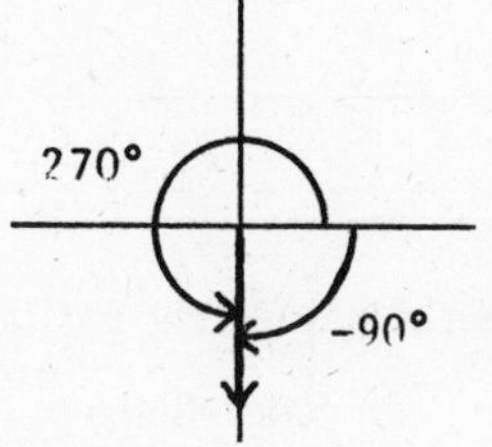

A positive angle between 0° and 360° and a negative angle between 0° and -360° have the same terminal side <u>if the sum of their absolute values is 360°</u>. That is:

$$210° + 150° = 360°$$
$$304° + 56° = 360°$$
$$270° + 90° = 360°$$

Name the <u>positive angle</u> between 0° and 360° which has the same terminal side as each of these negative angles:

(a) -50° __________ (b) -130° __________ (c) -180° __________ (d) -320° __________

Answers: a) 310° b) 230° c) 180° d) 40°

164. In the figure on the right, the 200° angle and the -160° angle have the same terminal side. Therefore, they have the same reference angle and trig ratios.

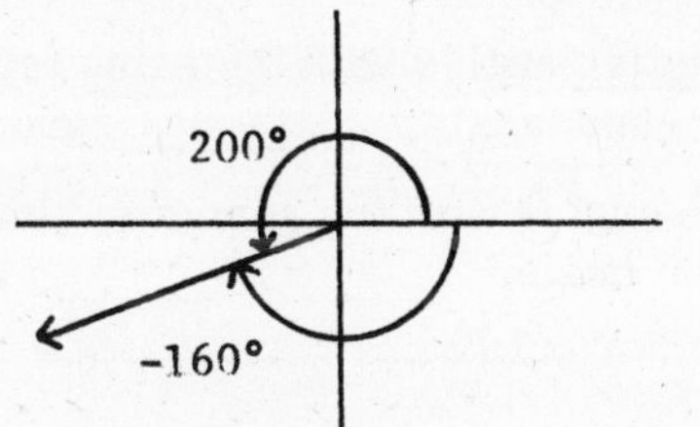

Name a <u>positive angle</u> between 0° and 360° which has <u>the same trig ratios</u> as each of the following angles:

(a) -80° __________ (b) -290° __________ (c) -270° __________ (d) -360° __________

Answers: a) 280° b) 70° c) 90° d) 0°

165. Since a -90° angle has the same trig ratios as a 270° angle:

sin(-90°) = sin 270° = -1

(a) cos(-90°) = ________ = ________

(b) tan(-90°) = ________ = ________

---

166. Since a -180° angle has the same trig ratios as a 180° angle:

sin(-180°) = sin 180° = 0

(a) cos(-180°) = ________ = ________

(b) tan(-180°) = ________ = ________

a) cos 270° = 0

b) tan 270° = -----

---

167. Since a -270° angle has the same trig ratios as a 90° angle:

sin(-270°) = sin 90° = +1

(a) cos(-270°) = ________ = ________

(b) tan(-270°) = ________ = ________

a) cos 180° = -1

b) tan 180° = 0

---

168. Since a -360° angle has the same trig ratios as a 0° angle:

(a) sin(-360°) = _____ (b) cos(-360°) = _____ (c) tan(-360°) = _____

a) cos 90° = 0

b) tan 90° = -----

---

169. Any negative angle between 0° and -360° has the same terminal side and trig ratios that many negative angles beyond -360° have. For example, a -100° angle has the same terminal side and trig ratios as each of the following:

-460° = (-100°) + (-360°)
-820° = (-100°) + (-720°)

Any negative angle between 0° and -360° also has the same terminal side and trig ratios that various positive angles have. For example, a -100° angle has the same terminal side and trig ratios as each of the following:

+260°
+620° = 260° + 360°
+980° = 260° + 720°

(a) Name two other negative angles with the same terminal side and trig ratios as a -200° angle. ________ and ________

(b) Name three positive angles with the same terminal side and trig ratios as a -200° angle. _______, _______, and _______

a) 0

b) +1

c) 0

---

a) -560° and -920°

b) 160°, 520°, 880°

170. Any positive angle between 0° and 360° has the same trig ratios that other positive angles greater than 360° have. For example, a 300° angle has the same trig ratios as each of these:

$$660° = 300° + 360°$$
$$1020° = 300° + 720°$$

Any positive angle between 0° and 360° has the same trig ratios that many negative angles have. For example, a 300° angle has the same trig ratios as each of these:

$$-60°$$
$$-420° = (-60°) + (-360°)$$
$$-780° = (-60°) + (-720°)$$

(a) Name two other positive angles with the same trig ratios as a 70° angle.

__________ and __________

(b) Name three negative angles with the same trig ratios as a 70° angle.

__________, __________, and __________

---

171. Which of the following angles have the same terminal side? __________

(a) 150° (b) -150° (c) -210° (d) 410°

Answer to 170: a) 430°, 790° b) -290°, -650°, -1010°

---

172. Which of the following angles have the same reference angle? __________

(a) -50° (b) 50° (c) 310° (d) -410°

Answer to 171: Only (a) and (c)

---

173. Which of the following angles have the same trig ratios? __________

(a) 90° (b) 450° (c) -90° (d) -270°

Answer to 172: (a), (c), and (d)

Answer to 173: (a), (b), and (d)

SELF-TEST 9 (Frames 154-173)

List the reference angle for:

1. -317° α = ________ 2. -791° α = ________

Find the numerical values of each of the following:

3. sin(-210°) = ________ 4. cos(-100°) = ________ 5. tan(-720°) = ________

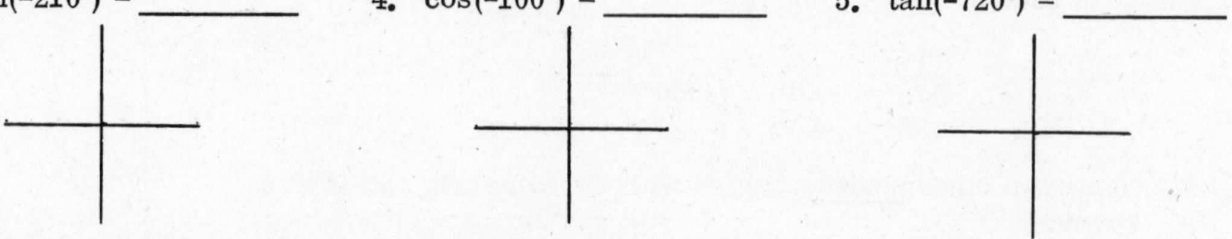

6. For the angle -140°, list three positive angles which have the same terminal side:

________, ________, ________

7. For the angle 330°, list three negative angles which have the same terminal side:

________, ________, ________

ANSWERS:

1. 43° (Q1)
2. 71° (Q4)
3. +0.5000
4. -0.1736
5. 0
6. 220°, 580°, 940°
7. -30°, -390°, -750°

3-10 OTHER CYCLES OF THE SINE-WAVE GRAPH

In an earlier section, we graphed $y = \sin \theta$ for values of θ between 0° and 360°. This part of the graph is called the "basic cycle" of the sine wave. In this section, we will extend the graph of this equation to the right (for angles greater than 360°) and to the left (for negative angles). We will show that the sine-wave graph repeats itself in cycles or periods.

174. To graph $\boxed{y = \sin \theta}$ for values of θ greater than 360°, we must make up a table of values for "θ" and "$\sin \theta$" for values of θ greater than 360°. For example:

(a) If θ = 420°, $\sin \theta = \sin 60°$ (Q1) = ________

(b) If θ = 520°, $\sin \theta = \sin 20°$ (Q2) = ________

(c) If θ = 570°, $\sin \theta = \sin 30°$ (Q3) = ________

(d) If θ = 710°, $\sin \theta = \sin 10°$ (Q4) = ________

a) +0.8660
b) +0.3420
c) -0.5000
d) -0.1736

175. The following pairs of values of θ and sin θ are especially useful when graphing the sine wave.

(a) If $\theta = 450°$, $\sin\theta = \sin\ 90° =$ ________

(b) If $\theta = 540°$, $\sin\theta = \sin 180° =$ ________

(c) If $\theta = 630°$, $\sin\theta = \sin 270° =$ ________

(d) If $\theta = 720°$, $\sin\theta = \sin 360° =$ ________

a) +1
b) 0
c) -1
d) 0

---

176. Here is a table of pairs of values which can be used to graph the sine wave $y = \sin\theta$ from 360° to 720°.

θ	sin θ	θ	sin θ	θ	sin θ	θ	sin θ
360°	0.0000	450°	+1.0000	540°	0.0000	630°	-1.0000
390°	+0.5000	480°	+0.8660	570°	-0.5000	660°	-0.8660
420°	+0.8660	510°	+0.5000	600°	-0.8660	690°	-0.5000
450°	+1.0000	540°	0.0000	630°	-1.0000	720°	0.0000

Plot these values on the graph below and draw a smooth curve through them to extend the graph of $y = \sin\theta$ to 720°.

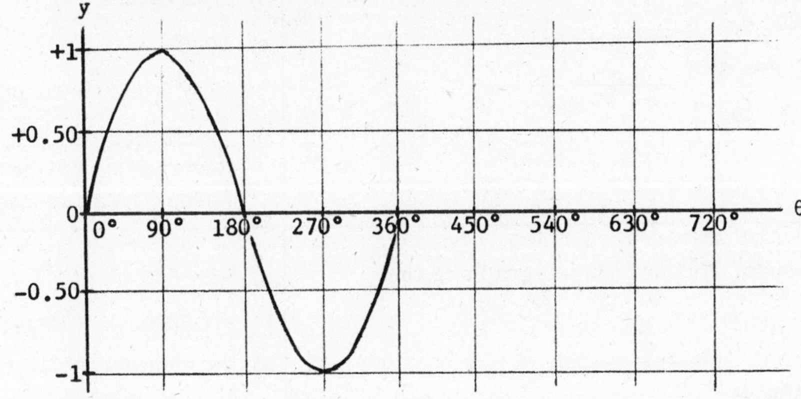

See the next frame.

---

177. Here is the completed graph of $y = \sin\theta$:

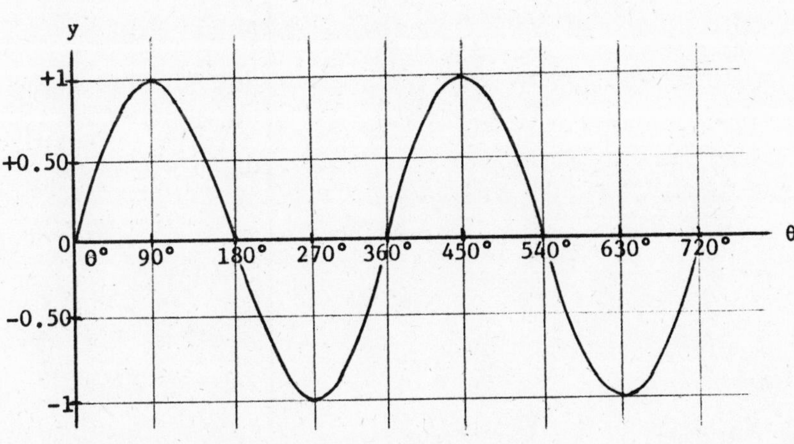

(Continued on following page.)

177. (Continued)

The new section of the graph, lying between 360° and 720°, is a duplicate of the section lying between 0° and 360°.

The complete "loop" lying between 0° and 360° is called a cycle (or period) of the sine-wave graph.

(a) How many degrees are there in 1 cycle (or 1 period) of the sine-wave graph? ________

(b) How many cycles (or periods) are shown in the graph on the preceding page? ________

(c) Since angles greater than 720° have sines, the graph can be extended to the right. The third cycle of the graph would begin at 720° and end at ______°.

(d) How far could the graph be extended to the right?

____

Answers:

a) 360°

b) Two cycles

c) 1080°

(720° + 360° = 1080°)

d) Indefinitely far. It never ends, but just keeps repeating itself.

---

178. To graph $y = \sin\theta$ for negative values of θ, we must make up a table of values for "θ" and "$\sin\theta$" for negative values of θ. For example:

(a) If $\theta = -60°$, $\sin\theta = \sin 60°$ (Q4) = ________

(b) If $\theta = -150°$, $\sin\theta = \sin 30°$ (Q3) = ________

(c) If $\theta = -210°$, $\sin\theta = \sin 30°$ (Q2) = ________

Answers:

a) −0.8660

b) −0.5000

c) +0.5000

---

179. The following pairs of values are especially useful when graphing the sine-wave.

(a) If $\theta = -90°$, $\sin\theta = \sin 270°$ = ________

(b) If $\theta = -180°$, $\sin\theta = \sin 180°$ = ________

(c) If $\theta = -270°$, $\sin\theta = \sin 90°$ = ________

(d) If $\theta = -360°$, $\sin\theta = \sin 0°$ = ________

Answers:

a) −1

b) 0

c) +1

d) 0

180. Here is a table of pairs of values which can be used to graph the sine wave from 0° to -360°.

θ	sin θ	θ	sin θ	θ	sin θ	θ	sin θ
0°	0.0000	-90°	-1.0000	-180°	0.0000	-270°	+1.0000
-30°	-0.5000	-120°	-0.8660	-210°	+0.5000	-300°	+0.8660
-60°	-0.8660	-150°	-0.5000	-240°	+0.8660	-330°	+0.5000
-90°	-1.0000	-180°	0.0000	-270°	+1.0000	-360°	0.0000

Plot these points on the graph below and draw the curve. This will extend the sine-wave graph (y = sin θ) to the left of 0° for negative angles.

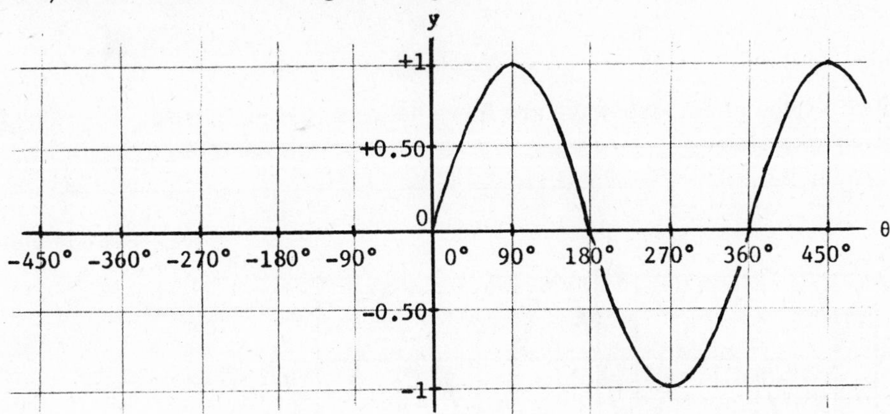

Answer to Frame 180: See the next frame.

181. Here is the completed graph of y = sin θ:

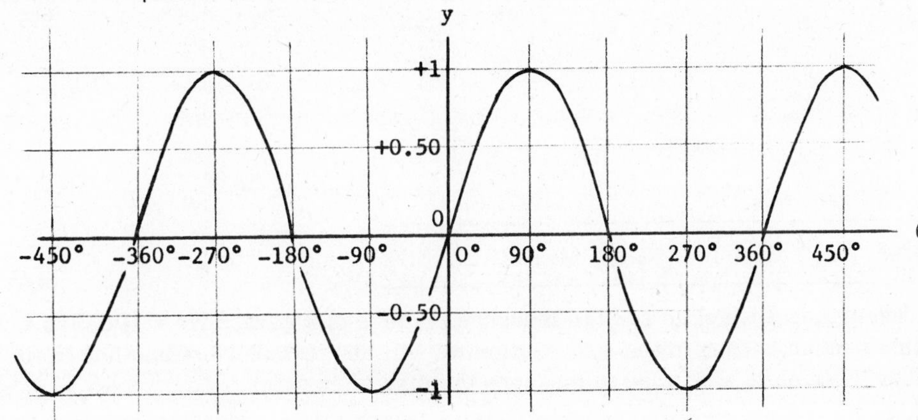

(a) How many cycles of the sine-wave graph lie between -360° and +360? ____________

(b) As you know, the graph can be extended indefinitely far to the right. How far can the graph be extended to the left? ____________

Answer to Frame 181: a) Two cycles b) It can be extended indefinitely far to the left.

SELF-TEST 10 (Frames 174-181)

1. On the axes at the right, sketch the graph of:

$y = \sin \theta$

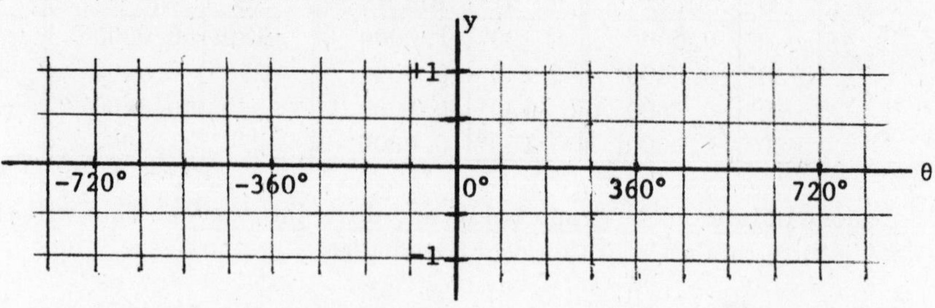

2. How many complete cycles of the sine-wave are there between $\theta = -1080°$ and $\theta = +1080°$?

____

ANSWERS:

1.

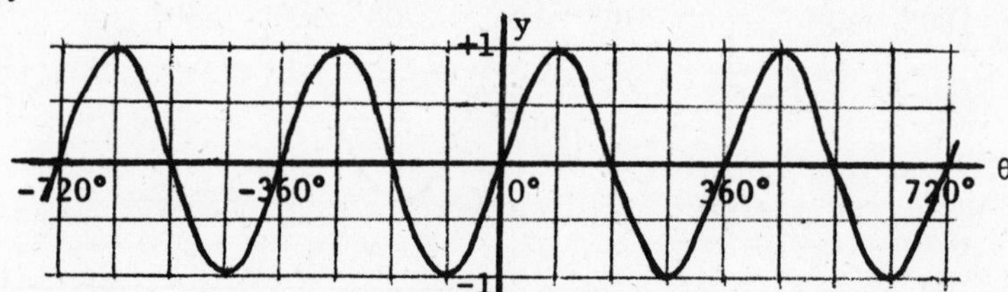

2. Six complete cycles.

3-11 DEFINITION OF A "DEGREE" IN ANGLE MEASUREMENT

A "degree" is a basic measuring unit which is used to measure the size of angles. We will define what is meant by a "degree" in this section. Then in the next section we will see that there is another basic measuring unit, the "radian", which is also used to measure the size of angles.

182. A "degree" is defined in terms of an arc-length of a circle. Therefore, before giving its definition, we must briefly review some facts about circles.

A central angle of a circle is an angle formed by two radii. Angle A0C, for example, is a central angle of the circle on the right.

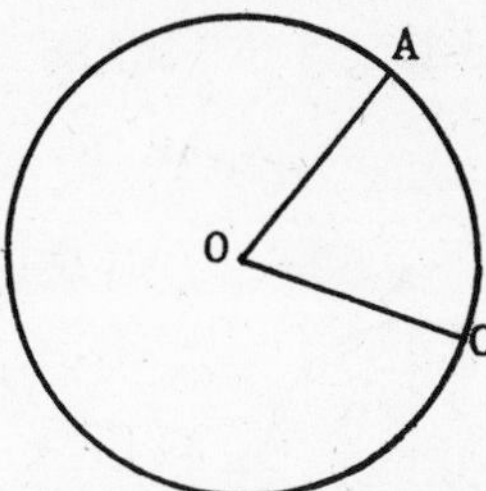

An arc of a circle is the curved line between two points on the circle. For example, the curved line between A and C on the circle above is an arc. We call it "arc AC" or $\widehat{AC}$. (Note: The curved line over AC means "arc".)

In the circle on the right:

(a) Name a central angle. ________

(b) Name an arc. ________

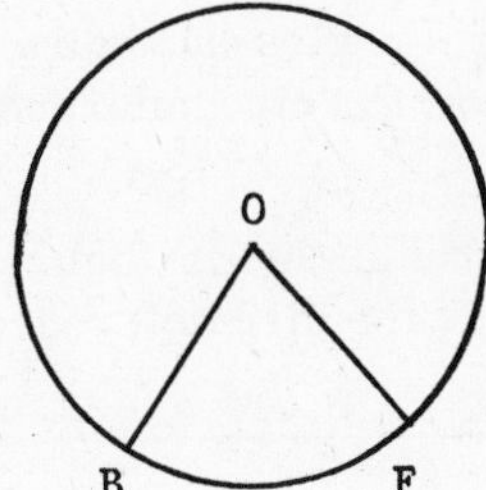

---

183. Any central angle of a circle cuts off an arc on the circle. For example:

There are four central angles in the circle on the right. They are angles A0B, B0C, C0D, and A0D.

Name the arc cut off by:

(a) Angle A0B ________

(b) Angle C0D ________

(c) Angle A0D ________

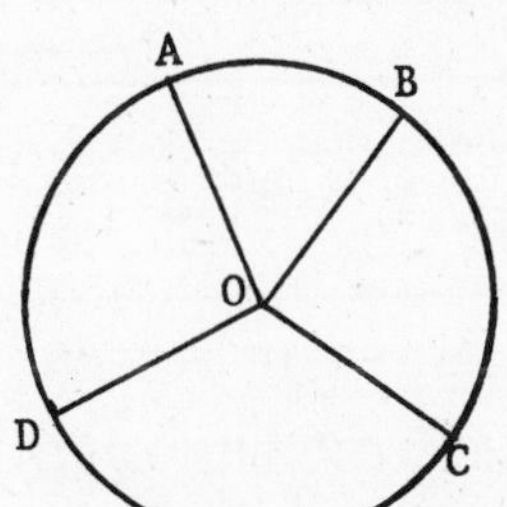

a) Angle B0F
b) arc BF or $\widehat{BF}$

---

184. In any given circle, equal central angles cut off arcs of the same length.

There are three equal central angles in the circle on the right. Name the three arcs which have the same length.

________, ________, ________

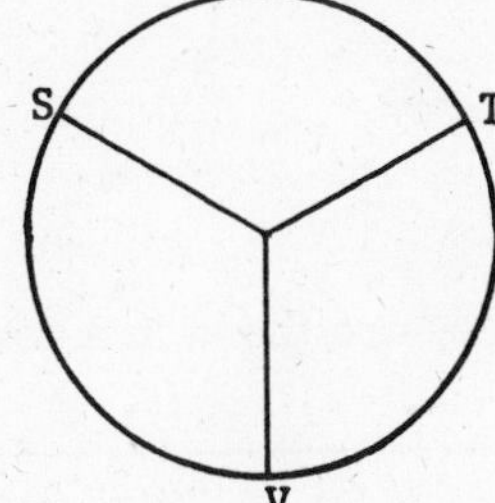

a) $\widehat{AB}$
b) $\widehat{CD}$
c) $\widehat{AD}$

$\widehat{ST}$, $\widehat{TV}$, $\widehat{SV}$

185. The circumference of a circle is the "distance around it". The length of an arc (or arc-length) can be compared to the length of the circumference of a circle. For example, since there are three equal central angles in Figure 1 below, each arc-length is $\frac{1}{3}$ of the circumference of the circle.

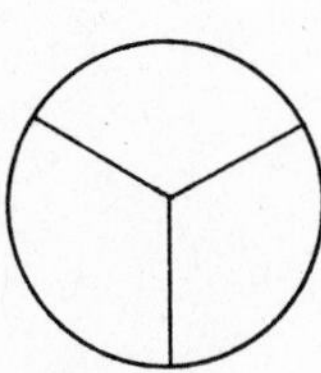

Figure 1

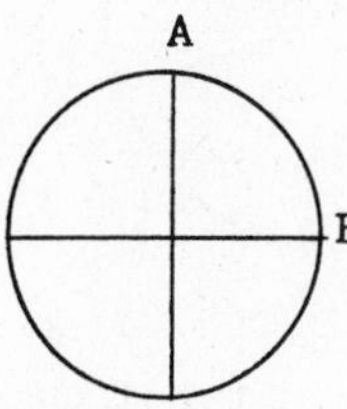

Figure 2

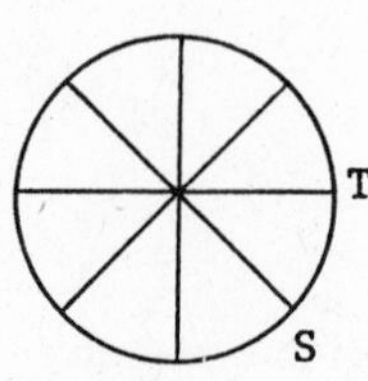

Figure 3

(a) There are four equal central angles in Figure 2. Therefore, $\overset{\frown}{AB}$ is what fractional part of the circumference of the circle? ______

(b) There are eight equal central angles in Figure 3. Therefore, $\overset{\frown}{ST}$ is what fractional part of the circumference of the circle? ______

---

186. (a) If a circle is divided into 50 equal central angles, each arc-length is what fractional part of its circumference? ______

(b) If a circle is divided into 360 equal central angles, each arc-length is what fractional part of its circumference? ______

Answers to 185: a) $\frac{1}{4}$ b) $\frac{1}{8}$

---

187. "One degree" can be defined in terms of a central angle of a circle which cuts off a definite arc-length.

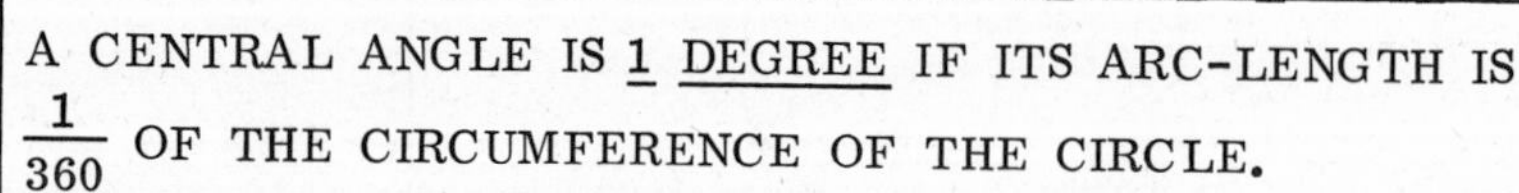

A CENTRAL ANGLE IS 1 DEGREE IF ITS ARC-LENGTH IS $\frac{1}{360}$ OF THE CIRCUMFERENCE OF THE CIRCLE.

The central angle on the right is 1° because its arc-length is $\frac{1}{360}$ of the circumference of the circle.

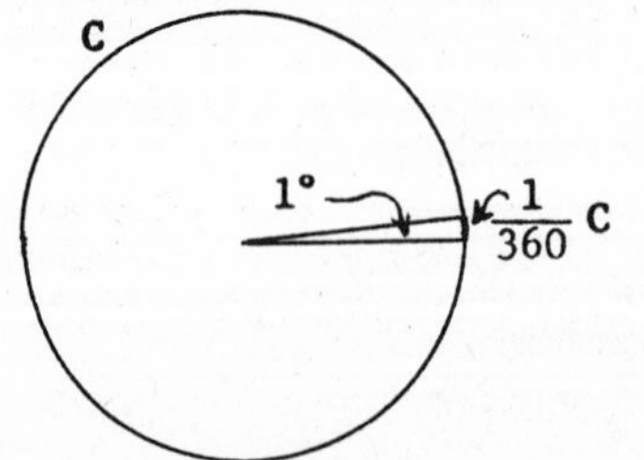

(a) How many central angles of 1° could be drawn in the circle above? ______

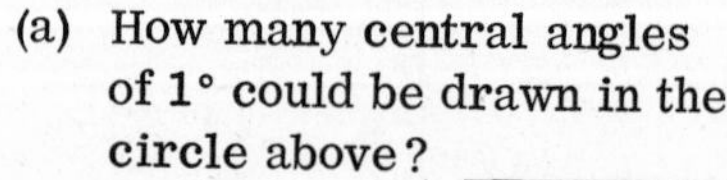

(b) Therefore, how many degrees are there in a completed circle? ______

Answers to 186: a) $\frac{1}{50}$ b) $\frac{1}{360}$

Answers to 187: a) 360 b) 360°

188. In any given circle, central angles are equal if their arc-lengths are equal.

The four arc-lengths ($\overset{\frown}{RS}$, $\overset{\frown}{SM}$, $\overset{\frown}{MT}$, and $\overset{\frown}{TR}$) are equal in the circle on the right. Each is $\frac{1}{4}$ or $\frac{90}{360}$ of the circumference of the circle.

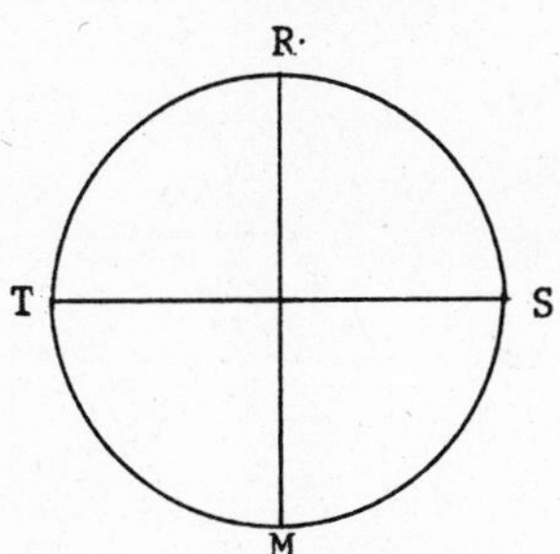

Since a complete circle contains 360°, each of the four equal central angles contains _______°.

90°

189. Since the three arc-lengths are equal in the circle on the right, the three central angles are equal. Each arc-length is $\frac{1}{3}$ or $\frac{120}{360}$ of the circumference of the circle.

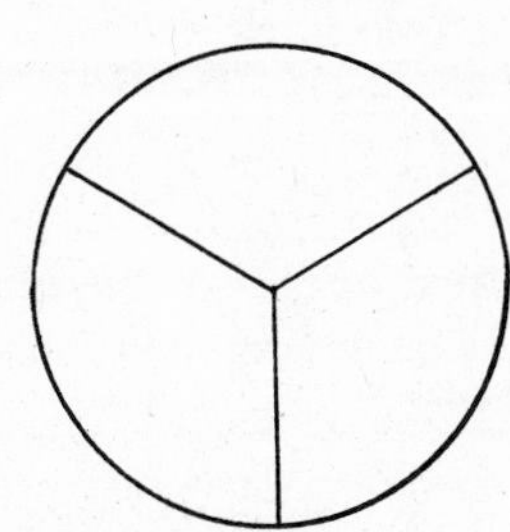

How many degrees are there in each central angle? _______°

120°

190. Each arc-length in the circle on the right is $\frac{1}{12}$ or $\frac{30}{360}$ of the circumference. How many degrees are there in each of the central angles?

_______°

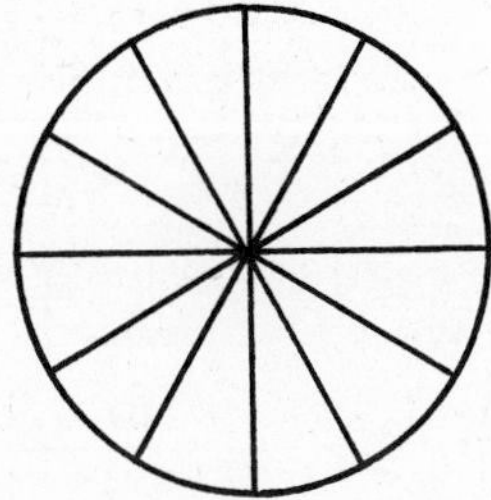

30°

191. (a) If the arc of a central angle is $\frac{1}{2}$ of the circumference of the circle, the central angle is _______°.

(b) If the arc of a central angle is $\frac{1}{6}$ of the circumference of the circle, the central angle is _______°.

(c) If the arc of a central angle is $\frac{1}{360}$ of the circumference of the circle, the central angle is _______°.

---

a) 180°
b) 60°
c) 1°

192. In order to make the definition of a "degree" meaningful for standard-position angles, we can define 1° as a standard-position angle, as we have done on the right. Notice these points:

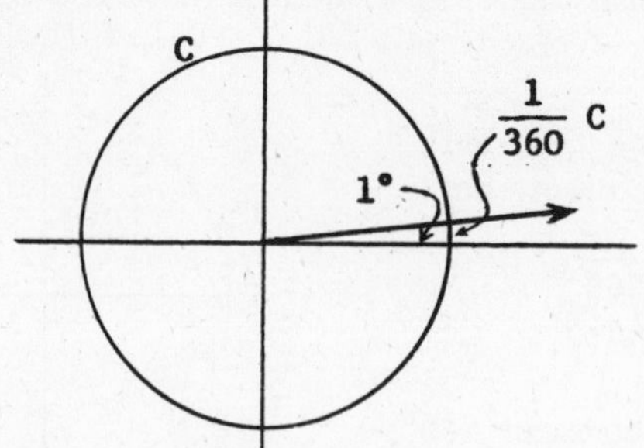

(1) A circle whose center is at the origin has been included on the coordinate system.

(2) The arc-length of the 1° angle is $\frac{1}{360}$ of the circumference of this circle.

(a) If the arc-length of an angle is $\frac{7}{360}$ of the circumference of the circle, the angle is _______°.

(b) If the arc-length of an angle is $\frac{2}{3}$ or $\frac{240}{360}$ of the circumference of the circle, the angle is _______°.

193.

a) 7°

b) 240°

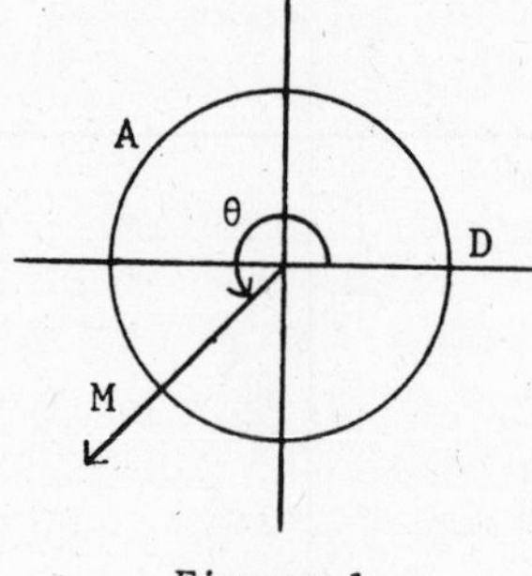

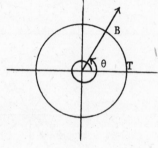

Figure 1

Figure 2

(a) In Figure 1 above, the arc-length $\widehat{DAM}$ of angle θ is $\frac{5}{8}$ or $\frac{225}{360}$ of the circumference of the circle. Therefore, θ = _________°.

(b) In Figure 2 above, angle θ is greater than 360°. Its arc-length is the circumference (one complete revolution) plus $\widehat{TB}$. Therefore, its arc-length is greater than the circumference. If its arc-length is $\frac{7}{6}$ or $\frac{420}{360}$ of the circumference of the circle, θ = _________°.

---

194. In the figure on the right, $\widehat{FK}$ is $\frac{1}{3}$ of the circumference of the circle.

How large is angle θ? _________

a) 225°

b) 420°

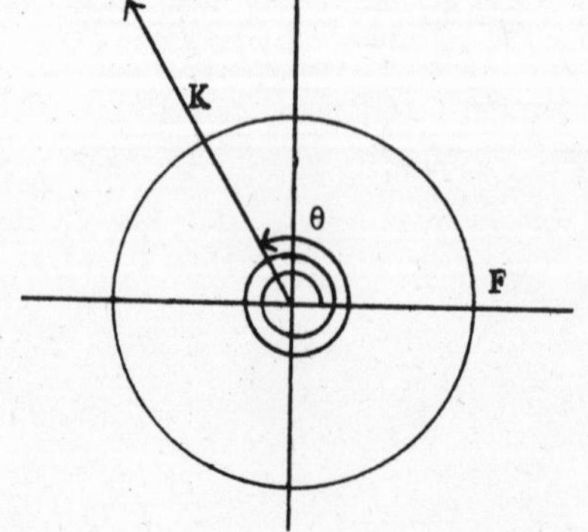

195. Negative angles are also measured in terms of their arc-lengths. They are negative because they generated in a clockwise direction. Here are two examples:

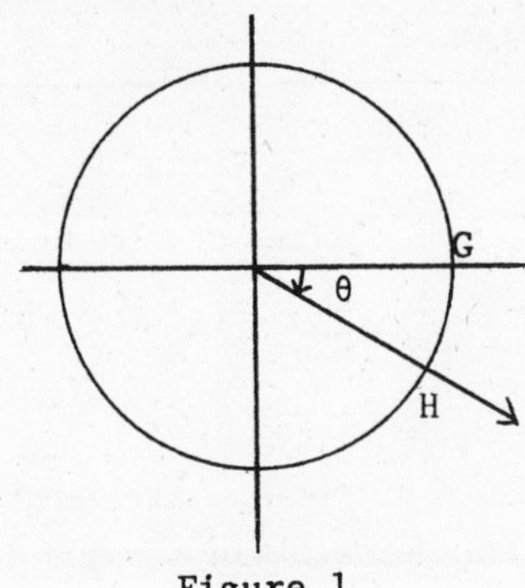

Figure 1

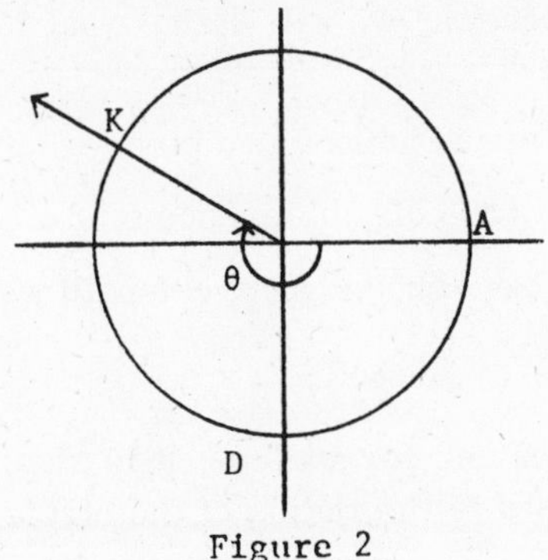

Figure 2

(a) In Figure 1, $\widehat{GH}$ is $\frac{1}{12}$ of the circumference of the circle. θ = ________°

(b) In Figure 2, $\widehat{ADK}$ is $\frac{7}{12}$ of the circumference of the circle. θ = ________°

840°

(2 revolutions + 120°)

Answer to Frame 195: a) -30° b) -210°

SELF-TEST 11 (Frames 182-195)

1. An angle is one degree in size if, when placed as the central angle of a circle, it cuts off an arc-length which is what fractional part of the circumference of the circle? ________

2. What is the size of an angle, in degrees, which cuts off an arc-length which is $\frac{1}{20}$ of the circumference of the circle? ________

3. If the circumference of a circle is divided into nine equal arcs, what is the size of the central angle of each arc? ________

4. If the arc cut off by a central angle is $\frac{1}{4}$ of the circumference of the circle, what is the size of the central angle? ________

5. On the coordinate system at the right, θ is a standard-position angle. Also, $\widehat{KMN}$ is $\frac{2}{5}$ of the circumference of the circle. Find the size of angle θ. θ = ________

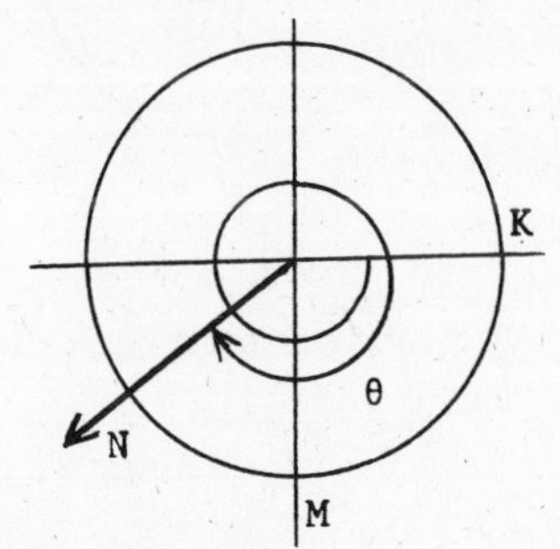

ANSWERS: 1. $\frac{1}{360}$ 2. 18° 3. 40° 4. 90° 5. θ = -504°

3-12 DEFINITION OF A "RADIAN" IN ANGLE MEASUREMENT

Though the "degree" is the most commonly used measuring unit for angles, angles are sometimes measured with a unit called a "radian". In this section, we will define what is meant by a "radian".

---

196. The degree is the most commonly used unit of angular measurement. When defining one degree, the basic arc-length chosen was $\frac{1}{360}$ of the circumference of a circle. This choice of a basic arc-length was purely arbitrary. We could have chosen an arc-length which is $\frac{1}{10}$, $\frac{1}{20}$, $\frac{1}{36}$, etc., of the circumference of a circle.

A radian is an alternate measuring unit for angles. It is also defined in terms of arc-length of a circle. However, the basic arc-length used is the radius of the circle. The definition of a radian is:

AN ANGLE OF 1 RADIAN IS A CENTRAL ANGLE WHOSE ARC-LENGTH EQUALS THE RADIUS OF THE CIRCLE. Note: The word "radian" is related to the word "radius".

A diagram of an angle of 1 radian is shown at the right. Angle $\theta = 1$ radian.

Note: (1) The radius of the circle is "r".

(2) The length of $\widehat{AB}$ is "r".

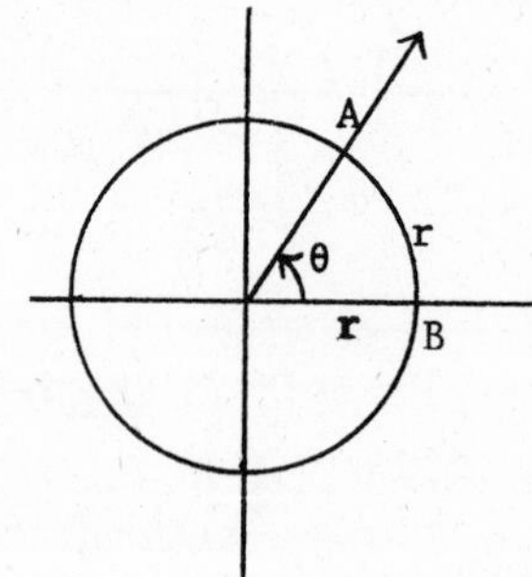

197.

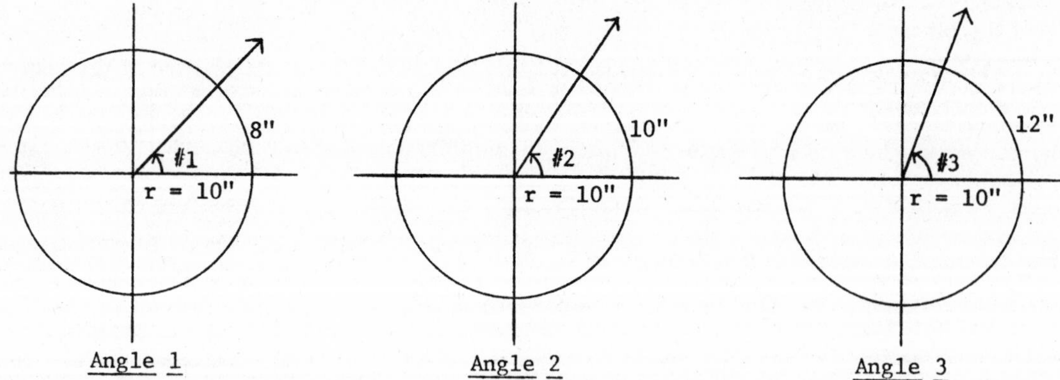

Of the three standard-position angles above, only Angle #2 contains exactly 1 radian. Why? ______________________

Because it is the only angle whose arc-length (10") equals the length of the radius (10").

198. To determine the number of radians in a central angle, we divide its arc-length by the radius of the circle, as shown in this equation:

$$\boxed{\theta \text{ (in radians)} = \frac{\text{arc-length}}{\text{radius}}}$$

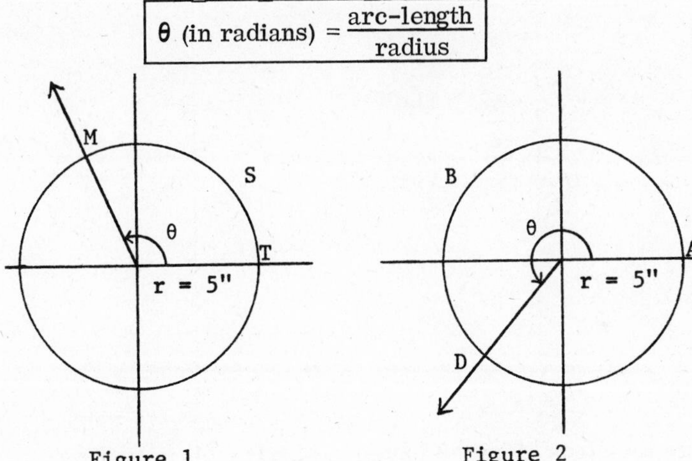

In each circle above, the radius is 5".

(a) In Figure 1, $\overparen{TSM}$ is 10". Therefore, θ = _____ radians.

(b) In Figure 2, $\overparen{ABD}$ is 20". Therefore, θ = _____ radians.

---

a) $\theta = 2$ radians (from $\frac{10''}{5''} = 2$)

b) $\theta = 4$ radians (from $\frac{20''}{5''} = 4$)

---

199. The number of radians in an angle is not always a whole number. Here are two examples:

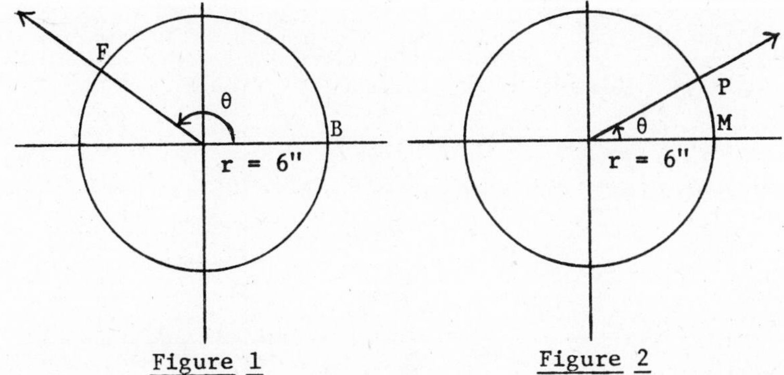

Figure 1 Figure 2

In each circle above, the radius is 6".

(a) In Figure 1, $\overparen{BF} = 15''$. Therefore, θ = _____ radians.

(b) In Figure 2, $\overparen{MP} = 3''$. Therefore, θ = _____ radians.

---

a) $\theta = 2.5$ radians (from $\frac{15''}{6''} = 2.5$)

b) $\theta = 0.5$ radians (from $\frac{3''}{6''} = 0.5$)

200. Angles greater than 360° can also be measured in radians. Here are two examples. In both figures, the radius of the circle is 5.00".

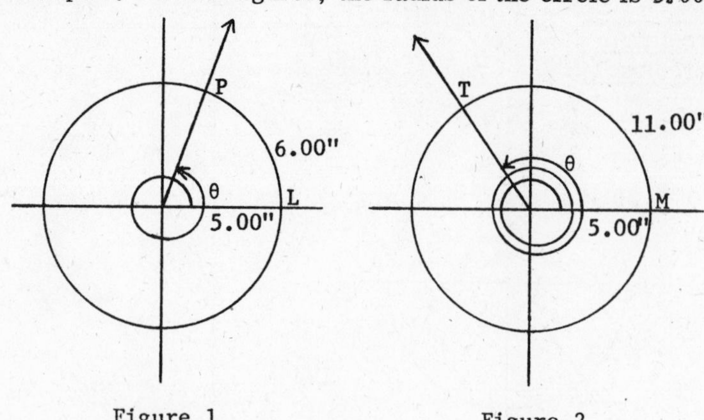

Figure 1 Figure 2

In Figure 1, the arc-length of θ equals the circumference (one revolution) plus $\widehat{LP}$.

Since $C = 2\pi r = 2(3.14)(5.00")$ or 31.4" and $\widehat{LP} = 6.00"$, the arc-length of $\theta = 31.4" + 6.00" = 37.4"$.

Therefore, $\theta \text{ (in radians)} = \dfrac{\text{arc-length}}{\text{radius}} = \dfrac{37.4"}{5.00"} = 7.48$ radians

In Figure 2, C is also 31.4" and $\widehat{MT} = 11.0"$.

(a) What is the arc-length of θ? ________ inches

(b) How many radians are there in θ? ________ radians

a) 73.8"
(2 revolutions + $\widehat{MT}$)

b) $\theta = \dfrac{73.8"}{5.00"}$
$= 14.8$ radians

---

201. Just as we can have negative angles when measuring with degrees, we can have negative angles when measuring with radians. Here is an example:

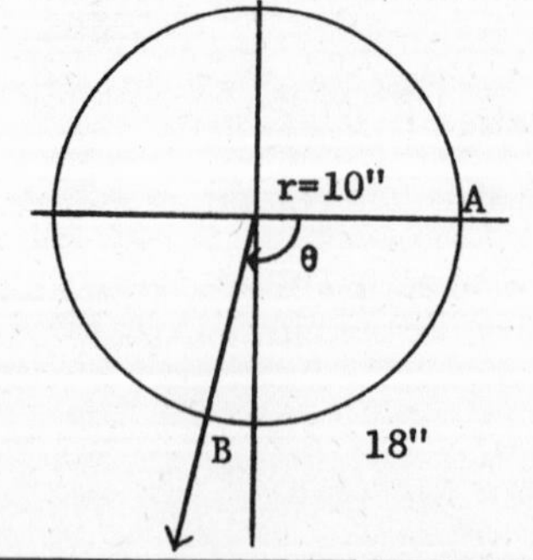

The radius of the circle is 10".
$\widehat{AB} = 18"$.

Angle θ contains ________ radians.

-1.8 radians

202. (a) To determine the number of degrees in an angle, we compare its arc-length to $\frac{1}{360}$ of the _________________ of the circle.

(b) To determine the number of radians in an angle, we compare its arc-length to the _______________ of the circle.

Answer to Frame 202: a) circumference b) radius

SELF-TEST 12 (Frames 196-202)

1. In a circle of radius 8.00", a central angle θ cuts off an arc-length of 2.60" on the circumference. The size of angle θ is __________ radian.

2. In a circle of radius 30", a central angle cuts off an arc-length equal to the diameter of the circle. The size of the central angle is _______ radians.

3. In the diagram at the right, the radius of the circle is 500 centimeters, and the length of $\overset{\frown}{EFG}$ is 1,860 centimeters. The size of angle θ is _______ radians.

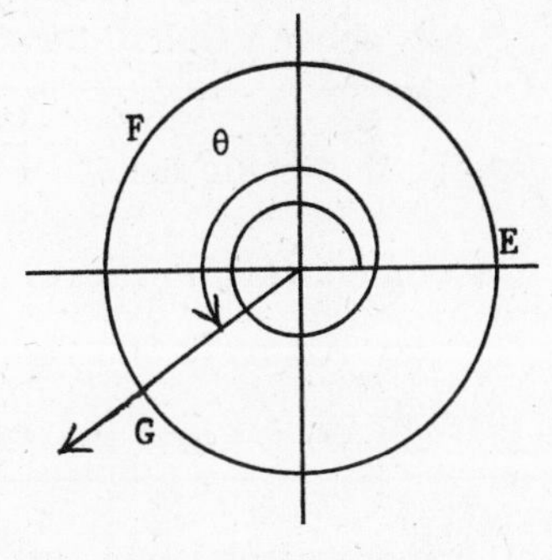

ANSWERS: 1. 0.325 radian 2. 2 radians 3. 10 radians

3-13 THE RELATIONSHIP BETWEEN DEGREES AND RADIANS

Most students are more familiar with angles measured in degrees than with angles measured in radians. The major purpose of this section is to show the relationship between the degree and the radian scales. This relationship will be used to convert radians to degrees, and degrees to radians.

203. In this frame, we will demonstrate the following basic fact: **1 radian = 57.3°**

We will use the 1-radian angle on the right to do so:

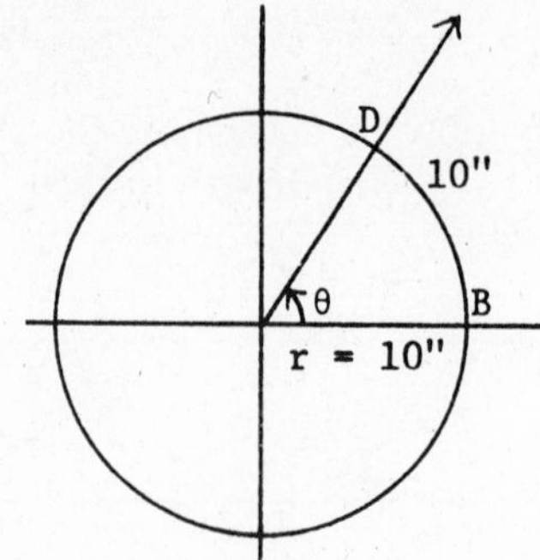

Step 1: The circumference of the circle is:

$$2\pi r = 2(3.14)(10'') = 62.8''$$

Step 2: $\overset{\frown}{BD}$ is a fractional part of the circumference.

It is $\frac{10''}{62.8''}$ or 0.1592 of the circumference.

Step 3: Since there are 360° in a complete circle (the total circumference):

$$\theta = (0.1592)(360°) \text{ or } 57.3°$$

Therefore: 1 radian = 57.3°

Note: By using a more exact value of π, mathematicians have found that:

1 radian = 57.2957795131...°

In actual problems involving slide rule calculations, however, it is satisfactory to round the above value to three significant digits, or 57.3°.

---

204. Using this basic fact: **1 radian = 57.3°**

we can convert any angle from radians to degrees.

(a) An angle of 3 radians = ________°.

(b) An angle of 5.1 radians = ________°.

Go to the next frame.

---

205. Using the same basic fact: **1 radian = 57.3°**

we can convert from degrees to radians by dividing the number of degrees by 57.3°. For example:

An angle of 57.3° = $\frac{57.3°}{57.3°}$ = 1 **radian.**

(a) An angle of 114.6° = $\frac{114.6°}{57.3°}$ = ___ radians.

(b) An angle of 229.2° = $\frac{229.2°}{57.3°}$ = ___ radians.

a) 171.9° or 172°

b) 292.2° or 292°

---

206. Convert each of the following from degrees to radians. (Report the answers to three significant digits, which is slide rule accuracy.)

(a) 142° = _______ radians (c) 35° = _______ radians

(b) 427° = _______ radians (d) -117° = _______ radians

a) 2 radians

b) 4 radians

---

a) 2.48 radians

b) 7.45 radians

c) 0.611 radian

d) -2.04 radians

207. In order to get a feel for the relationship between "number of degrees" and "number of radians", you can use this basic fact:

1 radian is approximately 60°

Since 1 radian is approximately 60°, approximately how many degrees are contained in:

(a) An angle of 2 radians? ________°

(b) An angle of 4 radians? ________°

(c) An angle of $\frac{1}{2}$ radian ? ________°

---

208. Similarly, we can roughly estimate the number of radians in an angle given in degrees by dividing by 60°. For example:

An angle of 180° ≐ 3 radians (from $\frac{180°}{60°}$)

(a) An angle of 360° ≐ _____ radians.

(b) An angle of 90° ≐ _____ radians.

(c) An angle of -120° ≐ _____ radians.

Answers to Frame 207: a) 120° b) 240° c) 30°

---

209. Using this basic fact: 1 radian ≐ 60°

give an approximate answer for each of the following:

(a) An angle of 720° ≐ __________ radians.

(b) An angle of 5 radians ≐ __________°.

(c) An angle of -90° ≐ __________ radians.

(d) An angle of -2.5 radians ≐ __________°.

Answers to Frame 208: a) 6 radians b) $1\frac{1}{2}$ or 1.5 radians c) -2 radians

---

Answer to Frame 209: a) 12 radians b) 300° c) -1.5 radians d) -150°

---

210. To get a quick, approximate conversion from radians to degrees, or vice-versa, you can use the fact that: 1 radian ≐ 60°

In the tables at the right, we have given some exact conversions from degrees to radians, and vice-versa. Examine the tables so that you become more familiar with the relationship between these two measuring units for angles.

Degrees	Radians
720°	12.57
540°	9.42
360°	6.28
270°	4.71
180°	3.14
90°	1.57
0°	0.00
-90°	-1.57
-180°	-3.14

Radians	Degrees
6	343.8°
5	286.5°
4	229.2°
3	171.9°
2	114.6°
1	57.3°
0	0.0°
-1	-57.3°
-2	-114.6°

SELF-TEST 13 (Frames 203-210)

In making the following conversions, use this fact: 1 radian = 57.3°

Convert to degrees:		Convert to radians:	
1. 0.750 radian	______	4. 106°	______
2. 6.28 radians	______	5. 1°	______
3. -2.81 radians	______	6. -180°	______

ANSWERS:

1. 43°
2. 360°
3. -161°
4. 1.85 radians
5. 0.01745 radian
6. -3.14 radians

3-14 RADIANS EXPRESSED IN TERMS OF "π"

Up to this point, we have always expressed radians in terms of ordinary numbers. In mathematics, however, radians are frequently expressed in terms of "π". Generally, "π" is only used when expressing some specific angles (like 30°, 45°, 60°, 90°, etc.) in radians. We will discuss how angles are expressed in terms of "π" in this section.

211. By dividing by 57.3°, we can convert 360° and 180° to radians:

$360° = \frac{360°}{57.3°} = 6.28$ radians $180° = \frac{180°}{57.3°} = 3.14$ radians

Since $3.14 = \pi$, we can express 360° and 180° in radians as follows:

360° = 6.28 radians or $\underline{2\pi}$ radians

180° = 3.14 radians or ____ radians

Answer: π radians

212. Here are the facts we found in the last frame:

$180° = \pi$ radians
$360° = 2\pi$ radians

We can demonstrate these facts by returning to our original definition of radians:

$$\theta \text{ (in radians)} = \frac{\text{Arc-Length}}{\text{Radius}}$$

Here are angles of 360° and 180° in standard position:

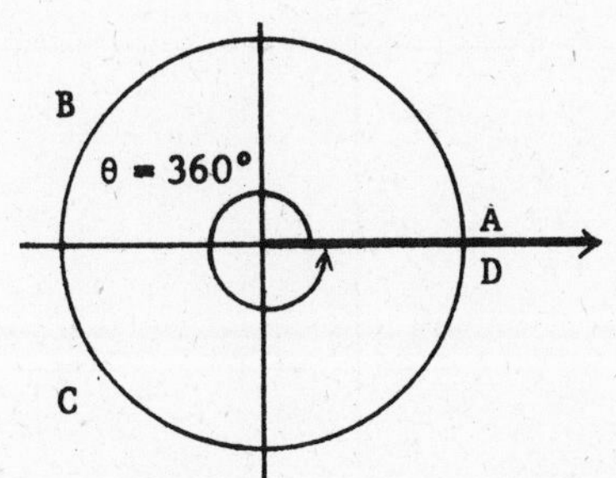

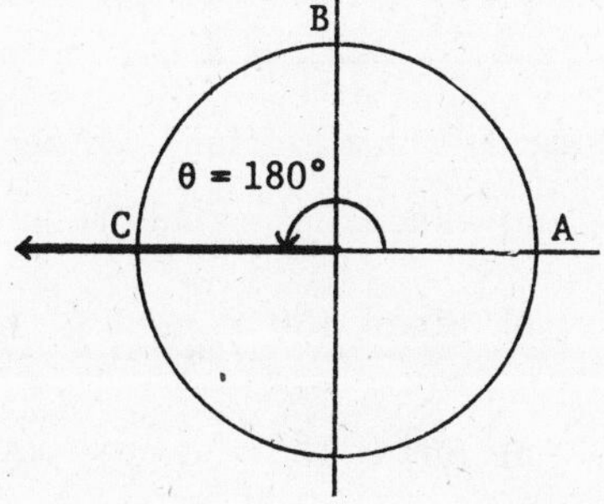

The arc of the 360° angle is $\widehat{ABCD}$, which equals the circumference of the circle (one complete revolution). The circumference = $2\pi r$. Therefore:

$$360° \text{ (in radians)} = \frac{\text{Arc-Length}}{\text{Radius}} = \frac{2\pi r}{r} = 2\pi\left(\frac{r}{r}\right) \text{ or } \underline{2\pi \text{ radians}}$$

The arc of the 180° angle is $\widehat{ABC}$, which equals $\frac{1}{2}$ of the circumference of the circle. Since the circumference is $2\pi r$, $\frac{1}{2}$ the circumference is $\frac{1}{2}(2\pi r)$ or πr. Therefore:

$$180° \text{ (in radians)} = \frac{\text{Arc-Length}}{\text{Radius}} = \frac{\pi r}{r} = \pi\left(\frac{r}{r}\right) \text{ or } \underline{\pi \text{ radians}}$$

Go to the next frame.

---

213. Complete: (a) An angle of π radians = ________°.

(b) An angle of 2π radians = ________°.

a) 180°

b) 360°

---

214. We have seen these two facts: $180° = \pi$ radians
$360° = 2\pi$ radians

To convert from degrees to radians expressed in terms of "π", it is easier to use the first fact:

$180° = \pi$ radians

Complete these conversions:

$$90° = \frac{1}{2}(180°) = \frac{1}{2}(\pi \text{ radians}) = \frac{\pi}{2} \text{ radians.}$$

(a) $60° = \frac{1}{3}(180°) = \frac{1}{3}(\pi \text{ radians}) = \underline{\quad}$ radians.

(b) $45° = \frac{1}{4}(180°) = \frac{1}{4}(\pi \text{ radians}) = \underline{\quad}$ radians.

(c) $30° = \frac{1}{6}(180°) = \frac{1}{6}(\pi \text{ radians}) = \underline{\quad}$ radians.

215. Complete these conversions:

$$120° = \frac{2}{3}(180°) = \frac{2}{3}(\pi \text{ radians}) = \frac{2\pi}{3} \text{ radians.}$$

(a) $135° = \frac{3}{4}(180°) = \frac{3}{4}(\pi \text{ radians}) =$ ____ radians.

(b) $150° = \frac{5}{6}(180°) = \frac{5}{6}(\pi \text{ radians}) =$ ____ radians.

a) $\frac{\pi}{3}$
b) $\frac{\pi}{4}$
c) $\frac{\pi}{6}$

---

216. Using $180° = \pi$ radians, you should be able to complete these:

(a) Since $30° = \frac{1}{6}(180°)$, $30° =$ ______ radians.

(b) Since $150° = \frac{5}{6}(180°)$, $150° =$ ______ radians.

(c) Since $45° = \frac{1}{4}(180°)$, $45° =$ ______ radians.

(d) Since $90° = \frac{1}{2}(180°)$, $90° =$ ______ radians.

a) $\frac{3\pi}{4}$
b) $\frac{5\pi}{6}$

---

217. Do these mentally. Express the radians in terms of "π":

(a) $60° =$ ____ radians (c) $120° =$ ____ radians

(b) $135° =$ ____ radians (d) $-18° =$ ____ radians

a) $\frac{\pi}{6}$ c) $\frac{\pi}{4}$
b) $\frac{5\pi}{6}$ d) $\frac{\pi}{2}$

---

218. To convert from radians expressed in terms of "π" back to degrees, we use the same fact, namely, π radians = 180°.

$$\frac{\pi}{3} \text{ radians} = \frac{1}{3}(\pi \text{ radians}) = \frac{1}{3}(180°) = 60°$$

(a) $\frac{\pi}{4}$ radians $= \frac{1}{4}(\pi \text{ radians}) = \frac{1}{4}(180°) =$ ______

(b) $\frac{2\pi}{3}$ radians $= \frac{2}{3}(\pi \text{ radians}) = \frac{2}{3}(180°) =$ ______

(c) $\frac{5\pi}{6}$ radians $= \frac{5}{6}(\pi \text{ radians}) = \frac{5}{6}(180°) =$ ______

a) $\frac{\pi}{3}$ c) $\frac{2\pi}{3}$
b) $\frac{3\pi}{4}$ d) $-\frac{\pi}{10}$

---

219. You should be able to do these conversions mentally:

(a) $\frac{\pi}{3}$ radians = ______ ° (c) $\frac{\pi}{2}$ radians = ______ °

(b) $\frac{3\pi}{4}$ radians = ______ ° (d) $\frac{5\pi}{6}$ radians = ______ °

a) 45°
b) 120°
c) 150°

---

a) 60° c) 90°
b) 135° d) 150°

220. We use the same fact to convert angles greater than 180° from degrees to radians expressed in terms of "π":

$$270° = \frac{3}{2}(180°) = \frac{3}{2}(\pi \text{ radians}) = \frac{3\pi}{2} \text{ radians.}$$

(a) $450° = \frac{5}{2}(180°) = \frac{5}{2}(\pi \text{ radians}) =$ ____ radians.

(b) $210° = \frac{7}{6}(180°) = \frac{7}{6}(\pi \text{ radians}) =$ ____ radians.

(c) $540° = 3(180°) = 3(\pi \text{ radians}) =$ ____ radians.

Answers: a) $\frac{5\pi}{2}$ b) $\frac{7\pi}{6}$ c) 3π

221. Ordinarily, we are not as interested in converting from degrees to radians expressed in terms of "π" as we are in converting from radians expressed in terms of "π" to degrees. Complete these:

(a) $\frac{5\pi}{3}$ radians $= \frac{5}{3}(\pi \text{ radians}) = \frac{5}{3}(180°) =$ ________

(b) 3π radians $= 3(\pi \text{ radians}) = 3(180°) =$ ________

(c) $\frac{7\pi}{4}$ radians $= \frac{7}{4}(\pi \text{ radians}) = \frac{7}{4}(180°) =$ ________

Answers: a) 300° b) 540° c) 315°

222. Complete these conversions:

(a) 4π radians = ______°

(b) $\frac{13\pi}{6}$ radians = ______°

(c) $-\frac{\pi}{2}$ radians = ______°

(d) $-\frac{4\pi}{3}$ radians = ______°

Answers: a) 720° b) 390° c) −90° d) −240°

223. Complete:

(a) 90° = ____ radians

(b) 270° = ____ radians

(c) $\frac{\pi}{6}$ radians = ______°

(d) $\frac{7\pi}{6}$ radians = ______°

Answers: a) $\frac{\pi}{2}$ b) $\frac{3\pi}{2}$ c) 30° d) 210°

224. Complete:

(a) $\frac{\pi}{4}$ radians = ______°

(b) $\frac{7\pi}{3}$ radians = ______°

(c) −60° = ____ radians

(d) 135° = ____ radians

Answers: a) 45° b) 420° c) $-\frac{\pi}{3}$ d) $\frac{3\pi}{4}$

225. When converting from degrees to radians, we usually express the number of radians in terms of "π" for only some specific angles. Even for these angles, we can express the number of radians as an ordinary number. At the right is a table which shows both ways of expressing "number of radians" for some typical angles:

Degrees	Radians (π)	Radians (Ordinary Numbers)
360°	2π	6.28
270°	$\frac{3\pi}{2}$	4.71
180°	π	3.14
150°	$\frac{5\pi}{6}$	2.62
135°	$\frac{3\pi}{4}$	2.36
120°	$\frac{2\pi}{3}$	2.09
90°	$\frac{\pi}{2}$	1.57
60°	$\frac{\pi}{3}$	1.05
45°	$\frac{\pi}{4}$	0.79
30°	$\frac{\pi}{6}$	0.52
0°	0	0.00

SELF-TEST 14 (Frames 211-225)

In making the following conversions, use this fact: π radians = 180°

Convert to degrees:

1. 3π radians ________
2. $\frac{\pi}{2}$ radians ________
3. $\frac{2\pi}{3}$ radians ________

Convert to radians: (Express each answer in terms of π.)

4. 45° ________
5. 150° ________
6. 270° ________

ANSWERS:

1. 540°
2. 90°
3. 120°
4. $\frac{\pi}{4}$ radians
5. $\frac{5\pi}{6}$ radians
6. $\frac{3\pi}{2}$ radians

3-15 TRIG RATIOS OF ANGLES EXPRESSED IN RADIANS

Sometimes it is necessary to find the trig ratios of angles expressed in radians. Since angles in a trig table are usually given in degrees, it is necessary to convert the angle from radians to degrees before using the table. We will briefly show the method in this section.

226. The symbol ° is used as an abbreviation for the word "degree". When an angle is measured in degrees, this symbol is always used. For example:

$$\sin 2^\circ \text{ means: } \sin 2 \text{ degrees}$$

There is no comparable symbol which is used as an abbreviation for the word "radian". Instead of writing the word "radian", usually just the number is written. For example:

$$\sin 2 \text{ means: } \sin 2 \text{ radians}$$

$$\tan \frac{\pi}{2} \text{ means: } \tan \frac{\pi}{2} \text{ radians}$$

Complete these with either "degrees" or "radians":

(a) $\cos 37^\circ = \cos 37$ ______________

(b) $\tan \pi = \tan \pi$ ______________

(c) $\cos 1.3 = \cos 1.3$ ______________

227. Whenever the symbol ° is not written after the number, you know that the angle is given in radians. To use the trig table, you must convert to degrees first. When doing so, round to the nearest degree.

Complete these, using your trig table:

(a) $\sin 1 = \sin 57^\circ =$ ________ (c) $\tan \frac{\pi}{4} = \tan 45^\circ =$ ________

(b) $\cos \frac{\pi}{2} = \cos 90^\circ =$ ________

Answers to 226:
a) degrees
b) radians
c) radians

228. Complete: (a) $\cos 1.5 = \cos$ ______° = ____________

(b) $\sin 4 = \sin$ ______° = ____________

(c) $\tan \frac{\pi}{6} = \tan$ ______° = ____________

Answers to 227:
a) 0.8387
b) 0.0000
c) 1.0000

229. Complete: (a) $\cos \pi = \cos$ ______° = ____________

(b) $\sin -2 = \sin$ ______° = ____________

(c) $\tan 1.59 = \tan$ ______° = ____________

Answers to 228:
a) $\cos 86^\circ = 0.0698$
b) $\sin 229^\circ = -0.7547$
c) $\tan 30^\circ = 0.5774$

Answers to 229:
a) $\cos 180^\circ = -1.0000$
b) $\sin -115^\circ = -0.9063$
c) $\tan 91^\circ = -57.29$

230. Occasionally you will see a sine wave in which the horizontal axis is scaled in radians rather than in degrees.

(1) The word "radian" will not appear.

(2) The number of radians in an angle will be expressed in terms of "π".

Here is an example of such a graph for $y = \sin \theta$. We have given the angles in degrees below each radian measure.

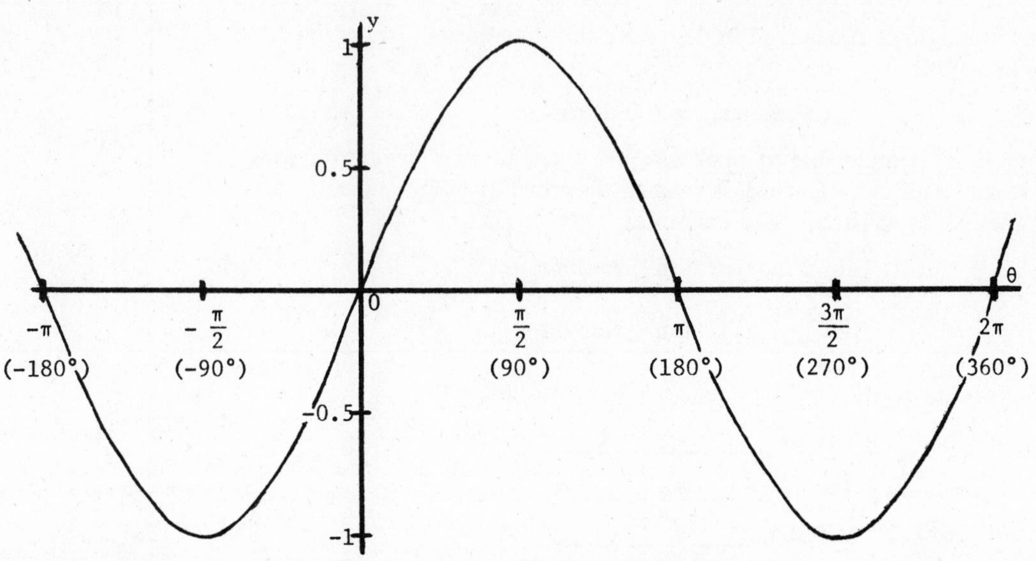

SELF-TEST 15 (Frames 226-230)

Using the trig table, find the numerical value of each of the following:

1. $\sin\frac{\pi}{6} = \sin$ ______° = ________
2. $\cos 2\pi = \cos$ ______° = ________
3. $\tan(-\pi) = \tan$ ______° = ________
4. $\sin 1.17 = \sin$ ______° = ________

ANSWERS:

1. $\sin 30° = 0.5000$
2. $\cos 360° = 1.0000$
3. $\tan(-180°) = 0.0000$
4. $\sin 67° = 0.9205$

FINDING TRIGONOMETRIC RATIOS ON THE SLIDE RULE

THE REMAINING SECTIONS IN THIS CHAPTER ARE DEVOTED TO "FINDING TRIG RATIOS ON THE SLIDE RULE". THESE SECTIONS ARE OPTIONAL. NO ITEMS FROM THEM WILL BE INCLUDED IN THE CHAPTER TEST.

3-16 FINDING SINE RATIOS ON THE SLIDE RULE

The numerical values of the sine, cosine, and tangent of an angle are usually found by referring to a trig-ratio table. However, these ratios can also be obtained from the slide rule. In this section, we will show how the S and A scales can be used to find the sine of any angle.

231. To find the sine of any angle, we use the S (or Sine) scale together with the A scale.

Locate the S scale on your slide rule. Since slide-rule manufacturers differ somewhat in their layout of the trig scales, your S scale may be either on the front or back of the slide, or on the frame. If your S scale is on the back of the slide, remove the slide from the frame and then reinsert it into the frame so that the S scale is on the front.

Though slide rules differ, one common position of the A and S scales is shown on the right.

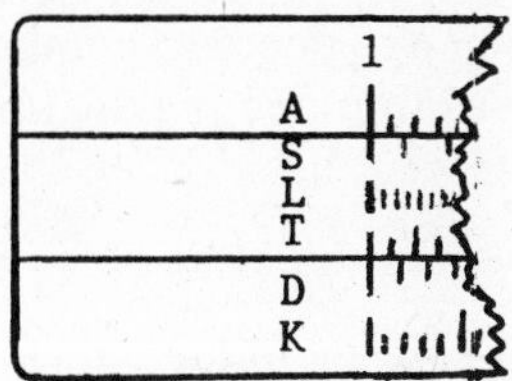

Note: If you have any difficulty in the instruction which follows because of the positioning of your S scale, either consult the manufacturer's instruction booklet or confer with your instructor.

232. Here is a diagram showing the general layout of the S and A scales. A few of the angles on the S scale are labeled:

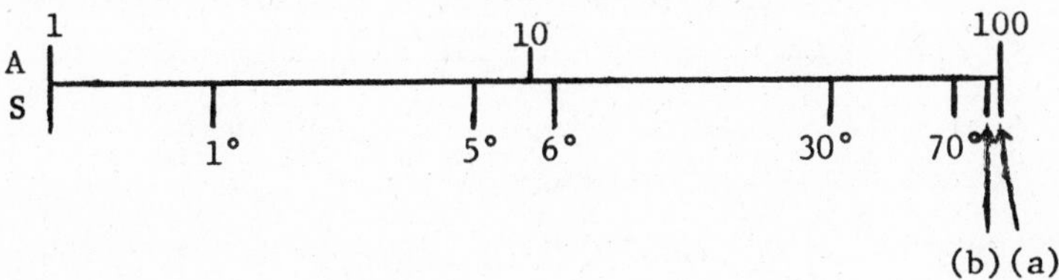

Note these characteristics of the S scale:

1. It is a scale of angles.
2. The angles near the right end are very crowded.
3. The left half of the scale includes angles up to about 6°.
4. The right half of the scale includes angles from about 6° to 90°.

The right end of the S scale, labeled point (a), represents a 90° angle. Point (b) represents what angle? _______

80°

233. Adjust the slide so that the left ends of S and A are together. Having done so, it is easy to find the sine of an angle:

> To find the sine of an angle, find the angle on the S scale and move the hairline to it. Then read the sine of the angle on the A scale under the hairline.

Follow this procedure to find sin 20°:

Step 1: Move the hairline to 20° on S.
Step 2: Read the answer "342" on A. Therefore, sin 20° = "342".
Step 3: Place the decimal point. Then, sin 20° = 0.342.
(The decimal-point placement is discussed in the next frame.)

Find the "sequence of digits" for sin 25°: sin 25° = "________"

---

234. To place the decimal point in the sines of angles from 6° to 90°, we can use the following facts from the trig-ratio table:

sin 6° = 0.1045
sin 90° = 1.0000

It is evident from the values above that sines of angles lying between 6° and 90° are decimal numbers lying between 0.1 and 1.0. This fact can be used to place the decimal point in the "slide rule digits". For example:

From the slide rule, sin 25° = "423". Therefore, sin 25° = 0.423

From the slide rule, sin 60° = "866". Therefore, sin 60° = 0.866

Do these problems on your slide rule:

(a) sin 10° = ________ (d) sin 50° = ________

(b) sin 30° = ________ (e) sin 70° = ________

(c) sin 40° = ________ (f) sin 80° = ________

Answer to 233: "423"

---

235. When sine ratios are read on the A scale, we usually obtain only three significant digits. Find each of these on your slide rule:

(a) sin 18° = ________ (c) sin 7° = ________

(b) sin 45° = ________ (d) sin 67° = ________

Answers to 234: a) 0.174 b) 0.500 c) 0.643 d) 0.766 e) 0.940 f) 0.985

---

236. Do these on your slide rule:

(a) sin $9\frac{1}{2}$° = ________ (b) sin $33\frac{1}{2}$° = ________

Answers to 235: a) 0.309 b) 0.707 c) 0.122 d) 0.920

Answers to 236: a) 0.165 b) 0.552

237. If the sine of an angle is known, the angle can easily be found on the slide rule. For example, if $\sin\varphi = 0.407$, then angle φ is found as follows:

Step 1: Move hairline to "407" on the RIGHT HALF of A.
Step 2: Read angle φ on S under the hairline. $\varphi = 24°$.

Find these angles using your slide rule:

(a) If $\sin A = 0.695$, then $A =$ _______°.

(b) If $\sin B = 0.156$, then $B =$ _______°.

(c) If $\sin C = 0.810$, then $C =$ _______°.

---

238. Up to this point, we have found the sines of angles lying between 6° and 90°. The answers always appeared on the right half of A. For angles less than 6°, the sines will appear on the LEFT HALF of A.

On your slide rule, move the hairline to 3° on S. Read the answer on A under the hairline: $\sin 3° =$ "_________" (sequence of digits)

a) $A = 44°$

b) $B = 9°$

c) $C = 54°$

---

239. To place the decimal point in $\sin 3° =$ "523", examine these values from a trig-ratio table:

$$\sin 1° = 0.0175$$
$$\sin 5° = 0.0872$$

From the above ratios, it is clear that the sine of any angle between 1° and 5° will be a number having one zero immediately to the right of the decimal point. Therefore:

$$\sin 3° = 0.0523$$

(a) $\sin 2° =$ _________ (b) $\sin 4° =$ _________

"523"

---

240. If the sine of an angle is read on the left half of A, the sine ratio is a decimal number which has one zero immediately to the right of the decimal point. For example:

$$\sin 3° = 0.0523$$
$$\sin 5° = 0.0872$$
$$\sin 1° = 0.0175$$

If the sine of an angle is read on the right half of A, it is a decimal number which has a non-zero digit immediately to the right of the decimal point. For example:

$$\sin 27° = 0.454$$
$$\sin 79° = 0.982$$
$$\sin 11° = 0.191$$

Do these on your slide rule:

(a) $\sin 1° =$ _________

(b) $\sin 37° =$ _________

(c) $\sin 5° =$ _________

(d) $\sin 13° =$ _________

(e) $\sin 2\frac{1}{2}° =$ _________

a) 0.0349

b) 0.0698

241. If the numerical value of the sine of an angle is known, we can use the A scale and the S scale to find the size of the angle.

For example, if $\sin \theta = 0.259$, the size of angle θ can be found as follows:

Step 1: Move the hairline to "259" on the right half of the A scale.
Step 2: Read the angle on the S scale, under the hairline. $\theta = 15°$

Using your slide rule, find these angles to the nearest degree:

(a) If $\sin B = 0.455$, B = ______° (b) If $\sin G = 0.880$, G = ______°

a) 0.0175 b) 0.602 c) 0.0872 d) 0.225 e) 0.0436

---

242. If the known numerical value of the sine ratio is a number between 0.010 and 0.100, the angle is found by setting the sine value on the left half of the A scale.

For example, if $\sin \varphi = 0.0350$, the size of angle φ is found by moving the hairline to "350" on the left half of A, and then reading the angle ($\varphi = 2°$) on S under the hairline.

Using your slide rule, find these angles to the nearest degree:

(a) If $\sin F = 0.0532$, F = ______° (b) If $\sin H = 0.0865$, H = ______°

a) B = 27° b) G = 62°

---

243. If the known numerical value of the sine:

(1) Lies between 0.010 and 0.100, set the value on the LEFT HALF of the A scale, and read the angle on the S scale.

(2) Lies between 0.100 and 1.000, set the value on the RIGHT HALF of the A scale, and read the angle on the S scale.

Using your slide rule, find these angles to the nearest degree:

(a) If $\sin M = 0.0705$, M = ______° (b) If $\sin P = 0.705$, P = ______°

a) F = 3° b) H = 5°

---

244. We can also use the slide rule to find the sines of angles greater than 90°. To do so, we simply find the sine of their reference angles. When doing so, use a sketch of the angle so that you do not make a "sign" error.

(a) sin 151° = ________ (b) sin 299° = ________

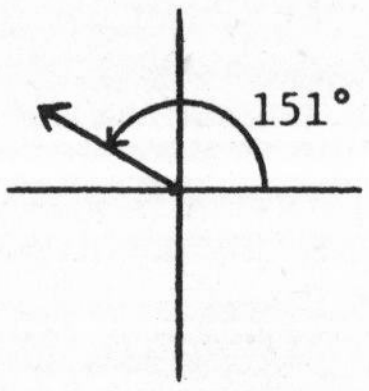

151°

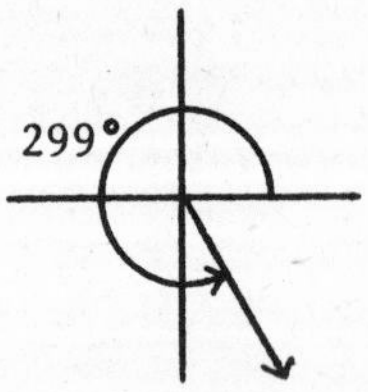

299°

a) M = 4° b) P = 45°

Answer to Frame 244: a) 0.485 From: sin 29° (Q2) b) -0.875 From: sin 61° (Q4)

SELF-TEST 16 (Frames 231-244)

Do these problems on your slide rule:

1. $\sin 27° =$ ______
2. $\sin 4\frac{1}{2}° =$ ______
3. $\sin 58° =$ ______
4. $\sin 7° =$ ______
5. $\sin 178° =$ ______
6. $\sin 210° =$ ______
7. If $\sin A = 0.176$, $A =$ ______°
8. If $\sin C = 0.0176$, $C =$ ______°

ANSWERS:

1. 0.454
2. 0.0785
3. 0.848
4. 0.122
5. 0.0349
6. -0.500
7. A = 10°
8. C = 1°

3-17 FINDING COSINE RATIOS ON THE SLIDE RULE

Though the slide rule contains a special "sine scale", it does not contain a special "cosine scale". The S scale is also used to find the cosines of angles. The process involves converting a cosine expression into an equivalent sine expression. We will show the method in this section.

---

245. The cosine of any angle between 0° and 90° is equal to the sine of some angle between 0° and 90°. This fact is illustrated by the following pairs of values taken from a trig table:

#1 $\begin{cases} \cos 10° = 0.9848 \\ \sin 80° = 0.9848 \end{cases}$ #2 $\begin{cases} \cos 39° = 0.7771 \\ \sin 51° = 0.7771 \end{cases}$ #3 $\begin{cases} \cos 73° = 0.2924 \\ \sin 17° = 0.2924 \end{cases}$

In Pair #1, the sum of the two angles (10° and 80°) is 90°.

Find the sum of the two angles in: (a) Pair #2 ______

(b) Pair #3 ______

---

246. The cosine of any angle is equal to the sine of another angle if the sum of the two angles is 90°. Therefore:

$\cos 19° = \sin 71°$

(a) $\cos 12° = \sin$ ____ (b) $\cos 30° = \sin$ ____ (c) $\cos 64° = \sin$ ____

Answers to 245: a) 90° b) 90°

---

247. The general principle is: **If A + B = 90° then cos A = sin B**

(a) $\cos 6° = \sin$ ______°
(b) $\cos 45° = \sin$ ______°
(c) $\cos 89° = \sin$ ______°
(d) $\cos 90° = \sin$ ______°

Answers to 246: a) 78° b) 60° c) 26°

Answers to 247: a) 84° b) 45° c) 1° d) 0°

248. This general principle can easily be verified by examining the relationships in the right triangle shown at the right.

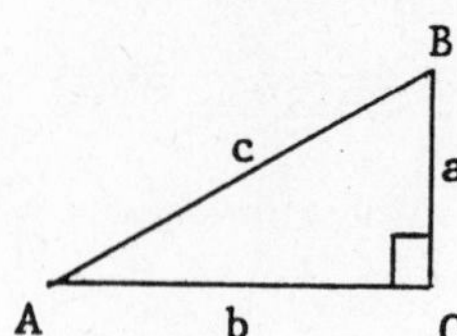
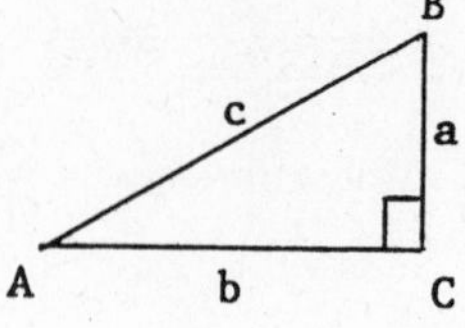

Since the triangle is a right triangle, A + B = 90°.

(a) $\cos A = \frac{b}{c}$ and $\sin B =$ ______

(b) Does $\cos A = \sin B$? ______

---

249. We can use this relationship to find the cosine of an angle on the slide rule. To find cos 60°, the steps are:

<u>Step 1</u>: Convert cos 60° to its equivalent sine relation: cos 60° = sin 30°

<u>Step 2</u>: Find sin 30° on the S and A scales of the slide rule: sin 30° = 0.500

<u>Step 3</u>: Therefore: cos 60° = <u>0.500</u> cos 60° = <u>0.500</u>

On your slide rule, find the numerical value of:

(a) cos 34° = sin 56° = ______

(b) cos 81° = sin 9° = ______

(c) cos 15° = ______

Answers (248): a) $\frac{b}{c}$ b) Yes

---

250. Find these on your slide rule:

(a) cos 68° = ______ (d) cos 45° = ______

(b) cos 88° = ______ (e) cos 85° = ______

(c) cos 9° = ______

Answers (249): a) 0.829 b) 0.156 c) 0.966

---

251. If the numerical value of the cosine of an angle is known, we can use the S scale to find the size of the angle. To do so, we use this fact:

$$\boxed{\cos A = \sin(90° - A)}$$

Let's find A when cos A = 0.829.

If cos A = 0.829, sin(90° − A) = 0.829.

Therefore: 90° − A = 56°

and A = 34°

Let's find B when cos B = 0.391.

If cos B = 0.391, sin(90° − B) = 0.391.

Therefore: (a) 90° − B = ______

(b) B = ______

Answers (250): a) 0.375 b) 0.0349 c) 0.988 d) 0.707 e) 0.0872

Answers (251): a) 23° b) 67°

252. (a) If $\cos B = 0.743$, $90° - B =$ _______

$B =$ _______

(b) If $\cos A = 0.407$, $90° - A =$ _______

$A =$ _______

Answers to Frame 252:
a) 48°
42°

b) 24°
66°

253. We can also use the slide rule to find the cosines of angles greater than 90°. Here are the steps needed to find cos 120°:

(1) Sketch the angle, as we have done on the right.

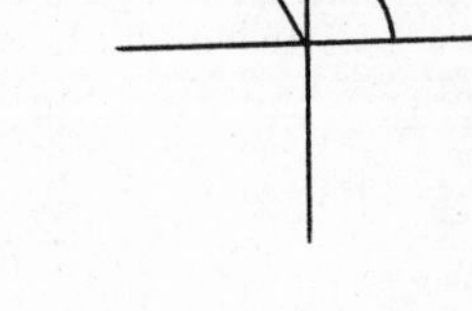

(2) Identify its reference angle. In this case, it is 60° (Q2).

(3) To find the <u>absolute value</u> of the cosine, use the S scale to find $\sin(90° - 60°)$ or sin 30°. It is 0.500.

(4) Determine the <u>sign</u> of the cosine. Since 120° lies in the second quadrant, its cosine is <u>negative</u>.

(5) Therefore, $\cos 120° = -0.500$

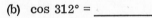

Following the steps above, find each cosine below on your slide rule.

(a) $\cos 160° =$ _________ (b) $\cos 312° =$ _________

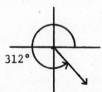

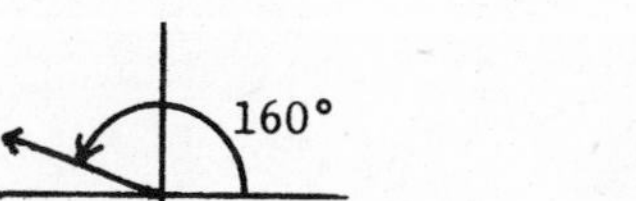

312°

<u>Answer to Frame 253</u>: a) -0.940 b) $+0.669$

<u>SELF-TEST 17 (Frames 245-253)</u>

Complete the following by writing in an angle between 0° and 90°:

1. cos 4° = sin _______
2. cos 72° = sin _______
3. cos 45° = sin _______

Do these problems on your slide rule:

4. cos 8° = _________
5. cos 51° = _________
6. cos 87° = _________
7. cos 153° = _________
8. If $\cos A = 0.630$, $A =$ _______°
9. If $\cos B = 0.225$, $B =$ _______°

<u>ANSWERS</u>:

1. sin <u>86°</u>
2. sin <u>18°</u>
3. sin <u>45°</u>
4. 0.990
5. 0.630
6. 0.0523
7. -0.891
8. A = 51°
9. B = 77°

3-18 FINDING TANGENT RATIOS ON THE SLIDE RULE

Tangent ratios can also be found on the slide rule. To do so, we use the T and D scales. We will show the method in this section.

---

254. Locate the T (or Tangent) scale on your slide rule. Like the S scale, it can be located on either the front or back of the slide or on the frame. Since the T scale is used in conjunction with the D scale, the slide must be reversed from its usual position if the T scale is on the back of it.

When the ends of T and D are lined up, the following procedure is used to find the tangent of an angle:

> To find the tangent of an angle, move the hairline to the angle on T. Then read the tangent of the angle on D under the hairline.

Follow this procedure to find tan 20°:

1. Move the hairline to 20° on T.
2. Read the answer "364" on D. Therefore, tan 20° = "364"
3. Place the decimal point. Then, tan 20° = 0.364
   (Decimal-point placement is discussed in the next frame.)

Find the "sequence of digits" for tan 25°: tan 25° = "________"

---

255. To place the decimal point in the tangents of angles from 6° to 45°, we can use the following facts from the trig table:

"466"

$$\tan 6° = 0.1051$$
$$\tan 45° = 1.0000$$

It is evident from the values above that tangents of angles lying between 6° and 45° will be decimal numbers lying between 0.1 and 1.0. This fact is used in placing the decimal point in tangent "sequences of digits". For example:

From the slide rule, tan 30° = "577"
Therefore, tan 30° = 0.577

From the slide rule, tan 9° = "158"
Therefore, tan 9° = 0.158

Do these on your slide rule:

(a) tan 7° = ________ (d) tan 44° = ________

(b) tan 37° = ________ (e) tan 26° = ________

(c) tan 17° = ________ (f) tan $15\frac{1}{2}°$ = ________

---

a) 0.123 d) 0.966

b) 0.754 e) 0.488

c) 0.306 f) 0.277

256. Examine the T scale. The smallest angle on it is about 6° and the largest angle on it is 45°. However, the tangents of angles less than 6° or between 45° and 90° can also be found on the slide rule.

Let's discuss the method of finding the tangents of angles less than 6° first. From a trig table:

tan 5° = 0.0875 sin 5° = 0.0872	tan 3° = 0.0524 sin 3° = 0.0523	tan 1° = 0.0175 sin 1° = 0.0175

As you can see, for small angles (less than 6°), the tangent ratio is approximately equal to the sine ratio. That is:

(a) $\tan 4° \doteq \sin$ ______° (b) $\tan 2\frac{1}{2}° \doteq \sin$ ______°

---

257. To find the tangent of an angle less than 6°, we can find the sine of the angle using the S and A scales. The sine value is approximately equal to the tangent value to two or three significant digits.

Do these on your slide rule:

(a) $\tan 4° \doteq \sin 4° \doteq$ __________ (c) $\tan 3° \doteq$ __________

(b) $\tan 2° \doteq$ __________ (d) $\tan 1\frac{1}{2}° \doteq$ __________

Answers to 256: a) 4° b) $2\frac{1}{2}°$

---

258. To find the tangents of angles between 45° and 90°, a conversion is necessary. Here are some examples of the conversion which is used.

Example #1: $\tan 70° = \frac{1}{\tan 20°}$ Note: (70° + 20° = 90°)

Since $\tan 70° = 2.75$ and $\frac{1}{\tan 20°} = \frac{1}{0.364} = 2.75$

Example #2: $\tan 58° = \frac{1}{\tan 32°}$ Note: (58° + 32° = 90°)

Since $\tan 58° = 1.60$ and $\frac{1}{\tan 32°} = \frac{1}{0.625} = 1.60$

The general relationship is:

$$\tan A = \frac{1}{\tan B}, \text{ where } A + B = 90°$$

Using this relationship, complete these:

(a) $\tan 80° = \frac{1}{\tan \_\_°}$ (b) $\tan 48° = \frac{1}{\tan \_\_°}$ (c) $\tan \_\_° = \frac{1}{\tan 30°}$

Answers to 257: a) 0.0698 b) 0.0349 c) 0.0523 d) 0.0262

Answers to 258: a) 10° b) 42° c) 60°

259. The general relationship can also be written in this form:

$$\boxed{\tan A = \frac{1}{\tan(90° - A)}}$$

Note that the sum of the two angles is 90°. That is:

$$\begin{aligned} A + (90° - A) &= A + 90° - A \\ &= A - A + 90° \\ &= \quad 0 \quad + 90° = 90° \end{aligned}$$

Examples: $\tan 71° = \frac{1}{\tan(90° - 71°)} = \frac{1}{\tan 19°}$

$\tan 47° = \frac{1}{\tan(90° - 47°)} = \frac{1}{\tan 43°}$

Write each of these as an expression involving the tangent of an angle between 0° and 45°:

(a) tan 81° = ______ (b) tan 64° = ______ (c) tan 75° = ______

---

a) $\frac{1}{\tan 9°}$

b) $\frac{1}{\tan 26°}$

c) $\frac{1}{\tan 15°}$

---

260. To find tan 70° on the slide rule, follow these steps:

(1) $\tan 70° = \frac{1}{\tan 20°}$

(2) Using the T scale, tan 20° = 0.364

(3) Therefore: $\tan 70° = \frac{1}{0.364} = 2.75$

(Either divide 1 by 0.364 on the C and D scales or use the C and CI scales to find the reciprocal of 0.364.)

Using a similar procedure, find the numerical value of these:

(a) tan 83° = ______ (b) tan 49° = ______

---

a) 8.13, from: $\frac{1}{\tan 7°} = \frac{1}{0.123}$

b) 1.15, from: $\frac{1}{\tan 41°} = \frac{1}{0.869}$

---

261. Use your slide rule for these:

(a) tan 139° = tan 41° (Q2) = ______

(b) tan 210° = tan 30° (Q3) = ______

(c) tan 177° = tan 3° (Q2) = ______

(d) tan 355° = tan 5° (Q4) = ______

---

a) −0.869

b) +0.577

c) −0.0523

d) −0.0872

262. To find tan 108° on the slide rule, we use the following steps:

(1) Sketch the angle.

(2) Its reference angle is 72° (Q2).

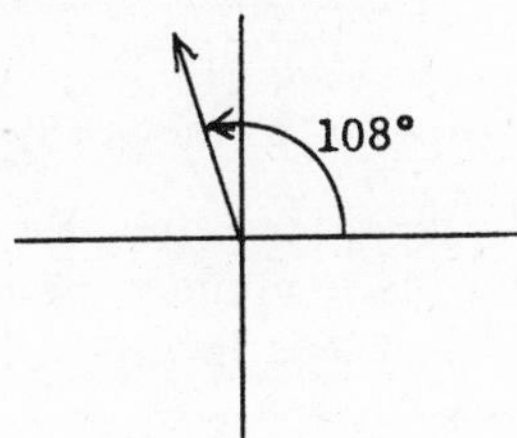

(3) To find the absolute value of the tangent, we evaluate

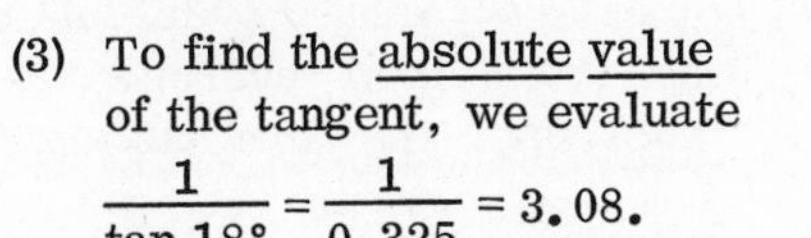

$$\frac{1}{\tan 18°} = \frac{1}{0.325} = 3.08.$$

(4) Since tangents are negative in the second quadrant, the sign of the tangent is negative.

(5) Therefore, tan 108° = -3.08.

Using the steps above, complete this one:

tan 130° = ________

---

263. Here is an exercise involving all three ratios. Use your slide rule to find the numerical value of each.

(a) sin 14° = ________ (d) cos 27° = ________

(b) cos 62° = ________ (e) tan 39° = ________

(c) sin $4\frac{1}{2}$° = ________ (f) tan 52° = ________

-1.19

Answer to Frame 263: a) 0.242 b) 0.470 c) 0.0785 d) 0.891 e) 0.810 f) 1.28

SELF-TEST 18 (Frames 254-263)

On your slide rule, find the numerical value of each tangent ratio:

1. tan 13° = ________ 2. tan $2\frac{1}{2}$° = ________ 3. tan 66° = ________

Do these problems on your slide rule:

4. sin 45° = ________ 5. cos 74° = ________ 6. tan 150° = ________

ANSWERS:

1. 0.231
2. 0.0436
3. 2.25
4. 0.707
5. 0.276
6. -0.577

3-19 AN ALTERNATE PROCEDURE FOR FINDING TRIG RATIOS ON THE SLIDE RULE

This section applies only to those students whose slide rules have the S and T scales on the back of the slide. If your slide rule has the S and T scales on the face of the slide or on the frame, you can skip this section.

When the S and T scales are on the back of the slide, the slide must be removed, turned over, and reinserted before they can be used. This procedure is clumsy. Furthermore, when the slide is reversed, the C scale is not visible and cannot be used in operations involving the C scale. Therefore, the slide has to be reversed again in order to use the C scale in operations.

Some slide rules are designed so that the S and T scales can be used while they are on the underside of the slide. By using them in this way:

(1) The slide does not have to be reversed.
(2) The C scale remains in position for operations on the face of the slide.

We will discuss this new procedure in this section.

---

264. Let's find sin 30° using the alternate procedure. Follow these steps:

Step 1: Insert the slide so that the C scale is on the face of the rule. The S (and T) scales are then on the underside of the slide.

Step 2: Turn the entire slide rule over. At the right end you will note a hairline. We will call this the right back hairline. It is fixed on the frame and cannot be moved.

Step 3: Move the slide so that 30° on S is under the right back hairline.

Step 4: Turn the slide rule over. Read the answer on B under the right end of A. The answer is "500".

Step 5: Place the decimal point in "500". Therefore: sin 30° = 0.500

Using this procedure, find the numerical value of these sines:

(a) sin 11° = ______ (b) sin 58° = ______ (c) sin 3° = ______ (d) sin 8° = ______

---

265. A similar procedure is used for finding tangents. For example, to find tan 30°, follow these steps:

Step 1: With the slide rule turned over, move the slide so that 30° on T is under the right back hairline.

Step 2: Turn the slide rule over. Read the answer on C at the right end of D. The answer is "577".

Step 3: Place the decimal point, and obtain tan 30° = 0.577

Using this procedure, find these tangents:

(a) tan 34° = ______ (c) tan 42° = ______

(b) tan 15° = ______ (d) tan 7° = ______

Answers to 264:
a) 0.191
b) 0.848
c) 0.0523
d) 0.139

Answers to 265:
a) 0.675
b) 0.268
c) 0.900
d) 0.123

266. The procedure of the last frame makes it very easy to find the tangents of angles lying between 45° and 90°. Here are the steps needed to find tan 60°:

Step 1: $\tan 60° = \frac{1}{\tan 30°}$

Step 2: Find tan 30° by moving 30° on T to the right back index. Then turn the slide rule over; the value of tan 30° is on the C scale under the right end of D. It actually is 0.577, but it does not have to be read. See the next step.

Step 3: The value of tan 30° (0.577) appears on C over the "1" on the right end of D. This position is actually the set-up needed for the division: $\frac{1}{0.577}$

The answer to this division appears on D under the left end of C. It is "173". Therefore: $\frac{1}{0.577} = 1.73$

Step 4: Therefore: $\tan 60° = \frac{1}{\tan 20°} = \frac{1}{0.577} = 1.73$

Work these, using the same procedure:

(a) tan 64° = ______ (b) tan 49° = ______ (c) tan 73° = ______

---

267. Cosines can also be quickly found using the back hairline. For example, to find cos 68° follow these steps:

Step 1: Write cos 68° as an equivalent sine expression:

$$\cos 68° = \sin(90° - 68°) = \sin 22°$$

Step 2: Using the back hairline (S scale) and the B scale, find sin 22°:

$$\sin 22° = 0.375$$

Step 3: Therefore: cos 68° = 0.375

Find these: (a) cos 60° = ______ (c) cos 74° = ______

(b) cos 20° = ______ (d) cos 88° = ______

Answers to Frame 266:

a) 2.05
b) 1.15
c) 3.27

Answer to Frame 267: a) 0.500 b) 0.940 c) 0.276 d) 0.0349

Since trig ratios are not found on the slide rule frequently, it is easy to forget that the S scale is used with the A scale and the T scale is used with the D scale. To reorient yourself to the scales, it is helpful to remember these two facts:

sin 30° = 0.500 tan 30° = 0.577

Then if you set 30° on the S scale, you can look for the scale which reads "500". And if you set 30° on the T scale, you can look for the scale which reads "577".

Chapter 4 OBLIQUE TRIANGLES

Oblique triangles are triangles which do not contain a right angle. In this chapter, we will discuss the solution of oblique triangles. After a brief introduction to the right-triangle method of solving oblique triangles, the Law of Sines and the Law of Cosines will be introduced and used to solve them. The strategies used to solve oblique triangles will be emphasized. Solutions involving obtuse angles are delayed until all types of solutions involving acute angles are examined. After the methods of solving oblique triangles are completely discussed, the oblique-triangle method of adding vectors (both on and off the coordinate system) will be discussed.

4-1 THE DEFINITION OF AN OBLIQUE TRIANGLE

A triangle is called a "right" triangle if it contains a 90° (or right) angle. Triangles which do not contain a 90° angle are called "oblique" triangles. In this section, we will briefly discuss "oblique" triangles.

1. If a triangle does not contain a 90° angle, it is called an oblique triangle. (The word "oblique" is pronounced "ō-bleek".)

   The triangle on the right is an oblique triangle because it does not contain a right angle.

   85°   45°   50°

   The sum of the three angles is _______ degrees.

2. The angle-sum principle applies to both right and oblique triangles. Therefore, the sum of the three angles in any triangle is _______°.

   Answer (frame 1): 180°

3. (a) In triangle ABC, how large is angle A? _______

   (b) Is triangle ABC a right or an oblique triangle? _________

   B 75°   C 25°   A

   Answer (frame 2): 180°

Answers (frame 3):

a) 80°

b) Oblique

4. If one angle of a triangle is 65° and a second is 30°: (a) The third angle contains _______°. (b) The triangle is a ___________ (right/oblique) triangle.	
5. Triangle #1 (55°, 45°); Triangle #2 (50°, 40°) After using the angle-sum principle to find the third angle in each triangle, answer these: (a) Triangle #1 is a ___________ (right/oblique) triangle. (b) Triangle #2 is a ___________ (right/oblique) triangle.	a) 85° b) oblique
6. Any angle between 0° and 90° is called an "acute" angle. Any angle between 90° and 180° is called an "obtuse" angle. (There is no special name for angles greater than 180°.) (a) Is a 40° angle an acute or obtuse angle? ___________ (b) Is a 97° angle an acute or obtuse angle? ___________ (c) Is a 240° angle an acute or obtuse angle? ___________	a) oblique b) right
7. Which of the following are acute angles? ___________ (a) 79° (b) 178° (c) 54° (d) 210°	a) Acute b) Obtuse c) Neither. There is no special name for angles greater than 180°.
8. Which of the following are obtuse angles? ___________ (a) 209° (b) 146° (c) 192° (d) 17°	(a) and (c)
9. The sum of the three angles of a triangle is always 180°. (a) Since angle D is 120°, it is an ___________ (acute/obtuse) angle. (b) The sum of angles C and F must be _______°. (c) Both angle C and angle F must be ___________ (acute/obtuse) angles. (Triangle C, D, F with 120° at D)	Only (b)

10. One angle in a triangle is an obtuse angle of 91°.

(a) The sum of the other two angles must be ______°.

(b) If the sum of two angles in a triangle is 89°, can either of them be an obtuse angle? ______

(c) Is it possible to have two obtuse angles (for example, 94°, 91°) in the same triangle? ______

a) obtuse

b) 60°

c) acute, since their sum is only 60°.

---

11. Both triangles below are oblique triangles since neither contains a right angle.

Triangle #1 Triangle #2

In Triangle #1, all three angles are acute angles. We call such a triangle an ACUTE oblique triangle.

In Triangle #2, one angle is obtuse. We call such a triangle an OBTUSE oblique triangle.

(a) If a triangle contains angles of 13°, 29°, and 138°, it is an __________ (acute/obtuse) oblique triangle.

(b) If a triangle contains angles of 47°, 51°, and 82°, it is an __________ (acute/obtuse) oblique triangle.

(c) How many obtuse angles can one triangle contain? __________

a) 89°

b) No

c) No, since the sum of two obtuse angles is greater than 180°.

Answer to Frame 11: a) obtuse b) acute c) Only one

SELF-TEST 1 (Frames 1-11)

1. In triangle ABC, angle A = ______°.
2. Angle A is an __________ (acute/obtuse) angle.
3. Triangle ABC is a __________ (right/oblique) triangle.
4. Triangle ABC is an __________ (acute/obtuse) oblique triangle.

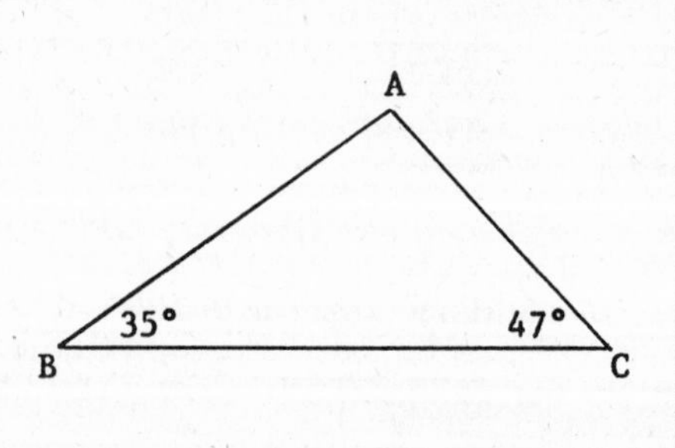

ANSWERS: 1. 98° 2. obtuse 3. oblique 4. obtuse

4-2 SOLVING OBLIQUE TRIANGLES BY RIGHT-TRIANGLE METHODS

"Solving triangles" means finding the lengths of unknown sides and the sizes of unknown angles. Right triangles can be solved directly by means of either the angle-sum principle, the Pythagorean Theorem, or the three basic trigonometric ratios. The angle-sum principle also applies directly to oblique triangles. However, since the Pythagorean Theorem and the three basic trigonometric ratios apply directly only to right triangles, they cannot be used to solve oblique triangles directly.

In this section, however, we will show that oblique triangles can be solved indirectly by right-triangle methods. To do so, the oblique triangle must be divided into two right triangles by drawing one of its altitudes. We will see that solutions by this method require a two-step process. Only some of the many possible types of solutions will be shown.

12. If two angles in either a right triangle or an oblique triangle are known, we can use the angle-sum principle to find the size of the third angle. For example:

(a) If two angles in a right triangle are 30° and 90°, the third angle is ______°.

(b) If two angles in an oblique triangle are 50° and 60°, the third angle is ______°.

---

13. Though the angle-sum principle applies to all triangles, the Pythagorean Theorem applies only to right triangles. It does not apply to oblique triangles.

Triangle MPF below is a right triangle. Triangle DHR is an oblique triangle. The lengths of two sides in each triangle are known.

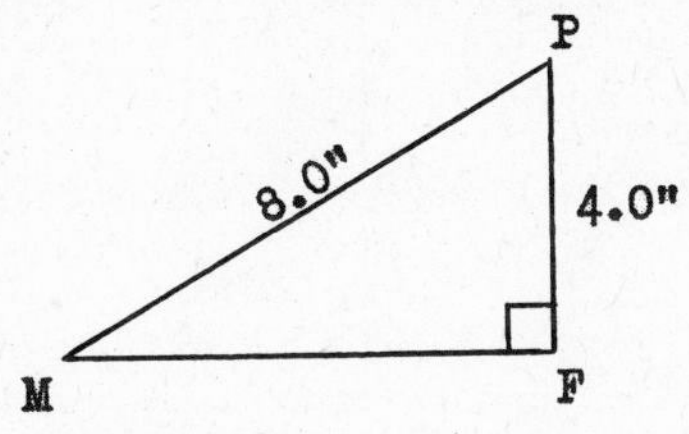

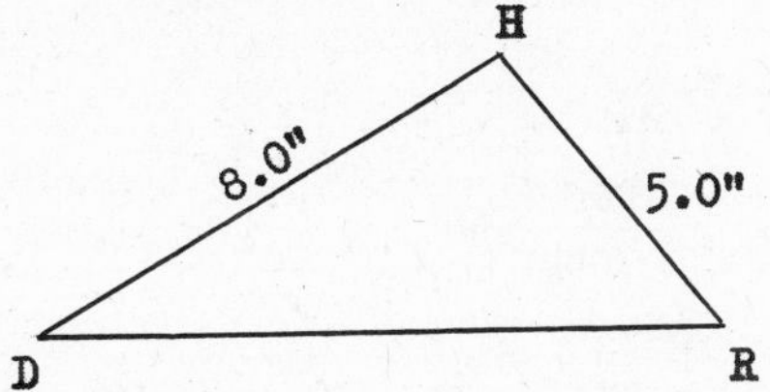

(a) Can we use the Pythagorean Theorem to find the length of MF in triangle MPF? ______

(b) Can we use the Pythagorean Theorem to find the length of DR in triangle DHR? ______

Answers to 12:
a) 60°
b) 70°

Answers to 13:
a) Yes, since it is a right triangle.
b) No, since it is an oblique triangle.

14. The three basic trigonometric ratios (sine, cosine, and tangent) are comparisons of the sides of right triangles. They are not comparisons of the sides of oblique triangles.

Triangle CDE below is a right triangle. Triangle GHK is an oblique triangle.

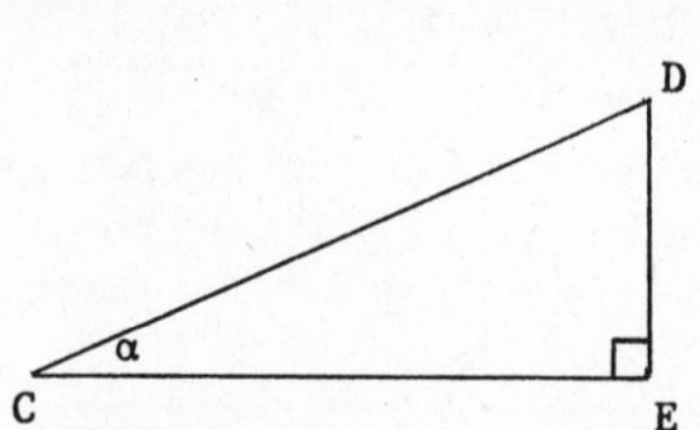

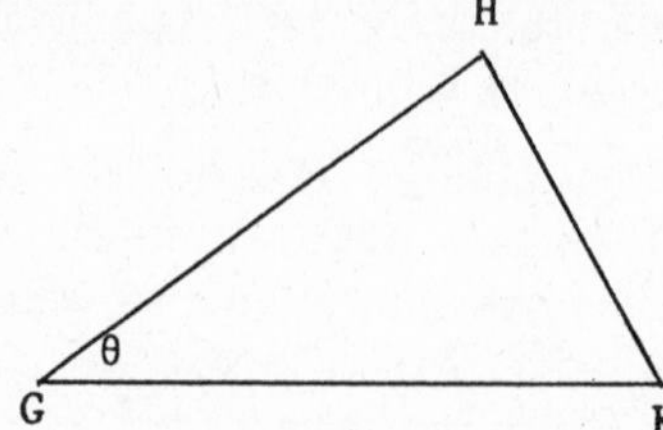

Using the proper sides of the triangles, complete these:

(a) In triangle CDE, sin α = _______

(b) In triangle GHK, sin θ = _______

---

15. Triangle MPQ is a right triangle. Triangle STV is an oblique triangle.

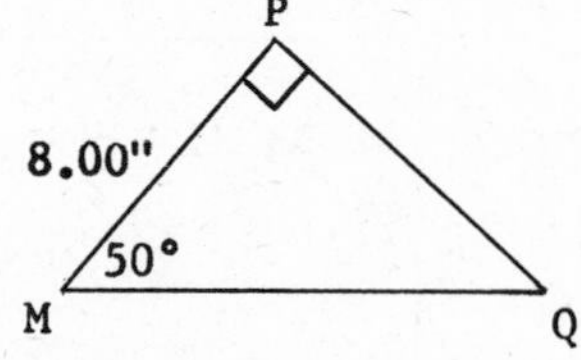

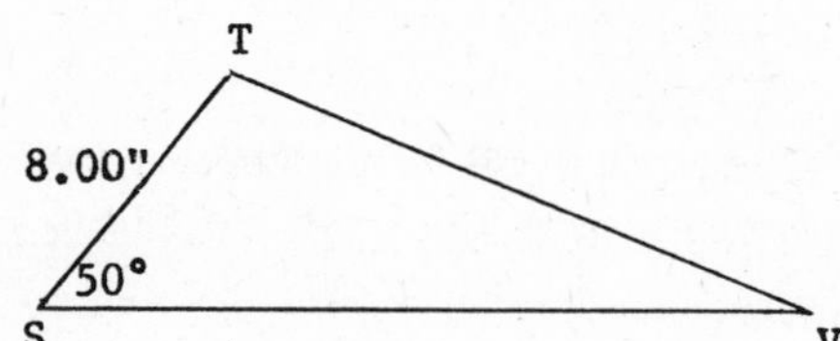

In triangle MPQ, we can use the tangent ratio to find the length of PQ. We get:

$$\tan 50° = \frac{PQ}{MP}$$

$$1.192 = \frac{PQ}{8.00''} \quad \text{and} \quad PQ = 9.54 \text{ inches}$$

Can we use the tangent ratio in a similar manner to find the length of TV in triangle STV? _______

Answers to 14:

a) $\frac{DE}{CD}$

b) Since GHK is an oblique triangle, sin θ is not a comparison of any of the sides of GHK.

---

16. Any oblique triangle can be divided into two right triangles by drawing an altitude from one angle to the opposite side. We have done so in the two oblique triangles below.

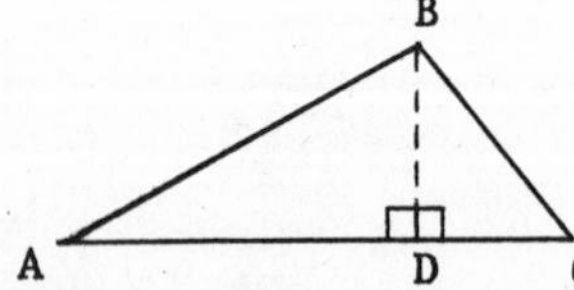

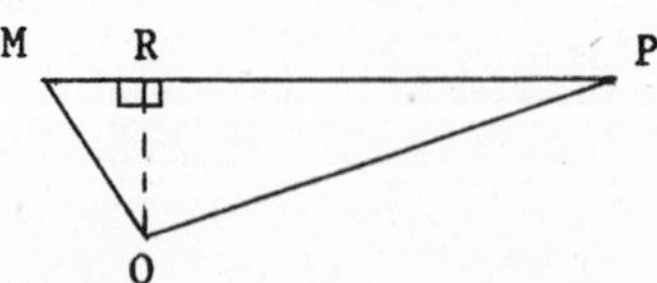

The altitude BD divides oblique triangle ABC into the two right triangles ABD and BCD.

The altitude RQ divides oblique triangle MPQ into the two right triangles _________ and _________.

Answer to 15:

No. Since STV is an oblique triangle, tan S (or tan 50°) is not a comparison of TV and ST.

MRQ and QRP

17. When an altitude divides an oblique triangle into two right triangles, the three basic trigonometric ratios hold for each right triangle.

Below we have drawn oblique triangle LMP twice. On the right, we have drawn altitude MN which divides triangle LMP into the two right triangles LMN and MNP. MN is 6.00 inches long.

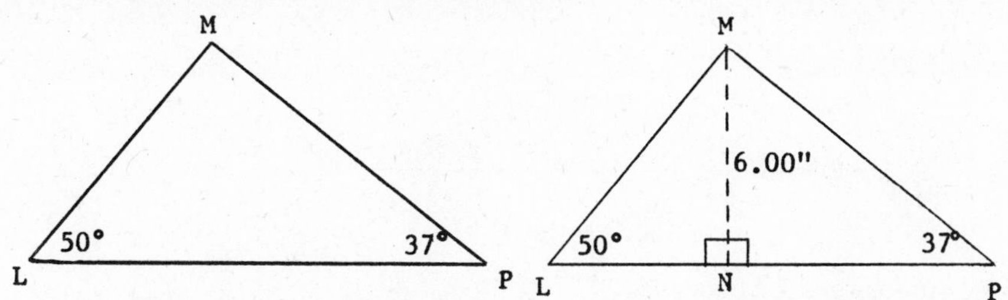

(a) In right triangle MNL, we can use the sine ratio to find the length of LM, as follows:

Since $\sin L = \dfrac{MN}{LM}$, $\sin 50° = \dfrac{6.00}{LM}$, and LM = ________

(b) In right triangle MNP, we can use the sine ratio to find the length of MP, as follows:

Since $\sin P = \dfrac{MN}{MP}$, $\sin 37° = \dfrac{6.00}{MP}$, and MP = ________

---

a) 7.83"

b) 9.97"

18. Oblique triangle BDF on the right is divided into two right triangles by altitude DH. The lengths of DH, BD, and DF are given.

By using the Pythagorean Theorem twice to find the lengths of BH and HF, we can find the length of BF.

(a) The length of BH is ______ inches.

(b) The length of HF is ______ inches.

(c) The length of BF is ______ inches.

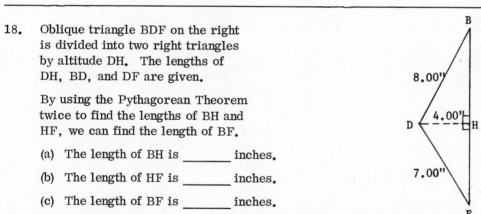

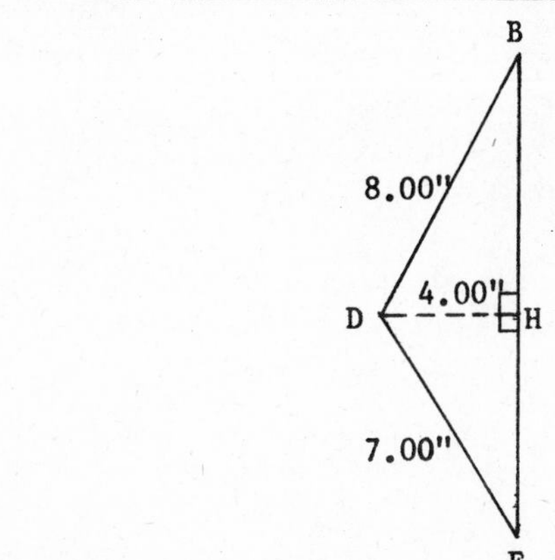
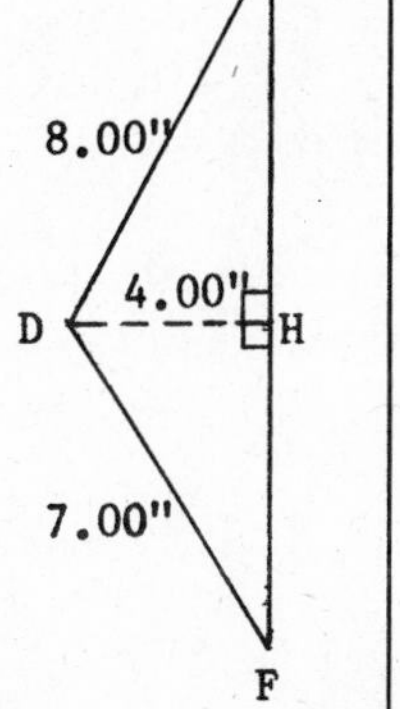

---

a) 6.93 inches

b) 5.74 inches

c) 12.67 inches

19. Oblique triangle QRS is divided into two right triangles by altitude TR. The lengths of TR, QR, and RS are given.

Using these lengths and the sine ratio, we can find the size of angles Q and S, as follows:

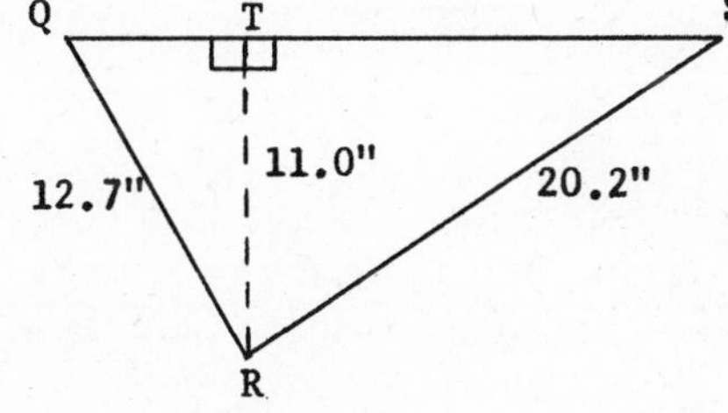

(a) Since $\sin Q = \dfrac{TR}{QR} = \dfrac{11.0}{12.7} = 0.865$, angle Q is ______°.

(b) Since $\sin S = \dfrac{TR}{RS} = \dfrac{11.0}{20.2} = 0.545$, angle S is ______°.

20. When we draw an altitude to divide an oblique triangle into two right triangles, we ordinarily do not know the length of the altitude. Therefore, we need a two-step process to find the length of a side or the size of an angle. The first step is finding the length of the altitude. Here is an example:

Oblique triangle ABC is divided into two right triangles by altitude BD. The length of AB and the sizes of angles A and C are given.

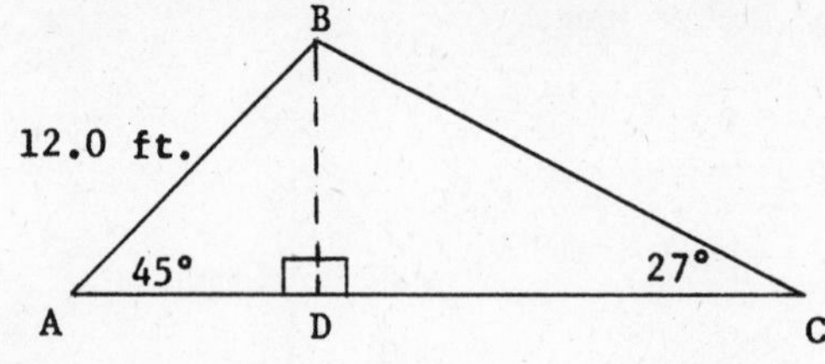

We can find the length of BC by a two-step process:

Step 1: Finding the length of BD, the altitude.

$$\frac{BD}{12.0} = \sin 45° \qquad BD = (12.0)(\sin 45°) = 8.49 \text{ ft.}$$

Step 2: Using the length of altitude BD to find the length of BC.

$$\frac{BD}{BC} = \sin 27° \qquad BC = \frac{BD}{\sin 27°} = \frac{8.49}{\sin 27°} = \underline{\hspace{3cm}}$$

a) 60°

b) 33°

---

21. In oblique triangle STV, altitude WV is drawn but its length is not given. Therefore, we need a two-step process to find the length of side SV.

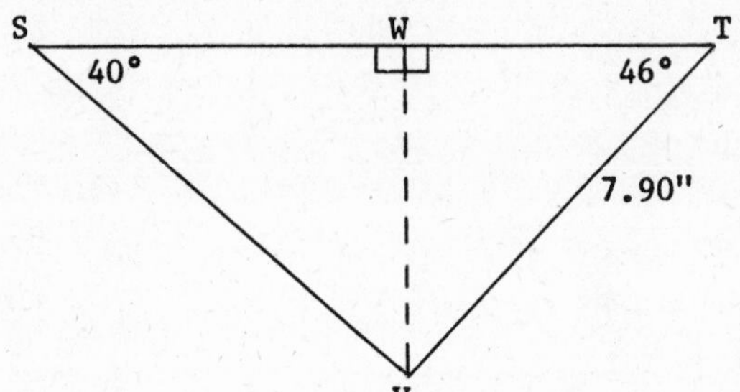

(a) Find the length of the altitude WV.
WV = ________

(b) Now find the length of side SV. SV = ________

18.7 ft.

---

a) WV = 5.68"

b) SV = 8.84"

22. Oblique triangle BKQ is divided into two right triangles by altitude KR, whose length is not given. Therefore, we need a two-step process to find the size of angle Q.

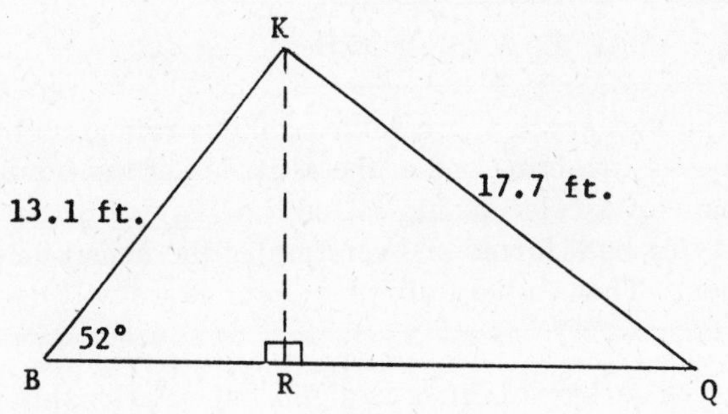

(a) Find the length of altitude KR.
KR = ________

(b) Since $\sin Q = \frac{KR}{KQ}$, Q contains ______°

Answer to Frame 22: a) 10.3 ft. b) 36° (from sin Q = 0.582)

SELF-TEST 2 (Frames 12-22)

In oblique triangle EFG, FH is an altitude. Find the following:

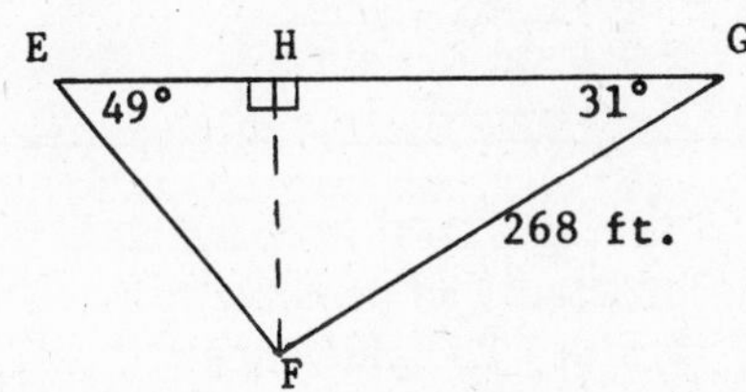

1. The length of altitude FH. ________
2. The length of side EF. ________
3. The size of angle EFG. ________

ANSWERS: 1. FH = 138 ft. 2. EF = 183 ft. 3. 100°

4-3 THE LAW OF SINES

In the last section, we solved oblique triangles by right-triangle methods. The solutions required a two-step process involving one of the altitudes of the triangle. Many oblique triangles can be solved more directly by means of a relationship called the "Law of Sines". In this section, we will introduce the Law of Sines. After showing both forms and confirming the sensibleness of each by a numerical example, we will prove the Law of Sines. Then in the following sections we will use the law to solve oblique triangles.

---

23. The Law of Sines is a general relationship which applies to any triangle. It involves the ratios of the three sides to the sines of the angles opposite them. There are three ratios of this type in any triangle. The Law of Sines says that these three ratios are equal.

For oblique triangle ABC on the right, the Law of Sines says this:

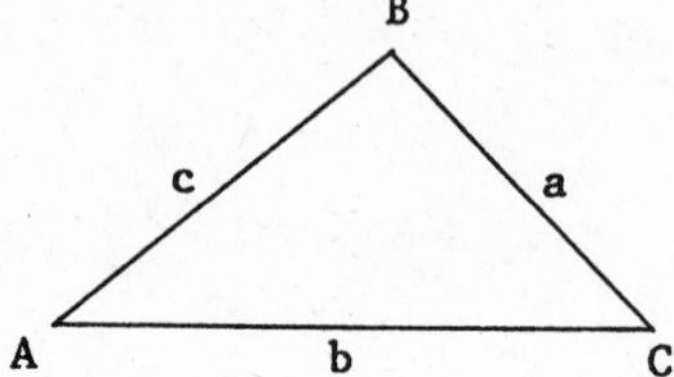

$$\frac{\text{side opposite angle A}}{\text{sine of A}} = \frac{\text{side opposite angle B}}{\text{sine of B}} = \frac{\text{side opposite angle C}}{\text{sine of C}}$$

or, more simply,

$$\frac{a}{\sin A} = \frac{b}{\sin B} = \frac{c}{\sin C}$$

Since the Law of Sines says that all three ratios are equal, we can write the following three proportions:

$$\frac{a}{\sin A} = \frac{b}{\sin B} \qquad \frac{a}{\sin A} = \frac{c}{\sin C} \qquad \frac{b}{\sin B} = \frac{c}{\sin C}$$

---

24. Triangle DFH is drawn as precisely as possible to scale. The sides are measurements reported precisely to "hundredths". The reported measurements include some rounding. We will use this triangle to confirm the Law of Sines.

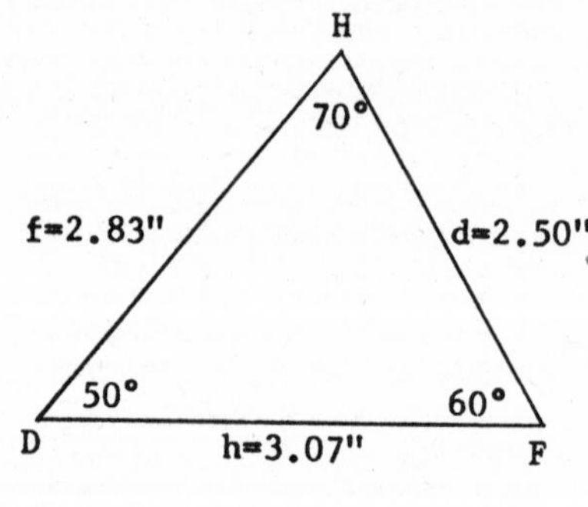

The Law of Sines says: $\frac{h}{\sin H} = \frac{d}{\sin D} = \frac{f}{\sin F}$

To confirm the Law of Sines, use your slide rule to evaluate each ratio:

(a) $\frac{h}{\sin H} = \frac{3.07}{\sin 70°} = \frac{3.07}{0.9397} =$ ________

(b) $\frac{d}{\sin D} = \frac{2.50}{\sin 50°} = \frac{2.50}{0.7660} =$ ________

(c) $\frac{f}{\sin F} = \frac{2.83}{\sin 60°} = \frac{2.83}{0.8660} =$ ________

(d) Are all three ratios equal? ________

25. The Law of Sines says that the ratios of sides to the sines of the opposite angles are equal in any triangle.

According to the Law of Sines, in triangle BDR:

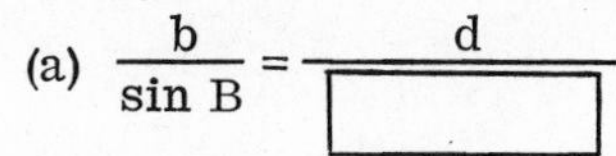
(a) $\dfrac{b}{\sin B} = \dfrac{d}{\boxed{}}$

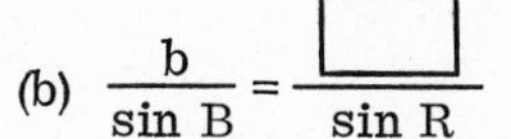
(b) $\dfrac{b}{\sin B} = \dfrac{\boxed{}}{\sin R}$

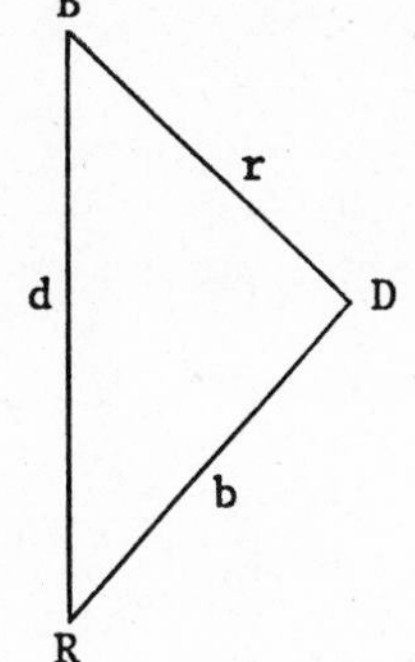

a) 3.27

b) 3.26

c) 3.27

d) Yes, given the amount of rounding involved in reporting the length of the sides to "hundredths".

---

26. Triangle BFH is an oblique triangle.

According to the Law of Sines:

(a) $\dfrac{BH}{\sin F} = \dfrac{BF}{\boxed{}}$

(b) $\dfrac{BH}{\sin F} = \dfrac{\boxed{}}{\sin B}$

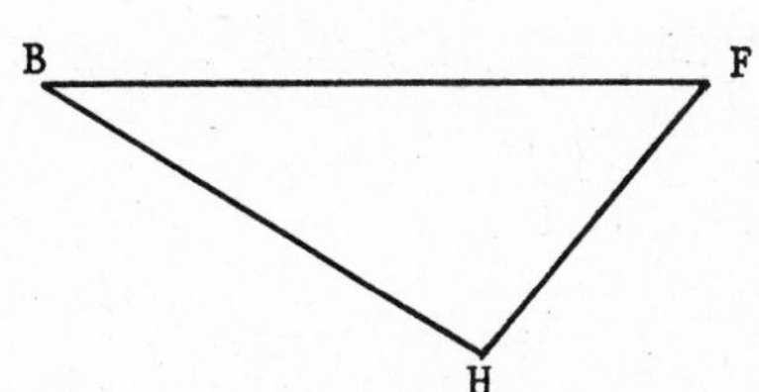

a) $\boxed{\sin D}$

b) $\boxed{r}$

---

27. Here is another oblique triangle.

According to the Law of Sines:

(a) $\dfrac{m}{\sin \beta} = \dfrac{\boxed{}}{\sin \theta}$

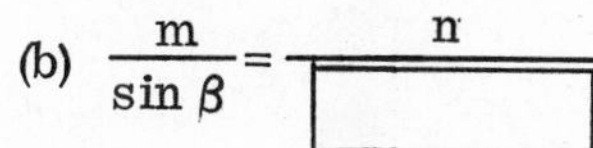
(b) $\dfrac{m}{\sin \beta} = \dfrac{n}{\boxed{}}$

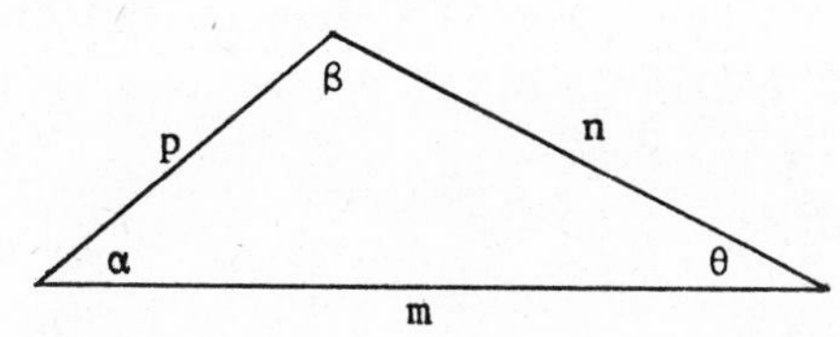

a) $\boxed{\sin H}$

b) $\boxed{HF}$

---

28. The Law of Sines is a proportion which contains three ratios. Up to this point, we have written these ratios with the "sides" as the numerators and the "sines" as the denominators. However, if the ratios are inverted (turned upside down), the new equation is also a proportion containing three equal ratios.

Therefore, for triangle ABC on the right, we can write the Law of Sines in either of two ways:

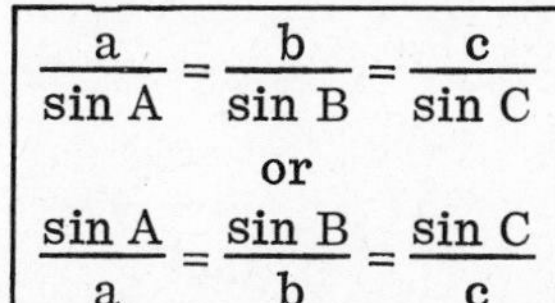
$$\frac{a}{\sin A} = \frac{b}{\sin B} = \frac{c}{\sin C}$$
or
$$\frac{\sin A}{a} = \frac{\sin B}{b} = \frac{\sin C}{c}$$

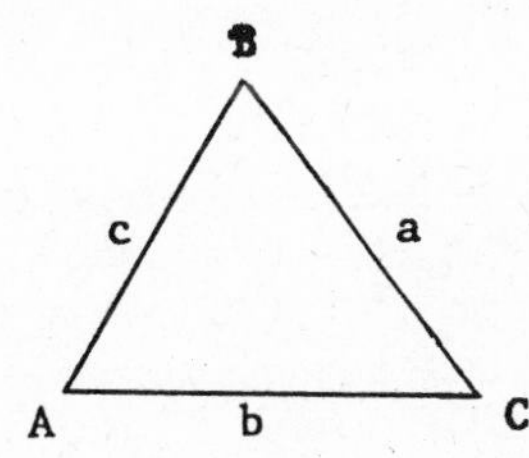

Write the Law of Sines in two different ways for triangle MPQ:

_____ = _____ = _____

_____ = _____ = _____

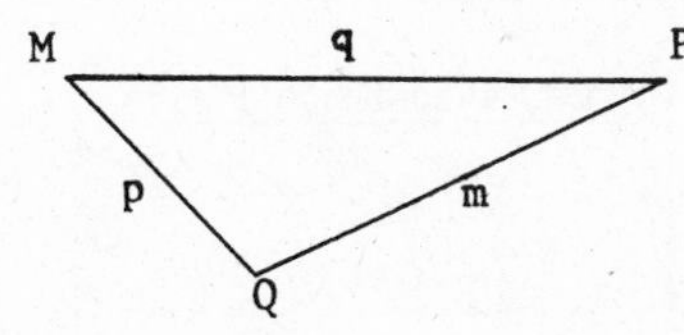

a) $\boxed{p}$

b) $\boxed{\sin \alpha}$

$$\frac{m}{\sin M} = \frac{p}{\sin P} = \frac{q}{\sin Q}$$

$$\frac{\sin M}{m} = \frac{\sin P}{p} = \frac{\sin Q}{q}$$

29. In an earlier frame, we confirmed the Law of Sines numerically by using triangle DFH on the right. In that frame, we confirmed the form with the "sides" as numerators. In this frame, we will confirm the form with the "sines" as numerators. That is, we will confirm:

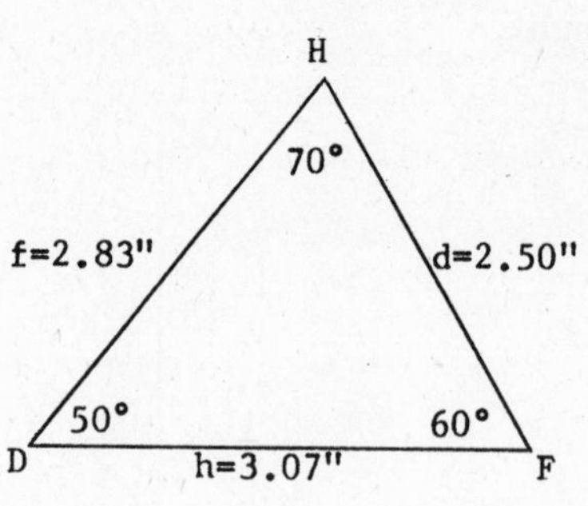

$$\boxed{\frac{\sin H}{h} = \frac{\sin D}{d} = \frac{\sin F}{f}}$$

Use your slide rule to evaluate each ratio:

(a) $\frac{\sin H}{h} = \frac{\sin 70°}{3.07} = \frac{0.9397}{3.07} =$ ________

(b) $\frac{\sin D}{d} = \frac{\sin 50°}{2.50} = \frac{0.7660}{2.50} =$ ________

(c) $\frac{\sin F}{f} = \frac{\sin 60°}{2.83} = \frac{0.8660}{2.83} =$ ________

(d) Are all three ratios equal? ________

a) 0.306

b) 0.306

c) 0.306

d) Yes

30. Using the form of the Law of Sines with the "sines" as numerators, complete these for oblique triangle PDR:

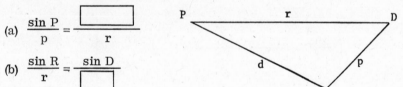

(a) $\frac{\sin P}{p} = \frac{\boxed{}}{r}$

(b) $\frac{\sin R}{r} = \frac{\sin D}{\boxed{}}$

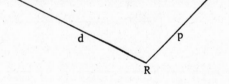

a) 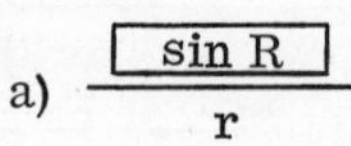

r

b) sin D
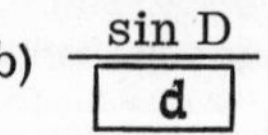

31. When writing a proportion based on the Law of Sines, you must be consistent. That is, the numerators of both ratios must either be "sides" or "sines".

Using triangle AMT, complete each proportion with one of the two possible ratios:

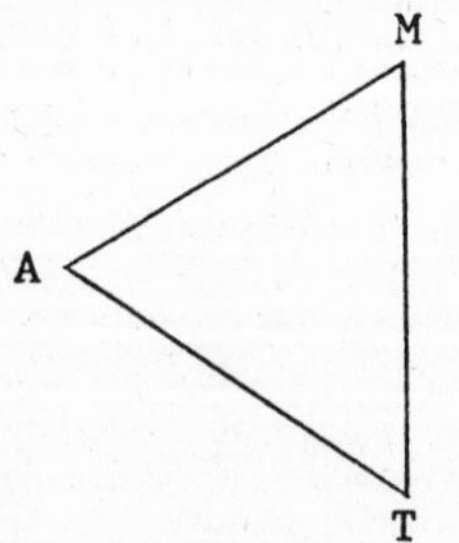

(a) $\frac{AM}{\sin T} =$ ________

(b) $\frac{\sin M}{AT} =$ ________

Answer to Frame 31: a) $\frac{MT}{\sin A}$ or $\frac{AT}{\sin M}$ b) $\frac{\sin A}{MT}$ or $\frac{\sin T}{AM}$

32. In the next few frames, we will prove the Law of Sines. To do so, we will use the general oblique triangle MPS. Notice that we have divided MPS into two right triangles by drawing the altitude "h".

In this frame, we will prove that: $\boxed{\dfrac{s}{\sin S} = \dfrac{m}{\sin M}}$

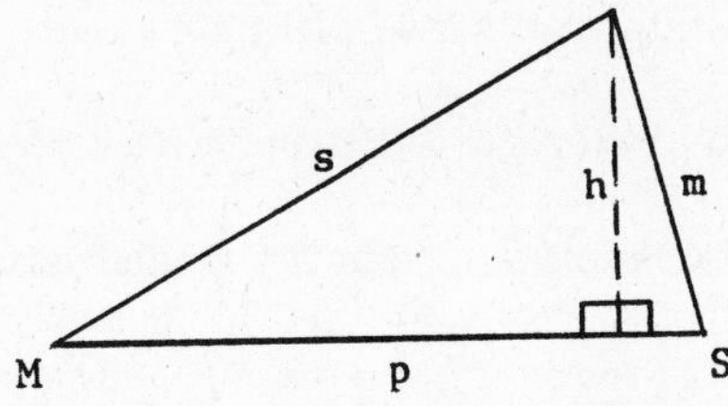

Step 1: Each equation below can be verified by examining the two right triangles on the right:

$$\sin M = \frac{h}{s} \quad \text{and} \quad \sin S = \frac{h}{m}$$

Step 2: Each equation below can be obtained by solving for "h" in the equation above, and equating the two solutions:

$$h = s(\sin M) \quad \text{and} \quad h = m(\sin S)$$

Therefore: $\boxed{s(\sin M) = m(\sin S)}$

Step 3: Each part below is an algebraic manipulation of the last equation:

(a) Multiplying both sides of the boxed equation by $\dfrac{1}{\sin M}$, we get:

$$\boxed{s(\sin M) = m(\sin S)}$$

$$\left(\frac{1}{\sin M}\right)s(\sin M) = \left(\frac{1}{\sin M}\right)m(\sin S)$$

$$s\left(\frac{\sin M}{\sin M}\right) = \frac{m}{\sin M}(\sin S)$$

$$s = \frac{m}{\sin M}(\sin S)$$

(b) Multiplying both sides of this last equation by $\dfrac{1}{\sin S}$, we get:

$$\left(\frac{1}{\sin S}\right)s = \frac{m}{\sin M}(\sin S)\left(\frac{1}{\sin S}\right)$$

$$\frac{s}{\sin S} = \frac{m}{\sin M}\left(\frac{\sin S}{\sin S}\right)$$

$$\boxed{\frac{s}{\sin S} = \frac{m}{\sin M}}$$

(This is the proportion we set out to prove.)

33. Here's the same triangle again. This time we have drawn the altitude from M to side m and labeled it "g".

In this frame, we will prove that: $\boxed{\dfrac{s}{\sin S} = \dfrac{p}{\sin P}}$

To prove that this proportion is true, we use the same procedure:

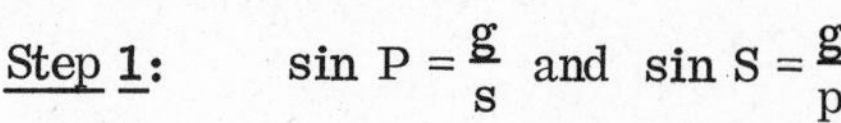

Step 1: $\sin P = \dfrac{g}{s}$ and $\sin S = \dfrac{g}{p}$

Step 2: $g = s(\sin P)$ and $g = p(\sin S)$

Therefore: $\boxed{s(\sin P) = p(\sin S)}$

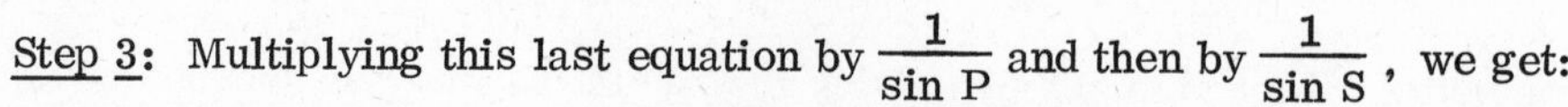

Step 3: Multiplying this last equation by $\dfrac{1}{\sin P}$ and then by $\dfrac{1}{\sin S}$, we get:

$$\boxed{\frac{s}{\sin S} = \frac{p}{\sin P}}$$

---

34. In the last two frames, we proved the following two proportions:

$$\frac{s}{\sin S} = \frac{m}{\sin M} \quad \text{and} \quad \frac{s}{\sin S} = \frac{p}{\sin P}$$

Since both of the other ratios are equal to $\dfrac{s}{\sin S}$, they must also be equal. Therefore, we have proven the general Law of Sines:

$$\boxed{\frac{s}{\sin S} = \frac{m}{\sin M} = \frac{p}{\sin P}}$$

The form of the Law of Sines which we have proven has the "sides" in the numerators. It can also be shown algebraically that the Law of Sines can be written in the following form, with the "sines" in the numerators:

$$\boxed{\frac{\sin S}{s} = \frac{\sin M}{m} = \frac{\sin P}{p}}$$

Note: In this course, we will not expect you to memorize this proof in order to reproduce it. The proof was inserted only to show that a general proof is possible.

SELF-TEST 3 (Frames 23-34)

Complete the following equations by applying the Law of Sines to oblique triangle FGH:

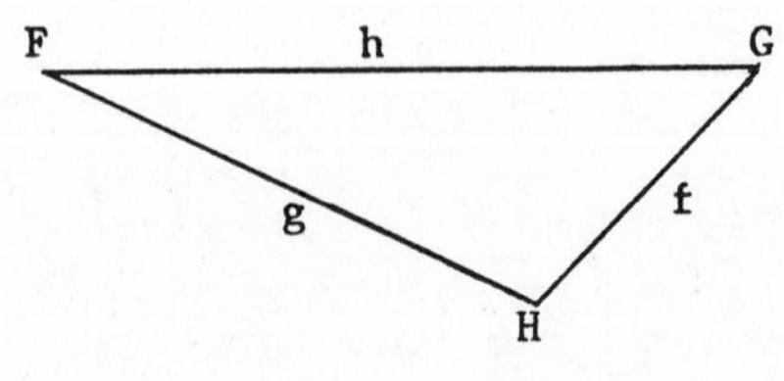

1. $\dfrac{h}{\boxed{}} = \dfrac{f}{\boxed{}}$

2. $\dfrac{\sin G}{\boxed{}} = \dfrac{\sin H}{\boxed{}}$

ANSWERS:

1. $\dfrac{h}{\boxed{\sin H}} = \dfrac{f}{\boxed{\sin F}}$

2. $\dfrac{\sin G}{\boxed{g}} = \dfrac{\sin H}{\boxed{h}}$

4-4 USING THE LAW OF SINES TO FIND UNKNOWN SIDES OF OBLIQUE TRIANGLES

In this section, we will use the Law of Sines to solve for unknown sides in oblique triangles. The strategy involved in using the Law of Sines will be emphasized. In order to simplify the solutions, proportions which contain obtuse angles will be avoided.

---

35. In an earlier section, we found the length of BC in triangle ABC by the two-step, right-triangle method. Using that method, we found that BC = 18.7 feet.

The same problem can be solved more directly by using the Law of Sines. Here is the method.

According to the Law of Sines: $\boxed{\dfrac{\text{length of AB}}{\sin C} = \dfrac{\text{length of BC}}{\sin A}}$

Substituting the known values, we get: $\dfrac{12.0 \text{ feet}}{0.4540} = \dfrac{BC}{0.7071}$

$$BC = \frac{0.7071(12.0 \text{ feet})}{0.4540}$$

$$BC = 18.7 \text{ feet}$$

Did we get the same answer by the "Law of Sines" method? ________

Answer: Yes

---

36. In an earlier section, we also found the length of SV by the two-step, right-triangle method. We found that SV = 8.84".

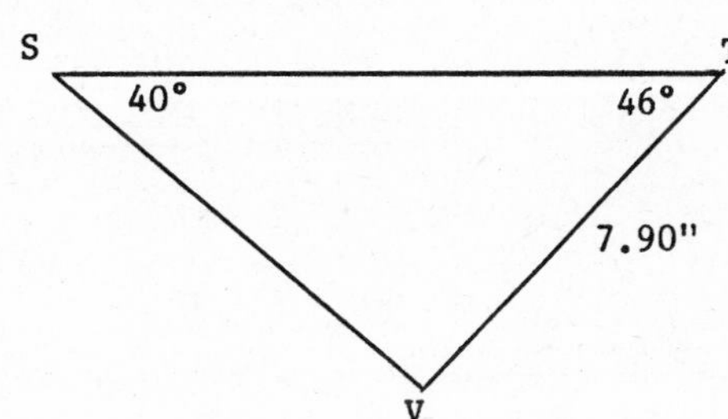

By using the Law of Sines, we can solve for SV in one step. Here is the method:

$$\frac{VT}{\sin S} = \frac{SV}{\sin T}$$

$$\frac{7.90''}{\sin 40°} = \frac{SV}{\sin 46°}$$

$$\frac{7.90''}{0.6428} = \frac{SV}{0.7193}$$

(a) Complete the solution. $SV = \dfrac{(0.7193)(7.90'')}{0.6428} =$ ________

(b) Is this the same answer we obtained earlier by the two-step right-triangle method? ________

Answers:

a) SV = 8.84"

b) Yes

37. When using the Law of Sines to find unknown sides of triangles, we must set up and solve proportions like the one below. Since a proportion is easier to solve if the unknown is in the numerator, the sides are put in the numerators.

$$\frac{c}{0.7071} = \frac{40.0''}{0.8660}$$

One way to solve this proportion on the slide rule is to isolate "c" and then perform a combined multiplication and division on the C and D scales. We get:

$$c = \frac{(0.7071)(40.0'')}{0.8660} = 32.7''$$

Using the method above, solve the following proportion:

$$\frac{14.8'}{0.5299} = \frac{t}{0.9703}$$

t = ________

---

38. Though the complete Law of Sines contains three ratios, the proportion used to find an unknown side of a triangle contains only two ratios. The strategy for selecting the proper two ratios is discussed in the next few frames.

Answer to 37: t = 27.1'

We want to use the Law of Sines to find the length of side "c" in triangle ABC.

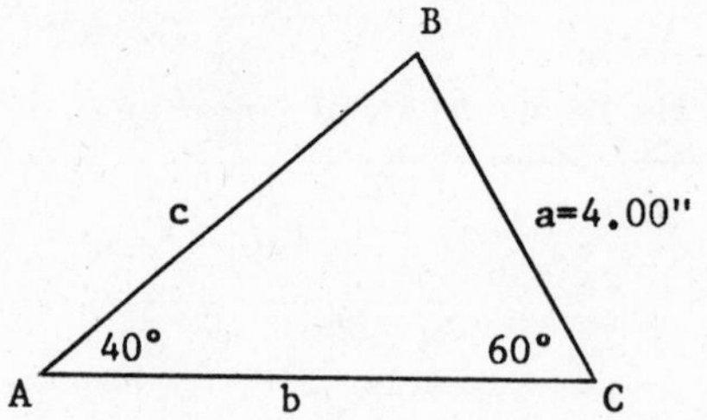

B

c a=4.00"

40° 60°

A b C

The complete Law of Sines for triangle ABC is:

$$\frac{a}{\sin A} = \frac{b}{\sin B} = \frac{c}{\sin C}$$

$$\text{or} \quad \frac{4.0''}{\sin 40°} = \frac{b}{\sin 80°} = \frac{c}{\sin 60°}$$

(Note: We used the angle-sum principle to find the size of angle B.)

If we want to solve for "c", we must use the ratio in which "c" appears, plus one other ratio. The two possibilities are:

$$\frac{c}{\sin 60°} = \frac{4.0''}{\sin 40°} \quad \text{and} \quad \frac{c}{\sin 60°} = \frac{b}{\sin 80°}$$

A proportion can be solved only if it contains a single unknown.

The proportion on the left contains one unknown, "c".
The proportion on the right contains two unknowns, "c" and "b".

Therefore, which proportion must we use to solve for "c", the one on the left or the one on the right? ________________

Answer to 38: The one on the left.

39. When using the Law of Sines to find an unknown side of a triangle, one ratio must contain that side. And since a proportion cannot be solved if it contains more than one unknown, <u>the other ratio must be one in which we know both the length of the side and the size of the angle opposite it.</u>

Let's set up the proportion needed to solve for "q" in triangle RTQ.

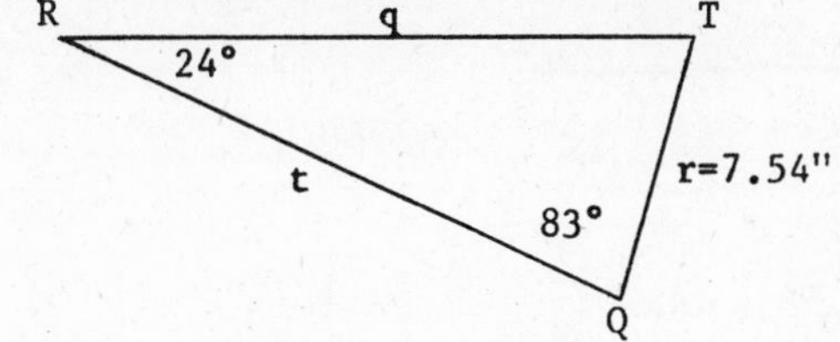

(a) Side "q" appears in what ratio? ________

(b) Which other ratio should we use? ________

(c) The proportion is: ________ = ________

---

40. Sometimes we have to use the angle-sum principle to find the size of the angle opposite the side we are trying to find. Here is an example:

Let's set up the proportion needed to solve for PQ in triangle PDQ.

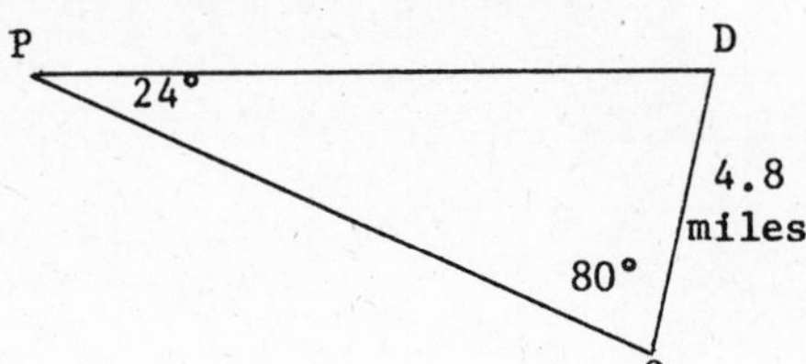

(a) How large is angle D? ________

(b) Side "PQ" appears in what ratio? ________

(c) What other ratio should we use? ________

(d) The proportion is: ________ = ________

a) $\frac{q}{\sin Q}$ or $\frac{q}{\sin 83°}$

b) $\frac{r}{\sin R}$ or $\frac{7.54''}{\sin 24°}$ (Since we know both "r" and "sin R".)

c) $\frac{q}{\sin 83°} = \frac{7.54''}{\sin 24°}$

---

41. Sometimes we have to use the angle-sum principle to obtain a second ratio in which we know both the side and the angle. Here is an example:

Let's set up the proportion needed to solve for side "FE" in triangle CFE.

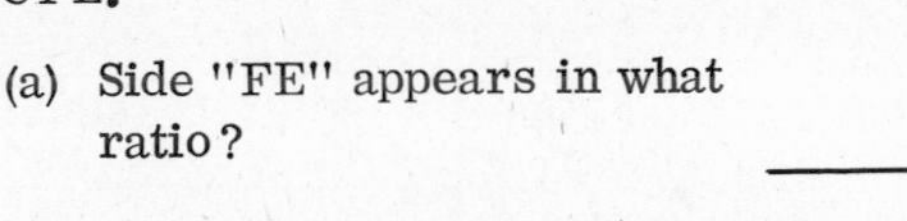

(a) Side "FE" appears in what ratio? ________

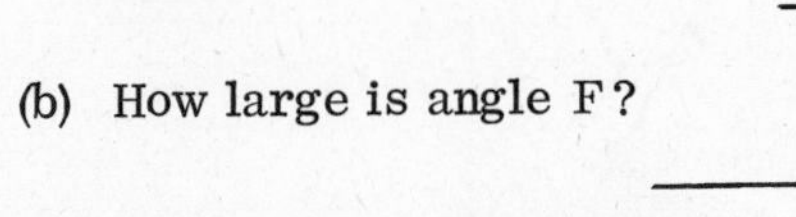

(b) How large is angle F? ________

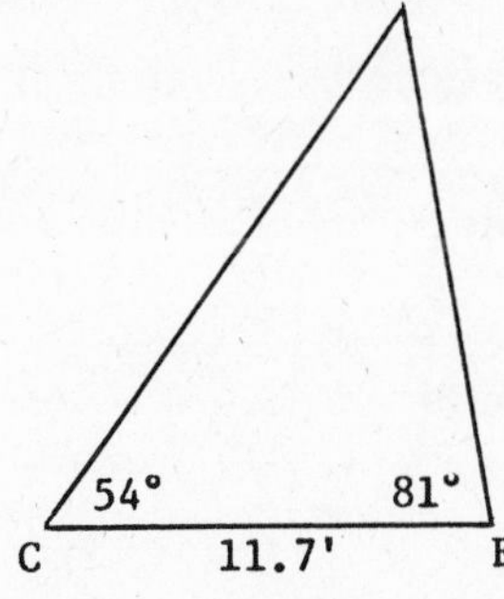

(c) What other ratio should we use? ________

(d) The proportion is: ________ = ________

a) 76°

b) $\frac{PQ}{\sin D}$ or $\frac{PQ}{\sin 76°}$

c) $\frac{DQ}{\sin P}$ or $\frac{4.8 \text{ miles}}{\sin 24°}$

d) $\frac{PQ}{\sin 76°} = \frac{4.8 \text{ miles}}{\sin 24°}$

42. (a) To solve for "k", what proportion would you use?

____

(b) To solve for "j", what proportion would you use?

____

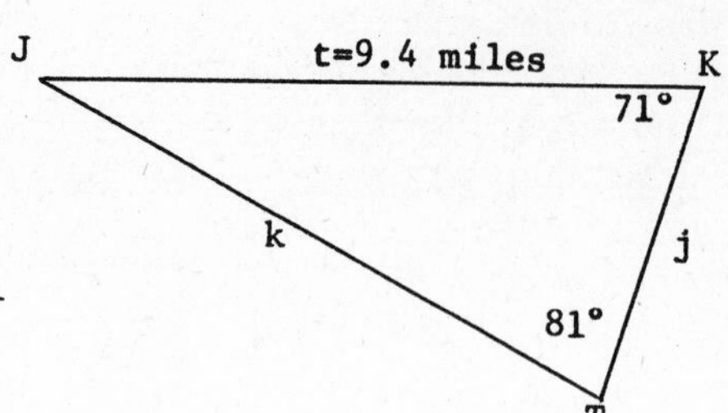

a) $\frac{FE}{\sin C}$ or $\frac{FE}{\sin 54°}$

b) 45°

c) $\frac{CE}{\sin F}$ or $\frac{11.7'}{\sin 45°}$

d) $\frac{FE}{\sin 54°} = \frac{11.7'}{\sin 45°}$

---

43. (a) To solve for VS, what proportion would you use?

____

(b) To solve for VC, what proportion would you use?

____

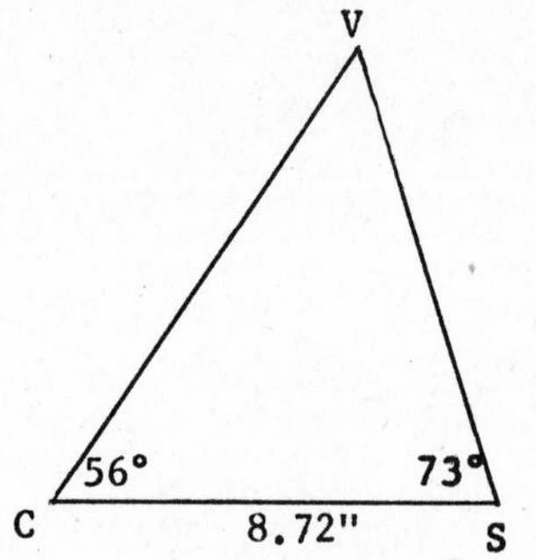

a) $\frac{k}{\sin K} = \frac{t}{\sin T}$

or

$\frac{k}{\sin 71°} = \frac{9.4 \text{ miles}}{\sin 81°}$

b) $\frac{j}{\sin J} = \frac{t}{\sin T}$

or

$\frac{j}{\sin 28°} = \frac{9.4 \text{ miles}}{\sin 81°}$

---

44. There is a general relationship between the sides and angles of any triangle which should be kept in mind when using the Law of Sines so that you do not compute blindly. The general relationship is this:

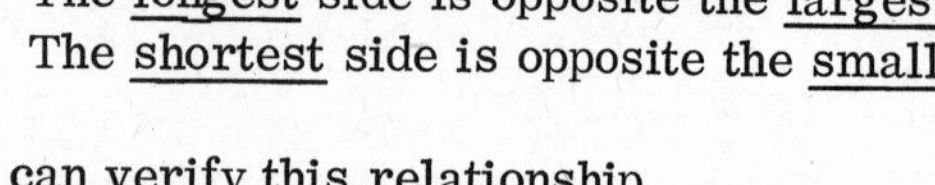

The <u>longest</u> side is opposite the <u>largest</u> angle.
The <u>shortest</u> side is opposite the <u>smallest</u> angle.

We can verify this relationship in triangle DHF:

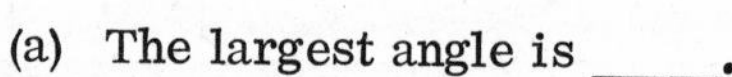

(a) The largest angle is _____.

(b) The longest side is _____.

(c) The smallest angle is _____.

(d) The shortest side is _____.

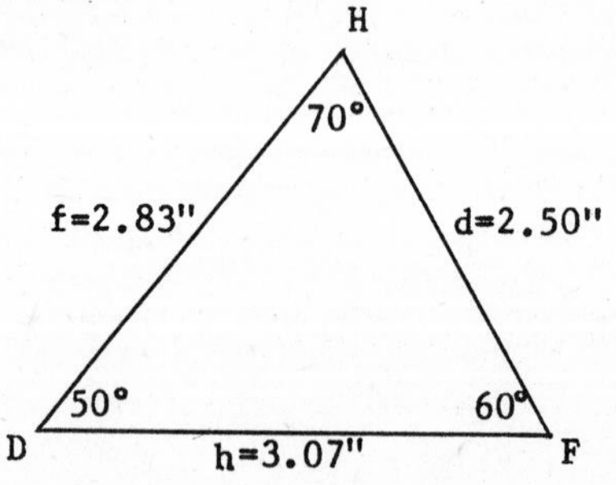

a) $\frac{VS}{\sin C} = \frac{CS}{\sin V}$

or

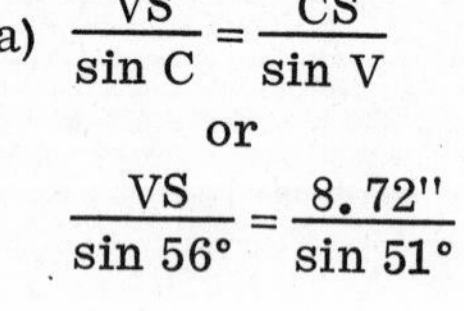

$\frac{VS}{\sin 56°} = \frac{8.72''}{\sin 51°}$

b) $\frac{VC}{\sin S} = \frac{CS}{\sin V}$

or

$\frac{VC}{\sin 73°} = \frac{8.72''}{\sin 51°}$

---

a) H (or 70°)

b) h (or 3.07")

c) D (or 50°)

d) d (or 2.50")

45. In this triangle:

(a) The largest side would have to be _____.

(b) The smallest side would have to be _____.

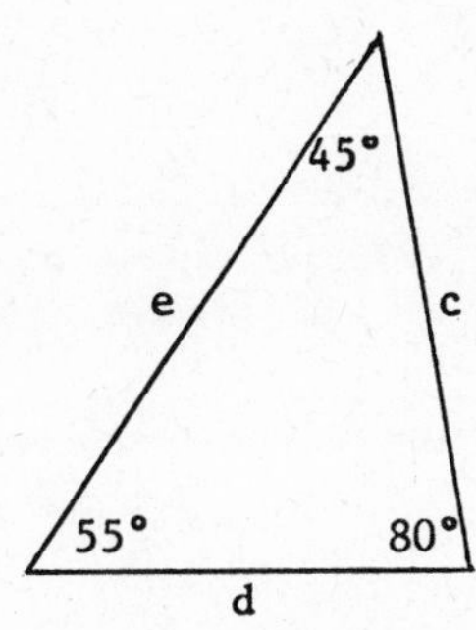

46. In triangle PTV, which side would be larger?

(a) PT or VT _______

(b) PT or PV _______

(c) PV or VT _______

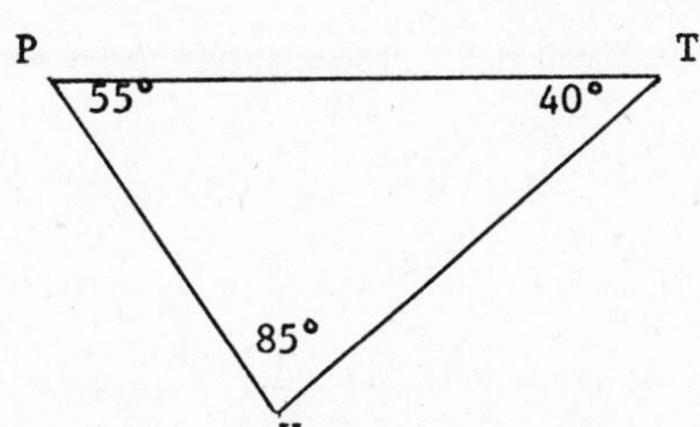

a) e

b) d

47. Let's solve for side "TV" in triangle TVW.

(a) Is TV longer or shorter than 61.0'? ___________

(b) What proportion must you use to solve for TV?

____

(c) Complete the solution. TV = __________ feet.

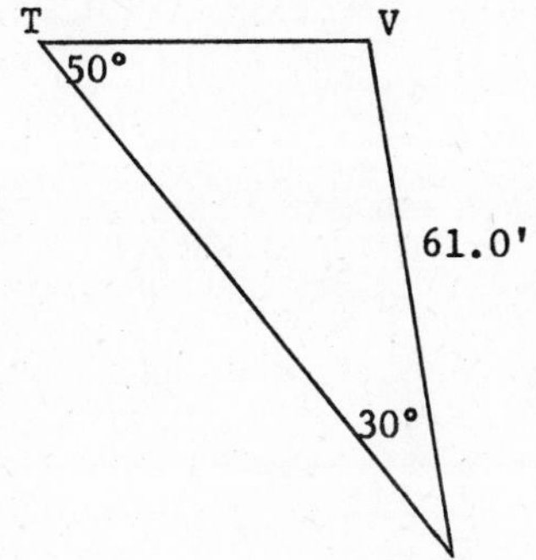

a) PT

b) PT

c) VT

a) Shorter, since VW is 61.0' and angle W is smaller than angle T.

b) $\frac{TV}{\sin W} = \frac{VW}{\sin T}$

or

$\frac{TV}{\sin 30°} = \frac{61.0'}{\sin 50°}$

c) TV = 39.8 feet

48. Let's solve for "b" in triangle ABC.

(a) How large is angle B? ________

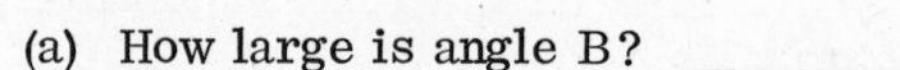

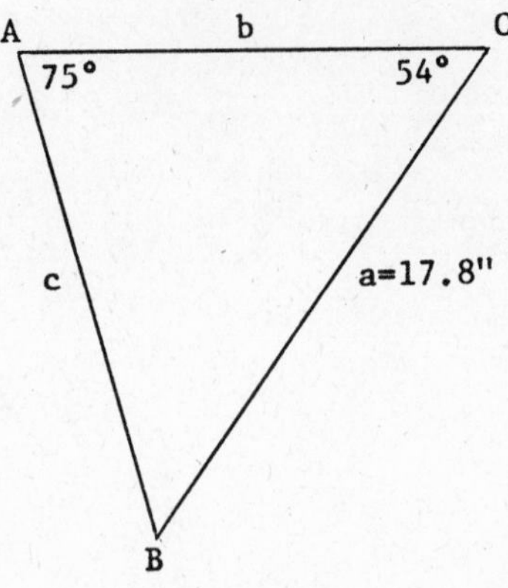

(b) Is "b" longer or shorter than 17.8"? ____________

(c) What proportion must you use to solve for "b"?

____

(d) Complete the solution: b = ________ inches.

---

49. Let's solve for "t" in triangle DFT.

(a) How large is angle F? ______

(b) Is "t" longer or shorter than 27.8'? ________

(c) What proportion must you use to solve for "t"?

____

(d) Complete the solution. t = ________ feet.

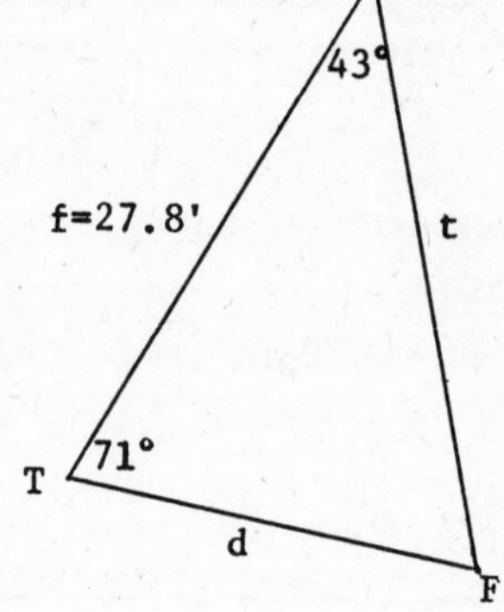

a) 51°

b) Shorter, since "a" is 17.8", and angle B is smaller than angle A.

c) $\frac{b}{\sin B} = \frac{a}{\sin A}$

or

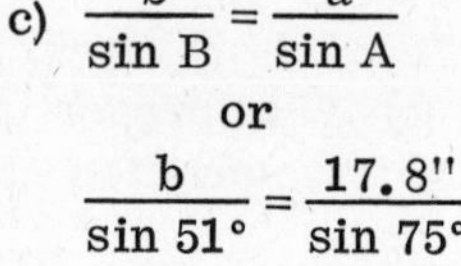

$\frac{b}{\sin 51°} = \frac{17.8''}{\sin 75°}$

d) b = 14.3 inches

---

a) 66°

b) Longer, since "f" is 27.8', and angle T is larger than angle F.

c) $\frac{t}{\sin 71°} = \frac{27.8'}{\sin 66°}$

d) t = 28.8 feet

SELF-TEST 4 (Frames 35-49)

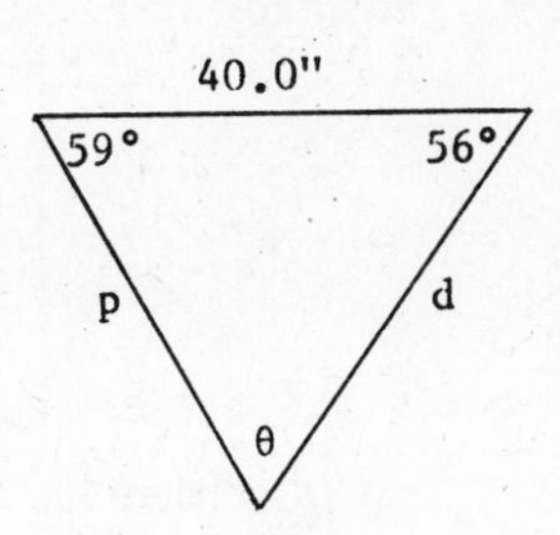

1. Find angle θ. θ = ________
2. Is side d greater or less than 40.0"? ________
3. Using the Law of Sines, find side d. d = ________
4. Is side p greater or less than side d? ________
5. Using the Law of Sines, find side p. p = ________

ANSWERS: 1. θ = 65° 2. Less 3. d = 37.8" 4. Less 5. p = 36.6"

4-5 USING THE LAW OF SINES TO FIND UNKNOWN ANGLES IN OBLIQUE TRIANGLES

In this section, we will use the Law of Sines to find unknown angles in oblique triangles. In order to simplify the solutions, proportions which contain obtuse angles (angles between 90° and 180°) will be avoided.

50. In an earlier section, we found the size of angle Q by the two-step, right-triangle method. We found that angle Q contains 36°.

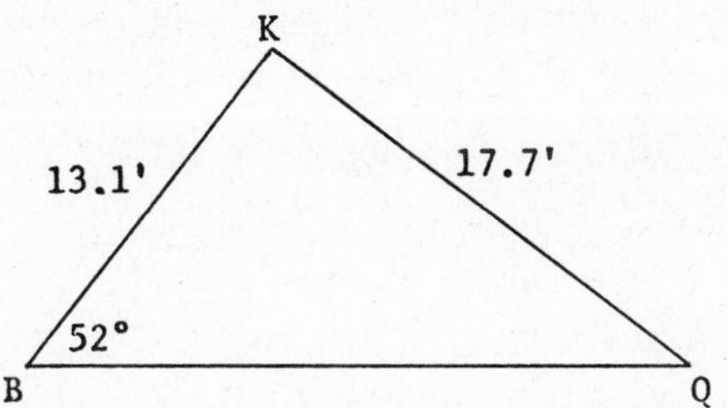

By using the Law of Sines, we can find the size of angle Q in one step. The steps are given below. Notice that we put the sines of the angles in the numerators of the ratios.

$$\frac{\sin Q}{BK} = \frac{\sin B}{KQ}$$

$$\frac{\sin Q}{13.1'} = \frac{\sin 52°}{17.7'}$$

$$\frac{\sin Q}{13.1'} = \frac{0.7880}{17.7'}$$

Now we can solve this last equation for "sin Q", and then use the "Trig Ratios" table to find the size of angle Q.

(a) $\sin Q = \frac{(13.1')(0.7880)}{17.7'} =$ ________

(b) Angle Q contains ______°.

(c) Is this the same answer we obtained earlier by the right-triangle method? ________

51. As we saw earlier, there are two equivalent forms of the Law of Sines. They are:

$$\frac{a}{\sin A} = \frac{b}{\sin B} = \frac{c}{\sin C}$$

$$\frac{\sin A}{a} = \frac{\sin B}{b} = \frac{\sin C}{c}$$

That is, either the sides or the sines of the angles can be written as the numerators of the ratios. Since it is easier to solve a proportion when the unknown is a numerator:

(1) We put "sides" in the numerator when solving for sides.
(2) We put "sines" in the numerator when solving for angles.

Which proportion below would we use to solve for angle D in this triangle? ________

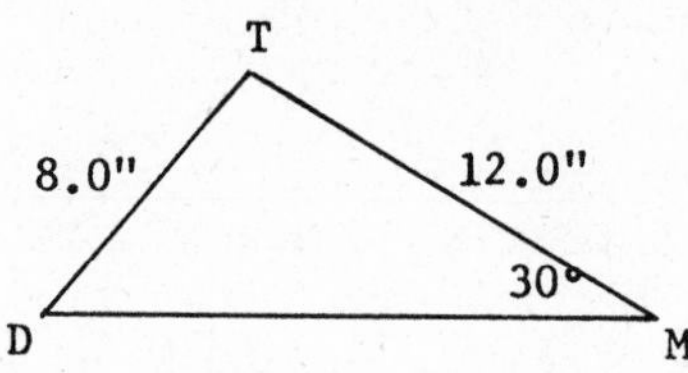

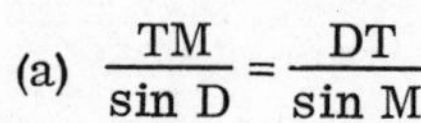

(a) $\frac{TM}{\sin D} = \frac{DT}{\sin M}$

(b) $\frac{\sin D}{TM} = \frac{\sin M}{DT}$

a) 0.583

b) 36°

c) Yes

---

52. When using the Law of Sines to solve for an unknown angle, we must set

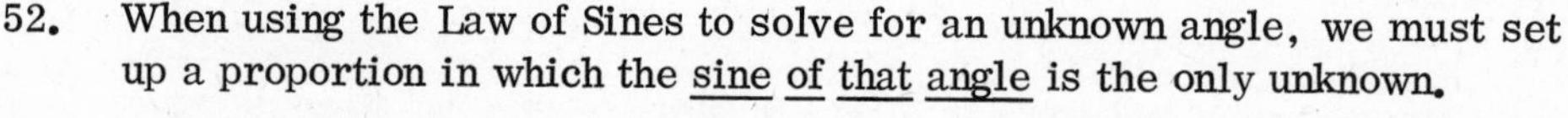

up a proportion in which the sine of that angle is the only unknown.

Below each triangle, write the proportion needed to solve for angle B:

(a)

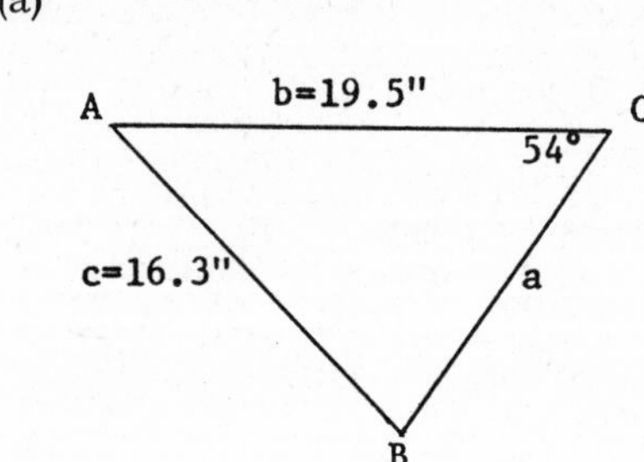

(b)

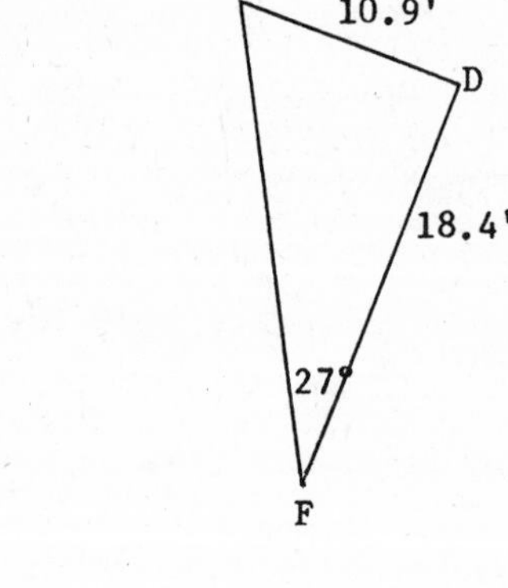

_______ = _______   _______ = _______

(b) The proportion with the "sines" in the numerator.

---

a) $\frac{\sin B}{b} = \frac{\sin C}{c}$ or $\frac{\sin B}{19.5''} = \frac{\sin 54°}{16.3''}$

b) $\frac{\sin B}{DF} = \frac{\sin F}{BD}$ or $\frac{\sin B}{18.4'} = \frac{\sin 27°}{10.9'}$

53. So that you do not compute blindly when solving for an angle in a triangle, remember these facts:

(1) The largest angle is opposite the longest side.
(2) The smallest angle is opposite the shortest side.

Therefore, in triangle CDM:

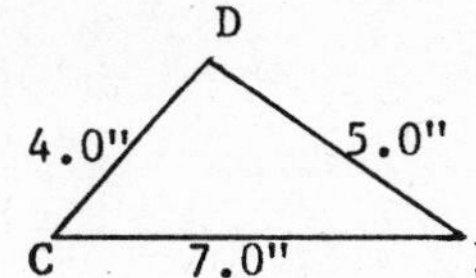

(a) The largest angle is angle _____.

(b) The smallest angle is angle _____.

---

54. In triangle PDK, angle D contains 50°.

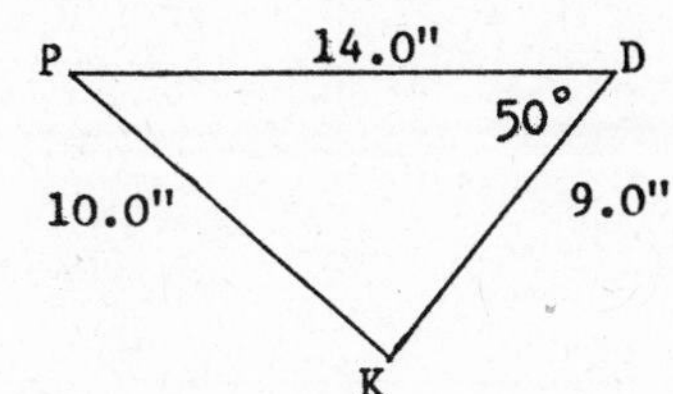

(a) Is angle P larger or smaller than 50°? ___________

(b) Is angle K larger or smaller than 50°? ___________

Answers to 53:
a) D
b) M

---

55. Let's use the Law of Sines to solve for angle T in this triangle.

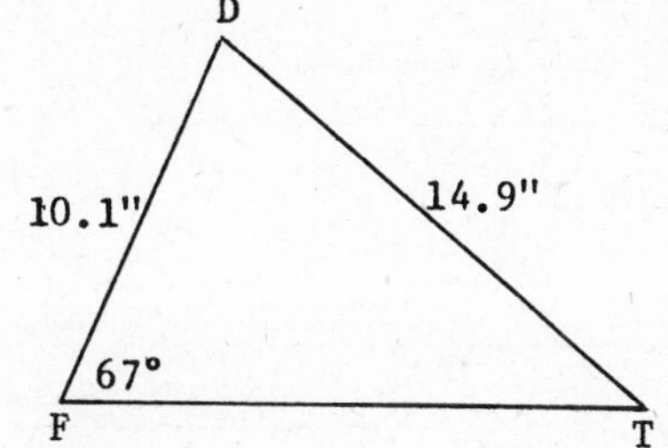

(a) Is angle T larger or smaller than 67°? ___________

(b) What proportion must we use to find "sin T"?

____

(c) sin T = __________

(d) Angle T contains _______°.

Answers to 54:
a) Smaller, since DK is shorter than PK.
b) Larger, since PD is longer than PK.

Answers to 55:
a) Smaller, since DF is shorter than DT.
b) $\frac{\sin T}{10.1''} = \frac{\sin 67°}{14.9''}$
c) 0.623
d) 39°

56. Let's solve for angle V in this triangle.

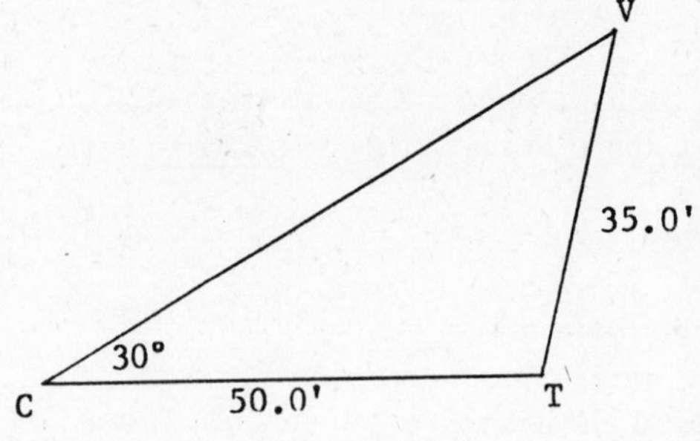

(a) Is angle V larger or smaller than 30°? ________

(b) What proportion is used to solve for "sin V"?

____

(c) sin V = ________

(d) Angle V contains ______°.

Answer to Frame 56:

a) Larger, since CT is longer than VT.

b) $\frac{\sin V}{50.0'} = \frac{\sin 30°}{35.0'}$

c) 0.714

d) 46°

57. (a) We do not need the Law of Sines to find angle T in this triangle. Why not? ________________

____

(b) Angle T contains ______°.

T

11.5"

65°　52°

M　11.3"　R

Answer to Frame 57: a) Since we know two angles, we can use the angle-sum principle to find angle T. (b) 63°

SELF-TEST 5 (Frames 50-57)

1. Is angle G greater or less than 29°? ________
2. Using the Law of Sines, find angle G. G = ______
3. Without using the Law of Sines, find Angle H. H = ______

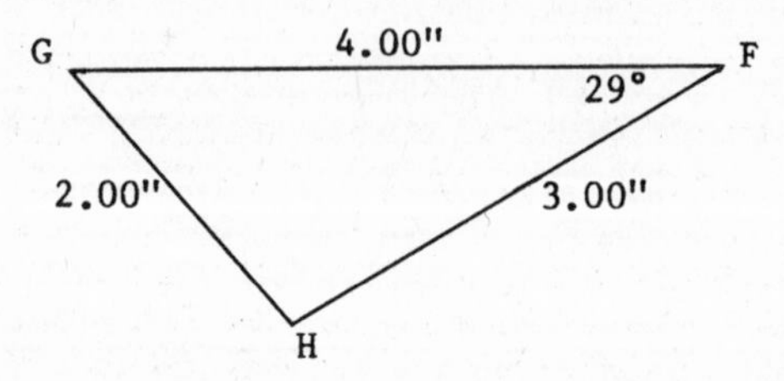

ANSWERS: 1. Greater 2. G = 47° 3. H = 104°

4-6 LIMITATIONS IN USING THE LAW OF SINES

The Law of Sines cannot be used for all triangle solutions. In this section, we will discuss the cases in which the Law of Sines can be used only indirectly. Then we will discuss the cases in which the Law of Sines cannot be used at all.

58. In order to solve for a side or angle of an oblique triangle by means of the Law of Sines, we must be able to write a proportion in which that side or the sine of that angle is the only unknown, since proportions with more than one unknown cannot be solved. Up to this point, we have seen two types of solutions:

(1) Those in which the proportion can be set up immediately. For example:

To solve for angle A in triangle ABT, we can set up this proportion immediately:

$$\frac{\sin A}{a} = \frac{\sin T}{t}$$

or

$$\frac{\sin A}{13.5''} = \frac{\sin 70°}{12.9''}$$

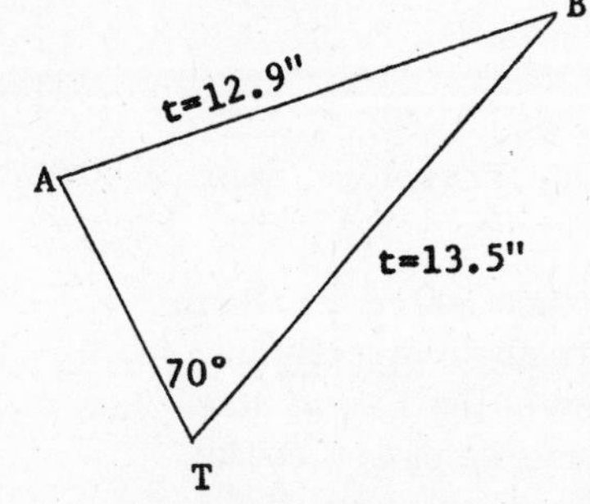

(2) Those in which the proportion can be set up only after the angle-sum principle is used. For example:

To solve for side "d" in triangle DRT, we must use the angle-sum principle to find the size of angle T. Then we can set up this proportion:

$$\frac{d}{\sin D} = \frac{t}{\sin T} \quad \text{or} \quad \frac{d}{\sin 83°} = \frac{10.0''}{\sin 60°}$$

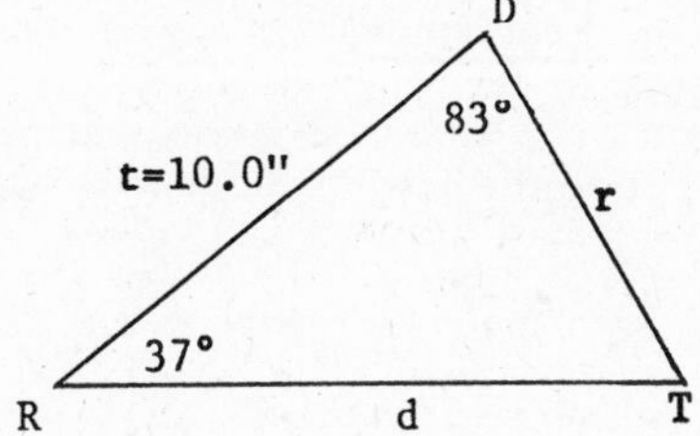

59. Sometimes, the Law of Sines cannot be used directly to find a side or angle even though it is part of the solution. Here is an example:

To solve for angle A in triangle ADF, we cannot directly set up a proportion in which "sin A" is the only unknown. The two possibilities are:

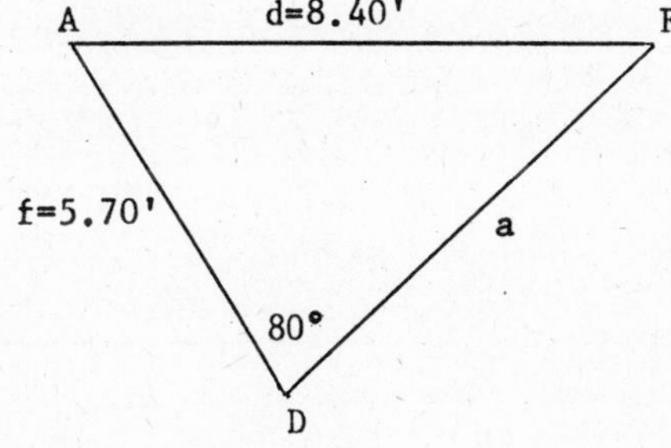

$$\frac{\sin A}{a} = \frac{\sin D}{d} \quad \text{or} \quad \frac{\sin A}{a} = \frac{\sin 80°}{8.40'}$$

(where sin A and side a are unknowns)

$$\frac{\sin A}{a} = \frac{\sin F}{f} \quad \text{or} \quad \frac{\sin A}{a} = \frac{\sin F}{5.70'}$$

(where sin A, side a, and sin F are unknowns)

(Continued on following page.)

59. (Continued)

Though we cannot solve for "sin A" directly, we can use the Law of Sines to solve for angle A indirectly. The two steps are:

(1) Using the Law of Sines to find angle F.
(2) Then using the angle-sum principle to find angle A.

Let's use this two-step process to find the size of angle A.

(a) sin F = ________

(b) Therefore, angle F contains ______°.

(c) Therefore, angle A contains ______°.

Answers: a) sin F = 0.668 b) 42° c) 58°

---

60. We cannot solve for GK in this triangle directly by means of the Law of Sines because we cannot set up a proportion in which GK is the only unknown. However, we can solve for GK in three steps. They are:

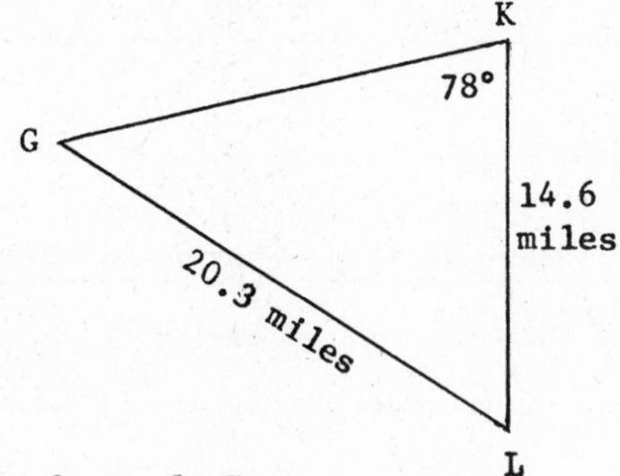

(a) Use the Law of Sines to solve for angle G.

Angle G = ______°

(b) Use the angle-sum principle to solve for angle L.

Angle L = ______°

(c) Now use the Law of Sines to find the length of GK.

GK = ________ miles

Answers: a) G = 45° b) L = 57° c) GK = 17.4 miles

61. The Law of Sines cannot be used for all triangle solutions. Here is an example of a case in which it does not work.

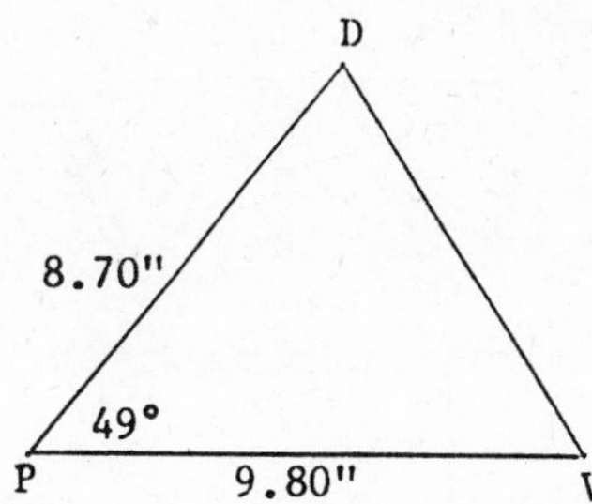

We want to find the length of DV in triangle DPV. We know two sides (DP and PV) and the angle between them (angle P).

Here are the three ratios we can set up:

$$\frac{DV}{\sin P} = \frac{DP}{\sin V} = \frac{PV}{\sin D} \quad \text{or} \quad \frac{DV}{\sin 49°} = \frac{8.70''}{\sin V} = \frac{9.80''}{\sin D}$$

Notice these points: (1) There is one unknown in each ratio.
(2) We cannot use the angle-sum principle to find the size of either unknown angle.

Here are the two proportions we can set up that include DV:

$$(1)\ \frac{DV}{\sin 49°} = \frac{8.70''}{\sin V} \qquad (2)\ \frac{DV}{\sin 49°} = \frac{9.80''}{\sin D}$$

Since there are two unknowns in each proportion, we cannot solve for DV in either. To solve for DV, either "sin V" or "sin D" must be known.

When each of the three ratios contains an unknown, and if we cannot use the angle-sum principle to find one of the unknown angles, the Law of Sines cannot be used.

---

62. Here is a triangle in which we know the length of the three sides. Obviously, we cannot use the angle-sum principle to find any of the three angles.

Here are the three ratios:

$$\frac{\sin P}{AK} = \frac{\sin A}{KP} = \frac{\sin K}{AP}$$

or

$$\frac{\sin P}{16.6''} = \frac{\sin A}{10.0''} = \frac{\sin K}{16.9''}$$

(a) What angle could we solve for by using the Law of Sines?

____

(b) Why not? ________________

____

a) None of them.

b) Because each proportion that can be set up will have two unknowns, and such a proportion cannot be solved.

63. The three ratios in triangle QRS are:

$$\frac{QS}{\sin R} = \frac{QR}{\sin S} = \frac{SR}{\sin Q}$$

or

$$\frac{QS}{\sin 46°} = \frac{15.7'}{\sin S} = \frac{16.9'}{\sin Q}$$

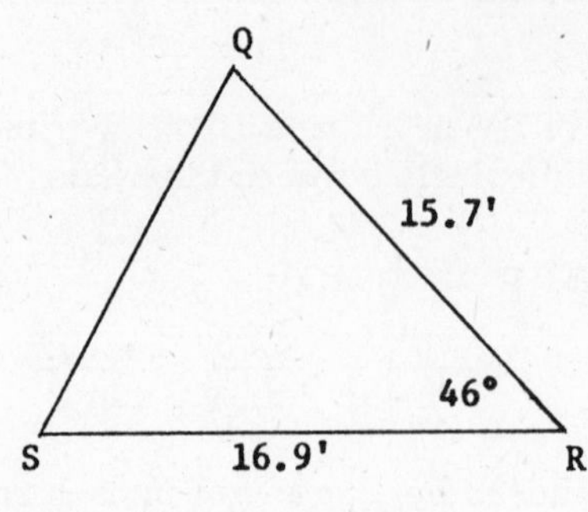

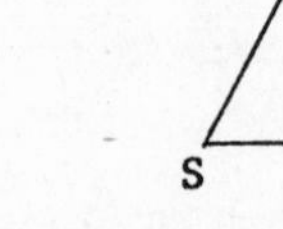

Using the Law of Sines, could we solve for:

(a) Side QS? _____ (b) Angle S? _____ (c) Angle Q? _____

---

a) No

b) No

c) No

---

64. Given the information in triangle ABC:

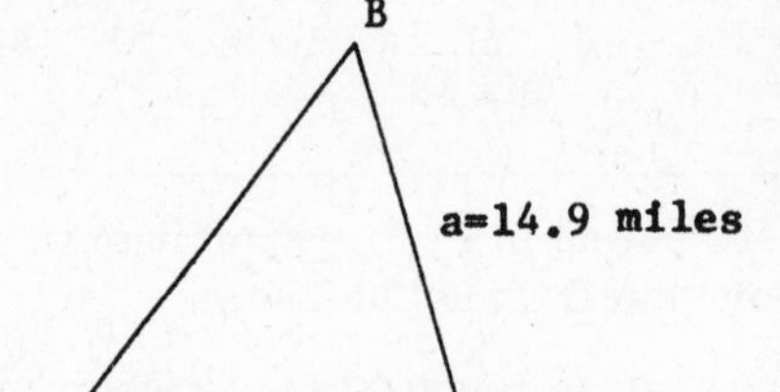

(a) For which one of the three ratios is both numerator and denominator known?

_____

(b) Could you use the Law of Sines to solve directly for angle B? _____

(c) How could you solve for angle C? _____

_____

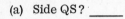

a) $\frac{a}{\sin A}$ or $\frac{14.9}{\sin 52°}$

b) Yes, since "b" is known.

c) First find angle B and then use the angle-sum principle.

---

65. In triangle DEF, the three sides are known.

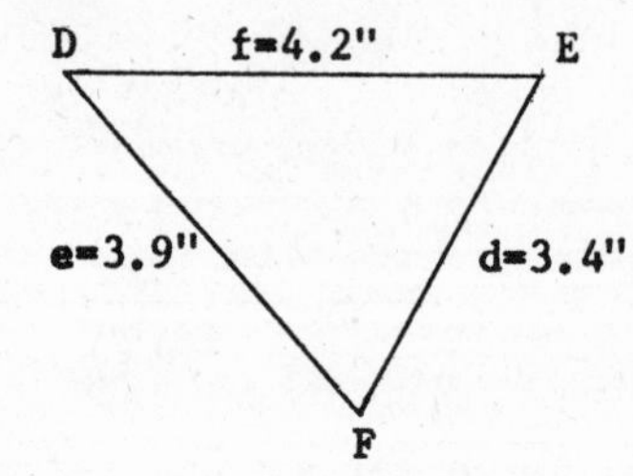

Using the Law of Sines, you cannot solve for any of the three angles. Why not?

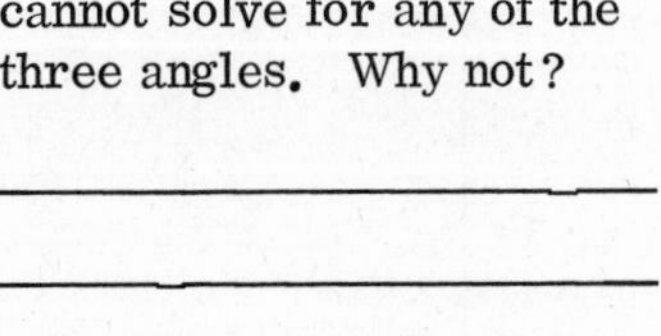

_____

_____

_____

Because the "sine" in each ratio is unknown, and we cannot use the angle-sum principle to find any of the angles.

66. Here is a triangle in which we know two angles and the side between them. Superficially, it looks as if we cannot use the Law of Sines because we do not know an angle plus the side opposite it.

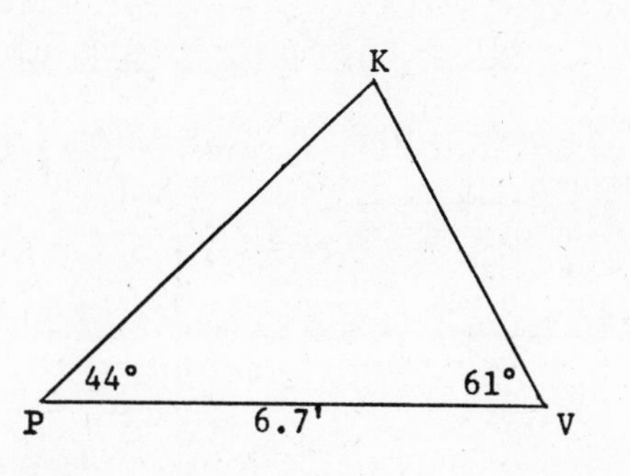

(a) Even though it is not listed on the triangle, angle K contains _______°.

(b) Could we use the Law of Sines to solve for KV? _________

Answer to Frame 66: a) 75° b) Yes, since we know angle K and side PV.

67. In order to use the Law of Sines, we must know both parts of at least one ratio. There are two possibilities:

(1) A side and the angle opposite it are given.
(2) A side is given, and the angle opposite it can be found immediately by using the angle-sum principle.

There are only two cases in which the Law of Sines cannot be used. Here are examples of each case:

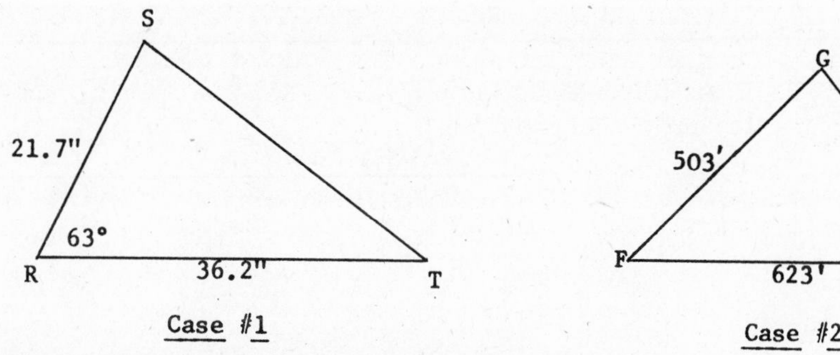

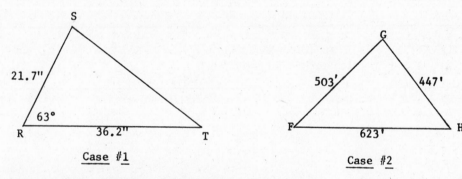

In Case #1, we know two sides and the angle between them.
In Case #2, we know three sides but no angles.

In each of these two cases, we do not know both parts of at least one ratio, and we cannot find both parts of one ratio by an immediate use of the angle-sum principle. Therefore, the Law of Sines cannot be used to find unknown parts in either triangle.

SELF-TEST 6 (Frames 58-67)

1. Can the Law of Sines be used to find angle Q in this triangle? ________

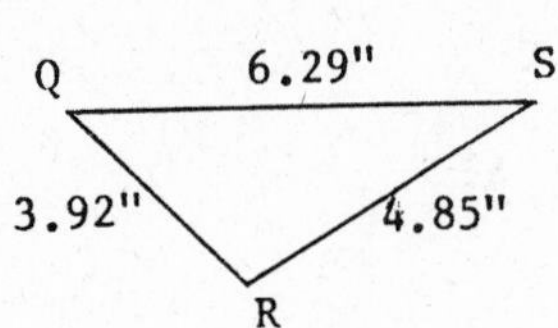

2. Can the Law of Sines be used to find side p in this triangle? ________

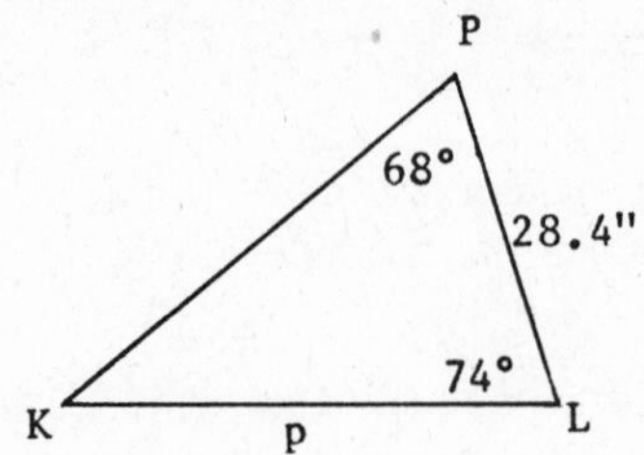

3. Can the Law of Sines be used to find angle A in this triangle? ________

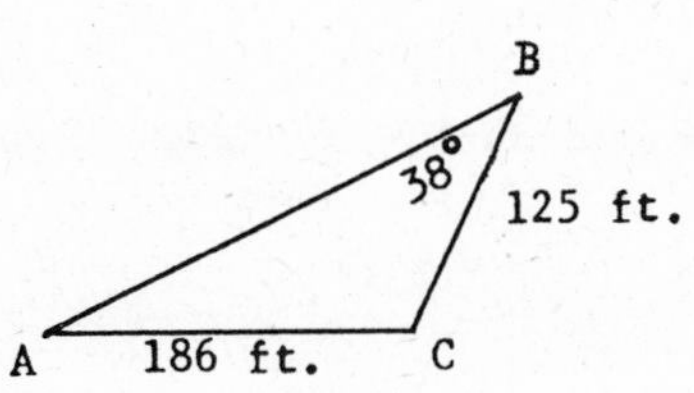

4. Can the Law of Sines be used to find side h in this triangle? ________

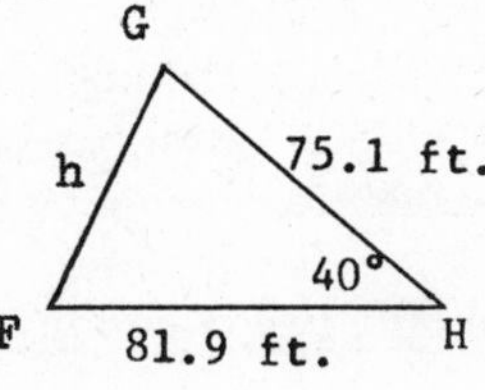

ANSWERS: 1. No 2. Yes. (First find angle K by the angle-sum principle.) 3. Yes 4. No

4-7 THE LAW OF COSINES

When the Law of Sines cannot be used to solve an oblique triangle, we can solve it by means of a relationship called the "Law of Cosines". In this section, we will discuss and prove the Law of Cosines. Then in the next section, we will use the Law of Cosines to find unknown sides and angles of oblique triangles.

68. In triangle CFK on the left below (Figure 1), we are given the lengths of two sides ("c" and "f") and the size of the angle (K) between them. We want to find the length of side "k". As we saw in the last section, the Law of Sines cannot be used to find "k" in this triangle.

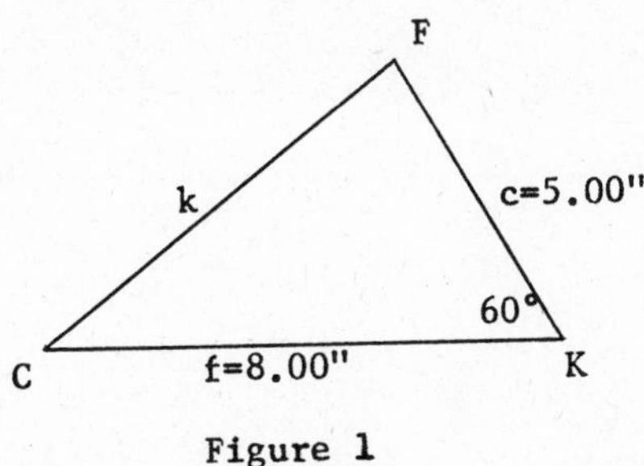

Figure 1

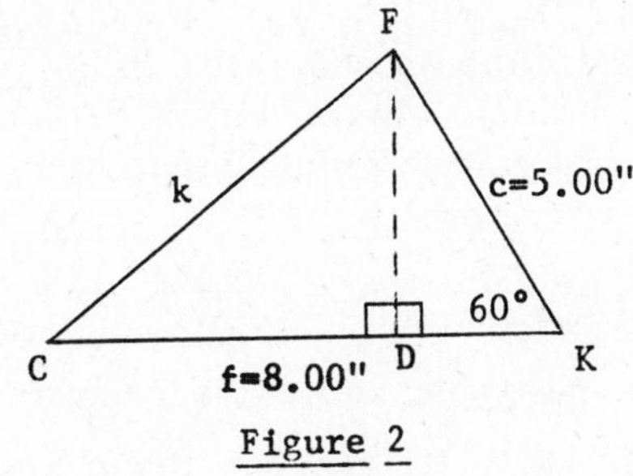

Figure 2

One way to find "k" is by the right-triangle method. In order to use this method, we have divided triangle CFK into two right triangles by drawing altitude FD in the figure on the right above (Figure 2). Now we can find "k" by the series of steps below:

Step 1: Finding the length of FD. $\sin 60° = \frac{FD}{c}$

$$0.8660 = \frac{FD}{5.00''}$$

$$FD = 4.33''$$

Step 2: Finding the length of DK. $\cos 60° = \frac{DK}{c}$

$$0.5000 = \frac{DK}{5.00''}$$

$$DK = 2.50''$$

Step 3: Finding the length of CD. Since $CK = CD + DK$, $CD = CK - DK$

$$CD = 8.00'' - 2.50''$$

$$CD = 5.50''$$

Step 4: Finding the length of "k".

Since "k" is the hypotenuse of right triangle CFD and we know the length of both legs ($CD = 5.50''$ and $FD = 4.33''$), we can use the Pythagorean Theorem to find "k", as follows:

$$k^2 = (CD)^2 + (FD)^2$$

$$k^2 = (5.50)^2 + (4.33)^2$$

$$k = \sqrt{(5.50)^2 + (4.33)^2} = \sqrt{30.2 + 18.8} = \sqrt{49.0}$$

$k = 7.00''$ (This is the solution of the problem.)

69. In the last frame, we solved for "k" in triangle CFK by the right-triangle method. The solution was: k = 7.00". Since the right-triangle method required four steps, it was rather tedious.

There is a relationship called the "Law of Cosines" which can be used to solve for "k" in one step. The specific "Law-of-Cosine" formula needed to solve for "k" is given in the box below.

$$k^2 = c^2 + f^2 - 2cf \cos K$$

Notice these points about the formula:

(1) "k" is the side opposite angle K, the known angle.

(2) "c" and "f" are the two known sides which form the known angle.

(3) "2cf cos K" is merely the product of four factors, including the two known sides and the cosine of the known angle. We could write it with parentheses around each factor. It then becomes: (2)(c)(f)(cos K)

Let's plug in the formula and see if we get 7.00" for "k":

$$k^2 = (5.00)^2 + (8.00)^2 - (2)(5.00)(8.00)(0.5000)$$

$$= 25.0 + 64.0 - 10.0(8.00)(0.5000)$$

$$= 25.0 + 64.0 - (80.0)(0.5000)$$

$$= 25.0 + 64.0 - 40.0$$

$$= 89.0 - 40.0$$

$$= 49.0$$

If $k^2 = 49.0$, then k = 7.00".

Is this the same solution we obtained by the right-triangle method in the last frame? ________

Yes

---

70. The Law of Cosines is a rather complicated relationship. Not only is the formula itself complicated, but for any given triangle, we can state three formulas based on the Law of Cosines. It is important that you understand what is meant by this law.

In this frame, we will discuss the Law of Cosines as it applies to oblique triangle DMT.

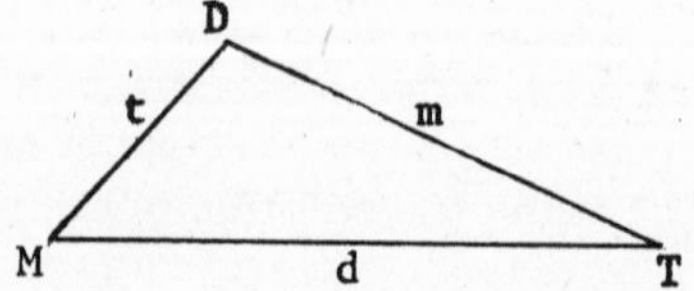

(Continued on following page.)

70. (Continued)

One formula based on the Law of Cosines is this:

$$t^2 = m^2 + d^2 - 2md \cos T$$

Note: (1) "t" is the side opposite angle T, and "cos T" appears in the formula.

(2) "m" and "d" are the other two sides.

A second formula based on the Law of Cosines is this:

$$m^2 = t^2 + d^2 - 2td \cos M$$

Note: (1) "m" is the side opposite angle M, and "cos M" appears in the formula.

(2) "t" and "d" are the other two sides.

Complete the third formula based on the Law of Cosines:

$d^2 = t^2 + m^2 -$ [ ]

Answer: $d^2 = t^2 + m^2 -$ $\boxed{2tm \cos D}$

71. Using triangle SVT as an example, let's translate one of the formulas based on the Law of Cosines into words:

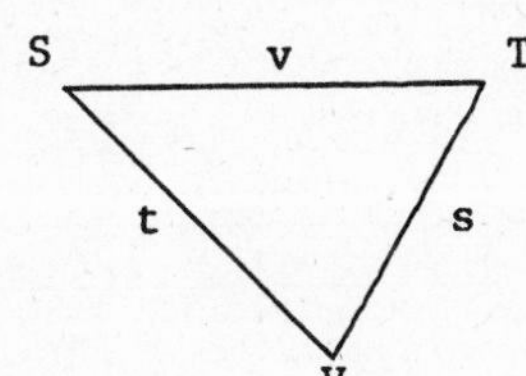

t^2	=	$v^2 + s^2$	-	$2vs$	$\cos T$
a side squared	=	sum of the squares of the other two sides	-	twice the product of the other two sides	times the cosine of the opposite angle

Complete the other two "Law-of-Cosine" formulas for triangle SVT:

(a) $v^2 = t^2 + s^2 -$ [ ]

(b) $s^2 =$ ________

Answers:

a) $v^2 = t^2 + s^2 -$ $\boxed{2ts \cos V}$

b) $s^2 = t^2 + v^2 - 2tv \cos S$

72. Given any triangle, we can state three formulas based on the Law of Cosines.

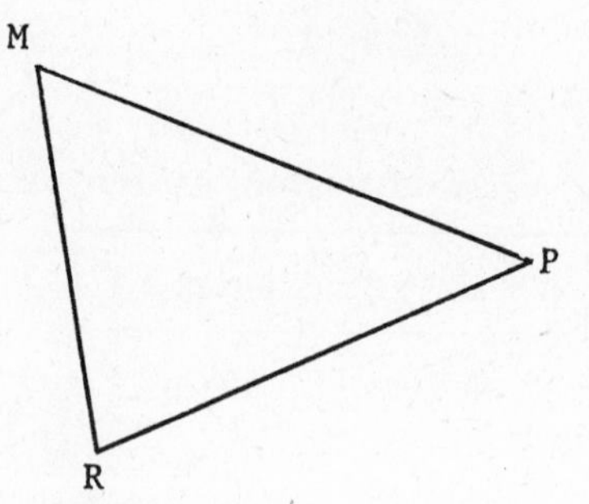

Here is one of the three formulas for triangle MPR:

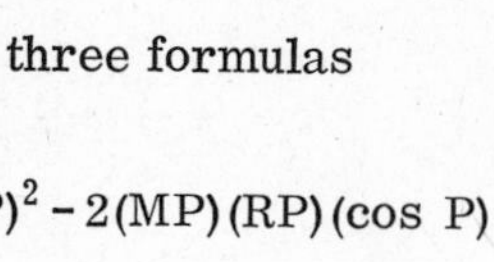

$(MR)^2 = (MP)^2 + (RP)^2 - 2(MP)(RP)(\cos P)$

Complete the two other formulas for triangle MPR:

(a) $(MP)^2 = (MR)^2 + (RP)^2 -$ [ ]

(b) $(RP)^2 =$ [ ] $- 2(MR)(MP)(\cos M)$

---

73. State the three formulas based on the Law of Cosines for triangle BDF:

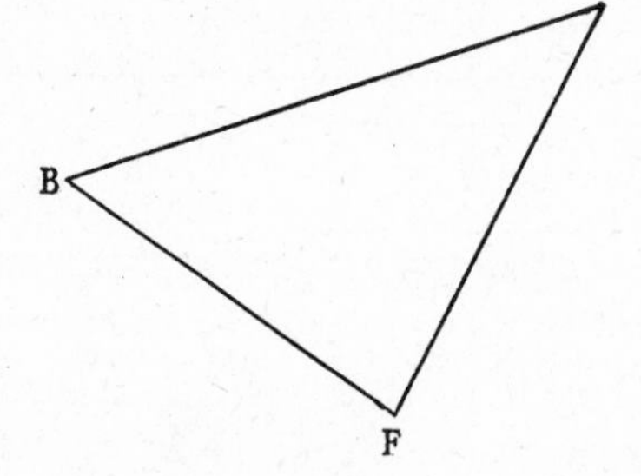

(a) $(BF)^2 =$ ____________________

____

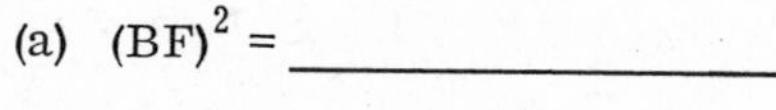

(b) $(BD)^2 =$ ____________________

____

(c) $(FD)^2 =$ ____

Answers to Frame 72:

a) $2(MR)(RP)(\cos R)$

b) $(MR)^2 + (MP)^2$

Answer to Frame 73:

a) $(BF)^2 = (BD)^2 + (FD)^2 - 2(BD)(FD)(\cos D)$

b) $(BD)^2 = (BF)^2 + (FD)^2 - 2(BF)(FD)(\cos F)$

c) $(FD)^2 = (BF)^2 + (BD)^2 - 2(BF)(BD)(\cos B)$

---

74. We will prove the Law of Cosines in this frame. Though we will not expect you to memorize or reproduce this proof, we want you to see that a general proof is possible.

There are four preliminary points you must know in order to understand the proof:

(1) The symbol for squaring "sin A" is $\sin^2 A$.
The symbol for squaring "cos A" is $\cos^2 A$.

Therefore: $\sin^2 47° = (\sin 47°)(\sin 47°)$
$\sin^2 C = (\sin C)(\sin C)$
$\cos^2 47° = (\cos 47°)(\cos 47°)$

(2) Given any angle A, $\sin^2 A + \cos^2 A = 1$

For example: Since $\sin 47° = 0.7314$ and $\cos 47° = 0.6820$,

$$\sin^2 47° + \cos^2 47° = (0.7314)^2 + (0.6820)^2$$
$$= 0.535 + 0.465 \quad \text{(slide rule)}$$
$$= 1$$

(3) $(ab)^2 = (ab)(ab) = a^2b^2$
$(a \cos C)^2 = (a \cos C)(a \cos C) = a^2\cos^2 C$

(Continued on following page.)

74. (Continued)

(4) $(a - b)^2 = (a - b)(a - b) = a^2 - 2ab + b^2$

$(b - a \cos C)^2 = (b - a \cos C)(b - a \cos C)$

$= b^2 - 2ab \cos C + a^2\cos^2 C$

To prove the Law of Cosines, we will use triangle ABC as an example. We will prove the following formula based on the Law of Cosines for that triangle.

$$\boxed{c^2 = a^2 + b^2 - 2ab \cos C}$$

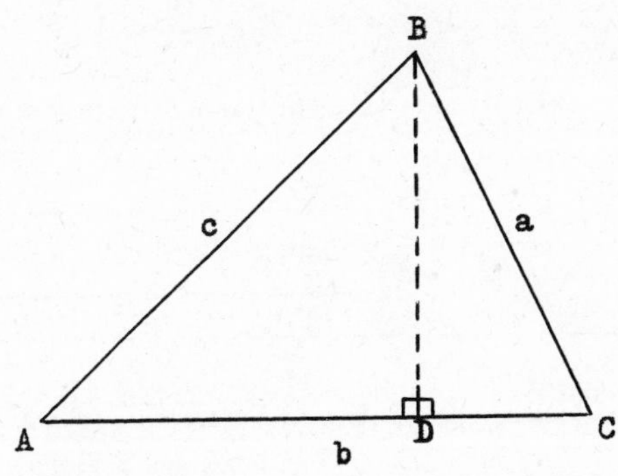

We have drawn altitude BD. Since triangle ABD is a right triangle, the Pythagorean Theorem gives us the following formula:

$$\boxed{c^2 = (BD)^2 + (AD)^2}$$

We will now show that the two boxed equations are equivalent by expressing $(BD)^2$ and $(AD)^2$ in terms of "a", "b" and angle C.

Finding $(BD)^2$: Since $\sin C = \frac{BD}{a}$,

$BD = a \sin C$

and $(BD)^2 = \boxed{a^2\sin^2 C}$

Finding $(AD)^2$: $AD = b - DC$

But $DC = a \cos C$, since $\cos C = \frac{DC}{a}$

Therefore: $AD = b - a \cos C$

and $(AD)^2 = (b - a \cos C)^2 = \boxed{b^2 - 2ab \cos C + a^2\cos^2 C}$

Now we will substitute the boxed expressions for $(BD)^2$ and $(AD)^2$ into the Pythagorean relationship.

If $c^2 = (BD)^2 + (AD)^2$: $c^2 = \boxed{a^2\sin^2 C} + \boxed{b^2 - 2ab \cos C + a^2\cos^2 C}$

By rearranging we get: $c^2 = a^2\sin^2 C + a^2\cos^2 C + b^2 - 2ab \cos C$

By factoring the first two terms we get: $c^2 = a^2(\sin^2 C + \cos^2 C) + b^2 - 2ab \cos C$

Since $\sin^2 C + \cos^2 C = 1$,

we get: $c^2 = a^2(1) + b^2 - 2ab \cos C$

or

$$\boxed{c^2 = a^2 + b^2 - 2ab \cos C}$$ (Law of Cosines)

SELF-TEST 7 (Frames 68-74)

For triangle RST, one formula based on the Law of Cosines is:

$$t^2 = r^2 + s^2 - 2rs \cos T$$

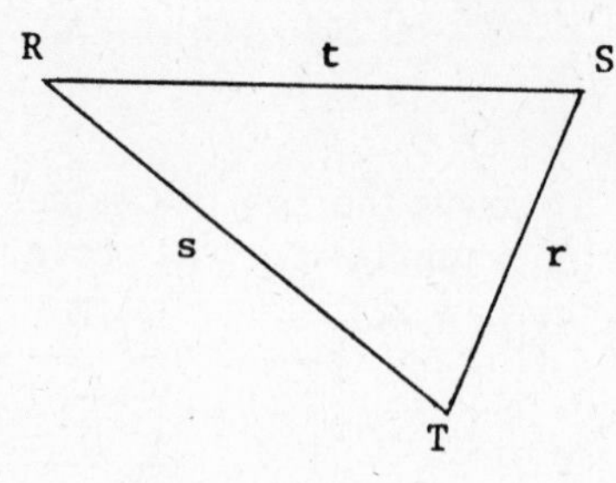

Similarly, use the Law of Cosines to complete these formulas:

1. $r^2 =$ ______________________

2. $s^2 =$ ______________________

ANSWERS:

1. $r^2 = s^2 + t^2 - 2st \cos R$
2. $s^2 = r^2 + t^2 - 2rt \cos S$

4-8 USING THE LAW OF COSINES TO FIND UNKNOWN SIDES AND ANGLES OF OBLIQUE TRIANGLES

In this section, we will use the Law of Cosines to find unknown sides and angles of oblique triangles. In order to simplify the solutions, obtuse angles will be avoided.

75. We can use the Law of Cosines to find the length of KD in triangle DKL.

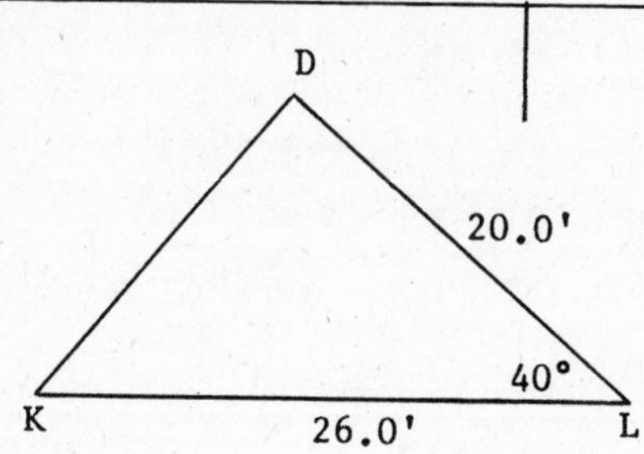

From the Law of Cosines, we know that:

$$(KD)^2 = (DL)^2 + (KL)^2 - 2(DL)(KL)(\cos L)$$

Plugging in the known values, we get:

$$(KD)^2 = (20.0)^2 + (26.0)^2 - 2(20.0)(26.0)(0.7660)$$

Finish the computations:

(a) $(KD)^2 =$ __________ (b) $KD =$ __________

a) 279 or 280

b) 16.7'

76. In triangle FGT, two sides and their included angle are known. The Law of Sines <u>cannot</u> be used to find the length of side FT. Therefore, what formula would we use to find the length of side FT?

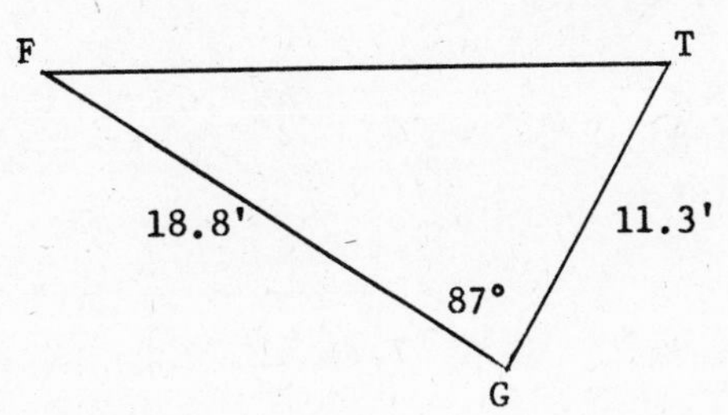

____

$(FT)^2 = (FG)^2 + (GT)^2 - 2(FG)(GT)(\cos G)$

or

$(FT)^2 = (18.8)^2 + (11.3)^2 - 2(18.8)(11.3)(\cos 87°)$

---

77. Triangle DMV cannot be solved by the Law of Sines because all of the angles are unknown. We can, however, use the Law of Cosines to find the size of angle D. To do so, we use the following formula:

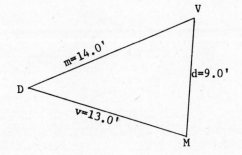

$$d^2 = m^2 + v^2 - 2mv \cos D$$

or

$$(9.0)^2 = (14.0)^2 + (13.0)^2 - 2(14.0)(13.0)(\cos D)$$

Notice these points:

(1) Of the three possible relationships based on the Law of Cosines, we picked the one in which "cos D" appears. (This is the one in which "d" appears alone on the left side of the equation. "d" is the side opposite angle D, the angle we are solving for.)

(2) Since we know the length of the three sides, "cos D" is the only unknown in the equation.

Here is the equation again. Let's complete the solution.

$$(9.0)^2 = (14.0)^2 + (13.0)^2 - 2(14.0)(13.0)(\cos D)$$

$$81 = \underbrace{196 + 169}_{365} - 364(\cos D)$$

$$81 = 365 - 364(\cos D)$$

Adding (-365) to both sides, we get:

$$-284 = -364(\cos D)$$

Taking the opposite of both sides, we get:

$$284 = 364(\cos D)$$

Multiplying both sides by the reciprocal of 364, we get:

$$\frac{284}{364} = \cos D \quad \text{or} \quad \cos D = \frac{284}{364}$$

(a) Using your slide rule, cos D = __________

(b) Therefore: D = _______°

---

a) cos D = 0.780

b) D = 39°

78. We are given only the three sides in triangle FGH. Therefore, we must use the Law of Cosines to find the size of angle H. What formula would we use?

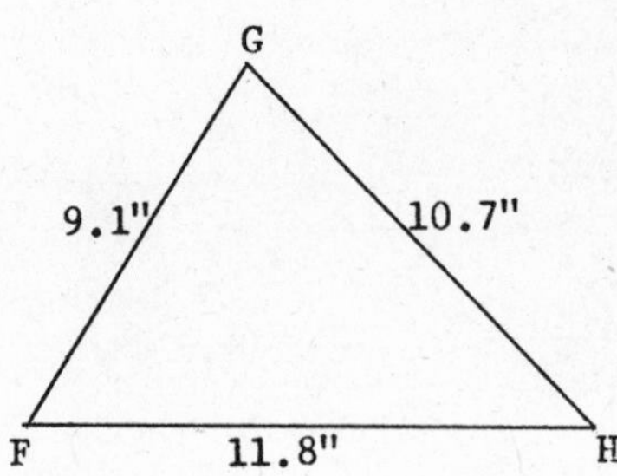

____

Answer to 78:

$(FG)^2 = (FH)^2 + (GH)^2 - 2(FH)(GH)(\cos H)$

$(9.1)^2 = (11.8)^2 + (10.7)^2 - 2(11.8)(10.7)(\cos H)$

---

79. The three sides are given in triangle DKS.

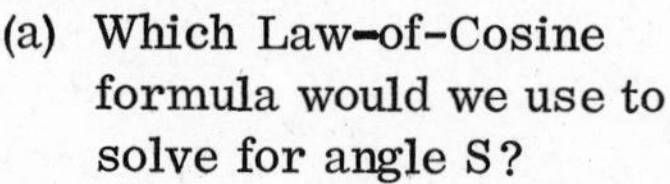

(a) Which Law-of-Cosine formula would we use to solve for angle S?

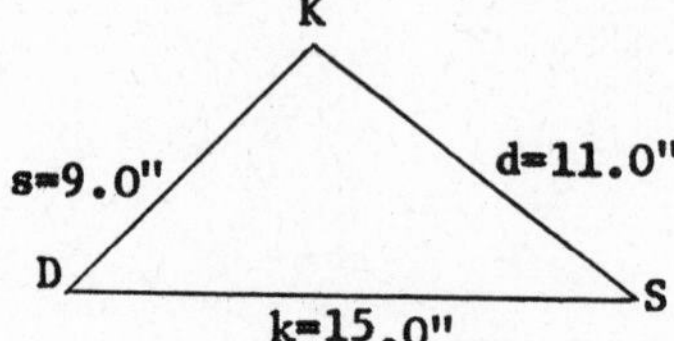

____

(b) Plug in the known values and solve for "cos S". The numerical value of cos S is: cos S = ________

(c) Therefore, angle S contains ______°.

Answer to 79:

a) $s^2 = d^2 + k^2 - 2dk \cos S$

or

$(9.0)^2 = (11.0)^2 + (15.0)^2 - 2(11.0)(15.0)(\cos S)$

b) $\cos S = 0.803$

c) 37°

80. In the last frame, we found that angle S in triangle DKS contains 37°. In that frame, we solved for "cos S" by plugging the known values into the following formula:

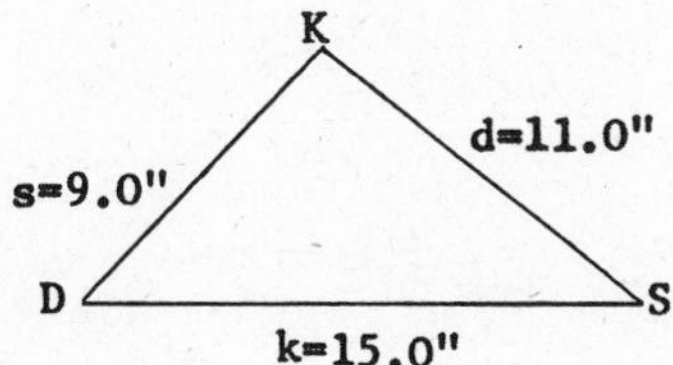

$$\boxed{s^2 = d^2 + k^2 - 2dk \cos S}$$

Instead of plugging the known values into the formula above as it stands, by algebraic manipulations we could rearrange for "cos S" <u>before plugging in the known values</u>. Let's do so.

(1) By adding $(-d^2) + (-k^2)$ to both sides, we get:

$$s^2 + (-d^2) + (-k^2) = -2dk \cos S$$

(2) Taking the opposite of both sides of the equation, we get:

$$(-s^2) + d^2 + k^2 = 2dk \cos S$$

or

$$d^2 + k^2 - s^2 = 2dk \cos S$$

(3) Multiplying both sides by the reciprocal of "2dk", we get:

$$\left(\frac{1}{2dk}\right)(d^2 + k^2 - s^2) = \underbrace{\left(\frac{1}{2dk}\right)2dk}_{(1)} \cos S$$

$$\frac{d^2 + k^2 - s^2}{2dk} = (1)\cos S$$

or

$$\boxed{\cos S = \frac{d^2 + k^2 - s^2}{2dk}}$$

(4) Plugging in the known values, we get:

$$\cos S = \frac{(11.0)^2 + (15.0)^2 - (9.0)^2}{2(11.0)(15.0)}$$

Let's complete the solution by this new method:

(a) cos S = ___________

(b) S = _______°

(c) Do we get the same solution we obtained in the last frame? _______

---

a) cos S = 0.803

b) S = 37°

c) Yes

81. We cannot use the Law of Sines to find angle C in triangle VCT because we are given only the three sides.

We can, however, find angle C by means of the Law of Cosines, as follows:

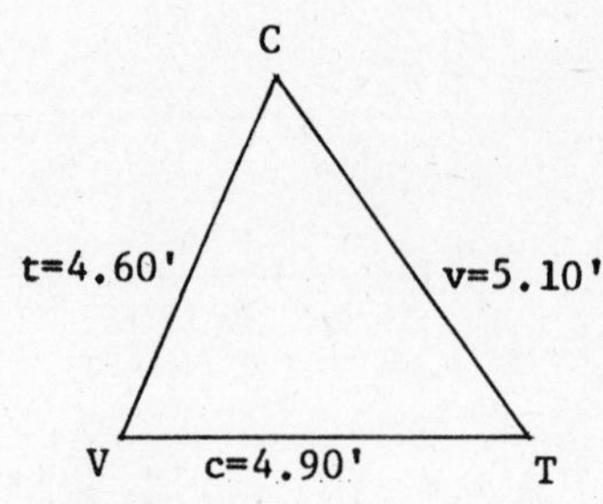

$$\boxed{c^2 = t^2 + v^2 - 2tv \cos C}$$

To find the numerical value of angle C, we can use either of two strategies:

(1) Plug the known values of "c", "t", and "v" into the equation above, and then solve for "cos C".

(2) Or, first rearrange the equation above, solving it for "cos C" by algebraic manipulations. Then, plug in the known values of "c", "t", and "v".

Rearrange the equation above, solving it for "cos C":

$$c^2 = t^2 + v^2 - 2tv \cos C$$

$$\cos C = \frac{t^2 + v^2 - c^2}{2tv}$$

82. The triangle at the right is the same triangle shown in the last frame.

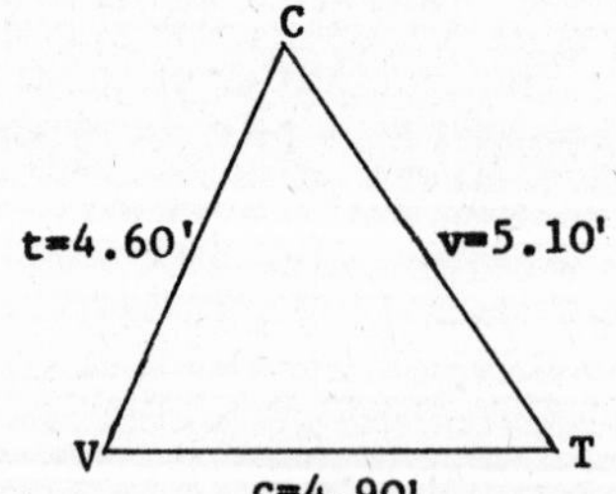

(a) To solve for angle T, what basic equation should be used? (Use letters.)

____

(Continued on following page.)

82. (Continued)

(b) Rearrange the equation, solving it for "cos T".

(c) Next, plug in the known values and find the numerical value of "cos T". cos T = ________

(d) Finally, find angle T. T = ______°

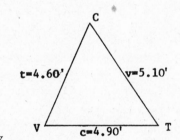

a) $t^2 = c^2 + v^2 - 2cv \cos T$

b) $\cos T = \dfrac{c^2 + v^2 - t^2}{2cv}$

c) $\cos T = 0.576$

d) $T = 55°$

---

83. Here is the same triangle again.

In the last frame, we found that angle T is 55°.

To solve for angle V, we can plug the known values into either of the following two equations based on the Law of Cosines:

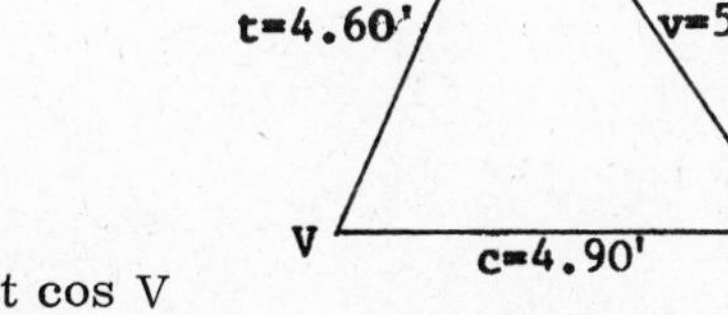

(1) $v^2 = c^2 + t^2 - 2ct \cos V$

(2) $\cos V = \dfrac{c^2 + t^2 - v^2}{2ct}$

Because angle T is known to be 55° (from the last frame), angle V can also be found by using the Law of Sines.

(a) Set up the proportion for finding angle V by means of the Law of Sines.

____

(Continued on following page.)

83. (Continued)

(b) Solve the proportion: Angle V = ______°

Answer to Frame 83: a) $\frac{\sin V}{5.10} = \frac{\sin 55°}{4.60}$ b) V = 65° (from sin V = 0.908)

SELF-TEST 8 (Frames 75-83)

1. In the oblique triangle at the right, find side p.

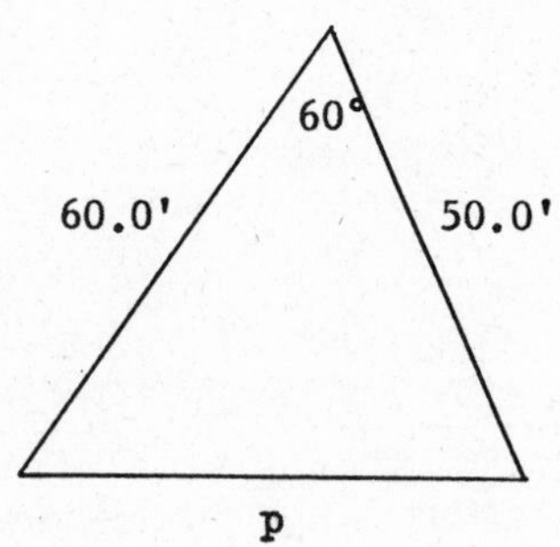

2. In the oblique triangle at the right, the three sides are known. Find angle G.

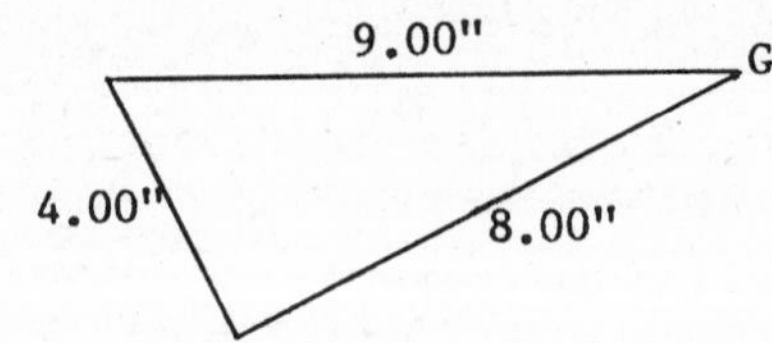

ANSWERS: 1. p = 55.7' (from p^2 = 3100) 2. G = 26° (from cos G = 0.896)

4-9 RIGHT TRIANGLES - LAW OF SINES AND LAW OF COSINES

The Law of Sines and the Law of Cosines are general relationships which apply to all triangles, including right triangles. In this section, we will show that these two laws apply also to right triangles. We will also show, however, that they are not used to solve right triangles because simpler methods are available.

84. The Law of Sines also holds for right triangles. To show that this fact is true, we will use right triangle ABC as an example.

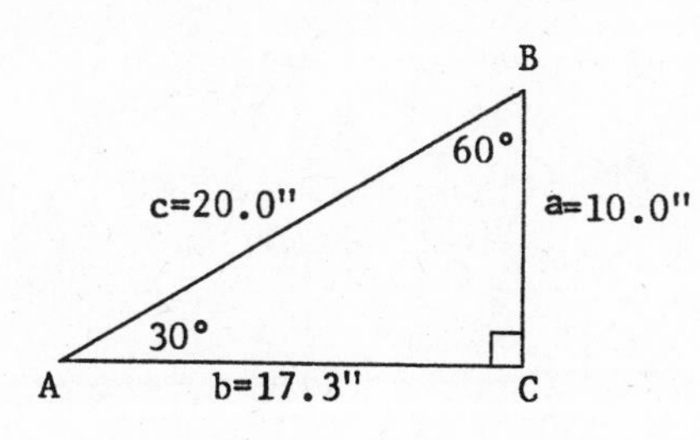

The Law of Sines for triangle ABC is: $\frac{a}{\sin A} = \frac{b}{\sin B} = \frac{c}{\sin C}$

Plugging in the known values, we get: $\frac{10.0}{\sin 30°} = \frac{17.3}{\sin 60°} = \frac{20.0}{\sin 90°}$

or $\frac{10.0}{0.5000} = \frac{17.3}{0.8660} = \frac{20.0}{1.0000}$

Let's evaluate the three ratios to show that they are equal:

(a) $\frac{10.0}{0.5000} =$ ________ (b) $\frac{17.3}{0.8660} =$ ________ (c) $\frac{20.0}{1.0000} =$ ________

(d) Are the three ratios equal? ________

Answer to Frame 84: a) 20.0 b) 20.0 c) 20.0 d) Yes

85. The Law of Cosines also holds for right triangles. To show this fact, we will use right triangle DKT as an example.

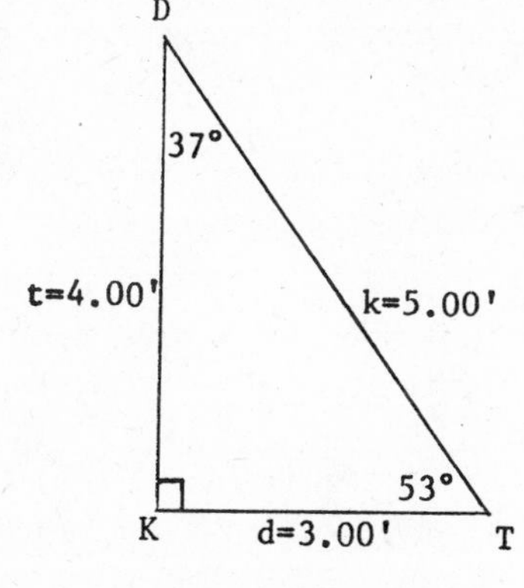

One formula based on the Law of Cosines is this:

$$k^2 = d^2 + t^2 - 2dt \cos K$$

Plugging in the known values and simplifying, we get:

$$(5.00)^2 = (3.00)^2 + (4.00)^2 - 2(3.00)(4.00)(\cos 90°)$$

or $25.0 = \underbrace{9.00 + 16.0}_{25.0} - \underbrace{2(3.00)(4.00)(0)}_{0}$

or $25.0 = 25.0 - 0$

or $25.0 = 25.0$

Note: Since $\cos 90° = 0$, the value of the term on the right is "0".

A second formula based on the Law of Cosines is this:

$$t^2 = d^2 + k^2 - 2dk \cos T$$

Plug in the known values and simplify to show that the sides are equal. (Note: In this case, the value of the last term will not be "0" since cos 53° is not "0".

Answer to Frame 85: $(4.00)^2 = (3.00)^2 + (5.00)^2 - 2(3.00)(5.00)(\cos 53°)$

$16.0 = \underbrace{9.00 + 25.0}_{34.0} - \underbrace{2(15.0)(0.6018)}_{18.0}$

$16.0 = 34.0 - 18.0$

$16.0 = 16.0$

86. The ordinary method for solving for "b" in triangle BDQ is to use the tangent ratio. For example:

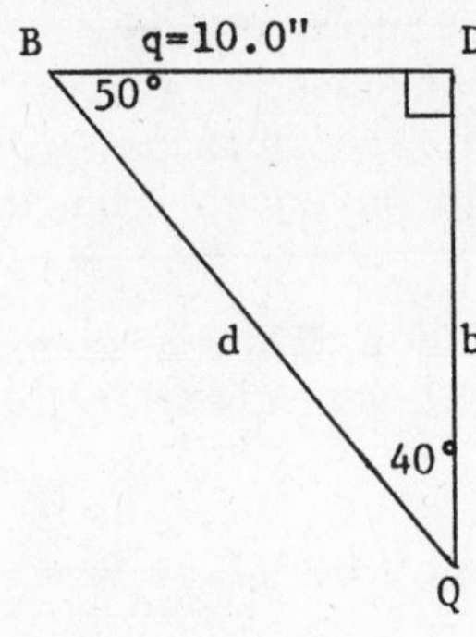

Since $\tan 50° = \frac{b}{10.0"}$,

$$b = (10.0")(1.192) = 11.9"$$

However, we can also solve for "b" by means of the Law of Sines.

(a) Using the Law-of-Sines formula on the right, solve for "b". $\frac{b}{\sin B} = \frac{q}{\sin Q}$

b = ____________

(b) Is this the same value we obtained by using the tangent ratio above? ________

---

a) $\frac{b}{\sin 50°} = \frac{10.0"}{\sin 40°}$

$b = \frac{(\sin 50°)(10.0")}{\sin 40°}$

$= \frac{(0.7660)(10.0")}{0.6428}$

$= 11.9"$

b) Yes

---

87. The ordinary method of solving for angle C in right triangle CDF is to use one of the trig ratios. For example:

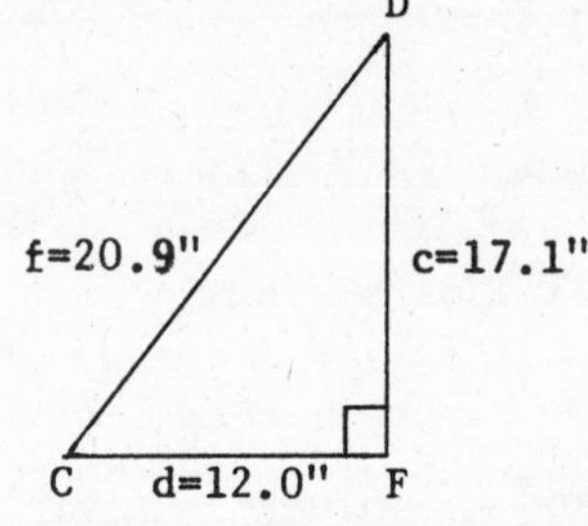

Since $\sin C = \frac{c}{f} = \frac{17.1"}{20.9"} = 0.819$,

C contains 55°.

However, we could also solve for angle C by means of the Law of Cosines. Here are the steps:

$$c^2 = d^2 + f^2 - 2df \cos C$$

$$(17.1)^2 = (12.0)^2 + (20.9)^2 - 2(12.0)(20.9)\cos C$$

$$292 = 144 + 437 - 2(251)\cos C$$

$$292 = 581 - 502 \cos C$$

$$-289 = -502 \cos C$$

$$289 = 502 \cos C$$

$$\cos C = \frac{289}{502} = 0.576$$

Therefore, angle C contains 55°.

(a) Did we get the same answer by the Law-of-Cosine method as we did by the trig-ratio method? ____________

(b) Is the "trig-ratio" or "Law-of-Cosine" method easier? ______________

---

a) Yes

b) Trig-ratio method

88. The Law of Sines and the Law of Cosines are general relationships which apply to both oblique and right triangles. However, though these two laws are used to solve oblique triangles, they are a clumsy way to solve right triangles. When solving right triangles, we use the trig ratios or the Pythagorean Theorem.

SELF-TEST 9 (Frames 84-88)

In the right triangle shown, side "d" can be found by either of these methods:

20.0"
d
31°
59°

Trig-Ratio Method	Law-of-Sines Method
$\tan 31° = \frac{d}{20.0''}$	$\frac{d}{\sin 31°} = \frac{20.0''}{\sin 59°}$

1. Which method is easiest to use? ______________________

2. Find the length of side "d". d = __________

ANSWERS: 1. Trig-Ratio Method 2. d = 12.0"

4-10 STRATEGIES FOR SOLVING TRIANGLES

Various strategies (or methods) are used to solve right and oblique triangles. In this section, we will briefly review these strategies. Then exercises will be given in which you must outline the strategy you would use for various types of solutions.

89. Different strategies (or methods) are used to solve right and oblique triangles.

For right triangles, we use one of the three trig ratios or the Pythagorean Theorem.

For oblique triangles, we use either the Law of Sines or the Law of Cosines.

The Law of Sines is used in all cases except these two:

(1) When only two sides and the angle between them are known.
(2) When only the three sides are known.

In these two cases, we use the Law of Cosines because for such cases we cannot set up a Law-of-Sines proportion which contains only one unknown.

If you have trouble remembering the two special cases of oblique triangles, you can always use the following general strategy:

TRY TO USE THE LAW OF SINES FIRST. IF YOU CANNOT SET UP AN EQUATION CONTAINING ONLY ONE UNKNOWN, THEN USE THE LAW OF COSINES.

---

90. In right triangle ABC, which trig ratio would you use:

(a) To solve for "c"? ____________

(b) To solve for "b"? ____________

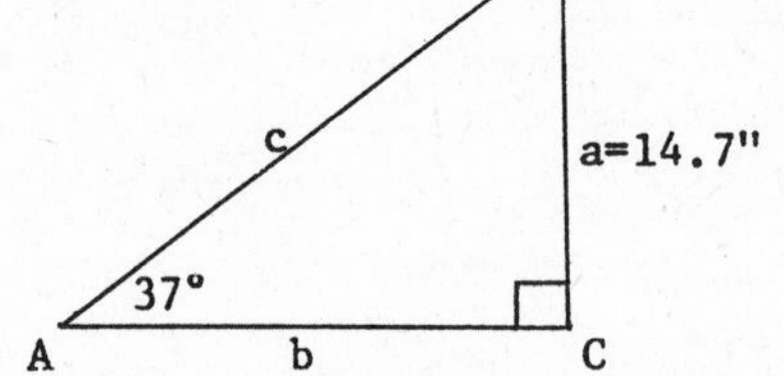

---

91. In triangle DFK, sides "d" and "f" are given.

(a) How would you solve for side "k"?

____

____

(b) How would you solve for angle D? ____________________

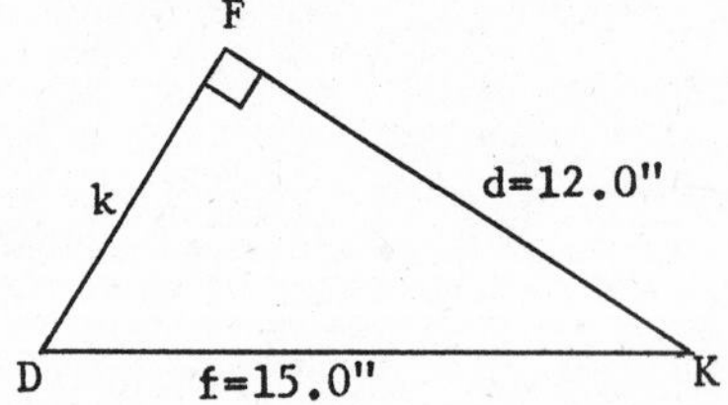

a) sin A or sin 37°

b) tan A or tan 37°

---

92. (a) To solve for AT in oblique triangle ACT, would you use the Law of Sines or the Law of Cosines?

____

(b) What specific equation would you use?

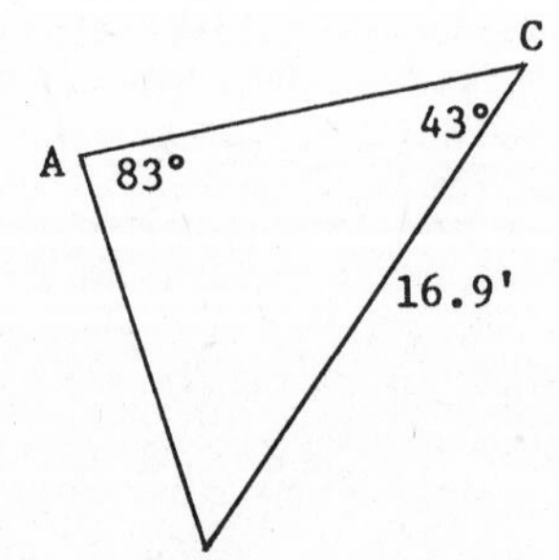

____

a) Use the Pythagorean Theorem.

b) Use the sine ratio, since:

$$\sin D = \frac{d}{f}$$

93. (a) To solve for side "b" in oblique triangle BDF, would you use the Law of Sines or the Law of Cosines?

____

(b) What specific equation would you use?

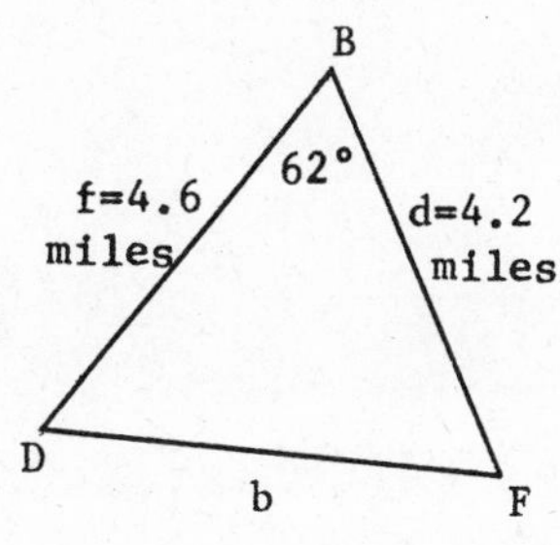

____

a) Law of Sines

b) $\dfrac{AT}{\sin C} = \dfrac{CT}{\sin A}$

or

$\dfrac{AT}{\sin 43°} = \dfrac{16.9'}{\sin 83°}$

---

94. (a) To solve for angle T, which law would you use?

____

(b) What specific equation would you use?

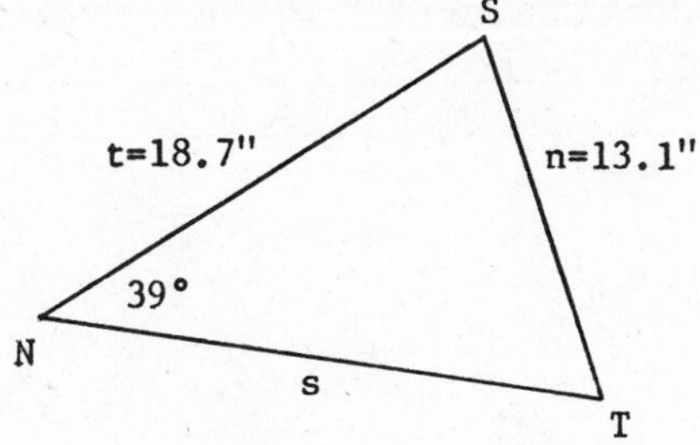

____

a) Law of Cosines

b) $b^2 = d^2 + f^2 - 2df \cos B$

or

$b^2 = (4.2)^2 + (4.6)^2 - 2(4.2)(4.6)(\cos 62°)$

---

95. (a) To solve for angle H, which law would you use?

____

(b) What equation would you use?

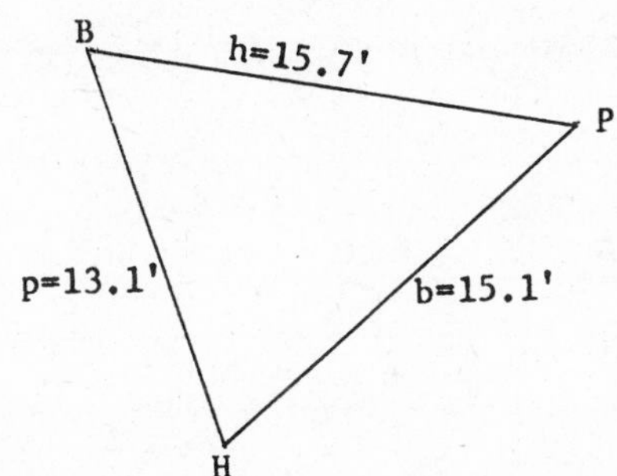

____

a) Law of Sines

b) $\dfrac{\sin T}{t} = \dfrac{\sin N}{n}$

or

$\dfrac{\sin T}{18.7''} = \dfrac{\sin 39°}{13.1''}$

(Remember, when solving for the sine of an angle, we put the sines in the numerators.)

---

a) Law of Cosines

b) $h^2 = b^2 + p^2 - 2bp \cos H$

or

$\cos H = \dfrac{b^2 + p^2 - h^2}{2bp}$

or

$\cos H =$

$\dfrac{(15.1)^2+(13.1)^2-(15.7)^2}{2(15.1)(13.1)}$

96. (a) To solve for angle H, two steps are required. What are they?

(1) ______________________

(2) ______________________

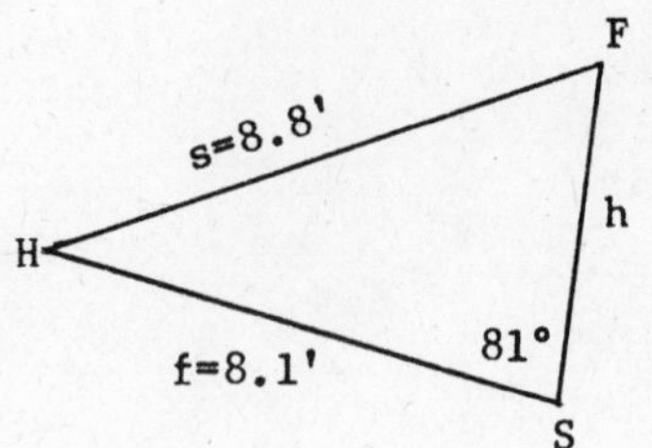

(b) To solve for angle F, what equation would you use?

____

---

97. (a) To solve for side "b", three steps are required. They are:

(1) ______________________

(2) ______________________

(3) ______________________

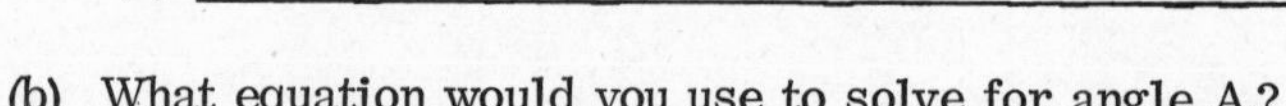

(b) What equation would you use to solve for angle A?

____

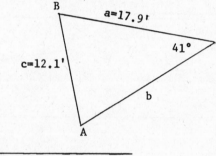

(c) What equation would you use to solve for "b" after the three angles are known?

____

---

a) (1) Solve for angle F.

(2) Solve for angle A by means of the angle-sum principle.

b) $\frac{\sin F}{f} = \frac{\sin S}{s}$

or

$\frac{\sin F}{8.1'} = \frac{\sin 81°}{8.8'}$

98. In triangle DJK, we want to find the size of two angles, D and K.

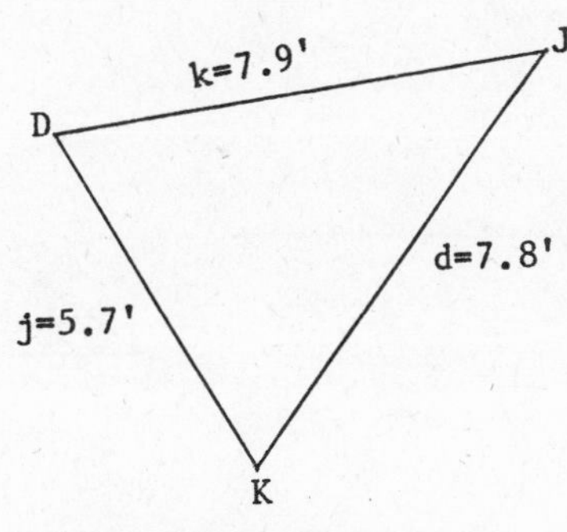

(a) Suppose we find angle D first. What equation should you use?

____

(b) Having found angle D, you can solve for angle K in either of two ways. What are the two equations?

(1) ______________________________

(2) ______________________________

a) (1) Solve for angle A.

(2) Find angle B by the angle-sum principle.

(3) Solve for "b".

b) $\frac{\sin A}{a} = \frac{\sin C}{c}$ or $\frac{\sin A}{17.9'} = \frac{\sin 41°}{12.1'}$

c) $\frac{b}{\sin B} = \frac{c}{\sin C}$ or $\frac{b}{\sin B} = \frac{12.1'}{\sin 41°}$

<u>Answer to Frame 98:</u>

a) $d^2 = j^2 + k^2 - 2jk \cos D$ or $\cos D = \frac{j^2 + k^2 - d^2}{2jk}$

b) (1) Use the Law of Sines: $\frac{\sin K}{k} = \frac{\sin D}{d}$

(2) Or, use the Law of Cosines:

$k^2 = d^2 + j^2 - 2dj \cos K$ or $\cos K = \frac{d^2 + j^2 - k^2}{2dj}$

<u>Note:</u> It is preferable to use the Law of Sines since it involves easier computation.

99. To find side "v" in triangle VSD, two steps are required. Name them.

(1) ______________________________

(2) ______________________________

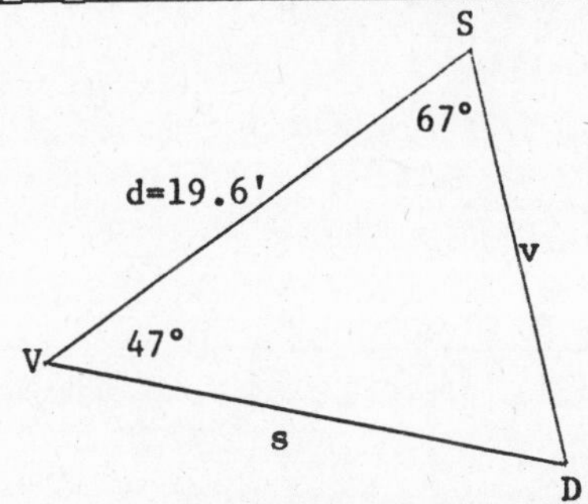

(1) Find angle D by the angle-sum principle.

(2) Use the Law of Sines to find "v":

$$\frac{v}{\sin V} = \frac{d}{\sin D}$$

100. Would you use the Law of Sines or the Law of Cosines to find CT in this triangle?

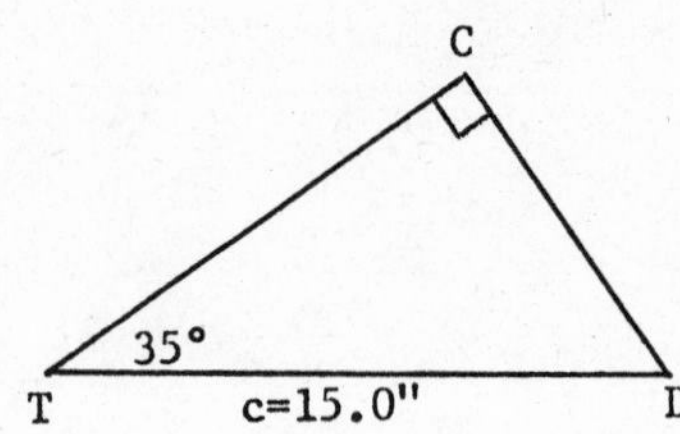

101. To find angle F in triangle FPT, two steps are required.

(a) What is the first step?

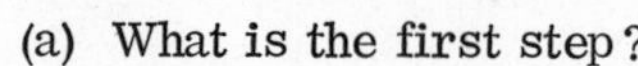

(b) What equation would you use for the first step?

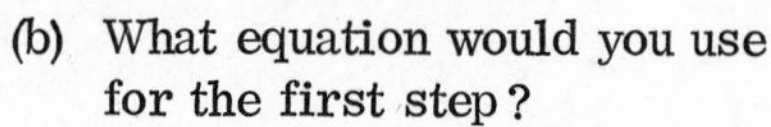

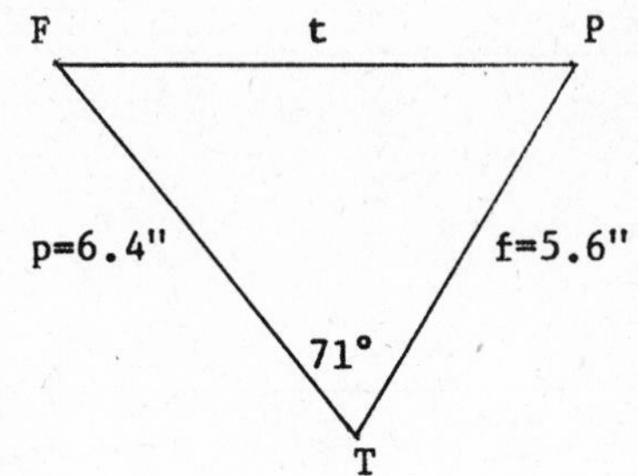

(c) Knowing side "t", you can solve for angle F in either of two ways. Name them.

(1) ____________________

(2) ____________________

(d) In (c), which of the two ways is easier? ____________

Neither, since it is a right triangle, we would use the cosine ratio.

Answer to Frame 101:

a) Find side "t".

b) $t^2 = f^2 + p^2 - 2fp \cos T$

c) (1) Use Law of Sines: $\dfrac{\sin F}{f} = \dfrac{\sin T}{t}$

(2) Use Law of Cosines: $f^2 = p^2 + t^2 - 2pt \cos F$

or

$$\cos F = \frac{p^2 + t^2 - f^2}{2pt}$$

d) The Law-of-Sines equation is easier computationally.

SELF-TEST 10 (Frames 89-101)

For each right triangle in Problems 1 to 3, state which one of the following principles would be used directly to find the unknown side or angle (designated by a single letter in each triangle):

Angle-Sum Principle Sine Ratio Cosine Ratio Tangent Ratio Pythagorean Theorem

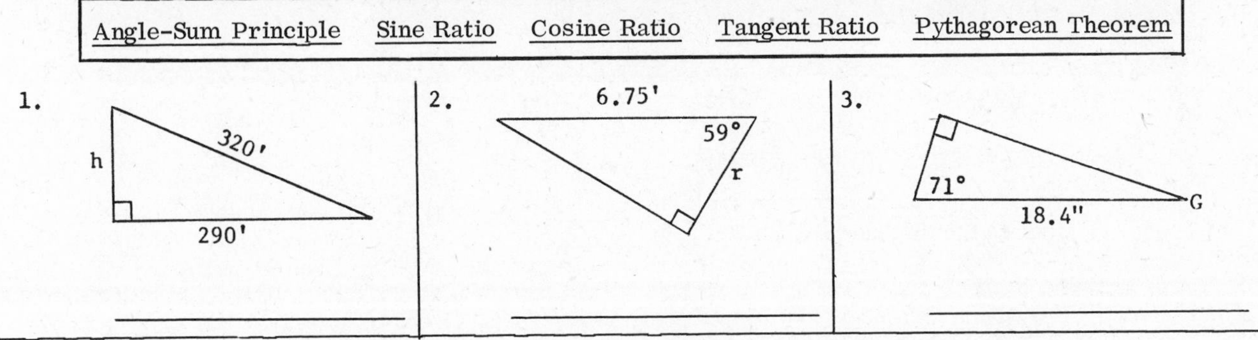

For each oblique triangle in Problems 4 to 7, state which one (or two) of the following principles would be used to find the unknown side or angle (designated by a single letter in each triangle):

Angle-Sum Principle Law of Sines Law of Cosines

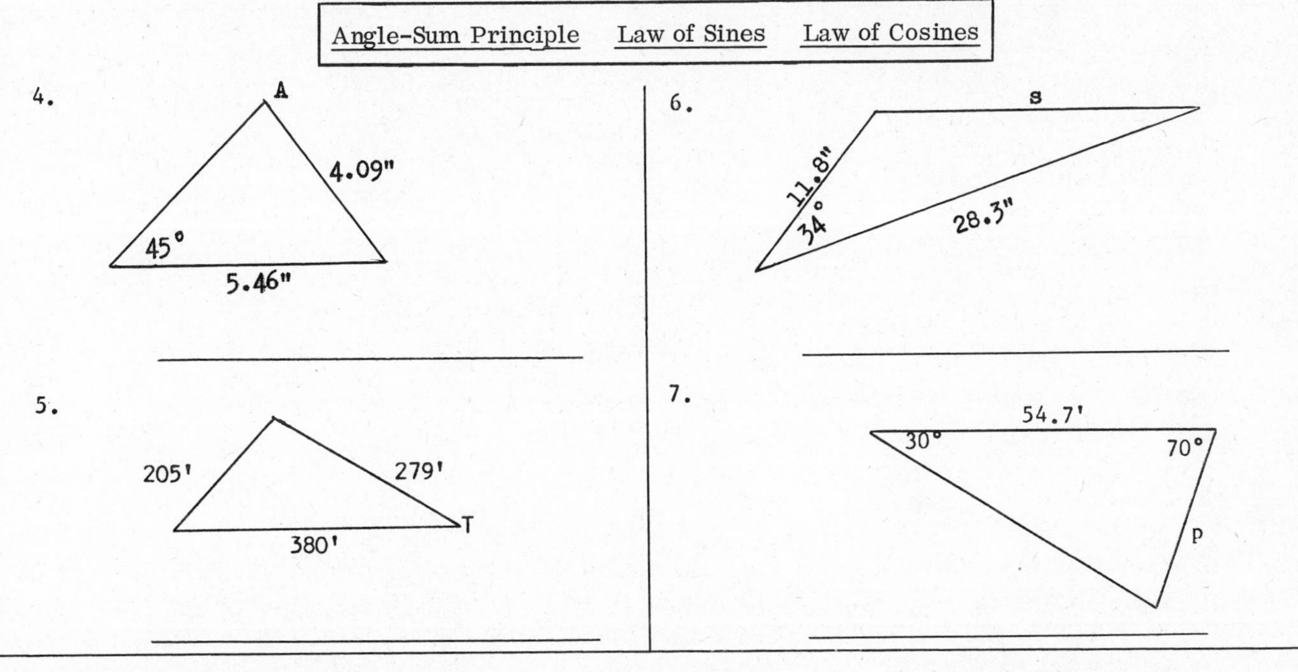

ANSWERS:

1. Pythagorean Theorem
2. Cosine Ratio
3. Angle-Sum Principle
4. Law of Sines
5. Law of Cosines
6. Law of Cosines
7. Angle-Sum Principle and Law of Sines

4-11 SOME APPLIED PROBLEMS

In this section, we will examine some applied problems which involve oblique triangles. Of course, either the Law of Sines or the Law of Cosines is needed in order to solve them.

102. A surveyor is required to measure the distance (PR) across a river.

He first set up line PQ = 100 ft. He then measured angle P and angle Q by means of a transit (a surveyor's instrument for measuring angles), and found that angle P = 107° and angle Q = 52°.

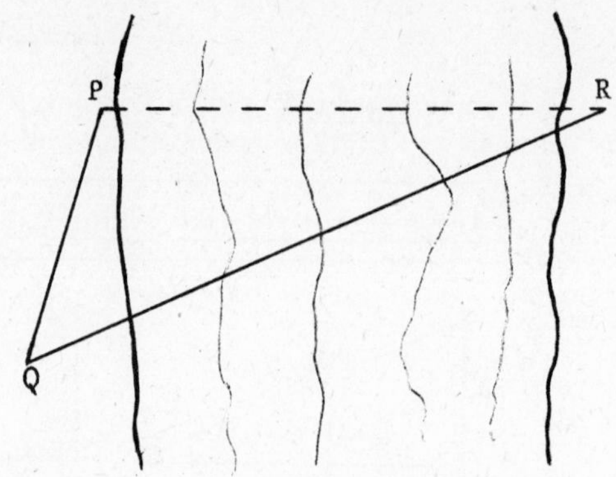

(a) The size of angle R is _______°.

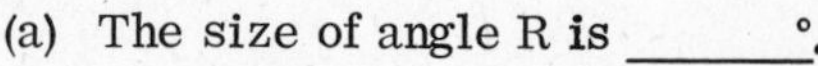

(b) Complete: $\dfrac{PQ}{\sin R} = \dfrac{PR}{\boxed{}}$

(c) Calculate the distance PR. PR = __________

---

103. In this diagram, distance GH cannot be measured directly. However, since it is a side of oblique triangle FGH, GH can be calculated. Do so.
GH = ______________

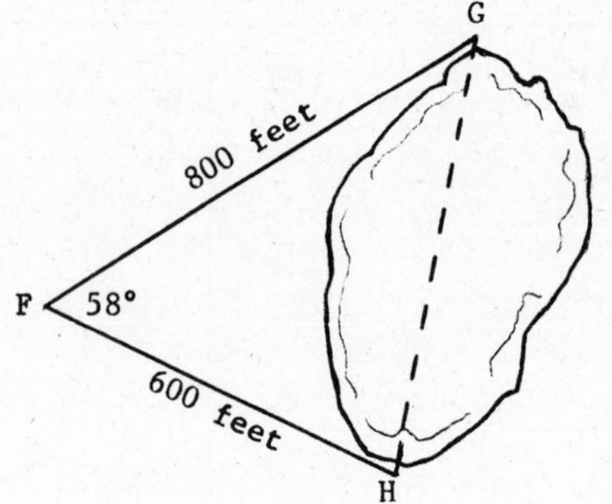

a) 21°

b) $\boxed{\sin Q}$

c) PR = 220 ft.

From Law of Sines:

$$PR = \frac{100 \sin 52°}{\sin 21°}$$

---

104. In this diagram, calculate the inaccessible distance MP. MP = ______________

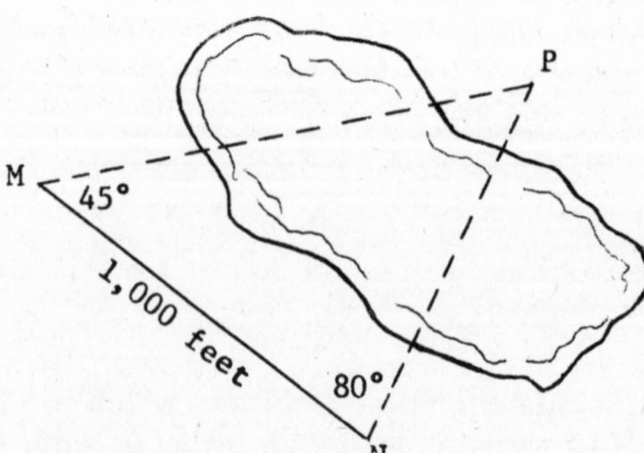

GH = 701 feet

From Law of Cosines:

$$(GH)^2 = (600)^2 + (800)^2 - 2(600)(800)(\cos 58°)$$

$$(GH)^2 = 491{,}000$$

$$GH = 701 \text{ ft.}$$

105. A diagram of a structural steel shape is shown on the right.

(a) Calculate distance "d".
d = ____________

e d d e
70° 52° 52° 70°
20.0 ft. 20.0 ft.

(b) Calculate distance "e".
e = ____________

MP = 1200 ft.

From Law of Sines:

$$\frac{MP}{\sin 80°} = \frac{1000}{\sin 55°}$$

---

106. A metal part contains three holes. The center-to-center distances between the holes are 2.50", 2.00", and 1.50". Find angle φ, the angle between the 2.50" and 1.50" lengths.
φ = ________°

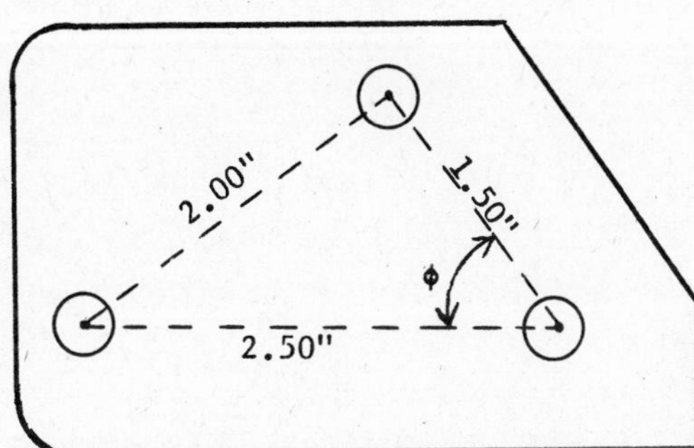

a) d = 22.2 ft.

From Law of Sines:

$$\frac{d}{\sin 70°} = \frac{20.0}{\sin 58°}$$

b) e = 18.6 ft.

From Law of Sines:

$$\frac{e}{\sin 52°} = \frac{20.0}{\sin 58°}$$

$\varphi = 53°$

From Law of Cosines:

$\cos \varphi = 0.600$

107. A diagram of a metal plate is given on the left below. It has the shape of a trapezoid. The 2.00" and 6.00" sides are parallel. We want to find the length of side "w".

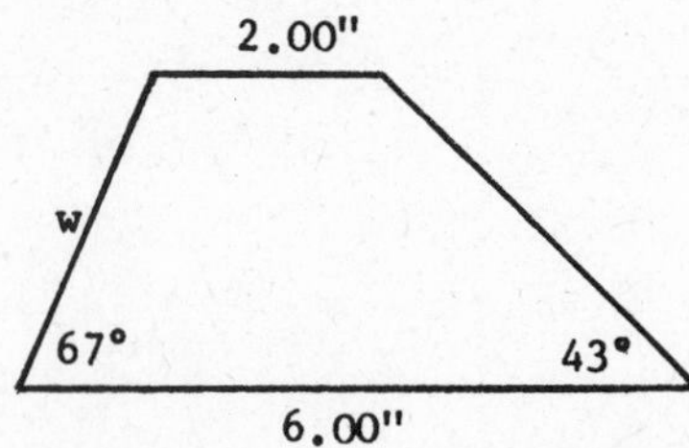

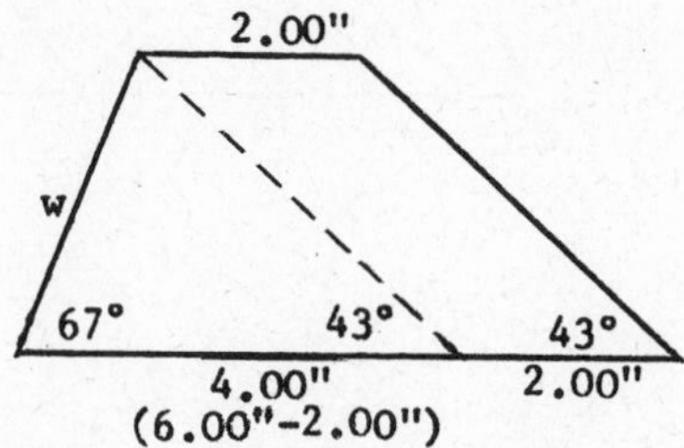

Since the figure on the left does not contain a triangle, we must draw an extra line so that oblique-triangle methods can be used. The extra line has been drawn in the figure on the right. It was drawn parallel to the slant side on the right.

Find side "w" by solving the oblique triangle. w = ________

---

w = 2.90"

From Law of Sines:

$$\frac{w}{\sin 43°} = \frac{4.00}{\sin 70°}$$

---

108. In this diagram, PQ is the height of a tall television tower.

To calculate PQ, follow these steps:

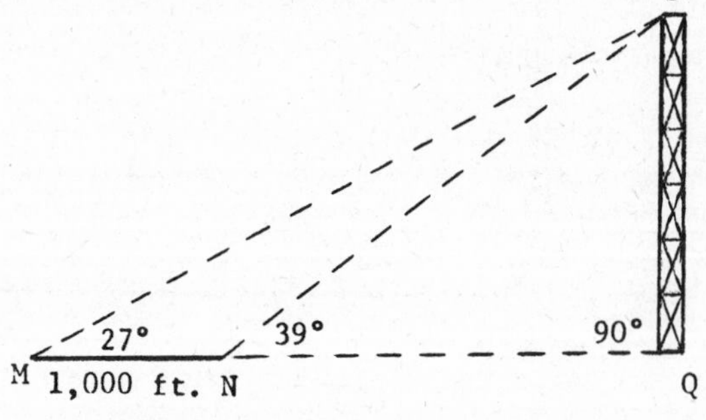

(a) In oblique triangle MNP:

Angle MNP = 180° - 39° = ______°

Also, angle MPN = ______°

(b) In oblique triangle MNP, using the Law of Sines:

$$\frac{PN}{\sin 27°} = \frac{1{,}000}{\sin 12°}$$ Therefore: PN = ________

(c) Finally, in right triangle PQN, find PQ. PQ = ________

Answer to Frame 108: a) Angle MNP = 141° b) PN = 2,180 ft. c) PQ = 1,370 ft.
Angle MPN = 12°

SELF-TEST 11 (Frames 102-108)

A metal plate contains three holes, located as shown in the diagram on the right.

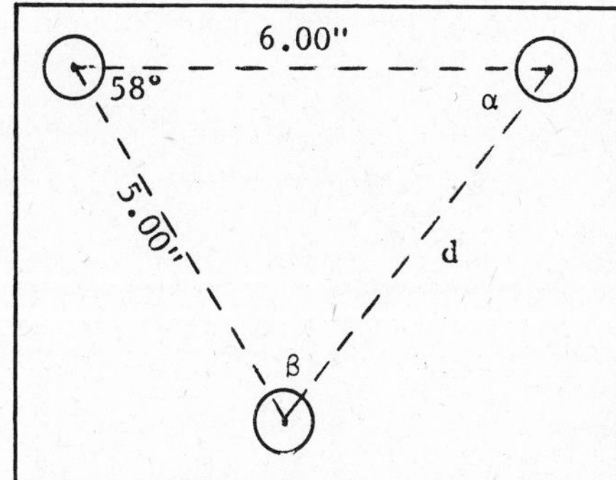

1. Using the Law of Cosines, find distance "d".
   d = ________
2. Using the Law of Sines, find angle α.
   α = ________
3. Using the angle-sum principle, find angle β.
   β = ________

ANSWERS: 1. d = 5.40" 2. $\alpha = 52°$ 3. $\beta = 70°$

4-12 THE LAW OF SINES AND OBTUSE ANGLES

Obtuse angles are angles between 90° and 180°. Up to this point, we have avoided obtuse angles when solving oblique triangles by means of the Law of Sines. In this section, we will discuss the solution of oblique triangles by means of the Law of Sines when an obtuse angle is involved.

109. Angles between 0° and 90° are called acute angles. Angles between 90° and 180° are called obtuse angles.

(a) If an oblique triangle contains three acute angles, it is called an ____________ oblique triangle.

(b) If an oblique triangle contains an obtuse angle, it is called an ____________ oblique triangle.

a) acute

b) obtuse

110. Oblique triangle ABC contains an obtuse angle.

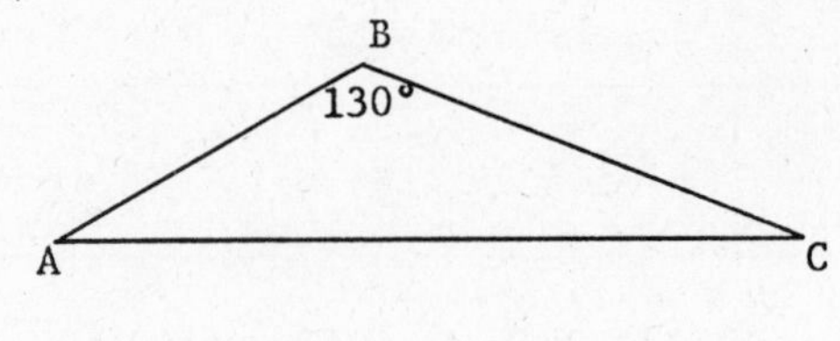

(a) Because of the angle-sum principle, the sum of angles A and C must be ______°.

(b) Can a triangle contain more than one obtuse angle? ________

---

111. Let's briefly review the "sines" of obtuse angles. Since all obtuse angles are angles between 90° and 180°:

(a) They all lie in the ____________ quadrant.

(b) Their sines are ____________ (positive/negative).

Answers to 110: a) 50° b) No, because the sum of the angles would then be more than 180°.

---

112. In the figure on the right:

θ is an obtuse angle (between 90° and 180°).

α is its reference angle (Quadrant 2).

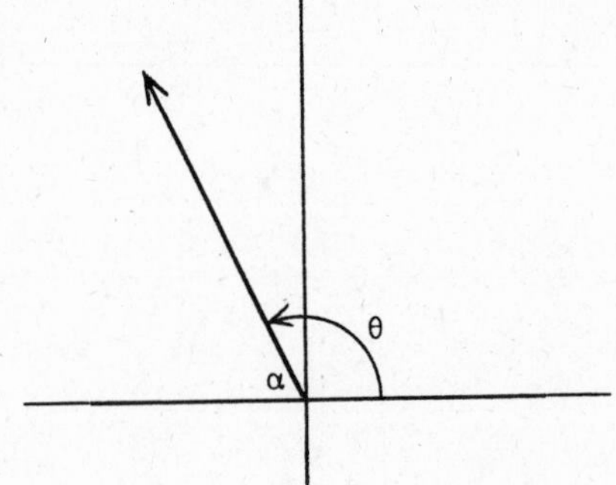

(a) $\alpha + \theta =$ ______°

(b) What is the relationship between $\sin\theta$ and $\sin\alpha$?

____

Answers to 111: a) second b) positive

---

113. The following facts should be obvious:

(1) The sum of an obtuse angle and its reference angle is 180°.

(2) The sine of an obtuse angle is positive, and is numerically identical to the sine of its reference angle.

That is: $\sin 160° = \sin 20°$

(a) $\sin 130° = \sin$ ____°

(b) $\sin 121° = \sin$ ____°

(c) $\sin 97° = \sin$ ____°

Answers to 112: a) 180° b) They are equal.

Answers to 113: a) 50° b) 59° c) 83°

114. The Law of Sines still holds when a triangle contains an obtuse angle. We will use triangle BDT as an example:

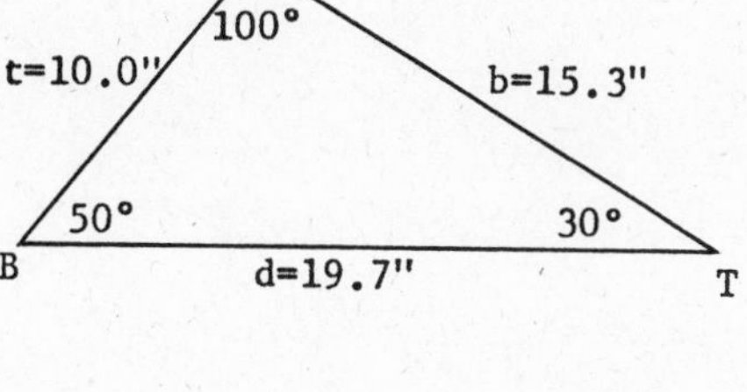

The Law of Sines is:

$$\frac{b}{\sin B} = \frac{d}{\sin D} = \frac{t}{\sin T}$$

Plugging in the known values, we get:

$$\frac{15.3}{\sin 50°} = \frac{19.7}{\sin 100°} = \frac{10.0}{\sin 30°}$$

$$\frac{15.3}{0.7660} = \frac{19.7}{0.9848} = \frac{10.0}{0.5000}$$

(Note: $\sin 100° = \sin 80° = 0.9848$)

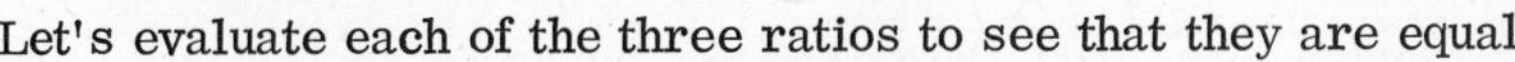

Let's evaluate each of the three ratios to see that they are equal:

(a) $\frac{15.3}{0.7660} =$ ______ (b) $\frac{19.7}{0.9848} =$ ______ (c) $\frac{10.0}{0.5000} =$ ______

(d) Are the three ratios equal? ______

---

a) 20.0

b) 20.0

c) 20.0

d) Yes

115. The use of the Law of Sines to find an unknown side of a triangle is straightforward even when the proportion involves an obtuse angle. Here is an example:

M, K, L; 34°, 125°, 15.0"

To solve for ML in triangle MKL, we use the following proportion based on the Law of Sines.

$$\frac{ML}{\sin K} = \frac{KL}{\sin M}$$

$$\frac{ML}{\sin 125°} = \frac{15.0''}{\sin 34°}$$

$$\frac{ML}{0.8192} = \frac{15.0''}{0.5592}$$ (Note: $\sin 125° = \sin 55° = 0.8192$)

Complete the solution. ML = ______ inches

22.0"

116. We can use the Law of Sines to solve for PD in triangle PDN. However, we must find the size of angle D first.

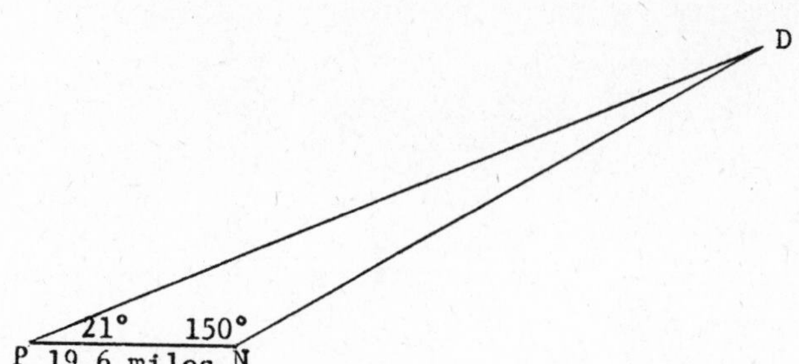

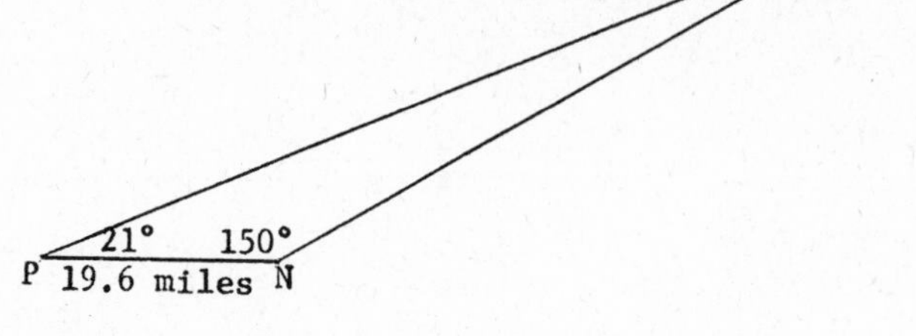

(a) How large is angle D? _____°

(b) What equation would you use to solve for PD?

____

(c) sin 150° = _______

(d) Complete the solution. PD = _________

Answers:

a) 9°

b) $\frac{PD}{\sin N} = \frac{PN}{\sin D}$

or

$\frac{PD}{\sin 150°} = \frac{19.6 \text{ mi.}}{\sin 9°}$

c) 0.5000

d) PD = 62.7 miles

---

117. The use of the Law of Sines to find an unknown angle presents a slight problem since any positive "sine" could be the "sine" of either an acute angle or an obtuse angle. Here is an example:

If we determine that sin B = 0.9063 in triangle ABC, angle B could be either 65° or 115°.

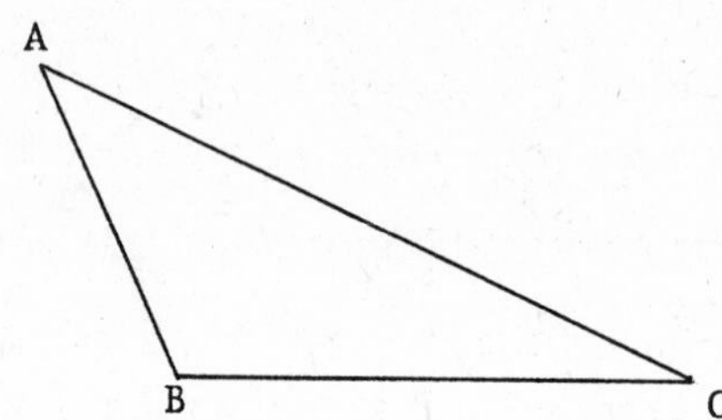

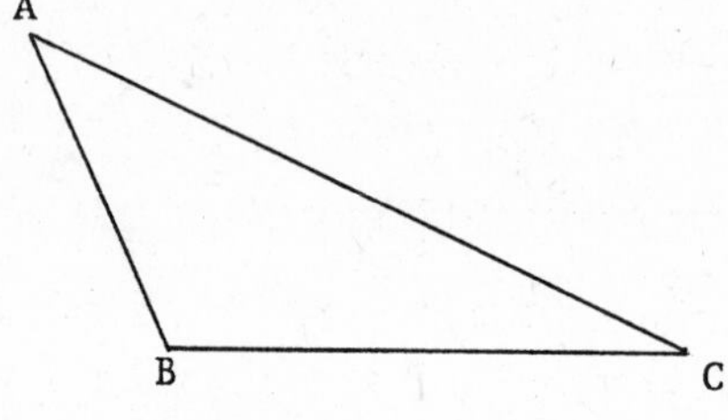

In order to make the correct choice, we must look at the triangle.

(a) Is angle B an acute or obtuse angle? _______

(b) Therefore, angle B must contain _____°.

Answers:

a) obtuse

b) 115°

---

118. When using the Law of Sines to solve for an angle in an obtuse oblique triangle:

(1) You find the sine of the angle first.

If sin B = 0.4384, angle B can be either 26° or 154°.

(2) You then check the diagram to see whether angle B is an acute or obtuse angle.

If angle B is acute, it is a 26° angle.
If angle B is obtuse, it is a 154° angle.

(Continued on following page.)

118. (Continued)

Let's apply this principle to triangle DPQ.

D P Q

If sin Q = 0.9816:

(a) Angle Q could be either ______° or ______°.

(b) Since angle Q is an obtuse angle, it must be ______°.

If sin D = 0.7314:

(c) Angle D could be either ______° or ______°.

(d) Since angle D is an acute angle, it must be ______°.

---

119. We want to solve for angle K in this triangle.

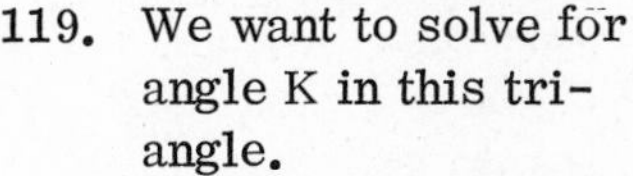

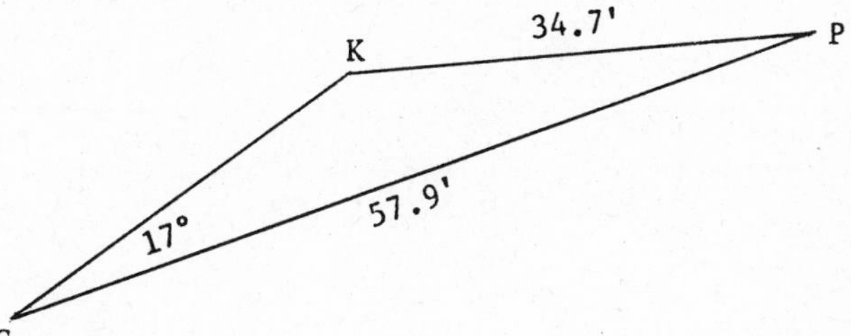

(a) What proportion would you use to solve for sin K?

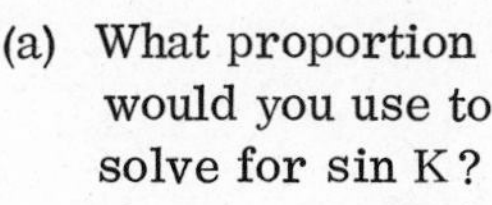

____

(b) Solve the proportion for sin K, and find the numerical value of sin K. sin K = ________

(c) Angle K is an __________ (acute/obtuse) angle.

(d) Therefore, angle K contains ______°.

Answers to 118:

a) 79° or 101°

b) 101°

c) 47° or 133°

d) 47°

Answers to 119:

a) $\frac{\sin K}{GP} = \frac{\sin G}{KP}$

or

$\frac{\sin K}{57.9'} = \frac{\sin 17°}{34.7'}$

b) 0.488

c) obtuse

d) 151°

120. We want to find the size of angle G in obtuse triangle GLB.

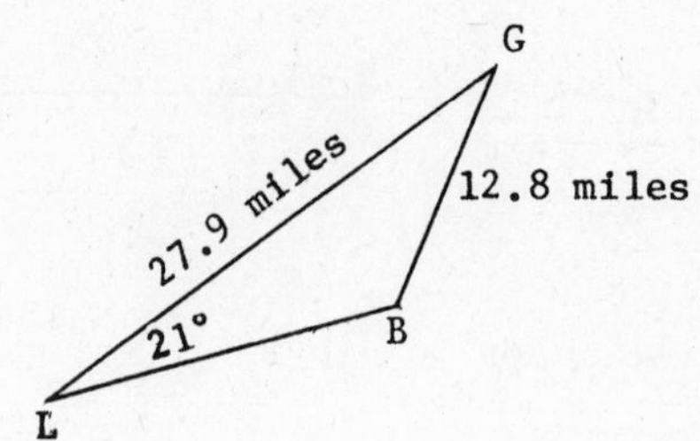

Angle G cannot be found directly by the Law of Sines or the Law of Cosines, but must instead be found indirectly.

(a) To find angle G, what procedure should be used?

____

(b) Find the numerical value of sin B. sin B = ________

(c) Angle B contains ______°.

(d) Angle G contains ______°.

---

a) Find angle B, using the Law of Sines. Then use the angle-sum principle to find angle G.

b) 0.781

c) 129°

d) 30°

SELF-TEST 12 (Frames 109-120)

1. Find the numerical value of:

sin 162° = ________

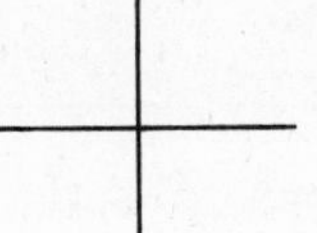

2. Given:

$\sin \theta = 0.9643$

(a) α = ________

(b) θ = ________

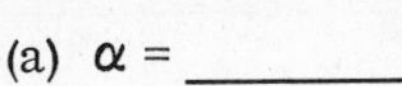

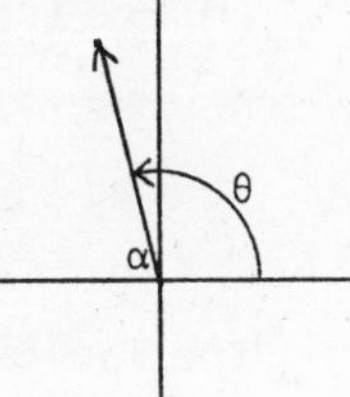

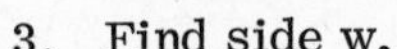

3. Find side w.

w = ________

4. Find angle φ.

φ = ________

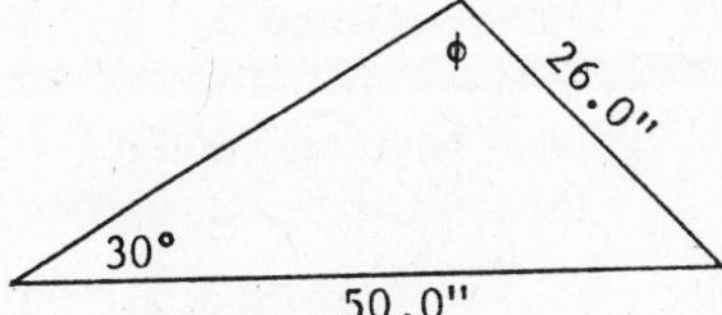

ANSWERS: 1. 0.3090 2. (a) $\alpha = 75°$ (Q2) (b) $\theta = 105°$ 3. w = 9.40" 4. $\varphi = 106°$

4-13 THE LAW OF COSINES AND OBTUSE ANGLES

In this section, we will show that the Law of Cosines also applies to obtuse oblique triangles. Then we will use it to solve oblique triangles when an obtuse angle is involved.

121. Since all obtuse angles lie in the second quadrant, the graph of angle θ on the right can stand for any obtuse angle.

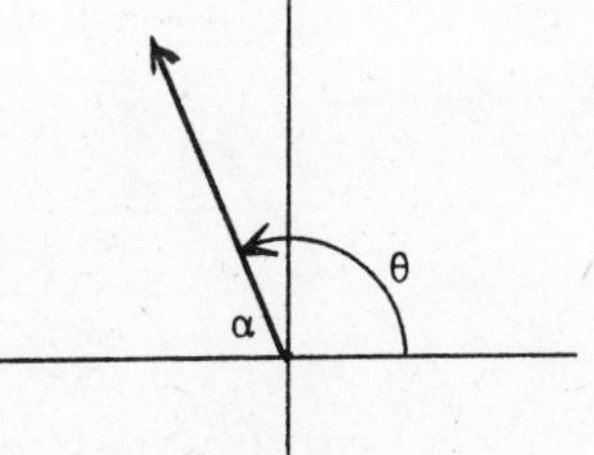

The cosine of any angle (θ) is equal to the cosine of its reference angle (α).

(a) The sign of the cosine of any reference angle in the second quadrant is ________.

(b) Therefore, the sign of the cosine of any obtuse angle is ________.

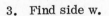

122. $\cos 120° = \cos 60°$ (Q2) $= -0.5000$

(a) $\cos 130° = \cos 50°$ (Q2) = ________

(b) $\cos 175° = \cos 5°$ (Q2) = ________

(c) $\cos 95° = \cos 85°$ (Q2) = ________

a) negative

b) negative

---

123. The Law of Cosines also applies to obtuse oblique triangles. We will use triangle ABC on the right as an example. The lengths shown for sides "a", "b", and "c" are measurements which have been rounded to three significant digits.

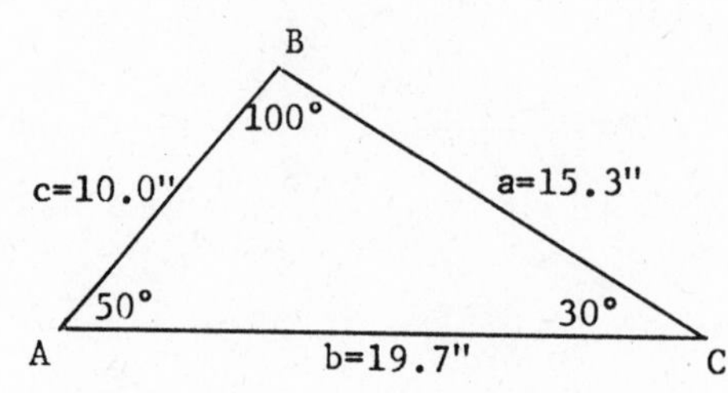

We will show that the following formula based on the Law of Cosines is true for triangle ABC:

$$b^2 = a^2 + c^2 - 2ac \cos B$$

Plugging in the known values and simplifying, we get:

$$(19.7)^2 = (15.3)^2 + (10.0)^2 - 2(15.3)(10.0)(\cos 100°)$$

$$388 = 234 + 100 - 2(153)(-0.1736)$$

$$388 = 334 - (306)(-0.1736)$$

$$388 = 334 - (-53.2)$$

$$388 = 334 + (+53.2)$$

$$388 = 387.2$$

Note: Examine the last few steps carefully. In them, we converted the subtraction of a negative term to an addition.

Taking into consideration the amount of rounding involved in this calculation and in the original triangle, does the Law of Cosines hold for this obtuse triangle? ________

a) -0.6428

b) -0.9962

c) -0.0872

Note: All are negative.

Yes

124. To solve for PW in this obtuse triangle, we must use the Law of Cosines. To do so, we use the following formula:

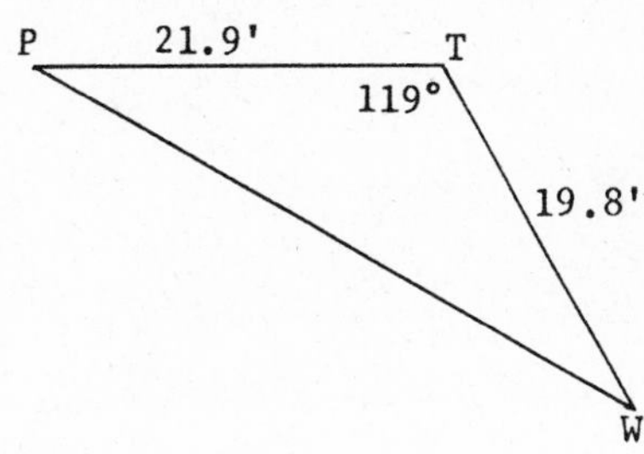

$$(PW)^2 = (PT)^2 + (TW)^2 - 2(PT)(TW)(\cos T)$$

Note: $\cos T = \cos 119° = \cos 61°$ (Q2) $= -0.4848$

Plugging in the known values, we get:

$$(PW)^2 = (21.9)^2 + (19.8)^2 - [2(21.9)(19.8)(-0.4848)]$$
$$= 480 + 392 - [(43.8)(19.8)(-0.4848)]$$
$$= 872 - [(867)(-0.4848)]$$
$$= 872 - [-420]$$
$$= 872 + (+420)$$
$$= 1292$$

Note: Examine the last few steps carefully. In them, we converted the subtraction of a negative term to an addition.

If $(PW)^2 = 1292$, then PW = ___________

Answer: 35.9 ft.

125. When solving for the size of an angle in a triangle with the Cosine Law,

(1) If the angle is acute, its cosine is positive.

(2) If the angle is obtuse, its cosine is negative.

Here is a triangle in which θ is an obtuse angle. If we find that $\cos \theta = -0.4540$:

(a) The reference angle for θ is _________.

(b) θ = _________

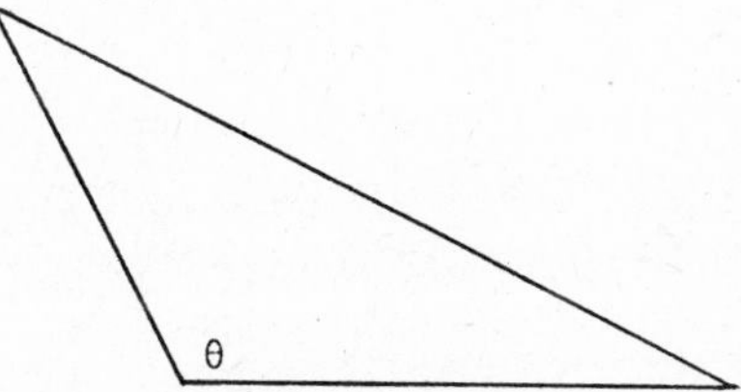

Answer: a) 63° (Q2) b) 117°

126. If θ is an angle in a triangle and its cosine is -0.6820:

(a) Specify the reference angle for θ. ___________

(b) How large is angle θ? _______°

Answer: a) 47° (Q2) b) 133°

127. If the cosine of angle B in a triangle is -0.8910:

(a) The reference angle for B is ________.

(b) How large is angle B? ______°

a) 27° (Q2)

b) 153°

---

128. Let's solve for angle J in this triangle. Since only the three sides are given, we must use the Law of Cosines to find angle J, as follows:

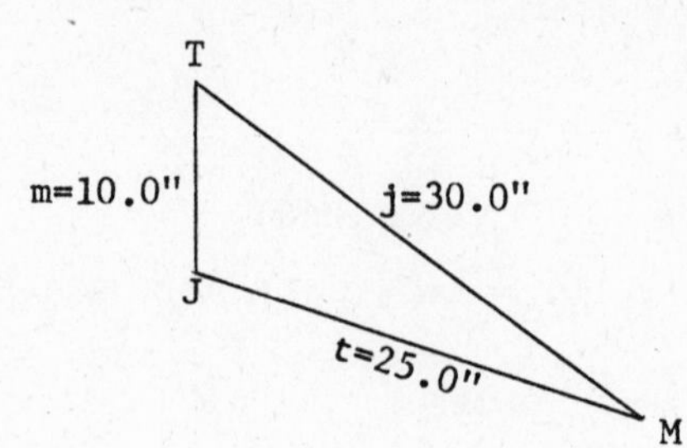

$$j^2 = m^2 + t^2 - 2mt \cos J$$

(a) Rearrange the above equation, solving it for "cos J".

(b) Find the numerical value of "cos J". cos J = ________

(c) Angle J contains ______°.

a) $\cos J = \dfrac{m^2 + t^2 - j^2}{2mt}$

b) $\cos J = -0.350$

c) 110°

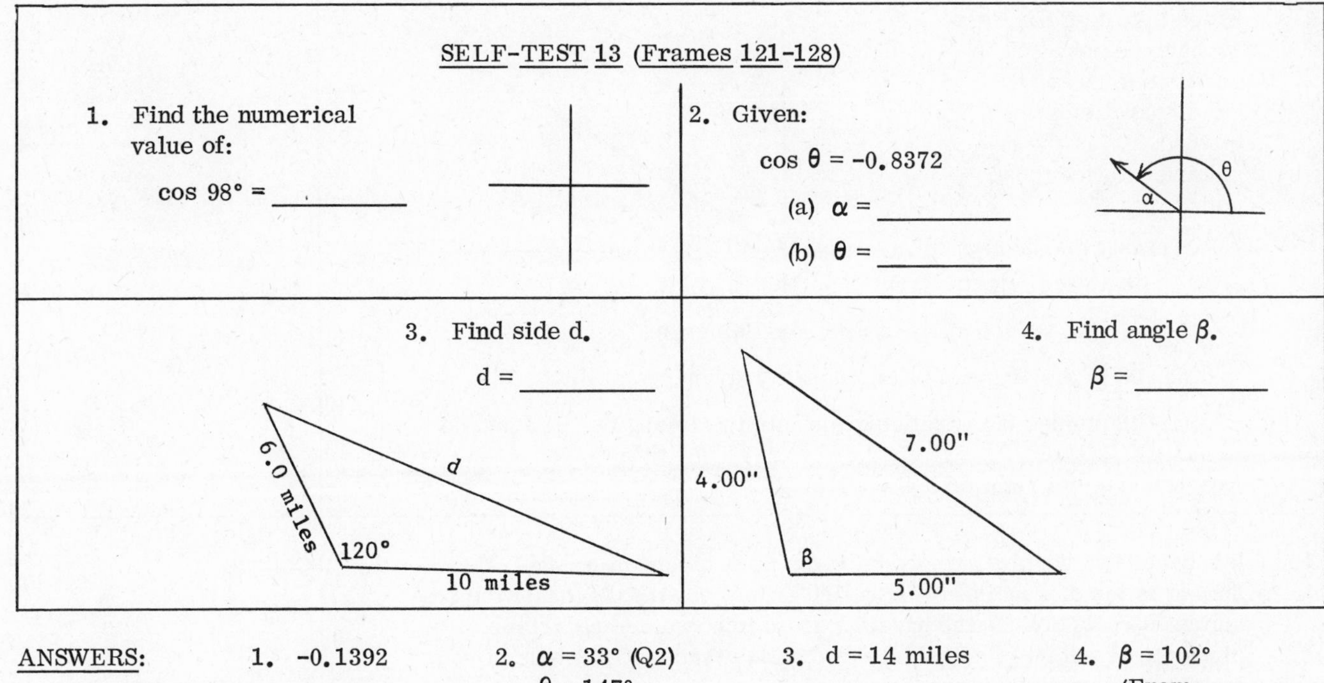

ANSWERS: 1. -0.1392 2. $\alpha = 33°$ (Q2) $\theta = 147°$ 3. d = 14 miles 4. $\beta = 102°$ (From: $\cos \beta = -0.200$)

4-14 VECTOR-ADDITION BY THE OBLIQUE-TRIANGLE METHOD

In an earlier chapter, we discussed vector-addition. A typical vector-addition problem involves the following conditions:

(1) You are given the lengths and directions of two or more vectors.
(2) You must find the length and direction of the resultant (the sum of the given vectors).

In that earlier chapter, we discussed vector-addition by both the parallelogram method and the component method. The parallelogram method is ordinarily avoided because it requires a precise scale-drawing. The component method is rather tedious because it requires all of the following steps:

(1) Finding the horizontal and vertical components of each original vector.
(2) Adding the components of the original vectors to find the components of the resultant.
(3) Using the components of the resultant to find both its length and its direction.

Though the component method must be used to add three or more vectors, there is a simpler method which can be used to add two vectors. This simpler method is called the "oblique-triangle method". We will discuss the oblique-triangle method of vector-addition in this section.

129. In the figure on the right, we have sketched the addition of $\overrightarrow{0F}$ and $\overrightarrow{0T}$ by the parallelogram method. $\overrightarrow{0S}$ is the resultant. Notice these points:

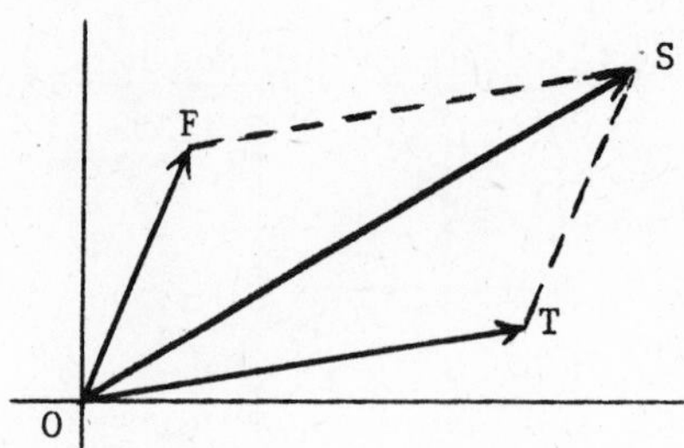

(1) The figure 0FST is a parallelogram.

(2) $\overrightarrow{0F}$ and $\overrightarrow{0T}$ are sides of this parallelogram.

(3) $\overrightarrow{0S}$ is the diagonal of this parallelogram.

(4) $\overrightarrow{0S}$ divides the parallelogram into two triangles, 0FS and 0ST.

Are 0FS and 0ST right or oblique triangles? ______________

Answer: Oblique

130. Whenever two vectors are added by the parallelogram method, the resultant is the diagonal of the parallelogram. As the diagonal, the resultant usually divides the parallelogram into two oblique triangles. Since the resultant is one side of two oblique triangles, we can find its length and direction by solving one of the two triangles.

Before discussing the oblique-triangle method of vector-addition, however, we must review some facts about parallelograms.

As you can see by examining parallelogram KPQM on the right:

(1) Opposite sides of parallelograms are equal.

(2) Opposite angles of parallelograms are equal.

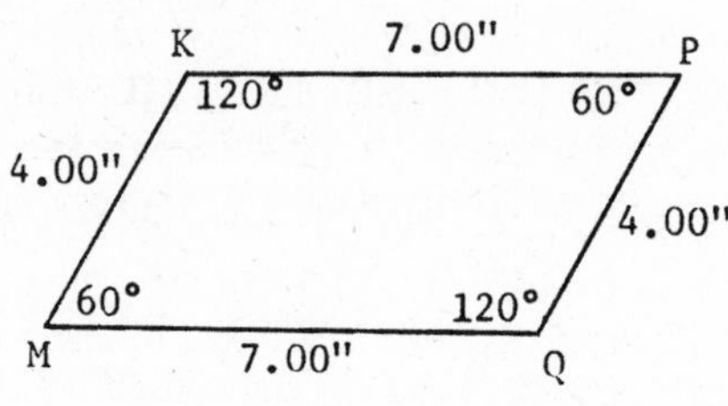

That is: (a) Side KP has the same length as side ________.

(b) Angle P is the same size as angle ________.

Answers: a) MQ b) M

131. The four angles are given in parallelogram LVFA on the right. As you can see by examining this parallelogram:

(1) The sum of the four angles of a parallelogram is 360°

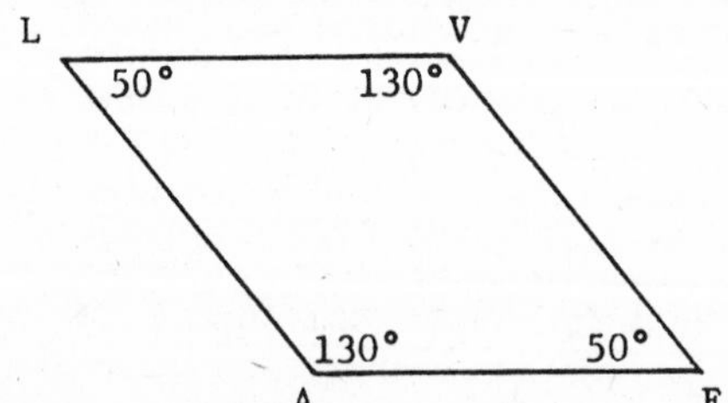

(2) The sum of any two unequal angles in a parallelogram is 180°. That is:

The sum of L and V = 180°.
The sum of A and F = 180°.
The sum of L and A = 180°.
The sum of F and V = 180°.

(Continued on following page.)

131. (Continued)

When the sum of two angles is 180°, we call them "supplementary angles". How many pairs of supplementary angles are there in parallelogram LVFA on the preceding page? __________

Four. They are:

L and V
A and F
L and A
F and V

---

132. None of the dimensions of parallelogram ABCD are given. However, you should be able to answer the following questions:

(a) Side AB is equal in length to side _______.

(b) The sum of angles A, B, C, and D is _______°.

(c) The sum of angles A and D is _______°.

(d) If angle B is 125°, angle D is _______°.

(e) If angle B is 125°, angle C is _______°.

a) CD
b) 360°
c) 180°
d) 125°
e) 55°

---

133. In parallelogram ACFT, if angle A = 49°:

(a) How large is angle T? _______°

(b) How large is angle F? _______°

(c) How large is angle C? _______°

a) 131°
b) 49°
c) 131°

---

134. In parallelogram HJSR, if angle J = 143°:

(a) Angle H = _______°

(b) Angle R = _______°

(c) Angle S = _______°

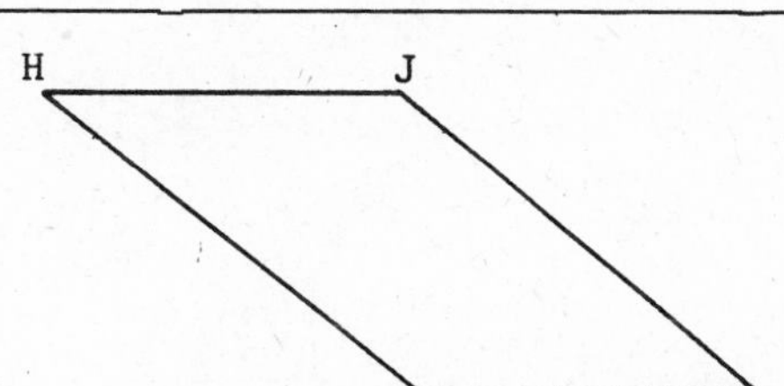

a) 37°
b) 143°
c) 37°

135. In parallelogram ABCD, both angle A and angle C contain 60°. The diagonal of the parallelogram is drawn. As you can see from the diagram:

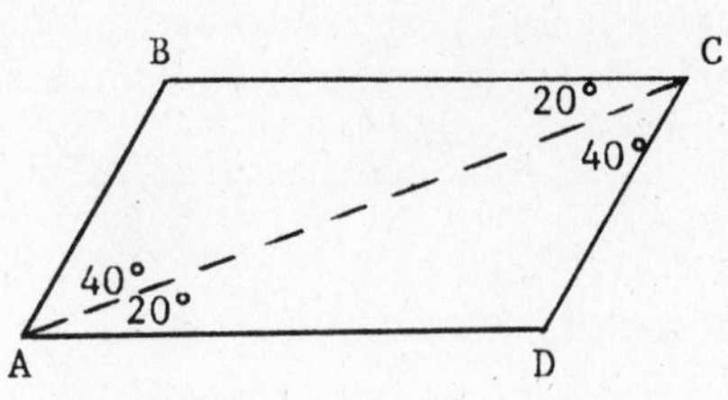

The diagonal of a parallelogram does not necessarily divide angles A and C into two equal angles.

Note: Both angles A and C are divided into two angles, but one is 40° and one is 20° in each case.

A diagonal has been drawn in parallelogram CDFT at the right. It divides angle T into angles #1 and #2.

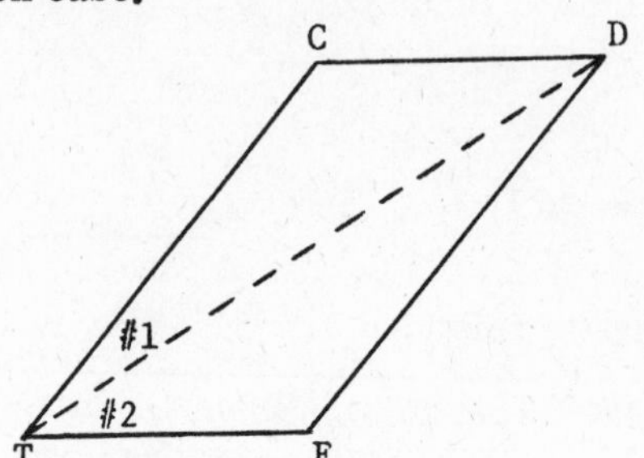

If angle T contains 50°:

(a) How large is angle F? _______°

(b) Can you tell the size of angles #1 and #2? __________

a) 130°

b) No. The diagonal does not necessarily divide angle T into two equal angles.

---

136. When vectors are added, they are frequently not drawn on the coordinate system. For example, $\overrightarrow{0A}$ and $\overrightarrow{0B}$ in Figure 1 below are not drawn on the coordinate system. The angle between them (angle 0) is 40°.

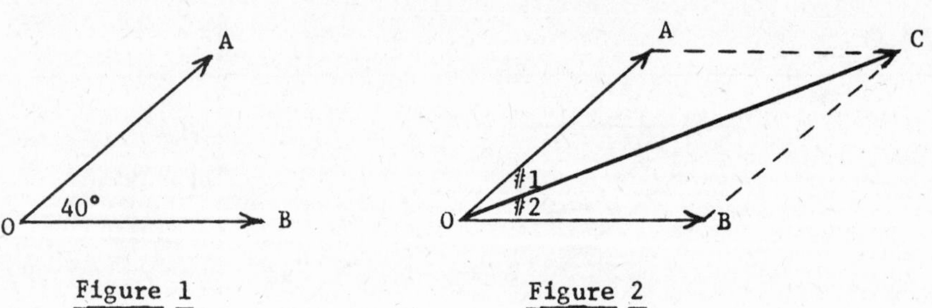

Figure 1 Figure 2

In Figure 2, we have completed the parallelogram and drawn the resultant $\overrightarrow{0C}$. Given the fact that angle 0 contains 40°:

(a) How large is angle B? _______°

(b) How large are angles #1 and #2? ____________________

a) 140°

b) You can't tell.

137. In Figure 1 below, we have drawn vectors 0S and 0T. The angle between them is 33°. In Figure 2, we have completed the parallelogram 0SRT. In 0SRT:

(1) Angle T must contain 147° since angles 0 and T are supplementary angles.

(2) $\overrightarrow{TR}$ must be 18.0 units, since $\overrightarrow{0S}$ and $\overrightarrow{TR}$ are opposite sides of the parallelogram.

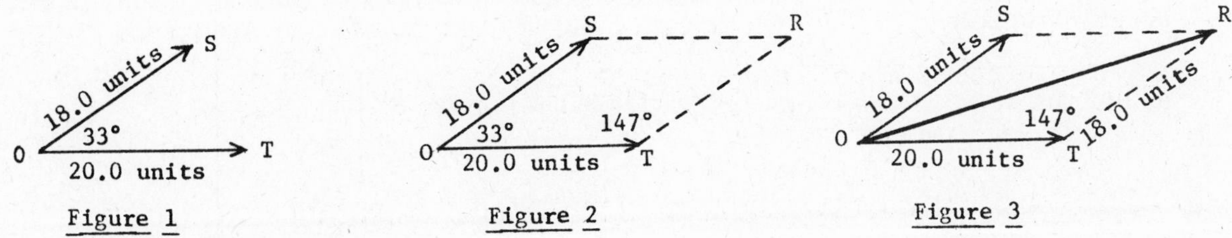

In Figure 3, we have drawn the resultant $\overrightarrow{0R}$. It divides the parallelogram into two oblique triangles. We know the following facts about oblique triangle 0RT which is drawn alone below.

(1) $\overrightarrow{0T}$ is 20.0 units.

(2) $\overrightarrow{TR}$ is 18.0 units.

(3) Angle T is 147°.

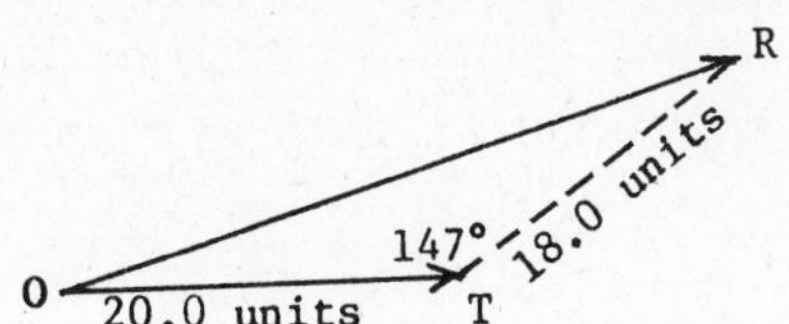

From the diagram, note that in triangle 0RT we know two sides and the included angle. Therefore, we can use the Law of Cosines to solve for the length of the resultant, $\overrightarrow{0R}$. To do so, we use the following Law-of-Cosine formula:

$$\left(\overrightarrow{0R}\right)^2 = \left(\overrightarrow{0T}\right)^2 + \left(\overrightarrow{TR}\right)^2 - 2\left(\overrightarrow{0T}\right)\left(\overrightarrow{TR}\right)(\cos T)$$

or $$\left(\overrightarrow{0R}\right)^2 = (20.0)^2 + (18.0)^2 - 2(20.0)(18.0)(-0.8387)$$

Note: $\cos T = \cos 147° = -0.8387$

Complete the solution. (Note: Be careful of the sign when evaluating the last term.)

$\overrightarrow{0R}$ = ____________

$\overrightarrow{0R}$ = 36.4 units

From: $\left(\overrightarrow{0R}\right)^2 = 1328$

138. Here is a similar problem. We want to add $\overrightarrow{0A}$ and $\overrightarrow{0B}$. The resultant is $\overrightarrow{0C}$.

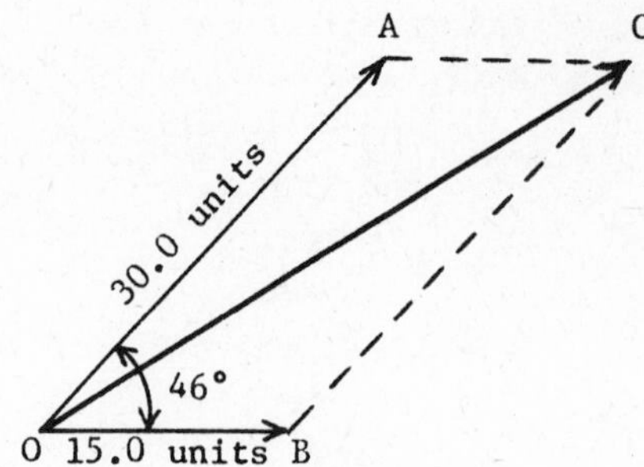

(a) If angle 0 (the angle formed by $\overrightarrow{0A}$ and $\overrightarrow{0B}$) is 46°, how large is angle B? ______°

(b) How long is $\overrightarrow{BC}$? ______

(c) Using triangle 0CB, what equation would you use to solve for the length of $\overrightarrow{0C}$, the resultant?

______

---

a) 134°

b) 30.0 units

(The same as $\overrightarrow{0A}$.)

c) $\left(\overrightarrow{0C}\right)^2 = \left(\overrightarrow{0B}\right)^2 + \left(\overrightarrow{BC}\right)^2 - 2\left(\overrightarrow{0B}\right)\left(\overrightarrow{BC}\right)(\cos B)$

or

$\left(\overrightarrow{0C}\right)^2 = (15.0)^2 + (30.0)^2 - 2(15.0)(30.0)(\cos 134°)$

---

139. Two vectors, $\overrightarrow{0F}$ and $\overrightarrow{0D}$, are drawn in Figure 1. The angle between them is 124°. In Figure 2, we have completed the parallelogram and drawn the resultant, $\overrightarrow{0S}$.

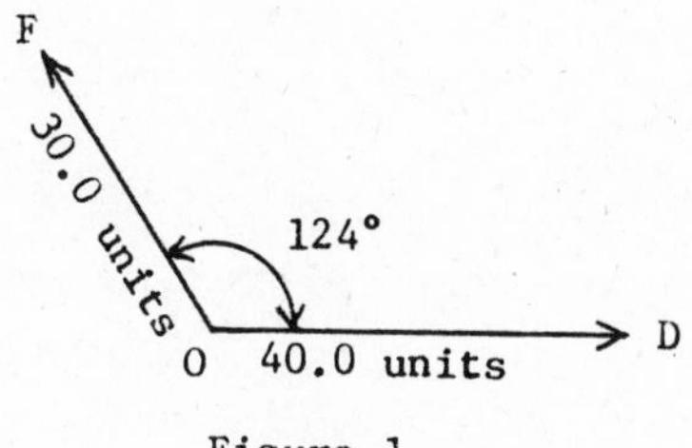

Figure 1

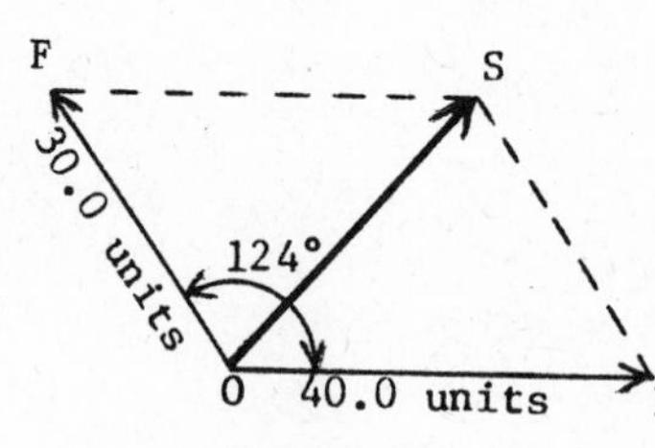

Figure 2

(a) If angle 0 is 124°, angle D is ______°.

(b) How long is $\overrightarrow{DS}$? ______ units.

(c) Using triangle 0DS, what equation would you use to solve for the length of $\overrightarrow{0S}$?

______

---

a) 56°

b) 30.0 units

c) $\left(\overrightarrow{0S}\right)^2 = \left(\overrightarrow{0D}\right)^2 + \left(\overrightarrow{DS}\right)^2 - 2\left(\overrightarrow{0D}\right)\left(\overrightarrow{DS}\right)(\cos D)$

or

$\left(\overrightarrow{0S}\right)^2 = (40.0)^2 + (30.0)^2 - 2(40.0)(30.0)(\cos 56°)$

---

140. We want to add $\overrightarrow{0R}$ and $\overrightarrow{0P}$. The resultant is $\overrightarrow{0Q}$.

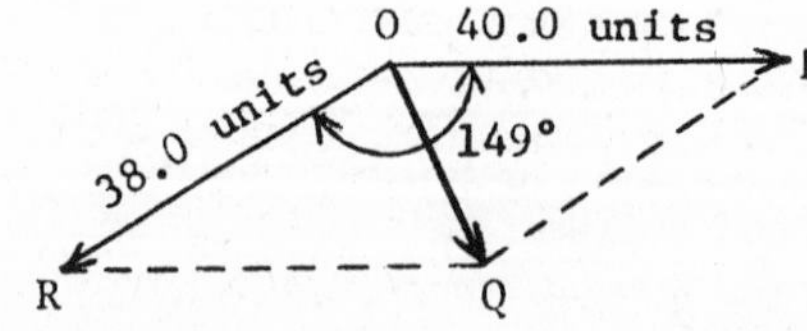

(a) Angle 0 is 149°. How large is angle R? ______°

(b) How long is $\overrightarrow{RQ}$? ______

(c) Using triangle 0RQ, what equation would you use to find the length of $\overrightarrow{0Q}$?

______

141. In the last few frames, we set up the equation for finding the lengths of various resultants by the oblique-triangle method. When using this method, we are also interested in determining the angle between the resultant and each of the original vectors. We will discuss this procedure in the next few frames.

In Figure 1 below, the angle between the two vectors, $\overrightarrow{0S}$ and $\overrightarrow{0T}$, is 42°. In Figure 2, we have completed the parallelogram and drawn the resultant, $\overrightarrow{0V}$. Using the Law of Cosines, we found that $\overrightarrow{0V}$ is 28.2 units. We want to find the size of angle #1 (the angle between $\overrightarrow{0S}$ and $\overrightarrow{0V}$) and the size of angle #2 (the angle between $\overrightarrow{0T}$ and $\overrightarrow{0V}$).

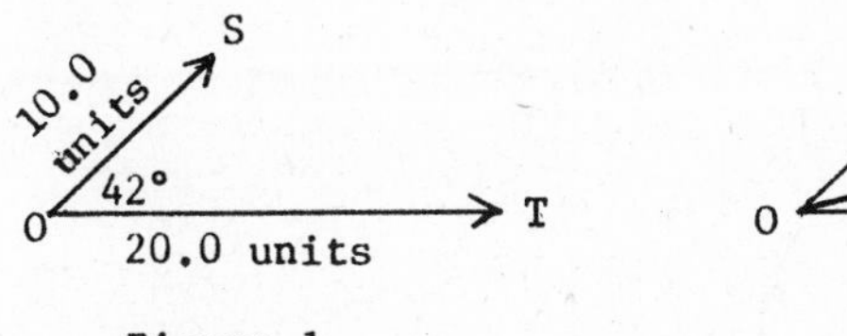

Figure 1

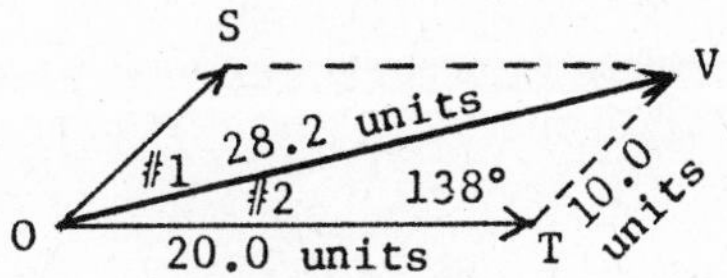

Figure 2

Since the sum of angle #1 and angle #2 is 42°, we can easily find the size of the second angle if we can determine the size of one of them. We will find the size of angle #2.

Let's examine triangle 0VT alone. We know the lengths of the three sides and the size of angle T. Therefore, we can use the Law of Sines to determine the size of angle #2.

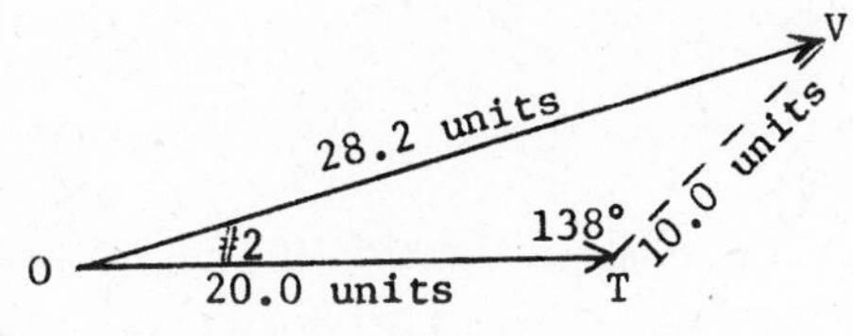

(a) What specific Law-of-Sines equation can we use to find the size of angle #2?

(b) Angle #2 contains _______°.

(c) Look back at Figure 2. Knowing the size of angle #2, how large is angle #1? _______°

a) 31°

b) 40.0 units

c) $\left(\overrightarrow{0Q}\right)^2 = \left(\overrightarrow{0R}\right)^2 + \left(\overrightarrow{RQ}\right)^2 - 2(\overrightarrow{0R})(\overrightarrow{RQ})(\cos R)$

or

$\left(\overrightarrow{0Q}\right)^2 = (38.0)^2 + (40.0)^2 - 2(38.0)(40.0)(\cos 31°)$

142. The diagram on the right shows the addition of $\overrightarrow{0A}$ and $\overrightarrow{0B}$. The angle between them is 80°. Using the Law of Cosines, we have determined that the length of the resultant $\overrightarrow{0C}$ is 33.2 units. We want to find the size of angles #2 and #1.

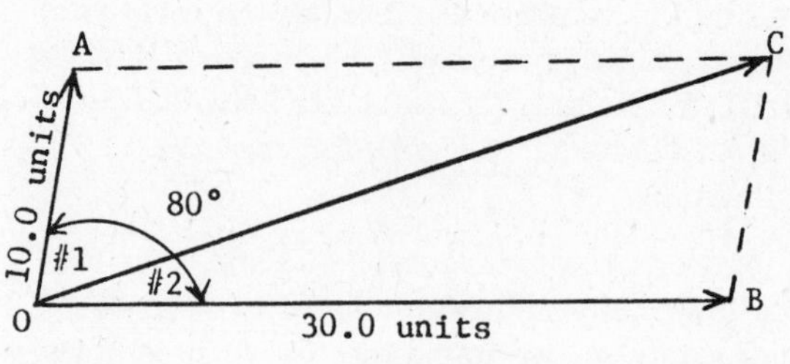

(a) Triangle 0CB is drawn at the right. Write in the information you know about this triangle.

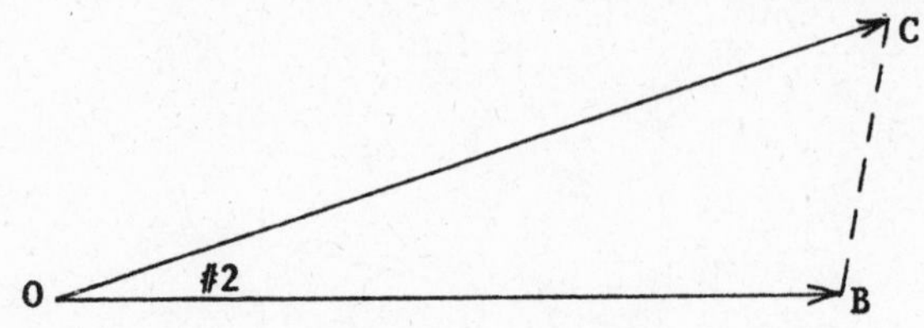

(b) Using the Law of Sines, find the size of angle #2. ______°

(c) Since angle 0 (the angle between $\overrightarrow{0A}$ and $\overrightarrow{0B}$) is 80°, how large is angle #1 ? ______°

a) $\dfrac{\sin \#2}{10.0} = \dfrac{\sin 138°}{28.2}$

b) 14°

c) 28°, since:
42° - 14° = 28°

143.

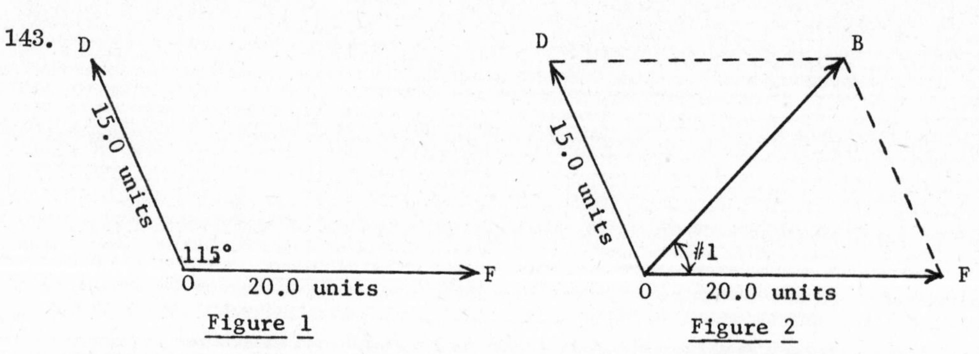

We want to add $\overrightarrow{0D}$ and $\overrightarrow{0F}$ in Figure 1. The angle between them is 115°. In Figure 2, we have completed the parallelogram and drawn the resultant $\overrightarrow{0B}$.

(a) How large is angle F? ______°

(b) How long is the resultant $\overrightarrow{0B}$? __________

(Continued on following page.)

a)

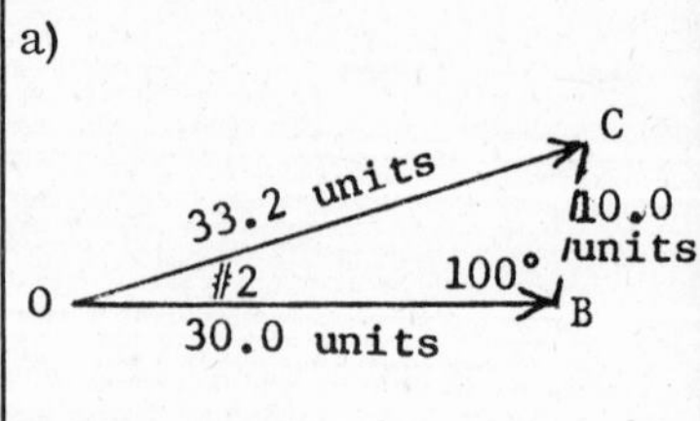

b) 17°

c) 63°

143. (Continued)

(c) How large is angle #1, the angle formed by $\overrightarrow{0B}$ and $\overrightarrow{0F}$? ______°

a) 65°

b) 19.3 units

From: $\overrightarrow{0B} = \sqrt{371}$

c) 45°

From: sin #1 = 0.705

144.

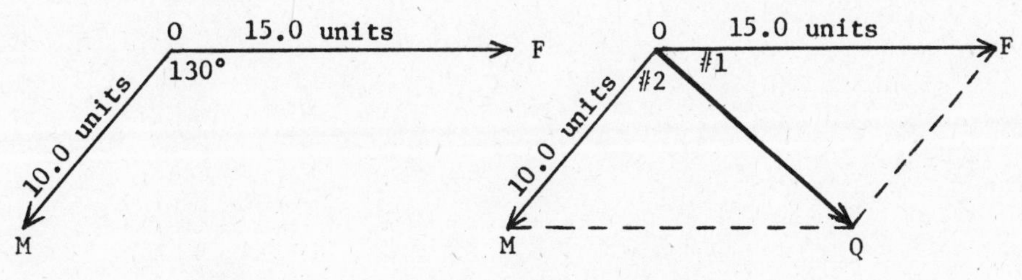

Figure 1 Figure 2

We want to add $\overrightarrow{0F}$ and $\overrightarrow{0M}$ in Figure 1. The angle between them is 130°. We have completed the parallelogram in Figure 2.

(a) How large is angle F? ______°

(b) How long is the resultant $\overrightarrow{0Q}$? ____________

(c) How large is angle #1? ______°

(d) How large is angle #2? ______°

a) 50°

b) 11.5 units

From: $\overrightarrow{0Q} = \sqrt{132}$

c) 42°

d) 88°

SELF-TEST 14 (Frames 129-144)

As shown on the right, two vectors of 30.0 units and 50.0 units act at an angle of 60°.

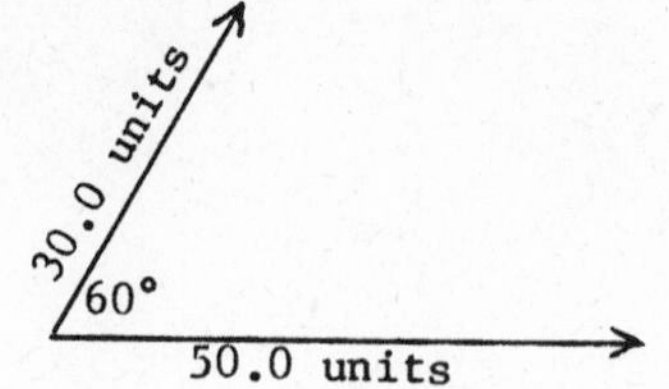

1. Find the length of their resultant. __________

2. Find the angle between the 50.0 unit vector and the resultant. __________

ANSWERS: 1. 70.0 units 2. 22°

4-15 VECTOR-ADDITION ON THE COORDINATE SYSTEM BY THE OBLIQUE-TRIANGLE METHOD

In the last section, we discussed the oblique-triangle method for the addition of two vectors. In that section, the vectors which were added were not on the coordinate system. The oblique-triangle method can also be used to add two vectors on the coordinate system. We will discuss the method in this section.

---

145. Before attempting to add vectors on the coordinate system by the oblique-triangle method, we will review a few facts about specifying their "directions".

The direction of a vector on the coordinate system can be specified in two ways:

(1) By its reference angle, or

(2) By its standard-position angle.

For example, $\overrightarrow{OM}$ is a vector in the second quadrant. Its direction is either:

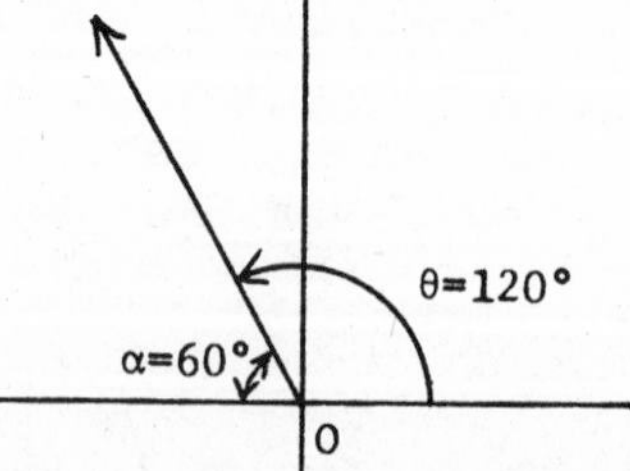

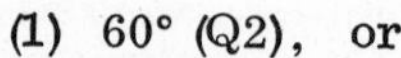

(1) 60° (Q2), or
(2) 120°

(a) If the reference angle of a vector is 30° (Q2), what is its direction in terms of its standard-position angle? __________

(b) If the standard-position angle of a vector is 155°, what is its direction in terms of its reference angle? __________

146. $\overrightarrow{0T}$ is a vector in the third quadrant.

Its direction is either:

(1) 50° (Q3), its reference angle, or
(2) 230° , its standard-position angle.

θ=230°
0
α=50°
T

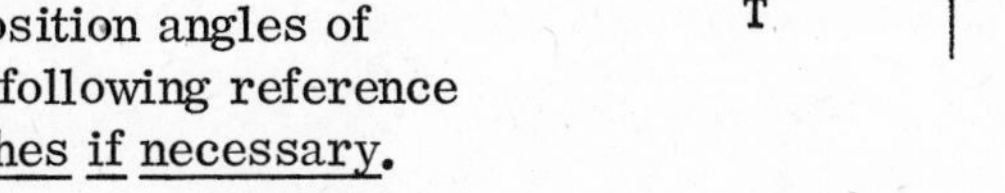

Give the standard-position angles of the vectors with the following reference angles. Draw sketches if necessary.

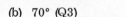

(a) 30° (Q3) ______ (b) 70° (Q3) ______ (c) 40° (Q2) ______

Give the reference angles of the vectors with the following standard-position angles.

(d) 260° ______ (e) 190° ______ (f) 135° ______

a) 150°

b) 25° (Q2)

---

147. Here is a vector in the fourth quadrant.

Its direction is either:

(1) 55° (Q4), or
(2) 305°

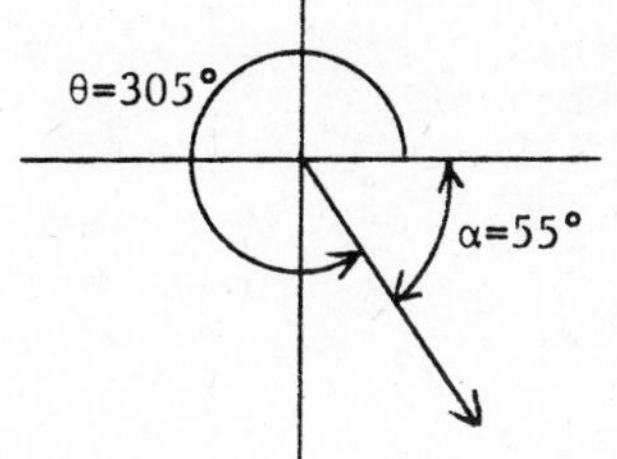

Give the standard-position angles of the vectors with the following reference angles.

(a) 30° (Q4) ______

(b) 70° (Q4) ______

(c) 70° (Q3) ______

(d) 70° (Q2) ______

Give the reference angles of the vectors with the following standard-position angles.

(e) 310° ______

(f) 285° ______

(g) 151° ______

(h) 211° ______

a) 210°
b) 250°
c) 140°

d) 80° (Q3)
e) 10° (Q3)
f) 45° (Q2)

---

a) 330° e) 50° (Q4)
b) 290° f) 75° (Q4)
c) 250° g) 29° (Q2)
d) 110° h) 31° (Q3)

148. If a vector is in the first quadrant, what is the relationship between its reference angle and its standard-position angle?

____

They are identical.

149. $\overrightarrow{0M}$ and $\overrightarrow{0F}$ are vectors in the first quadrant. Their direction is given by standard-position angles, 70° for $\overrightarrow{0M}$ and 20° for $\overrightarrow{0F}$.

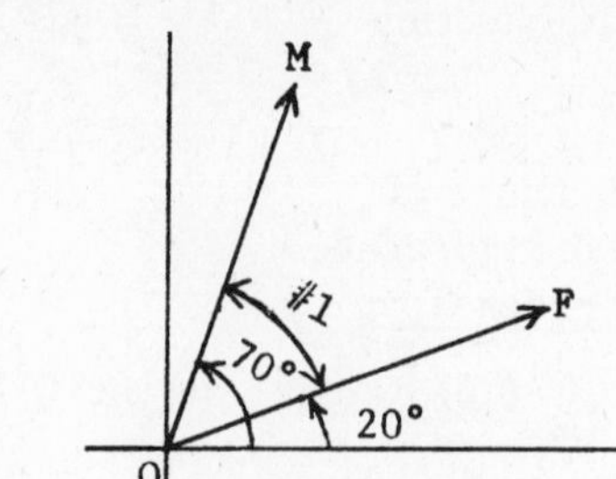

In order to use the oblique-triangle method, we must be able to determine the size of the angle formed by $\overrightarrow{0M}$ and $\overrightarrow{0F}$. We have labeled this angle #1.

To find the size of angle #1, we simply find the difference between the two standard-position angles. Angle #1 is a 50° angle, since 70° - 20° = 50°.

In the figure on the right, the angle formed by $\overrightarrow{0T}$ and $\overrightarrow{0V}$ has been labeled angle #1. How large is angle #1? ______°

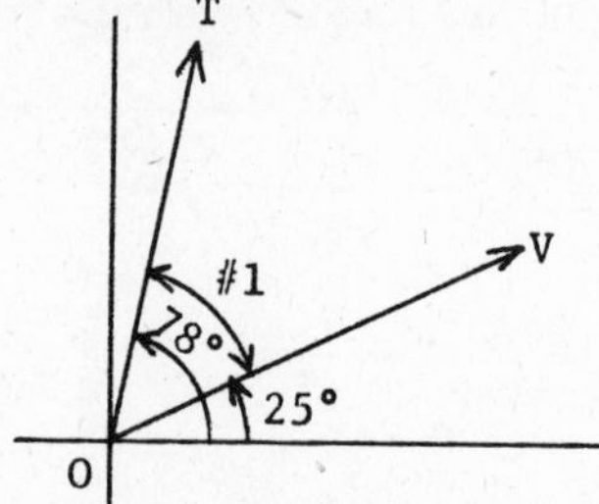

53°

150. Here is a case where one of the vectors is in the second quadrant.

The standard-position angles are:

140° for $\overrightarrow{0B}$.
20° for $\overrightarrow{0C}$.

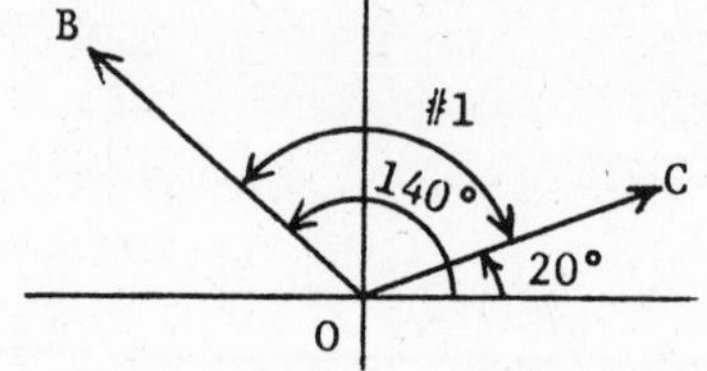

The angle formed by $\overrightarrow{0B}$ and $\overrightarrow{0C}$ has been labeled angle #1. How large is angle #1? ______°

120°

Since:

140° - 20° = 120°

151. Here is a case where one of the vectors is in the fourth quadrant. The angle between them is labeled angle #1.

(a) What is the reference angle (α) for $\overrightarrow{0K}$? ____________

(b) How large is angle #1? ______°

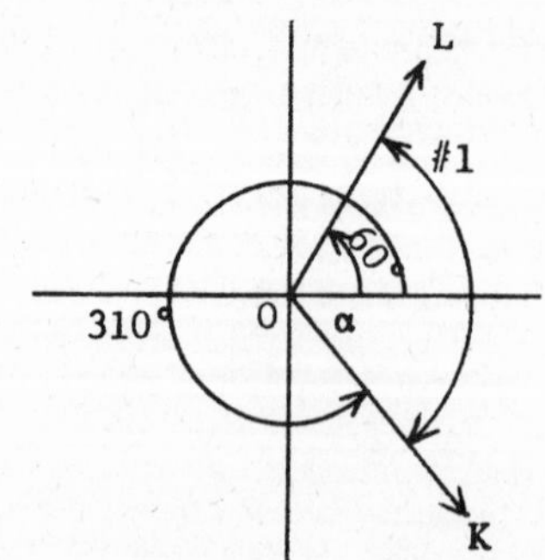

---

a) 50° (Q4)

b) 110°

Since:

60° + 50° = 110°

152. Having reviewed some facts about the angles of vectors on the coordinate system, we can now find the length of the resultant of $\overrightarrow{0D}$ and $\overrightarrow{0M}$ by the oblique-triangle method. In the diagram at the right, we have completed the parallelogram. $\overrightarrow{0P}$ is the resultant. Angle #1 is the angle between $\overrightarrow{0D}$ and $\overrightarrow{0M}$.

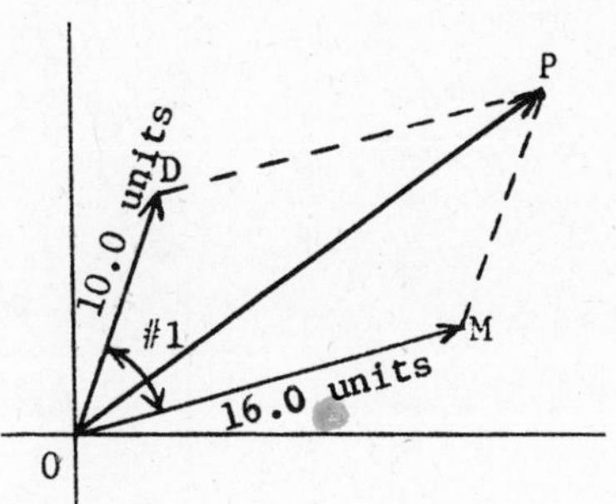

(a) If the direction of $\overrightarrow{0D}$ is 70° and the direction of $\overrightarrow{0M}$ is 15°, how large is angle #1? _______°

(b) How large is angle M? _______°

(c) cos M = __________

(d) How long is $\overrightarrow{MP}$? __________

(e) Given what you know about triangle 0PM, what equation would you use to find the length of the resultant, $\overrightarrow{0P}$?

____

---

a) 55°
Since:
70° - 15° = 55°

b) 125° Since:
Angle #1 + Angle M = 180°

c) cos M = -0.5736

d) 10.0 units
Since:
$\overrightarrow{MP} = \overrightarrow{0D} = 10.0$ units

e) $\left(\overrightarrow{0P}\right)^2 = (16.0)^2 + (10.0)^2 - 2(16.0)(10.0)(\cos 125°)$

---

153. We want to add $\overrightarrow{0B}$ and $\overrightarrow{0D}$. We have completed the parallelogram and drawn their resultant, $\overrightarrow{0R}$.

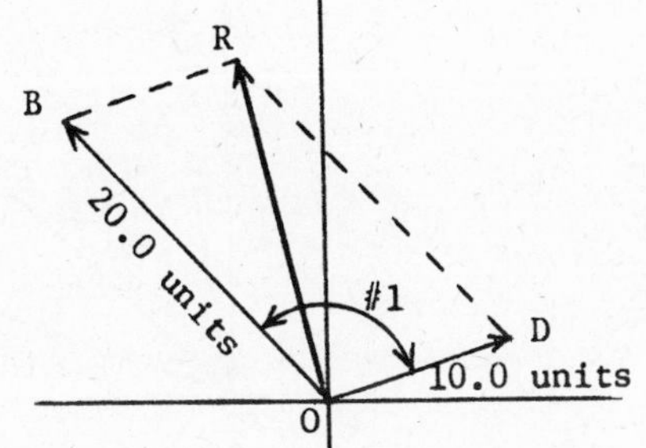

(a) If the direction of $\overrightarrow{0B}$ is 135° and the direction of $\overrightarrow{0D}$ is 20°, how large is angle #1? _______°

(b) How large is angle D? _______°

(c) How long is $\overrightarrow{DR}$? __________

(d) Using triangle 0DR, what equation would you use to find the length of the resultant $\overrightarrow{0R}$?

____

---

a) 115°

b) 65°

c) 20.0 units

d) $\left(\overrightarrow{0R}\right)^2 = (10.0)^2 + (20.0)^2 - 2(10.0)(20.0)(\cos 65°)$

154. In the figure at the right, $\overrightarrow{0S}$ is the resultant of $\overrightarrow{0L}$ and $\overrightarrow{0N}$.

The direction of $\overrightarrow{0L}$ is 65°.

The direction of $\overrightarrow{0N}$ is 330°.

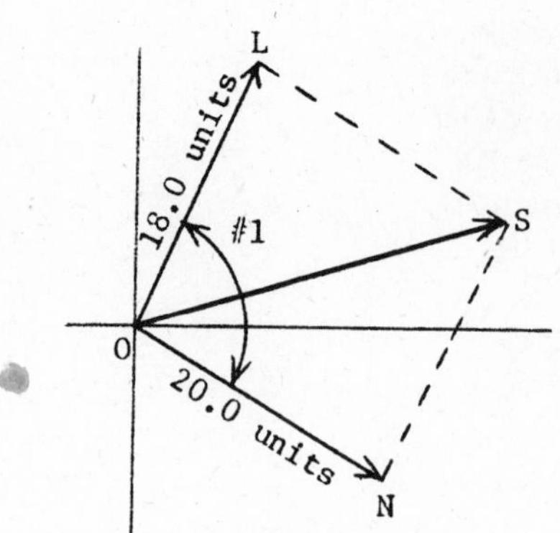

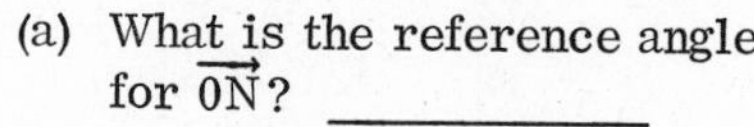
(a) What is the reference angle for $\overrightarrow{0N}$? ________

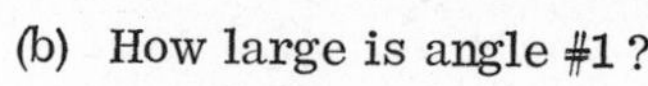
(b) How large is angle #1? ______°

(c) How large is angle L? ______°

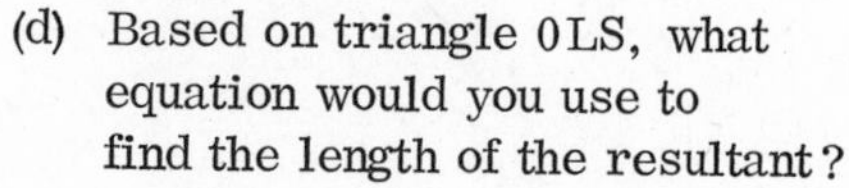
(d) Based on triangle 0LS, what equation would you use to find the length of the resultant?

____

---

155. In the next few frames, we will show you how to find the <u>direction</u> of a resultant after you have determined its length by the oblique-triangle method.

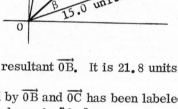

In this case, $\overrightarrow{0B}$ is the resultant of $\overrightarrow{0A}$ and $\overrightarrow{0C}$.

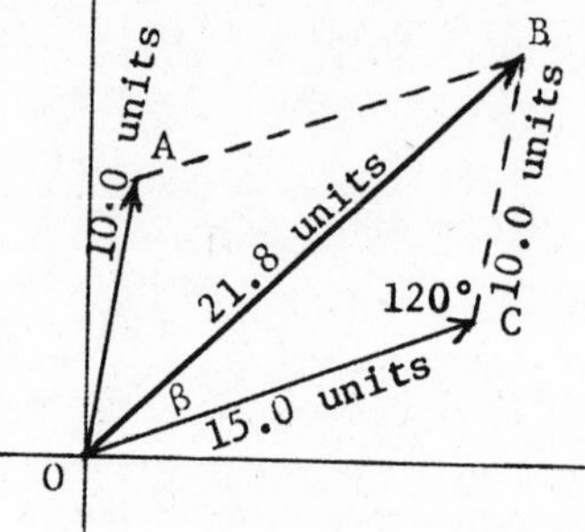

(1) The direction of $\overrightarrow{0A}$ is 80°.
The direction of $\overrightarrow{0C}$ is 20°.

(2) Since the angle formed by $\overrightarrow{0A}$ and $\overrightarrow{0C}$ is 80° − 20° = 60°, angle C is 120°.

(3) Using the Law of Cosines, we have calculated the length of the resultant $\overrightarrow{0B}$. It is 21.8 units.

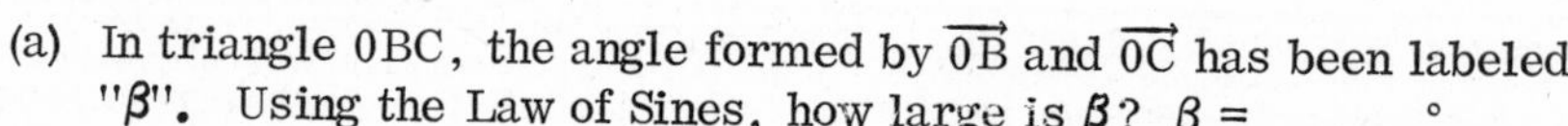
(a) In triangle 0BC, the angle formed by $\overrightarrow{0B}$ and $\overrightarrow{0C}$ has been labeled "β". Using the Law of Sines, how large is β? β = ______°

(b) The direction of the resultant is the sum of angle β + the standard-position angle of $\overrightarrow{0C}$. Therefore, the direction of $\overrightarrow{0B}$ is ______°.

a) 30° (Q4)

b) 95°
Since:
65° + 30° = 95°

c) 85°
Since:
Angle #1 + Angle L = 180°

d) $(\overrightarrow{0S})^2 = (18.0)^2 + (20.0)^2 - 2(18.0)(20.0)(\cos 85°)$

156. In the figure on the right, $\overrightarrow{0M}$ is the resultant of $\overrightarrow{0S}$ and $\overrightarrow{0T}$.

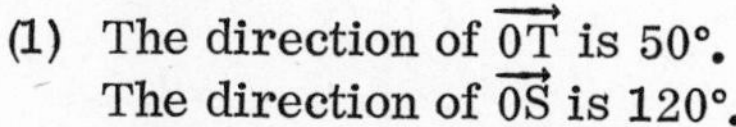

(1) The direction of $\overrightarrow{0T}$ is 50°.
The direction of $\overrightarrow{0S}$ is 120°.

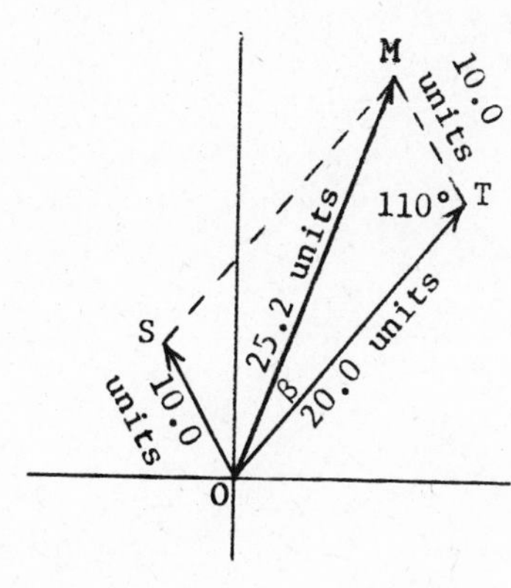

(2) Since the angle formed by $\overrightarrow{0S}$ and $\overrightarrow{0T}$ is 120° - 50° = 70°, angle T contains 110°.

(3) Using the Law of Cosines in triangle 0MT, we have calculated the length of the resultant $\overrightarrow{0M}$. It is 25.2 units.

(4) Angle β is the angle formed by the resultant $\overrightarrow{0M}$ and $\overrightarrow{0T}$.

(a) In triangle 0MT, how large is angle β? β = ______°

(b) The standard-position angle of the resultant is the sum of angle β and the standard-position angle of $\overrightarrow{0T}$. Therefore, the direction of $\overrightarrow{0M}$ is ______°.

a) 23°
(sin β = 0.397)

b) 43°
Since:
23° + 20° = 43°

---

157. In this diagram, $\overrightarrow{0F}$ is the resultant of $\overrightarrow{0D}$ and $\overrightarrow{0G}$.

(1) The direction of $\overrightarrow{0D}$ is 55°.
The direction of $\overrightarrow{0G}$ = 295°. Its reference angle is 65° (Q4).

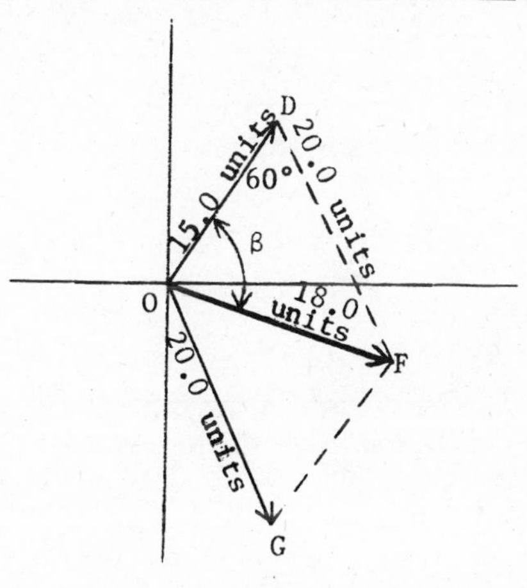

(2) The angle formed by $\overrightarrow{0D}$ and $\overrightarrow{0G}$ is the sum of the standard-position angle of $\overrightarrow{0D}$ (55°) and the reference angle of $\overrightarrow{0G}$ (65°). It is a 120° angle.

(3) Since the angle formed by $\overrightarrow{0D}$ and $\overrightarrow{0G}$ is 120°, angle D is 60°.

(4) Using the Law of Cosines in triangle 0DF, we calculated the length of the resultant $\overrightarrow{0F}$. It is 18.0 units.

(5) The angle formed by $\overrightarrow{0D}$ and the resultant $\overrightarrow{0F}$ has been labeled "β".

(Continued on following page.)

a) 22°
(sin β = 0.373)

b) 72°
Since:
22° + 50° = 72°

157. (Continued)

(a) Using the Law of Sines in triangle 0DF, how large is angle β?
β = ______°

(b) The resultant $\overrightarrow{0F}$ is in the 4th quadrant. To find its reference angle, we subtract the standard-position angle of $\overrightarrow{0D}$ (55°) from angle β (74°). The reference angle of $\overrightarrow{0F}$ is ____________.

(c) The standard-position angle of the resultant is ______°.

Answer to Frame 157: a) 74° ($\sin \beta = 0.962$) b) 19° (Q4) c) 341°

SELF-TEST 15 (Frames 145-157)

Two vectors on the coordinate system have these lengths and standard-position angles:

10.0 units at 40°
20.0 units at 150°

Sketch these vectors on the coordinate system, and then calculate the following:

1. Length of their resultant. __________
2. Standard-position angle of the resultant. ______°

Two vectors on the coordinate system have these lengths and standard-position angles:

6.00 units at 22°
4.00 units at 308°

Sketch these vectors, and calculate the following:

3. Length of their resultant. __________
4. Standard-position angle of the resultant. ______°

ANSWERS: 1. 19.1 units 2. 120° 3. 8.08 units 4. 354°

Chapter 5 APPLIED GEOMETRIC PROBLEMS

In this chapter, we will review many geometric principles and facts, and will then use them to solve applied problems. Most of the problems will involve "circle" concepts, such as central angles, arcs, chords, sectors, segments, rotational speed, circumferential speed, tangents, and half-tangents. A few, however, will involve triangles. Many of the problems will require the use of right-triangle or oblique-triangle trigonometry for their solution.

5-1 CENTRAL ANGLES AND ARCS OF CIRCLES

In this section, we will review some basic facts about central angles and arcs of circles. These facts will be used to solve problems in later sections.

---

1. A radius is a line which connects the center of a circle with any point on the circle. For example, 0A, 0B, and 0C are radii of the circle on the right. (Note: The plural of radius is radii, pronounced "ray-dee-eye".)

   Since all radii of the same circle are equal, if 0A is 4.0 inches, 0B and 0C are also ________ inches.

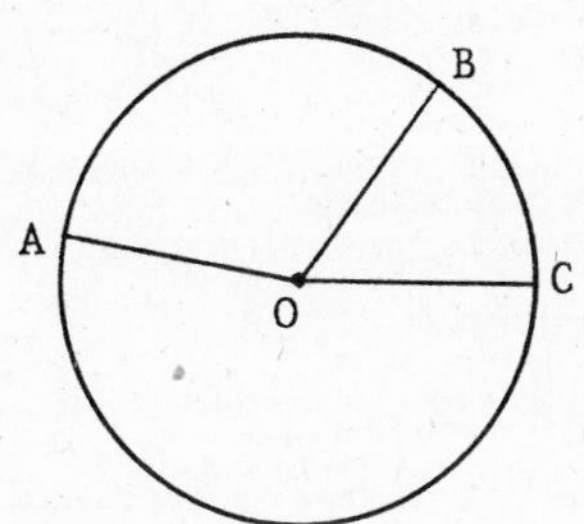

4.0

2. The angle formed by two radii is called a central angle. For example, angle M0T in the circle on the right is a central angle since it is formed by radii 0M and 0T.

   What do we know about the lengths of 0M and 0T? ________________

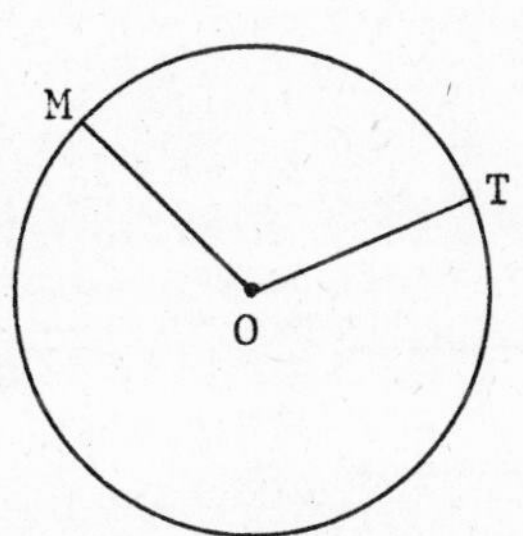

They are equal.

3. The circle on the right is divided into four equal central angles.

Since there are 360° in a complete circle, how large is each of the four central angles? _______°

---

4. How many degrees are there in each central angle of a circle if the circle is divided into:

(a) 3 equal central angles? _______°

(b) 10 equal central angles? _______°

(c) 12 equal central angles? _______°

Answer to 3: 90°

---

5. An arc of a circle is a piece of the circumference of the circle.

For example, in the circle on the right, the curved lines between BD and between MS are "arcs" of the circle.

Instead of writing "arc BD", we usually write $\overset{\frown}{BD}$.

(Note: The curved line over BD means "arc".)

Instead of writing "arc MS", we usually write _______.

Answers to 4: a) 120° b) 36° c) 30°

---

6. Any central angle of a circle cuts off an arc of the circle. For example, in the circle on the right:

Angle TOV cuts off $\overset{\frown}{TV}$.

(a) Angle VOD cuts off _______.

(b) Angle COD cuts off _______.

Answer to 5: $\overset{\frown}{MS}$

Answers to 6: a) $\overset{\frown}{VD}$ b) $\overset{\frown}{CD}$

7. If two central angles in the same circle are equal, they cut off arcs of equal length.

In the circle on the right, central angles M0T and S0V are both 45°. Therefore, what two arcs are equal? ______ and ______

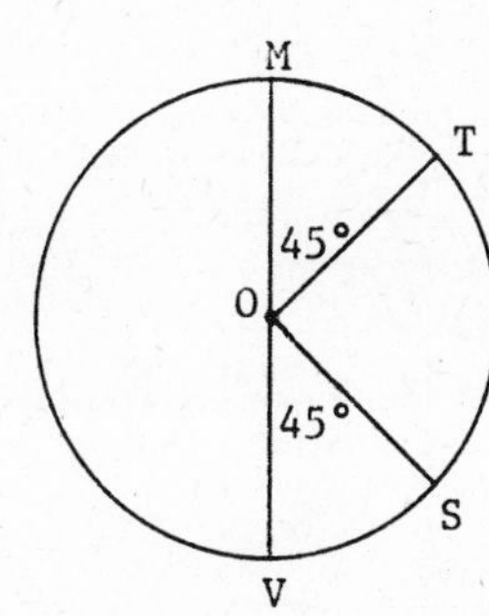

8. If two central angles in a circle cut off arcs of equal length, the central angles are equal.

In the circle on the right, $\overset{\frown}{AB}$ and $\overset{\frown}{CD}$ are the same length. Therefore, central angles ______ and ______ are equal.

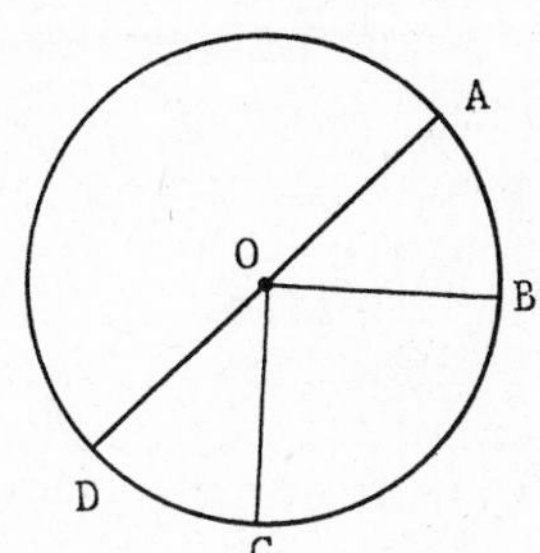

$\overset{\frown}{MT}$ and $\overset{\frown}{SV}$

9. Just as angles are measured in <u>angle-degrees</u>, arcs are measured in <u>arc-degrees</u>. The number of <u>arc-degrees</u> in an arc is the same as the number of <u>angle-degrees</u> in the central angle which cuts off that arc. For example, in the circle below:

Since there are 60° in central angle A0B, there are 60 arc-degrees in $\overset{\frown}{AB}$.

Since there are 45° in central angle B0C, there are _____ arc-degrees in $\overset{\frown}{BC}$.

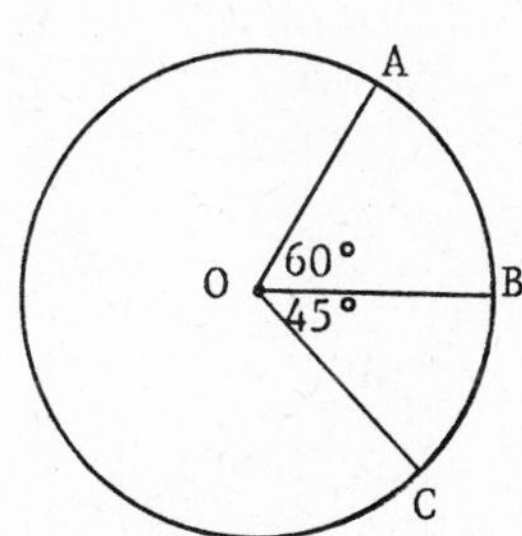

A0B and C0D

10. Just as there are 360 <u>angle-degrees</u> in a circle, there are 360 <u>arc-degrees</u> in a circle.

In the circle on the right, the sizes of two central angles are given.

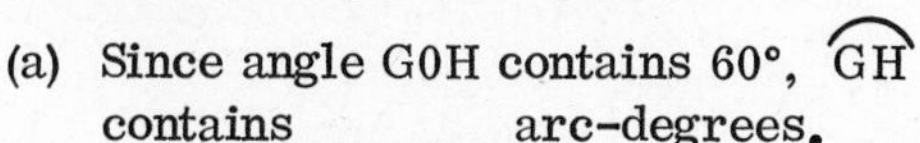

(a) Since angle G0H contains 60°, $\overset{\frown}{GH}$ contains ______ arc-degrees.

(b) Angle G0F contains ______°.

(c) $\overset{\frown}{FG}$ contains ______ arc-degrees.

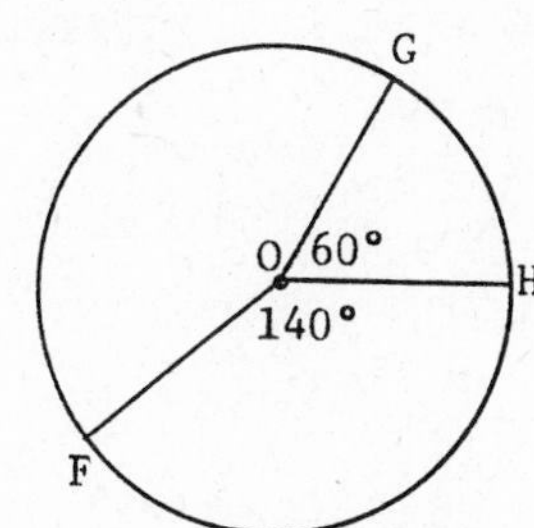

45

11. If there are an equal number of arc-degrees in $\widehat{MN}$, $\widehat{NF}$, and $\widehat{MF}$:

(a) Each arc contains _______ arc-degrees.

(b) Each corresponding central angle contains _______ angle-degrees.

M N θ F

a) 60

b) 160°
From: 360° - 200°

c) 160

---

12. The six central angles in the circle on the right are equal.

(a) There are _______ degrees in each of the six central angles.

(b) There are _______ arc-degrees in each of the six arcs.

a) 120

b) 120

---

13. How many degrees are there in each arc if a circle is divided into:

(a) 4 equal arcs? _______°

(b) 5 equal arcs? _______°

(c) 8 equal arcs? _______°

(d) 9 equal arcs? _______°

(e) 10 equal arcs? _______°

(f) 20 equal arcs? _______°

a) 60

b) 60

Answer to Frame 13:

a) 90° d) 40°
b) 72° e) 36°
c) 45° f) 18°

SELF-TEST 1 (Frames 1-13)

1. If a circle is divided into 9 equal central angles, how many degrees are there in each central angle? _______°

2. If a circle is divided into 30 equal arcs, how many degrees are there in each arc? _______ arc-degrees

A circle is divided into central angles, as shown on the right.

3. Find central angle θ. θ = _______°

4. Two equal arcs are _______ and _______.

5. $\widehat{AD}$ contains _______ arc-degrees.

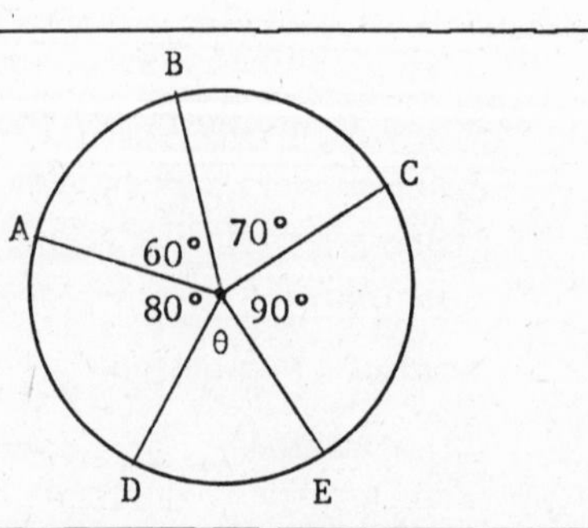

ANSWERS: 1. 40° 2. 12 arc-degrees 3. θ = 60° 4. $\widehat{AB}$ and $\widehat{DE}$ 5. 80 arc-degrees

5-2 CHORDS OF CIRCLES

In this section, we will review some basic facts about chords of circles. These facts will be used to solve problems in later sections.

---

14. A chord (pronounced "cord") is a straight line which joins any two points on a circle. For example, both AB and CD are chords of the circle below.

A diameter is a chord which passes through the center of a circle. For example, CD is a diameter of the circle below.

(a) Is 0T a chord of the circle? _______

(b) The length of the diameter CD is the sum of the lengths of the two radii _______ and _______.

(c) If C0 is 4 inches, CD is _______ inches.

(d) If CD is 12 inches, 0D is _______ inches.

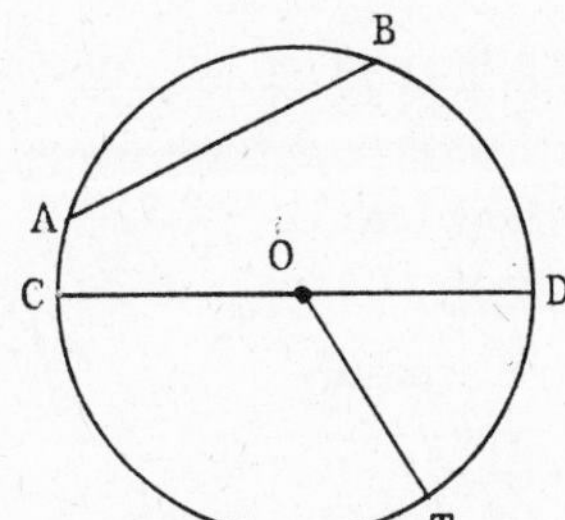

---

15. In the circle on the right, the two radii 0P and 0T are drawn to the ends of chord PT. The two radii and the chord form triangle 0PT. Since the two radii are equal:

(1) 0PT is an isosceles triangle.
(2) Angles #1 and #2 (the angles opposite the radii) are equal.

Therefore, if angle #1 contains 50°:

(a) Angle #2 contains _______°.

(b) Central angle P0T contains _______°.

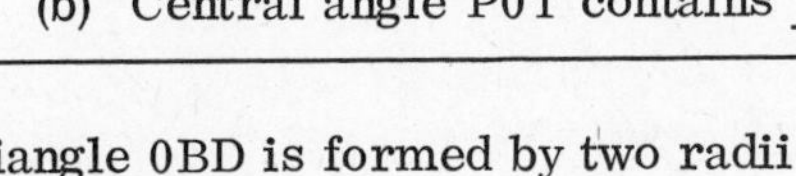

Answers to 14:

a) No, since it does not join two points on the circle.

b) 0C and 0D

c) 8 inches

d) 6 inches

---

16. Triangle 0BD is formed by two radii and a chord.

If central angle B0D contains 50°:

(a) Angle #1 contains _______°.

(b) Angle #2 contains _______°.

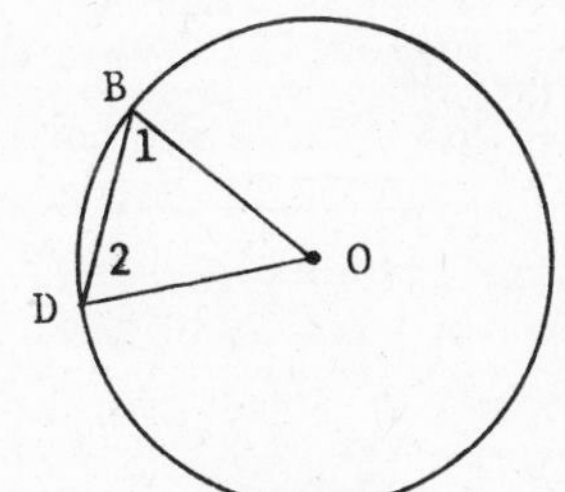

Answers to 15:

a) 50°

b) 80°

Answers to 16:

a) 65°

b) 65°

17. If $\overset{\frown}{RPS}$ contains 140 arc-degrees:

(a) Angle ROS contains ______°.

(b) Angle ORS contains ______°.

(c) Angle OSR contains ______°.

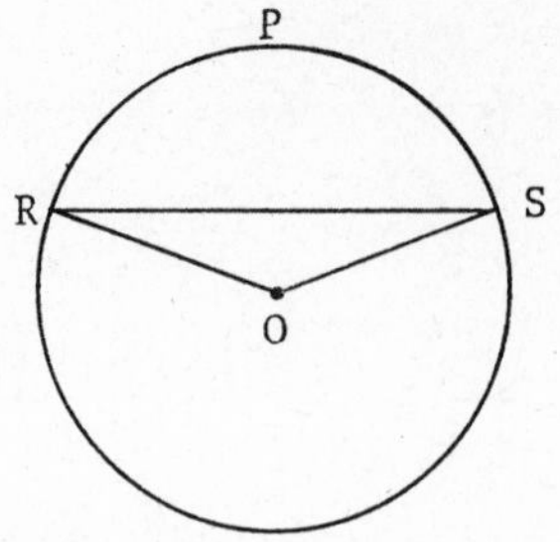

18. In any given circle, <u>equal arcs have equal chords.</u> All five arcs are equal in the circle below.

(a) If chord CD is 1.5", how long is chord FE? ________

(b) <u>Since</u> the five arcs are equal, $\overset{\frown}{CD}$ contains ______ arc-degrees.

(c) Angle COD contains ______°.

(d) Angle #1 and angle #2 are equal. Each contains ______°.

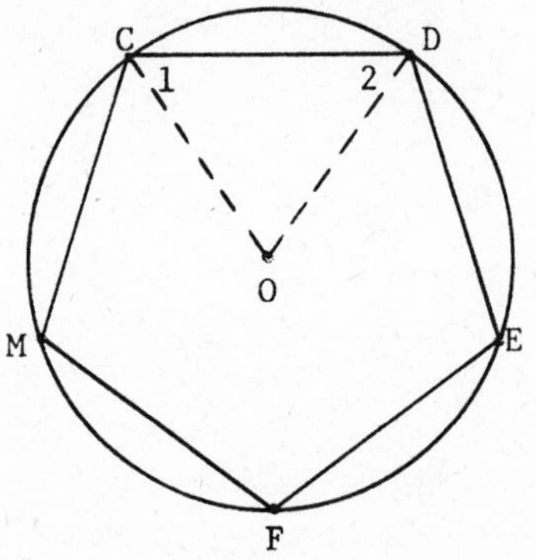

a) 140°
b) 20°
c) 20°

---

19. The circle on the right is divided into 8 equal arcs. Three chords (GH, AB, and CE) are drawn.

(a) Chord GH is equal to chord ______.

(b) $\overset{\frown}{CD}$ contains ______ arc-degrees.

(c) $\overset{\frown}{CDE}$ contains ______ arc-degrees.

(d) Angle COE contains ______°.

(e) Angles 1 and 2 each contain ______°.

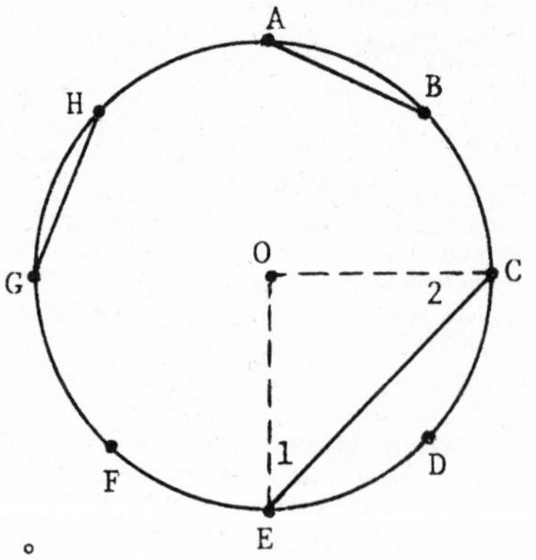

a) 1.5"
b) 72 arc-degrees
c) 72°
d) 54°

---

a) AB
b) 45
c) 90
d) 90°
e) 45°

20. If a radius (or diameter) of a circle is drawn perpendicular to a chord, it bisects (cuts into two equal parts) the chord.

In this circle, radius 0V is drawn perpendicular to chord GH.

(a) Since 0V is perpendicular to GH, angle 0FG is a right angle. One other right angle is angle ______.

(b) Since 0V is perpendicular to GH, we know that GF is equal to ______.

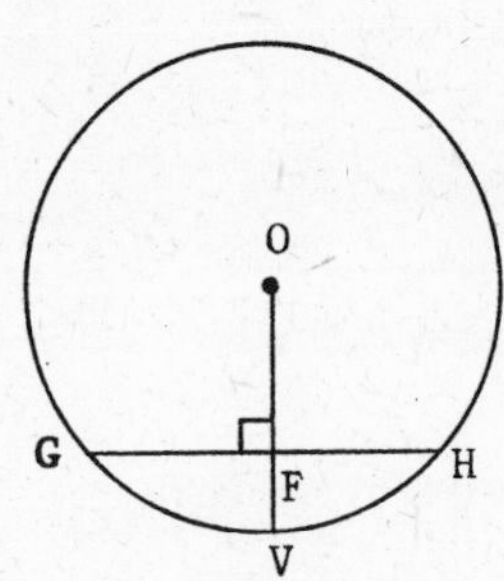

---

a) Angle 0FH

b) FH

21. In this circle, radius 0D is perpendicular to chord ST.

(a) If ST is 10.0", how long is BT? ________

(b) If 0D is 8.0" and 0B is 6.2", how long is BD? ________

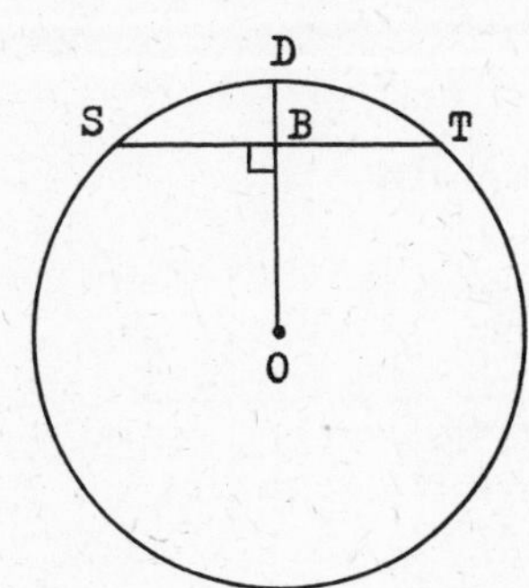

---

a) 5.0"

b) 1.8"
Since:
BD = 0D - 0B

22. In this circle, radius 0R is perpendicular to chord MP. 0M and 0P are also radii.

(a) If MP is 2.0", how long are MA and AP? ________

(b) If 0R is 3.0", how long are 0M and 0P? ________

(c) If 0R is 3.0" and AR is 0.2", how long is 0A? ________

(d) Because 0M = 0P, triangle 0MP is an ________ triangle.

(e) Triangles 0MA and 0AP are ________ triangles.

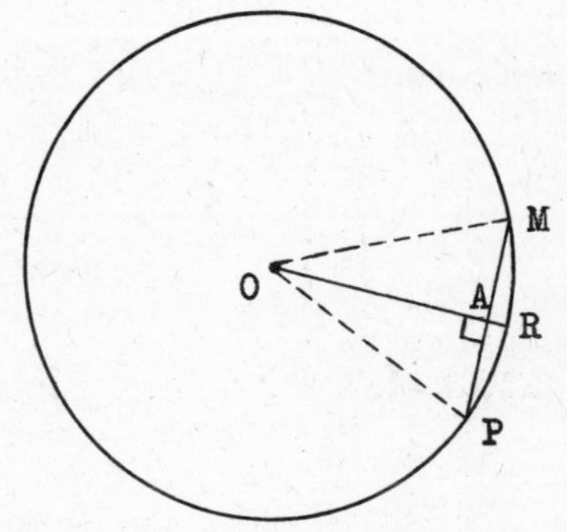

---

a) 1.0"

b) 3.0", since all three are radii.

c) 2.8"

d) isosceles

e) right

23. In the circle on the right, 0A, 0B, and 0C are radii. 0B is perpendicular to chord AC.

(a) Name the isosceles triangle in this figure. ________

(b) Name the two right triangles in this figure. ________ and ________

(c) 0A is the hypotenuse of right triangle ________.

(d) 0D and CD are the legs of right triangle ________.

---

24. If a radius is drawn perpendicular to a chord, it bisects:

(1) the chord
(2) the arc of the chord
(3) the central angle of the chord

In the circle on the right, radius 0M is perpendicular to chord FG.

(a) If $\overset{\frown}{FG}$ contains 60 arc-degrees, how many arc-degrees are there in $\overset{\frown}{FM}$ and $\overset{\frown}{MG}$? ______ arc-degrees

(b) Angle F0G contains ______°.

(c) Angles #1 and #2 each contain ______°.

a) 0AC

b) 0AD and 0CD

c) 0AD

d) 0CD

---

25. Radius 0P is drawn perpendicular to chord AB. $\overset{\frown}{APB}$ contains 150 arc-degrees.

(a) $\overset{\frown}{AP}$ contains ______ arc-degrees.

(b) Angle A0P contains ______°.

(c) Angle 0AC contains ______°.

(d) The hypotenuse of right triangle A0C is ______.

(e) The legs of right triangle 0BC are ______ and ______.

(f) If AC is 5", how long is chord AB? ______

a) 30 arc-degrees

b) 60°

c) 30°

---

a) 75

b) 75°

c) 15°

d) 0A

e) 0C and BC

f) 10"

SELF-TEST 2 (Frames 14-25)

In the circle shown, radius 0E is perpendicular to chord DF, and $\widehat{DEF}$ contains 120°.

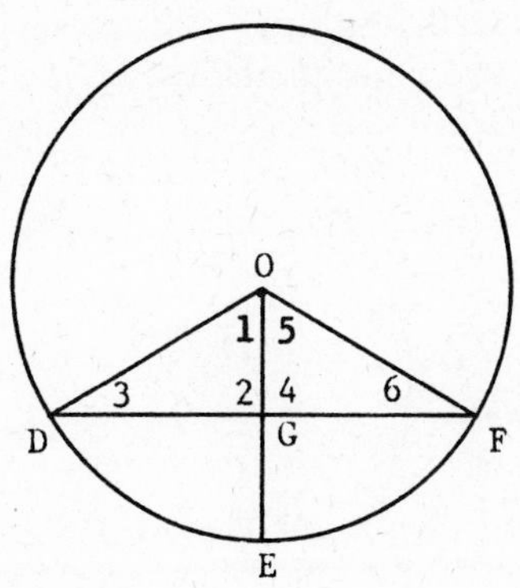

1. Angle #1 = ______°
2. Angle #2 = ______°
3. Angle #3 = ______°
4. Angle #4 = ______°
5. Angle #5 = ______°
6. Angle #6 = ______°

Radius 0F = 5.00", chord DF = 8.66", and 0G = 2.50".

7. DG = ________
8. GE = ________

ANSWERS:

1. 60°	4. 90°	7. 4.33"
2. 90°	5. 60°	8. 2.50"
3. 30°	6. 30°	

5-3 BASIC PROBLEMS INVOLVING CHORDS OF CIRCLES

There are various types of basic problems involving chords of circles. The major ones are finding their lengths, distances from the center of the circle, depths, or central angles. All of these problems can be solved by drawing appropriate radii to form two right triangles, and then using one of the right triangles as the key step in the solution. In this section, we will discuss this method and the various strategies involved. The solution of one type of problem by means of the Law of Sines will also be discussed.

26. Chord problems involve finding lengths and angles. Since it is easiest to find a length or an angle if it is part of a triangle, especially a right triangle, we solve chord problems by means of a right triangle. Therefore, the first step in any problem of this type is drawing radii to form two right triangles. Here is an example:

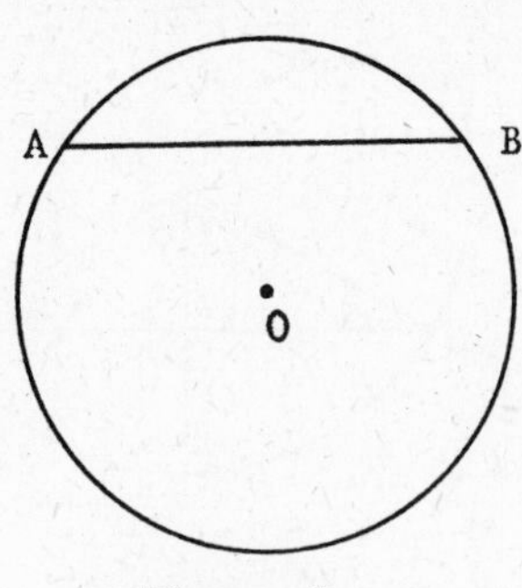

Figure 1

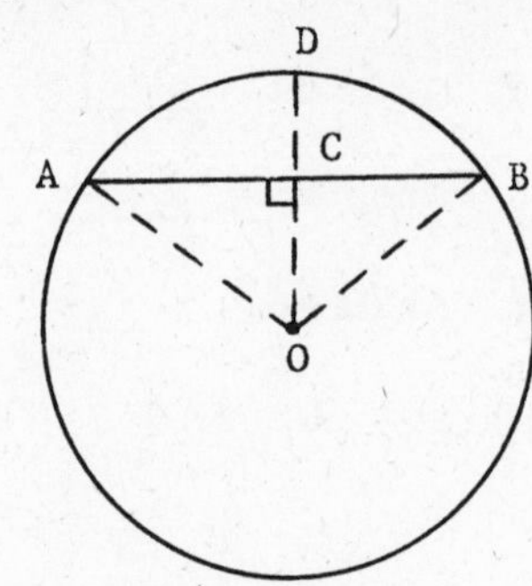

Figure 2

In Figure 1, chord AB alone is drawn.

In Figure 2, radii 0A, 0B, and 0D are also drawn.

Note: (1) 0A and 0B are drawn to the ends of chord AB.
(2) 0D is drawn perpendicular to chord AB.

Name the two right triangles which are formed by drawing these three radii. __________ and __________

0AC and 0BC

27. After drawing radii to form two right triangles, the solution of one of the right triangles is the key step in the solution of any chord problem. Of course, to solve the right triangle, we use the Pythagorean Theorem or one of the trigonometric ratios. Here is an example:

In the figure on the right, three radii are drawn to form two right triangles. 0M is the distance of chord CD from the center of the circle.

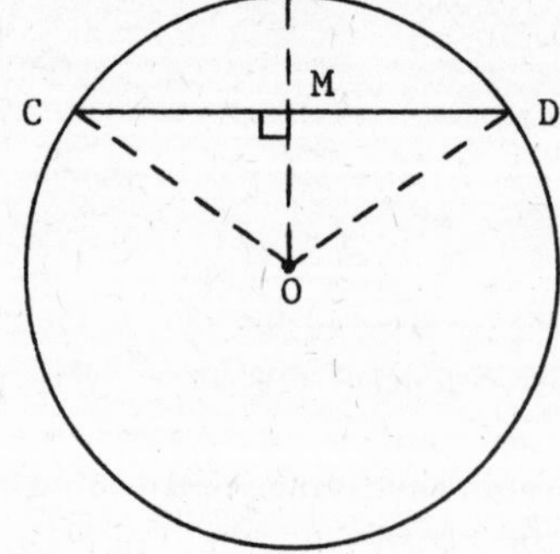

In right triangle 0MD, we know that radius 0D is 7.0" and half-chord MD is 5.0". Since 0D is the hypotenuse of right triangle 0MD, and since MD is one of its legs, how can we find the length of 0M?

____

By means of the Pythagorean Theorem, since:

$$(0M)^2 = (0D)^2 - (MD)^2$$

28. Here is another figure of the same type. 0R is the distance of chord BT from the center of the circle. BR is half of chord BT.

Radius 0B is the hypotenuse of right triangle 0BR. If we know that 0B is 10.0" and angle #1 contains 27°:

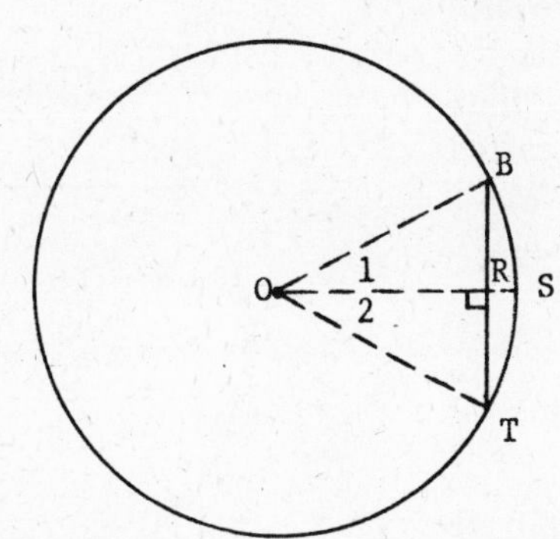

(a) What trig ratio could we use to find BR? ____________

(b) What trig ratio could we use to find 0R? ____________

---

29. In the figure on the right, angle A0C is the central angle of chord AC. Angles #1 and #2 are each half of angle A0C.

Radius 0C is the hypotenuse of right triangle 0TC. Half-chord TC is one of its legs. If we know that 0C is 10.0" and TC is 8.1", what trig ratio could we use to find the size of angle #2? ____________

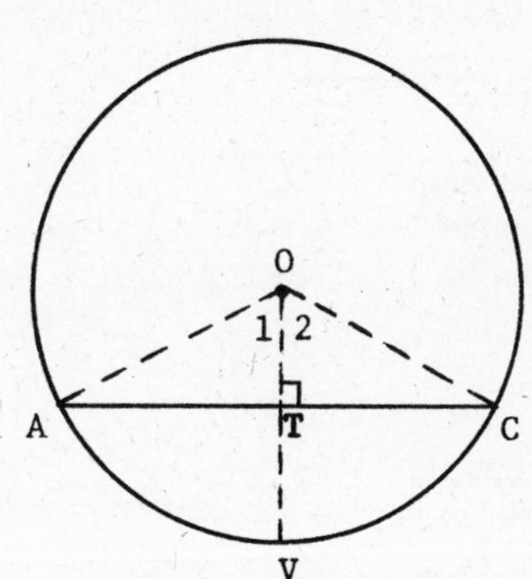

a) The sine ratio, since:

$$\sin \#1 = \frac{BR}{0B}$$

b) The cosine ratio, since:

$$\cos \#1 = \frac{0R}{0B}$$

---

30. In the figure on the right, radius 0P is perpendicular to chord CD. 0P is 2.0" and CD is 3.0". We want to find the length of 0F, the distance of the chord from the center of the circle.

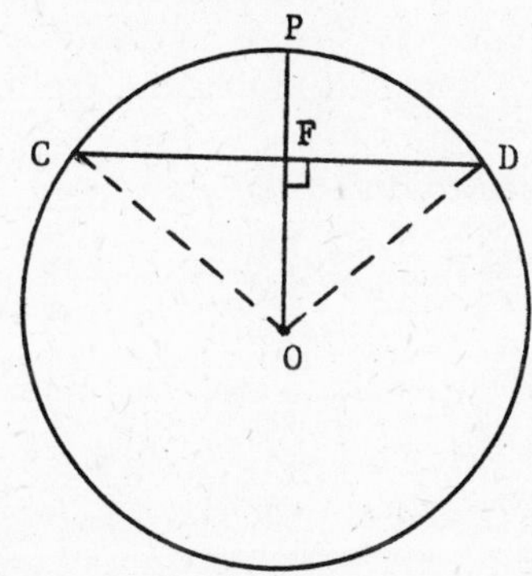

We can use right triangle 0FD to find 0F. In that triangle:

0D is the hypotenuse. Since it is a radius, 0D is also 2.0".

FD is a leg. Since 0P bisects chord CD (which is 3.0"), FD is 1.5".

Using the Pythagorean Theorem, find the length of 0F. 0F = ____________

The sine ratio, since:

$$\sin \#2 = \frac{TC}{0C}$$

31. Here is the same figure we saw in the last frame. PF is the distance from the center of the chord to the circle. It is called the "<u>depth</u>" of the chord. We want to find this "<u>depth</u>".

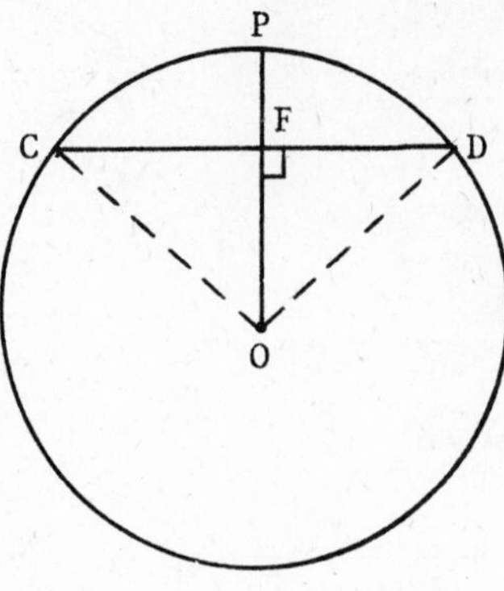

We know these facts:

(1) The length of radius 0P is 2.0".

(2) The length of 0F is 1.3". (From the Pythagorean Theorem)

Therefore, the "<u>depth</u>" (or PF) can be found by a subtraction. The "depth" of the chord is ______________.

1.3" (two significant digits)

Since:

$$0F = \sqrt{(0D)^2 - (FD)^2}$$

---

32. In the circle on the right, we want to find 0D, the distance of chord TS from the center of the circle. The radius of the circle is 8.0". Central angle T0S is 120°.

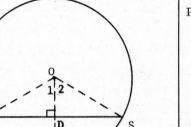

Since 0D is a leg of right triangle 0DT, we can use that triangle to find its length. We know these facts:

(1) Radius 0T is 8.0".
(2) Angle #1 is 60° (since it is half of central angle T0S).

(a) What trig ratio can be used to find 0D? ______________

(b) Find the length of 0D. 0D = __________

PF = 0.7", since:

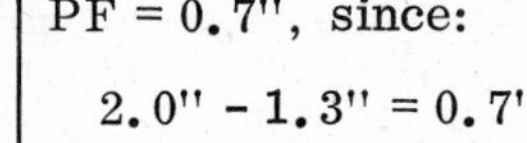
2.0" - 1.3" = 0.7"

---

a) The cosine ratio, since:

$$\cos \#1 = \frac{0D}{0T}$$

$$\text{or } \cos 60° = \frac{0D}{8.0''}$$

b) 0D = 4.0"

33. We want to find the length of chord VS in the circle on the right. The radius of the circle is 20.0". Central angle V0S is 120°.

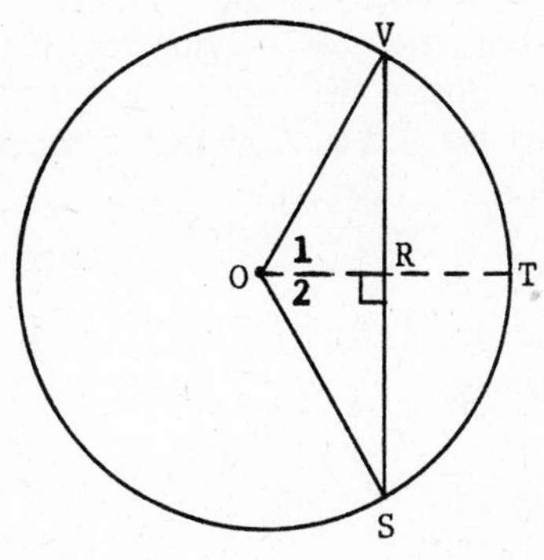

Since VR and RS are half of chord VS, we can find the length of the chord if we can find the length of either VR or RS. Let's find the length of RS. To do so, we will use right triangle 0RS. In that triangle:

(1) 0S is 20.0" since it is a radius.
(2) Angle #2 is 60° since it is half of central angle V0S.

(a) Which trig ratio can be used to find RS? ____________________

(b) Find the length of RS. RS = __________

(c) Now find the length of chord VS. VS = __________

---

34. We want to find the size of central angle D0F in the circle on the right. The radius of the circle is 2.5". Chord DF is 3.0".

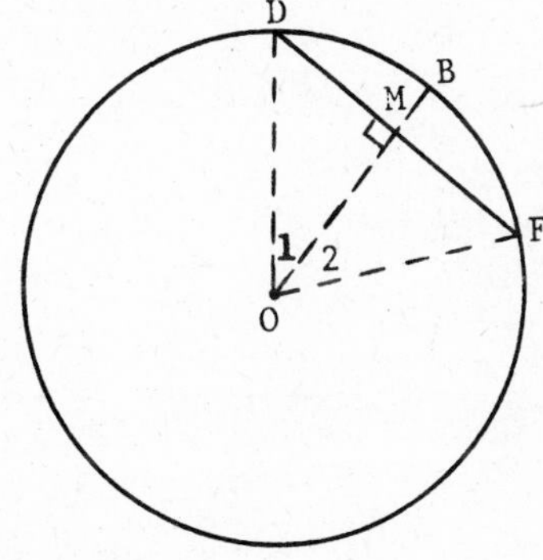

Since angles #1 and #2 are half of central angle D0F, we can find the size of the central angle if we can find the size of either angle #1 or angle #2. Let's find the size of angle #1. To do so, we will use right triangle 0DM. In that triangle:

(1) 0D is 2.5" since it is a radius.
(2) DM is 1.5" since it is half of chord DF (which is 3.0").

(a) Which trig ratio can be used to find angle #1? ____________________

(b) How large is angle #1? _______°

(c) How large is central angle D0F? _______°

a) The sine ratio, since:

$$\sin \#2 = \frac{RS}{0S}$$

or $$\sin 60° = \frac{RS}{20.0''}$$

b) RS = 17.3"

c) VS = 2(17.3") or 34.6"

35. Though the numerical solutions of problems involving chords are usually quite simple, the strategies for solving them frequently include a number of steps. It is important that you be able to recognize the required steps. Therefore, we will review these strategies in the next few frames.

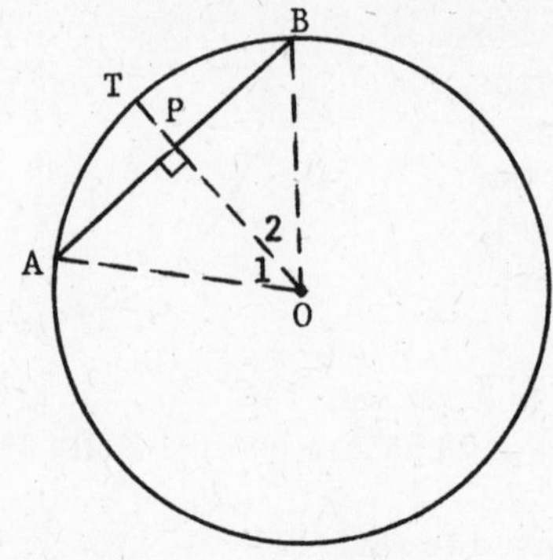

We want to find 0P, the distance of chord AB from the center of the circle. We are given these facts:

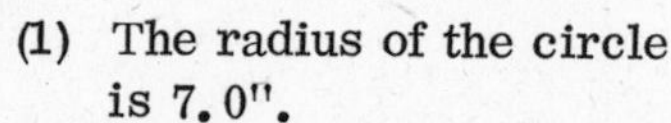

(1) The radius of the circle is 7.0".
(2) Chord AB is 8.0".

Let's use right triangle 0AP to find the length of 0P.

(a) What two facts do we know about triangle 0AP?

(1) ______________ (2) ______________

(b) Given these two known facts, what method can we use to find 0P?

____

a) The sine ratio, since:

$$\sin \#1 = \frac{DM}{0D}$$

b) 37°

c) 74°

---

36. We want to find the length of chord CD in this circle. We are given this information:

(1) The radius of the circle is 10.0".
(2) Central angle C0D is 110°.

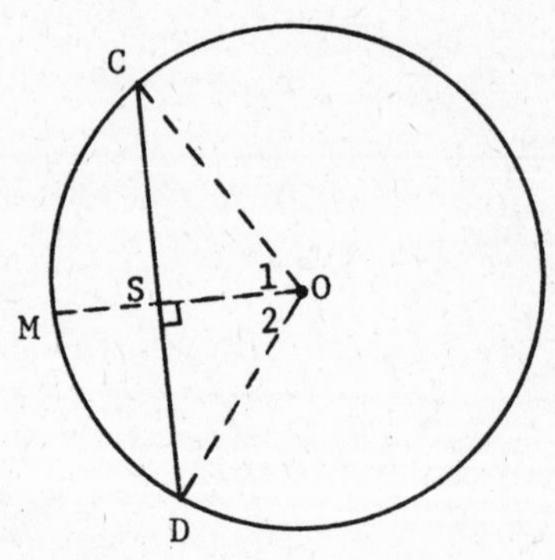

To find chord CD, the general strategy is to find the length of one of the half-chords (CS or SD) and then double it. Let's find the length of SD. We can use right triangle 0SD to do so.

(a) What two facts do we know about triangle 0SD?

(1) ______________ (2) ______________

(b) Given these facts, what method can we use to find SD?

____

(c) Is the problem completed when we have found the length of SD? ________

a) (1) A0 = 7.0"
(2) AP = 4.0"

b) The Pythagorean Theorem

---

a) (1) 0D = 10.0"
(2) Angle #2 = 55°

b) The sine ratio, since:

$$\sin \#2 = \frac{SD}{0D}$$

c) No. We must double SD to find chord CD.

37. We want to find the "depth" of chord RV in this circle. The "depth" is the distance MT. We are given this information:

(1) The radius of the circle is 12.0".

(2) $\widehat{RTV}$ contains 100 arc-degrees.

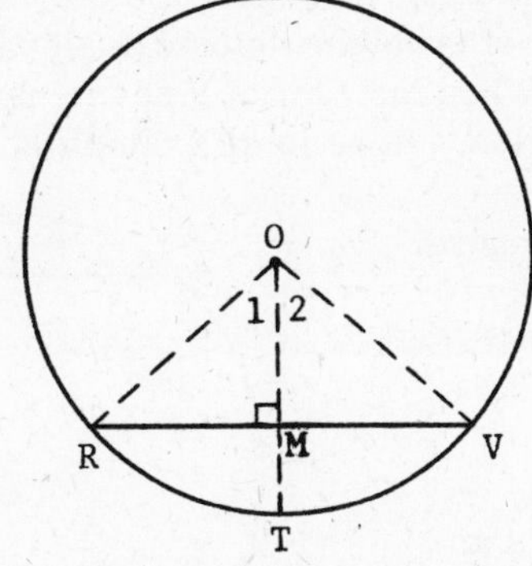

The general strategy for finding the "depth" (or MT) is to find the length of 0M and subtract it from the known length of radius 0T. To find the length of 0M, we can use right triangle 0RM.

(a) What two facts do we know about triangle 0RM?

(1) ________________ (2) ________________

(b) Given these two facts, what method can we use to find 0M?

____

(c) Is the problem solved when we have found the length of 0M? __________

---

38. We want to find the size of central angle E0F in this circle. We are given this information:

(1) The radius of the circle is 6.0".

(2) Chord EF is 8.0".

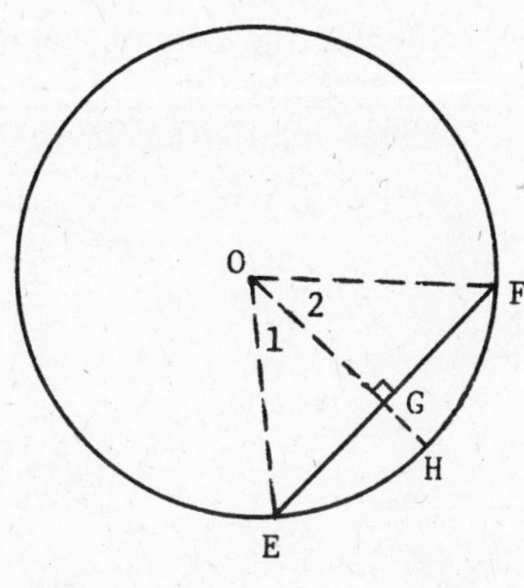

To find central angle E0F, the general strategy is to find either angle #1 or angle #2 and then double it. Let's use right triangle 0FG to find the size of angle #2.

(a) What two facts do we know about triangle 0FG?

(1) ________________ (2) ________________

(b) Given these facts, what method can we use to find the size of angle #2? ________________________

(c) After finding angle #2, what must we do to find central angle E0F? ________________________________

Answers to 37:

a) (1) 0R = 12.0"
(2) Angle #1 = 50°

b) The cosine ratio, since:

$$\cos \#1 = \frac{0M}{0R}$$

c) No. We must subtract 0M from 0T to find MT.

Answers to 38:

a) (1) 0F = 6.0"
(2) GF = 4.0"

b) The sine ratio, since:

$$\sin \#2 = \frac{GF}{0F}$$

c) Double the size of angle #2.

39. Up to this point, we have solved all chord-problems by means of right triangles. There is one type of problem which can also be solved by means of the Law of Sines. When the Law of Sines is used, we do not have to construct right triangles. Here is an example:

We want to find the length of chord AB in this circle. The radius of the circle is 10.0", and central angle A0B is 80°.

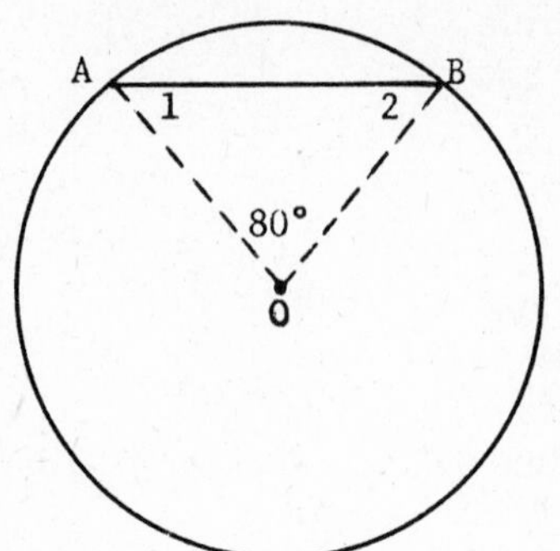

We have drawn only two radii, 0A and 0B. Since triangle A0B is an isosceles triangle, angles #1 and #2 each contain 50°. To find AB, we can use the following Law-of-Sines formula:

$$\frac{AB}{\sin A0B} = \frac{0B}{\sin \#1} \quad \text{or} \quad \frac{AB}{\sin 80°} = \frac{10.0''}{\sin 50°}$$

Solving for AB: $AB = \frac{(10.0'')(\sin 80°)}{\sin 50°}$

Complete the solution. AB = ________

AB = 12.9"

40. Of the various types of chord-problems, the Law of Sines can be used only to find the length of a chord when the radius and the central angle are known. Let's use the Law of Sines for this problem:

We want to find the length of chord CD. The radius of the circle is 20.0". $\overset{\frown}{CD}$ contains 140 arc-degrees.

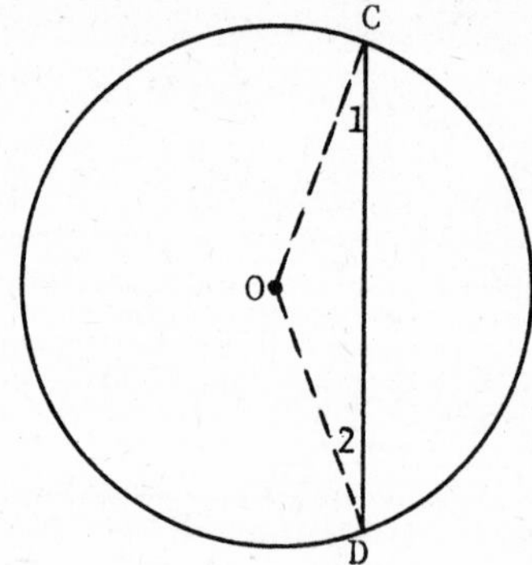

(a) How large is central angle C0D? ________°

(b) How large are angles #1 and #2? ________°

(c) What Law-of-Sines formula can be used to find CD?

____

(d) Using this formula, CD = ________.

Answer to Frame 40: a) 140° b) 20° in each. c) $\frac{CD}{\sin 140°} = \frac{20.0''}{\sin 20°}$ d) CD = 37.6"

SELF-TEST 3 (Frames 26-40)

In the circle at the right, the radius is 5.00" and the length of the chord is 7.40".

1. Find "d", the depth of the chord. d = __________

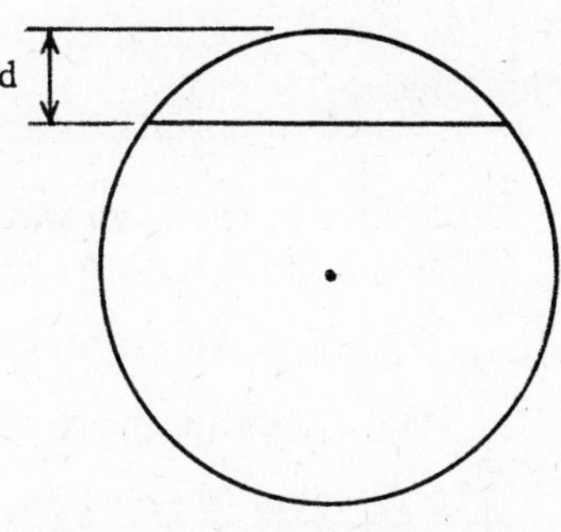

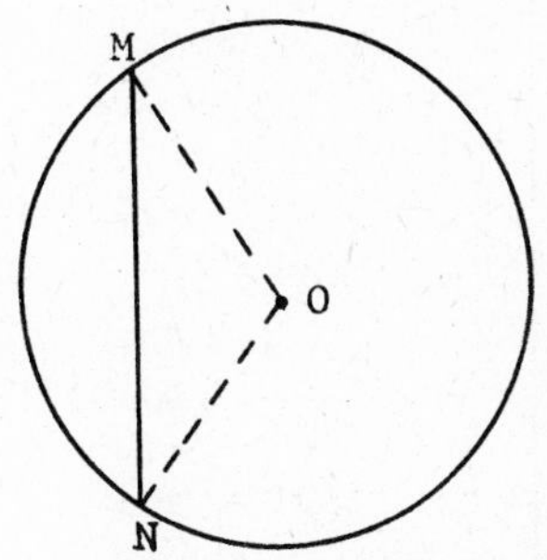

In the circle at the left, radius ON = 15.5" and central angle MON = 110°.

2. Find the length of chord MN. MN = __________

In the circle at the right, radius OR = 25.0" and chord PR = 42.8".

3. Find central angle POR. ______°

4. Find "d", the depth of the chord. d = __________

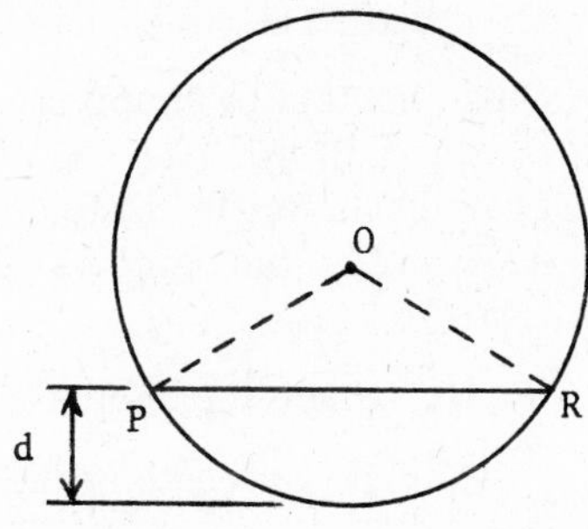

ANSWERS: 1. d = 1.64" 2. MN = 25.4" 3. 118° 4. d = 12.1"

5-4 APPLIED PROBLEMS INVOLVING CHORDS OF CIRCLES

In the last section, we solved the basic types of problems involving chords of circles. In this section, we will solve some applied problems involving chords of circles. The same methods, of course, are used for the solutions.

41. In Figure 1, we have shown the end-view of a steel shaft on which a "flat" is to be cut. As you can see from the given dimensions:

(1) The diameter of the circle is 2.50".

(2) We want the length of the chord to be 2.00".

<u>The problem is to determine "d", the depth of the cut.</u>

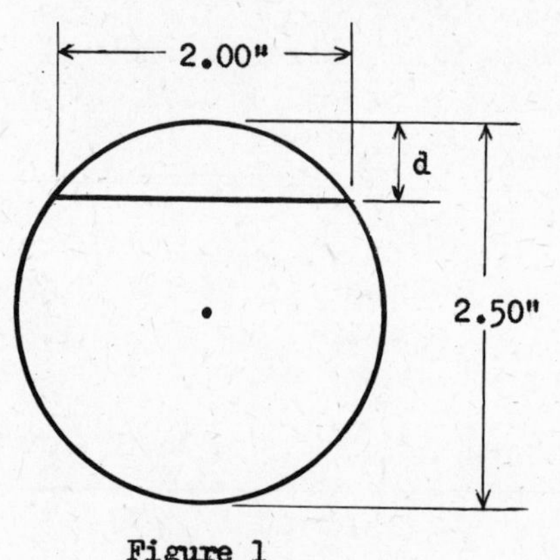

Figure 1

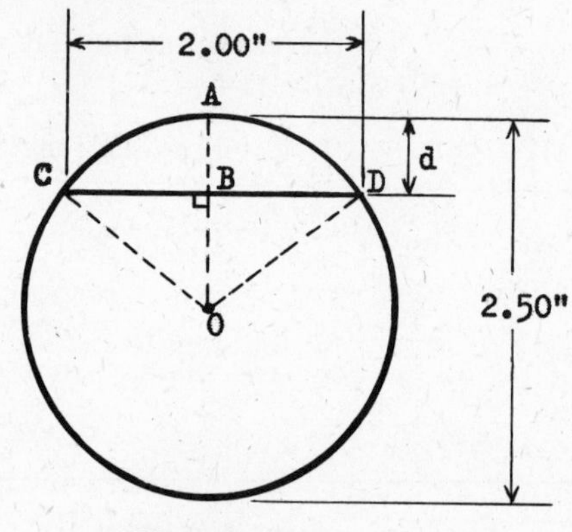

Figure 2

This problem is simply a basic chord-problem. Therefore, we have drawn radii 0C, 0A, and 0D in Figure 2. Now we can use one of the two right triangles to find the length of 0B, and then subtract to find AB which is the same as "d", the depth of the cut. We will use right triangle 0BC.

(a) Since the diameter is 2.50", how long is radius 0C? ________

(b) If the length of the chord is to be 2.00", how long is BC? ________

(c) Using the Pythagorean Theorem, how long is 0B? ________

(d) How long is AB? ________

(e) How long is the depth "d"? ________

a) 1.25"

b) 1.00"

c) 0.75"

d) 0.50"

e) 0.50"

42. A draftsman must give all of the dimensions of the metal part in Figure 1. Various dimensions are given. He must include the value of dimension "x" on the drawing.

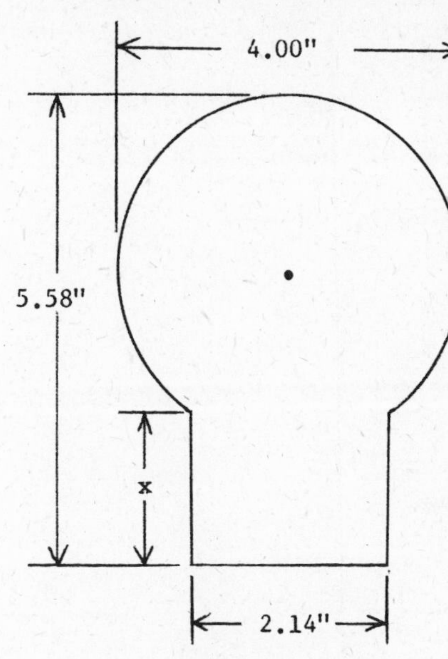

Figure 1

Figure 2

Various additional lines have been drawn in Figure 2. As you can see from Figure 2, "x" is equal to CF. Therefore, we can find its length by means of the following equation:

$$x \text{ (or CF)} = HF - H0 - 0C$$

From the given dimensions, we know these facts:

(1) HF is 5.58".
(2) H0 is 2.00" (since it is a radius of the circle whose diameter is 4.00").

Therefore: $x \text{ (or CF)} = 5.58'' - 2.00'' - 0C$

We can find "x" if we can find 0C. Since 0C is the distance of chord AB from the center of the circle, finding 0C is a familiar problem. To find 0C, we will use right triangle 0AC. We know these facts about triangle 0AC:

(1) 0A is 2.00", since it is a radius.
(2) AC is 1.07", since it is half of AB which is 2.14".

(a) Using the Pythagorean Theorem, how long is 0C? __________

(b) Now plugging this value into the equation above, how long is "x" or CF? __________

43. The diagram in Figure 1 shows the end of a shaft which has a "flat" (a chord) cut on it. We want to find the length of the chord and its central angle.

a) 1.69"

b) 1.89"

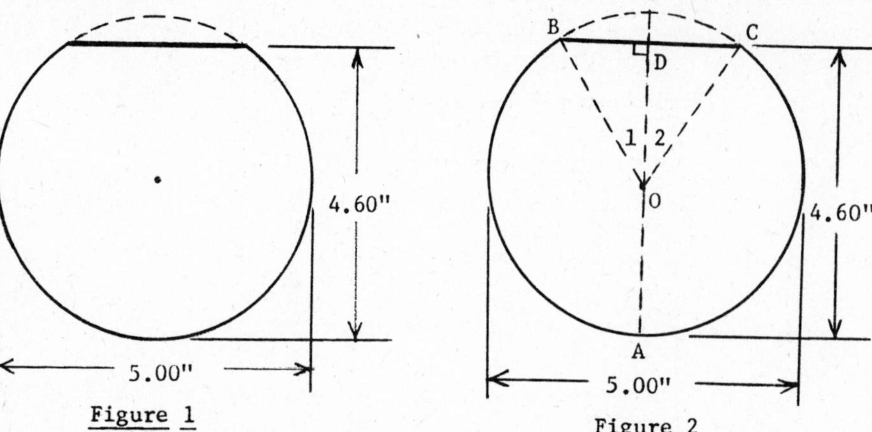

Figure 1

Figure 2

In Figure 2, we have drawn four radii. From the given dimensions, we know these facts:

(1) The length of each radius is 2.50", since the given diameter is 5.00".

(2) The length of AD is 4.60".

To find the length and central angle of chord BC, we can use right triangle 0BD. We know two facts about triangle 0BD:

(1) Radius 0B is 2.50".

(2) 0D is 2.10", since 0D = AD - 0A = 4.60" - 2.50".

(a) Using the Pythagorean Theorem, find the length of half-chord BD. BD is __________.

(b) The length of chord BC is __________.

(c) Using the cosine ratio (0B and 0D were given), find the size of angle #1. Angle #1 is ______°.

(d) Therefore, the central angle of the chord (angle B0C) is ______°.

a) 1.36"

b) 2.72"

c) 33° (cos #1 = 0.8400)

d) 66°

44. In Figure 1 below, we have drawn the end of a metal part which is called a flange. On this flange, 5 equally-spaced holes are drilled. The center of these holes lies on the dotted circle (called a "hole-circle"). If the radius of the hole-circle is 3.00", how far apart are the centers of two successive holes?

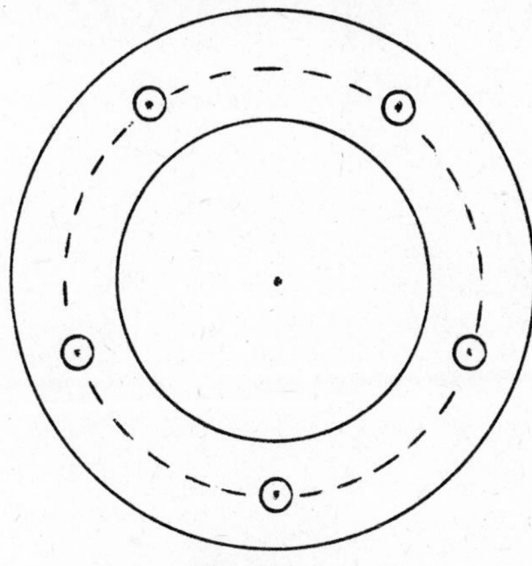

Figure 1

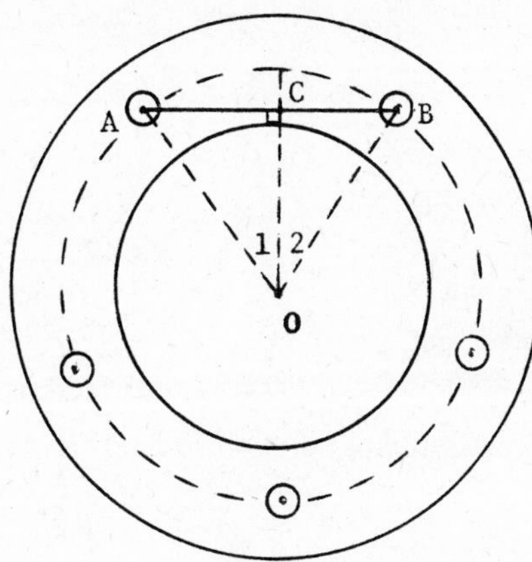

Figure 2

We have drawn some extra lines in Figure 2. As you can see, finding the distance between the centers of two successive holes is the same as finding the length of chord AB.

To find the length of AB, we can find the length of either AC or BC and then double it. Let's find the length of AC. We will use right triangle 0AC to do so.

(a) How long is 0A? __________

(b) Since each arc is one-fifth of the circle, how large is central angle A0B? ______°

(c) How large is angle #1? ______°

(d) Using the proper trig ratio, find the length of AC.
AC = __________

(e) How long is AB, the distance between the centers of two successive holes? AB = __________

---

a) 3.00", since it is a radius.

b) 72°

c) 36°

d) 1.76"
From: $\sin 36° = \frac{AC}{3.00"}$

e) 3.52"

45. In the last frame, we found the distance between the centers of two successive holes on a "hole-circle" by solving a right triangle. Let's solve the same problem by means of the Law of Sines.

Here is the same diagram. We have drawn radii 0A and 0B which are both 3.00". Let's find the length of AB.

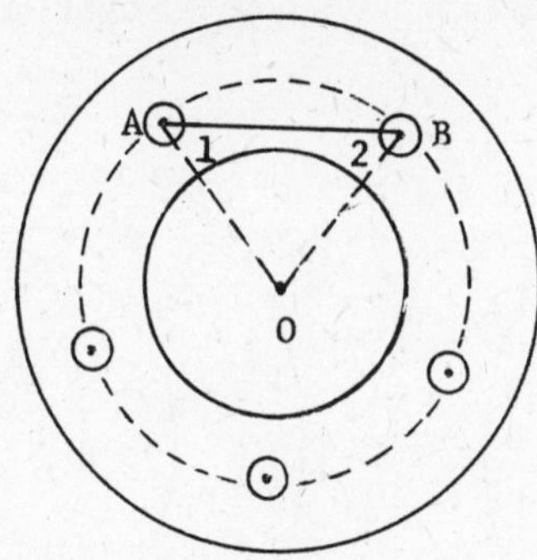

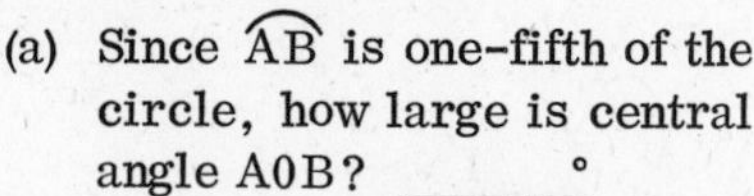

(a) Since $\widehat{AB}$ is one-fifth of the circle, how large is central angle A0B? _______°

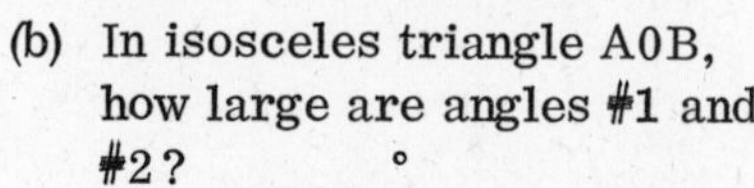

(b) In isosceles triangle A0B, how large are angles #1 and #2? _______°

(c) What Law-of-Sines equation can be used to find the length of AB?

_______ = _______

(d) Find the length of AB. AB = _______

(e) Is this the same solution we obtained in the last frame? _______

---

46. Here is another flange. We have drilled 10 equally-spaced holes whose centers lie on the "hole-circle". The radius of the "hole-circle" is 9.85".

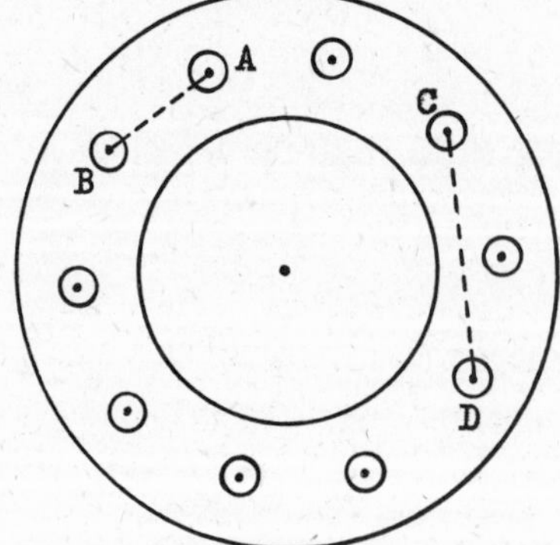

Using either the "right-triangle" method or the "Law-of-Sines" method:

(a) Find the distance between the centers of two successive holes, like A and B.

a) 72°

b) 54° each

c) $\frac{AB}{\sin 72°} = \frac{3.00''}{\sin 54°}$

d) 3.52"

e) Yes

(Continued on following page.)

46. (Continued)

(b) Find the distance between the centers of two alternate holes, like C and D. _______________

a) 6.09"

b) 11.58"

---

47. Holes A, B, and D were drilled in the hole-circle. The holes are not spaced equally. The radius of the hole-circle is 1.50". Chord AB is 2.00". We want to find the number of arc-degrees in $\widehat{AB}$.

We can find the number of arc-degrees in $\widehat{AB}$ if we can find the size of central angle A0B. To do so, we will find the size of angle #2 and then double it.

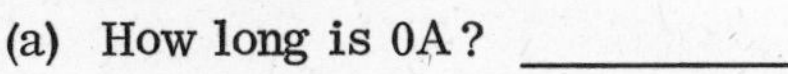

(a) How long is 0A? __________

(b) How long is AC? __________

(c) Using the sine ratio, find the size of angle #2. Angle #2 = _______°

(d) Angle A0B = _______°

(e) $\widehat{AB}$ contains _______ arc-degrees.

a) 1.50", since it is a radius.

b) 1.00", since it is a half-chord.

c) 42°

d) 84°

e) 84 arc-degrees

SELF-TEST 4 (Frames 41-47)

1. In the circle diagram at the right, find "h". h = ________

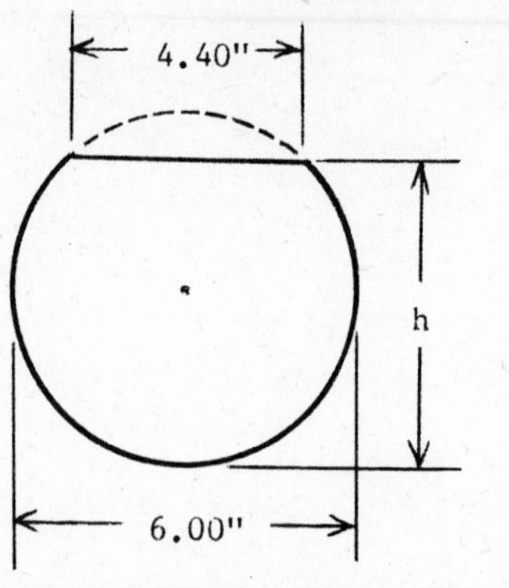

Nine equally-spaced holes are placed on the circumference of a circle whose diameter is 20.0", as shown in the diagram at the right.

2. Find "d", the distance between two successive holes.
   d = ________

3. Find "s", the distance between two alternate holes.
   s = ________

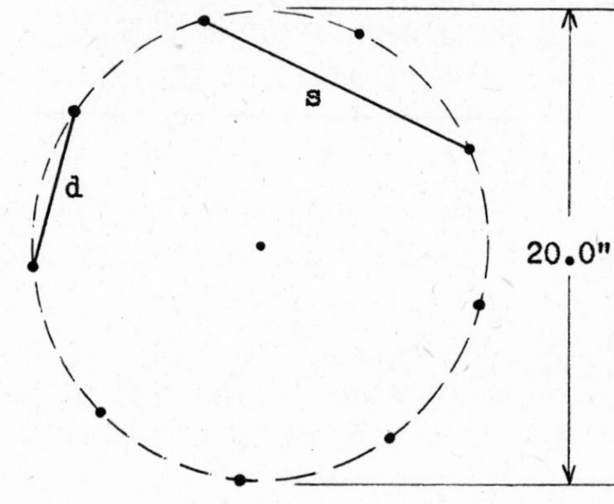

ANSWERS: 1. h = 5.04" 2. d = 6.84" 3. s = 12.9"

5-5 ARC-LENGTHS OF CIRCLES

In this section, we will discuss the method for finding the lengths of arcs of circles. Since any arc-length is a fractional part of the circumference of a circle, we will begin by reviewing "circumference" formulas. A few applied problems will be included.

48. The circumference of a circle is the distance around it. The basic formula for computing the circumference of a circle is:

$\boxed{C = \pi d}$ or $\boxed{C = 3.14d}$

where: "C" is the circumference.
"d" is the diameter of the circle.
"π" (pronounced "pie") is an unending number which is usually rounded to 3.14.

As you can see from the formula on the right above, the circumference of any circle is slightly more than __________ times as long as its diameter.

---

49. Using the formula $\boxed{C = 3.14d}$, find the circumferences of the circles with the following diameters.

(a) If d = 4.26 feet , C = __________ feet.

(b) If d = 11.2 inches, C = __________ inches.

Answer (48): three

---

50. Since the diameter of a circle is composed of two radii, we know that: d = 2r. By substituting "2r" for "d", we can obtain a circumference formula expressed in terms of the radius of the circle. That is:

Since $C = \pi d$, $C = \pi(2r)$ or $\boxed{C = 2\pi r}$

or $C = 3.14d$, $C = 3.14(2r)$ or $\boxed{C = 6.28r}$

As you can see from the formula C = 6.28r, the circumference of any circle is slightly more than __________ times as long as its radius.

Answers (49):
a) 13.4 feet
b) 35.2 inches

---

51. Using the formula $\boxed{C = 6.28r}$, find the circumferences of the circles with the following radii.

(a) If r = 53.5", C = __________ inches.

(b) If r = 1.88', C = __________ feet.

Answer (50): six

---

52. To find the <u>diameter</u> of a circle whose circumference is known, we can solve $C = \pi d$ for "d". We get:

$\boxed{d = \frac{C}{\pi}}$ or $\boxed{d = \frac{C}{3.14}}$

Using this formula, find the diameters of the circles with the following circumferences:

(a) If C = 72.3", d = __________ inches.

(b) If C = 4.21", d = __________ inches.

Answers (51):
a) 336 inches
b) 11.8 feet

Answers (52):
a) 23.0 inches
b) 1.34 inches

53. To find the radius of a circle whose circumference is known, we can solve $C = 2\pi r$ for "r". We get:

$$\boxed{r = \frac{C}{2\pi}} \quad \text{or} \quad \boxed{r = \frac{C}{6.28}}$$

Using this formula, find the radii of the circles with the following circumferences:

(a) If C = 214' , r = _________ feet.

(b) If C = 7.60", r = _________ inches.

---

a) 34.1 feet

b) 1.21 inches

54. The length of an arc of a circle is always a fractional part of the circumference of the circle. Here is an example:

The circle on the right is divided into four equal arcs. Therefore, each arc is $\frac{1}{4}$ of the circumference of the circle.

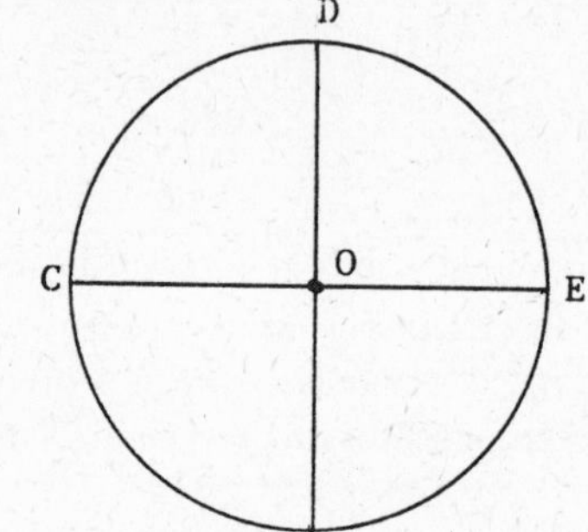

If the circumference is 24":

(a) How long is $\overparen{CD}$? ______

(b) How long is $\overparen{CDE}$? ______

---

a) 6"

b) 12"

55. The circle on the right is divided into 6 equal arcs. Therefore, each arc is $\frac{1}{6}$ of the circumference of the circle.

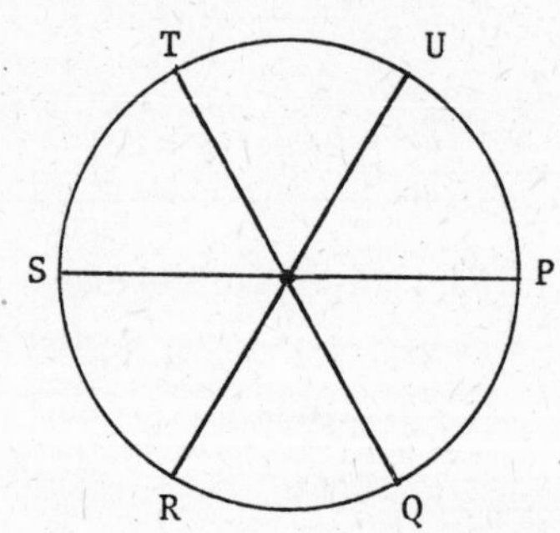

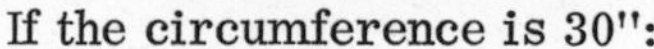

If the circumference is 30":

(a) How long is $\overparen{UP}$? ______

(b) How long is $\overparen{STU}$? ______

---

a) 5"

b) 10"

56. Here is a semi-circle (half-circle). The semi-circle is divided into 5 equal arcs.

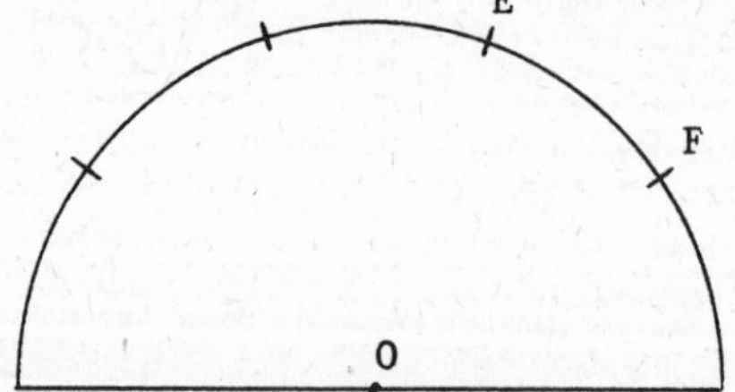

If the diameter of the circle is 20.0":

(a) The circumference of the circle is _________ inches.

(b) The length of the semi-circle is _________ inches.

(c) The length of $\overparen{EF}$ is _________ inches.

57. Here is a semi-circle divided into 10 equal arcs.

If the radius 0R is 5.00 feet:

(a) The circumference of the circle is ________ feet.

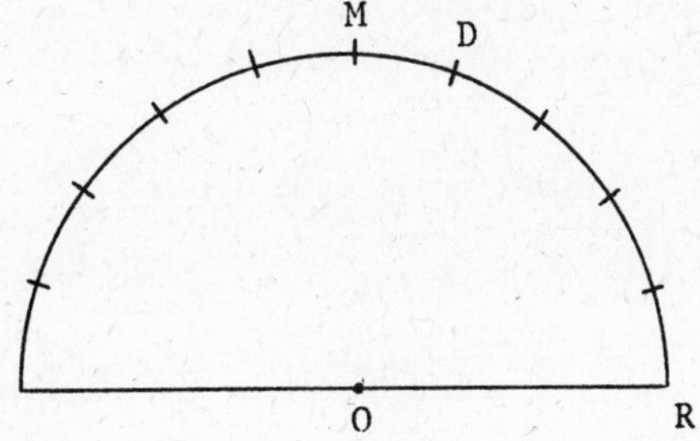

(b) The length of the semi-circle is ________ feet.

(c) The length of $\overset{\frown}{MD}$ is ________ feet.

a) 62.8 inches

b) 31.4 inches

c) 6.28 inches

---

58. Here is a semi-circular arch. It is constructed with seven equal-sized parts. 0R is the radius of the inner circle; 0S is the radius of the outer circle.

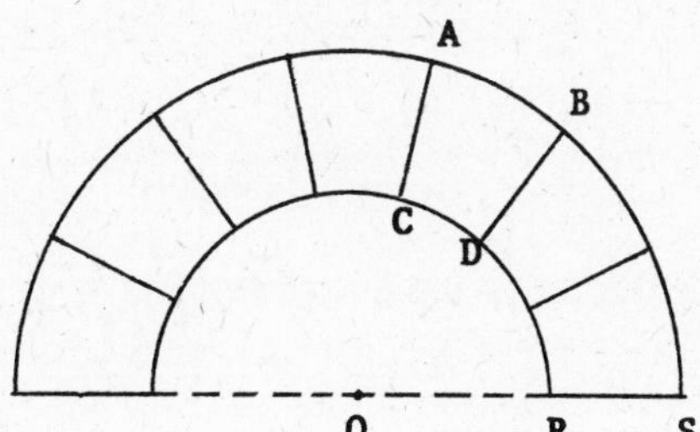

If 0R = 5.00 ft.:

(a) How long is the inner semi-circle? ________ feet

(b) How long is $\overset{\frown}{CD}$? ________ feet

If 0S = 8.00 ft.:

(c) How long is the outer semi-circle? ________ feet

(d) How long is $\overset{\frown}{AB}$? ________ feet

a) 31.4 feet

b) 15.7 feet

c) 1.57 feet

---

a) 15.7 feet

b) 2.24 feet

c) 25.1 feet

d) 3.59 feet

59. If the circumference of a circle is divided into equal arcs, it is easy to determine the "fractional part of the circumference" for each arc. For example:

If a circle is divided into 5 equal arcs, each arc is "$\frac{1}{5}$ of the circumference" or "$\frac{1}{5}$(C)".

In terms of C, how long is each arc if the circle is divided into:

(a) 10 equal arcs? ______ (b) 20 equal arcs? ______

---

a) $\frac{1}{10}$(C) b) $\frac{1}{20}$(C)

60. To determine what fractional part of the circumference a single arc-length is, we can compare the number of arc-degrees in the arc with the number of arc-degrees in the complete circle. For example:

If an arc contains 90 arc-degrees, its length is $\frac{90}{360}$(C) or $\frac{1}{4}$(C).

If an arc contains 60 arc-degrees, its length is $\frac{60}{360}$(C) or $\frac{1}{6}$(C).

(a) If an arc contains 40 arc-degrees, its length is $\frac{40}{360}$(C) or ______.

(b) If an arc contains 36 arc-degrees, its length is $\frac{36}{360}$(C) or ______.

---

a) $\frac{1}{9}$(C)

b) $\frac{1}{10}$(C)

61. When determining the fractional part of the circumference which a single arc is by comparing arc-degrees, frequently the ratio cannot be reduced to lower terms. For example:

If an arc contains 29 arc-degrees, its length is $\frac{29}{360}$(C).

If an arc contains 73 arc-degrees, its length is ______.

$\frac{73}{360}$(C)

62. Since the number of arc-degrees in an arc is the same as the number of degrees in its central angle, we can also compare an arc-length to the circumference by means of central angles. For example:

If the central angle of an arc contains 120°, its arc-length is $\frac{120}{360}$(C) or $\frac{1}{3}$(C).

If the central angle of an arc contains 45°, its arc-length is $\frac{45}{360}$(C) or ______.

$\frac{1}{8}$(C)

63. (a) If an arc contains 45 arc-degrees in a circle whose circumference is 80":

Its arc-length is $\frac{45}{360}(80'') = \frac{1}{8}(80'') = $ ______

(b) If an arc has a central angle of 60° in a circle whose circumference is 66":

Its arc-length is $\frac{60}{360}(66'') = \frac{1}{6}(66'') = $ ______

---

64. If the circumference of a circle is 100":

(a) The arc-length of a 67° arc is $\frac{67}{360}(100'') = $ ______

(b) The arc-length of a 149° arc is $\frac{149}{360}(100'') = $ ______

a) 10"

b) 11"

---

65. If the <u>circumference</u> of a circle is 2.79":

(a) The arc-length of an arc whose central angle is 47° is:

$\frac{47}{360}(2.79'') = \frac{(47)(2.79'')}{360} = $ ______ inches

(b) The arc-length of an arc whose central angle is 111° is:

$\frac{111}{360}(2.79'') = \frac{111(2.79'')}{360} = $ ______ inches

a) 18.6 inches

b) 41.4 inches

---

66. If the <u>radius</u> of a circle is 5.70":

(a) Its circumference is ______ inches.

(b) The length of an arc with an 89° central angle is ______ inches.

a) 0.364 inches

b) 0.860 inches

---

67. If the <u>diameter</u> of a circle is 15.0":

(a) Its circumference is ______ inches.

(b) The length of a 48° arc is ______ inches.

a) 35.8 inches

b) 8.85 inches

---

a) 47.1 inches

b) 6.28 inches

68. A flat metal strip is to be bent to form an arc of a circle, as shown. The central angle is 78°, and the radius is 4.25".

How long must the flat strip be? __________

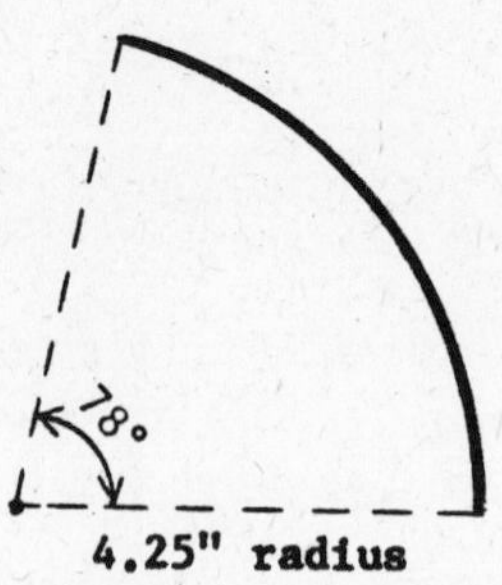

5.78"

---

69. The radius of this circle is 1.50". Angle POQ is 100°.

(a) The circumference of the circle is __________ inches.

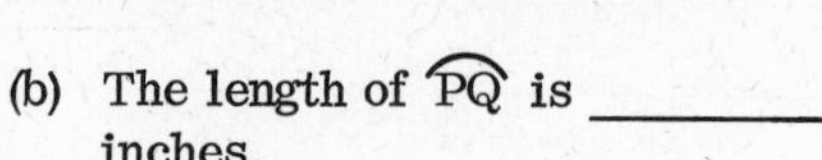

(b) The length of $\overset{\frown}{PQ}$ is __________ inches.

(c) How long is chord PQ? ________ inches

a) 9.42 inches

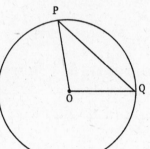

b) 2.62 inches

c) 2.30 inches

---

70. The arc of a circle is 20.0". It has a central angle of 72°. We want to find the radius of the circle.

We know this fact:

$$20.0'' = \frac{72}{360}(C)$$

$$\text{or } 20.0'' = \frac{72C}{360}$$

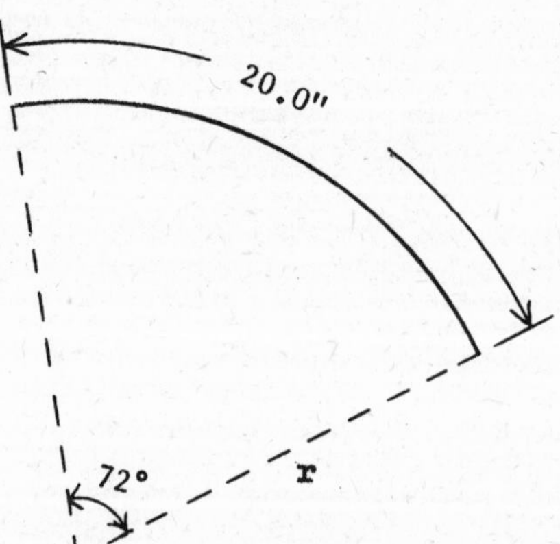

(a) Find "C", the circumference of the circle. C = __________

(b) Now find "r", the radius of the circle. r = __________

71. In the surveying problem diagramed at the right, chord EF = 460 feet, and $\overset{\frown}{EF}$ contains 80 arc-degrees.

(a) Find radius 0F.
0F = ________

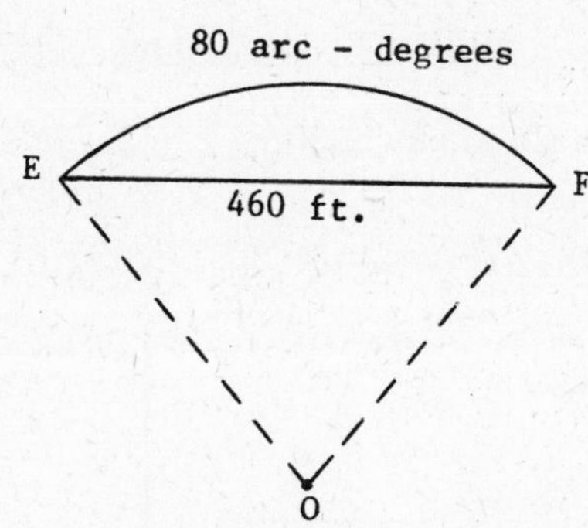

(b) Find the circumference of the circle. C = ________

(c) Find the length of $\overset{\frown}{EF}$. $\overset{\frown}{EF}$ = ________

a) C = 100"

b) r = 15.9"

From: $r = \frac{C}{2\pi}$

---

a) 0F = 358 feet

b) C = 2,250 feet

c) $\overset{\frown}{EF}$ = 500 feet

SELF-TEST 5 (Frames 48-71)

1. Find the radius of a circle whose circumference is 64.8".

Radius = __________

2. In a circle of 1.86" diameter, an arc lies on a central angle of 82°. Find the length of the arc.

Arc-length = __________

The circumference of a circle of 45.0" diameter is divided into nine equal arcs. For one of the arcs, find:

3. The central angle: __________

4. The arc length: __________

5. The chord length: __________

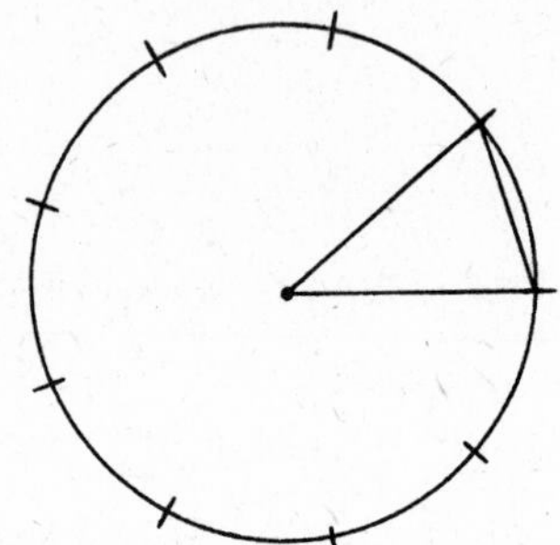

6. Arc PQ is 1.70" long and lies on the circumference of a circle of 1.91" radius. Find the central angle θ on which the arc lies.

θ = __________

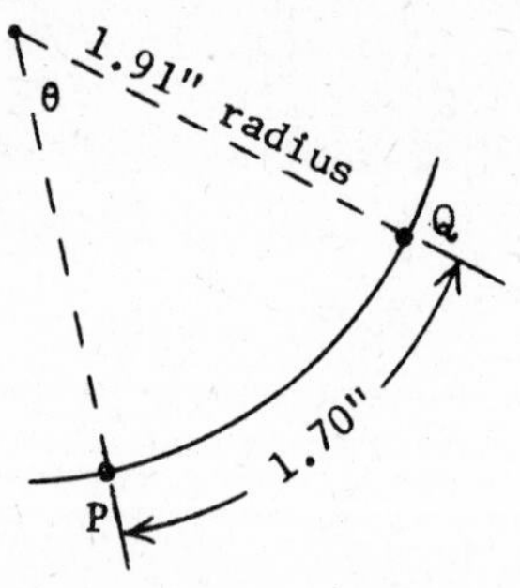

ANSWERS:

1. 10.3"
2. 1.33"
3. 40°
4. 15.7"
5. 15.4"
6. θ = 51°

5-6 SECTORS AND SEGMENTS OF CIRCLES

In this section, we will discuss the methods for finding the areas of sectors and segments of circles. Since the area of a sector is a fractional part of the total area of the circle, we will begin by reviewing the formulas for finding the area of a circle.

---

72. The basic formula for the area of a circle is:

$\boxed{A = \pi r^2}$ Where: "r" is the radius, and
"π" is 3.14, approximately.

For example, the area of a circle whose radius is 8.47" is:

$$A = (3.14)(8.47)^2$$

The actual calculation can be quickly done on the slide rule. Follow these steps on your slide rule:

<u>Step 1</u>: Move the hairline to 8.47 on the D scale.

<u>Step 2</u>: Move the right end of the B scale to the hairline.

<u>Step 3</u>: Move the hairline to π (3.14) on the B scale.

<u>Step 4</u>: Read the digits of the answer on the A scale under the hairline. The digits are "225".

<u>Step 5</u>: Place the decimal point by estimation. A = <u>225. sq. in.</u>

In this process, you first squared 8.47, and then multiplied this square by π using the A and B scales.

Using your slide rule, find the areas of these circles:

(a) r = 1.97 ft. A = ________ (b) r = 0.234" A = ________

---

a) A = 12.2 sq. ft.

b) A = 0.172 sq. in.

73. The formula for the area of a circle in terms of its diameter is:

$$\boxed{A = 0.785d^2}$$

This formula is easily derived as follows:

Since $A = \pi r^2$ and $r = \frac{d}{2}$ (from $d = 2r$),

$$A = \pi\left(\frac{d}{2}\right)^2 = \pi\left(\frac{d^2}{4}\right) = \frac{\pi}{4}(d^2) = \frac{3.14}{4}(d^2) = 0.785d^2$$

For example, the area of a circle whose diameter is 4.40" is:

$$A = (0.785)(4.40)^2$$

To do the actual calculation, follow these steps on your slide rule:

Step 1: Move the hairline to 4.40 on the D scale.

Step 2: Move the right end of the B scale to the hairline.

Step 3: Move the hairline to 0.785 on the B scale. (Note: Some slide rules have a special mark for 0.785 on their A and B scales.)

Step 4: Read the digits of the answer on the A scale under the hairline. The digits are "152".

Step 5: Place the decimal point by estimation. A = 15.2 sq. in.

In this process, you first squared 4.40, and then multiplied this square by 0.785 using the A and B scales.

Using your slide rule and the above procedure, find the area of a circle whose diameter is 61.3 ft. A = ______________

A = 2,950 sq. ft.

74. Using your slide rule and the formula $A = 0.785d^2$, find the areas of these circles:

(a) d = 7.31" A = __________ (b) d = 18.3 ft. A = __________

---

a) A = 42.0 sq. in.

b) A = 263 sq. ft.

75. A <u>sector</u> of a circle is the part of the circle that lies between the sides of a central angle. It has the shape of a "piece of pie" as shown in the figure on the right.

The shaded area in this figure is called <u>sector 0AB</u>. Angle A0B is the angle of the sector. It is, of course, a central angle.

The circle on the right is divided into four equal sectors. The area of each sector is $\frac{1}{4}$ of the area of the complete circle.

If the area of the complete circle is 40 sq. in., the area of each sector is ______________.

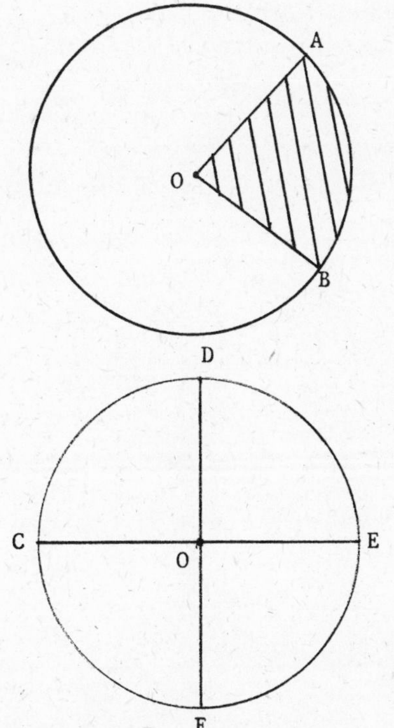

---

10 sq. in.

76. The area of a sector is a fractional part of the total area of the circle. We can determine how much of the total area of the circle is contained in the sector <u>by comparing the size of the central angle to 360°</u>. For example:

If the central angle is 60°,

the area of the sector = $\left(\frac{60}{360}\right)$(the total area).

If the central angle is 79°,

the area of the sector = $\left(\frac{\quad}{\quad}\right)$(the total area).

$\left(\frac{79}{360}\right)$

77. If the total area of a circle is 90.0 sq. in.:

(a) The area of a sector whose central angle is 120° is:

$\frac{120}{360}(90.0) =$ __________ sq. in.

(b) The area of a sector whose central angle is 37° is:

$\frac{37}{360}(90.0) =$ __________ sq. in.

---

a) 30.0 sq. in.

b) 9.25 sq. in.

78. The radius of this circle is 5.00". Therefore:

(a) The total area is _______ sq. in.

(b) The area of sector 0FP is _______ sq. in.

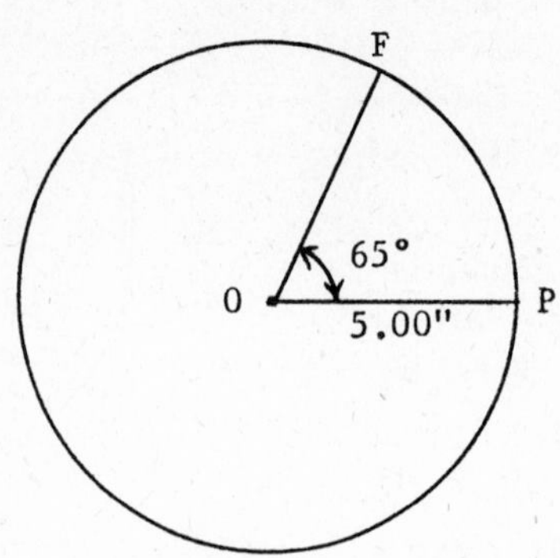

---

a) 78.5 sq. in.

b) 14.2 sq. in.

79. The radius of this circle is 7.50". Therefore:

(a) The total area is _______ sq. in.

(b) The area of sector 0TS is _______ sq. in.

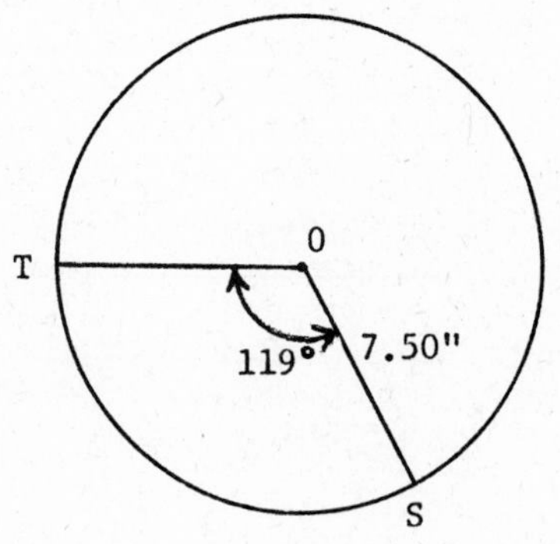

---

a) 177 sq. in.

b) 58.5 sq. in.

80. A segment of a circle is the part of a circle that lies between a chord and its arc. The shaded area in this figure is segment AB.

The area of segment AB can be found by subtracting the area of triangle 0AB from the area of sector 0AB. That is:

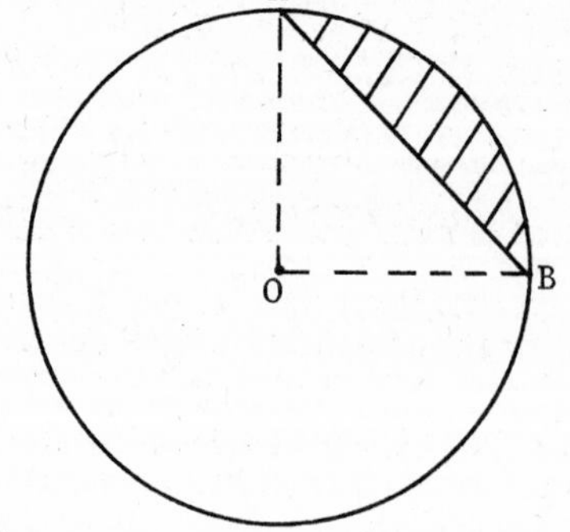

If the area of sector 0AB is 24.0 sq. in. and the area of triangle 0AB is 16.0 sq. in.,

the area of segment AB is ______________.

8.0 sq. in.

81. If radius $0V = 10.0''$:

(a) The area of the total circle is ________ sq. in.

(b) The area of sector 0VM is ________ sq. in.

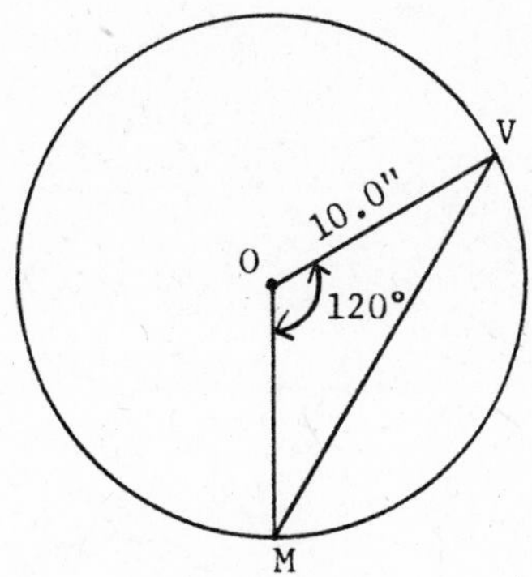

If the area of triangle 0VM is 43 sq. in.:

(c) The area of segment VM is ________ sq. in.

a) 314 sq. in.

b) 105 sq. in.

c) 62 sq. in.

---

82. In order to find the area of a segment, we subtract the area of its related triangle from the area of its sector. Let's examine the method for finding the area of its related triangle.

In the circle on the right, triangle M0T is the related triangle for segment MT.

M D T
0

The formula for the area of any triangle is:

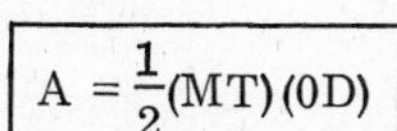

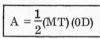

where "h" is the altitude drawn to base "b".

Since we have drawn 0D perpendicular to chord MT, 0D is the altitude to base "MT" in triangle M0T. Therefore, by substitution, the area of triangle M0T is:

$$A = \frac{1}{2}(MT)(0D)$$

But since both MD and DT equal $\frac{1}{2}(MT)$, we can make a further substitution. We get:

$$A = (MD)(0D) \quad \text{or} \quad A = (DT)(0D)$$

That is: The area of the related triangle is equal to the half-chord times the distance of the chord from the center of the circle.

Using this "half-chord" formula, write one of the two possible formulas for the area of the triangle in each circle below.

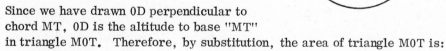

(a)

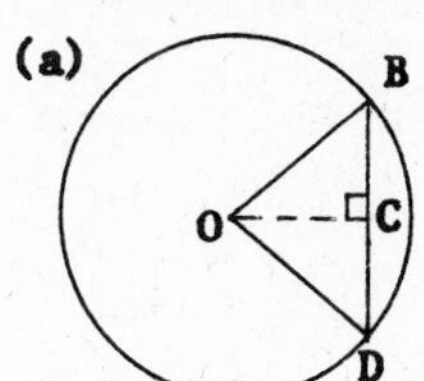

(b)

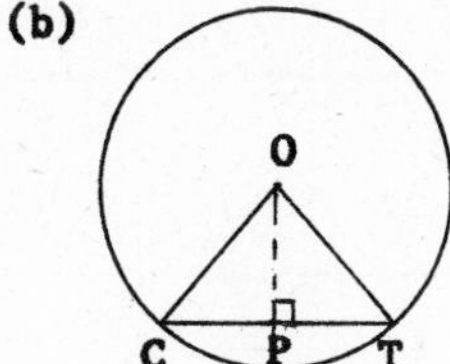

(c)

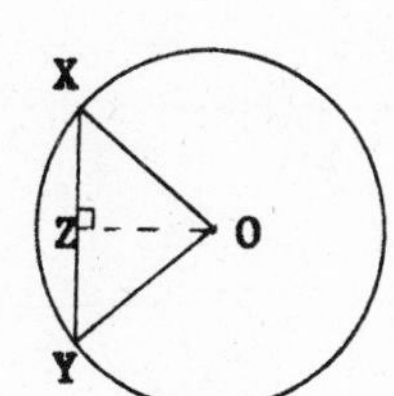

A = ____________ A = ____________ A = ____________

83. Let's find the area of triangle C0F in the circle on the right. The radius is 10.0". Central angle C0F contains 120°. Since central angle C0F is 120°, angles #1 and #2 each contain 60°.

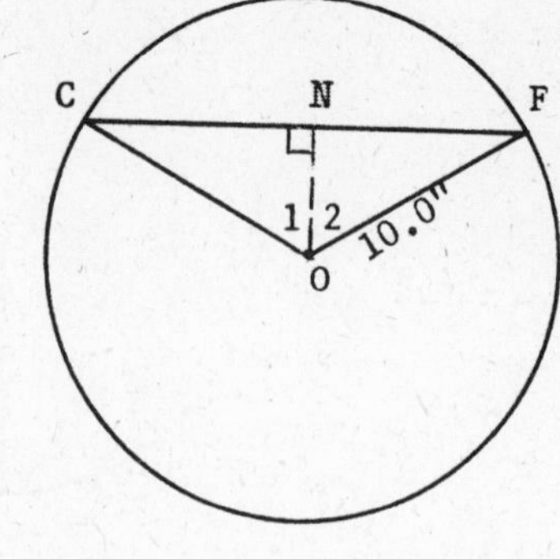

One of the "half-chord" formulas for the area of triangle C0F is:

A = (NF)(0N)

We can use right triangle 0NF to find NF and 0N.

(a) Using the sine ratio, find NF. NF = ________

(b) Using the cosine ratio, find 0N. 0N = ________

(c) Plugging these values into the formula above, find the area of triangle C0F. A = ________

a) A = (BC)(0C) or A = (CD)(0C)

b) A = (CP)(0P) or A = (PT)(0P)

c) A = (XZ)(0Z) or A = (ZY)(0Z)

---

84. The radius of this circle is 10.0". Central angle B0D contains 84°.

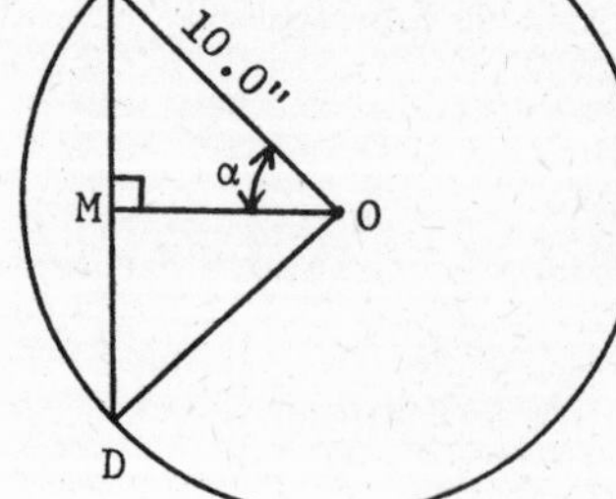

(a) How large is angle α? ______°

(b) How long is BM? ________

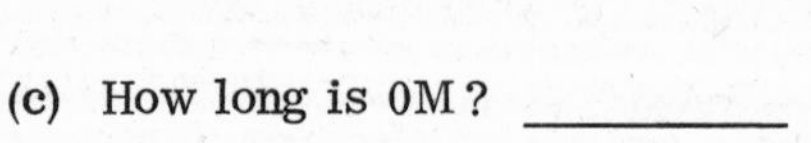

(c) How long is 0M? ________

(d) What is the area of triangle 0BD? ________

a) 8.66", since $\sin 60° = \frac{NF}{10.0''}$

b) 5.00", since $\cos 60° = \frac{0N}{10.0''}$

c) 43.3 sq. in.

---

a) 42°
b) 6.69"
c) 7.43"
d) 49.7 sq. in.

85. The radius of this circle is 20.0". Central angle P0V equals 100°.

(a) Find the area of triangle P0V. ____________

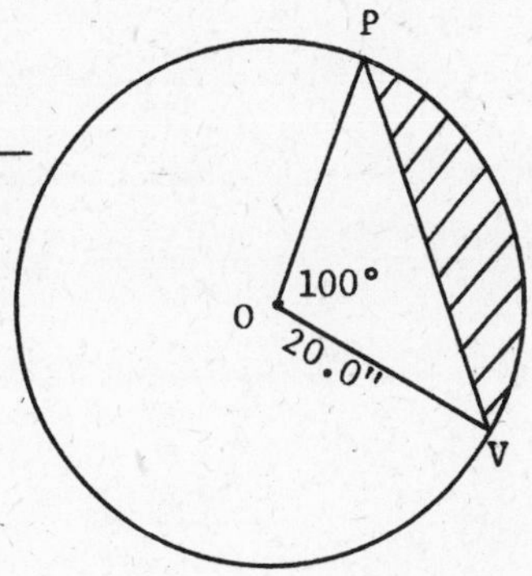

(b) The area of sector 0PV is 349 sq. in. Find the area of segment PV (the shaded area). ____________

---

a) 197 sq. in.

b) 152 sq. in.

86. The radius of this circle is 10.0". Central angle D0T equals 126°.

(a) Find the area of the circle. ____________

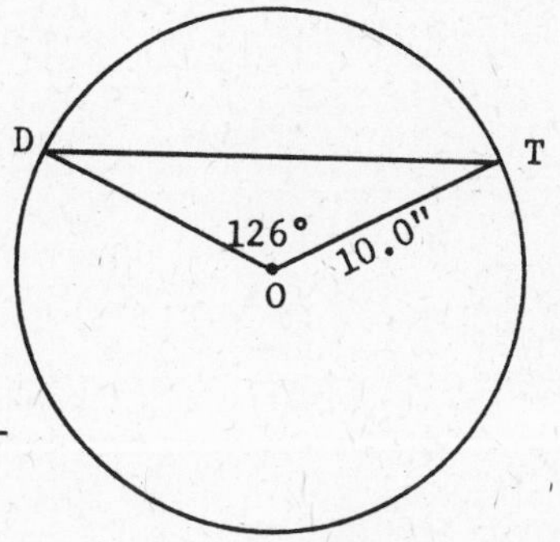

(b) Find the area of sector D0T. ____________

(c) Find the area of triangle D0T. ____________

(d) Find the area of segment DT. ____________

---

a) 314 sq. in.

b) 110 sq. in.

c) 40.4 sq. in.

d) 69.6 sq. in.

SELF-TEST 6 (Frames 72-86)

1. Find the area of sector 0GH. ______________

2. Find the area of segment GH. ______________

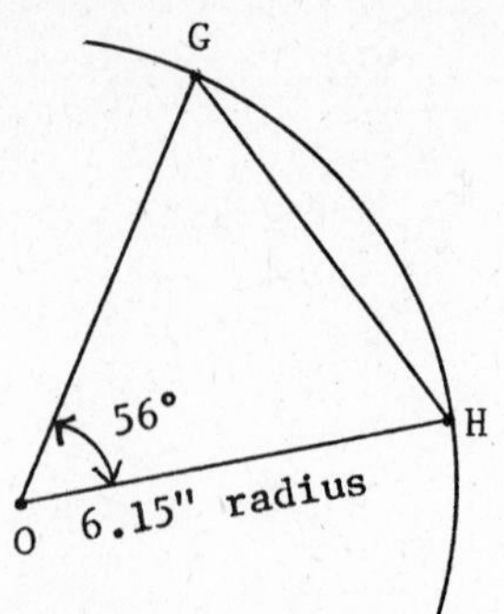

3. Find the area of the shaded portion of the circle shown at the right. ______________

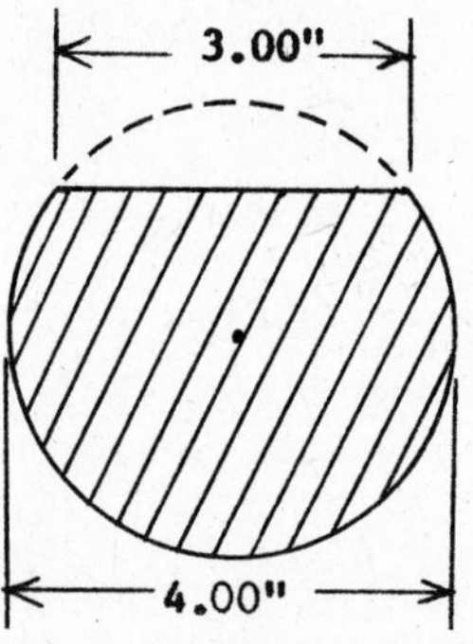

ANSWERS:

1. Sector 0GH = 18.5 sq. in.
2. Segment GH = 2.8 sq. in.
3. 11.1 sq. in.

5-7 REVOLUTIONS AND ROTATIONAL SPEED

When an object (like a wheel, gear, shaft, or drill) is turning with a circular motion, we say that it is rotating. The speed of this rotation can be measured in terms of the number of revolutions the object makes in a unit of time. This measure is called "rotational speed". We will discuss rotational speed in this section.

87. When an object turns with a circular motion, we say that it is "rotating". Rotations are frequently measured in terms of "revolutions".

The figure on the right shows what is meant by one revolution. Vector 0A, with one end fastened at point 0, has rotated until it ended up where it started.

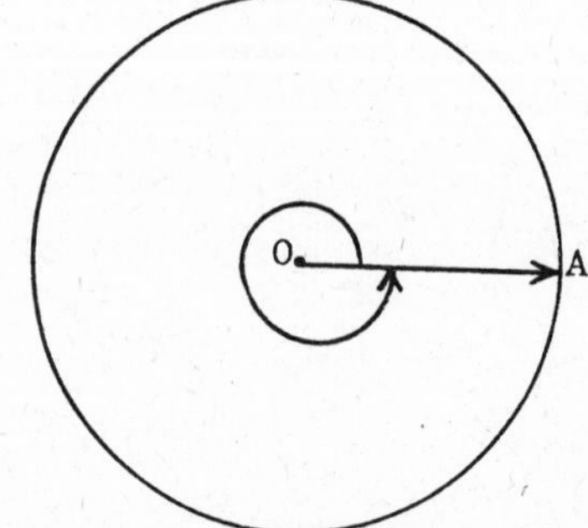

How many degrees are there in one revolution? ______ °

360°

88. Ordinarily we are interested in the speed of a rotating object. This speed is called its "rotational speed". "Rotational speed" is the number of revolutions in a fixed period of time. The fixed period of time is usually "1 minute" or "1 second". Therefore:

$$\text{Rotational Speed} = \frac{\text{Number of Revolutions}}{\text{1 Unit of Time}}$$

When reporting a rotational speed of "10 revolutions in a period of 1 minute":

Instead of writing: $\frac{\text{10 revolutions}}{\text{1 minute}}$,

we usually write: 10 revolutions per minute
or
10 revolutions/minute

Though the number "1" is not explicitly written:

"10 revolutions per minute" means "10 revolutions per 1 minute".

"10 revolutions/minute" means "10 revolutions/1 minute".

Write the two usual ways of reporting a rotational speed of "35 revolutions in a period of 1 second":

____

____

---

89. When reporting rotational speeds, the following abbreviations are used:

"revs." for "revolutions"
"min." for "minutes"
"sec." for "seconds"

Therefore: 48 revs./min. means 48 revolutions per minute.

25 revs./sec. means 25 ____________________.

Answer to 88: 35 revolutions per second and 35 revolutions/second

---

90. Instead of "revolutions per minute" or "revs./min.", the abbreviation "rpm" is frequently used.

Instead of "revolutions per second" or "revs./sec.", the abbreviation "rps" is sometimes used.

(a) 330 rpm means 330 ____________________

(b) 18 rps means 18 ____________________

Answer to 89: revolutions per second

Answer to 90:

a) revolutions per minute or revs./min.

b) revolutions per second or revs./sec.

91. The following formula can be used to compute the rotational speed of a rotating object:

$$\text{Rotational Speed} = \frac{\text{Total Number of Revolutions}}{\text{Total Amount of Time}}$$

Let's use this formula to compute the rotational speed of an object which rotates 100 times in a period of 5 minutes.

$$\text{Rotational Speed} = \frac{100}{5} \text{ or } \frac{20}{1} \text{ or } 20 \text{ rpm}$$

Note: By division, the fraction $\frac{100}{5}$ is reduced to 20 which is really $\frac{20}{1}$ or $\frac{20 \text{ revolutions}}{1 \text{ minute}}$.

Compute the rotational speed of an object which rotates 50 times in a period of 10 seconds.

Rotational Speed = ______________________

---

92. To convert from "revs./min." to "revs./sec." and vice versa, we use the following fact:

1 minute = 60 seconds

Since there are 60 seconds in 1 minute, we multiply by 60 when converting from "revs./sec." to "revs./min.". For example:

4 revs./sec. = 4(60) or 240 revs./min.

(a) 30 revs./sec. = ________ revs./min.

(b) 12 rps = ________ rpm

Answer to 91: 5 revs./sec. or 5 rps

From: $\frac{50}{10}$ or $\frac{5}{1}$

---

93. To convert from "revs./min." to "revs./sec.", we simply reverse the procedure. That is, we divide by 60. For example:

$$120 \text{ revs./min.} = \frac{120}{60} \text{ or } 2 \text{ revs./sec.}$$

(a) 600 revs./min. = ______ revs./sec.

(b) 8700 rpm = ______ rps

Answers to 92:

a) 1,800

b) 720

---

94. When converting from "revs./sec." to "revs./min." and vice versa, remember that a minute is much longer than a second. Therefore:

If the same rotational speed is measured "per second" and "per minute", there will be many more revolutions in 1 minute.

(a) If 100 revs./sec. is converted to "revs./min.", will the number of "revs./min." be larger or smaller than 100? __________

(b) If 120 revs./min. is converted to "revs./sec.", will the number of "revs./sec." be larger or smaller than 120? __________

Answers to 93:

a) 10

b) 145

Answers to 94:

a) Larger

b) Smaller

95. Complete: (a) 78 revs. /sec. = ________ revs. /min.

(b) 965 revs. /min. = ________ revs. /sec.

(c) 12,300 rpm = ________ rps

(d) 32.2 rps = ________ rpm

Answer to Frame 95: a) 4,680 b) 16.1 c) 205 d) 1,930

SELF-TEST 7 (Frames 87-95)

1. Convert 215 revolutions per minute to revolutions per second.

____

2. Convert 56.5 revolutions per second to revolutions per minute.

____

3. Complete: 7,920 rpm = ________ rps

4. Complete: 18.7 rps = ________ rpm

ANSWERS: 1. 3.58 rps 2. 3,390 rpm 3. 132 rps 4. 1,120 rpm

5-8 CIRCUMFERENTIAL SPEED

The speed of a rotating object can be measured in two different ways. One measure is rotational speed, which we discussed in the last section. A second measure is circumferential speed, which measures the distance traveled in 1 minute or in 1 second by a point on the circumference of a rotating object. We will discuss circumferential speed in this section.

96. A point on the circumference of a rotating object travels a definite distance during each revolution of the object. In this diagram, point A lies on the circumference of a rotating object whose center is at point 0.

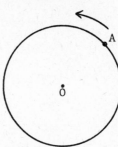

In one revolution, point A travels around the circle and ends up where it started. Therefore, the distance traveled by point A in one revolution is equal to the circumference of the circle.

A

0

If the circumference is 10.0", point A travels ________ in one revolution.

---

97. If an object makes more than one revolution, a point on its circumference keeps going around the same circular path. In each revolution, the distance traveled equals the circumference of the circle.

Answer: 10.0"

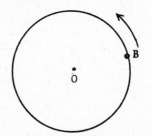

Point B lies on the circumference of a rotating object whose center is at point 0. If the circumference of the circle is 10.0", how far does point B travel in 5 revolutions? ________

B

0

---

98. The distance traveled by a point on the circumference of a rotating object is calculated by this formula:

Answer: 50.0"

$$\text{Distance Traveled} = \begin{pmatrix}\text{Number of} \\ \text{Revolutions}\end{pmatrix} \times \text{Circumference}$$

A rotating circle has a circumference of 3.77 feet. How far does a point on its circumference move:

(a) In 20 revolutions? ____________

(b) In 65.5 revolutions? ____________

---

99. Write the formula for the circumference of a circle:

Answer: a) 75.4 feet
b) 247 feet

(a) In terms of its diameter: C = ________

(b) In terms of its radius: C = ________

Answer: a) $C = \pi d$ or $C = 3.14d$
b) $C = 2\pi r$ or $C = 6.28r$

100. If the diameter of a circle is 10.0":

(a) The circumference of the circle is _________.

(b) How far does a point on the circumference travel in 3 revolutions? _________

---

101. If the radius of a circle is 0.500 feet:

(a) The circumference of the circle is _________.

(b) How far does a point on the circumference travel in 7.40 revolutions? _________

Answers to 100: a) 31.4" b) 94.2"

---

102. A point on the circumference of a circle traveled 120 feet during 48 revolutions. What is the length of the circumference? _____________

Answers to 101: a) 3.14 feet b) 23.2 feet

---

103. The circumference of a circle is 1.97 feet long. How many revolutions of the circle are required to cause a point on its circumference to travel 82.5 feet? _____________

Answer to 102: 2.50 feet

From: $\frac{120}{48} = 2.50$

---

104. Ordinarily we are interested in the speed at which a point on the circumference travels. This speed is called "circumferential speed". "Circumferential speed" is the distance traveled by a point on the circumference in 1 unit of time. That is:

$$\text{Circumferential Speed} = \frac{\text{Distance Traveled}}{\text{1 Unit of Time}}$$

Since: the distance traveled is measured in "feet" or "inches,

and the unit of time is either "1 minute" or "1 second",

the possible units for circumferential speed are:

feet per minute (or ft./min.)
feet per second (or ft./sec.)

inches per minute (or in./min.)
inches per second (or in./sec.)

Though the number "1" is not written explicitly:

"10 ft./min." means "10 ft./1 min."

"20 in./sec." means "20 in./__ sec."

Answer to 103: 41.9 revolutions

From: $\frac{82.5}{1.97} = 41.9$

Answer to 104: 20 in./1 sec.

105. The circumferential speed of a point can be computed by means of the following formula:

$$\text{Circumferential Speed} = \frac{\text{Total Distance Traveled}}{\text{Total Amount of Time}}$$

That is, we can find the circumferential speed if we <u>divide</u> the "total distance traveled" by the "total amount of time" taken to do so. For example:

If a point on a circumference travels 10 feet in 2 seconds:

$$\text{Its circumferential speed} = \frac{10}{2} \text{ or } \frac{5}{1} \text{ or } 5 \text{ ft./sec.}$$

(a) If a point on a circumference travels 400 feet in 2 minutes, its circumferential speed is ________________.

(b) If a point on a circumference travels 1,810 inches in 17.1 seconds, its circumferential speed is ________________.

---

106. The <u>diameter</u> of the circle on the right is 5.00". The circle revolves 10 times in 5.31 seconds. We want to find the circumferential speed of point P.

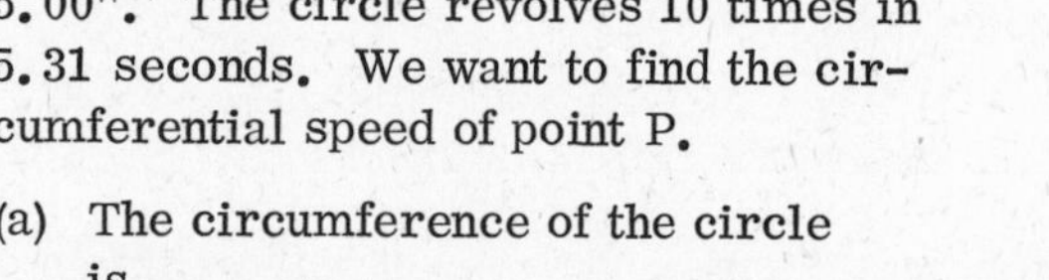

(a) The circumference of the circle is __________.

(b) The distance traveled by point P in 10 revolutions is __________.

(c) The circumferential speed of point P is __________.

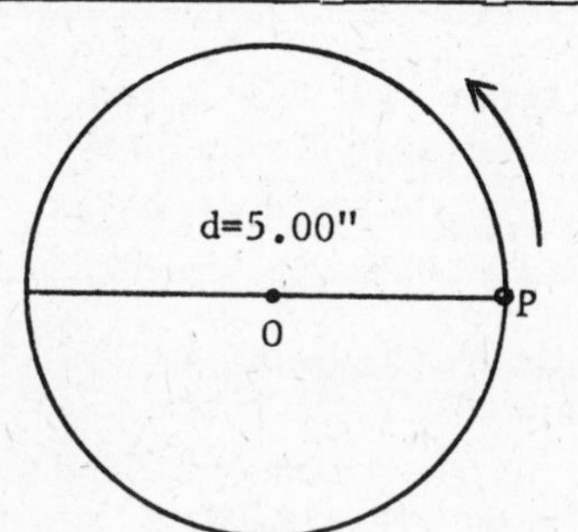

a) 200 ft./min.

b) 106 in./sec.

---

107. Don't confuse rotational speed and circumferential speed.

<u>Rotational Speed</u> measures how fast the <u>rotating object is turning</u>. Therefore, it measures "<u>revolutions</u>" in a unit of time.

<u>Circumferential Speed</u> measures how fast <u>a point on the circumference is moving</u>. Therefore, it measures "<u>feet</u>" or "<u>inches</u>" in a unit of time.

Which of the following are measures of <u>circumferential</u> speed? __________

(a) 17.9 rpm (b) 6.43 in./sec. (c) 800 ft./min. (d) 479 rps

a) 15.7"

b) 157"

c) 29.6 in./sec.

From: $\frac{157}{5.31}$

Only (b) and (c)

108. The radius of the circle on the right is 5.00". The circle revolves 100 times in 2.51 minutes. We want to find the circumferential speed of point T.

(a) The circumference of the circle is __________.

(b) The distance traveled by point T in 100 revolutions is __________.

(c) The circumferential speed of point T is ______________.

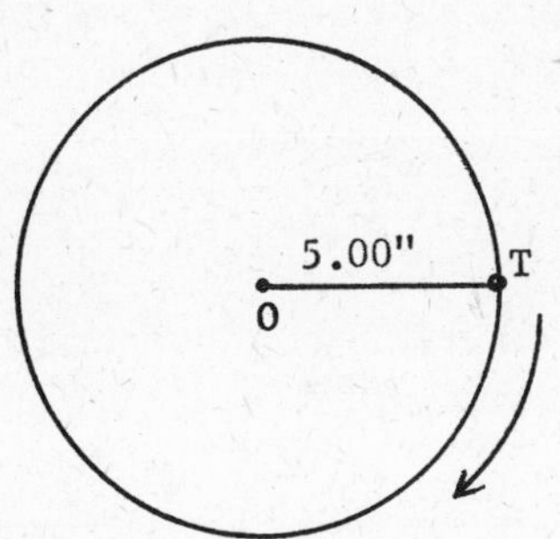

(d) The rotational speed of the circle is ______________.

---

a) 31.4"

b) 3,140"

c) 1,250 in./min.

From: $\frac{3,140}{2.51}$

d) 39.8 rpm

From: $\frac{100}{2.51}$

---

109. If a circular object rotates at a speed of 125 rpm and its circumference is 10.0":

(1) We can find the distance traveled by a point on its circumference by multiplying the circumference by the number of revolutions. That is:

Distance Traveled = (125)(10.0") or 1,250"

(2) And since 125 rpm means "125 revolutions in 1 minute", the circumferential speed of the point has the same numerical value as the distance traveled. That is:

$$\text{Circumferential Speed} = \frac{1,250}{1} \text{ or } 1,250 \text{ in./min.}$$

Therefore, if we know the circumference of a circle and its rotational speed in RPM, we can obtain the circumferential speed of a point on the circle by means of the following formula:

Circumferential Speed = Rotational Speed x Circumference

Using this formula, find the circumferential speed of a point on the circles with the following rotational speeds and circumferences:

(a) 250 rpm with a circumference of 15.0 inches.

Circumferential Speed = ______________________

(b) 400 rpm with a circumference of 2.75 feet.

Circumferential Speed = ______________________

110. The diameter of a circle is 7.50". The circle rotates at a speed of 100 rpm. Let's find the circumferential speed of a point on the circle.

(a) The circumference of the circle is __________.

(b) The circumferential speed of the point is ____________________.

a) 3,750 in./min.

b) 1,100 ft./min.

---

111. The radius of a circle is 1.25". The circle rotates at a speed of 200 rpm. Find the circumferential speed of a point on the circle.

(a) The circumference of the circle is __________.

(b) The circumferential speed is ____________________.

a) 23.6"

b) 2,360 in./min.

From: (100)(23.6)

---

112. When computing the circumferential speed of a point on a circle whose dimensions are given in inches, frequently the numerical value of the circumferential speed in "inches per minute" is quite large. To make this number smaller, we can convert "inches per minute" to "feet per minute". To do so, we divide by 12, since there are 12 inches in 1 foot. For example:

$$1{,}200 \text{ in./min.} = \frac{1{,}200}{12} \text{ or } 100 \text{ ft./min.}$$

Convert each of these to "feet per minute":

(a) 2,400 in./min. = __________ ft./min.

(b) 32,900 in./min. = __________ ft./min.

a) 7.85"

b) 1,570 in./min.

---

a) 200

b) 2,740

113. A circular object with a circumference of 28.5 inches rotates at a speed of 175 rpm. Let's find the circumferential speed of a point on the circle.

(a) In "inches per minute", the circumferential speed is ___________________.

(b) In "feet per minute", the circumferential speed is ___________________.

---

114. The diameter of a circular object is 10.0". It rotates at a speed of 650 rpm. Find the circumferential speed of a point on the circle in "feet per minute".

Circumferential Speed = ___________________

a) 4,990 in./min.

b) 416 ft./min.

---

115. The principle of "circumferential speed" is important in metal-cutting operations. The principle is used in the operation of such common machines as the lathe, drill press, milling machine, and grinder. In such applications, circumferential speed is called "cutting speed" or "rim speed".

To illustrate, the diagram at the right shows the typical set-up for a metal-cutting machine called a lathe. In this set-up, a rotating metal shaft (shaded circle) is being cut by a stationary cutting tool. The circumferential speed of a point on the circumference of the shaft (such as point A) is called the "cutting speed". Cutting speed is usually expressed in "feet per minute".

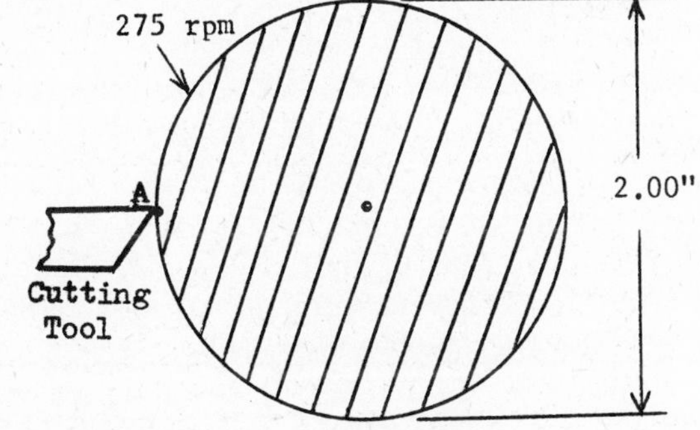

In the diagram, the diameter of the shaft is 2.00", and the rotational speed is 275 rpm.

(a) Calculate the circumferential speed of point A in "feet per minute". ___________________

(b) Therefore, the "cutting speed" is ___________________.

1,700 ft./min.

Answer to Frame 115: a) 144 ft./min. b) 144 ft./min.

SELF-TEST 8 (Frames 96-115)

1. A wheel of 27.0" diameter makes 100 complete revolutions. Find the total distance, in feet, traveled by a point on the circumference. ________

Do these conversions:

2. 1,620 in./min. = ________ ft./min.

3. 21.8 in./sec. = ________ in./min.
   = ________ ft./min.

4. Complete this formula:

   Circumferential Speed = Rotational Speed x ________________

A circle of 4.00" radius rotates at a speed of 220 rpm.

5. Find its circumference, in feet.
   ________ ft.

6. Find the circumferential speed of a point on its circumference, in feet per minute. ________ ft./min.

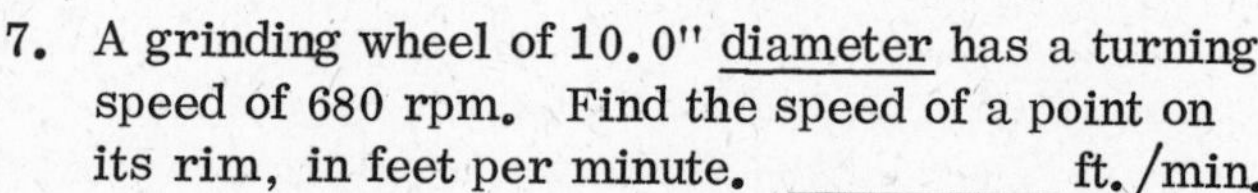

7. A grinding wheel of 10.0" diameter has a turning speed of 680 rpm. Find the speed of a point on its rim, in feet per minute. ________ ft./min.

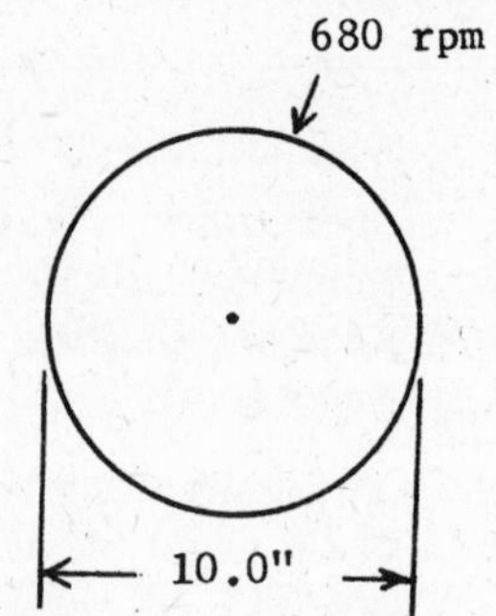

ANSWERS:

1. 707 ft.
2. 135 ft./min.
3. 1,308 in./min.
   109 ft./min.
4. Circumference
5. 2.09 ft.
6. 460 ft./min.
7. 1,780 ft./min.

5-9 ANGULAR MEASURES OF ROTATIONAL SPEED

Though rotational speed is frequently measured in terms of "revolutions" in a unit of time, it can also be measured in terms of the size of the angle generated in a unit of time. When rotational speed is measured in terms of "angle size", it is called "angular velocity". Since there are two alternate ways of measuring angle size (degrees and radians), there are two alternate ways of measuring "angular velocity". We will discuss both ways in this section, and then show the methods of converting from one way to the other.

116. Since a rotation can be measured in terms of the size of the angle generated, rotational speed can also be measured in terms of degrees rotated in a unit of time. That is:

$$\text{Rotational Speed} = \frac{\text{Number of Degrees Rotated}}{\text{1 Unit of Time}}$$

Of course, to compute a measure of rotational speed in terms of "degrees", we simply divide the number of degrees by the amount of time. For example:

If an object rotates through 500° in 10 seconds,

its rotational speed $= \frac{500}{10}$ or $\frac{50}{1}$ or 50°/sec.

If an object rotates through 30,000° in 5 minutes,

its rotational speed = ________°/min.

Answer: 6,000°/min.

117. To convert rotational speeds reported in "revolutions per unit time" to "degrees per unit time" (and vice versa), you must be able to convert revolutions to degrees (and vice versa). The basic fact used in these conversions is:

1 revolution = 360°

Therefore: (1) To convert from revolutions to degrees, we multiply by 360°. For example:

5 revolutions = 5(360°) or 1,800°

(2) To convert from degrees to revolutions, we divide by 360°. For example:

$720° = \frac{720°}{360°}$ or 2 revolutions

Using the procedure above, complete these conversions:

(a) 10 revolutions = ________°

(c) 1.5 revolutions = ________°

(b) 1,080° = _____ revolutions

(d) 450° = ________ revolutions

Answers: a) 3,600° b) 3 revs. c) 540° d) 1.25 revs.

118. Complete these conversions:

(a) 180° = _______ revolution

(c) 1,674° = _______ revolutions

(b) 6.5 revolutions = ________°

(d) 0.7 revolution = ________°

119. To make the following conversions from one measure of rotational speed to the other, we simply convert the revolutions to degrees (or vice versa). Do so in these problems:

(a) 3.5 rpm = _______°/min. (c) 0.8 rev./sec. = _______°/sec.

(b) 1,760°/min. = _______revs./min. (d) 498°/sec. = _______ revs./sec.

a) $\frac{1}{2}$ or 0.5 rev.

b) 2,340°

c) 4.65 revs.

d) 252°

---

120. Since the size of the angle generated can be measured in radians, a third measure of rotational speed is the number of radians rotated in a unit of time. That is:

$$\boxed{\text{Rotational Speed} = \frac{\text{Number of Radians Rotated}}{\text{1 Unit of Time}}}$$

To compute a measure of rotational speed in terms of "radians", we simply divide the number of radians by the amount of time. For example:

If an object rotates through 14 radians in 2 seconds,

its rotational speed = $\frac{14}{2}$ or $\frac{7}{1}$ or 7 radians/sec.

If an object rotates through 480 radians in 10 minutes,

its rotational speed = _______ radians/min.

a) 1,260°/min.

b) 4.89 revs./min.

c) 288°/sec.

d) 1.38 revs./sec.

---

121. Let's briefly review the concept of "radians".

Radians are an alternate way of measuring the size of angles. A central angle contains 1 radian if its arc-length equals the radius of the circle. Angle α in the figure on the right contains 1 radian.

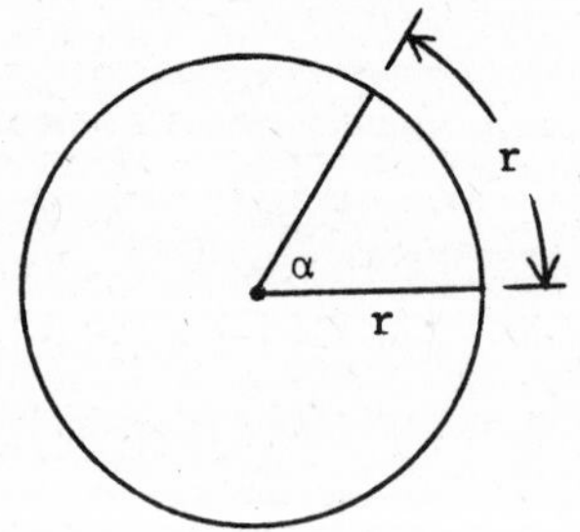

In each figure below, the radius of the circle and the arc-length of the central angle are given. The radius in each case is 5.0".

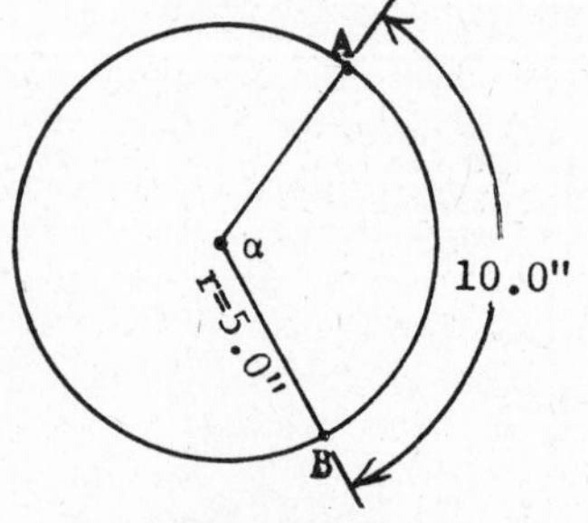

Figure 1

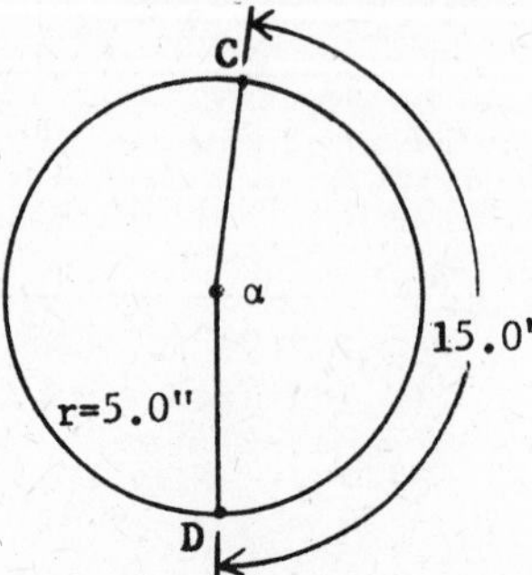

Figure 2

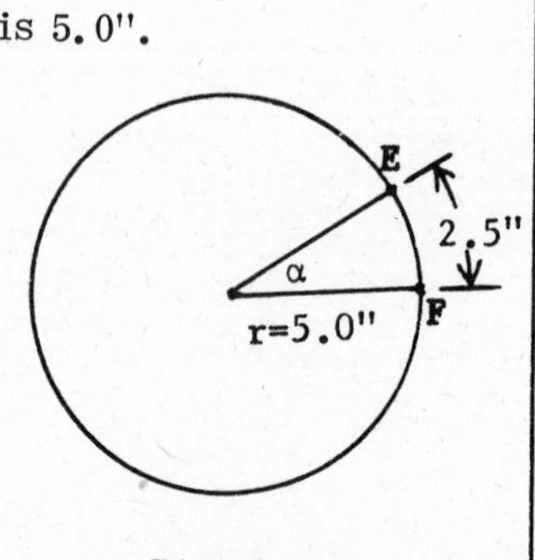

Figure 3

(Continued on following page.)

48 radians/min.

121. (Continued)

(a) In Figure 1, $\widehat{AB}$ is 10.0". Therefore, α contains _______ radians.

(b) In Figure 2, $\widehat{CD}$ is 15.0". Therefore, α contains _______ radians.

(c) In Figure 3, $\widehat{EF}$ is 2.5". Therefore, α contains _______ radians.

---

122. As you can see from the figure on the right, there are 6.28 radians in a complete circle. This fact can be obtained from the circumference formula.

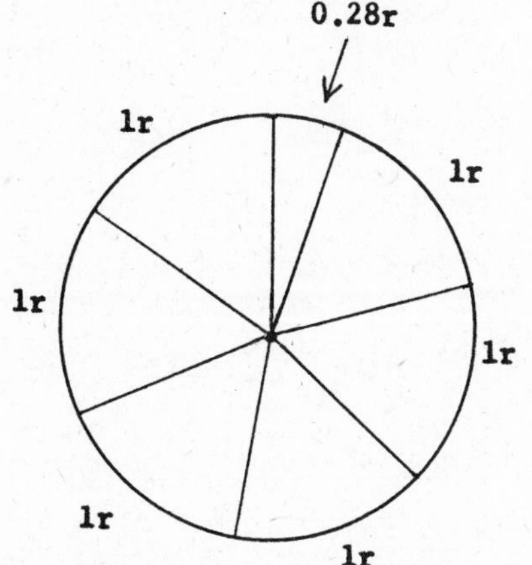

$$C = 2\pi r = 2(3.14)r = 6.28r$$

Since there are 6.28 radians in a complete circle, there are 57.3° in 1 radian. To get the number 57.3°, we divide 360° by 6.28. That is:

$$\frac{360°}{6.28} = 57.3°$$

(a) How many radians are there in a complete circle? _________ radians

(b) How many degrees are there in 1 radian? _________°

Answers to 121: a) 2 radians b) 3 radians c) $\frac{1}{2}$ or 0.5 radian

---

123. To convert from radians to degrees (and vice versa), we use the following fact:

1 radian = 57.3°

When converting from radians to degrees, we multiply by 57.3°. For example:

3 radians = 3(57.3°) or 171.9° (or 172°)

When converting from degrees to radians, we divide by 57.3°.

$$114.6° = \frac{114.6}{57.3} \text{ or 2 radians.}$$

Complete the following conversions:

(a) 4 radians = _________°

(b) 720° = _______ radians

(c) 0.45 radian = _________°

(d) 41° = _______ radian

Answers to 122: a) 6.28 radians b) 57.3°

Answers to 123: a) 229.2° or 229° b) 12.6 radians c) 25.8° d) 0.716 radian

124. To convert radians to revolutions (and vice versa), we use the following fact:

1 revolution = 6.28 radians

When converting from revolutions to radians, we multiply by 6.28. For example:

2 revolutions = 2(6.28) or 12.56 radians

When converting from radians to revolutions, we divide by 6.28. For example:

$$12.56 \text{ radians} = \frac{12.56}{6.28} \text{ or } 2 \text{ revolutions}$$

Complete the following conversions:

(a) 3 revolutions = _____ radians (c) 1.8 revolutions = _____ radians

(b) 25.7 radians = _____ revolutions (d) 147 radians = _____ revolutions

---

125. Here are the two facts needed to convert from radians to the other two measures (or vice versa):

1 radian = 57.3°
1 revolution = 6.28 radians

Using the facts above, complete these conversions:

(a) 497° = _____ radians (c) 18.7 radians = _______ revolutions

(b) 3.56 radians = ______° (d) 3.1 revolutions = _______ radians

Answers to 124:
a) 18.8 radians
b) 4.09 revolutions
c) 11.3 radians
d) 23.4 revolutions

---

126. We use the same two facts to convert from one measure of rotational speed to another when they have the same time-unit. Complete these conversions:

(a) 7.40 rpm = _________ radians/min.

(b) 14 radians/sec. = _________°/sec.

(c) 196 radians/sec. = _________ revs./sec.

(d) 2,400°/min. = _________ radians/min.

Answers to 125:
a) 8.36 radians
b) 204°
c) 2.98 revolutions
d) 19.5 radians

127. To make a conversion from "revolutions per minute" to "degrees per second", we must convert both the measure of rotation and the time-unit. Such a conversion requires two steps. Here is an example:

To convert 5 revs./min. to "degrees/sec.", follow these steps:

Step 1: First, convert "revolutions" to "degrees":

$$5 \text{ revs./min.} = 5(360°) \text{ or } 1{,}800°/\text{min.}$$

Step 2: Then, convert "minutes" to "seconds":

$$1{,}800°/\text{min.} = \frac{1800}{60} \text{ or } 30°/\text{sec.}$$

Therefore: 5 revs./min. = 30°/sec.

To convert 37 revs./min. to "radians/sec.", complete these steps:

(a) Convert "revolutions" to "radians":

37 revs./min. = ________ radians/min.

(b) Now convert "minutes" to "seconds":

37 revs./min. = ________ radians/sec.

a) 46.5 radians/min.

b) 802°/sec.

c) 31.2 revs./sec.

d) 41.9 radians/min.

---

128. Each of the following requires a two-step conversion:

(a) 1570 radians/min. = ________ revs./sec.

(b) 2 rpm = ______ °/sec.

(c) 1.30 radians/sec. = ________ °/min.

a) 232 radians/min.

b) 3.87 radians/sec.

---

a) 4.17 revs./sec.

b) 12°/sec.

c) 4,470°/min.

129. Rotational speed (or velocity) can be reported in any of three ways:

(1) revolutions per minute (or second)
(2) degrees per minute (or second)
(3) radians per minute (or second)

Since "degrees" and "radians" are measures of angle-size, both "degrees per minute (or second)" and "radians per minute (or second)" are called angular velocity.

(a) Does the abbreviation "rpm" stand for "radians per minute" or "revolutions per minute"? ______________

(b) "Angular velocity" is another term for ______________.

(c) Is the term "angular velocity" used in conjunction with "revolutions per minute"? ________

Answer to Frame 129: a) Revolutions per minute b) rotational speed c) No

SELF-TEST 9 (Frames 116-129)

Complete these basic relationships:

1. 1 radian = ________°
2. 1 revolution = ________°
3. 1 revolution = ________ radians

Do the following conversions:

4. 25 rpm = ________°/min.
5. 400 rpm = ________ radians/min.
6. 1,350°/sec. = ________ revs./sec.
7. 73.5 radians/sec. = ________ revs./sec.

Using the two-step process, do these conversions:

8. 720 rpm = ________ radians/sec.
9. 600 radians/sec. = ________ rpm
10. 600°/sec. = ________ rpm
11. 4,800 radians/min. = ________ revs./sec.

ANSWERS:

1. 57.3°
2. 360°
3. 6.28 radians
4. 9,000°/min.
5. 2,510 radians/min.
6. 3.75 revs./sec.
7. 11.7 revs./sec.
8. 75.4 radians/sec.
9. 5,730 rpm
10. 100 rpm
11. 12.7 revs./sec.

5-10 CIRCUMFERENTIAL SPEED AND ANGULAR VELOCITY (IN RADIANS)

When a rotation is measured in revolutions, the distance traveled by a point on the circle can be easily computed by using the circumference of the circle. Therefore, circumferential speed can be computed by multiplying the revolutions per minute (rpm) times the circumference.

However, when a rotation is measured in radians, the distance traveled by a point on the circle can be computed more easily by using the radius of the circle. Therefore, circumferential speed can be computed by multiplying the angular velocity (in radians) times the radius. We will discuss the method in this section.

---

130. In order to compute the circumferential speed of a point on the circumference of a rotating circle, we must know the "distance traveled" by the point, since:

$$\text{Circumferential Speed} = \frac{\text{Distance Traveled By a Point}}{\text{1 Unit of Time}}$$

When the rotation is measured in revolutions, the "distance traveled by a point" is computed in terms of the circumference of the circle. For example:

If an object whose diameter is 10.0" makes 3 revolutions, the distance traveled by a point on its circumference is:

3(Circumference) = 3(31.4") or 94.2"

If an object whose radius is 2.0" makes 10 revolutions, the distance traveled by a point on its circumference is:

10(Circumference) = 10(______) or ________

Answer: 10(12.6") or 126"

---

131. When the rotation is measured in "radians", it is easier to compute the distance traveled by a point on the circumference in terms of the "radius" of the circle. It is easier because the point travels the length of 1 radius for each radian of rotation. To show this fact, we will examine a few cases.

In this diagram, the central angle is 1 radian, and the radius is 0.80 feet.

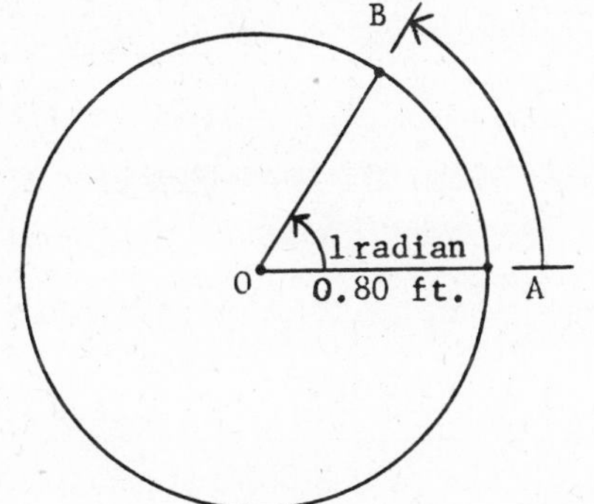

(a) What is the length of arc AB?

____

(b) If the circle rotates through an angle of 1 radian, point A moves to point B. Through what distance does point A move?

____

Answers:

a) 0.80 ft.

(The definition of 1 radian is an angle whose arc equals the radius of the circle.)

b) 0.80 ft.

132. The radius of this circle is 1.25 ft. If the circle rotates through an angle of 2 radians, point P moves to point Q. Through what distance does point P move? ____________

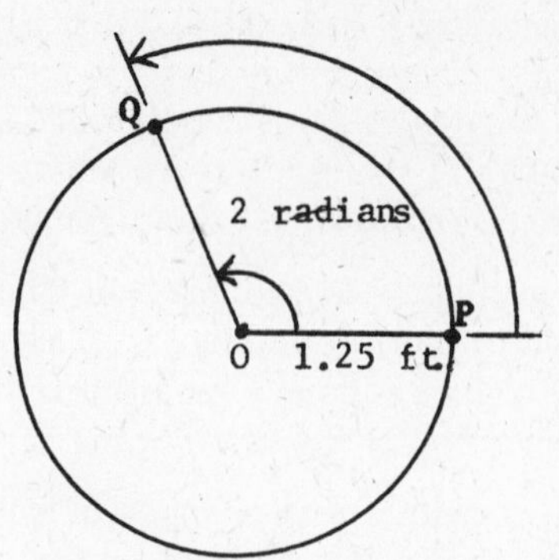

---

133. The radius of this circle is 2.58 ft. If the circle rotates through an angle of 4.50 radians, point G moves to point H. Through what distance does point G move?

____

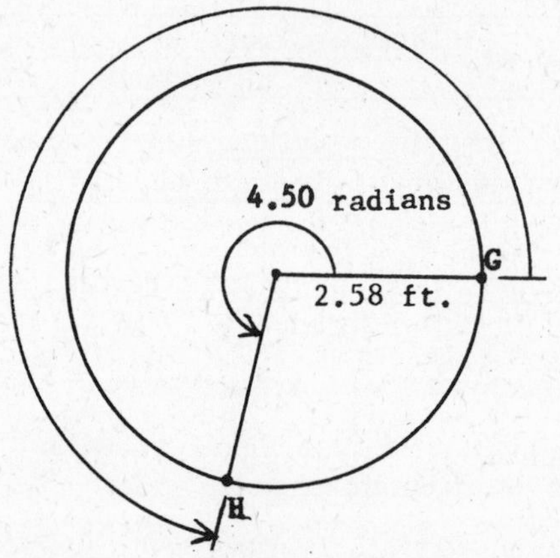

2.50 ft.

From: $2 \times 1.25 = 2.50$

---

134. If the angle of rotation is measured in radians, we can find the distance traveled by a point on the circumference, in terms of the radius, by means of the following formula:

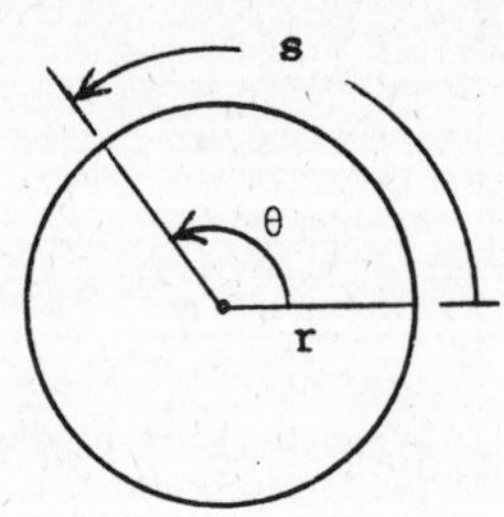

$$s = \theta r$$

Where: "s" is the distance traveled

"θ" is the angle of rotation in radians

"r" is the radius of the circle

Find "s", the distance traveled, if:

(a) $\theta = 15$ radians and $r = 0.40$ feet. s = ____________

(b) $\theta = 0.327$ radian and $r = 1.80$ feet. s = ____________

11.6 ft.

From: $4.50 \times 2.58 = 11.6$

---

a) $s = 6.0$ feet

b) $s = 0.589$ feet

135. Using the formula $s = \theta r$ to find the "distance traveled" by a point on the circumference, we can find its circumferential speed if we know the "amount of time" needed to travel that distance. The circumferential speed is then found by dividing the "distance traveled" by the "amount of time".

Find the circumferential speed of a point when the angle generated is 20 radians, the radius of the circle is 0.85", and the amount of time needed is 3.6 seconds.

(a) The distance traveled is ______________.

(b) The amount of time is ______________.

(c) The circumferential speed of the point is ______________.

---

a) 17 inches
From: 20×0.85

b) 3.6 seconds

c) 4.7 in./sec.
From: $\frac{17}{3.6}$

136. If the angle of rotation is measured in degrees, we must first convert the degrees to either revolutions or radians in order to find the distance traveled by a point on the circumference. Here is an example:

If a circular object whose diameter is 10.0" rotates through 720°, we can find the distance traveled in either of two ways:

(1) We can convert 720° to revolutions and then use the circumference of the circle.

$$\text{Since } 720° = \frac{720}{360} \text{ or 2 revolutions,}$$

$$\text{and } C = \pi d = 3.14(10.0'') = 31.4'', \text{ then the}$$

$$\text{Distance Traveled} = (2)(C) = 2(31.4'') = 62.8''$$

(2) We can convert 720° to radians and then use the radius of the circle.

$$\text{Since } 720° = \frac{720}{57.3} \text{ or 12.6 radians,}$$

$$\text{and } r = \frac{10.0''}{2} \text{ or 5.00'', then the}$$

$$\text{Distance Traveled} = \theta r = (12.6)(5.00'') = 63.0''$$

Considering the amount of rounding involved in the two methods, did we get the same value for the distance traveled? __________

Yes

137. A circular object whose radius is 1.25" rotates through 1,476° in 4.50 seconds. Let's find the circumferential speed of a point on its circumference by converting the degrees to revolutions.

(a) 1,476° = ________ revolutions

(b) The circumference of the circle is ________.

(c) The distance traveled by the point is ________.

(d) The amount of time is ____________.

(e) The circumferential speed of the point is ____________.

---

a) 4.10 revolutions

b) 7.85"

c) 32.2"

d) 4.50 seconds

e) 7.16 in./sec.

From: $\frac{32.2}{4.50}$

138. A circular object whose radius is 0.750" rotates through an angle of 2,260° in 1.25 minutes. Let's find the circumferential speed of a point on its circumference by converting the degrees to radians.

(a) 2,260° = ________ radians

(b) The distance traveled by the point is ________.

(c) The amount of time is ____________.

(d) The circumferential speed of the point is ____________.

---

a) 39.4 radians

From: $\frac{2,260}{57.3}$

b) 29.6"

From:

s = (39.4)(0.750)

c) 1.25 minutes

d) 23.7 in./min.

From: $\frac{29.6}{1.25}$

139. When rotational speed (or velocity) is measured in revolutions per minute, circumferential speed can be computed by means of the following formula:

Circumferential Speed = Revolutions Per Minute x Circumference

When rotational speed (or velocity) is measured in "radians per minute (or second)", it is called "angular velocity". When this is the case, circumferential speed (or velocity) can be computed by means of this formula:

Circumferential Speed = Angular Velocity x Radius

To show that the latter formula makes sense, we will discuss the following example:

A circle whose radius is 1.59 feet has an angular velocity of 27.4 radians per minute. Let's find the circumferential speed of a point on the circle.

(1) We can find the distance traveled by the point by multiplying the radius by the number of radians. That is:

$$s = \theta r = (27.4)(1.59) = 43.6 \text{ ft.}$$

(2) And since the angular velocity is 27.4 radians in 1 minute, the circumferential speed of the point has the same numerical value as the distance traveled. That is:

$$\text{Circumferential Speed} = \frac{43.6}{1} \text{ or } 43.6 \text{ feet/minute}$$

Go to next frame.

---

140. Here is the same formula:

Circumferential Speed = Angular Velocity x Radius

This formula is frequently expressed in symbol form as we have done below:

$\boxed{v = \omega r}$, where v = circumferential speed (or velocity)
ω = angular velocity
r = radius

Let's use the formula to compute the circumferential speed when the angular velocity is 16.5 radians/min. and the radius is 2.40 feet.

(a) ω = __________

(b) r = __________

(c) $v = \omega r$ = __________

---

a) 16.5 radians/min.

b) 2.40 ft.

c) 39.6 ft./min.

141. When using the formula $v = \omega r$, be sure to check the "units" which are given. "r" can be given in either "feet" or "inches". "ω" can be given in either radians "per minute" or radians "per second". The units given determine the units of the circumferential speed. For example:

If "r" is given in "feet" and "ω" is given in radians "per second", "v" is reported in "ft./sec.".

Watch the units when doing these:

(a) Find v if $r = 3.92$ in. and $\omega = 500$ radians per minute.
v = ________________

(b) Find v if $r = 0.790$ ft. and $\omega = 18.5$ radians per second.
v = ________________

---

a) 1,960 in./min.

b) 14.6 ft./sec.

142. In the formula $v = \omega r$, "r" is the radius of the circle. Be sure that you plug in the radius and not the diameter of the circle.

The diameter of a circle is 1.50 feet. It rotates at an angular velocity of 400 radians per minute. Find the circumferential speed of a point on the circle. v = ________________

---

v = 300 ft./min.

143. In the formula $v = \omega r$, "ω" is the angular velocity in terms of radians. If the angular velocity is given in terms of degrees, you have to convert to radians first. Do this problem:

The angular velocity of a rotating object is 71.6°/sec. Its diameter is 3.00 in. Find the circumferential speed of a point on the circle.

(a) In terms of radians, ω = ________________

(b) r = __________

(c) $v = \omega r$ = ________________

---

a) 1.25 radians/sec.

b) 1.50"

c) 1.88 in./sec.

144. If the radius of a circular object is 10.8" and its angular velocity is 16.0 radians/sec., its circumferential speed is:

$$v = \omega r = (16.0)(10.8) \text{ or } 173 \text{ in./sec.}$$

To convert "v" to "ft./min.", we need two steps:

(a) Convert the "inches" to "feet". v = ________ ft./sec.

(b) Convert the "seconds" to "minutes". v = ________ ft./min.

Answer to Frame 144: a) 14.4 ft./sec. b) 864 ft./min.

SELF-TEST 10 (Frames 130-144)

1. A rotating circle has a radius of 1.25 feet. How far does a point on its circumference move in 7.40 revolutions? ________

2. A circle of radius 0.82 feet rotates through an angle of 2.50 radians. Through what distance does a point on its circumference move? ________

3. A circle of radius 0.482 feet has an angular velocity of 128 radians per minute. Find the velocity, in feet per minute, of a point on its circumference. ________

4. A point on the circumference of a circle of radius 6.00 inches has an angular velocity of 32 radians per second. Find the velocity of the point, in feet per minute. ________

ANSWERS: 1. 58.1 ft. 2. 2.05 ft. 3. 61.7 ft./min. 4. 960 ft./min.

5-11 AREAS OF TRIANGLES

Trigonometry frequently has to be used in order to find the areas of triangles. Trigonometry is used to find the legs of right triangles and the altitudes of oblique triangles, so that the formula for triangle area can be used. We will discuss the steps needed to find the areas of both types of triangles in this section.

145. The formula for the area of a triangle is:

$$A = \frac{1}{2}bh$$

Where "b" is any base (or side), and "h" is the altitude drawn to that base (or side).

State whether or not the area of each triangle can be computed using the information given:

(a) ________

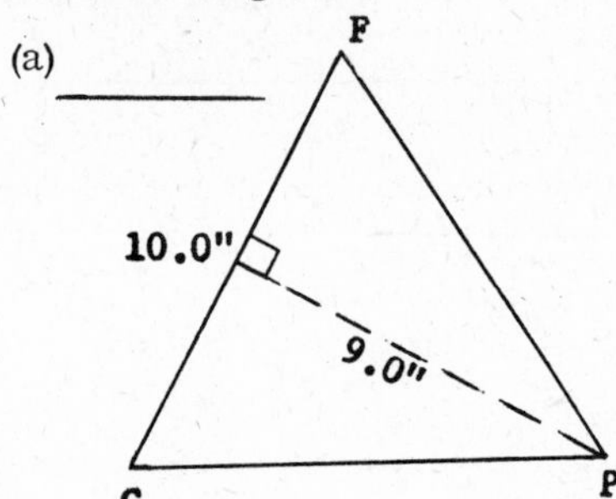

(b) ________

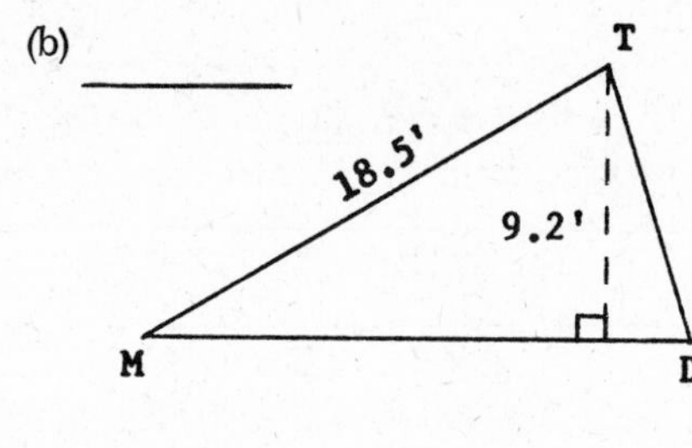

146. In a right triangle, each leg is the altitude to the other leg. For example:

In right triangle NS0:

If 0S is the base, NS is its altitude.

If NS is the base, 0S is its altitude.

Therefore, the area of a right triangle is $\frac{1}{2}$ times the product of its two legs.

In terms of 0S and NS, the formula for the area of right triangle NS0 above is: A = ________

Answers to 145:

a) Yes, since both the altitude and the base to which it is drawn are known. (Its area is 45 sq. in.)

b) No, since the altitude is drawn to the base MD, whose length is not known.

147. (a) The lengths of the two legs of this right triangle are ________ and ________.

(b) Therefore, the area of the triangle is ______ sq. ft.

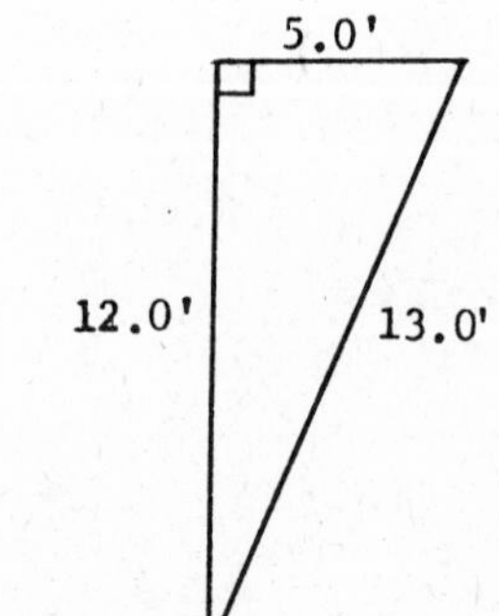

Answer to 146: $A = \frac{1}{2}(0S)(NS)$

148. To find the area of right triangle TWV, we must find the lengths of legs VW and TW first. To do so, we use trig ratios.

(a) The length of VW is ________.

V

20.0"

30°

T W

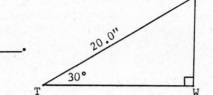

(b) The length of TW is ________.

(c) The area of the triangle is ____________.

a) 5.0' and 12.0'

b) 30 sq. ft.

---

149. To find the area of this right triangle, we must find the length of leg "s_1".

(a) The length of s_1 is ________.

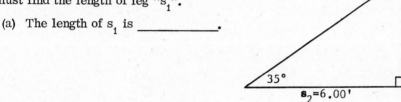

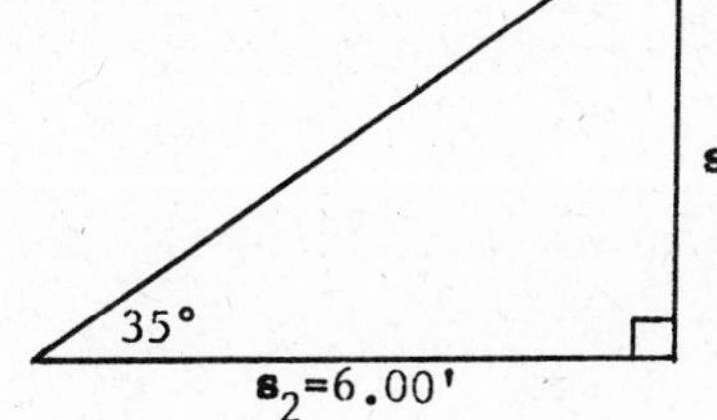

(b) The area of the triangle is ____________.

a) 10.0"

From: $\sin 30° = \frac{VW}{20.0''}$

b) 17.3"

From: $\cos 30° = \frac{TW}{20.0''}$

c) 86.5 sq. in.

---

150. To find the area of an <u>oblique</u> <u>triangle</u>, it is usually necessary to first calculate the length of an altitude.

For example, in oblique triangle EFG, we draw altitude FH to base EG. Base EG is 50.0".

Altitude FH can be found by trigonometry, since it is a side of right triangle EHF.

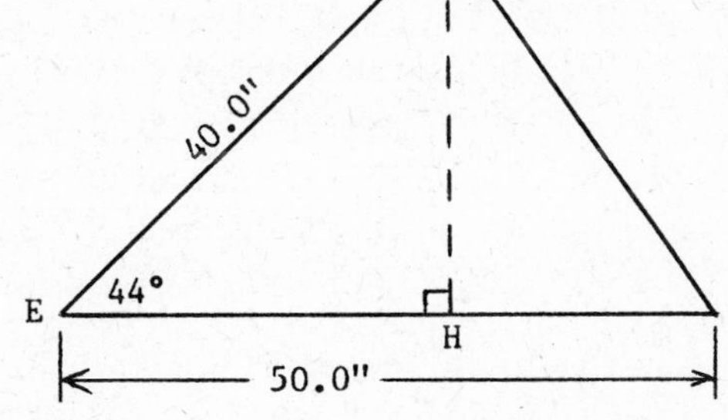

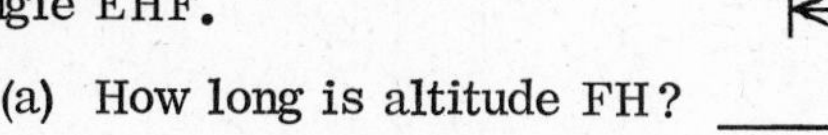

(a) How long is altitude FH? ________

(b) Find the area of oblique triangle EFG. ____________

a) 4.20 feet

From: $\tan 35° = \frac{s_1}{6.00}$

b) 12.6 sq. ft.

151. To find the area of an oblique triangle, draw the altitude to a base whose length is known. Then you don't have to calculate the length of the base, too.

In oblique triangle PQR, you can draw an altitude either to PR or to QR.

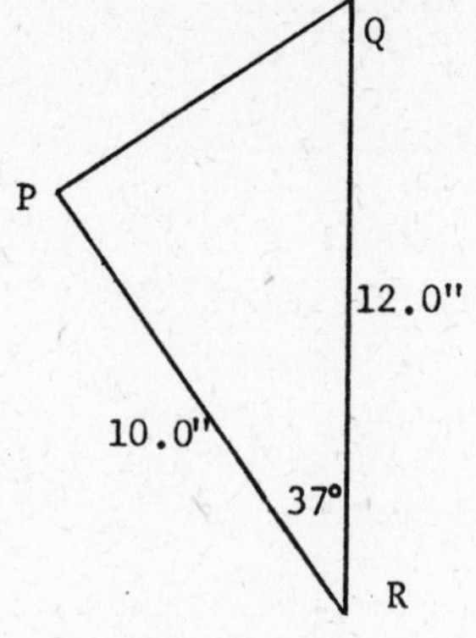

(a) Draw the altitude to QR. The length of this altitude is __________.

(b) The area of oblique triangle PQR is ______________.

a) FH = 27.8"

From: $\frac{FH}{40.0''} = \sin 44°$

b) 695 sq. in.

---

152. In oblique triangle AFT:

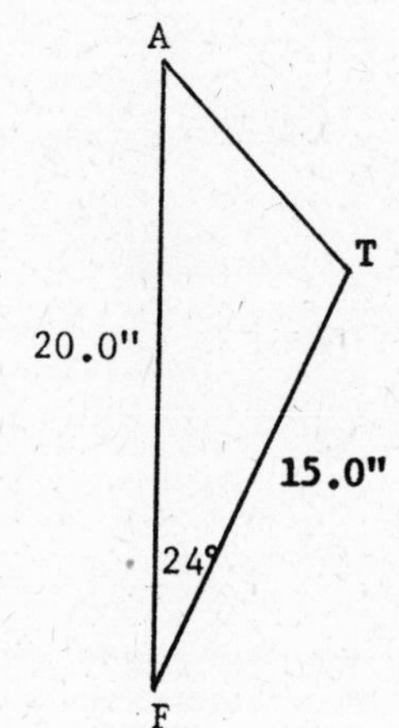

(a) The length of the altitude drawn to AF is __________.

(b) The area of the triangle is ______________.

a) 6.02"

From: $\frac{\text{Altitude}}{10.0''} = \sin 37°$

b) 36.1 sq. in.

---

a) 6.10"

b) 61.0 sq. in.

153. In the oblique triangle on the right, we drew altitude MA to the only known side, VT. In order to find the length of MA, we can use either right triangle MVA or right triangle MAT. However, to use a trig ratio with one of those triangles, we must find the length of either MV or MT.

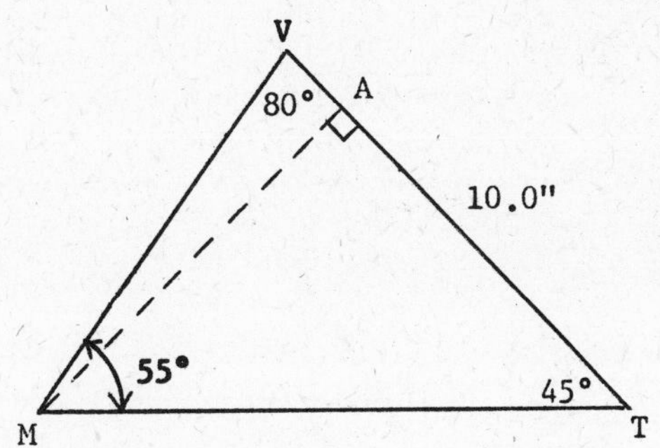

Let's find the length of MV. To do so, we can use the Law of Sines with the original oblique triangle MVT.

$$\frac{MV}{\sin 45°} = \frac{10.0''}{\sin 55°} \quad \text{or} \quad MV = \frac{(10.0'')(\sin 45°)}{\sin 55°} = 8.63 \text{ in.}$$

Now, using right triangle MVA, we can find the length of altitude MA.

(a) How long is MA? ______________

(b) The area of triangle MVT is ______________.

---

a) 8.50 in.

b) 42.5 sq. in.

154. In this oblique triangle, we know two bases. We can compute the area by drawing an altitude to either AC or FC. We have drawn the altitude to AC.

To find the length of this altitude, we must determine the size of angle C.

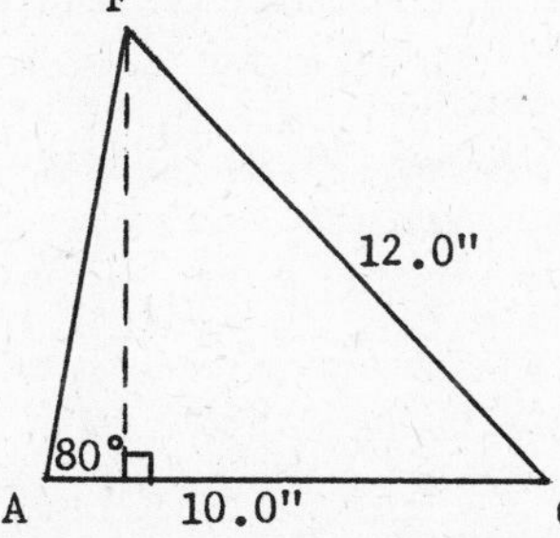

(a) To find the size of angle C, we need two steps. Name them:

(1) ____

(2) ____

(b) Find Angle C. Angle C = ______°.

(c) The length of the altitude to AC is ________.

(d) The area of oblique triangle AFC is ______________.

155. In this oblique triangle, the three sides are known. We want to find the area. As the first step, we have drawn the altitude to side "t".

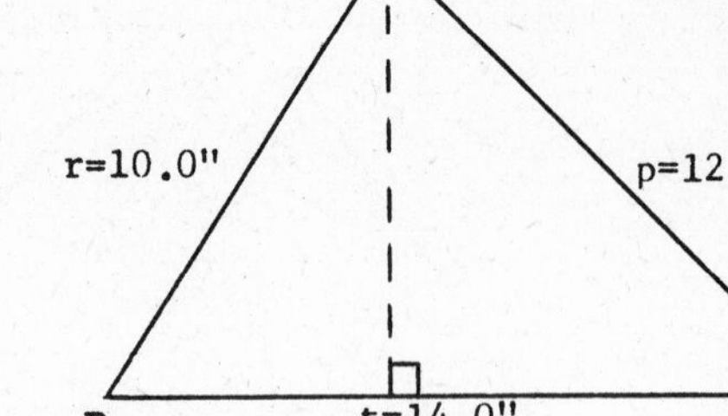

(a) How can we find the size of angle P?

____

____

(b) Find angle P. Angle P = ______°.

(c) The length of the altitude is ________.

(d) The area is ____________.

a) (1) Use the Law of Sines to find angle F in oblique triangle AFC.
(2) Subtract the sum of angles A and F from 180°, to obtain angle C.

b) Angle C = 45°
(Since: Angle F = 55°)

c) 8.48"

d) 42.4 sq. in.

---

156. When a triangle includes an obtuse angle, sometimes a base has to be extended when the altitude is drawn to it. Here is such a case.

Base MT is extended to P. FP is the altitude to base MT.

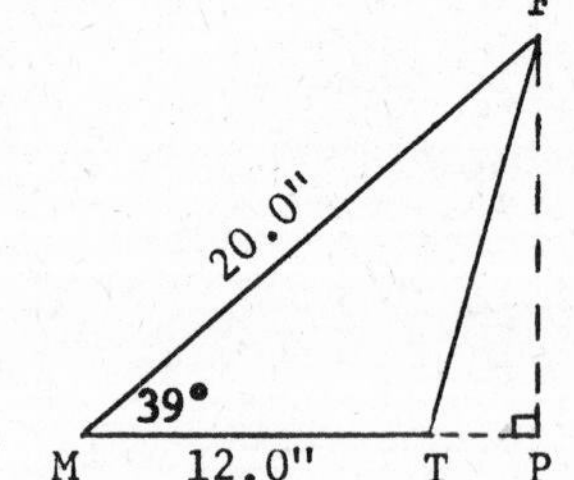

(a) How long is FP? ________

(b) The area of the triangle MFT is ____________.

a) Use the Law of Cosines:

$$p^2 = r^2 + t^2 - 2rt \cos P$$

b) Angle P = 57°
From: $\cos P = 0.543$

c) 8.39"

d) 58.7 sq. in.

---

a) 12.6"
From: $\frac{FP}{20.0''} = \sin 39°$

b) 75.6 sq. in.

157. In this oblique triangle, we have extended base TS and drawn the altitude to it. Angle T contains 25°.

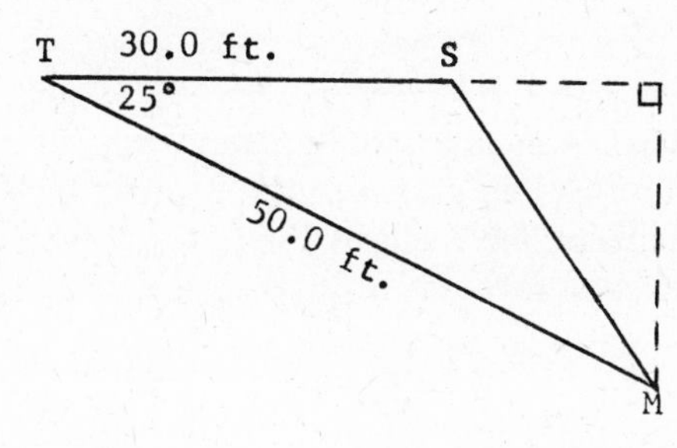

(a) How long is the altitude? __________

(b) The area of triangle TSM is __________.

Answer to Frame 157: a) 21.1 ft. b) 317 sq. ft.

SELF-TEST 11 (Frames 145-157)

Find the area of each of these right triangles.

1. Area = __________

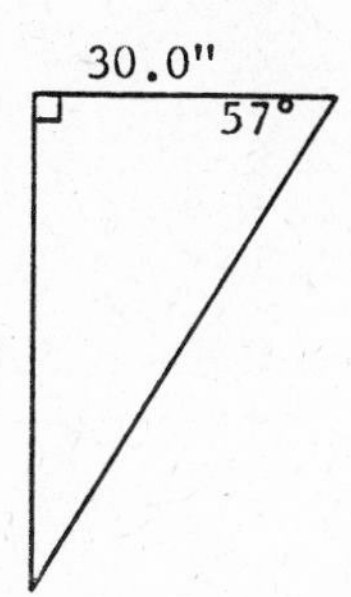

2. Area = __________

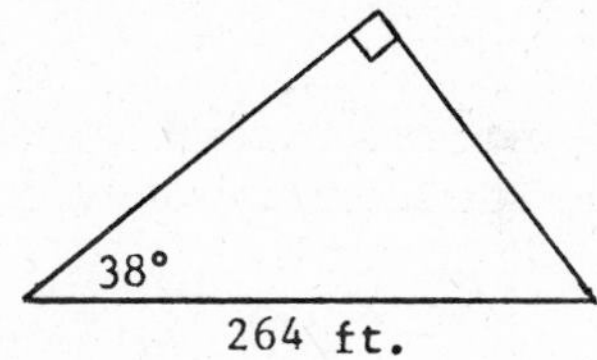

Find the area of each of these oblique triangles.

3. Area = __________

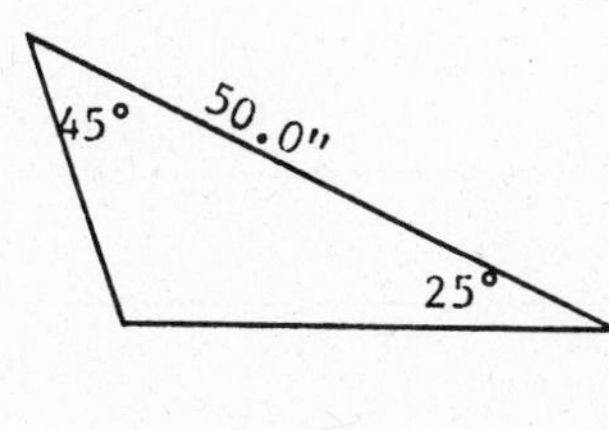

4. Area = __________

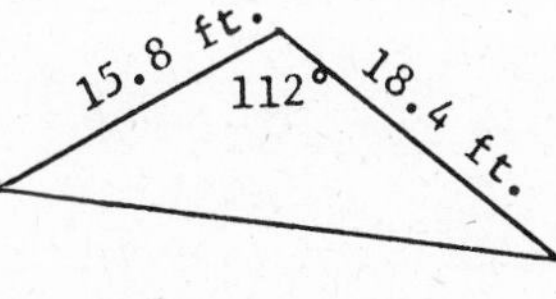

ANSWERS: 1. 693 sq. in. 2. 16,900 sq. ft. 3. 398 sq. in. 4. 135 sq. ft.

5-12 TANGENTS, HALF-TANGENTS AND RELATED ANGLES

In this section, we will define what is meant by a "tangent" and a "half-tangent" of a circle. We will show that a tangent is perpendicular to a radius drawn to the point of tangency. We will discuss the relationship between the angle formed by a tangent and a chord and the central angle of the chord. We will also discuss various angle relationships which are formed by drawing two half-tangents to a circle from the same external point.

158. A tangent to a circle is a line which touches the circle at only one point.

Line BD is a tangent. Point C is called the point of tangency.

Line MN is not a tangent. Why not?

____

____

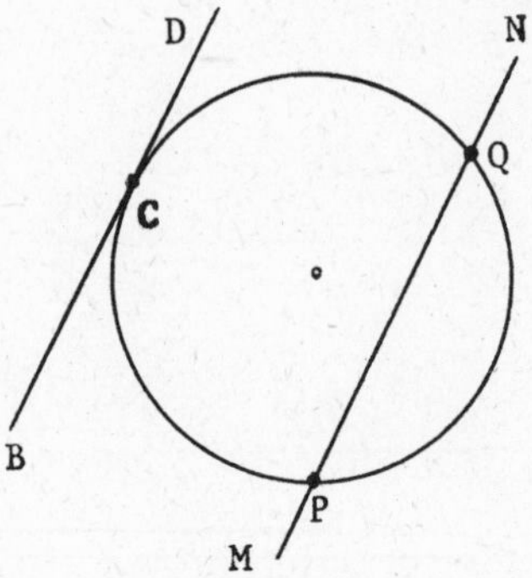

159. If a radius is drawn to the point of tangency, it is perpendicular to the tangent.

In this figure:

MR is a tangent.
P is the point of tangency.
0P is a radius.

Name two right angles:
__________ and __________

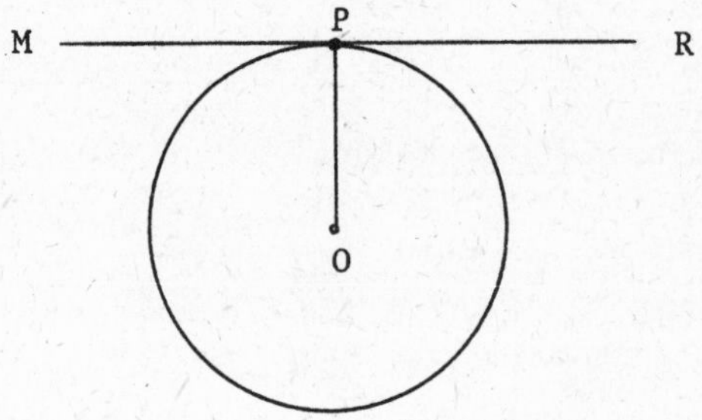

Answer (158): Because it touches the circle at two points, P and Q.

160. In the figure on the right:

(a) Radius 0A is perpendicular to tangent _______.

(b) Radius 0B is perpendicular to tangent _______.

(c) Radius 0C is perpendicular to tangent _______.

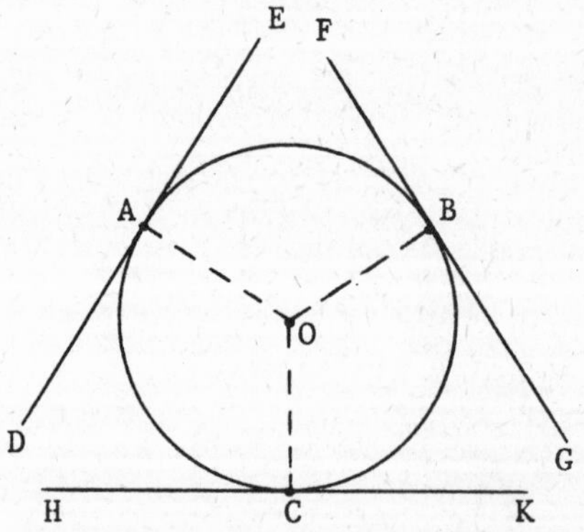

Answer (159): Angles MP0 and 0PR

Answer (160):

a) DE

b) FG

c) HK

161. In this figure, AC is a tangent. Notice that tangent line AC is drawn on both sides of the point of tangency.

DE is called a "half-tangent", since only one-half of the tangent line is drawn.

How large is angle 0ED? _______°

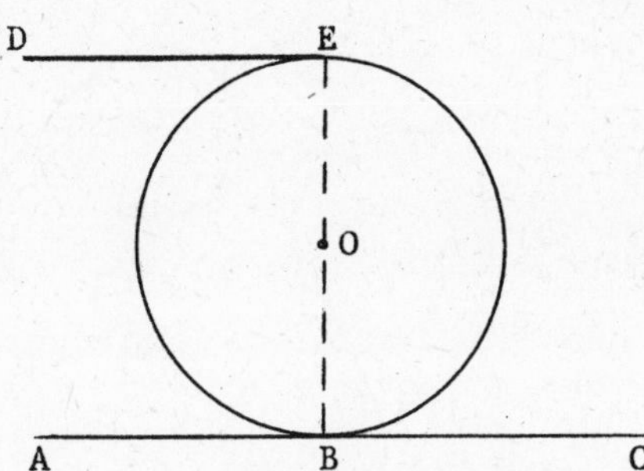

---

162. In practical situations, most of the "tangents" that you meet are really half-tangents.

90°

Two pulleys and a V-belt are shown in the figure on the right.

MP and QT are the straight parts of the V-belt.

MP is a half-tangent

(1) to the large pulley at point P, and
(2) to the small pulley at point M.

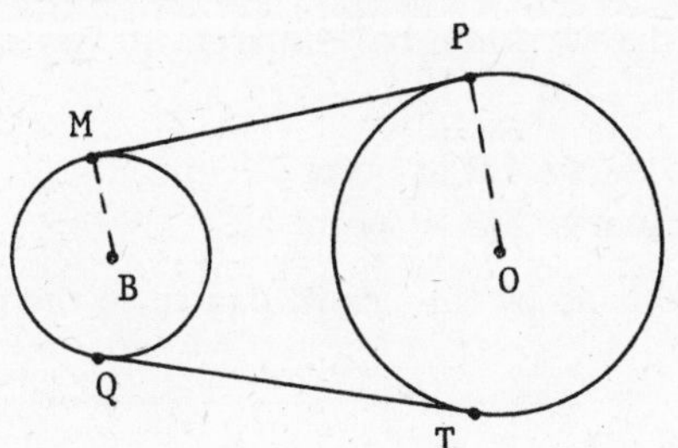

How large are angles MP0 and PMB? _______°

163.

90°

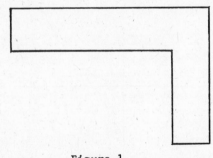

Figure 1

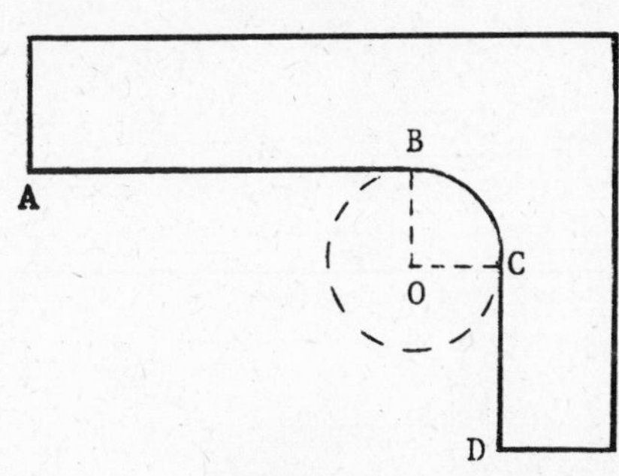

Figure 2

Figure 1 shows an L-shaped metal piece.

Figure 2 shows the same L-shaped metal piece with the inner sharp corner replaced by a circular corner from B to C. This circular corner is commonly called a "radius" or "fillet". A dash-line circle has been drawn at this corner, with radii 0B and 0C drawn to the points of tangency. Note the following:

AB is a half-tangent to the curve BC at point B.
DC is a half-tangent to the curve BC at point C.

Angles AB0 and DC0 each contain _______°.

90°

164. A triangular steel plate with circular corners is shown at the right. In practical work, curved corners are usually <u>circular</u>.

MN, PT, and LR are circular curves.

LM, NT, and PR are straight lines.

The complete circles are shown in dashed lines, and two radii have been drawn in each circle.

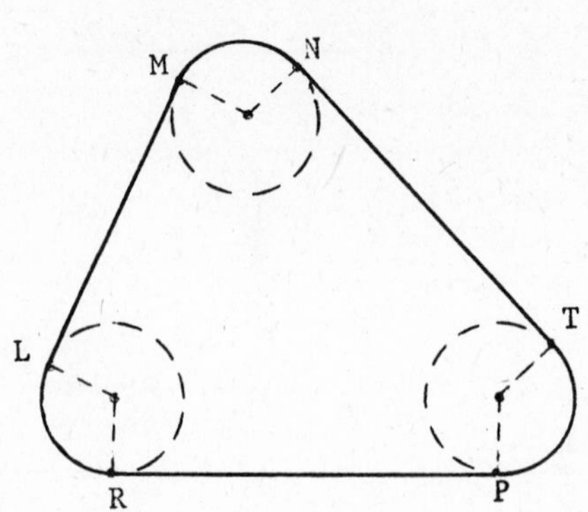

Each straight line on the drawing is a half-tangent to two circular corners. For example:

LM is tangent (1) to curve LR at point L, and
(2) to curve MN at point M.

Angles of what size are formed by the radii drawn to the points of tangency? _______°

---

165. In the figure on the right:

CT is a tangent to the circle.
Point F is the point of tangency.
FS is a chord of the circle.

(a) Angle 0FT = _______°

(b) Angle SFT = 44°
Therefore, angle 0FS = _______°

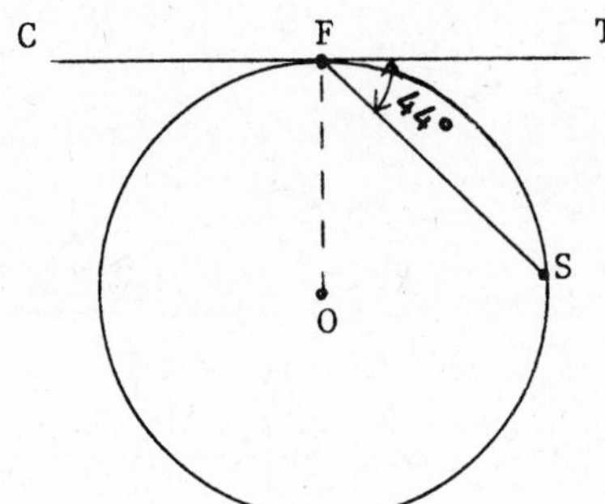

90° (right angles)

---

166. In the figure on the right:

MD is a half-tangent of the circle.
Point D is the point of tangency.
SD is a chord of the circle.

(a) Angle 0DM contains _______°.

(b) If angle SDM (the angle formed by the half-tangent and the chord) contains 30°, angle 0DS contains _______°.

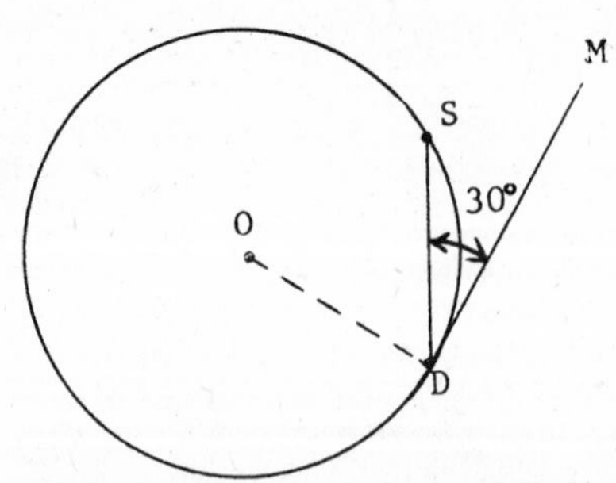

a) 90°
b) 46°

---

a) 90°
b) 60°

167. In Figure 1 below, AB is a chord of the circle. Since 0A and 0B are radii:

(1) Triangle 0AB is an isosceles triangle.
(2) Angles #1 and #2 are equal since they are opposite the two equal sides of the triangle.

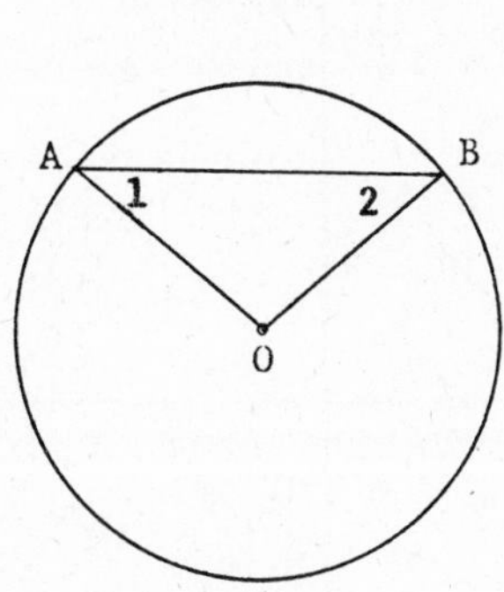

Figure 1

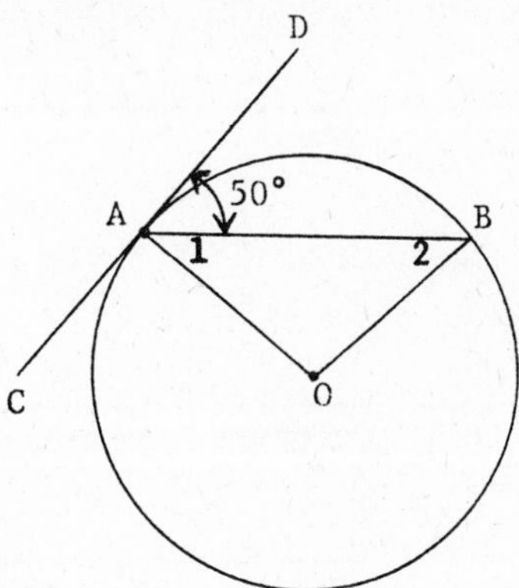

Figure 2

In Figure 2, we have drawn the same triangle. However, we have added tangent CD whose point of tangency is at A. Angle DAB, which is formed by the tangent and chord AB, contains 50°.

(a) Angle 0AD contains ______°.

(b) Angle #1 contains ______°.

(c) Angle #2 contains ______°.

(d) Central angle A0B contains ______°.

(e) How large is angle DAB compared to central angle A0B? ______________________

---

168. In the figure at the right:

MF is a chord. Angle M0F is its central angle.

(a) Angle 0MA contains ______°.

(b) Why are angles #1 and #2 equal? ______________________

(c) Angle #1 and angle #2 each contain ______°.

(d) How large is central angle M0F? ______°

(e) How large is central angle M0F compared to angle AMF (the angle formed by the tangent and the chord)? ______________________

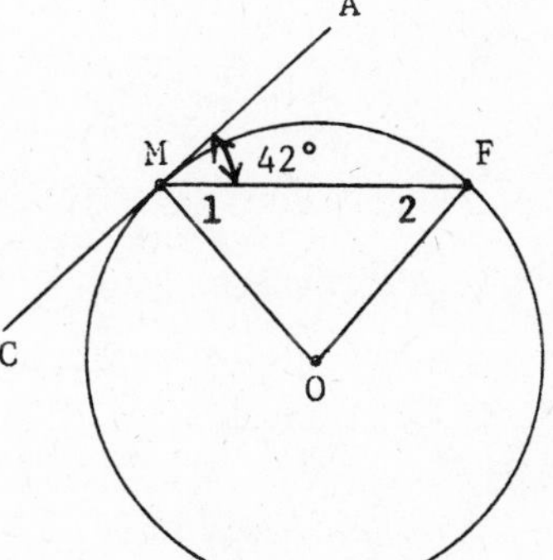

a) 90°
b) 40°
c) 40°
d) 100°
e) One-half as large.

169. The point of the last few frames was to show this fact:

> The angle formed by a tangent and a chord is $\frac{1}{2}$ as large as the central angle of the chord.

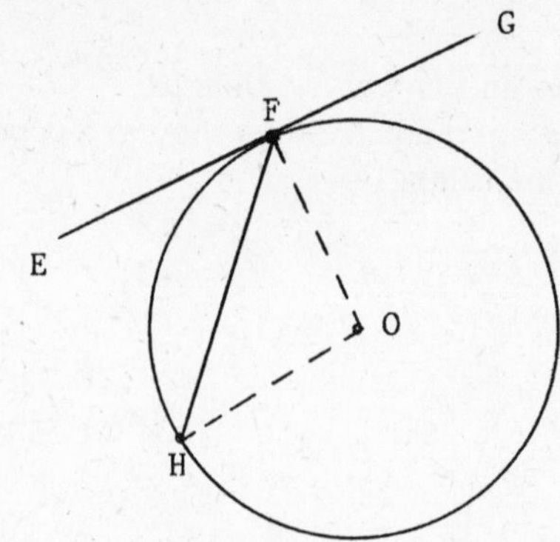

(a) If central angle F0H contains 96°, how large is angle EFH? ______°

(b) If angle EFH contains 46°, how large is central angle F0H? ______°

a) 90°

b) Because they are opposite the equal sides (radii) in an isosceles triangle.

c) 48°

d) 84°

e) It is twice as large.

---

170. In the figure at the right, 0E is perpendicular to chord BD.

How many degrees are there in the following angles:

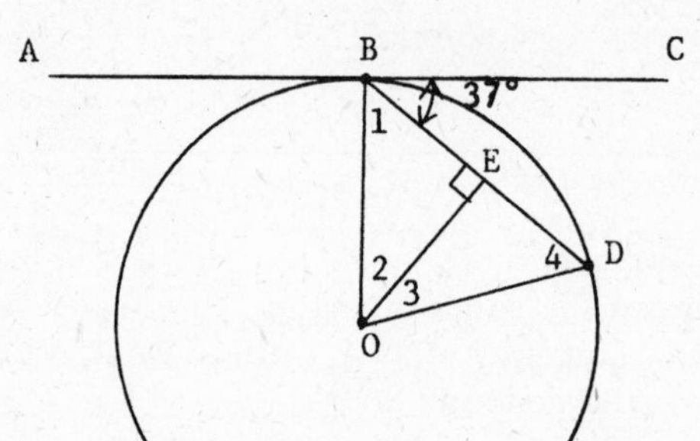

(a) Angle 0BC = ______°

(b) Angle #1 = ______°

(c) Angle #2 = ______°

(d) Angle #3 = ______°

(e) Angle #4 = ______°

(f) Central angle B0D = ______°

a) 48°

b) 92°

---

171. In the circle at the right, 0M is perpendicular to chord VS, angle VTS = 90°, and VT is tangent to the circle at point V.

If central angle V0S = 58°, find the size of each of the following angles:

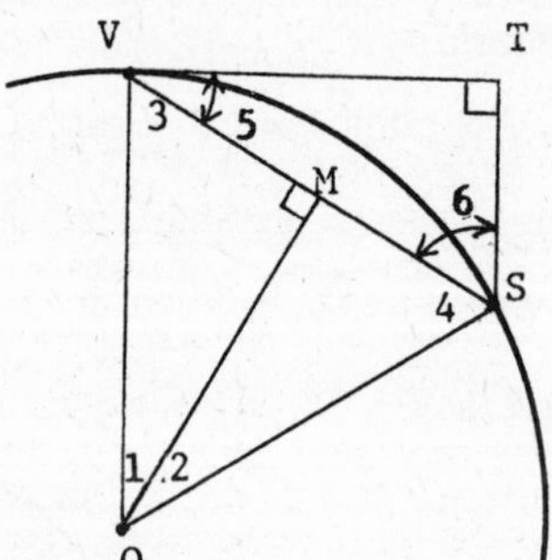

(a) Angle #1 = ______°

(b) Angle #2 = ______°

(c) Angle #3 = ______°

(d) Angle #4 = ______°

(e) Angle #5 = ______°

(f) Angle #6 = ______°

a) 90°

b) 53°

c) 37°

d) 37°

e) 53°

f) 74°

172. In the figure on the right:

CD and CM are half-tangents to the circle from the same external point.

CO connects the external point and the center of the circle.

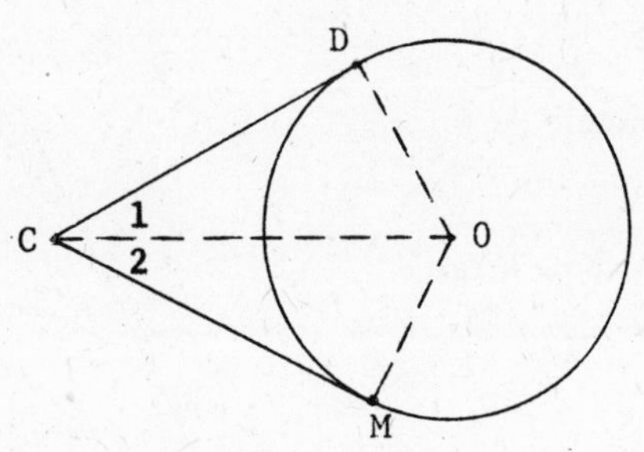

(a) 0D and 0M are ___________ of the circle.

(b) Angles CD0 and CM0 each contain ______°.

(c) Triangles CD0 and CM0 are both ___________ triangles.

(d) The hypotenuse of both right triangles is _________.

a) 29° d) 61°
b) 29° e) 29°
c) 61° f) 61°

---

173. DM and DV are half-tangents to the circle from the same external point. Whenever a line is drawn from the external point to the center of the circle, it bisects the external angle. That is:

Angle #1 equals angle #2.

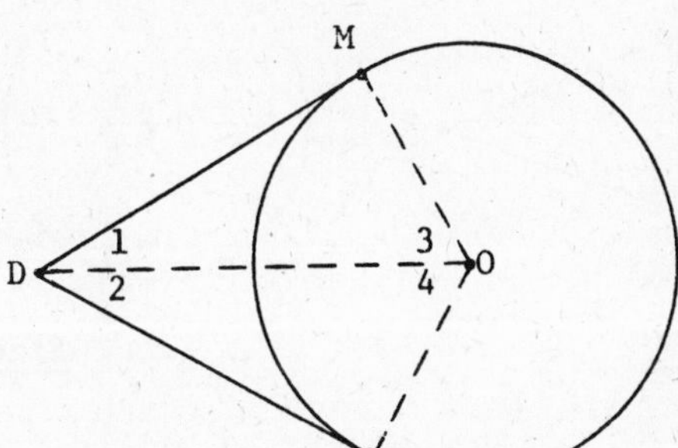

(a) Name two right triangles.
_______ and _______

(b) If angles #1 and #2 each contain 30°, angles #3 and #4 each contain ______°.

a) radii
b) 90°
c) right
d) C0

---

174. If angle #1 contains 35°, how large are the following angles?

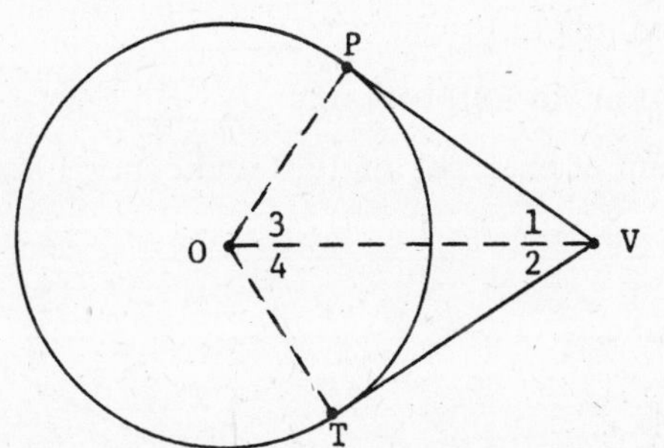

(a) Angle #2 = ______°
(b) Angle #3 = ______°
(c) Angle #4 = ______°
(d) Angle PVT = ______°
(e) Central angle P0T = ______°

a) DM0 and DV0
b) 60°

---

a) 35°
b) 55°
c) 55°
d) 70°
e) 110°

175. (a) The hypotenuse of triangle AB0 is _______.

(b) The hypotenuse of triangle AC0 is _______.

(c) If angle #2 equals 25°, how large is central angle B0C? _______

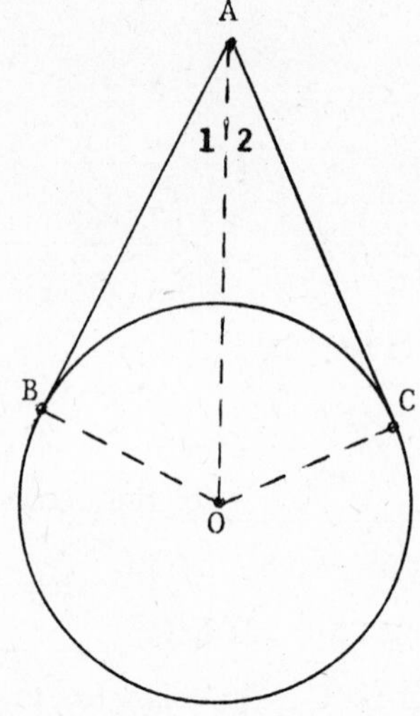

---

a) A0

b) A0

c) 130°

176.

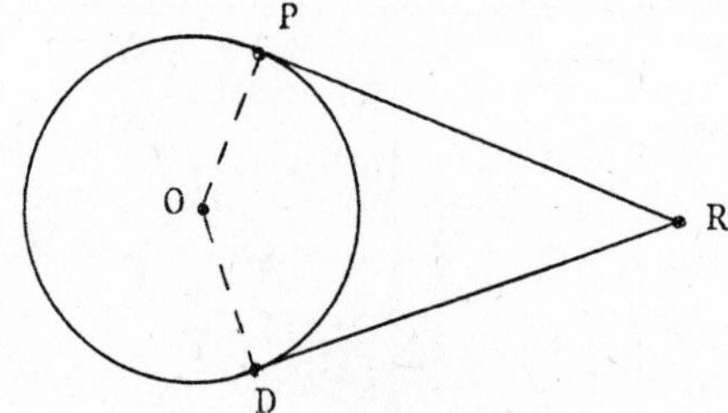

Figure 1

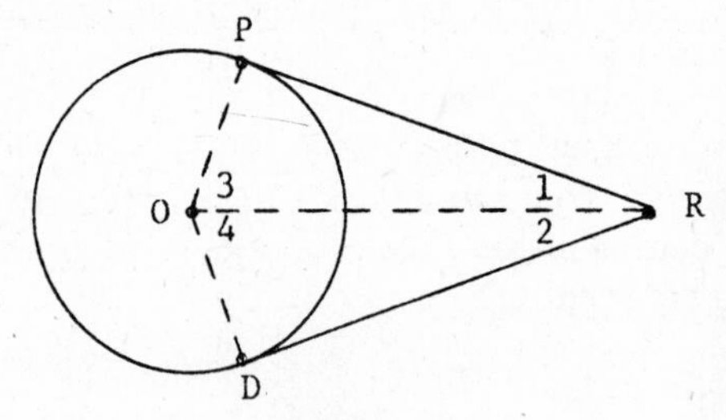

Figure 2

We want to show that the sum of external angle R and central angle P0D in Figure 1 is 180°. To do so, we have drawn 0R in Figure 2. Angle #1 = Angle #2 = 20°.

Since angles #1 and #2 each contain 20°, angles #3 and #4 each contain 70°.

Therefore: (a) External angle R contains _______°.

(b) Central angle P0D contains _______°.

(c) The sum of external angle R and central angle P0D is _________°.

---

a) 40°

b) 140°

c) 180°

---

177. The point of the last frame was to show you that the sum of external angle M and central angle N0A is 180°.

(a) If angle M contains 42°, angle N0A contains _________°.

(b) If angle N0A contains 136°, angle M contains _______°.

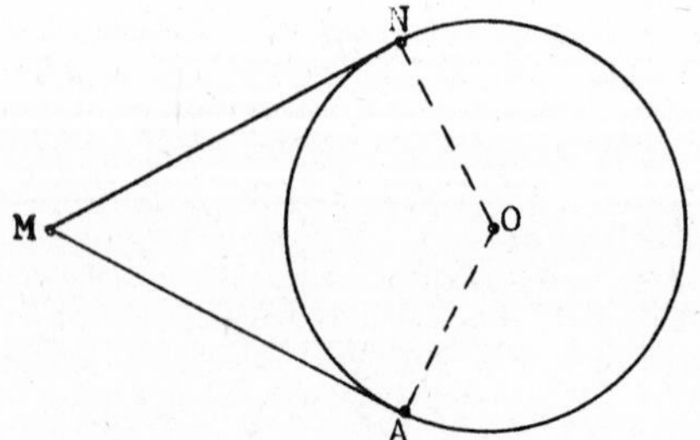

---

a) 138°

b) 44°

178. In the figure on the right, two half-tangents are drawn from point A to points B and C. The chord from B to C is drawn. The following two facts about A0 and BC are true:

(1) A0 is perpendicular to BC.
(2) A0 bisects BC.

(a) Since A0 is perpendicular to BC, angles ADB and ADC are ____________ angles.

(b) Triangles ADB and ADC are ____________ triangles.

(c) If angles #1 and #2 each contain 25°, angles #3 and #4 each contain ________°.

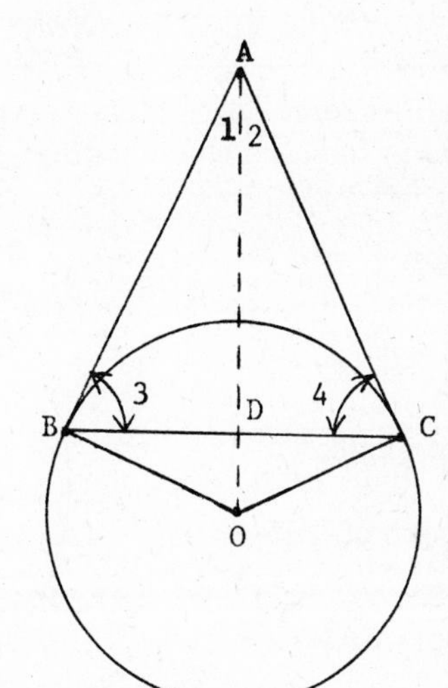

179. In the figure at the right, TM and TS are half-tangents. Also, angle #1 = 40°.

(a) Name six right triangles in the figure:

_____ _____ _____

_____ _____ _____

(b) Angle #2 = ______°

(c) Angle #3 = ______°

(d) Angle #4 = ______°

(e) Angle #5 = ______°

(f) Angle #6 = ______°

(g) Angle #7 = ______°

(h) Angle #8 = ______°

a) right
b) right
c) 65°

Answer to Frame 179: a) 0MT and 0ST, 0MV and 0SV, MVT and SVT b) 40° c) 50° d) 50° e) 40° f) 40° g) 50° h) 50°

SELF-TEST 12 (Frames 158-179)

In the figure at the right, GB and GA are half-tangents. Central angle A0B is 136°.

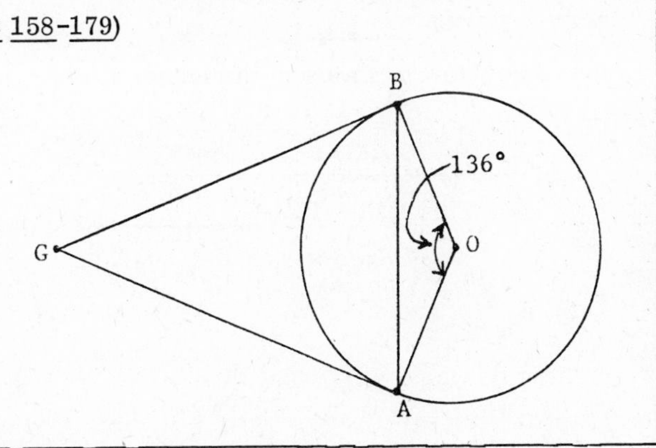

Find the following angles:

1. Angle 0AB = ______°
2. Angle 0BA = ______°
3. Angle GBA = ______°
4. Angle GAB = ______°
5. Angle AGB = ______°

ANSWERS: 1. 22° 2. 22° 3. 68° 4. 68° 5. 44°

5-13 APPLIED PROBLEMS INVOLVING TANGENTS AND HALF-TANGENTS

In the last section, we examined some angle facts related to tangents and half-tangents of circles. In this section, we will use these facts to solve some applied problems.

180. In the figure on the right:

AC is a half-tangent.
Central angle A0B = 64°.
Chord AB = 20.0".
Angle ACB is a right angle.

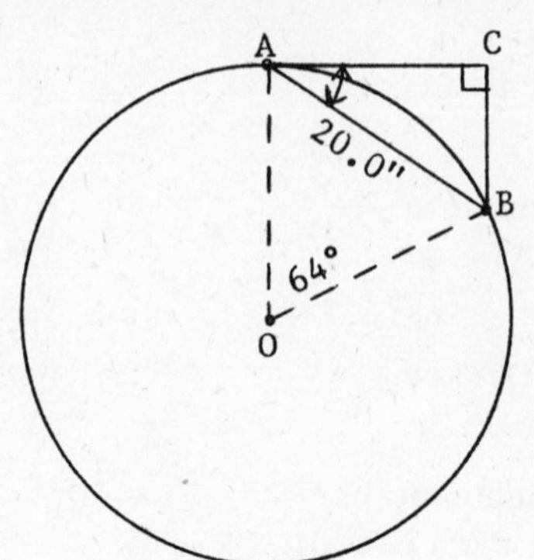

We want to find the length of half-tangent AC.

(a) How large is angle CAB? ________°

(b) Using right triangle ACB, what trig ratio can be used to find the length of half-tangent AC? ____________

(c) Find AC. AC = ________

181. In the figure on the right:

The radius of the circle is 10.0".
The angle formed by half-tangent DT and chord DM is 50°.

We want to find the length of chord DM.

(a) How large is central angle D0M? ________°

(b) How large are angles #1 and #2? ________°

(c) How can we find the length of DM? ____________________________

(d) Find DM. DM = ________

a) 32° It is half of central angle A0B.

b) Cosine ratio

c) AC = 17.0"

From: $\frac{AC}{20.0''} = \cos 32°$

182. In the figure on the right:

Chord ST is 15.0". The angle formed by half-tangent TR and chord ST is 38°.

We want to find the radius of the circle.

(a) How large are angles #1 and #2? __________

(b) How large is central angle T0S? __________

(c) How can we find the length of 0T?

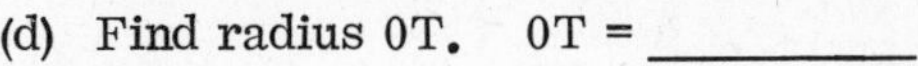

(d) Find radius 0T. 0T = __________

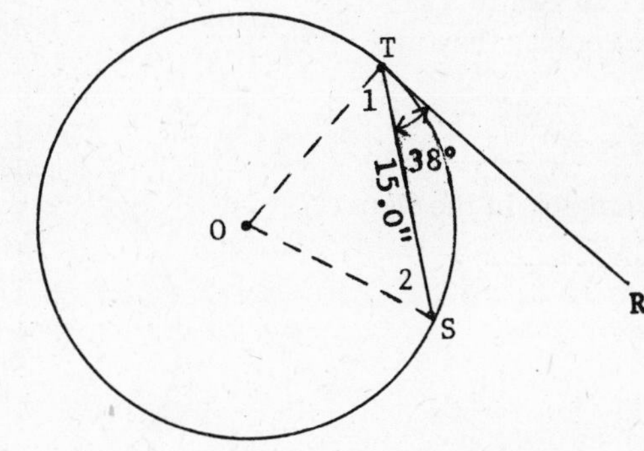

a) 100° It is twice as large as angle TDM.

b) 40°

c) Use the Law of Sines in oblique triangle D0M. Or, draw a perpendicular from center 0 to chord DM, and then solve one of the resulting right triangles for one-half of chord DM.

d) DM = 15.3"

---

183. In the figure on the right:

AB is a half-tangent.
Angle ABC is 75°.
The radius of the circle is 10.0".

(a) The circumference of the circle is __________.

(b) How large is central angle C0B? __________

(c) How many arc-degrees are in $\overset{\frown}{CB}$? __________

(d) How long is $\overset{\frown}{CB}$? __________

(e) The area of the circle is __________.

(f) The area of sector C0B is __________.

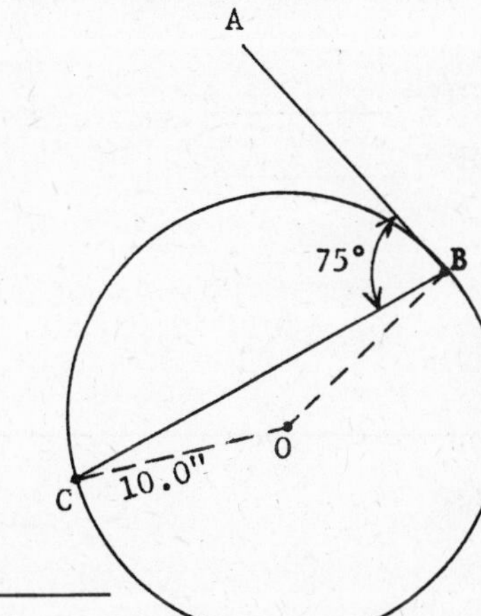

a) 52°

b) 76°

c) Use the Law of Sines in oblique triangle T0S. Or, draw a perpendicular from center 0 to chord ST, and then solve one of the resulting right triangles for the radius.

d) Radius 0T = 12.2"

184. If two half-tangents are drawn to a circle from the same external point, they are equal. Therefore:

(a) If PR = 7.5", PS = ________.

(b) Name two right triangles in the figure.
________ and ________

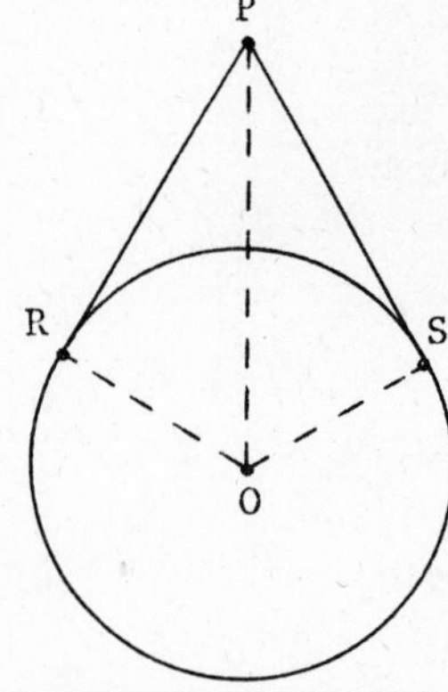

a) 62.8"

b) 150°

c) 150 arc-degrees

d) 26.2"
From: $\frac{150}{360}(62.8)$

e) 314 sq. in.

f) 131 sq. in.
From: $\frac{150}{360}(314)$

---

185. In the figure at the right:

PA and PB are half-tangents.
Radius 0A = 5.00".
Angle AP0 = 25°.

Calculate the following:

(a) How long is AP? AP = ________

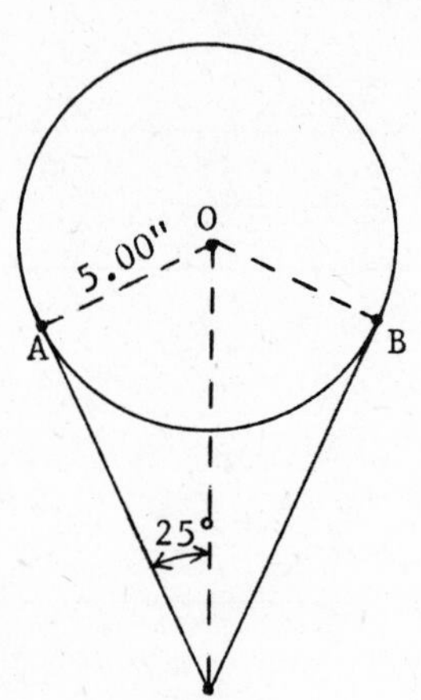

(b) How long is BP? BP = ________

(c) How long is 0P? 0P = ________

a) 7.5"

b) PR0 and PS0

---

186. In the figure at the right:

Radius 0C = 15.0".
Half-tangent CM = 25.0".

Calculate the following:

(a) Angle #1 = ________

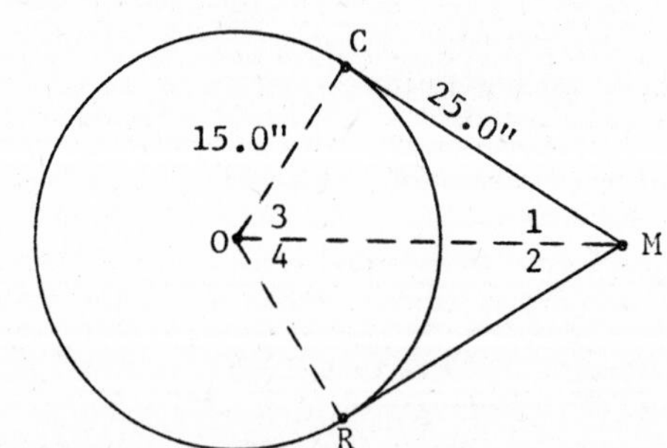

(b) Angle #2 = ________

(c) Angle #3 = ________

(d) Angle #4 = ________

(e) Distance 0M = ________

a) AP = 10.7"
From: $\frac{5.00''}{AP} = \tan 25°$

b) BP = 10.7"

c) 0P = 11.8"
From: $\frac{5.00''}{0P} = \sin 25°$

187. In the figure, AB and AC are half-tangents. Also, external angle A contains 50°, and radius 0C is 10.0".

(a) How large is central angle B0C? ________

(b) The circumference of the circle is ________.

(c) The length of $\widehat{BC}$ is ________.

(d) The area of the circle is ________.

(e) The area of sector 0BC is ________.

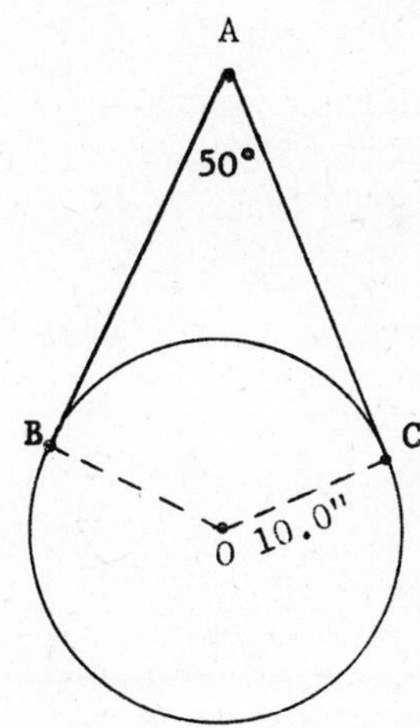

a) 31°

From: tan #1 $= \frac{15.0}{25.0}$

$= 0.600$

b) 31°

c) 59°

d) 59°

e) 0M = 29.2"

---

188. ABCD is a steel strap. It is bent so that BC is a circular arc.

AB and CD are each 10.00" long. The radius of the circular arc is 1.00". Also, external angle BVC = 60°.

(a) How large is central angle B0C? ________

(b) The circumference of the circle is ________.

(c) The length of $\widehat{BC}$ is ________.

(d) The total length of the steel strap is ________.

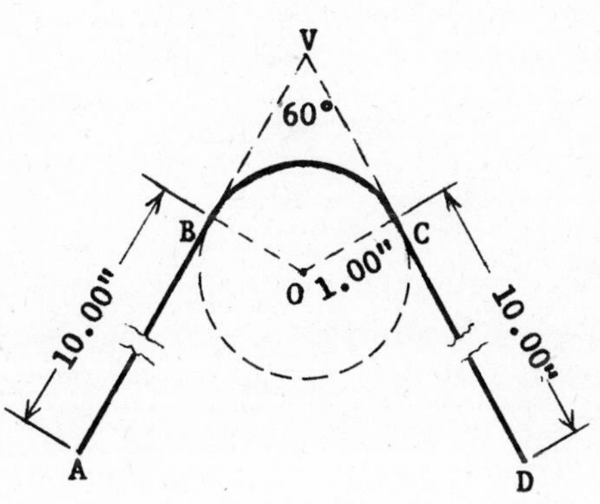

a) 130°

b) 62.8 in.

c) 22.7 in.

d) 314 sq. in.

e) 113 sq. in.

189. Here is another steel strap.

PF is a circular arc.
The radius of the circular arc is 2.00".
External angle PRF contains 95°.
MP is 5.00" and CF is 7.00".

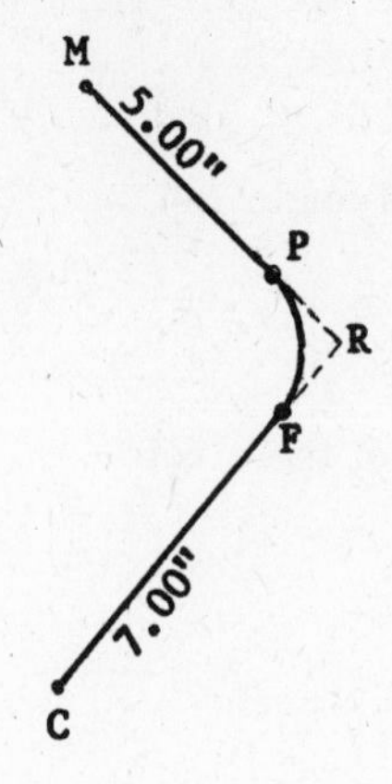

(a) The length of $\widehat{PF}$ is ________.

(b) The total length of the steel strap is ________.

a) 120°
b) 6.28"
c) 2.09"
d) 22.09"

Answer to Frame 189: a) 2.96" b) 14.96"

SELF-TEST 13 (Frames 180-189)

In the figure at the right:

PD and PE are half-tangents.
The radius of the circle is 2.00".
External angle DPE = 76°.

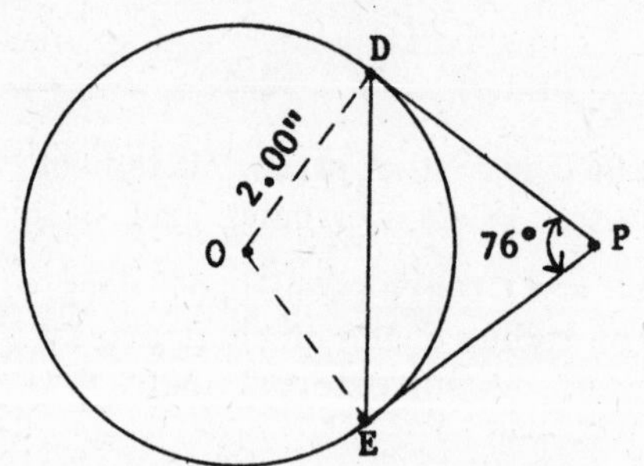

Find the following angles:

1. Angle PDE = ______ 2. Angle DOE = ______

Find the following lengths:

3. Half-tangent PD = ________ 4. Chord DE = ________ 5. Distance OP = ________ 6. Arc-length $\widehat{DE}$ = ________

ANSWERS:

1. 52°
2. 104°
3. 2.56"
4. 3.15"
5. 3.25"
6. 3.63"

Chapter 6 IDENTITIES AND INTERPOLATION

The two major topics covered in this chapter are identities and interpolation. Though the eight basic identities are discussed and used to show the process of verifying more-complicated identities, the treatment of identities is not exhaustive. After introducing the principles of interpolation in the context of interpolating from a degree table to minutes, these principles are extended to interpolations from a degree-minute table to seconds, and from a degree table to decimal subdivisions of a degree.

The chapter also includes sections on the following topics: (1) definitions of the three remaining trigonometric ratios - the cosecant, secant, and cotangent of an angle, (2) solving trigonometric equations, and (3) inverse trigonometric notation.

6-1 DEFINITIONS OF THE COSECANT, SECANT, AND COTANGENT OF AN ANGLE

There are six trigonometric ratios for any angle. Up to this point, we have defined and used only the three most basic ones - the sine, cosine, and tangent of an angle. In this section, we will define the other three ratios - the cosecant, secant, and cotangent of an angle. We will show that these other ratios are the reciprocals of the sine, cosine, and tangent of an angle.

---

1. We have seen that two quantities are a pair of reciprocals if their product is +1. That is:

   "8" and "$\frac{1}{8}$" are a pair of reciprocals, since $(8)\left(\frac{1}{8}\right) = 1$

   "$\frac{3}{2}$" and "$\frac{2}{3}$" are a pair of reciprocals, since $\left(\frac{3}{2}\right)\left(\frac{2}{3}\right) = 1$

   Write the reciprocal of each of the following:

   (a) $\frac{2}{9}$ ______ (b) $\frac{5}{d}$ ______ (c) $\frac{m}{q}$ ______

---

a) $\frac{9}{2}$

b) $\frac{d}{5}$

c) $\frac{q}{m}$

2. Trigonometric ratios are comparisons of the lengths of the sides of a right triangle. Since six comparisons of this type are possible, there are six trigonometric ratios. Up to this point we have defined and used only three of them - the sine, cosine, and tangent of an angle.

Using the triangle on the right, we have defined the six trigonometric ratios below. The three familiar ratios are defined on the left. The three new ratios are defined on the right.

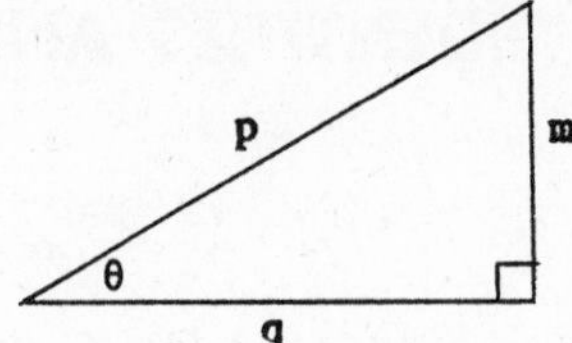

$$\text{sine } \theta = \frac{\text{side opposite}}{\text{hypotenuse}} = \frac{m}{p} \qquad \text{cosecant } \theta = \frac{\text{hypotenuse}}{\text{side opposite}} = \frac{p}{m}$$

$$\text{cosine } \theta = \frac{\text{side adjacent}}{\text{hypotenuse}} = \frac{q}{p} \qquad \text{secant } \theta = \frac{\text{hypotenuse}}{\text{side adjacent}} = \frac{p}{q}$$

$$\text{tangent } \theta = \frac{\text{side opposite}}{\text{side adjacent}} = \frac{m}{q} \qquad \text{cotangent } \theta = \frac{\text{side adjacent}}{\text{side opposite}} = \frac{q}{m}$$

As you can see from the definitions above, "sine θ" and "cosecant θ" are a pair of reciprocals, since:

$$\text{sine } \theta = \frac{m}{p}, \quad \text{cosecant } \theta = \frac{p}{m}, \quad \text{and} \quad \left(\frac{m}{p}\right)\left(\frac{p}{m}\right) = +1$$

Of the three new ratios, which one is the reciprocal of:

(a) cosine θ ______________ (b) tangent θ ______________

Answers: a) secant θ b) cotangent θ

---

3. Notice carefully which pairs of ratios are reciprocals.

sine θ	←--------→	cosecant θ
cosine θ	←--------→	secant θ
tangent θ	←--------→	cotangent θ

Even though cosine θ and cosecant θ both begin with the letters "co", are they a pair of reciprocals? __________

Answer: No

---

4. Write the names of the ratios which are the reciprocals of each of the following:

(a) sine θ ______________ (d) cotangent θ ______________

(b) tangent θ ______________ (e) cosine θ ______________

(c) secant θ ______________ (f) cosecant θ ______________

Answers:
a) cosecant θ
b) cotangent θ
c) cosine θ

d) tangent θ
e) secant θ
f) sine θ

5. In terms of "side opposite", "side adjacent", and "hypotenuse", define each of the following:

(a) cosecant θ = ____________________

(b) secant θ = ____________________

(c) cotangent θ = ____________________

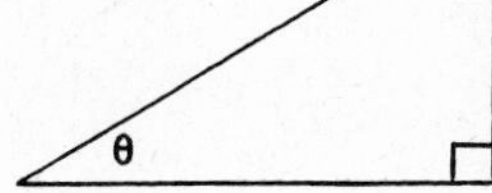

---

6. In this right triangle, define:

(a) sine φ = ____

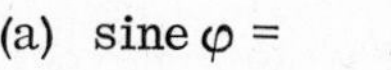

(b) cosecant φ = ____

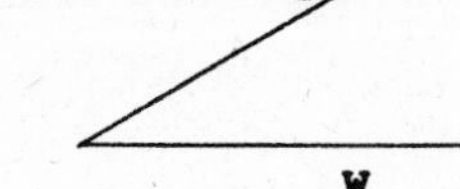

(c) cosine φ = ____

(d) secant φ = ____

(e) tangent φ = ____

(f) cotangent φ = ____

a) $\frac{\text{hypotenuse}}{\text{side opposite}}$

b) $\frac{\text{hypotenuse}}{\text{side adjacent}}$

c) $\frac{\text{side adjacent}}{\text{side opposite}}$

---

7. Just as there are abbreviations for sine (sin), cosine (cos), and tangent (tan), there are abbreviations for cosecant, secant, and cotangent.

Full Name	Abbreviation
cosecant θ	csc θ
secant θ	sec θ
cotangent θ	cot θ

In this right triangle:

(a) csc γ = ____

(b) sec γ = ____

(c) cot γ = ____

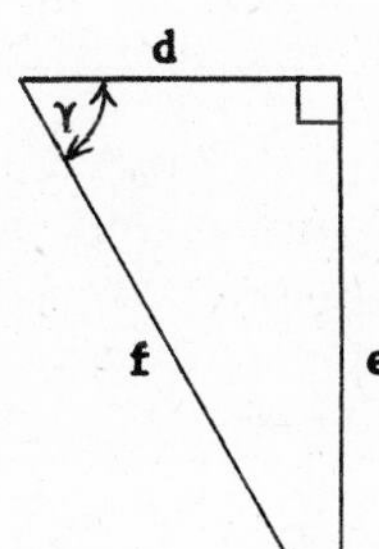

a) $\frac{w}{s}$

b) $\frac{s}{w}$

c) $\frac{t}{s}$

d) $\frac{s}{t}$

e) $\frac{w}{t}$

f) $\frac{t}{w}$

---

a) $\frac{f}{e}$

b) $\frac{f}{d}$

c) $\frac{d}{e}$

8. In this right triangle:

(a) $\cot \varphi =$ ____

(b) $\sec \varphi =$ ____

(c) $\csc \varphi =$ ____

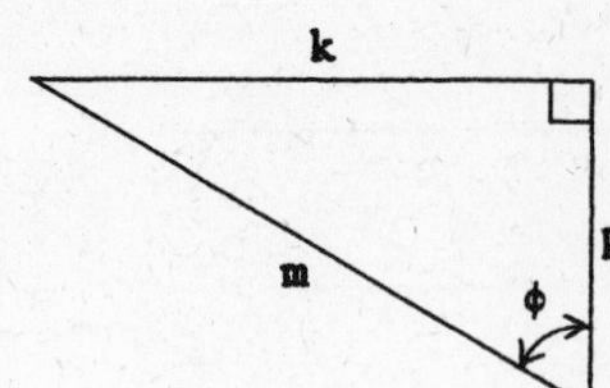

9. In this right triangle:

(a) $\sec A =$ ____

(b) $\cot A =$ ____

(c) $\csc A =$ ____

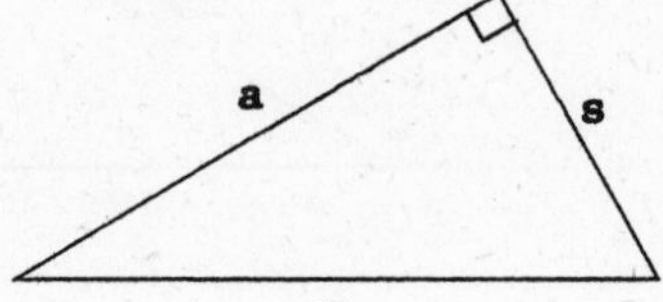

a) $\frac{p}{k}$

b) $\frac{m}{p}$

c) $\frac{m}{k}$

10. In each blank, write in the appropriate reciprocal trig ratio in abbreviated form:

(a) If $\sin \theta = \frac{3}{5}$, then ________ $= \frac{5}{3}$

(b) If $\tan \theta = \frac{7}{4}$, then ________ $= \frac{4}{7}$

(c) If $\cos \theta = \frac{5}{8}$, then ________ $= \frac{8}{5}$

(d) If $\csc \theta = \frac{4}{3}$, then ________ $= \frac{3}{4}$

a) $\frac{w}{s}$

b) $\frac{s}{a}$

c) $\frac{w}{a}$

a) $\csc \theta$

b) $\cot \theta$

c) $\sec \theta$

d) $\sin \theta$

SELF-TEST 1 (Frames 1-10)

Name the trig ratio which is the reciprocal of:

1. $\tan\theta$ ________ 2. $\sin\theta$ ________ 3. $\cos\theta$ ________

Using the right triangle shown, define the following ratios in terms of sides "b", "h", and "p":

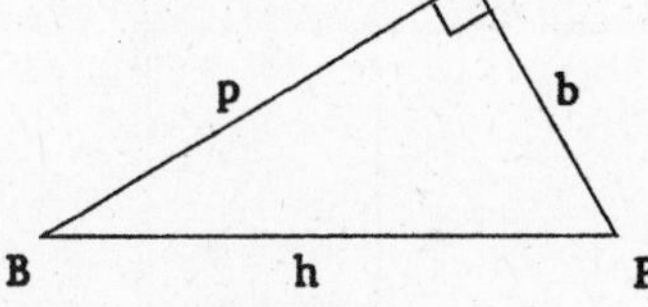

4. sin B = ____	7. cot B = ____	10. sec P = ____
5. cos B = ____	8. sec B = ____	11. csc P = ____
6. tan B = ____	9. csc B = ____	12. cot P = ____

ANSWERS:

1. $\cot\theta$	4. $\frac{b}{h}$	7. $\frac{p}{b}$	10. $\frac{h}{b}$
2. $\csc\theta$	5. $\frac{p}{h}$	8. $\frac{h}{p}$	11. $\frac{h}{p}$
3. $\sec\theta$	6. $\frac{b}{p}$	9. $\frac{h}{b}$	12. $\frac{b}{p}$

6-2 FINDING THE COSECANT, SECANT, AND COTANGENT OF AN ANGLE

The cosecant, secant, and cotangent of an angle were defined in the last section. In this section, we will find the numerical values of the cosecant, secant, and cotangent of angles of various sizes. We will also find the size of angles when the numerical values of their cosecants, secants, or cotangents are given. Though tables of cosecants, secants, and cotangents are available, we will emphasize the use of the reciprocal relationships to solve these types of problems.

11. Tables of values of the cosecants, secants, and cotangents of angles are available. Here is part of such a table. (The sines, cosines, and tangents of the angles are also included.)

Angle θ	sin θ	cos θ	tan θ	csc θ	sec θ	cot θ
-----	------	------	------	-----	-----	-----
31°	0.5150	0.8572	0.6009	1.942	1.167	1.664
32°	0.5299	0.8480	0.6249	1.887	1.179	1.600
33°	0.5446	0.8387	0.6494	1.836	1.192	1.540
34°	0.5592	0.8290	0.6745	1.788	1.206	1.483
35°	0.5736	0.8192	0.7002	1.743	1.221	1.428
-----	------	------	------	-----	-----	-----

We can use the values in the table to verify the fact that pairs of these ratios are reciprocals. (Of course, since the values in the table are rounded to four significant digits and since we round further to use the slide rule, we do not necessarily get exactly "+1" as the product.)

$$(\sin 31°)(\csc 31°) = (0.5150)(1.942) = 1.00$$

$$(\cos 33°)(\sec 33°) = (0.8387)(1.192) = 1.00$$

$$(\tan 35°)(\cot 35°) = (0.7002)(1.428) = 1.00$$

Using the table above, find these angles:

(a) If $\csc \theta = 1.788$, $\theta =$ _______°

(b) If $\cot \theta = 1.540$, $\theta =$ _______°

---

a) 34° b) 33°

---

12. Though tables of values for cosecants, secants, and cotangents are available, we can also find their numerical values for any angle by using the "reciprocal" relationships and tables of values of sines, cosines, and tangents. For example:

csc 30° = the reciprocal of sin 30°

sec 45° = the reciprocal of cos 45°

cot 60° = the reciprocal of tan 60°

To use the reciprocal relationships, we must be able to find the reciprocals of numbers like 0.7892 or 1.732.

Since $(0.7892)\left(\frac{1}{0.7892}\right) = \frac{0.7892}{0.7892} = 1$,

the reciprocal of 0.7892 is $\frac{1}{0.7892}$, and

the reciprocal of $\frac{1}{0.7892}$ is 0.7892.

Write the <u>reciprocal</u> of each of the following:

(a) 1.732 _______

(b) $\frac{1}{0.0396}$ _______

(c) 3.6697 _______

---

a) $\frac{1}{1.732}$

b) 0.0396

c) $\frac{1}{3.6697}$

13. In order to express the reciprocal of 0.7892 as a regular number, we can use the slide rule. Two procedures are possible to find the reciprocal:

(1) Divide "1" by "0.7892" on the C and D scales.
(2) Or, use the CI and D scales.

Using your slide rule, write the reciprocals of each of the following as a regular decimal number:

(a) 0.7679 ________ (c) 3.680 ________

(b) 1.430 ________ (d) 0.0307 ________

---

14. Given the fact that cos 32° = 0.8480, let's find sec 32° by using the reciprocal relationship.

$$\sec 32° = \text{the reciprocal of } \cos 32° = \frac{1}{0.8480} = 1.18$$

Given the limits of slide rule accuracy, is this the same value of sec 32° which is given in the partial table in Frame 11? ________

Answers to 13:
a) 1.30
b) 0.699
c) 0.272
d) 32.6

---

15. Given the fact that sin 34° = 0.5592, let's find csc 34° by using the reciprocal relationship.

(a) $\csc 34° = \text{the reciprocal of } \sin 34° = \frac{1}{0.5592} =$ ____________

(b) Does this value generally agree with the value of csc 34° shown in the table in Frame 11? ________

Answer to 14: Yes

---

16. Given the fact that tan 31° = 0.6009, let's find cot 31° by using the reciprocal relationship.

(a) $\cot 31° = \text{the reciprocal of } \tan 31° = \frac{1}{0.6009} =$ ____________

(b) Does this value coincide with the table value for cot 31°? ______

Answers to 15:
a) csc 34° = 1.79
b) Yes, given slide rule accuracy.

---

17. We will not give a complete "csc-sec-cot" table in this book. To find the cosecant, secant, or cotangent of an angle, you can use the "reciprocal" relations and the "sin-cos-tan" table. For example:

$$\sec 29° = \text{reciprocal of } \cos 29° = \frac{1}{0.8746} = 1.14$$

Complete each of these using your slide rule:

(a) $\csc 66° = \text{the reciprocal of } \sin 66° = \frac{1}{0.9135} =$ ________

(b) sec 47° = the reciprocal of cos 47° = $\frac{\quad}{\_\_\_\_\_\_}$ = ________

(c) cot 81° = the reciprocal of tan 81° = $\frac{\quad}{\_\_\_\_\_\_}$ = ________

Answers to 16:
a) cot 31° = 1.66
b) Yes, given slide rule accuracy.

18. Use the "sin-cos-tan" table to complete these:

(a) csc 17° = ________ (b) cot 83° = ________ (c) sec 61° = ________

a) = 1.09

b) $\frac{1}{0.6820} = 1.47$

c) $\frac{1}{6.314} = 0.158$

---

19. The definitions of the cosecant, secant, and cotangent of an angle apply to angles of any size. And though we will not discuss the general proof, it can be shown that the "reciprocal" relationships hold for angles of any size. Therefore, we can use the same method to find the cosecant, secant, and cotangent of angles greater than 90°. For example:

sec 142° = the reciprocal of cos 142°

$$= \frac{1}{\cos 142°} = \frac{1}{\cos 38° \text{ (Q2)}} = \frac{1}{-0.7880}$$

= -1.27 (sec 142° is negative.)

Find the following. Be careful of the signs.

(a) csc 130° = ________

(b) sec 160° = ________

(c) cot 95° = ________

a) $\frac{1}{0.2924} = 3.42$

b) $\frac{1}{8.144} = 0.123$

c) $\frac{1}{0.4848} = 2.06$

---

20. We can also use the "sin-cos-tan" table to find the size of an angle whose cosecant, secant, or cotangent is known. To do so, we use the same "reciprocal" relationships. For example:

If csc θ = 1.89, sin θ = the reciprocal of 1.89.

If sec θ = 2.14, cos θ = the reciprocal of 2.14.

If cot θ = 1.32, tan θ = the reciprocal of 1.32.

Using a reciprocal relationship and the "sin-cos-tan" table, we can find the size of an angle whose cosecant is 2.19. That is:

If csc θ = 2.19, sin $\theta = \frac{1}{2.19}$ or 0.457

If θ is a first-quadrant angle, how large is it? ______°

a) $\frac{1}{0.7660} = 1.31$

b) $\frac{1}{-0.9397} = -1.06$

c) $\frac{1}{-11.43} = -0.0875$

$\theta = 27°$

21. Here is another example of finding the size of an angle:

If $\sec \theta = 3.76$, $\cos \theta = \frac{1}{3.76}$ or 0.266

If θ is a first-quadrant angle, how large is it? ______°

75°

22. Using the same procedure, solve the following for the size of the angle. In each case, the angle is a first-quadrant angle.

(a) If $\cot A = 0.569$, A contains ______°.

(b) If $\csc \theta = 1.43$, θ contains ______°.

(c) If $\sec \varphi = 5.71$, φ contains ______°.

---

a) $A = 60°$

(From: $\tan A = 1.76$)

b) $\theta = 44°$

(From: $\sin \theta = 0.699$)

c) $\varphi = 80°$

(From: $\cos \varphi = 0.175$

SELF-TEST 2 (Frames 11-22)

Using a "sin-cos-tan" table and your slide rule, calculate the numerical value of each of the following:

1. csc 74° = ____________
2. cot 17° = ____________
3. sec 42° = ____________
4. cot 120° = ____________
5. csc 240° = ____________
6. sec 310° = ____________

Find each of the following angles, using a "sin-cos-tan" table and your slide rule. In each case, the angle is a first-quadrant angle.

7. If csc θ = 4.16, θ = ______°
8. If cot B = 0.268, B = ______°
9. If sec P = 1.56, P = ______°

ANSWERS:

1. $\frac{1}{0.9613} = 1.04$
2. $\frac{1}{0.3057} = 3.27$
3. $\frac{1}{0.7431} = 1.35$
4. $\frac{1}{-1.732} = -0.577$
5. $\frac{1}{-0.8660} = -1.15$
6. $\frac{1}{0.6428} = 1.56$
7. $\theta = 14°$
8. B = 75°
9. P = 50°

6-3 THE RATIO IDENTITIES

A trigonometric "identity" is an equation with two properties:

(1) It contains more than one of the six trigonometric ratios.
(2) It is true for all angles, regardless of their size.

In this section, we will discuss the two basic "ratio" identities. They are called "ratio" identities because one side of the equation is a ratio or comparison of two trigonometric ratios.

23. The first ratio identity is given in the box below:

$$\boxed{\tan\theta = \frac{\sin\theta}{\cos\theta} \quad \text{or} \quad \frac{\sin\theta}{\cos\theta} = \tan\theta}$$

This identity says that "the tangent of any angle is equal to the ratio of its sine to its cosine".

We can easily prove this identity for any first-quadrant angle by plugging in the ratios from a right triangle.

In this right triangle:

$\sin\theta = \frac{m}{h}$, $\cos\theta = \frac{t}{h}$, $\tan\theta = \frac{m}{t}$

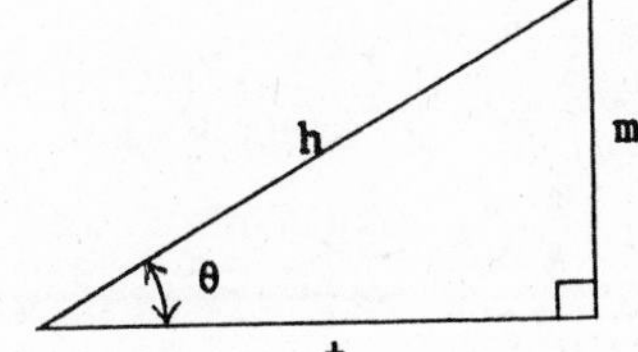

Plugging these ratios into

$$\boxed{\tan\theta = \frac{\sin\theta}{\cos\theta}}$$, we get:

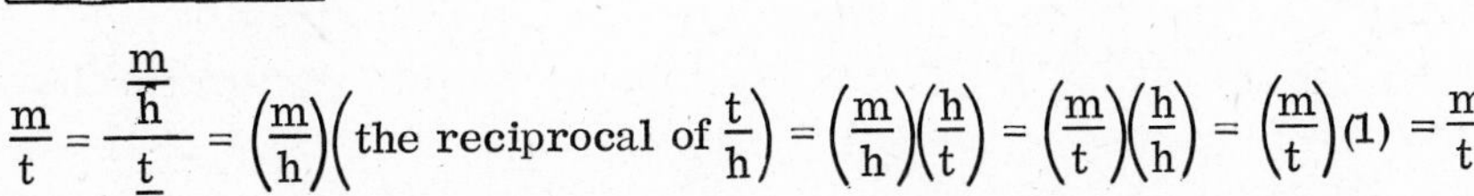

$$\frac{m}{t} = \frac{\frac{m}{h}}{\frac{t}{h}} = \left(\frac{m}{h}\right)\left(\text{the reciprocal of } \frac{t}{h}\right) = \left(\frac{m}{h}\right)\left(\frac{h}{t}\right) = \left(\frac{m}{t}\right)\left(\frac{h}{h}\right) = \left(\frac{m}{t}\right)(1) = \frac{m}{t}$$

Go to next frame.

24. Let's show that this identity is true for a 35° angle.

If $\theta = 35°$, $\sin\theta = 0.5736$
$\cos\theta = 0.8192$
$\tan\theta = 0.7002$

Plugging these values into the identity below on the left, we get the numerical equation on the right.

$\tan\theta = \frac{\sin\theta}{\cos\theta}$ $\qquad$ $0.7002 = \frac{0.5736}{0.8192}$

(a) Using your slide rule, $\frac{0.5736}{0.8192} =$ ________

(b) Does this quotient equal 0.7002? ________

---

a) 0.700
b) Yes, within the limits of slide rule accuracy.

25. Let's show that this identity is true for a 67° angle.

If $\theta = 67°$, $\sin\theta = 0.9205$
$\cos\theta = 0.3907$
$\tan\theta = 2.356$

Plugging these values into the identity, we get the numerical equation on the right.

$\tan\theta = \frac{\sin\theta}{\cos\theta}$ $\qquad$ $2.356 = \frac{0.9205}{0.3907}$

(a) $\frac{0.9205}{0.3907} =$ ________ $\qquad$ (b) Does this quotient equal 2.356? ________

---

a) 2.36
b) Yes, within the limits of slide rule accuracy.

26. We have given a proof of the first ratio identity for any first-quadrant angle. Though a general proof is possible for an angle of any size, we will not discuss this proof. However, we will show that the identity holds for some specific angles outside of the first quadrant in the next few frames.

Let's show that the identity is true for a 230° angle.

If $\theta = 230°$, $\sin\theta = -0.7660$
$\cos\theta = -0.6428$
$\tan\theta = +1.192$

Plugging in these values, we get: $+1.192 = \dfrac{-0.7660}{-0.6428}$

(a) $\dfrac{-0.7660}{-0.6428} =$ ________ (b) Does this quotient equal +1.192? ________

---

27. Let's show that the identity holds for a -55° angle.

If $\theta = -55°$, $\sin\theta = -0.8192$
$\cos\theta = +0.5736$
$\tan\theta = -1.428$

Plugging in these values, we get: $-1.428 = \dfrac{-0.8192}{+0.5736}$

(a) $\dfrac{-0.8192}{+0.5736} =$ ________ (b) Does this quotient equal -1.428? ________

Answers to frame 26:
a) +1.19
b) Yes, within the limits of slide rule accuracy.

---

28. The second ratio identity can be obtained from the first by means of this principle: If two expressions are equal, their reciprocals are equal. That is:

If $\dfrac{2}{3} = \dfrac{6}{9}$, $\dfrac{3}{2} = \dfrac{9}{6}$

Here is the first ratio identity: $\tan\theta = \dfrac{\sin\theta}{\cos\theta}$

(a) Using the "reciprocal" relationship, the reciprocal of $\tan\theta =$ ________.

(b) The reciprocal of $\dfrac{\sin\theta}{\cos\theta} =$ ________

(c) Therefore, the second ratio identity is: ________ = ________

Answers to frame 27:
a) -1.43
b) Yes, within the limits of slide rule accuracy.

Answers to frame 28:
a) $\cot\theta$
b) $\dfrac{\cos\theta}{\sin\theta}$
c) $\cot\theta = \dfrac{\cos\theta}{\sin\theta}$

29. The second ratio identity is given in the box below:

$$\boxed{\cot\theta = \frac{\cos\theta}{\sin\theta} \quad \text{or} \quad \frac{\cos\theta}{\sin\theta} = \cot\theta}$$

This identity says that "the cotangent of any angle is equal to the ratio of its cosine to its sine".

We can also prove this identity for any first-quadrant angle by using the ratios from a right triangle. Complete the following proof based on the ratios in the right triangle below.

$$\cot\theta = \frac{\cos\theta}{\sin\theta}$$

$$\frac{a}{p} = \frac{\frac{a}{r}}{\frac{p}{r}}$$

$$\frac{a}{p} = \left(\frac{a}{r}\right)\left(\frac{r}{p}\right) = \underline{\qquad}$$

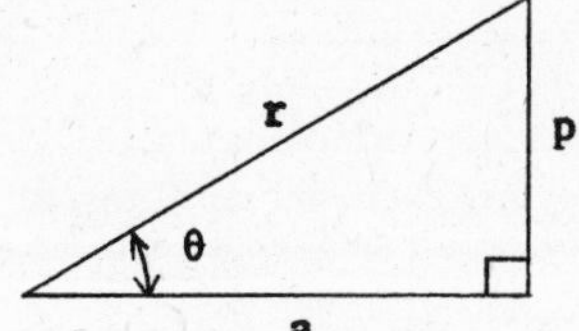

Answer: $\frac{a}{p}$

---

30. Let's show that this second ratio identity is true for a 61° angle.

If $\theta = 61°$, $\sin\theta = 0.8746$

$\cos\theta = 0.4848$

$\cot\theta = \frac{1}{1.804}$ or 0.554

(Since $\tan 61° = 1.804$)

Plugging these values into the identity below on the left, we get the numerical equation on the right.

$$\cot\theta = \frac{\cos\theta}{\sin\theta} \qquad 0.554 = \frac{0.4848}{0.8746}$$

(a) $\frac{0.4848}{0.8746} = \underline{\qquad}$

(b) Does this quotient equal 0.554? ______

Answers: a) 0.554 b) Yes

---

31. There is also a general proof of this second ratio identity for angles of any size. Though we will not discuss this general proof, we will show that the identity is true for one second-quadrant angle.

If $\theta = 140°$, $\sin\theta = +0.6428$

$\cos\theta = -0.7660$

$\cot\theta = \frac{1}{-0.8391} = -1.19$

(Since $\tan 140° = -0.8391$)

Plugging in these values, we get: $-1.19 = \frac{-0.7660}{+0.6428}$

(a) $\frac{-0.7660}{+0.6428} = \underline{\qquad}$

(b) Does this quotient equal -1.19? ______

Answers: a) -1.19 b) Yes

32. The cotangent of any angle can be obtained by finding the reciprocal of its tangent. For example:

$$\cot 51° = \text{the reciprocal of } \tan 51° = \frac{1}{1.235} = 0.810$$

The cotangent of any angle can also be found by means of the second ratio identity. That is:

(a) $\cot 51° = \dfrac{\cos 51°}{\sin 51°} = \dfrac{0.6293}{0.7771} =$ ________

(b) Is this the same numerical value we obtained above? ________

---

33. The "solved for" variables in the two ratio identities are "tan θ" and "cot θ". However, we can solve for "sin θ" or "cos θ" in either of them. For example:

To solve for "sin θ" in $\boxed{\tan\theta = \dfrac{\sin\theta}{\cos\theta}}$, we multiply both sides by "cos θ". We get:

$$\cos\theta(\tan\theta) = \cos\theta\left(\frac{\sin\theta}{\cos\theta}\right)$$

$$\cos\theta(\tan\theta) = \left(\frac{\cos\theta}{\cos\theta}\right)\sin\theta$$

$$\cos\theta(\tan\theta) = (1)\sin\theta$$

or

$$\sin\theta = \cos\theta\tan\theta$$

Note: Though there are no parentheses around "tan θ", "cos θ tan θ" means cos θ times tan θ.

Solve for cos θ in the same identity:

$$\tan\theta = \frac{\sin\theta}{\cos\theta}$$

Answers to 32: a) 0.810 b) Yes

Answer to 33: $\cos\theta = \dfrac{\sin\theta}{\tan\theta}$

34. The second ratio identity is: $\boxed{\cot\theta = \frac{\cos\theta}{\sin\theta}}$

(a) Solve for cos θ. $\cos\theta =$ ____________

(b) Solve for sin θ. $\sin\theta =$ ____________

Answer to Frame 34: a) $\cos\theta = \sin\theta\cot\theta$ b) $\sin\theta = \frac{\cos\theta}{\cot\theta}$

SELF-TEST 3 (Frames 23-34)

Write the two ratio identities. 1. ____________ 2. ____________

Using the ratio identities, complete the following:

3. $\frac{\sin 218°}{\cos 218°} = \boxed{}$ 4. $\cot 700° = \frac{\cos 700°}{\boxed{}}$

If $\sin 19° = 0.3256$ and $\cos 19° = 0.9455$, use the ratio identities to find the numerical value of tan 19° and cot 19°:

5. $\tan 19° =$ ________ 6. $\cot 19° =$ ________

7. One of the following expressions is not an identity. Which one is it? ______

(a) $\cos\theta\tan\theta = \sin\theta$ (b) $\frac{\cos\theta}{\sin\theta} = \tan\theta$ (c) $\cos\theta = \sin\theta\cot\theta$

ANSWERS:

1. $\tan\theta = \frac{\sin\theta}{\cos\theta}$
2. $\cot\theta = \frac{\cos\theta}{\sin\theta}$
3. tan 218°
4. sin 700°
5. $\tan 19° = 0.344$
6. $\cot 19° = 2.90$
7. (b)

6-4 THE PYTHAGOREAN IDENTITIES

There are three identities which are based on the Pythagorean Theorem. We will discuss these three Pythagorean identities in this section.

---

35. Before introducing the Pythagorean identities, we will review some symbolism related to "squaring".

Squaring a fraction is the equivalent of squaring both its numerator and its denominator. For example:

$$\left(\frac{a}{c}\right)^2 = \left(\frac{a}{c}\right)\left(\frac{a}{c}\right) = \frac{a\cdot a}{c\cdot c} = \frac{a^2}{c^2}$$

Notice the two equivalent expressions on the ends. There is a set of parentheses on the left, but none on the right.

Write an equivalent expression without parentheses for each of the following:

(a) $\left(\frac{b}{c}\right)^2 =$ ______ (b) $\left(\frac{m}{h}\right)^2 =$ ______

---

36. Write an equivalent expression with parentheses for each of the following:

(a) $\frac{c^2}{d^2} =$ ______ (b) $\frac{t^2}{s^2} =$ ______

Answers to 35: a) $\frac{b^2}{c^2}$ b) $\frac{m^2}{h^2}$

---

37. We have squared "sin θ" below:

$$(\sin\theta)^2 = (\sin\theta)(\sin\theta) = \sin^2\theta$$

Notice the two equivalent expressions on the ends. There is again a set of parentheses on the left, but none on the right. Both are expressions for "the square of sin θ".

Write an equivalent expression without parentheses for each of the following:

(a) $(\cos\theta)^2 =$ ________ (b) $(\tan\theta)^2 =$ ________

Answers to 36: a) $\left(\frac{c}{d}\right)^2$ b) $\left(\frac{t}{s}\right)^2$

---

38. Write an equivalent expression with parentheses for each of the following:

(a) $\cos^2\theta =$ ________ (b) $\tan^2\theta =$ ________

Answers to 37: a) $\cos^2\theta$ b) $\tan^2\theta$

---

39. Mathematicians like to avoid writing parentheses. When writing "the square of sin θ" in symbol form, they write $\sin^2\theta$ instead of $(\sin\theta)^2$, even though both are equivalent expressions.

How would mathematicians write each of these in symbol form?

(a) "the square of cos θ" ________

(b) "the square of tan θ" ________

Answers to 38: a) $(\cos\theta)^2$ b) $(\tan\theta)^2$

40. There are three identities based on the Pythagorean Theorem. The first Pythagorean identity is given in the box below. It says that "the sum of the squares of the sine and cosine of any angle is +1".

$$\boxed{\sin^2\theta + \cos^2\theta = 1}$$

We can use this right triangle to prove the identity for any first-quadrant angle.

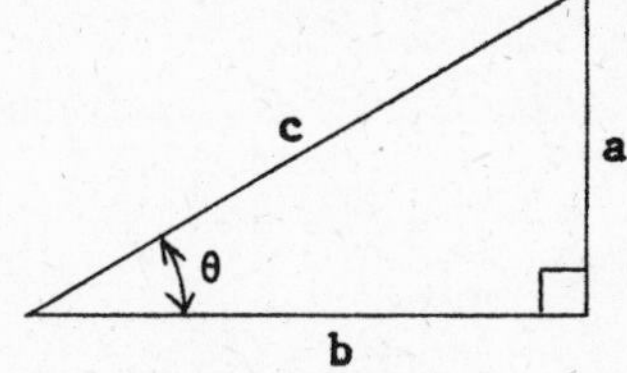

For this triangle, the Pythagorean Theorem is:

$$a^2 + b^2 = c^2$$

Multiplying both sides by $\frac{1}{c^2}$, the reciprocal of c^2, we can get a "1" on the right side:

$$\frac{1}{c^2}(a^2 + b^2) = \left(\frac{1}{c^2}\right)(c^2)$$

$$\frac{1}{c^2}(a^2) + \frac{1}{c^2}(b^2) = 1$$

$$\frac{a^2}{c^2} + \frac{b^2}{c^2} = 1$$

$$\left(\frac{a}{c}\right)^2 + \left(\frac{b}{c}\right)^2 = 1$$

But since $\frac{a}{c} = \sin\theta$ and $\frac{b}{c} = \cos\theta$,

$$\left(\frac{a}{c}\right)^2 = \sin^2\theta \text{ and } \left(\frac{b}{c}\right)^2 = \cos^2\theta.$$

By substituting these values in the last equation, we get the identity:

$$\sin^2\theta + \cos^2\theta = 1$$

a) $\cos^2\theta$

b) $\tan^2\theta$

---

41. Let's show that the first Pythagorean identity is true for a 52° angle.

If $\theta = 52°$, $\sin\theta = 0.7880$ and $\cos\theta = 0.6157$

Plugging these values into the identity and simplifying, we get:

$$\sin^2\theta + \cos^2\theta = 1$$

$$(0.7880)^2 + (0.6157)^2 = 1$$

$$0.621 + 0.379 = 1$$

$$1.000 = 1$$

Show that this identity is also true if $\theta = 32°$:

(sin 32° = 0.5299)
(cos 32° = 0.8480)

Go to next frame.

42. The second Pythagorean identity is given in the box below:

$$\boxed{\tan^2\theta + 1 = \sec^2\theta}$$

We can use this right triangle to prove the identity for any first-quadrant angle.

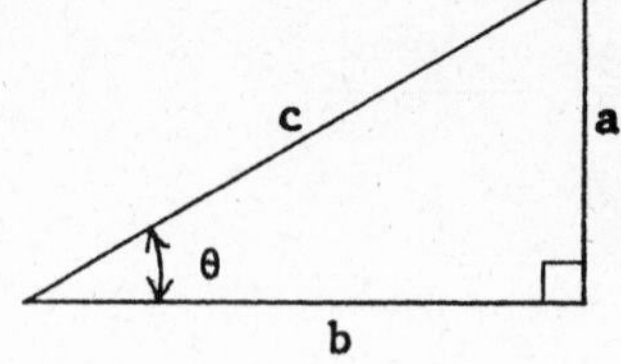

The Pythagorean Theorem is:

$$a^2 + b^2 = c^2$$

Multiplying both sides by $\frac{1}{b^2}$, the reciprocal of b^2, we can get a "1" as the second term on the left:

$$\left(\frac{1}{b^2}\right)(a^2 + b^2) = \left(\frac{1}{b^2}\right)c^2$$

$$\left(\frac{1}{b^2}\right)(a^2) + \left(\frac{1}{b^2}\right)(b^2) = \frac{c^2}{b^2}$$

$$\frac{a^2}{b^2} + \frac{b^2}{b^2} = \frac{c^2}{b^2}$$

$$\left(\frac{a}{b}\right)^2 + 1 = \left(\frac{c}{b}\right)^2$$

(a) Since $\frac{a}{b} = \tan\theta$, $\left(\frac{a}{b}\right)^2 =$ ________

(b) Since $\frac{c}{b} = \sec\theta$, $\left(\frac{c}{b}\right)^2 =$ ________

(c) By substitution, we get: ________ + 1 = ________

Answer (previous frame):

$(0.5299)^2 + (0.8480)^2 = 1$

$0.281 + 0.719 = 1$

$1.000 = 1$

---

43. Let's show that the second Pythagorean identity is true for a 29° angle.

If $\theta = 29°$, $\tan\theta = 0.5543$ and $\sec\theta = 1.143$

Substituting these values into the identity, we get:

$$\tan^2\theta + 1 = \sec^2\theta$$

$$(0.5543)^2 + 1 = (1.143)^2$$

$$0.307 + 1 = 1.307$$

$$1.307 = 1.307$$

Show that this identity is true if $\theta = 71°$:

(tan 71° = 2.904)
(sec 71° = 3.071)

Answers to 42:

a) $\tan^2\theta$

b) $\sec^2\theta$

c) $\underline{\tan^2\theta} + 1 = \underline{\sec^2\theta}$

Answer to 43:

$(2.904)^2 + 1 = (3.071)^2$

$8.43 + 1 = 9.43$

$9.43 = 9.43$

44. The third Pythagorean identity is given in the box below:

$$\boxed{1 + \cot^2\theta = \csc^2\theta}$$

We can use the same right triangle to prove this identity for any first-quadrant angle.

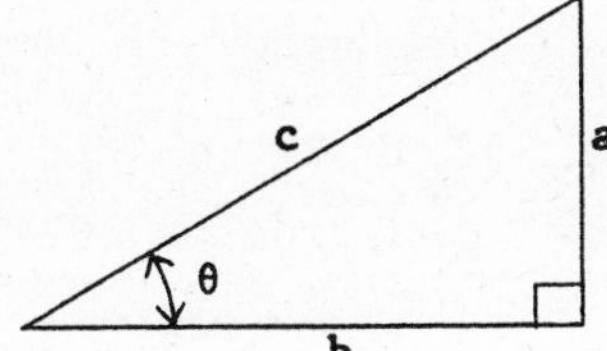

The Pythagorean Theorem is:

$$a^2 + b^2 = c^2$$

Multiplying both sides by $\frac{1}{a^2}$, the reciprocal of a^2, we can get a "1" as the first term on the left:

$$\left(\frac{1}{a^2}\right)(a^2 + b^2) = \left(\frac{1}{a^2}\right)(c^2)$$

$$\left(\frac{1}{a^2}\right)(a^2) + \left(\frac{1}{a^2}\right)(b^2) = \frac{c^2}{a^2}$$

$$\frac{a^2}{a^2} + \frac{b^2}{a^2} = \frac{c^2}{a^2}$$

$$1 + \left(\frac{b}{a}\right)^2 = \left(\frac{c}{a}\right)^2$$

By appropriate substitution, write the identity:

1 + ____________ = ____________

$1 + \cot^2\theta = \csc^2\theta$

45. Let's justify the third Pythagorean identity for a 43° angle.

If $\theta = 43°$, $\cot\theta = 1.072$ and $\csc\theta = 1.466$

Plugging these values into the identity, we get:

$$1 + \cot^2\theta = \csc^2\theta$$

$$1 + (1.072)^2 = (1.466)^2$$

$$1 + 1.15 = 2.15$$

$$2.15 = 2.15$$

Show that this identity is true for a 55° angle:

($\cot 55° = 0.700$)
($\csc 55° = 1.221$)

$1 + (0.700)^2 = (1.221)^2$

$1 + 0.490 = 1.49$

$1.490 = 1.49$

46. In the blanks on the right, write the three Pythagorean identities.

____

____

____

47. We have proven the three Pythagorean identities for all first-quadrant angles. Though a general proof is possible for angles of any size, we will not discuss this proof. However, we will show that the first Pythagorean identity is true for a few angles outside of the first quadrant.

Let's show that $\boxed{\sin^2\theta + \cos^2\theta = 1}$ is true for a 320° angle:

(a) $\sin 320° =$ ________ (b) $\cos 320° =$ ________

(c) Plugging these values into the identity and simplifying, we get:

$\sin^2\theta + \cos^2\theta = 1$

$\tan^2\theta + 1 = \sec^2\theta$

$1 + \cot^2\theta = \csc^2\theta$

---

48. Let's show that the first Pythagorean identity is true for a -217° angle:

(a) $\sin(-217°) =$ ________

(b) $\cos(-217°) =$ ________

(c) Plugging these values into the identity and simplifying, we get:

a) $\sin 320° = -0.6428$

b) $\cos 320° = +0.7660$

c) $(-0.6428)^2 + (0.7660)^2 = 1$
$(+0.413) + (0.587) = 1$
$1 = 1$

---

49. The first Pythagorean identity is $\boxed{\sin^2\theta + \cos^2\theta = 1}$. We can solve this identity for "$\sin^2\theta$" and "$\sin\theta$".

To solve for "$\sin^2\theta$", we add the opposite of "$\cos^2\theta$" to both sides and get:

$$\sin^2\theta = 1 - \cos^2\theta$$

To solve for "$\sin\theta$", we now take the square root of both sides of the last equation. We get:

$$\sin\theta = \sqrt{1 - \cos^2\theta}$$

Let's solve the same identity for "$\cos^2\theta$" and "$\cos\theta$":

(a) $\cos^2\theta =$ ____________________

(b) $\cos\theta =$ ____________________

a) $\sin(-217°) = +0.6018$

b) $\cos(-217°) = -0.7986$

c) $(+0.6018)^2 + (-0.7986)^2 = 1$
$(+0.362) + (+0.638) = 1$

---

a) $\cos^2\theta = 1 - \sin^2\theta$

b) $\cos\theta = \sqrt{1 - \sin^2\theta}$

50. The second Pythagorean identity is $\boxed{\tan^2\theta + 1 = \sec^2\theta}$. We can solve this identity for "$\tan^2\theta$" and "$\tan\theta$":

$$\tan^2\theta = \sec^2\theta - 1$$
$$\tan\theta = \sqrt{\sec^2\theta - 1}$$

The third Pythagorean identity is $\boxed{1 + \cot^2\theta = \csc^2\theta}$. Let's solve this identity for "$\cot^2\theta$" and "$\cot\theta$":

(a) $\cot^2\theta =$ ____________________

(b) $\cot\theta =$ ____________________

Answer to Frame 50: a) $\cot^2\theta = \csc^2\theta - 1$ b) $\cot\theta = \sqrt{\csc^2\theta - 1}$

SELF-TEST 4 (Frames 35-50)

Write the three Pythagorean identities:

1. ____________ 2. ____________ 3. ____________

Complete the following, using the Pythagorean identities:

4. $(\tan 13°)^2 + 1 =$ [ ]

5. [ ] $+ (\cos 225°)^2 = 1$

6. Verify the following identity for $A = 90°$:

$$\sin^2 A + \cos^2 A = 1$$

7. Verify the following identity for $\varphi = 45°$: ($\sec 45° = 1.414$)

$$\tan^2\varphi + 1 = \sec^2\varphi$$

ANSWERS:

1. $\sin^2\theta + \cos^2\theta = 1$
2. $\tan^2\theta + 1 = \sec^2\theta$
3. $1 + \cot^2\theta = \csc^2\theta$
4. $(\sec 13°)^2$
5. $(\sin 225°)^2$
6. $(\sin 90°)^2 + (\cos 90°)^2 = 1$
   $(1)^2 + (0)^2 = 1$
   $1 + 0 = 1$
   $1 = 1$
7. $(\tan 45°)^2 + 1 = (\sec 45°)^2$
   $(1.000)^2 + 1 = (1.414)^2$
   $1 + 1 = 2$
   $2 = 2$

6-5 THE RECIPROCAL IDENTITIES

In an earlier section, we discussed the three reciprocal relationships among the trigonometric ratios. There are three identities based on these relationships. They are called the "reciprocal" identities. We will briefly discuss them in this section.

51. Since the "sine" and "cosecant" of any angle are a pair of reciprocals, we can write the following identity:

$$\boxed{(\sin\theta)(\csc\theta) = 1}$$

We can solve this identity for either "sin θ" or "csc θ". To solve for "csc θ", we multiply both sides by $\frac{1}{\sin\theta}$, the reciprocal of sin θ. We get:

$$(\sin\theta)(\csc\theta) = 1$$

$$\underbrace{\left(\frac{1}{\sin\theta}\right)(\sin\theta)}_{(1)}(\csc\theta) = 1\left(\frac{1}{\sin\theta}\right)$$

$$(\csc\theta) = \frac{1}{\sin\theta}$$

$$\csc\theta = \frac{1}{\sin\theta}$$

Starting with the original identity, solve for "sin θ":

sin θ = ________

Answer: $\frac{1}{\csc\theta}$

52. Since the "cosine" and "secant" of any angle are a pair of reciprocals, we can write the following identity:

$$\boxed{(\cos\theta)(\sec\theta) = 1}$$

(a) Solve for "cos θ": (b) Solve for "sec θ":

cos θ = ________ sec θ = ________

Answers: a) $\frac{1}{\sec\theta}$ b) $\frac{1}{\cos\theta}$

53. Since the "tangent" and "cotangent" of any angle are a pair of reciprocals, we can write the following identity:

$$\boxed{(\tan\theta)(\cot\theta) = 1}$$

(a) Solve for "cot θ": (b) Solve for "tan θ":

$\cot\theta =$ ________ $\tan\theta =$ ________

Answers to 53: a) $\frac{1}{\tan\theta}$ b) $\frac{1}{\cot\theta}$

54. The three reciprocal identities are summarized in the table below:

Since: $(\sin\theta)(\csc\theta) = 1$
$(\cos\theta)(\sec\theta) = 1$
$(\tan\theta)(\cot\theta) = 1$,

$\sin\theta = \frac{1}{\csc\theta}$; $\csc\theta = \frac{1}{\sin\theta}$

$\cos\theta = \frac{1}{\sec\theta}$; $\sec\theta = \frac{1}{\cos\theta}$

$\tan\theta = \frac{1}{\cot\theta}$; $\cot\theta = \frac{1}{\tan\theta}$

Answer to 54: Go to next frame.

55. Using the right triangle on the right, we can show that the following identity is true.

$$\sin\theta = \frac{1}{\csc\theta}$$

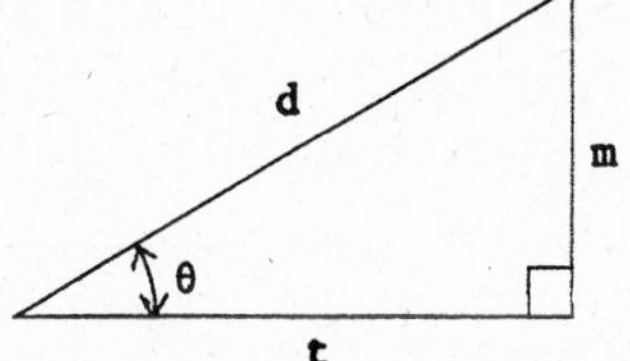

Since $\sin\theta = \frac{m}{d}$ and $\csc\theta = \frac{d}{m}$, we can substitute and get:

$$\frac{m}{d} = \frac{1}{\frac{d}{m}} = 1\left(\frac{m}{d}\right) = \frac{m}{d}$$

Using the same right triangle, show that the following identity is true:

$$\sec\theta = \frac{1}{\cos\theta}$$

56. Refer to the right triangle shown at the right. From this triangle:

(a) $\tan A =$ ______

(b) $\cot A =$ ______

(c) By plugging in these ratios, verify this identity:

$$\cot A = \frac{1}{\tan A}$$

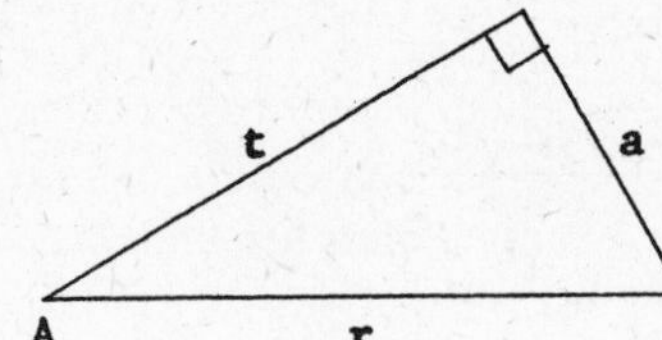

Since:

$\sec\theta = \frac{d}{t}$ and $\cos\theta = \frac{t}{d}$

we get:

$$\frac{d}{t} = \frac{1}{\frac{t}{d}} = 1\left(\frac{d}{t}\right) = \frac{d}{t}$$

Answer to Frame 56: a) $\frac{a}{t}$ b) $\frac{t}{a}$ c) $\frac{t}{a} = \frac{1}{\frac{a}{t}} = 1\left(\frac{t}{a}\right) = \frac{t}{a}$

SELF-TEST 5 (Frames 51-56)

Complete the following reciprocal identities:

1. $\sec A =$ ______
2. $\cot A =$ ______
3. $\csc A =$ ______

Using the reciprocal identities, complete the following:

4. $\frac{1}{\csc 58°} =$ ☐
5. $\frac{1}{\sec 115°} =$ ☐
6. $\frac{1}{\cot 5°} =$ ☐

7. What trig ratio is the reciprocal of $\sin\varphi$? ______
8. What trig ratio is the reciprocal of $\cos\varphi$? ______
9. What trig ratio is the reciprocal of $\tan\varphi$? ______

ANSWERS:

1. $\frac{1}{\cos A}$
2. $\frac{1}{\tan A}$
3. $\frac{1}{\sin A}$
4. $\sin 58°$
5. $\cos 115°$
6. $\tan 5°$
7. $\csc\varphi$
8. $\sec\varphi$
9. $\cot\varphi$

6-6 A SUMMARY OF THE EIGHT BASIC IDENTITIES

In this section, we will give a summary table of the eight basic identities. Then we will give some practice frames on these basic identities.

57. The eight basic identities are listed in the table below:

Reciprocal Identities	Ratio Identities	Pythagorean Identities
$\sin\theta\csc\theta = 1$	$\tan\theta = \dfrac{\sin\theta}{\cos\theta}$	$\sin^2\theta + \cos^2\theta = 1$
$\cos\theta\sec\theta = 1$		$\tan^2\theta + 1 = \sec^2\theta$
$\tan\theta\cot\theta = 1$	$\cot\theta = \dfrac{\cos\theta}{\sin\theta}$	$1 + \cot^2\theta = \csc^2\theta$

58. This frame and the following frames give practice in the basic identities. Using the reciprocal identities, complete these:

(a) The reciprocal of $\cos\theta$ is ________.

(b) The reciprocal of $\csc\theta$ is ________.

(c) $\dfrac{1}{\cot\theta} =$ ________ (d) $\dfrac{1}{\csc\theta} =$ ________

Go to next frame.

59. Using the ratio identities, complete these:

(a) $\tan\theta = \dfrac{\square}{\square}$ (b) $\cot\theta = \dfrac{\square}{\square}$

a) $\sec\theta$ c) $\tan\theta$
b) $\sin\theta$ d) $\sin\theta$

60. Using the Pythagorean identities, complete these:

(a) $\sin^2\theta +$ ________ $= 1$ (c) $1 +$ ________ $= \csc^2\theta$

(b) ________ $+ 1 = \sec^2\theta$

a) $\dfrac{\sin\theta}{\cos\theta}$ b) $\dfrac{\cos\theta}{\sin\theta}$

61. All three types of identities are needed to complete these:

(a) $\dfrac{\sin\theta}{\cos\theta} =$ ________ (c) $\dfrac{1}{\cot\theta} =$ ________

(b) $\sec^2\theta =$ ________

a) $\cos^2\theta$
b) $\tan^2\theta$
c) $\cot^2\theta$

62. (a) $\dfrac{1}{\sec\theta} =$ ________ (c) $\dfrac{\cos\theta}{\sin\theta} =$ ________

(b) $\csc^2\theta =$ ________

a) $\tan\theta$
b) $\tan^2\theta + 1$
c) $\tan\theta$

a) $\cos\theta$
b) $1 + \cot^2\theta$
c) $\cot\theta$

63. Using the basic identities, write two expressions for "$\tan\theta$":

$\tan\theta =$ ________ $\tan\theta =$ ________

---

64. Using the basic identities, write two expressions for "$\cot\theta$":

$\cot\theta =$ ________ $\cot\theta =$ ________

Answers to 63: $\tan\theta = \dfrac{1}{\cot\theta}$ $\tan\theta = \dfrac{\sin\theta}{\cos\theta}$

---

65. By rearranging the basic identities, we can write four different expressions which are equal to "$\sin\theta$":

$\sin\theta = \dfrac{1}{\csc\theta}$, from $(\sin\theta)(\csc\theta) = 1$

$\sin\theta = \cos\theta\tan\theta$, from $\tan\theta = \dfrac{\sin\theta}{\cos\theta}$

$\sin\theta = \dfrac{\cos\theta}{\cot\theta}$, from $\cot\theta = \dfrac{\cos\theta}{\sin\theta}$

$\sin\theta = \sqrt{1 - \cos^2\theta}$, from $\sin^2\theta + \cos^2\theta = 1$

By rearranging the basic identities, we can also write four different expressions which are equal to "$\cos\theta$". Do so:

$\cos\theta =$ ________ $\cos\theta =$ ________

$\cos\theta =$ ________ $\cos\theta =$ ________

Answers to 64: $\cot\theta = \dfrac{1}{\tan\theta}$ $\cot\theta = \dfrac{\cos\theta}{\sin\theta}$

---

66. By rearranging Pythagorean identities, we can write expressions which equal "$\tan\theta$" and "$\cot\theta$". Do so:

(a) $\tan\theta =$ ________ (b) $\cot\theta =$ ________

Answers to 65: $\cos\theta = \dfrac{1}{\sec\theta}$ $\cos\theta = \dfrac{\sin\theta}{\tan\theta}$ $\cos\theta = \sin\theta\cot\theta$ $\cos\theta = \sqrt{1 - \sin^2\theta}$

Answers to 66:

a) $\tan\theta = \sqrt{\sec^2\theta - 1}$

b) $\cot\theta = \sqrt{\csc^2\theta - 1}$

SELF-TEST 6 (Frames 57-66)

Complete the following identities:

1. $\frac{1}{\sec \varphi} = \boxed{}$

2. $\cot A = \frac{1}{\boxed{}}$

3. $\cot B = \frac{\boxed{}}{\boxed{}}$

4. $\boxed{} + \cos^2\theta = 1$

5. $\sin W = \frac{1}{\boxed{}}$

6. $1 + \cot^2\beta = \boxed{}$

7. $\tan^2 H + 1 = \boxed{}$

8. $\tan \alpha = \frac{\boxed{}}{\boxed{}}$

ANSWERS:

1. $\cos \varphi$
2. $\tan A$
3. $\frac{\cos B}{\sin B}$
4. $\sin^2\theta$
5. $\csc W$
6. $\csc^2\beta$
7. $\sec^2 H$
8. $\frac{\sin \alpha}{\cos \alpha}$

6-7 VERIFYING IDENTITIES

An identity is an equation which is true for all angles, regardless of their size. Besides the eight basic identities, the number of other possible identities is almost limitless. In this section, we will verify some of these other identities. By "verifying" an identity, we mean substituting by means of the basic identities on one side and then rearranging that side until we obtain an expression which is identical to the other side. The treatment of "verifying" identities in this section is in no way meant to be exhaustive of all types.

67. The equation in the box below is an identity:

$$\boxed{\sin \theta = \cos \theta \tan \theta}$$

To verify this identity, we can show that "cos θ tan θ" reduces to "sin θ" if we substitute $\frac{\sin \theta}{\cos \theta}$ for tan θ and simplify. Here are the steps:

$$\boxed{\sin \theta = \cos \theta \tan \theta}$$

$$= \cos \theta\left(\frac{\sin \theta}{\cos \theta}\right)$$

$$= \left(\frac{\cos \theta}{\cos \theta}\right)(\sin \theta)$$

$$= (1) \quad (\sin \theta)$$

$$= \sin \theta$$

(Continued on following page.)

67. (Continued)

By substituting $\frac{\cos\theta}{\sin\theta}$ for $\cot\theta$ and simplifying, verify the identity below:

$$\boxed{\cos\theta = \sin\theta\cot\theta}$$

68. We can verify the identity in the box below by substituting $\frac{\sin\theta}{\cos\theta}$ for $\tan\theta$ and simplifying the right side. Here are the steps:

$$\boxed{\cos\theta = \frac{\sin\theta}{\tan\theta}}$$

$$= \frac{\sin\theta}{\frac{\sin\theta}{\cos\theta}}$$

$$= \sin\theta\left(\frac{\cos\theta}{\sin\theta}\right)$$

$$= \left(\frac{\sin\theta}{\sin\theta}\right)(\cos\theta)$$

$$= \cos\theta$$

Verify the following identity by substituting $\frac{\cos\theta}{\sin\theta}$ for $\cot\theta$ and simplifying the right side:

$$\boxed{\sin\theta = \frac{\cos\theta}{\cot\theta}}$$

$\cos\theta = \sin\theta\cot\theta$

$$= \sin\theta\left(\frac{\cos\theta}{\sin\theta}\right)$$

$$= \left(\frac{\sin\theta}{\sin\theta}\right)(\cos\theta)$$

$$= \cos\theta$$

69. In order to verify the identity in the box below, we must make a double substitution on the right side before simplifying.

We substitute $\frac{\cos\theta}{\sin\theta}$ for $\cot\theta$.

We substitute $\frac{1}{\sin\theta}$ for $\csc\theta$.

Show the steps in the verification:

$$\boxed{\cos\theta = \frac{\cot\theta}{\csc\theta}}$$

$\sin\theta = \frac{\cos\theta}{\cot\theta}$

$$= \frac{\cos\theta}{\frac{\cos\theta}{\sin\theta}}$$

$$= \cos\theta\left(\frac{\sin\theta}{\cos\theta}\right)$$

$$= \left(\frac{\cos\theta}{\cos\theta}\right)(\sin\theta)$$

$$= \sin\theta$$

70. To verify the identity in the box below, we must make the following double substitution on the right side:

$\frac{1}{\cos\theta}$ for $\sec\theta$ and $\frac{\cos\theta}{\sin\theta}$ for $\cot\theta$

Show the steps in the verification of this identity:

$$\boxed{1 = \sin\theta\sec\theta\cot\theta}$$

$$\cos\theta = \frac{\cot\theta}{\csc\theta} = \frac{\frac{\cos\theta}{\sin\theta}}{\frac{1}{\sin\theta}} = \left(\frac{\cos\theta}{\sin\theta}\right)\left(\frac{\sin\theta}{1}\right) = \cos\theta\left(\frac{\sin\theta}{\sin\theta}\right) = \cos\theta$$

---

71. Here is a case in which we can simplify the right side by making the following substitution:

$(\tan^2\theta + 1)$ for $\sec^2\theta$ (from a Pythagorean identity)

The identity and the steps of the verification are given below:

$$\boxed{1 = \sec^2\theta - \tan^2\theta}$$
$$= (\tan^2\theta + 1) - \tan^2\theta$$
$$= \tan^2\theta + 1 - \tan^2\theta$$
$$= \underbrace{\tan^2\theta + (-\tan^2\theta)}_{0} + 1$$
$$= 0 + 1$$
$$= 1$$

The same general procedure can be used to verify the identity below. Do so:

$$\boxed{1 = \csc^2\theta - \cot^2\theta}$$

$$1 = \sin\theta\sec\theta\cot\theta = \sin\theta\left(\frac{1}{\cos\theta}\right)\left(\frac{\cos\theta}{\sin\theta}\right) = \frac{\sin\theta\cos\theta}{\sin\theta\cos\theta} = 1$$

$$1 = \csc^2\theta - \cot^2\theta$$
$$= (1 + \cot^2\theta) - \cot^2\theta$$
$$= 1 + \cot^2\theta - \cot^2\theta$$
$$= 1 + \underbrace{\cot^2\theta + (-\cot^2\theta)}_{0}$$
$$= 1 + 0$$
$$= 1$$

72. To verify the identity below, we must first factor by means of the distributive principle before substituting $\cos^2\theta$ for $(1 - \sin^2\theta)$. Notice that we make a further substitution for $\sec\theta$ in a later step.

$$\boxed{\cos\theta = \sec\theta - \sec\theta\sin^2\theta}$$
$$= \sec\theta(1 - \sin^2\theta)$$
$$= \sec\theta(\cos^2\theta)$$
$$= \left(\frac{1}{\cos\theta}\right)\cos^2\theta$$
$$= \frac{\cos^2\theta}{\cos\theta}$$
$$= \cos\theta$$

The same general procedure can be used to verify this identity. Do so:

$$\boxed{\sin\theta = \csc\theta - \csc\theta\cos^2\theta}$$

---

73. Frequently, a verification of an identity requires the addition or subtraction of two fractions. Here is such a case:

$$\boxed{\cos\theta = \sec\theta - \sin\theta\tan\theta}$$
$$= \frac{1}{\cos\theta} - \sin\theta\left(\frac{\sin\theta}{\cos\theta}\right)$$
$$= \frac{1}{\cos\theta} - \frac{\sin^2\theta}{\cos\theta}$$
$$= \frac{1 - \sin^2\theta}{\cos\theta}$$

Now by substituting for $(1 - \sin^2\theta)$, you can easily complete the verification. Do so:

$$\sin\theta = \csc\theta - \csc\theta\cos^2\theta$$
$$= \csc\theta(1 - \cos^2\theta)$$
$$= \csc\theta(\sin^2\theta)$$
$$= \left(\frac{1}{\sin\theta}\right)\sin^2\theta$$
$$= \frac{\sin^2\theta}{\sin\theta}$$
$$= \sin\theta$$

74. Here is a case in which we must subtract a non-fraction from a fraction:

$$\boxed{\cos\theta\cot\theta = \csc\theta - \sin\theta}$$

$$= \frac{1}{\sin\theta} - \sin\theta$$

$$= \frac{1}{\sin\theta} - \frac{\sin^2\theta}{\sin\theta}$$

$$= \frac{1 - \sin^2\theta}{\sin\theta}$$

Now by substituting for $(1 - \sin^2\theta)$, you should be able to complete the verification. Do so:

$$\cos\theta = \frac{1 - \sin^2\theta}{\cos\theta}$$

$$= \frac{\cos^2\theta}{\cos\theta}$$

$$= \cos\theta$$

Answer to Frame 74:

$$\cos\theta\cot\theta = \frac{1 - \sin^2\theta}{\sin\theta}$$

$$= \frac{\cos^2\theta}{\sin\theta}$$

$$= \cos\theta\left(\frac{\cos\theta}{\sin\theta}\right)$$

$$= \cos\theta\cot\theta$$

75. As you can see, the verification of identities can become quite complicated. Fortunately, the need to verify an identity does not occur often. The main use of identities is in the study of calculus where identities are used to make substitutions. In a calculus course, either the necessary identity is given, or it can be obtained from a "table of identities". A partial table of identities is given below and on the following page. It contains many of the identities which are most frequently used in calculus.

TRIGONOMETRIC IDENTITIES

Reciprocal Identities	Ratio Identities	Pythagorean Identities
$\sin\theta\csc\theta = 1$	$\tan\theta = \dfrac{\sin\theta}{\cos\theta}$	$\sin^2\theta + \cos^2\theta = 1$
$\cos\theta\sec\theta = 1$		$\tan^2\theta + 1 = \sec^2\theta$
$\tan\theta\cot\theta = 1$	$\cot\theta = \dfrac{\cos\theta}{\sin\theta}$	$1 + \cot^2\theta = \csc^2\theta$

(Continued on following page.)

75. (Continued)

TRIGONOMETRIC IDENTITIES

Double-Angle Identities

$\sin 2\theta = 2 \sin \theta \cos \theta$

$\cos 2\theta = \cos^2\theta - \sin^2\theta$

$\cos 2\theta = 2 \cos^2\theta - 1$

$\cos 2\theta = 1 - 2 \sin^2\theta$

$\tan 2\theta = \dfrac{2 \tan \theta}{1 - \tan^2\theta}$

Half-Angle Identities

$\sin \dfrac{\theta}{2} = \sqrt{\dfrac{1 - \cos \theta}{2}}$

$\cos \dfrac{\theta}{2} = \sqrt{\dfrac{1 + \cos \theta}{2}}$

$\tan \dfrac{\theta}{2} = \dfrac{\sin \theta}{1 + \cos \theta}$

Multiple-Angle Identities

$\sin 3\theta = 3 \sin \theta - 4 \sin^3\theta$

$\cos 3\theta = 4 \cos^3\theta - 3 \cos \theta$

$\sin 4\theta = 4 \sin \theta \cos \theta - 8 \sin^3\theta \cos \theta$

Power Identities

$\sin^2\theta = \dfrac{1}{2}(1 - \cos 2\theta)$

$\cos^2\theta = \dfrac{1}{2}(1 + \cos 2\theta)$

$\sin^3\theta = \dfrac{1}{4}(3 \sin \theta - \sin 3\theta)$

Addition and Subtraction Identities

$\sin(\alpha + \beta) = \sin \alpha \cos \beta + \cos \alpha \sin \beta$

$\cos(\alpha + \beta) = \cos \alpha \cos \beta - \sin \alpha \sin \beta$

$\sin(\alpha - \beta) = \sin \alpha \cos \beta - \cos \alpha \sin \beta$

$\cos(\alpha - \beta) = \cos \alpha \cos \beta + \sin \alpha \sin \beta$

$\sin \alpha + \sin \beta = 2 \sin \dfrac{1}{2}(\alpha + \beta) \cos \dfrac{1}{2}(\alpha - \beta)$

$\sin \alpha - \sin \beta = 2 \cos \dfrac{1}{2}(\alpha + \beta) \sin \dfrac{1}{2}(\alpha - \beta)$

$\cos \alpha + \cos \beta = 2 \cos \dfrac{1}{2}(\alpha + \beta) \cos \dfrac{1}{2}(\alpha - \beta)$

$\cos \alpha - \cos \beta = -2 \sin \dfrac{1}{2}(\alpha + \beta) \sin \dfrac{1}{2}(\alpha - \beta)$

SELF-TEST 7 (Frames 67-75)

Verify these identities:

1. $\dfrac{\sec \theta}{\csc \theta} = \tan \theta$

2. $\dfrac{\cot A}{\sin A} = \dfrac{\csc A}{\tan A}$

3. $1 + \sin \varphi = \sin \varphi(1 + \csc \varphi)$

ANSWERS:

1. On left side: $\dfrac{\sec\theta}{\csc\theta} = \dfrac{\dfrac{1}{\cos\theta}}{\dfrac{1}{\sin\theta}} = \left(\dfrac{1}{\cos\theta}\right)\left(\dfrac{\sin\theta}{1}\right) = \dfrac{\sin\theta}{\cos\theta} = \tan\theta$

2. On right side: $\dfrac{\csc A}{\tan A} = \dfrac{\dfrac{1}{\sin A}}{\dfrac{1}{\cot A}} = \left(\dfrac{1}{\sin A}\right)\left(\dfrac{\cot A}{1}\right) = \dfrac{\cot A}{\sin A}$

3. On right side: $\sin\varphi(1 + \csc\varphi) = \sin\varphi + \sin\varphi\csc\varphi = \sin\varphi + 1 = 1 + \sin\varphi$

6-8 TRIGONOMETRIC EQUATIONS

Trigonometric identities are equations which are true for all angles, regardless of their size. An example of an identity is: $\sin^2\theta + \cos^2\theta = 1$ Conditional trigonometric equations, usually called trigonometric equations, are equations which are true for only some angles. An example of a trigonometric equation is: $3 + 2\sin\theta = 4$ In this section, we will solve some simple trigonometric equations. "Solving" a trigonometric equation means finding the angle or angles, usually between 0° and 360°, which satisfy the equation.

---

76. The three equations below are trigonometric equations of the simplest type. We have already solved many equations like them.

$\boxed{\sin\theta = 0.7660}$ $\boxed{\cos\theta = 0.8660}$ $\boxed{\tan\theta = 0.3640}$

These equations are not identities, since they are not true for angles of any size. For example:

$\sin\theta = 0.7660$ is true for $\theta = 50°$, since $\sin 50° = 0.7660$

$\sin\theta = 0.7660$ is not true for $\theta = 60°$, since $\sin 60° \neq 0.7660$

Though a trigonometric equation is not true for angles of any size, there are many angles which satisfy any one of them. That is:

(1) Each equation above is satisfied by two angles between 0° and 360°. For example:

Both 50° and 130° satisfy $\sin\theta = 0.7660$

(2) Each equation above is satisfied by two angles in each revolution beyond 360°. For example:

$\sin\theta = 0.7660$ is satisfied by the following pairs of angles in the revolutions beyond 360°:

410° and 490°, 770° and 850°, 1130° and 1210°, etc.

Note: These pairs of angles were obtained by adding 360°, 720°, 1080° to 50° and 130°, the pair of angles between 0° and 360°.

(3) Each equation above is satisfied by various pairs of negative angles. For example:

$\sin\theta = 0.7660$ is satisfied by the following pairs of negative angles:

-230° and -310°, -590° and -670°, -950° and -1030°, etc.

Go to next frame.

77. Since there are so many angles which satisfy a trigonometric equation, the range of angles in which the desired solution (or solutions) lies is ordinarily specified, so that only one or two angles of the solution are required to be listed.

(a) Name the acute angle (between 0° and 90°) which satisfies:
$\sin\theta = 0.5000$ $\theta =$ ________

(b) Name the two angles between 0° and 360° which satisfy: $\cos\theta = 0.6428$
$\theta =$ ________ and ________

(c) Name the two negative angles between 0° and -360° which satisfy:
$\tan\theta = 0.3640$
$\theta =$ ________ and ________

---

a) 30°

b) 50° and 310°

c) -160° and -340°

---

78. (a) The acute angle which satisfies $\tan\theta = 3.277$ is ________.

(b) The obtuse angle (between 90° and 180°) which satisfies $\cos\theta = -0.5880$ is ________.

(c) The two angles between 0° and 360° which satisfy $\sin\theta = -0.9848$ are ________ and ________.

(d) The two negative angles between 0° and -360° which satisfy $\sin\theta = 0.7986$ are ________ and ________.

---

a) 73°

b) 126°

c) 260° and 280°

d) -233° and -307°

79. Name the angle between 0° and 360° which satisfies each of these equations:

(a) $\sin\theta = +1$ $\theta =$ ________

(b) $\sin\theta = 0$ $\theta =$ ________

(c) $\sin\theta = -1$ $\theta =$ ________

---

a) 90°

b) 180°

c) 270°

80. Name the angle (or angles) between 0° and 360° which satisfies each of these equations:

(a) $\cos\theta = -1$ $\theta =$ ________

(b) $\tan\theta = 0$ $\theta =$ ________

(c) $\cos\theta = 0$ $\theta =$ ________

---

a) 180°

b) 180°

c) 90° and 270°

81. The equations below are similar to the ones we have been solving:

$\sin\theta = 0.4692$ $-0.6837 = \cos\theta$ $\tan\theta = 1$

In each equation shown above, one side consists of the name of one trigonometric ratio, and the other side consists of the numerical value of that ratio. Since trigonometric equations of this type are the easiest to solve, they are called "basic trigonometric equations".

Though the following equations are trigonometric equations, they are not "basic" trigonometric equations:

$2\sin\theta = 1$ $2 - 3\tan\theta = 0$

$5 + \cos\theta = 4.3047$ $\cos\theta\tan\theta = 0.4839$

Which of the following are "basic" trigonometric equations? ________

(a) $5\cos\theta = 3$ (c) $3\tan\theta + 4 = 5\tan\theta$

(b) $\tan\theta = -1.96$ (d) $0.8759 = \sin\theta$

Only (b) and (d)

82. In order to solve "non-basic" trig equations, we must first reduce them to equivalent "basic" trig equations. To do so, we use ordinary algebraic principles and axioms. For example, to reduce the equation $2 \sin \theta = 1$ to an equivalent basic equation, we use the multiplication axiom, multiplying both sides by the reciprocal of "2". We get:

$$\boxed{2 \sin \theta = 1}$$

$$\frac{1}{2}(2 \sin \theta) = \frac{1}{2}(1)$$

$$1 \sin \theta = \frac{1}{2}$$

$$\sin \theta = \frac{1}{2} \text{ or } 0.5000$$

The acute-angle root of the last equation is $\theta = 30°$. Show that this root satisfies the original equation.

---

83. To reduce the equation below to a basic equation, we first apply the oppositing principle, and then multiply both sides by $\frac{1}{3}$. We get:

$$\boxed{-3 \sin \theta = -2}$$

$$3 \sin \theta = 2$$

$$\sin \theta = \frac{2}{3} \text{ or } 0.6667$$

What acute angle satisfies both the basic equation and the original equation? θ = ________

Answer to 82: $2(\sin 30°) = 1$; $2(0.5000) = 1$; $1.0000 = 1$

---

84. To reduce the equation below to a basic equation, we simply apply the oppositing principle.

$$-\cos \theta = 0.8660$$

$$\cos \theta = -0.8660$$

Name the obtuse angle which satisfies both equations. θ = ________

Answer to 83: 42°

---

85. Reduce each of the following to a basic equation:

(a) $8 \sin \theta = 4$ (b) $-5 \cos \theta = 2$ (c) $-\tan \theta = -1$

________ ________ ________

Answer to 84: 150°

86. In order to reduce the trig equation below to a "basic" equation, we use the addition axiom, adding (-5) to both sides. We get:

$$5 + \sin\theta = 4$$

$$\underbrace{(-5) + 5} + \sin\theta = \underbrace{4 + (-5)}$$

$$0 \quad + \sin\theta = \quad -1$$

$$\sin\theta = -1$$

What angle between 0° and 360° satisfies both the basic equation and the original equation? θ = ________

Answers to frame 85: a) $\sin\theta = 0.5000$ b) $\cos\theta = -0.4000$ c) $\tan\theta = 1$

---

87. In order to reduce the equation below to a "basic" equation, we must use both the addition axiom and the multiplication axiom.

$$\boxed{7 + 2\tan\theta = 5\tan\theta}$$

Adding "$-2\tan\theta$" to both sides, we get:

$$7 = 3\tan\theta$$

Multiplying both sides by $\frac{1}{3}$, we get:

$$\frac{7}{3} = \tan\theta$$

$$\text{or}\quad \tan\theta = 2.3333$$

Name two angles between 0° and 360° which satisfy both the basic equation and the original equation.

θ = ________ and θ = ________

Answer to frame 86: 270°

---

88. We must also use both axioms to reduce this equation to a "basic" equation.

$$\boxed{5\cos\theta + 2 = 0}$$

(a) Adding "-2" to both sides, we get:

____

(b) Multiplying both sides by $\frac{1}{5}$, we get:

____

(c) Name an <u>obtuse</u> angle which satisfies both the basic equation and the original equation. θ = ________

Answer to frame 87: 67° and 247°

89. Though the numerical coefficients of expressions such as "sin θ" and "-cos θ" are not explicitly shown, they are either "+1" or "-1". That is:

$$\sin\theta = 1\sin\theta$$

$$-\cos\theta = -1\cos\theta$$

The equation below has been rewritten so that the numerical coefficient of "sin θ" is explicit:

$$\boxed{3 + \sin\theta = 5\sin\theta}$$

$$3 + 1\sin\theta = 5\sin\theta$$

Reduce the equation to a basic equation: ______________

a) $5\cos\theta = -2$

b) $\cos\theta = -\frac{2}{5}$

or $\cos\theta = -0.4000$

c) 114°

---

90. The equation below has been rewritten so that the numerical coefficient of "tan θ" is explicit:

$$\boxed{5\tan\theta = 7 - \tan\theta}$$

$$5\tan\theta = 7 - 1\tan\theta$$

Reduce the equation to a basic equation: ______________

$\sin\theta = \frac{3}{4}$

or $\sin\theta = 0.7500$

---

91. The trig equation below contains two trigonometric ratios. In order to solve it, we must obtain a basic equation which contains only one trigonometric ratio. To do so, we can make a substitution based on one of the identities. We get:

$$\boxed{\cos\theta\tan\theta = 0.485}$$

$$\cos\theta\left(\frac{\sin\theta}{\cos\theta}\right) = 0.485$$

$$\sin\theta = 0.485$$

The acute angle which satisfies the basic equation is 29°. Show that this angle satisfies the original equation:

$\tan\theta = \frac{7}{6}$

or $\tan\theta = 1.1667$

92. Each equation below can be reduced to a "basic" equation by appropriate substitution using identities. Do so:

(a) $\sin\theta \cot\theta = 0.342$

(b) $3\cos\theta = \sin\theta$

(c) $\frac{\sec\theta}{\csc\theta} = 1$

$(\cos 29°)(\tan 29°) = 0.485$
$(0.8746)(0.5543) = 0.485$
$0.485 = 0.485$

Answer to Frame 92:

a) $\sin\theta\left(\frac{\cos\theta}{\sin\theta}\right) = 0.342$

$\cos\theta = 0.342$

b) $3 = \frac{\sin\theta}{\cos\theta}$

$3 = \tan\theta$

c) $\frac{\frac{1}{\cos\theta}}{\frac{1}{\sin\theta}} = 1$

$\frac{\sin\theta}{\cos\theta} = 1$

$\tan\theta = 1$

SELF-TEST 8 (Frames 76-92)

Each equation in this self-test is satisfied by two angles between 0° and 360°. Find them.

1. $\tan\theta = 1.600$

θ = ________
θ = ________

2. $\cos A = 0.454$

A = ________
A = ________

In solving Problems 3 and 4, first simplify each equation by algebraic rearrangement to obtain a "basic" equation:

3. $4 - 2\cos\varphi = 5$

φ = ________
φ = ________

4. $2 - \sin\alpha = 4\sin\alpha - 1$

α = ________
α = ________

In solving Problems 5 and 6, use substitution of trigonometric identities to obtain a "basic" equation:

5. $5\cos\theta\tan\theta = 4$

θ = ________
θ = ________

6. $2\sin B = 7\cos B$

B = ________
B = ________

ANSWERS:
1. $\theta = 58°$, $\theta = 238°$
2. A = 63°, A = 297°
3. $\varphi = 120°$, $\varphi = 240°$. From: $\cos\varphi = -0.500$
4. $\alpha = 37°$, $\alpha = 143°$. From: $\sin\alpha = 0.600$
5. $\theta = 53°$, $\theta = 127°$. From: $\sin\theta = 0.800$
6. B = 74°, B = 254°. From: $\tan B = 3.500$

6-9 INVERSE TRIGONOMETRIC NOTATION

Expressions like those below are called "inverse trigonometric expressions":

arcsin 0.7547	$\sin^{-1}$ 0.5446
arccos 0.8829	$\cos^{-1}$ 0.3090
arctan 3.487	$\tan^{-1}$ 0.9325

In this section, we will discuss the meaning of these expressions and solve some simple equations in which they occur.

---

93. The equation sin 30° = 0.5000 can be stated in words in these two ways:

"The sine of 30° is 0.5000".

"30° is the angle whose sine is 0.5000".

Similarly, the equation sin 45° = 0.7071 can be stated this way:

________ is the angle whose sine is ________.

---

94. As you have just seen, the equation sin 45° = 0.7071 can be stated this way:

45° is the angle whose sine is 0.7071.

Mathematicians use a special abbreviation to represent the phrase "the angle whose sine is". The abbreviation is "arcsin". Thus:

45° is [the angle whose sine is] 0.7071.

45° is [arcsin] 0.7071

This sentence is written in equation form as:

45° = arcsin 0.7071

Write the following in equation form using the arcsin notation:
60° is the angle whose sine is 0.8660. ________________

Answer to 93: 45° is the angle whose sine is 0.7071.

---

95. The meaning of "14° = arcsin 0.2419" is:

14° is the angle whose ________ is 0.2419.

Answer to 94: 60° = arcsin 0.8660

---

96. "87° is the angle whose sine is 0.9986" can be written in equation form as:

87° = ________ 0.9986

Answer to 95: sine

Answer to 96: arcsin

97. The equation: $\sin 36° = 0.5878$

can be written in arcsin notation as: $36° = \arcsin 0.5878$

Convert each of these equations to arcsin notation:

(a) $\sin 72° = 0.9511$ ______

(b) $\sin 5° = 0.0872$ ______

(c) $0.5592 = \sin 34°$ ______

(d) $\sin \theta = w$ ______

---

98. The equation "$90° = \arcsin 1.0000$" means this:

90° is the ______ 1.0000.

Answers to 97:

a) $72° = \arcsin 0.9511$

b) $5° = \arcsin 0.0872$

c) $34° = \arcsin 0.5592$

d) $\theta = \arcsin w$

---

99. The equation: $41° = \arcsin 0.6561$

can be written in regular form as: $\sin 41° = 0.6561$

Convert these arcsin equations to regular form:

(a) $79° = \arcsin 0.9816$ ______

(b) $8° = \arcsin 0.1392$ ______

(c) $\arcsin 0.7986 = 53°$ ______

(d) $\arcsin R = \alpha$ ______

90° is the angle whose sine is 1.0000.

---

100. Use a table of trig ratios for these:

(a) Find acute angle β, if: $\beta = \arcsin 0.2924$. $\beta =$ ______

(b) Find K, if: $22° = \arcsin K$. $K =$ ______

a) $\sin 79° = 0.9816$

b) $\sin 8° = 0.1392$

c) $\sin 53° = 0.7986$

d) $\sin \alpha = R$

---

101. Using a trig table, find the value of each letter. All angles are acute angles:

(a) $2° = \arcsin G$ $G =$ ______

(b) $T = \arcsin 0.9962$ $T =$ ______

(c) $\arcsin 0.3240 = A$ $A =$ ______

(d) $\arcsin R = 49°$ $R =$ ______

a) $\beta = 17°$

b) $K = 0.3746$

102. Mathematicians sometimes use another symbol for arcsin. The symbol used is $\sin^{-1}$. That is:

$\sin^{-1}$ means arcsin

arcsin and $\sin^{-1}$ both mean "the angle whose sine is"

Caution: In $\sin^{-1}$ the -1 is not an exponent.
The entire symbol means "the angle whose sine is".

Using the new symbol, the equation: $75° = \arcsin 0.9659$

can also be written as: $75° = \sin^{-1} 0.9659$

Write each of these equations using the $\sin^{-1}$ symbol:

(a) $46° = \arcsin 0.7193$ ________________

(b) $\theta = \arcsin N$ ________________

a) $G = 0.0349$

b) $T = 85°$

c) $A = 19°$

d) $R = 0.7547$

---

103. Complete these by referring to a table of trig ratios:

(a) Find acute angle B, if: $B = \sin^{-1} 0.8970$ B = ________

(b) Find P, if: $\sin^{-1} P = 27°$ P = ________

a) $46° = \sin^{-1} 0.7193$

b) $\theta = \sin^{-1} N$

---

104. Just as arcsin and $\sin^{-1}$ mean "the angle whose sine is", the following two symbols mean "the angle whose cosine is":

arccos and $\cos^{-1}$

Using these symbols, the equation: $\cos 30° = 0.8660$

can be written in these two forms: $30° = \arccos 0.8660$

and $30° = \cos^{-1} 0.8660$

Write each equation in two other forms:

(a) $\cos 77° = 0.2250$ ________________

____

(b) $y = \cos \theta$ ________________

____

a) $B = 64°$

b) $P = 0.4540$

---

a) $77° = \arccos 0.2250$

$77° = \cos^{-1} 0.2250$

b) $\theta = \arccos y$

$\theta = \cos^{-1} y$

105. Complete these by referring to a trig table. All angles are acute angles:

(a) Find M, if: M = arccos 0.3570   M = ________

(b) Find R, if: $\cos^{-1}R = 14°$ R = ________

(c) Find A, if: $\cos^{-1}0.0350 = A$ A = ________

(d) Find Q, if: arccos Q = 38°   Q = ________

---

106. The symbols [arctan] and [$\tan^{-1}$] mean "the angle whose tangent is".

Answers to 105:
a) M = 69°
b) R = 0.9703
c) A = 88°
d) Q = 0.7880

Using these symbols, the equation: tan 30° = 0.5774

can be written in these two forms: [30° = arctan 0.5774]

and [$30° = \tan^{-1}0.5774$]

Write each equation in two other forms:

(a) tan 81° = 6.314   ________________
____

(b) $m = \tan\alpha$ ________________
____

---

107. Complete these by referring to a trig table. All angles are acute angles:

Answers to 106:
a) 81° = arctan 6.314
$81° = \tan^{-1}6.314$
b) α = arctan m
$\alpha = \tan^{-1}m$

(a) Find E, if: E = arctan 2.050   E = ________

(b) Find N, if: $\tan^{-1}N = 88°$ N = ________

(c) Find P, if: $\tan^{-1}0.3852 = P$ P = ________

(d) Find H, if: arctan H = 45°   H = ________

---

108. In the formula below, find θ when R = 24.8 and Z = 43.2.

$$\theta = \cos^{-1}\left(\frac{R}{Z}\right)$$

θ = ________

Answers to 107:
a) E = 64°
b) N = 28.64
c) P = 21°
d) H = 1.000

Answer to 108:

$$\theta = \cos^{-1}\left(\frac{R}{Z}\right)$$
$$= \cos^{-1}\left(\frac{24.8}{43.2}\right)$$
$$= \cos^{-1}(0.574)$$
$$= 55°$$

109. In the formula below, find α when X = 2,860 and R = 1,270.

$$\alpha = \arctan\left(\frac{X}{R}\right)$$

α = ________

$$\alpha = \arctan\left(\frac{X}{R}\right) = \arctan\left(\frac{2860}{1270}\right) = \arctan(2.25) = 66°$$

110. To find X in the formula below when $\theta = 32°$ and Z = 515, we substitute as we have done on the right.

$$\boxed{\sin^{-1}\left(\frac{X}{Z}\right) = \theta} \qquad \sin^{-1}\left(\frac{X}{515}\right) = 32°$$

(a) Since $\sin^{-1}\left(\frac{X}{515}\right) = 32°$, $\frac{X}{515}$ = ________

(b) Since $\frac{X}{515} = 0.5299$, X = ________

Answer to Frame 110: a) $\frac{X}{515} = \sin 32°$ or $\frac{X}{515} = 0.5299$ b) X = 273

SELF-TEST 9 (Frames 93-110)

Write each equation using arcsin, arccos, or arctan notation:

1. $\sin 64° = 0.8988$ ________
2. $\tan 82° = 7.115$ ________
3. $0.9744 = \cos 13°$ ________

Write each equation using $\sin^{-1}$, $\cos^{-1}$, or $\tan^{-1}$ notation:

4. $1.192 = \tan 50°$ ________
5. $\sin \beta = N$ ________
6. $P = \cos \alpha$ ________

Using a trig table, find the value of each letter. All angles are acute angles.

7. F = arccos 0.890 F = ________
8. $\sin^{-1} K = 84°$ K = ________
9. arctan 4.00 = W W = ________

10. $\boxed{\theta = \cos^{-1}\left(\frac{E_r}{E_t}\right)}$ Find θ, if $E_r = 58.2$ and $E_t = 87.5$.

θ = ________

11. $\boxed{\arctan\left(\frac{R}{Z}\right) = \alpha}$ Find Z, if $\alpha = 70°$ and R = 4620.

Z = ________

ANSWERS:

1. 64° = arcsin 0.8988
2. 82° = arctan 7.115
3. 13° = arccos 0.9744
4. $50° = \tan^{-1} 1.192$
5. $\beta = \sin^{-1} N$
6. $\alpha = \cos^{-1} P$
7. F = 27°
8. K = 0.9945
9. W = 76°
10. $\theta = 48°$
11. Z = 1680

6-10 SUBDIVISIONS OF A DEGREE -- MINUTES AND SECONDS

Up to this point, we have generally used the "degree" as the basic unit of angle measurement. Sometimes more precise measurements of angles are needed. In such cases, the "degree" can be subdivided into "minutes" and "seconds". We will discuss subdivisions of a degree into minutes and seconds in this section.

111. A degree is frequently subdivided into "minutes". In this context, the term "minute" is a unit of angle measurement. It is not a unit of time measurement. By definition: 1 minute is $\frac{1}{60}$ of 1 degree Based on this definition, there are ______ minutes in 1 degree.	
112. The symbol ' is used as an abbreviation for the word "minutes". For example: 13 minutes is written as 13' When specifying the size of an angle in degrees and minutes, we write the minutes immediately after the degrees. For example: 35°10' means: 35° and 10' Using abbreviation symbols, write each of the following as concisely as possible: (a) 7 minutes = ________ (b) 14 degrees and 39 minutes = ________	60
113. Since there are 60 minutes in 1°, we simply multiply by 60 to convert degrees to minutes. That is: 5° = (5 x 60)' or 300' Convert each of the following to minutes: (a) 2° = ________ (b) 3° = ________ (c) 10° = ________	a) 7' b) 14°39'
114. If the number of minutes is a multiple of 60 (like 60', 120', 180', 240', etc.), we can easily convert the minutes to degrees by dividing by 60. For example: $120' = \left(\frac{120}{60}\right)^\circ = 2^\circ$ Convert each of the following to degrees: (a) 60' = ______ (b) 240' = ______ (c) 420' = ______	a) 120' b) 180' c) 600'
	a) 1° b) 4° c) 7°

115. When the number of minutes is greater than 60 but not a multiple of 60, we can convert it to a degree-minute expression. To do so, we separate the number of minutes into the largest possible multiple of 60 plus whatever is left over. For example:

$$87' = 60' + 27' = 1° + 27' = 1°27'$$
$$150' = 120' + 30' = 2° + 30' = 2°30'$$

Convert each of these to a degree-minute expression:

(a) 80' = ________ (c) 64' = ________

(b) 112' = ________ (d) 219' = ________

Answers to 115: a) 1°20' b) 1°52' c) 1°4' d) 3°39'

116. To add angles reported in degrees and minutes, we simply add the degrees separately and then the minutes separately. For example:

The addition of 17°24' and 20°7' is done on the right.

$$\begin{array}{r} 17°24' \\ \underline{20°\ 7'} \\ 37°31' \end{array}$$

If the sum of the minutes is more than 60', we convert that sum to a degree-minute expression and then simplify. For example:

$$\begin{array}{r} 19°47' \\ \underline{38°36'} \\ 57°83' \end{array} = 57° + 83' = 57° + 1°23' = 58°23'$$

(a) Find θ, if: $\theta = 74°22' + 18°58'$

(b) Find A, if: $A = 129°55' + 34°53'$

θ = ________ A = ________

Answers to 116: a) $\theta = 93°20'$ b) $A = 164°48'$

117. If the sum of the minutes is exactly 60', we convert the 60' to 1°. However, we write 0' in the sum so that the precision of the original angles is indicated. For example:

$$\begin{array}{r} 9°25' \\ \underline{10°35'} \\ 19°60' \end{array} = 20°0'$$

Find θ, if: $\theta = 7°29' + 3°31'$.

θ = ________

Answer to 117: $\theta = 11°0'$

118. To subtract angles reported to the nearest minute, we simply subtract the minutes separately and the degrees separately. For example:

We have subtracted 16°18' from 39°42' at the right.

$$\begin{array}{r} 39°42' \\ -\ 16°18' \\ \hline 23°24' \end{array}$$

Find θ, if: $\theta = 123°53' - 67°12'$.

θ = ____________

Answer: $\theta = 56°41'$

119. The subtraction on the left below cannot be done as it stands, since 24' is larger than 16'. In order to perform the subtraction, we must "borrow" 1° from 14°, convert it to 60', and add the 60' to 16', as we have done on the right:

$$\begin{array}{r} 14°16' \\ -\ \ 2°24' \\ \hline \end{array} \qquad \begin{aligned} 14°16' &= 13° + 1°16' \\ &= 13° + 76' \\ &= 13°76' \end{aligned}$$

By substituting 13°76' for 14°16', we can complete the subtraction:

$$\begin{array}{r} 14°16' \\ -\ \ 2°24' \\ \hline \end{array} \longrightarrow \begin{array}{r} 13°76' \\ -\ \ 2°24' \\ \hline 11°52' \end{array}$$

Find β, if: $\beta = 108°42' - 79°57'$.

β = ____________

Answer: 28°45'

120. To subtract 137°44' from 180°, we must convert 180° to a degree-minute expression. We can do so by borrowing 1° or 60' from 180° to get 179°60'. That is:

$$\begin{array}{r} 180° \\ -\ 137°44' \\ \hline \end{array} \longrightarrow \begin{array}{r} 179°60' \\ -\ 137°44' \\ \hline 42°16' \end{array}$$

Find θ, if: $\theta = 180° - 95°21'$.

θ = ____________

Answer: 84°39'

121. Let's find the third angle in a triangle if the other two angles are 38°49' and 75°33'.

(a) The sum of the two angles is ___________.   (b) The size of the third angle is ___________.

---

122. We can get a more precise measure of angles by further subdividing minutes into smaller parts called "seconds".

By definition: 1 second = $\frac{1}{60}$ of a minute

Based on the definition above, there are ______ seconds in 1 minute.

Answers to 121: a) 114°22' b) 65°38'

---

123. The symbol " is usually used as an abbreviation for the word "seconds". That is:

4 seconds is written as 4"

By using the abbreviations for degrees, minutes, and seconds, angle measurements which are precise to seconds can be written very concisely. That is:

34 degrees 27 minutes 55 seconds = 34°27'55"

18 degrees 47 minutes 13 seconds = ______________

Answer to 122: 60

---

124. (a) Since 1° = 60' and 1' = 60", how many seconds are there in one degree? 1 degree = __________ seconds

(b) Since $1'' = \frac{1'}{60}$ and $1' = \frac{1°}{60}$, a second is what fractional part of a degree? 1 second = __________ degree

Answer to 123: 18°47'13"

---

125. To convert from minutes to seconds, we simply multiply by 60. Therefore:

(a) 2' = __________"   (b) 5' = __________"   (c) 12' = __________"

Answers to 124: a) 3600 b) $\frac{1}{3600}$

An angle of 1 second is a very, very small angle.

---

126. 60 or more seconds can be easily converted to minutes if the number of seconds is a multiple of 60. To do so, we simply divide by 60. That is:

(a) 120 seconds = ______'   (b) 180" = ______'   (c) 300" = ______'

Answers to 125: a) 120" b) 300" c) 720"

Answers to 126: a) 2' b) 3' c) 5'

127. If the number of seconds is greater than 60 but not a multiple of 60, we can convert it to a minute-second expression. To do so, we separate the number of seconds into the largest possible multiple of 60 plus whatever is left over. For example:

75" = 60" + 15" = 1' + 15" = 1'15"
123" = 120" + 3" = 2' + 3" = 2'3"

Convert each of the following to a minute-second expression:

(a) 63" = ____________ (c) 147" = ____________

(b) 99" = ____________ (d) 184" = ____________

---

128. To add angles reported to the nearest second, we simply add the degrees, minutes, and seconds separately. For example:

The addition of 5°17'22" and 22°8'19" is done on the right.

```
 5°17'22"
22° 8'19"
---------
27°25'41"
```

Perform the addition on the right:

```
17°38'12"
 9°10' 3"
---------
```

Answers to Frame 127:

a) 1'3"
b) 1'39"
c) 2'27"
d) 3'4"

Answer to Frame 128: 26°48'15"

---

129. When adding angles, if the sum of the minutes or seconds is 60 or more, we do not leave the sum in that form.

If the sum of the minutes is 60' or more, we convert it to a degree-minute expression and simplify. For example:

```
10°52'11"
37°18' 9"
---------
47°70'20" = 48°10'20"  (Since 70' = 1°10'.)
```

If the sum of the seconds is 60" or more, we convert it to a minute-second expression and simplify. For example:

```
 7°19'39"
12° 5'42"
---------
19°24'81" = 19°25'21"  (Since 81" = 1'21".)
```

If the sum of both the minutes and seconds is 60 or more, we must make a double conversion. For example:

```
10°35'28"
14°42'51"
---------
24°77'79" = 24°78'19"  (Since 79" = 1'19".)
          = 25°18'19"  (Since 78' = 1°18'.)
```

Find α, if: α = 35°49'27" + 9°24'36" α = ____________

130. (a) Find θ, if: $\theta = 9°49'17'' + 12°35'58''$ θ = ___________	(b) Find A, if: $A = 74°59'47'' + 49°34'42''$ A = ___________	45°14'3''

131. If we get exactly 60' in the sum of two angles which are precise to seconds, we convert the 60' to 1°. However, we write 0' in the sum so that the number of seconds is not mistaken as the number of minutes. For example:

14°60'19'' is written 15°0'19''

If we get exactly 60'' in the sum of two angles, we convert the 60'' to 1'. However, we write 0'' so that the precision of the original angles is indicated. For example:

9°13'60'' is written 9°14'0''.

Write each of the following sums in the proper form:

(a) 10°60'18'' = ________________

(b) 27° 9'60'' = ________________

(c) 3°59'87'' = ________________

a) 22°25'15''

b) 124°34'29''

Answer to Frame 131: a) 11°0'18'' b) 27°10'0'' c) 4°0'27''

132. To subtract angles which are precise to seconds, we subtract the seconds, minutes, and degrees separately. Sometimes we have to "borrow" in order to complete the subtraction.

In this case, we have to borrow 1° (or 60'), since 42' is larger than 17':

```
 12°17'49''  ⟶   11°77'49''
- 5°42'10''     - 5°42'10''
                  6°35'39''
```

In this case, we have to borrow 1' (or 60''), since 46'' is larger than 33'':

```
 25°19'33''  ⟶   25°18'93''
-19° 3'46''     -19° 3'46''
                  6°15'47''
```

In this case, we have to borrow both 1' and 1°. Note that we get only 86' since 1' of the 27' was borrowed:

```
 14°27'14''  ⟶   13°86'74''
- 9°35'21''     - 9°35'21''
                  4°51'53''
```

(Continued on following page.)

132. (Continued)

Perform each of the subtractions below:

(a) $\begin{array}{r} 17°39'27'' \\ -\ 8°14'31'' \\ \hline \end{array}$

(b) $\begin{array}{r} 12°\ 6'17'' \\ -\ 9°41'\ 9'' \\ \hline \end{array}$

(c) $\begin{array}{r} 87°14'32'' \\ -\ 41°19'57'' \\ \hline \end{array}$

---

133. To find the third angle in a triangle, we must subtract the sum of the other two angles from 180°. If that sum is precise to seconds, we get a subtraction like the one below. Notice that we converted 180° to 179°59'60":

$$\begin{array}{r} 180° \\ -\ 105°14'27'' \\ \hline \end{array} \longrightarrow \begin{array}{r} 179°59'60'' \\ -\ 105°14'27'' \\ \hline 74°45'33'' \end{array}$$

Find θ, if: θ = 180° - 84°29'45".

θ = ____________

Answers to 132:

a) 9°24'56"

b) 2°25' 8"

c) 45°54'35"

---

134. Let's find the third angle in a triangle in which the other two angles are 67°14'29" and 45°51'40".

(a) The sum of the two angles is ____________.

(b) The size of the third angle is ____________.

Answer to 133: 95°30'15"

---

135. Two angles of a triangle are 32°25'14" and 27°13'46". Find the third angle. ____________

Answers to 134:

a) 113°6'9"

b) 66°53'51"

Answer to 135: 120°21'0"

SELF-TEST 10 (Frames 111-135)

1. A minute is what fractional part of a degree? ________

2. A second is what fractional part of a minute? ________

3. A second is what fractional part of a degree? ________

4. How many seconds are there in one degree? ________

5. Find A, if: A = 31°52' + 49°28'

6. Find φ, if: $\varphi = 180° - 71°18'$

Two angles of a triangle are 56°35' and 87°52'.

7. The sum of the two angles is ________.

8. The third angle is ________.

Two angles of a triangle are 73°38'45" and 62°20'51".

9. The sum of the two angles is ________.

10. The third angle is ________.

ANSWERS:

1. $\frac{1}{60}$
2. $\frac{1}{60}$
3. $\frac{1}{3600}$
4. 3600"
5. A = 81°20'
6. φ = 108°42'
7. 144°27'
8. 35°33'
9. 135°59'36"
10. 44°0'24"

6-11 SIMILAR RIGHT TRIANGLES

In this section, we will briefly discuss similar right triangles. We will emphasize the fact that corresponding sides of similar right triangles are proportional. The "proportionality" fact about similar right triangles is being introduced at this point because it is needed to understand the concept of interpolation which will be discussed in the next section.

136. Two right triangles are similar if the three angles of one are equal to the three angles of the other. For example, right triangles ABC and DEF below are similar because each contains a 30°, a 60°, and a 90° angle.

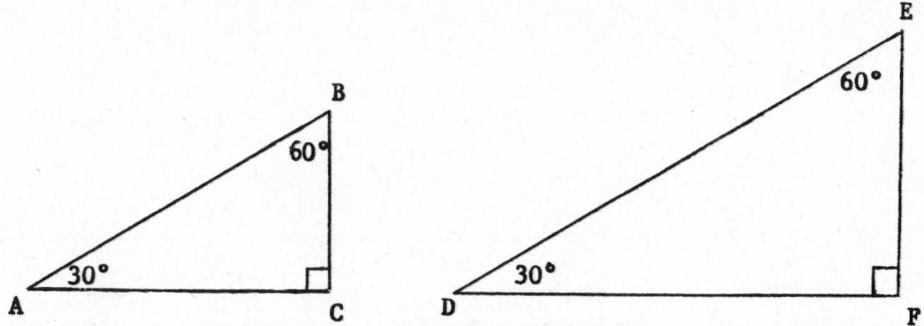

Though the two triangles above are similar, are they the same size? ________

No. Triangle DEF is obviously larger.

137. The corresponding sides of two similar right triangles are sides which have the same positions in the triangles. For example, side "a" in triangle ABC and side "d" in triangle DEF are "corresponding" sides because each is opposite the 30° angle.

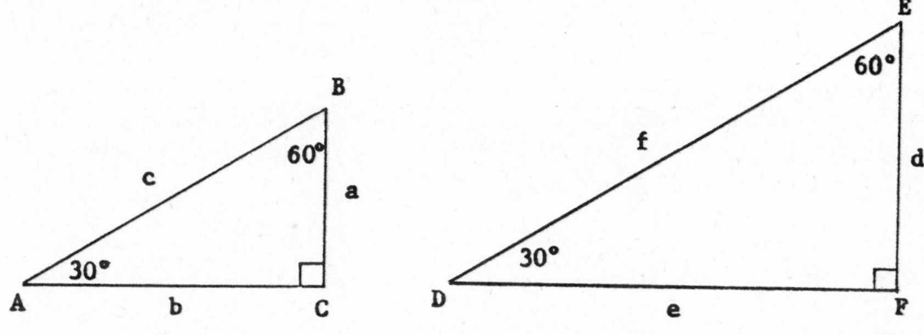

(a) Which side in triangle DEF corresponds to side "b" in triangle ABC? ________

(b) Which side in triangle ABC corresponds to side "f" in triangle DEF? ________

(c) Are the corresponding sides of the similar right triangles above equal? ________

a) Side "e"

b) Side "c"

c) No

138. Though the corresponding sides of similar right triangles are usually unequal, they are always proportional. That is, the ratio of any pair of corresponding sides is equal to the ratio of any other pair of corresponding sides. For example, in the two right triangles below:

$$\frac{t}{m} = \frac{r}{d} = \frac{p}{h}, \quad \text{since} \quad \frac{3.0''}{6.0''} = \frac{4.0''}{8.0''} = \frac{5.0''}{10.0''}$$

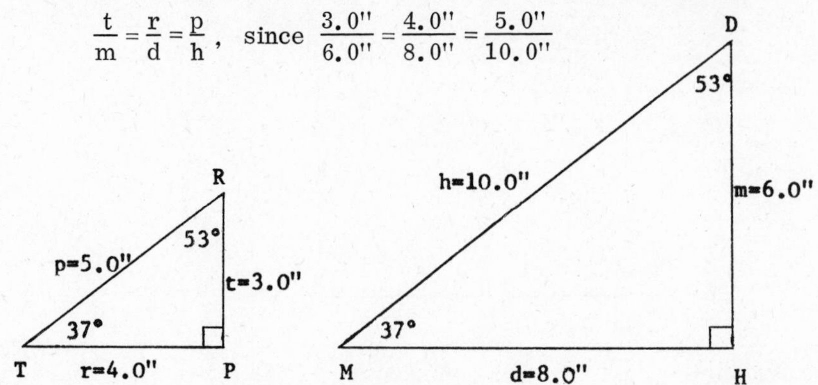

Below, we have written the ratios with the sides of triangle MDH as the numerators. Are the three ratios still equal? __________

$$\frac{m}{t} = \frac{d}{r} = \frac{h}{p}$$

Answer to Frame 138: Yes, since: $\frac{6.0''}{3.0''} = \frac{8.0''}{4.0''} = \frac{10.0''}{5.0''}$

139. In the last frame, we showed that the corresponding sides of a specific pair of similar right triangles were proportional. In this frame, we will show that the corresponding sides of any pair of similar right triangles are proportional. To do so, we will use the general pair of similar right triangles below. In them, angles A and D, B and E, and C and F are equal.

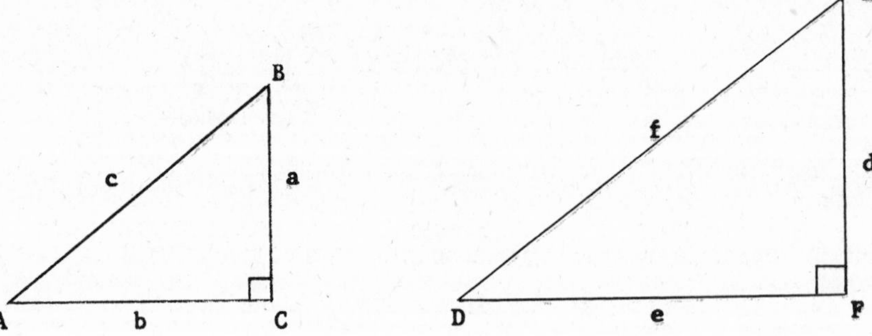

Our general strategy will be this: We will set up proportions based on trig ratios, and then rearrange them to get proportions involving ratios of "corresponding" sides.

Step 1: Showing that: $\boxed{\frac{a}{d} = \frac{b}{e}}$

Since angle A = angle D, tan A = tan D. Therefore: $\frac{a}{b} = \frac{d}{e}$

Clearing the fractions, we get: $ae = bd$

Multiplying both sides by $\frac{1}{de}$, we get: $\boxed{\frac{a}{d} = \frac{b}{e}}$

(Continued on following page.)

139. (Continued)

Step 2: Showing that: $\boxed{\frac{a}{d} = \frac{c}{f}}$

Since angle A = angle D, sin A = sin D. Therefore: $\frac{a}{c} = \frac{d}{f}$

Clearing the fractions, we get: af = cd

Multiplying both sides by $\frac{1}{df}$, we get: $\boxed{\frac{a}{d} = \frac{c}{f}}$

Step 3: Showing that: $\boxed{\frac{a}{d} = \frac{b}{e} = \frac{c}{f}}$

Since $\frac{a}{d} = \frac{b}{e}$ and $\frac{a}{d} = \frac{c}{f}$, $\frac{b}{e} = \frac{c}{f}$

Therefore: $\boxed{\frac{a}{d} = \frac{b}{e} = \frac{c}{f}}$

---

140. Based on the fact that the corresponding sides of similar right triangles are proportional, we can set up a number of proportions for any pair of similar right triangles.

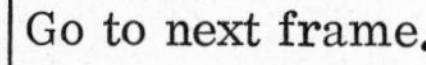
Go to next frame.

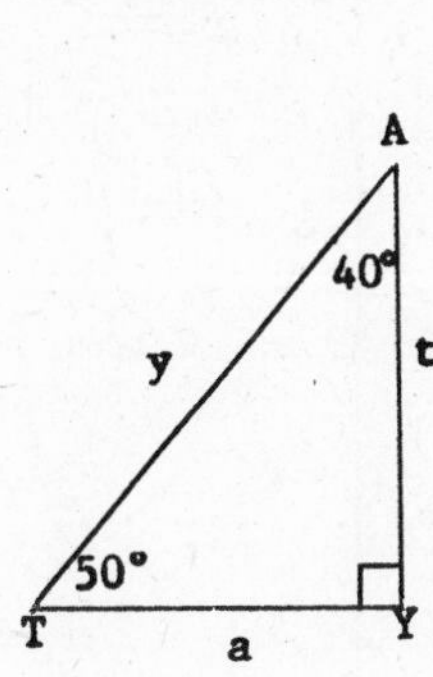

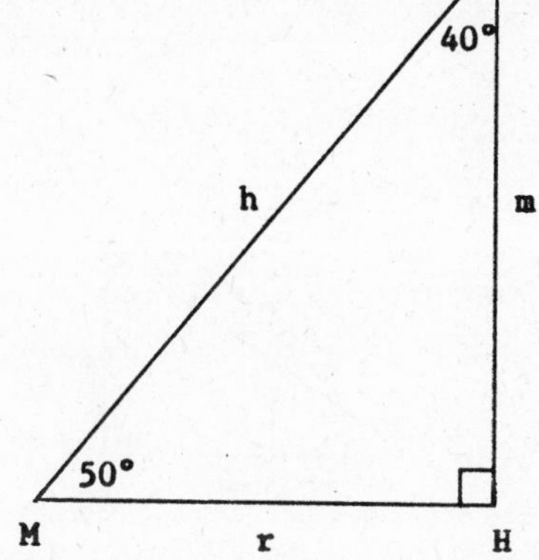

Complete the following proportions based on the pair of similar right triangles above:

(a) $\frac{r}{a} = \frac{m}{\square}$ (b) $\frac{t}{m} = \frac{y}{\square}$ (c) $\frac{h}{y} = \frac{\square}{a}$

---

a) $= \frac{m}{\boxed{t}}$

b) $= \frac{y}{\boxed{h}}$

c) $= \frac{\boxed{r}}{a}$

141. The following pair of right triangles are similar.

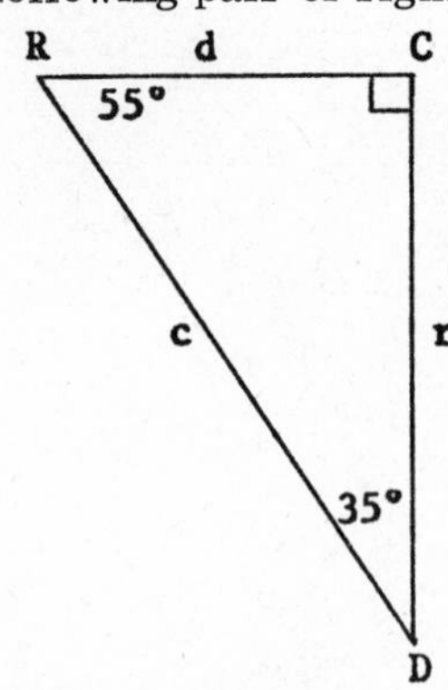

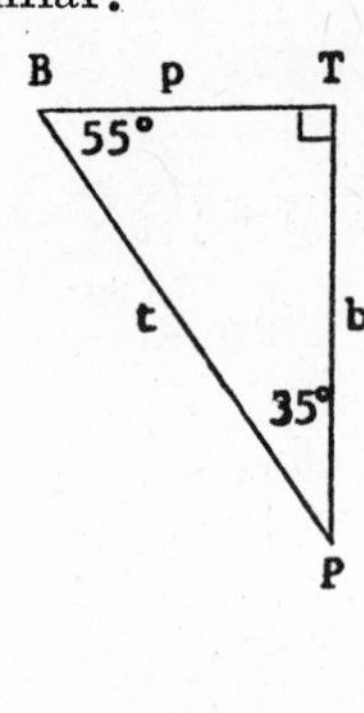

Complete the following proportions involving "corresponding" sides:

(a) $\frac{d}{p} = \frac{r}{\square}$ (b) $\frac{b}{r} = \frac{\square}{c}$ (c) $\frac{\square}{c} = \frac{p}{d}$

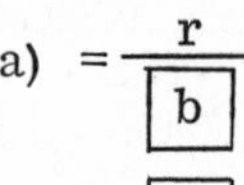

a) $= \frac{r}{\boxed{b}}$

b) $\frac{\boxed{t}}{c} =$

c) $\frac{\boxed{t}}{c} =$

---

142. We can use a proportion based on the corresponding sides of similar right triangles in order to find an unknown side of one of them. As an example, we will find the length of side "d" below. The two triangles are similar:

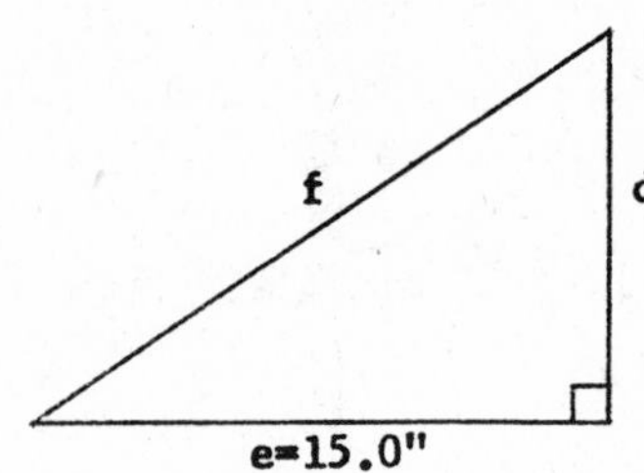

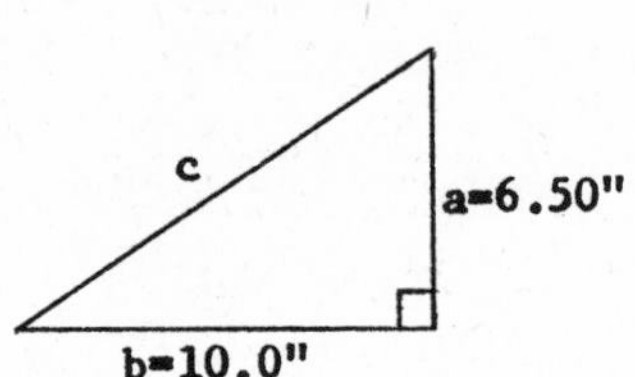

Because the triangles are similar: $\frac{d}{a} = \frac{e}{b}$

Substituting the known values, we get: $\frac{d}{6.5''} = \frac{15.0''}{10.0''}$

Solving for "d", we get: $d = \frac{(6.5'')(15.0'')}{10.0''}$

$d =$ ________

d = 9.75"

143. Let's use a proportion based on the corresponding sides of these two similar right triangles to find the length of side "h".

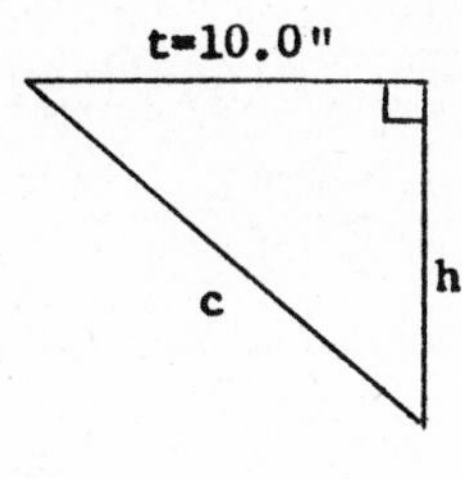

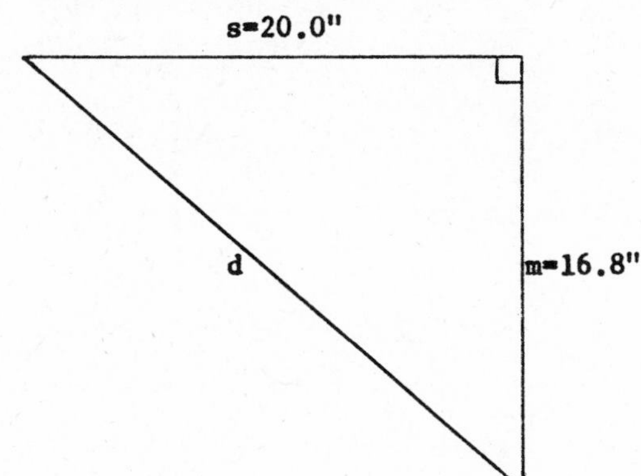

Because the triangles are similar: $\frac{h}{m} = \frac{t}{s}$

Substituting the known values, we get: $\frac{h}{16.8''} = \frac{10.0''}{20.0''}$

Solving for "h", we get: $h = \frac{(16.8'')(10.0'')}{20.0''}$

$h =$ ________

Answer: h = 8.4"

---

144. There are two right triangles (ACB and AED) in the figure on the right. Triangle ACB is inside triangle AED. Angle A, which contains 30°, is common to both triangles.

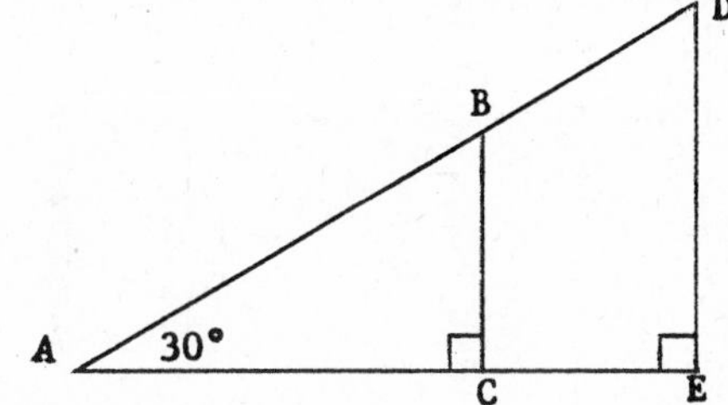

(a) How large is angle ABC? ________

(b) How large is angle ADE? ________

(c) The two right triangles are similar. Why?

____

Answers:

a) 60°

b) 60°

c) Their angles are equal. Each triangle contains a 30°, a 60°, and a 90° angle.

---

145. Whenever two right triangles contain a common acute angle, their other acute angles must also be equal. Therefore, they are similar triangles.

There are two right triangles in the figure on the right.

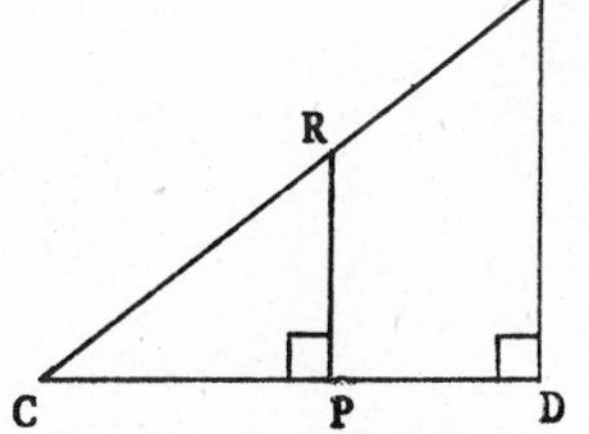

(a) The two right triangles are triangle ______ and triangle ______.

(b) The acute angle which is common to each is angle _____.

(c) The pair of equal acute angles is angle ______ and angle ______.

(d) The two triangles are a pair of ____________ right triangles.

146. There are two right triangles in this figure.

(a) The acute angle common to each is angle _______.

(b) The pair of equal acute angles is angle _______ and angle _______.

(c) The two similar right triangles are triangle _______ and triangle _______.

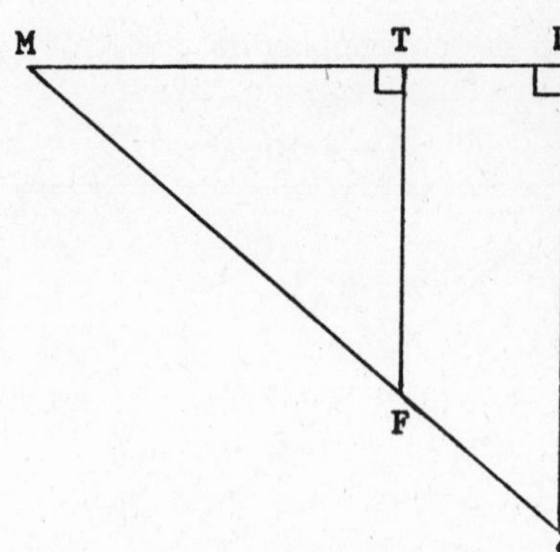

a) CPR and CDT
b) C
c) CRP and CTD
d) similar

---

147. In the figure on the right, triangles CHR and CTD are a pair of similar right triangles.

Complete the following proportions based on their corresponding sides:

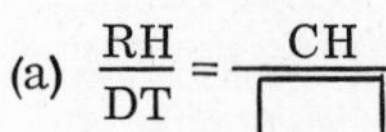
(a) $\frac{RH}{DT} = \frac{CH}{\square}$

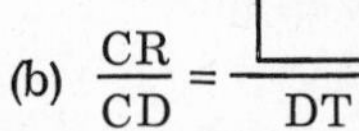
(b) $\frac{CR}{CD} = \frac{\square}{DT}$

(c) $\frac{CT}{CH} = \frac{\square}{CR}$

a) M
b) MFT and MSD
c) MTF and MDS

---

148. In the figure on the right, triangles AVS and ADK are a pair of similar right triangles.

(a) Name two ratios of corresponding sides which are equal to $\frac{AS}{AK}$: _____ and _____

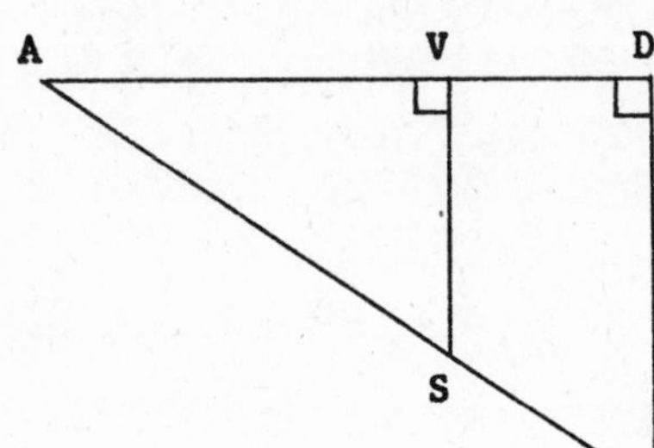

(b) Name two ratios of corresponding sides which are equal to $\frac{AD}{AV}$: _____ and _____

a) $= \frac{CH}{\boxed{CT}}$

b) $= \frac{\boxed{RH}}{DT}$

c) $= \frac{\boxed{CD}}{CR}$

---

a) $\frac{AV}{AD}$ and $\frac{VS}{DK}$

b) $\frac{AK}{AS}$ and $\frac{DK}{VS}$

149. In the figure on the right:

AE = 20.0", CE = 10.0", AD = 12.0".

Let's use a proportion based on corresponding sides to find the length of BD.

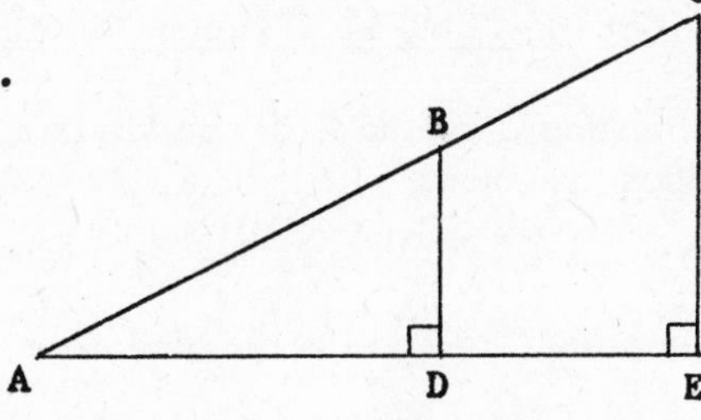

Since right triangles ADB and AEC are similar: $\frac{BD}{CE} = \frac{AD}{AE}$

Substituting the known values, we get: $\frac{BD}{10.0''} = \frac{12.0''}{20.0''}$

Solving for BD, we get: $BD = \frac{(10.0'')(12.0'')}{20.0''}$

BD = ______________

BD = 6.0 "

---

150. In the figure on the right:

TR = 20.0", TQ = 60.0", QM = 42.0".

Let's use a proportion based on corresponding sides to find the length of RV.

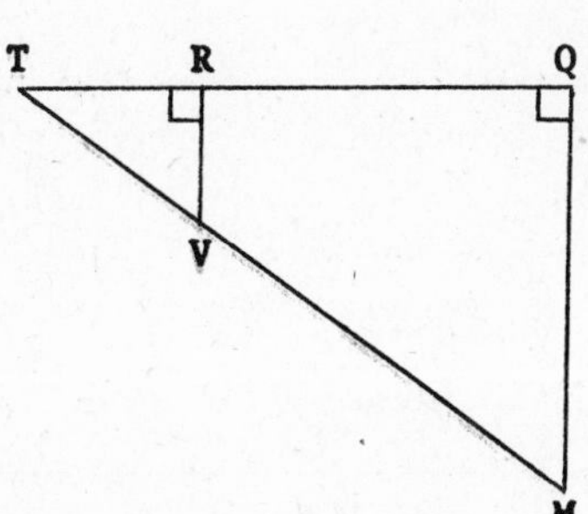

Since right triangles TRV and TQM are similar: $\frac{RV}{QM} = \frac{TR}{TQ}$

Substituting the known values, we get: $\frac{RV}{42.0''} = \frac{20.0''}{60.0''}$

Solving for RV, we get: $RV = \frac{(42.0'')(20.0'')}{60.0''}$

RV = ______________

RV = 14.0"

SELF-TEST 11 (Frames 136-150)

In the figure at the right, triangles GHK and GLP are similar right triangles. The following sides are known: KH = 256', PL = 525', and GL = 334'.

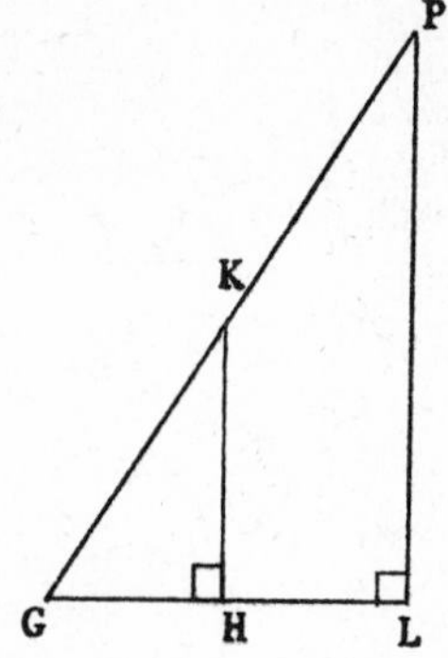

1. Find GH.

   GH = __________

In the figure at the right, triangle QTW and QRS are similar right triangles. The following sides are known: QR = 43.8", QT = 60.5", and WT = 37.3".

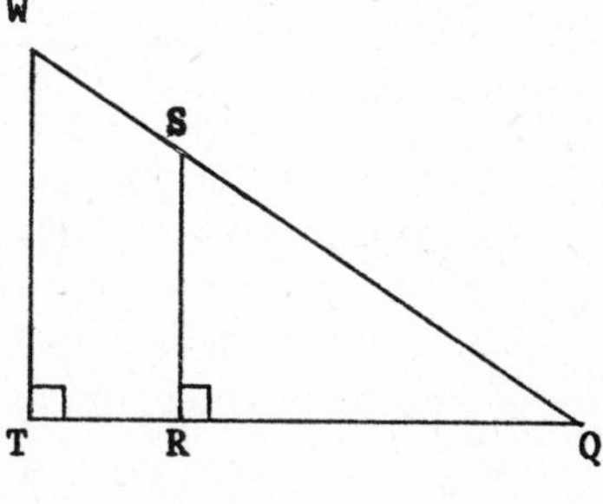

2. Find SR.

   SR = __________

ANSWERS: 1. GH = 163' From: $\frac{GH}{334'} = \frac{256'}{525'}$ 2. SR = 27.0" From: $\frac{SR}{37.3''} = \frac{43.8''}{60.5''}$

6-12 INTERPOLATING TO FIND SINE RATIOS OF ANGLES TO THE NEAREST MINUTE

Up to this point, our treatment of trigonometric ratios has been related to measurements of angles to the nearest degree. We can, however, approximate trigonometric ratios for angles measured to the nearest minute. In order to obtain these more precise values, we can use our ordinary "Table of Trig Ratios" and a process called "linear interpolation". In this section, we will discuss the process of interpolation and use it to find the sine ratios of angles reported to the nearest minute.

151. Mathematicians have a process by which they can compute the sine ratios of angles reported to the nearest minute. This process was used to compute the sines reported in the table below. Only the sines of some selected angles between 30° and 31°, 51° and 52°, and 75° and 76° are given.

θ	sin θ
30° 0'	0.5000
30°10'	0.5025
30°20'	0.5050
30°30'	0.5075
30°40'	0.5100
30°50'	0.5125
31° 0'	0.5150

θ	sin θ
51° 0'	0.7771
51°10'	0.7790
51°20'	0.7808
51°30'	0.7826
51°40'	0.7844
51°50'	0.7862
52° 0'	0.7880

θ	sin θ
75° 0'	0.9659
75°10'	0.9667
75°20'	0.9674
75°30'	0.9681
75°40'	0.9689
75°50'	0.9696
76° 0'	0.9703

Using the table, complete the following:

(a) sin 30°10' = ______ (b) sin 51°40' = ______ (c) If sin θ = 0.5075, θ = ______ (d) If sin θ = 0.9696, θ = ______

Answer to Frame 151: a) 0.5025 b) 0.7844 c) 30°30' d) 75°50'

152. The mathematicians' method of computing the sines of angles to the nearest minute is quite complicated. To avoid this complexity, we can use our ordinary "Table of Trig Ratios" and a process called "linear interpolation". Though interpolation gives only approximate values, the difference between these approximate values and the precise values is very small. Therefore, it is entirely acceptable to use interpolated values in basic science, technology, and practical work.

Linear interpolation is based on the following fact: Any very short segment of a curve is approximately a straight line. To show this fact, we have graphed the sines from the table in the last frame in the three figures below. The sines between the various pairs of angles (30° and 31°, 51° and 52°, 75° and 76°) are graphed in Figures 1, 2, and 3, respectively.

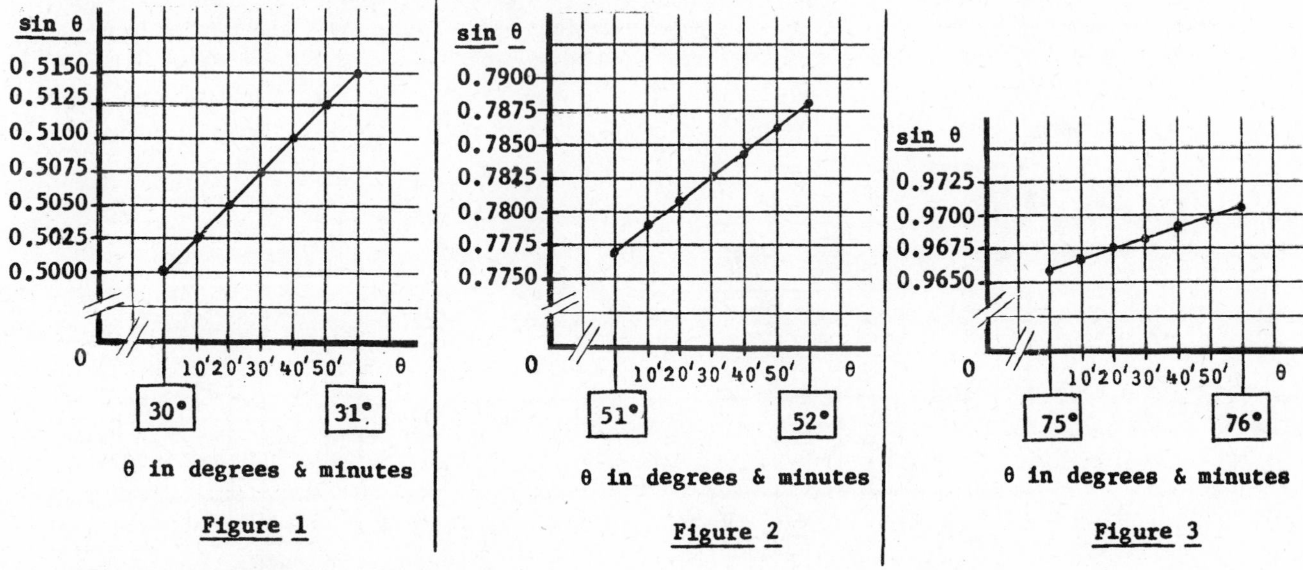

Figure 1 Figure 2 Figure 3

Notice these points about the graphs:

(1) The slash marks on the axes indicate that many values between the origin and the first listed value have been skipped.

(2) Though the graph of the sine wave is a curve, each of the three enlargements of a very short segment of that curve is approximately a straight line.

153. Since any very short segment of a curve is approximately a straight line, we can treat it as a straight line with only a very slight loss in accuracy.

Using the principle above, we have graphed the sine wave between 30° and 31° in Figures 1 and 2 below. In each case, we simply plotted sin 30° = 0.5000 and sin 31° = 0.5150, and then drew a straight line between them. These two sine values for 30° and 31° were obtained from our ordinary "Table of Trig Ratios".

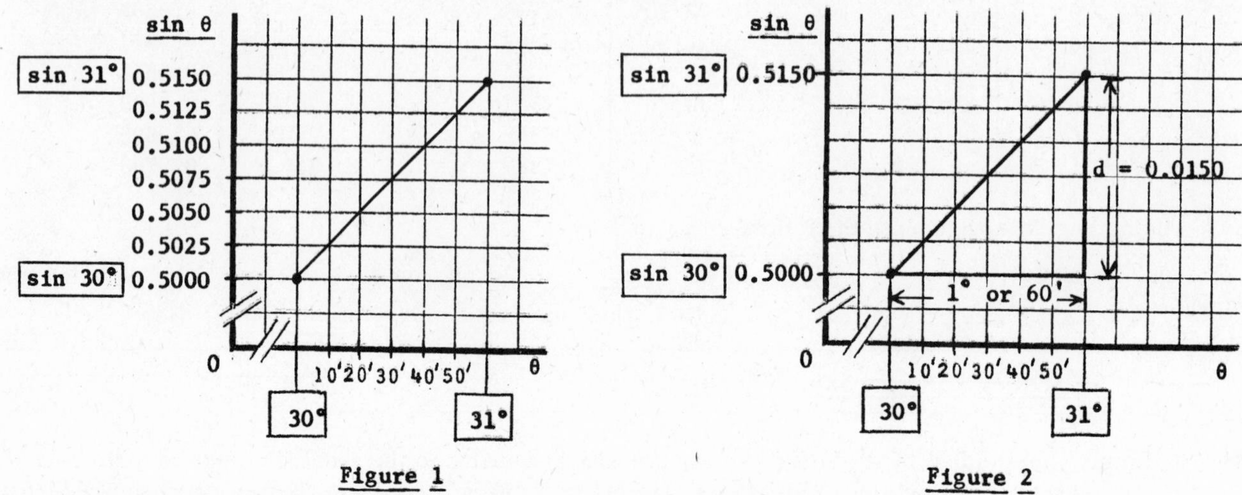

Figure 1 Figure 2

In Figure 2, we have drawn a horizontal and a vertical line to complete a right triangle.

The horizontal line represents the 1° (or 60') between 30° and 31°.

The vertical line (labeled "d") represents the "table difference" between sin 31° and sin 30°. That is:

d = sin 31° - sin 30° = 0.5150 - 0.5000 = 0.0150

Below we have used the same "two-point" method to graph the sine wave between 51° and 52° (Figure 3) and between 75° and 76° (Figure 4). In each case, the sines of the two exact-degree angles were obtained from our "Table of Trig Ratios".

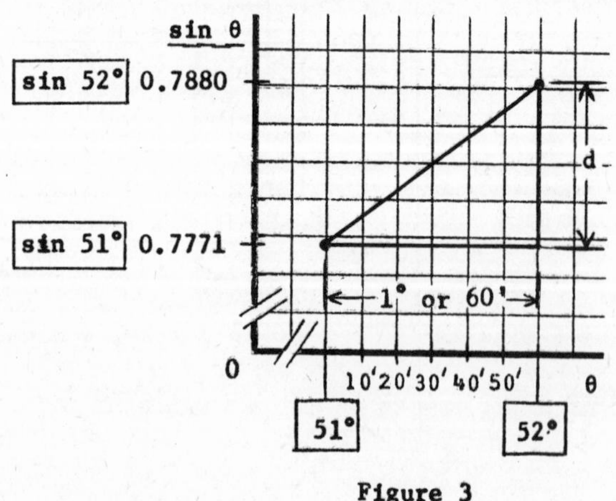

Figure 3

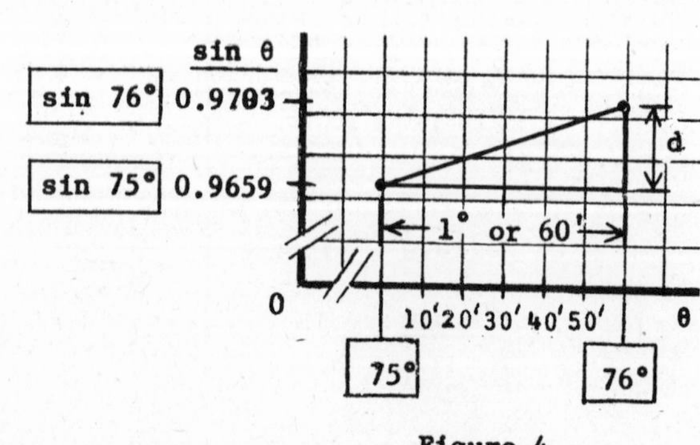

Figure 4

(Continued on following page.)

153. (Continued)

In each case, we have drawn a horizontal and a vertical line to complete a right triangle.

The horizontal line represents the 1° (or 60') between two adjacent exact-degree angles.

The vertical line (labeled "d") represents the table difference of the sines of the two exact-degree angles.

The numerical value of the table difference "d" is not given on either graph in Figures 3 and 4. Complete the subtractions below which are needed to find the table difference in each case:

(a) d = sin 52° - sin 51° = 0.7880 - 0.7771 = ________

(b) d = sin 76° - sin 75° = 0.9703 - 0.9659 = ________

---

a) 0.0109

b) 0.0044

154. As you saw in the last frame, we obtain the "table difference" between the sines of two angles by simply subtracting the sine of the smaller angle from the sine of the larger angle. For example, the "table difference" between sin 17° and sin 16° is:

sin 17° - sin 16° = 0.2924 - 0.2756 = 0.0168

We can find this "table difference" more easily by simply writing the two sines in vertical subtraction form, as we have done on the right.

sin 17° = 0.2924
sin 16° = 0.2756
0.0168

Using the vertical subtraction form, find the "table difference" between:

(a) sin 81° and sin 80° (b) sin 20° and sin 21° (c) sin 45° and 46°

sin 81° =
sin 80° = ________

Answer to Frame 154: a) 0.0029 b) 0.0164 c) 0.0122

---

155. Linear interpolation can be used to find the sine of an angle reported to the nearest minute. The interpolation process is called "linear" interpolation because it is based on treating a small segment of the sine curve as if it were a straight line. This process requires the solution of a proportion involving the corresponding sides of similar right triangles.

In this frame, we will explain the interpolation process for finding sin 30°30'. In the table given at the beginning of this section, we saw that sin 30°30' = 0.5075.

(Continued on following page.)

155. (Continued)

The interpolation process for finding sin 30°30' is based on the diagram on the right. Notice these points:

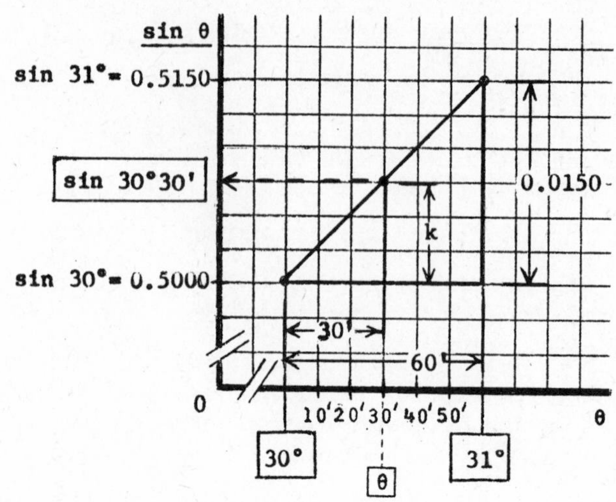

(1) Since sin 30°30' lies between sin 30° and sin 31°, its numerical value lies between 0.5000 and 0.5150. (See the dashed horizontal arrow.)

(2) Sin 30°30' = sin 30° + k = 0.5000 + k, where "k" is the amount we must add to sin 30° because of the extra 30'.

(3) To find "k", we can set up the following proportion based on the two similar right triangles in the figure:

$$\frac{k}{0.0150} = \frac{30'}{60'}$$

where: "k" is the amount we must add to sin 30°,

0.0150 is the "table difference" between sin 30° and 31°,

30' is the number of minutes beyond 30° in 30°30', and

60' is the number of minutes in the 1° between 30° and 31°.

Solving the proportion above for "k", we get:

$$k = \left(\frac{30'}{60'}\right)(0.0150) = \frac{1}{2}(0.0150) = 0.0075$$

Since sin 30°30' = 0.5000 + k,

sin 30°30' = 0.5000 + 0.0075 = 0.5075.

Using the interpolation process, we determined that sin 30°30' = 0.5075. Is this the same value of sin 30°30' which we gave in the table at the beginning of this section? __________

Yes

156. In the table at the beginning of this section, we saw that sin 51°20' = 0.7808. Let's use the interpolation process to find sin 51°20'. The interpolation process is based on the diagram below. Notice these points:

(1) Since sin 51°20' lies between sin 51° and sin 52°, sin 51°20' lies between 0.7771 and 0.7880. (See the dashed horizontal arrow.)

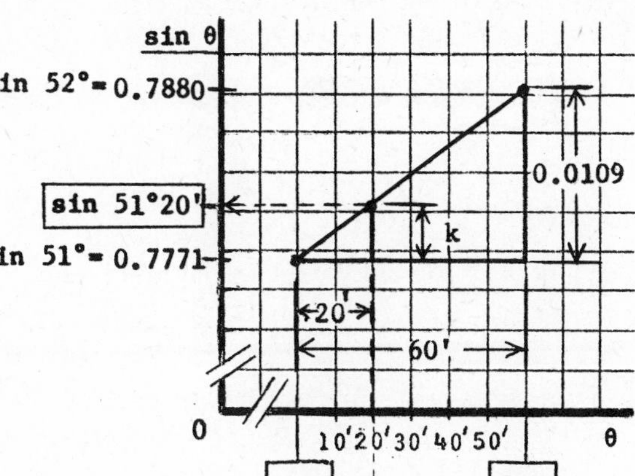

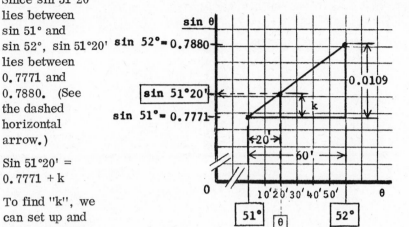

(2) Sin 51°20' = 0.7771 + k

(3) To find "k", we can set up and solve the following proportion based on the two similar right triangles in the figure:

$$\frac{k}{0.0109} = \frac{20'}{60'}$$ where 0.0109 is the "table difference".

Solving: $k = \left(\frac{20'}{60'}\right)(0.0109) = \left(\frac{1}{3}\right)(0.0109) = 0.0036$

(a) Since sin 51°20' = 0.7771 + k,
sin 51°20' = 0.7771 + 0.0036 = ________

(b) Is this the same value we gave earlier for sin 51°20' in the table? ________

---

a) 0.7807

b) Not quite the same, but it is very close.

---

157. In the table earlier, we saw that sin 75°40' = 0.9689. Let's use the interpolation process to find sin 75°40'. The process is based on the diagram below. Notice these points:

(1) Since sin 75°40' lies between sin 75° and sin 76°, sin 75°40' lies between 0.9659 and 0.9703. (See the dashed horizontal line.)

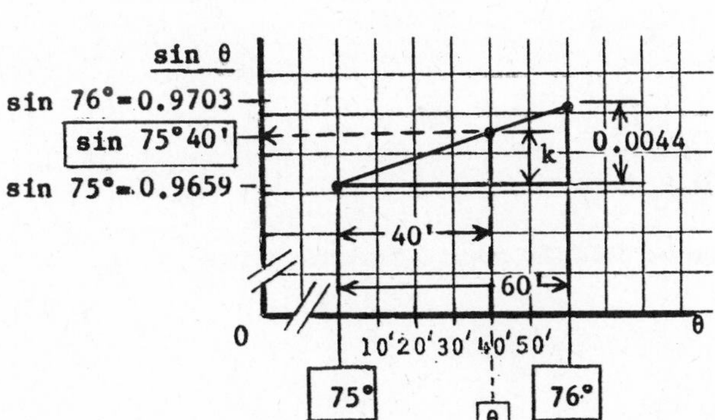

(2) Sin 75°40' = 0.9659 + k

(Continued on following page.)

157. (Continued)

(3) To find "k", we can set up and solve the following proportion based on the two similar right triangles:

$$\boxed{\frac{k}{0.0044} = \frac{40'}{60'}}$$ where 0.0044 is the "table difference".

Solving: $k = \left(\frac{40'}{60'}\right)(0.0044) = \frac{2}{3}(0.0044) = 0.0029$

(a) Since $\sin 75°40' = 0.9659 + k$,
$\sin 75°40' = 0.9659 + 0.0029 =$ ________

(b) Is this the same value we gave earlier for sin 75°40' in the table?

____

---

158. In the last three frames, we used diagrams in order to explain the interpolation process. We saw that the interpolated value of a sine is either identical to a more-accurate table value, or only slightly different from it. Since drawing diagrams for interpolations is tedious, let's examine the interpolations we have done in order to determine a method of interpolating which does not require diagrams:

$\sin 30°30' = \sin 30° + k$, where: $\frac{k}{0.0150} = \frac{30'}{60'}$

$\sin 51°20' = \sin 51° + k$, where: $\frac{k}{0.0109} = \frac{20'}{60'}$

$\sin 75°40' = \sin 75° + k$, where: $\frac{k}{0.0044} = \frac{40'}{60'}$

Note: (1) In each case, the interpolated sine is equal to the sine of the immediately smaller exact-degree angle, plus "k".

(2) "k" is obtained by solving a proportion which contains:

(a) The "table difference" between the immediately smaller and the immediately larger exact-degree angles.
(b) The number of minutes beyond the immediately smaller exact-degree angle.
(c) 60', the number of minutes in 1°.

Using the pattern above, we can find the sine of an angle by the interpolation method without using a diagram. As an example, we will find sin 42°18':

$\sin 42°18' = \sin 42° + k = 0.6691 + k$,

where: $\frac{k}{\text{Table Difference}} = \frac{18'}{60'}$

In order to solve for "k", we must find the table difference between sin 42° and sin 43°:

$\sin 43° - \sin 42° = 0.6820 - 0.6691 = 0.0129$

Substituting this value, we get: $\frac{k}{0.0129} = \frac{18'}{60'}$

or $k = \left(\frac{18'}{60'}\right)(0.0129) = \left(\frac{3}{10}\right)(0.0129) = 0.0039$

Therefore: $\sin 42°18' = 0.6691 + 0.0039 =$ ________

a) 0.9688

b) Not quite, but it is very close.

159. Let's use the same pattern to find sin 64°10':

$$\sin 64°10' = \sin 64° + k = 0.8988 + k,$$

$$\text{where: } \frac{k}{\text{Table Difference}} = \frac{10'}{60'}$$

(a) The table difference between sin 65° and sin 64° is ________.

(b) Substituting this value, solve for "k". k = ________

(c) Therefore: sin 64°10' = ________

0.6730

---

160. Let's use the same pattern to find sin 10°45':

$$\sin 10°45' = \sin 10° + k = 0.1736 + k,$$

$$\text{where: } \frac{k}{\text{Table Difference}} = \frac{45'}{60'}$$

(a) The table difference between sin 10° and sin 11° is ________.

(b) Solve for "k". k = ________

(c) Therefore: sin 10°45' = ________

a) 0.0075

b) 0.0013, from:

$$\frac{k}{0.0075} = \frac{10'}{60'}$$

$$\text{or } k = \left(\frac{1}{6}\right)(0.0075)$$

c) 0.9001

---

a) 0.0172

b) 0.0129, from:

$$\frac{k}{0.0172} = \frac{45'}{60'}$$

$$\text{of } k = \left(\frac{3}{4}\right)(0.0172)$$

c) 0.1865

161. The key step in the interpolation process is finding "k", the amount which must be added to the sine of the immediately smaller exact-degree angle. To find "k", we set up proportions like those on the left below, and then rearrange them to isolate "k" as we have done on the right:

$$\frac{k}{\text{Table Difference}} = \frac{10'}{60'} \quad \text{or} \quad k = \left(\frac{10'}{60'}\right)(\text{Table Difference})$$

$$\frac{k}{\text{Table Difference}} = \frac{45'}{60'} \quad \text{or} \quad k = \left(\frac{45'}{60'}\right)(\text{Table Difference})$$

In order to shorten this process, we can skip the formal proportions on the left and immediately write an equation like those on the right. For example, when interpolating to find sin 27°50', we can find "k" by writing:

$k = \left(\frac{50'}{60'}\right)(0.0155)$, where 0.0155 is the "table difference" between sin 27° and sin 28°.

Using the shortcut procedure above, complete each of these:

(a) To find sin 49°15', "k" = $\left(\frac{\quad}{\quad}\right)(\qquad)$

(b) To find sin 81°29', "k" = $\left(\frac{\quad}{\quad}\right)(\qquad)$

---

a) $\left(\frac{15'}{60'}\right)(0.0113)$

b) $\left(\frac{29'}{60'}\right)(0.0026)$

162. Here are examples of the type of equations we use in the shortcut procedure for finding "k":

$$k = \left(\frac{15'}{60'}\right)(0.0113) \qquad\qquad k = \left(\frac{29'}{60'}\right)(0.0026)$$

Since $\frac{15'}{60'}$ and $\frac{29'}{60'}$ are fractions, you can see that "k" is always some "fractional part" of the table difference between two successive exact-degree angles.

Let's find sin 18°30' by this shortcut procedure:

(a) The table difference between sin 18° and sin 19° is ________.

(b) $k = \left(\frac{\quad}{\quad}\right)(\qquad) =$ ________

(c) Therefore: sin 18°30' = sin 18° + k = ________

163. After setting up an equation to solve for "k", we ordinarily must use the slide rule to perform a combined multiplication and division. Since we are looking for some "fractional part" of the table difference, "k" must be less than the table difference. For example:

(a) If $k = \left(\frac{29'}{60'}\right)(0.0075)$, we know that "k" cannot be greater than ________.

(b) If $k = \left(\frac{59'}{60'}\right)(0.0105)$, we know that "k" cannot be greater than ________.

a) 0.0166

b) $\left(\frac{30'}{60'}\right)(0.0166) = 0.0083$

c) 0.3173, from (0.3090 + 0.0083)

---

164. When computing the value of "k", it is sometimes helpful to reduce the fraction to lowest terms if such a reduction is possible. One case in which such a reduction is especially helpful is when the numerator of the reduced fraction is "1". Here are some examples:

$$\left(\frac{30'}{60'}\right) = \frac{1}{2} \quad \left(\frac{20'}{60'}\right) = \frac{1}{3} \quad \left(\frac{15'}{60'}\right) = \frac{1}{4} \quad \left(\frac{12'}{60'}\right) = \frac{1}{5} \quad \left(\frac{10'}{60'}\right) = \frac{1}{6}$$

If the fraction can be reduced to one whose numerator is "1", we can find "k" by simply dividing the table difference by the denominator. For example:

$$k = \left(\frac{20'}{60'}\right)(0.0111) = \left(\frac{1}{3}\right)(0.0111) = \frac{0.0111}{3} = 0.0037$$

Complete each of these:

(a) $k = \left(\frac{30'}{60'}\right)(0.0057) =$ ____________________

(b) $k = \left(\frac{12'}{60'}\right)(0.0115) =$ ____________________

a) 0.0075

b) 0.0105

---

165. Another type of reduction which is very useful is one in which the fraction reduces to "tenths". This is true whenever the numerator is a multiple of "6". For example:

$$\left(\frac{6'}{60'}\right) = 0.1 \quad \left(\frac{18'}{60'}\right) = 0.3 \quad \left(\frac{36'}{60'}\right) = 0.6 \quad \left(\frac{54'}{60'}\right) = 0.9$$

Since these fractions reduce to "tenths", we can find "k" without using the slide rule. For example:

$$k = \left(\frac{24'}{60'}\right)(0.0069) = (0.4)(0.0069) = 0.00276 \text{ or } 0.0028$$

Complete each of the following:

(a) $k = \left(\frac{6'}{60'}\right)(0.0153) = (\quad)(\quad) =$ ________

(b) $k = \left(\frac{42'}{60'}\right)(0.0023) = (\quad)(\quad) =$ ________

a) $\frac{0.0057}{2} = 0.0029$

b) $\frac{0.0115}{5} = 0.0023$

---

a) (0.1)(0.0153) = 0.0015

b) (0.7)(0.0023) = 0.0016

166. Many other fractions which occur in the equations used to compute "k" can be reduced to lowest terms. For example:

$$\left(\frac{40'}{60'}\right) = \frac{2}{3} \quad \left(\frac{45'}{60'}\right) = \frac{3}{4} \quad \left(\frac{25'}{60'}\right) = \frac{5}{12} \quad \left(\frac{33'}{60'}\right) = \frac{11}{20}$$

In terms of actually computing the value of "k", we can make these statements about these reductions:

(1) Most of them do not eliminate the need to perform a combined multiplication and division on the slide rule.

(2) Some of them might help you to estimate the slide rule answer.

Therefore, whether you reduce fractions like those above to lowest terms or not is strictly a matter of personal choice. Use your personal choice to complete these:

(a) $\left(\frac{45'}{60'}\right)(0.0087) =$ ________ (b) $\left(\frac{21'}{60'}\right)(0.0152) =$ ________

---

167. Even if you don't reduce the fraction to lowest terms, it is helpful to mentally round it to a simple fraction in order to estimate the slide rule answer. For example:

$$\left(\frac{32'}{60'}\right) \doteq \frac{1}{2} \quad \left(\frac{18'}{60'}\right) \doteq \frac{1}{4} \text{ or } \frac{1}{3} \quad \left(\frac{7'}{60'}\right) \doteq \frac{1}{10}$$

One fact you should keep in mind when computing "k" is this:

Since "k" is a <u>fractional part</u> of the table difference, "k" cannot be larger than the ________________.

Answers to 166:
a) 0.0065
b) 0.0053

---

168. In general, then, we can make the following statements about reducing the fraction to lowest terms when computing the value of "k":

(1) A reduction is very useful if:

(a) the reduced numerator is "1", or if
(b) the fraction reduces to "tenths".

(2) Most other reductions do not eliminate the need to perform a combined operation on the slide rule. However, they may simplify the estimation.

(3) Since the value of "k" can always be obtained on the slide rule without a reduction to lowest terms, reducing to lowest terms is inefficient if it takes too much time.

Answer to 167: table difference

Go to next frame.

169. Let's review the complete interpolation process by finding sin 13°29':

(a) Find the table difference between sin 13° and sin 14°. ________

(b) $k = \left(\frac{\quad}{\quad}\right)(\qquad) =$ ________

(c) sin 13°29' = sin 13° + k = ________

---

170. Use interpolation to find the sines of the following angles:

(a) sin 2°30' = ________ (b) sin 20°15' = ________

Answers to 169:

a) 0.0169

b) $\left(\frac{29'}{60'}\right)(0.0169) = 0.0082$

c) 0.2332

---

171. Use interpolation to find these sines:

(a) sin 36°50' = ________ (b) sin 87°17' = ________

Answers to 170:

a) sin 2°30' = 0.0436

b) sin 20°15' = 0.3461

Answers to 171:

a) sin 36°50' = 0.5995

b) sin 87°17' = 0.9988

SELF-TEST 12 (Frames 151-171)

Find sin 8°30' by interpolation, as follows:

1. The table difference is ________.
2. k = ________
3. sin 8°30' = sin 8° + k = ________

Find sin 52°42' by interpolation, as follows:

4. The table difference is ________.
5. k = ________
6. sin 52°42' = sin 52° + k = ________

Find each of the following by interpolation:

7. sin 78°12' = ________
8. sin 24°35' = ________
9. sin 83°49' = ________

ANSWERS:

1. 0.0172
2. 0.0086
3. sin 8°30' = 0.1478
4. 0.0106
5. 0.0074
6. sin 52°42' = 0.7954
7. sin 78°12' = 0.9788
8. sin 24°35' = 0.4160
9. sin 83°49' = 0.9941

6-13 INTERPOLATING TO FIND TANGENT RATIOS OF ANGLES TO THE NEAREST MINUTE

In this section, we will use linear interpolation to find the tangents of angles reported to the nearest minute. The interpolation process for finding tangents is similar to the interpolation process for finding sines.

172. In the table below, we have listed the tangents of some angles reported to the nearest minute. The specific angles lie either between 22° and 23°, 46° and 47°, or 64° and 65°. These tangents were computed by a complicated mathematical procedure.

θ	tan θ
22° 0'	0.4040
22°10'	0.4074
22°20'	0.4108
22°30'	0.4142
22°40'	0.4176
22°50'	0.4211
23° 0'	0.4245

θ	tan θ
46° 0'	1.036
46°10'	1.042
46°20'	1.048
46°30'	1.054
46°40'	1.060
46°50'	1.066
47° 0'	1.072

θ	tan θ
64° 0'	2.050
64°10'	2.066
64°20'	2.081
64°30'	2.097
64°40'	2.112
64°50'	2.128
65° 0'	2.145

The pairs of values in the table above have been graphed below in Figures 1, 2, and 3:

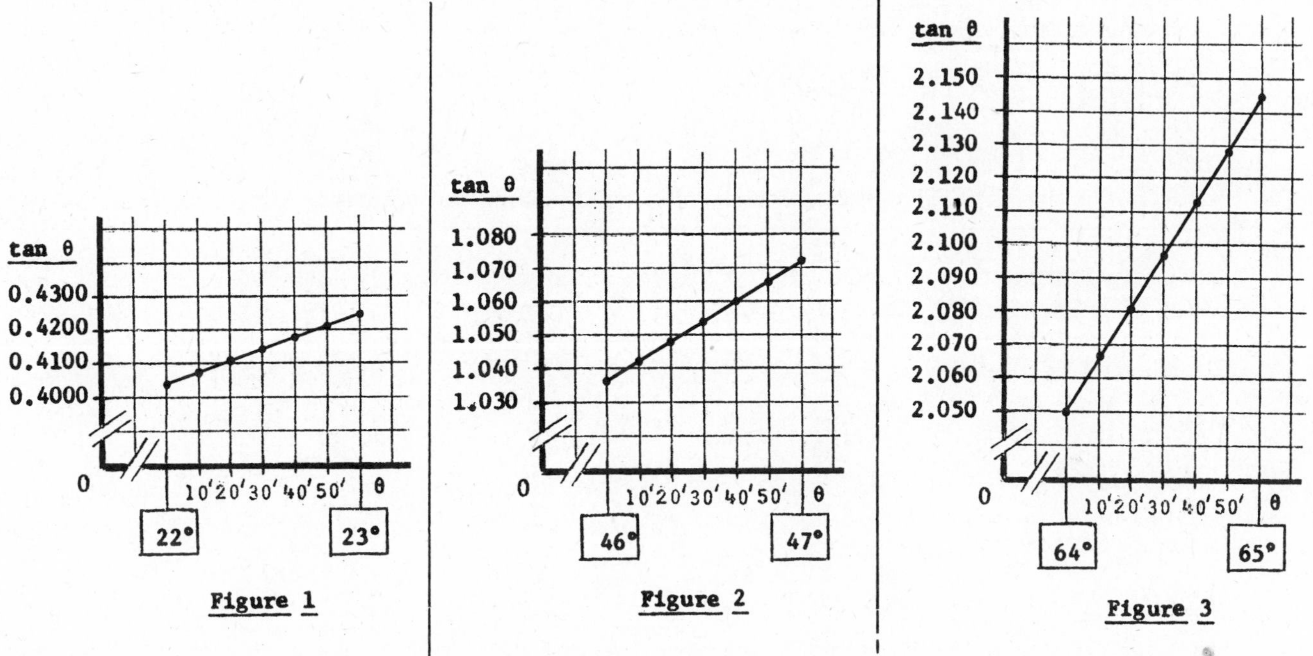

Though the graph of the tangent ratio is a curve, you can see that the three enlargements of short segments of that curve are approximately straight lines.

---

173. Since any short segment of the tangent curve is approximately a straight line, we can treat it as a straight line with only a slight loss of accuracy. Therefore, we can use <u>linear interpolation</u> to find the tangents of angles reported to the nearest minute. The values obtained by interpolation are very close to those generated by the mathematicians' complicated procedure.

The procedure for finding tangents by interpolation is similar to that for finding sines by interpolation. The key step is solving a proportion involving the corresponding sides of similar right triangles. The interpolation process for finding tan 46°40' is based on the diagram on the following page.

(Continued on following page.)

173. (Continued)

Notice these points:

(1) Since tan 46°40' lies between tan 46° and tan 47°, the numerical value of tan 46°40' lies between 1.036 and 1.072.

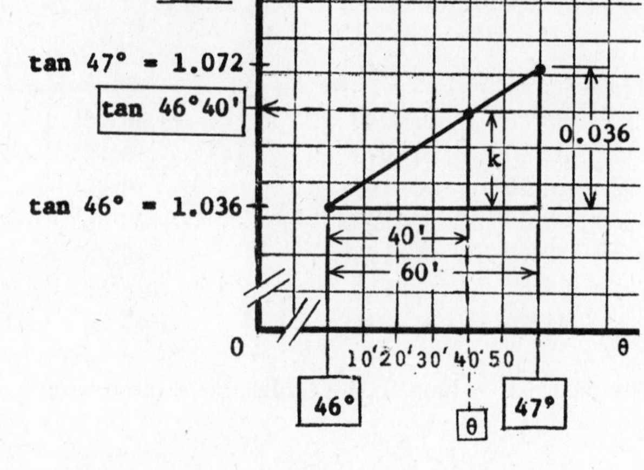

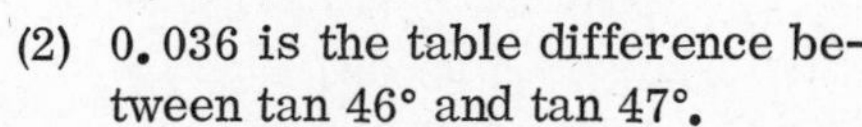

(2) 0.036 is the table difference between tan 46° and tan 47°.

(3) "k" is the amount which must be added to tan 46° in order to get tan 46°40'.

(4) To find "k", we set up and solve the following proportion based on the similar right triangles in the diagram:

$$\frac{k}{0.036} = \frac{40'}{60'} \quad \text{or} \quad k = \left(\frac{40'}{60'}\right)(0.036) = \left(\frac{2}{3}\right)(0.036) = 0.024$$

(a) tan 46°40' = tan 46° + k = 1.036 + 0.024 = ________

(b) Is this the same value given for tan 46°40' in the table in the last frame? ________

---

a) 1.060

b) Yes

174. When interpolating to find a tangent ratio, we can use the same shortcut we used when interpolating sine ratios. That is, we can immediately write a fraction times the table difference, instead of writing the formal proportion. For example:

When finding tan 20°15', $k = \left(\frac{15'}{60'}\right)(0.0199)$

(0.0199 is the table difference between tan 20° and tan 21°.)

Using the same principle, complete these:

(a) When finding tan 37°26', $k = \left(\frac{\quad}{\quad}\right)(\qquad)$

(Note: The table difference between tan 37° and tan 38° is 0.0277.)

(b) When finding tan 78°30', $k = \left(\frac{\quad}{\quad}\right)(\qquad)$

(Note: The table difference between tan 78° and tan 79° is 0.440.)

---

a) $\left(\frac{26'}{60'}\right)(0.0277)$

b) $\left(\frac{30'}{60'}\right)(0.440)$

175. Here is the complete procedure for finding tan 71°30' by interpolation:

(a) The table difference between tan 71° and tan 72° is ________.

(b) $k = \left(\frac{\quad}{\quad}\right)(\qquad) =$ ________

(c) tan 71°30' = ________

---

a) 0.174

b) $\left(\frac{30'}{60'}\right)(0.174) = 0.087$

c) 2.991

176. Use interpolation to find the following tangents:

(a) tan 17°20' = ________ (b) tan 84°45' = ________

---

a) tan 17°20' = 0.3121

b) tan 84°45' = 10.95

177. Use interpolation to find the following tangents:

(a) tan 44°10' = ________ (b) tan 61°47' = ________

---

a) tan 44°10' = 0.9714

b) tan 61°47' = 1.864

SELF-TEST 13 (Frames 172-177)

1. By interpolation, $\tan 7°36' =$ ________.

$\tan 8° = 0.1405$
$\tan 7° = 0.1228$

2. By interpolation, $\tan 72°50' =$ ________.

$\tan 73° = 3.271$
$\tan 72° = 3.078$

ANSWERS:

1. $\tan 7°36' = 0.1334$

From: $\tan 7°36' = \tan 7° + k$
$= 0.1228 + 0.0106$
$= 0.1334$

2. $\tan 72°50' = 3.239$

From: $\tan 72°50' = \tan 72° + k$
$= 3.078 + 0.161$
$= 3.239$

6-14 INTERPOLATING TO FIND COSINE RATIOS OF ANGLES TO THE NEAREST MINUTE

In this section, we will use linear interpolation to find the cosines of angles reported to the nearest minute. The interpolation process for finding cosines is similar to the interpolation process for finding sines and tangents. The only difference is due to the fact that the cosine ratio decreases from 0° to 90°.

178. Both the sine ratio and the tangent ratio increase in numerical value as the angle increases from 0° to 90°. To be more specific:

The sine ratio increases from 0 at 0° to +1 at 90°.

The tangent ratio increases from 0 at 0° to 57.29 at 89°.
(Note: tan 90° is indeterminate.)

We have given the cosines of a few angles from 0° to 90° in the table at the right. As you can see, the cosine ratio decreases as the angle increases from 0° to 90°. That is:

θ	$\cos \theta$
0°	1.0000
30°	0.8660
45°	0.7071
60°	0.5000
90°	0.0000

The cosine ratio decreases from ________ at 0° to ________ at 90°.

Answer to Frame 178: ... from 1.0000 (or +1) at 0° to 0.0000 (or 0) at 90°.

179. In the table below, we have listed the cosines of some angles reported to the nearest minute. The specific angles lie either between 14° and 15°, 48° and 49°, or 81° and 82°. These cosines were computed by a complicated mathematical procedure.

θ	cos θ
14° 0'	0.9703
14°10'	0.9696
14°20'	0.9689
14°30'	0.9682
14°40'	0.9674
14°50'	0.9667
15° 0'	0.9659

θ	cos θ
48° 0'	0.6691
48°10'	0.6670
48°20'	0.6648
48°30'	0.6626
48°40'	0.6604
48°50'	0.6583
49° 0'	0.6561

θ	cos θ
81° 0'	0.1564
81°10'	0.1536
81°20'	0.1507
81°30'	0.1478
81°40'	0.1449
81°50'	0.1421
82° 0'	0.1392

The pairs of values in the table above have been graphed below in Figures 1, 2, and 3:

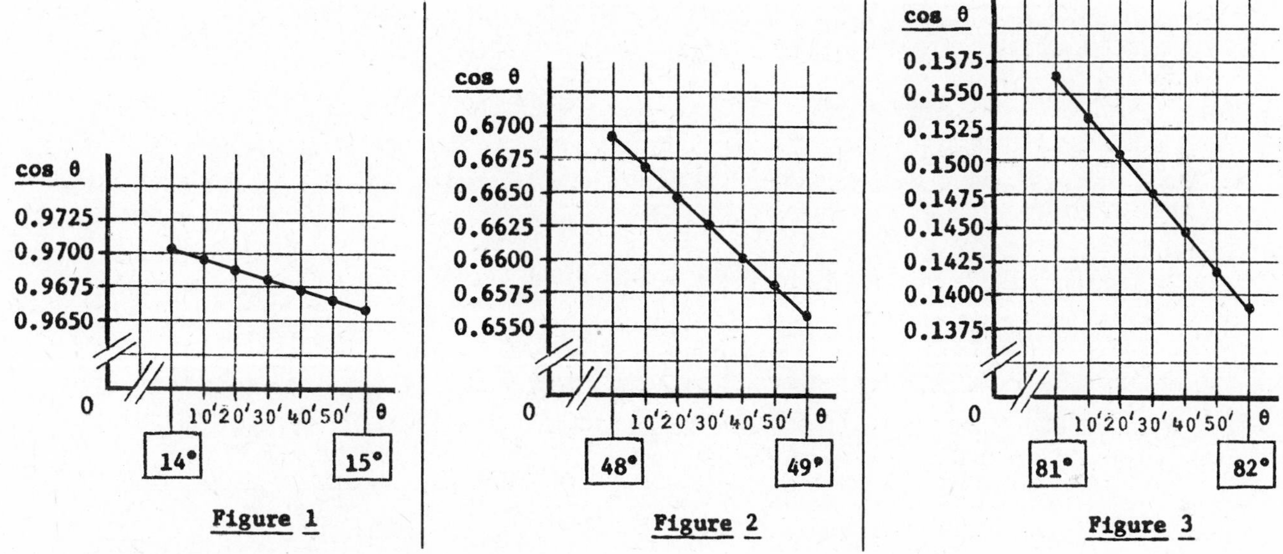

Figure 1 Figure 2 Figure 3

Notice these two points about the graphs:

(1) Though the graph of the cosine ratio is a curve, the three enlargements of short segments of that curve are approximately straight lines.

(2) The graphs slope downward to the right. That is, as the size of the angle increases, the numerical value of the cosine ratio decreases.

---

180. Since any short segment of the cosine curve is approximately a straight line, the cosine of angles reported to the nearest minute can be approximated quite closely by means of linear interpolation. The procedure for finding cosines by interpolation is similar to that for finding sines and tangents by interpolation. The only major difference is the fact that the cosine ratio decreases as the angle increases. This fact requires that "k" be subtracted.

(Continued on following page.)

180. (Continued)

The interpolation process for finding cos 48°30' is based on the diagram below.

Notice these points:

(1) Since cos 48°30' lies between cos 48° and cos 49°, the numerical value of cos 48°30' lies between 0.6691 and 0.6561.

(2) Since cos 48° = 0.6691 and cos 49° = 0.6561, the table difference between cos 48° and cos 49° is 0.0130.

(3) Since the cosine ratio <u>decreases</u>, cos 48°30' is <u>less</u> than cos 48°. That is, cos 48°30' is less than 0.6691.

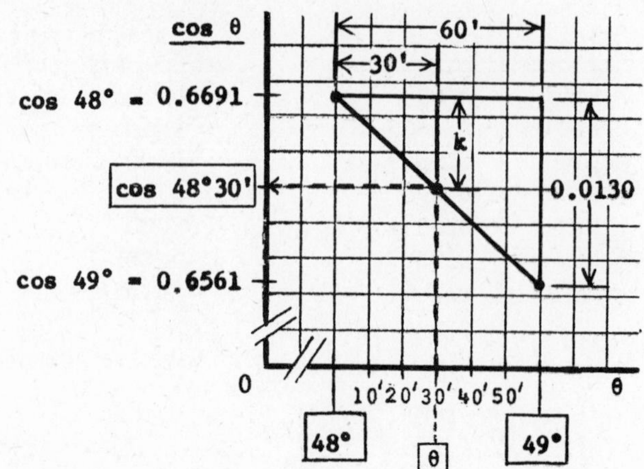

(4) "k" represents the amount we must <u>subtract</u> from cos 48° in order to get cos 48°30'. That is: cos 48°30' = cos 48° - k

(5) To find "k", we set up and solve the following proportion based on the two similar right triangles in the diagram:

$$\frac{k}{0.0130} = \frac{30'}{60'} \quad \text{or} \quad k = \left(\frac{30'}{60'}\right)(0.0130) = \left(\frac{1}{2}\right)(0.0130) = 0.0065$$

Complete the interpolation process:

(a) Therefore: cos 48°30' = cos 48° - k = 0.6691 - 0.0065 = ________

(b) In the table shown in the last frame, the listed table value for cos 48°30' was 0.6626. Is the interpolated value the same as the table value? ________

<u>Answer to Frame 180:</u> a) 0.6626 b) Yes

---

181. Let's use the interpolation process to find cos 14°40'. The process is based on the diagram below.

Notice these points:

(1) Since cos 14°40' lies between cos 14° and cos 15°, cos 14°40' lies between 0.9703 and 0.9659.

(2) The table difference between cos 14° and cos 15° is 0.0044.

(3) cos 14°40' is <u>less</u> than cos 14°.

(4) "k" is the amount we must subtract from cos 14° in order to get cos 14°40'. That is: cos 14°40' = cos 14° - k

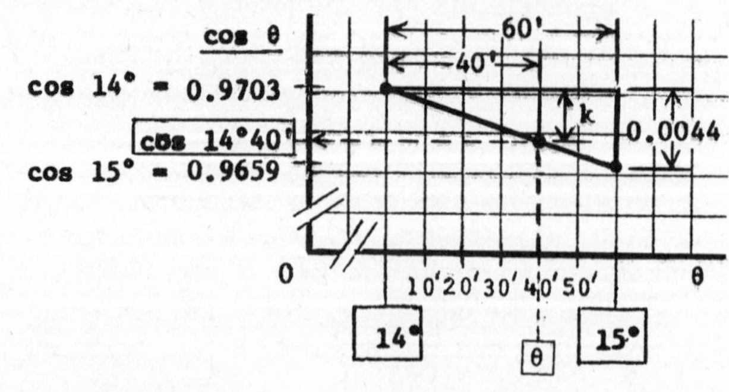

(5) To find "k", we set up and solve the following proportion:

$$\frac{k}{0.0044} = \frac{40'}{60'} \quad \text{or} \quad k = \left(\frac{40'}{60'}\right)(0.0044) = \left(\frac{2}{3}\right)(0.0044) = 0.0029$$

(Continued on following page.)

181. (Continued)

Complete the interpolation process:

(a) Therefore: cos 14°40' = cos 14° - k = 0.9703 - 0.0029 = __________

(b) Is this interpolated value the same as the table value shown earlier in this section? The table value for cos 14°40' was 0.9674. __________

Answer to Frame 181: a) 0.9674 b) Yes

182. Let's interpolate to find cos 81°20'. The interpolation is based on the diagram below.

Notice these points:

(1) Since cos 81°20' lies between cos 81° and cos 82°, cos 81°20' lies between 0.1564 and 0.1392.

(2) The table difference between cos 81° and cos 82° is 0.0172.

(3) cos 81°20' is <u>less</u> than cos 81°.

(4) "k" is the amount we must subtract from cos 81° in order to get cos 81°20'. That is: cos 81°20' = cos 81° - k

(5) To find "k", we set up and solve the following proportion:

$$\frac{k}{0.0172} = \frac{20'}{60'} \quad \text{or}$$

$$k = \left(\frac{20'}{60'}\right)(0.0172) = \left(\frac{1}{3}\right)(0.0172) = 0.0057$$

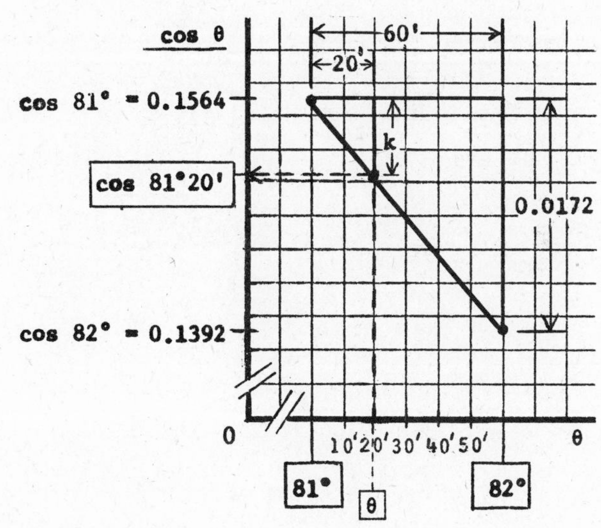

Complete the interpolation process:

(a) Therefore: cos 81°20' = cos 81° - k = 0.1564 - 0.0057 = __________

(b) Is this interpolated value the same as the table value shown earlier in this section? The table value for cos 81°20' was 0.1507. __________

183. The only difference between interpolating to find sines or tangents and interpolating to find cosines is this: <u>Because the cosine ratio is a decreasing quantity, we subtract "k" from the cosine of the immediately smaller exact-degree angle.</u>

Let's interpolate to find cos 56°15':

(a) The table difference between cos 56° and cos 57° is __________.

(b) $k = \left(\frac{\quad}{\quad}\right)(\qquad) =$ __________

(c) cos 56°15' = cos 56° - k = __________

a) 0.1507

b) Yes

184. Use interpolation to find these cosines:

(a) cos 21°30' = __________ (b) cos 74°10' = __________

a) 0.0146

b) $\left(\frac{15'}{60'}\right)(0.0146) = 0.0037$

c) cos 56°15' = 0.5555
From: 0.5592 - 0.0037

---

185. Interpolate to find these cosines:

(a) cos 63°48' = __________ (b) cos 37°31' = __________

a) cos 21°30' = 0.9304
From: 0.9336 - 0.0032

b) cos 74°10' = 0.2728
From: 0.2756 - 0.0028

---

186. In this frame and the next, we will interpolate to find sines, cosines, and tangents. Remember:

We add "k" to find sines and tangents.
We subtract "k" to find cosines.

Use interpolation for these:

(a) sin 45°20' = __________ (b) cos 8°40' = __________

a) cos 63°48' = 0.4415
From: 0.4540 - 0.0125

b) cos 37°31' = 0.7931
From: 0.7986 - 0.0055

187. Interpolate to find these:

(a) tan 77°15' = ________ (b) cos 87°36' = ________

a) sin 45°20' = 0.7112
From: 0.7071 + 0.0041

b) cos 8°40' = 0.9886
From: 0.9903 - 0.0017

Answer to Frame 187:

a) tan 77°15' = 4.425
From: 4.331 + 0.094

b) cos 87°36' = 0.0419
From: 0.0523 - 0.0104

SELF-TEST 14 (Frames 178-187)

Using interpolation, find the numerical value of each trig ratio:

1. cos 10°36' = ________

cos 10° = 0.9848
cos 11° = 0.9816

2. sin 26°45' = ________

sin 27° = 0.4540
sin 26° = 0.4384

3. tan 53°18' = ________

tan 54° = 1.376
tan 53° = 1.327

4. cos 68°21' = ________

cos 68° = 0.3746
cos 69° = 0.3584

ANSWERS:

1. cos 10°36' = 0.9829
From: 0.9848 - 0.0019

2. sin 26°45' = 0.4501
From: 0.4384 + 0.0117

3. tan 53°18' = 1.342
From: 1.327 + 0.015

4. cos 68°21' = 0.3689
From: 0.3746 - 0.0057

6-15 INTERPOLATING TO REPORT ANGLES TO THE NEAREST MINUTE

When reporting an angle with a given sine, cosine, or tangent up to this point, we have reported the nearest exact-degree angle. Since angles can be measured more precisely to the nearest minute, we can report an angle to the nearest minute for any given trigonometric ratio. Linear interpolation can be used to do so. We will discuss the method in this section.

188. Up to this point, we have reported the closest exact-degree angle when finding an angle with a specific sine. For example, when reporting the angle whose sine is 0.5100, we have reported 31°, since 0.5100 is closer to 0.5150 (which is sin 31°) than to 0.5000 (which is sin 30°). However, when reporting an angle with a specific sine, we can be more precise and report the angle to the nearest minute. In an earlier section, for example, we saw that 0.5100 is sin 30°40'.

Interpolation can be used to report an angle to the nearest minute. The interpolation process is again based on the solution of a proportion involving the corresponding sides of similar right triangles. The interpolation process for finding the "nearest-minute" angle whose sine is 0.5100 is based on the diagram below.

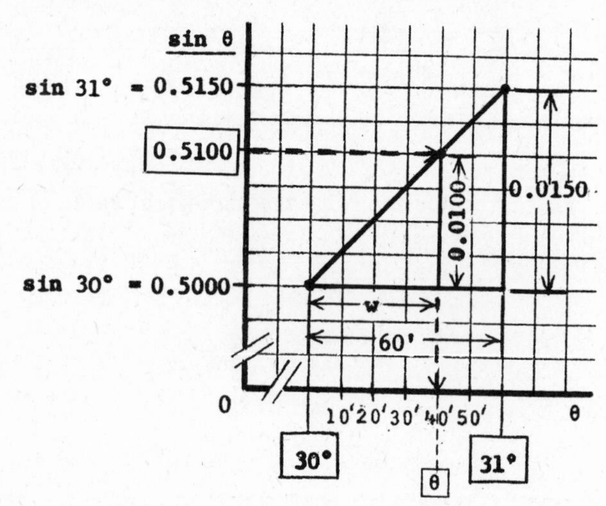

Notice these points:

(1) Since 0.5100 lies between 0.5000 and 0.5150, the angle whose sine is 0.5100 lies between 30° and 31°. (See the dashed arrows.)

(2) 0.0150 is the table difference between sin 30° and sin 31°.

(3) 0.0100 is the difference between 0.5100 and 0.5000. (0.5000 is the sine of 30°, the immediately smaller exact-degree angle.)

(4) "w" represents the number of minutes which must be added to 30° to get an angle whose sine is 0.5100.

(5) To find "w", we can set up and solve the following proportion based on the two similar right triangles in the diagram:

$$\boxed{\frac{w}{60'} = \frac{0.0100}{0.0150}} \quad \text{or} \quad w = \left(\frac{0.0100}{0.0150}\right)(60') = \left(\frac{2}{3}\right)(60') = 40'$$

Therefore, if $\sin \theta = 0.5100$, $\theta = 30° + w =$ ________

$\theta = 30° + 40'$ or $30°40'$

189. In an earlier section, we saw that $\theta = 51°20'$ when $\sin\theta = 0.7808$. Let's use interpolation to find the angle whose sine is 0.7808. The interpolation process is based on the diagram below.

Notice these points:

(1) Since 0.7808 lies between 0.7771 and 0.7880, the angle whose sine is 0.7808 lies between 51° and 52°.

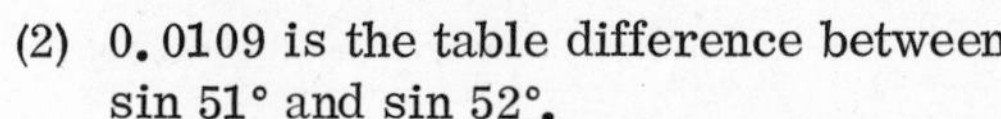

(2) 0.0109 is the table difference between sin 51° and sin 52°.

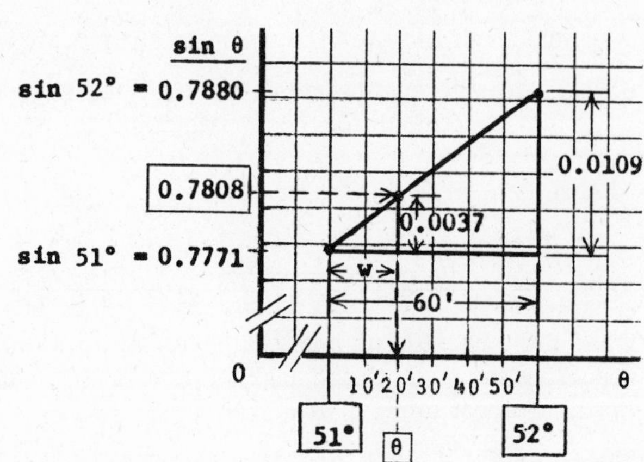

(3) 0.0037 is the difference between 0.7808 and 0.7771. (0.7771 is sin 51°.)

(4) "w" represents the number of minutes which must be added to 51° to get an angle whose sine is 0.7808.

(5) To find "w", we set up and solve the following proportion:

$$\boxed{\frac{w}{60'} = \frac{0.0037}{0.0109}} \quad \text{or} \quad w = \left(\frac{0.0037}{0.0109}\right)(60') = 20.4' \text{ or } 20'$$

Therefore, if $\sin\theta = 0.7808$, $\theta = 51° + w =$ ________

Answer: $\theta = 51° + 20'$ or $51°20'$

190. The proportions used in the interpolations in the last two frames are listed on the left below. On the right, we have solved for "w" in each proportion.

$$\frac{w}{60'} = \frac{0.0100}{0.0150} \quad \text{or} \quad w = \left(\frac{0.0100}{0.0150}\right)(60')$$

$$\frac{w}{60'} = \frac{0.0037}{0.0109} \quad \text{or} \quad w = \left(\frac{0.0037}{0.0109}\right)(60')$$

As you can see from the solutions on the right, "w" is always a fractional part of 60' (or 1°). In each case:

(1) The numerator of the fraction is the difference between the given sine and the sine of the immediately smaller exact-degree angle.

(2) The denominator is the "table difference" between the two adjacent exact-degree angles.

Using equations like those on the right above, we can find "w" without drawing a diagram. Here is an example:

Let's find θ (to the nearest minute) when $\sin\theta = 0.9537$.

Since $\sin 72° = 0.9511$ and $\sin 73° = 0.9563$, $\theta = 72° + w$.

The given sine minus $\sin 72° = 0.9537 - 0.9511 = 0.0026$.
The table difference between sin 72° and sin 73° is 0.0052.

Therefore: (a) $w = \left(\frac{0.0026}{0.0052}\right)(60') = \left(\frac{1}{2}\right)(60') =$ ________

(b) $\theta = 72° + w =$ ________

---

a) $w = 30'$

b) $\theta = 72°30'$

191. When using the slide rule to find "w", we round "w" to the nearest whole-number minute.

Let's find angle θ (to the nearest minute) when $\sin\theta = 0.2437$.

Since $\sin 14° = 0.2419$ and $\sin 15° = 0.2588$, $\theta = 14° + w$.

The given sine minus $\sin 14° = 0.2437 - 0.2419 = 0.0018$.
The table difference between $\sin 14°$ and $\sin 15°$ is 0.0169.

Therefore: (a) $w = \left(\frac{0.0018}{0.0169}\right)(60') =$ ________

(b) $\theta =$ ____________

a) 6'
b) 14°6'

192. When the tangent of an angle is given, the interpolation process for reporting the angle to the nearest minute is the same. Let's find θ (to the nearest minute) when $\tan\theta = 3.459$.

Since $\tan 73° = 3.271$ and $\tan 74° = 3.487$, $\theta = 73° + w$.

The given tangent minus $\tan 73°$ is 0.188.
The table difference between $\tan 73°$ and $\tan 74°$ is 0.216.

Therefore: (a) $w = \left(\frac{0.188}{0.216}\right)(60') =$ ________

(b) $\theta =$ __________

a) 52'
b) 73°52'

193. Use interpolation to find the following angles to the nearest minute:

(a) If $\sin\theta = 0.6375$, $\theta =$ _____. (b) If $\tan\theta = 1.616$, $\theta =$ _________.

a) $\theta = 39°36'$

Note:

$w = \left(\frac{0.0082}{0.0135}\right)(60') = 36'$

b) $\theta = 58°15'$

Note:

$w = \left(\frac{0.016}{0.064}\right)(60') = 15'$

194. When the cosine of an angle is given, the interpolation process for reporting the angle to the nearest minute is quite similar to the process for sine and tangent. The only difference is that the cosine ratio is a decreasing quantity between 0° and 90°. As an example, we will find θ (to the nearest minute) when $\cos\theta = 0.6613$. The interpolation process is based on the diagram below.

Notice these points:

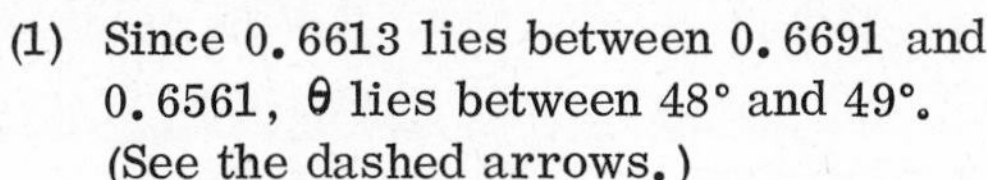

(1) Since 0.6613 lies between 0.6691 and 0.6561, θ lies between 48° and 49°. (See the dashed arrows.)

(2) The difference between cos 48° and 0.6613 is 0.0078.

(3) The table difference between cos 48° and cos 49° is 0.0130.

(4) "w" represents the number of minutes which must be added to 48° to get an angle whose cosine is 0.6613.

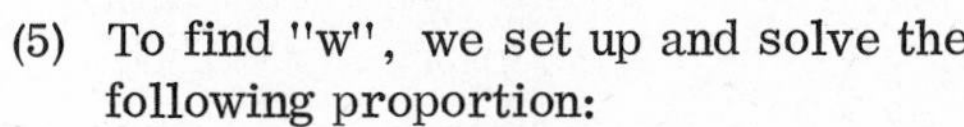

(5) To find "w", we set up and solve the following proportion:

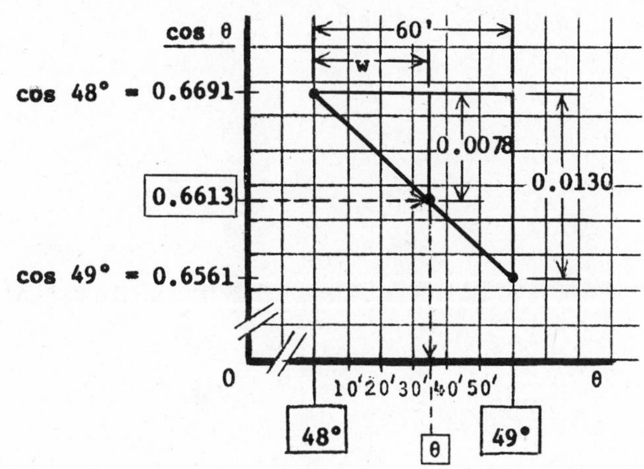

$$\frac{w}{60'} = \frac{0.0078}{0.0130} \quad \text{or} \quad w = \left(\frac{0.0078}{0.0130}\right)(60') = 36'$$

Therefore: $\theta = 48° + w$ or ____________

Answer: $\theta = 48°36'$

---

195. We can also find θ (to the nearest minute) without a diagram when $\cos\theta$ is given. Here is an example:

Let's find θ when $\cos\theta = 0.8888$.

Since $\cos 27° = 0.8910$ and $\cos 28° = 0.8829$, $\theta = 27° + w$.

Cos 27° minus the given cosine $= 0.8910 - 0.8888 = 0.0022$.
The table difference between cos 27° and cos 28° is 0.0081.

Therefore: (a) $w = \left(\frac{0.0022}{0.0081}\right)(60')$

(b) $\theta =$ ____________

Answer: a) $w = 16'$ b) $\theta = 27°16'$

---

196. Using the interpolation procedure in the last frame, report each angle to the nearest minute:

(a) If $\cos\theta = 0.1132$, $\theta =$ ______. (b) If $\cos\theta = 0.7466$, $\theta =$ ______.

197. When interpolating to find a trigonometric ratio, we add (or subtract) a fractional part of the table difference.

When interpolating to find an angle, we add a fractional part of 60'.

Use interpolation to complete these:

(a) $\sin 47°15' =$ ____________

(b) If $\cos \theta = 0.5075$, θ (to the nearest minute) = ____________

a) $\theta = 83°30'$

Note: $w = \left(\frac{0.0087}{0.0174}\right)(60')$

$w = 30'$

b) $\theta = 41°42'$

Note: $w = \left(\frac{0.0081}{0.0116}\right)(60')$

$w = 42'$

---

198. Use interpolation for these problems:

(a) If $\tan \theta = 2.025$, $\theta =$ __________. (b) $\cos 16°30' =$ __________

a) 0.7343

b) 59°30'

---

199. Report each angle to the nearest minute:

(a) If $\sin \theta = 0.1110$, $\theta =$ ______. (b) If $\cos \theta = 0.8640$, $\theta =$ ______.

a) 63°43'

b) 0.9588

---

a) $\theta = 6°22'$

b) $\theta = 30°14'$

SELF-TEST 15 (Frames 188-199)

By interpolation, find each angle to the nearest minute:

1. If $\sin A = 0.4014$,	2. If $\tan \varphi = 1.120$,	3. If $\cos \theta = 0.8534$,
$A =$ ________	$\varphi =$ ________	$\theta =$ ________
$\sin 24° = 0.4067$ $\sin 23° = 0.3907$	$\tan 49° = 1.150$ $\tan 48° = 1.111$	$\cos 31° = 0.8572$ $\cos 32° = 0.8480$

ANSWERS: 1. $A = 23°40'$ 2. $\varphi = 48°14'$ 3. $\theta = 31°25'$

6-16 INTERPOLATING WITH ANGLES REPORTED TO THE NEAREST SECOND

In the last four sections, we have introduced the principles of linear interpolation with angles reported to the nearest minute. Though the principles were introduced with angles to the nearest minute, interpolations of that type are not ordinarily needed since trig tables for angles to the nearest minute are readily available. The major use of interpolation is with angles reported to the nearest second, since trig tables for angles of that precision are not readily available. We will discuss this latter type of interpolation in this section.

200. Tables of trigonometric ratios for angles reported to the nearest minute are readily available. Since each degree contains 60 minutes, such a table is quite long. Though we will not give a complete table of that type in this book, some entries from such a table are given at the right. Notice that each trig ratio is carried to five significant digits, rather than the four significant digits we have been using.

Referring to the table at the right, answer the following:

(a) sin 17°5' = ________

(b) cos 44°29' = ________

(c) tan 76°59' = ________

(d) If tan θ = .30605, θ = ________

(e) If sin θ = .70132, θ = ________

(f) If cos θ = .22722, θ = ________

Angle θ	Sin θ	Cos θ	Tan θ
17° 0'	.29237	.95630	.30573
17° 1'	.29265	.95622	.30605
17° 2'	.29293	.95613	.30637
17° 3'	.29321	.95605	.30668
17° 4'	.29348	.95596	.30700
17° 5'	.29376	.95588	.30732
17° 6'	.29404	.95579	.30764
17° 7'	.29432	.95571	.30796
17° 8'	.29460	.95562	.30828
17° 9'	.29487	.95554	.30859
44°25'	.69987	.71427	.97984
44°26'	.70008	.71406	.98041
44°27'	.70029	.71386	.98098
44°28'	.70049	.71366	.98155
44°29'	.70070	.71345	.98212
44°30'	.70091	.71325	.98270
44°31'	.70112	.71305	.98327
44°32'	.70132	.71284	.98384
44°33'	.70153	.71264	.98441
44°34'	.70174	.71243	.98499
76°51'	.97378	.22750	4.2803
76°52'	.97384	.22722	4.2859
76°53'	.97391	.22693	4.2916
76°54'	.97398	.22665	4.2972
76°55'	.97404	.22637	4.3029
76°56'	.97411	.22608	4.3086
76°57'	.97417	.22580	4.3143
76°58'	.97424	.22552	4.3200
76°59'	.97430	.22523	4.3257
77° 0'	.97437	.22495	4.3315

---

201. Since our partial table contains entries for angles precise to the nearest minute, we can use it to interpolate the trig ratios of angles reported to the nearest second. The interpolation process, which is the same as the one we have been using, is based on the fact that there are 60 seconds in 1 minute.

Let's find sin 17°5'30" as an example:

(1) Since sin 17°5'30" lies between sin 17°5' and sin 17°6', it lies between .29376 and .29404.

(2) The table difference between sin 17°5' and sin 17°6' is .29404 - .29376 = .00028

(3) sin 17°5'30" = sin 17°5' + k, where:

$$k = \left(\frac{30''}{60''}\right)(.00028) = \frac{1}{2}(.00028) = .00014$$

Therefore: sin 17°5'30" = sin 17°5' + k

= .29376 + .00014 = ________

Answers to 200:

a) .29376

b) .71345

c) 4.3257

d) 17°1'

e) 44°32'

f) 76°52'

202. Let's find tan 44°32'15" by interpolation:

(1) Since tan 44°32'15" lies between tan 44°32' and tan 44°33', it lies between .98384 and .98441.

(2) The table difference between tan 44°32' and tan 44°33' is .98441 - .98384 = .00057.

(3) tan 44°32'15" = tan 44°32' + k, where

$$k = \left(\frac{15''}{60''}\right)(.00057) = \left(\frac{1}{4}\right)(.00057) = .00014$$

Therefore: tan 44°32'15" = tan 44°32' + k

= .98384 + .00014 = ___________

sin 17°5'30" = .29390

---

203. Let's find cos 76°52'20" by interpolation. Remember that we <u>subtract</u> "k" when interpolating to find cosines since the cosine ratio is a <u>decreasing</u> quantity from 0° to 90°.

(1) Since cos 76°52'20" lies between cos 76°52' and cos 76°53', it lies between .22722 and .22693.

(2) The table difference between cos 76°52' and cos 76°53' is .22722 - .22693 = .00029.

(3) cos 76°52'20" = cos 76°52' - k, where:

$$k = \left(\frac{20''}{60''}\right)(.00029) = \left(\frac{1}{3}\right)(.00029) = .00010$$

Therefore: cos 76°52'20" = cos 76°52' - k

= .22722 - .00010 = ___________

tan 44°32'15" = .98398

---

204. Using the procedures in the last three frames, complete these. Remember that "k" is only a fractional part of the table difference, and cannot be larger than the table difference.

(a) sin 44°26'12" = ___________

sin 44°27' = .70029
sin 44°26' = .70008

(b) cos 17°1'40" = ___________

cos 17°1' = .95622
cos 17°2' = .95613

(c) tan 76°55'37" = ___________

tan 76°56' = 4.3086
tan 76°55' = 4.3029

cos 76°52'20" = .22712

205. When given a specific trig ratio, we can also use interpolation to report an angle to the nearest second. As an example, let's find θ when $\sin\theta = .97414$.

(1) Since .97414 lies between .97411 and .97417, θ lies between 76°56' and 76°57'.

(2) The table difference between 76°56' and 76°57' is .97417 - .97411 = .00006.

(3) The difference between the given sine (.97414) and sin 76°56' = .97414 - .97411 = .00003.

(4) $\theta = 76°56' + w$, where:

$$w = \left(\frac{.00003}{.00006}\right)(60'') = \left(\frac{1}{2}\right)(60'') = 30''$$

Therefore, θ = ___________

a) sin 44°26'12" = .70012

b) cos 17° 1'40" = .95616

c) tan 76°55'37" = 4.3064

---

206. Let's use interpolation to find θ (to the nearest second) when $\tan\theta = .98057$.

(1) Since .98057 lies between .98041 and .98098, θ lies between 44°26' and 44°27'.

(2) The table difference between tan 44°26' and tan 44°27' is .00057.

(3) The difference between the given tangent (.98057) and tan 44°26' is .00016.

(4) $\theta = 44°26' + w$, where:

$$w = \left(\frac{.00016}{.00057}\right)(60'') = 17''$$

Therefore: $\theta = 44°26' + w$ = ___________

θ = 76°56'30"

---

207. Let's use interpolation to find θ (to the nearest second) when $\cos\theta = .22587$.

(1) Since .22587 lies between .22608 and .22580, θ lies between 76°56' and 76°57'.

(2) The table difference between cos 76°56' and cos 76°57' is .00028.

(3) The difference between cos 76°56' and the given cosine is .22608 - .22587 = .00021.

(4) $\theta = 76°56' + w$, where:

$$w = \left(\frac{.00021}{.00028}\right)(60'') = \left(\frac{3}{4}\right)(60'') = 45''$$

Therefore: θ = ___________

θ = 44°26'17"

θ = 76°56'45"

208. When interpolating to find an angle to the nearest second, remember that "w" is a fractional part of 60" (or 1'). It cannot be larger than 60".

Complete these, reporting each angle to the nearest second:

(a) If $\tan\theta = 4.3219$, $\theta =$ ________

tan 76°59' = 4.3257
tan 76°58' = 4.3200

(b) If $\cos\theta = .71320$, $\theta =$ ________

cos 44°30' = .71325
cos 44°31' = .71305

(c) If $\sin\theta = .29477$, $\theta =$ ________

sin 17°9' = .29487
sin 17°8' = .29460

---

a) θ = 76°58'20"

b) θ = 44°30'15"

c) θ = 17° 8'38"

209. Use interpolation to complete these. In (b), report the angle to the nearest second:

(a) sin 76°59'50" = ________

sin 77° 0' = .97437
sin 76°59' = .97430

(b) If $\cos\theta = .22566$, $\theta =$ ________

cos 76°57' = .22580
cos 76°58' = .22552

---

a) sin 76°59'50" = .97436

b) θ = 76°57'30"

210. Use interpolation to complete these. In (a), report the angle to the nearest second:

(a) If $\tan \theta = .30803$, $\theta =$ ________

tan 17°8' = .30828
tan 17°7' = .30796

(b) cos 44°28'49" = ________

cos 44°28' = .71366
cos 44°29' = .71345

Answer to Frame 210: a) θ = 17°7'13" b) cos 44°28'49" = .71349

SELF-TEST 16 (Frames 200-210)

Using interpolation, find the numerical value of each trig ratio:

1. sin 9°17'30" = ________

sin 9°18' = .16160
sin 9°17' = .16132

2. tan 58°31'44" = ________

tan 58°32' = 1.6340
tan 58°31' = 1.6329

3. cos 25°21'18" = ________

cos 25°21' = .90371
cos 25°22' = .90358

Using interpolation, find the size of each angle to the nearest second:

4. If $\tan \theta = 6.4770$, $\theta =$ ________

tan 81°14' = 6.4845
tan 81°13' = 6.4720

5. If sin A = 0.90381, A = ________

sin 64°40' = 0.90383
sin 64°39' = 0.90371

6. If $\cos \varphi = 0.76955$, $\varphi =$ ________

cos 39°41' = .76958
cos 39°42' = .76940

ANSWERS:

1. sin 9°17'30" = .16146
2. tan 58°31'44" = 1.6337
3. cos 25°21'18" = .90367
4. θ = 81°13'24"
5. A = 64°39'50"
6. φ = 39°41'10"

6-17 DECIMAL SUBDIVISIONS OF A DEGREE

We have seen that degrees can be subdivided into minutes and seconds. Degrees can also be subdivided into "decimal" subdivisions like tenths, hundredths, and so on. We will discuss these decimal subdivisions of degrees in this section.

211. A degree can be subdivided into decimal subdivisions. For example:

$0.7°$ means 7 tenths of a degree.

$0.52°$ means 52 hundredths of a degree.

Complete: (a) $0.03°$ means 3 ________________ of a degree.

(b) $0.9°$ means 9 ____________ of a degree.

(c) Write $\frac{1°}{2}$ in decimal form: ________

(d) Write $\frac{7°}{20}$ in decimal form: ________

---

a) hundredths

b) tenths

c) $0.5°$

d) $0.35°$

212. Since there are 60 minutes in $1°$, to convert a decimal part of a degree to minutes we simply multiply by 60. When doing so, we round to the nearest whole-number minute. For example:

$$0.6° = (0.6)(60') = 36'$$
$$0.14° = (0.14)(60') = 8.4' \text{ or } 8'$$

Convert each of these to minutes:

(a) $0.4°$ = ________ (b) $0.15°$ = ________

---

a) $24'$

b) $9'$

213. When the size of an angle is given in decimal subdivision form, we can convert it to degree-minute form by converting the decimal part to minutes. For example:

$$27.2° = 27° + 0.2° = 27° + 12' = 27°12'$$

Convert each of these to degree-minute form:

(a) $14.5°$ = ________ (b) $86.90°$ = ________ (c) $3.6°$ = ________

---

a) $14°30'$

b) $86°54'$

c) $3°36'$

214. To convert from minutes to a decimal part of a degree, we simply divide by 60. For example:

$$6' = \left(\frac{6}{60}\right)^\circ = \frac{1}{10}^\circ \text{ or } 0.1^\circ$$

$$45' = \left(\frac{45}{60}\right)^\circ = \frac{3}{4}^\circ \text{ or } 0.75^\circ$$

When converting from minutes to a decimal part of a degree in this section, round the decimal <u>to the nearest hundredth.</u> For example:

$$40' = \left(\frac{40}{60}\right)^\circ = \frac{2}{3}^\circ = 0.667^\circ \text{ or } 0.67^\circ$$

Convert each of the following to a decimal part of a degree:

(a) 48' = _____ (b) 15' = _____ (c) 20' = _____ (d) 29' = _____

---

215. To convert degree-minute expressions to degrees with decimal subdivisions, we simply convert the minutes to a decimal part of a degree. For example:

$$14°30' = 14° + 30' = 14° + 0.5° = 14.5°$$
$$39°10' = 39° + 10' = 39° + 0.17° = 39.17°$$

Convert each of the following to degrees with decimal subdivisions:

(a) 21°36' = __________ (b) 48°25' = __________

Answers to 214:
a) 0.8°
b) 0.25°
c) 0.33°
d) 0.48°

---

216. When an angle is reported with a decimal subdivision, we can interpolate to report a more precise trig ratio. To do so, we use our ordinary "exact-degree" trig table. As an example, let's find sin 31.5° by interpolation:

(1) Since 31.5° lies between 31° and 32°,
sin 31.5° lies between sin 31° (0.5150) and sin 32° (0.5299).

(2) The table difference between sin 31° and sin 32° is 0.0149.

(3) sin 31.5° = sin 31° + k, where:

$$k = (0.5)(0.0149) = 0.0075$$

Therefore: sin 31.5° = __________

Answers to 215:
a) 21.6°
b) 48.42°

Answer to 216: 0.5225

217. Let's find tan 49.76° by interpolation:

(1) Since 49.76° lies between 49° and 50°,
tan 49.76° lies between tan 49° (1.150) and tan 50° (1.192).

(2) The table difference between tan 49° and tan 50° is 0.042.

(3) tan 49.76° = tan 49° + k, where:

$$k = (0.76)(0.042) = 0.032$$

Therefore, tan 49.76° = ___________

tan 49.76° = 1.182

---

218. Let's find cos 77.3° by interpolation:

(1) Since 77.3° lies between 77° and 78°,
cos 77.3° lies between cos 77° (0.2250) and cos 78° (0.2079).

(2) The table difference between cos 77° and cos 78° is 0.0171.

(3) cos 77.3° = cos 77° − k, where:

$$k = (0.3)(0.0171) = 0.0051$$

Therefore, cos 77.3° = ___________

cos 77.3° = 0.2199

---

219. Find each of the following by interpolation:

(a) tan 6.7° = ___________

(b) cos 36.75° = ___________

(c) sin 84.19° = ___________

a) tan 6.7° = 0.1175

b) cos 36.75° = 0.8012

c) sin 84.19° = 0.9948

220. When given a specific trig ratio of an angle, we can also use interpolation to report the angle with a decimal subdivision. For example, let's find the angle (θ) whose sine is 0.9304:

(1) Since 0.9304 lies between 0.9272 (sin 68°) and 0.9336 (sin 69°), θ lies between 68° and 69°.

(2) The table difference between sin 68° and sin 69° is 0.0064.

(3) The difference between the given sine (0.9304) and sin 68° (0.9272) is 0.0032.

(4) Therefore: $\theta = 68° + w$, where:

$$w = \left(\frac{0.0032}{0.0064}\right)(1°) = (0.5)(1°) = 0.5°$$

How large is angle θ? ________

$\theta = 68.5°$

---

221. Let's use interpolation to find the angle (θ) whose tangent is 1.585.

(1) Since 1.585 lies between 1.540 (tan 57°) and 1.600 (tan 58°), θ lies between 57° and 58°.

(2) The table difference between tan 57° and tan 58° is 0.060.

(3) The difference between the given tangent (1.585) and tan 57° (1.540) is 0.045.

(4) Therefore: $\theta = 57° + w$, where:

$$w = \left(\frac{0.045}{0.060}\right)(1°) = (0.75)(1°) = 0.75°$$

How large is angle θ? ________

$\theta = 57.75°$

---

222. Let's use interpolation to find θ, when $\cos\theta = 0.9761$.

(1) Since 0.9761 lies between 0.9781 (cos 12°) and 0.9744 (cos 13°), θ lies between 12° and 13°.

(2) The table difference between cos 12° and cos 13° is 0.0037.

(3) The difference between cos 12° (0.9781) and the given cosine (0.9761) is 0.0020.

(4) $\theta = 12° + w$, where $w = \left(\frac{0.0020}{0.0037}\right)(1°) = 0.54°$

Therefore, θ = ________

$\theta = 12.54°$

223. Use interpolation to complete the following. Report each angle with a decimal subdivision, but round to the nearest hundredth when necessary:

(a) If $\cos\theta = 0.7518$, $\theta =$ ________

(b) If $\tan\theta = 1.018$, $\theta =$ ________

(c) If $\sin\theta = 0.0225$, $\theta =$ ________

---

a) $\theta = 41.25°$

b) $\theta = 45.5°$

c) $\theta = 1.29°$

224. Use interpolation to complete these. In (b), report angle θ to the nearest hundredth of a degree:

(a) $\sin 15.9° =$ ________

(b) If $\cos\theta = 0.2350$, $\theta =$ ________

---

a) $\sin 15.9° = 0.2739$

b) $\theta = 76.41°$

225. Use interpolation for these. In (b), report angle θ to the nearest hundredth of a degree:

(a) $\cos 31.46° =$ ____________

(b) If $\tan\theta = 4.450$, $\theta =$ ____________

Answer to Frame 225: a) $\cos 31.46° = 0.8530$ b) $\theta = 77.32°$

SELF-TEST 17 (Frames 211-225)

Using interpolation, find the numerical value of each trig ratio:

1. $\tan 30.62° =$ ____________

$\tan 31° = 0.6009$
$\tan 30° = 0.5774$

2. $\sin 7.75° =$ ____________

$\sin 8° = 0.1392$
$\sin 7° = 0.1219$

3. $\cos 23.2° =$ ____________

$\cos 23° = 0.9205$
$\cos 24° = 0.9135$

Using interpolation, find the size of each angle to the nearest hundredth of a degree:

4. If $\sin\alpha = 0.6978$, $\alpha =$ ____________

$\sin 45° = 0.7071$
$\sin 44° = 0.6947$

5. If $\tan\varphi = 2.278$, $\varphi =$ ____________

$\tan 67° = 2.356$
$\tan 66° = 2.246$

6. If $\cos B = 0.1753$, $B =$ ____________

$\cos 79° = 0.1908$
$\cos 80° = 0.1736$

ANSWERS:

1. $\tan 30.62° = 0.5920$
2. $\sin 7.75° = 0.1349$
3. $\cos 23.2° = 0.9191$
4. $\alpha = 44.25°$
5. $\varphi = 66.29°$
6. $B = 79.90°$

Chapter 7 SINE WAVE ANALYSIS

In an earlier chapter, we introduced the graph of the equation $y = \sin \theta$. Below, we have graphed this equation from $\theta = -360°$ to $\theta = 720°$. This graph is called the basic sine wave. The variables on the axes are "y" and "θ".

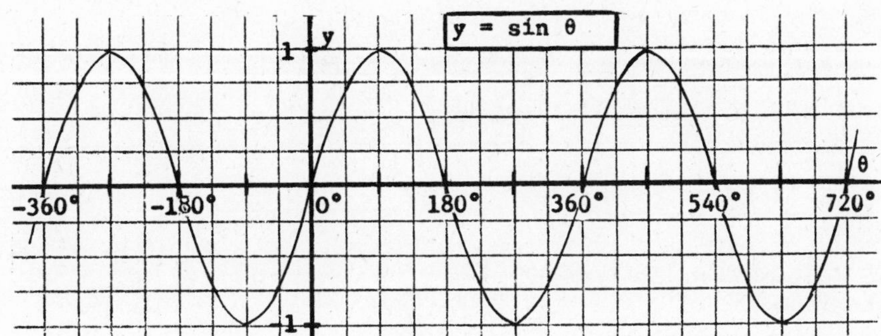

Though they differ somewhat from the basic sine wave, the graphs of many other equations are also sine waves. The graphs of all four equations below, for example, are sine waves. Notice that in each case the two variables are "y" and "θ".

$y = 2 \sin \theta$ $\quad\quad$ $y = \sin 2\theta$

$y = 3 \sin(\theta + 30°)$ $\quad\quad$ $y = 4 \sin(2\theta - 60°)$

In this chapter, we will discuss the following types of sine wave equations: fundamentals, non-fundamentals, and harmonics of the fundamentals and non-fundamentals. In doing so, we will discuss the following basic properties of sine waves: basic cycles, amplitude, phase difference and phase shifts. After discussing these types of equations and their properties, a procedure for sketching sine waves will be given. At the end of the chapter, the cosine wave will be briefly discussed.

7-1 A REVIEW OF THE SINES OF ANGLES OF ALL SIZES

In order to graph sine wave equations, a knowledge of the sines of angles of all sizes is necessary. Therefore, we will briefly review the sines of both positive and negative angles in this section.

1. In the four figures below, we have sketched a standard-position angle in each of the four quadrants. The sine of each standard-position angle (θ) is equal to the sine of its reference angle (α).

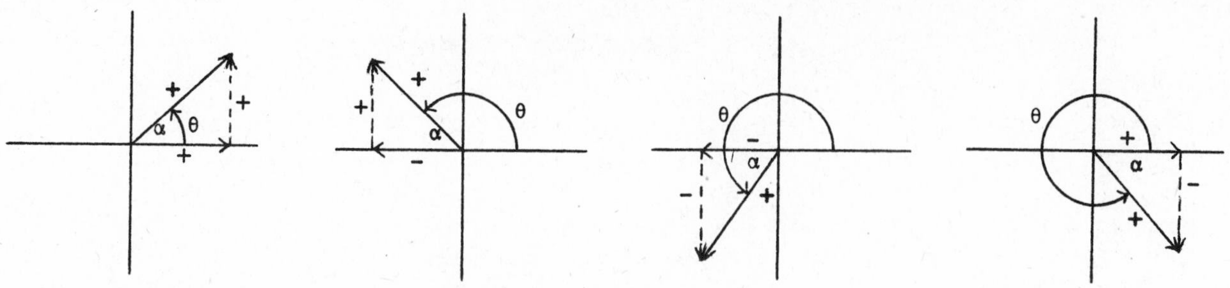

By examining the signs of the components, you can see that:

(a) The sines of angles are positive in Quadrants _____ and _____.

(b) The sines of angles are negative in Quadrants _____ and _____.

---

2. (a) The sine of any angle between 0° and 180° is __________________ (positive/negative).

(b) The sine of any angle between 180° and 360° is __________________ (positive/negative).

a) I and II

b) III and IV

---

3. Give the numerical values of the sines of each of the following standard-position angles:

(a) sin 120° = ________ (b) sin 210° = ________ (c) sin 300° = ________

a) positive

b) negative

---

4. Give the sines of each of the following special angles:

(a) sin 0° = _______ (d) sin 270° = _______

(b) sin 90° = _______ (e) sin 360° = _______

(c) sin 180° = _______

a) +0.8660

b) −0.5000

c) −0.8660

---

5. Any angle greater than 360° has the same terminal side as an angle between 0° and 360°. Therefore, any angle greater than 360° has the same sine as an angle between 0° and 360°.

If the angle lies between 360° and 720°, we can find the comparable angle by subtracting 360°.

A 500° angle has the same sine as a 140° angle, since 500° − 360° = 140°.

If the angle lies between 720° and 1080°, we can find the comparable angle by subtracting 720°.

A 900° angle has the same sine as a 180° angle, since 900° − 720° = 180°.

Each of the following angles has the same sine as an angle between 0° and 360°. Name the comparable angle for each:

(a) 450° _______ (b) 850° _______ (c) 640° _______ (d) 1000° _______

a) 0 d) −1

b) +1 e) 0

c) 0

Answer to Frame 5: a) 90° b) 130° c) 280° d) 280°

6. Any negative angle has the same terminal side as a positive angle between 0° and 360°. Therefore, any negative angle has the same sine as a positive angle between 0° and 360°.

Below we have sketched a negative angle in each quadrant:

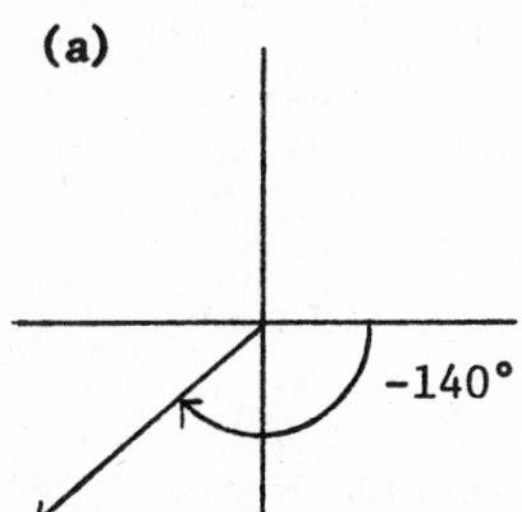

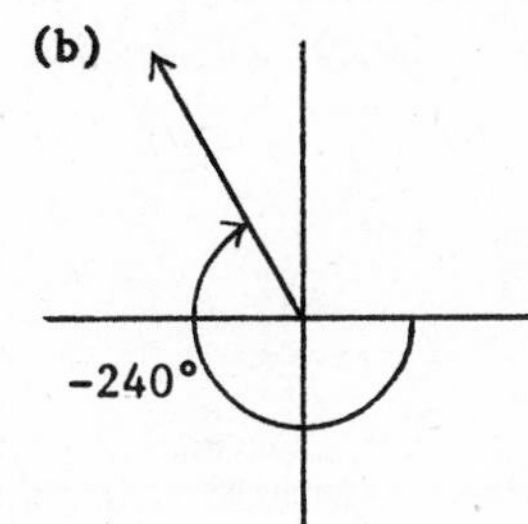

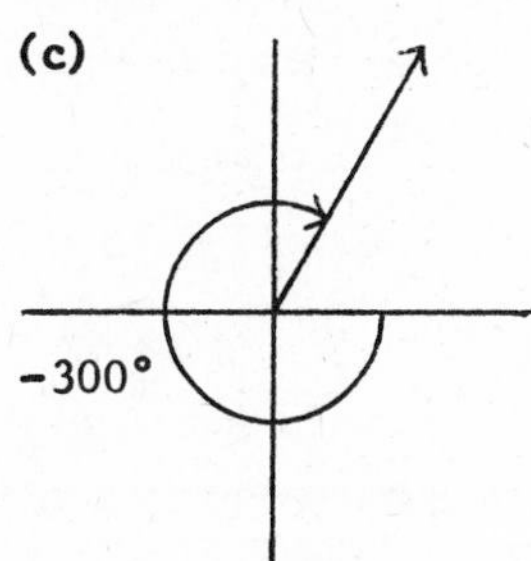

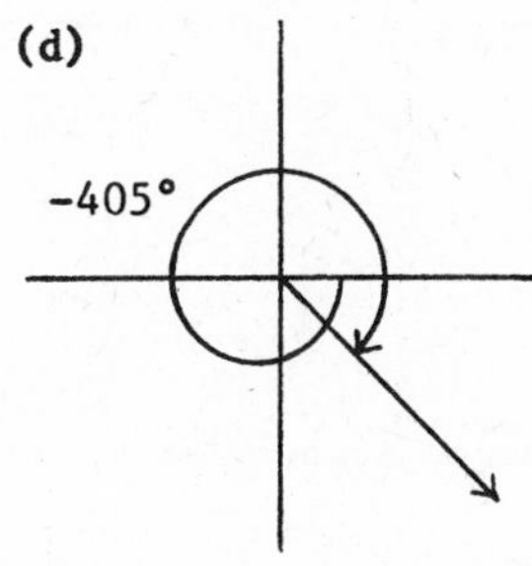

By examining each sketch, name the comparable positive angle between 0° and 360°:

(a) -140° ________ (b) -240° ________ (c) -300° ________ (d) -405° ________

Answer to Frame 6: a) 220° b) 120° c) 60° d) 315°

7. Name the positive angle between 0° and 360° that has the same sine as each of the following negative angles. To do so, first sketch each angle on the axes provided.

(a) -60° ________ (b) -180° ________ (c) -330° ________ (d) -500° ________

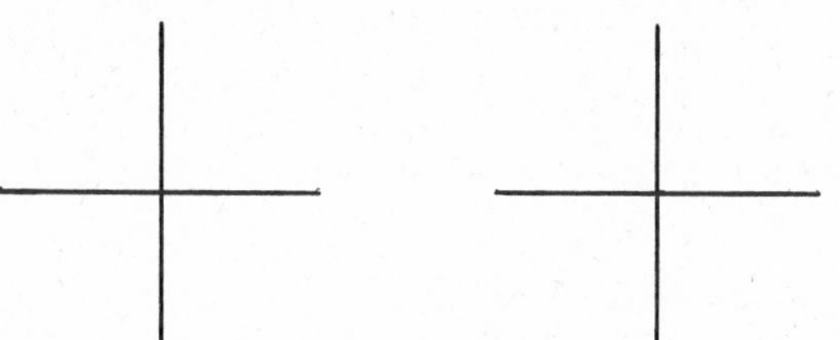

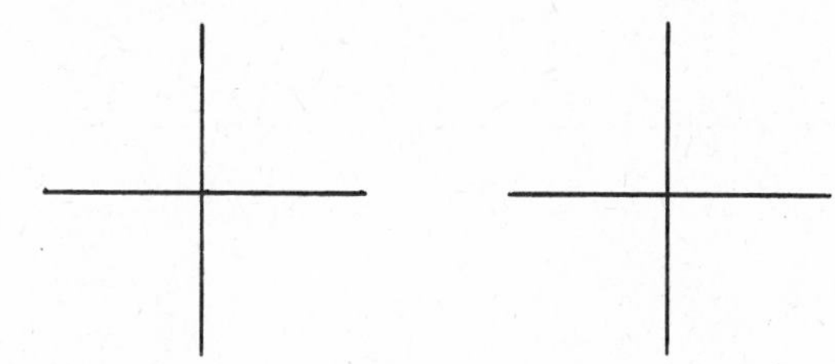

8. Name the angle between 0° and 360° that has the same sine as each of the following angles. Use sketches where necessary:

(a) 740° ________ (c) 560° ________ (e) -460° ________

(b) -80° ________ (d) -270° ________ (f) -540° ________

a) 300°

b) 180°

c) 30°

d) 220°

a) 20° c) 200° e) 260°

b) 280° d) 90° f) 180°

9. Complete each of the following:

(a) sin 390° = sin 30° = ________

(b) sin 450° = sin ____ = ________

(c) sin(−60°) = sin ____ = ________

(d) sin (−180°) = sin ____ = ________

a) = +0.5000
b) sin 90° = +1.0000
c) sin 300° = −0.8660
d) sin 180° = 0

10. Find the numerical value of each of the following:

(a) sin 540° = ________ (d) sin 870° = ________

(b) sin 600° = ________ (e) sin(−90°) = ________

(c) sin 720° = ________ (f) sin(−210°) = ________

Answer to Frame 10: a) 0 b) −0.8660 c) 0 d) +0.5000 e) −1 f) +0.5000

SELF-TEST 1 (Frames 1-10)

For each of the following angles, name the positive angle between 0° and 360° which has the same terminal side:

1. 408° ________
2. -214° ________
3. 840° ________
4. -408° ________

Find the numerical value of each of the following:

5. sin 500° = ________
6. sin(−300°) = ________
7. sin 990° = ________

ANSWERS:
1. 48°
2. 146°
3. 120°
4. 312°
5. +0.6428
6. +0.8660
7. -1

7-2 FUNDAMENTAL SINE WAVES

In this section, we will graph and discuss sine wave equations of the form: $\boxed{y = A \sin \theta}$, where A is a positive number. All sine waves of this type are called "fundamental" sine waves. Graphically, a "fundamental" sine wave can be identified by the fact that it has a basic cycle in the interval from 0° to 360°. When "fundamental" sine waves are graphed, the "maximum value of y" or "maximum height" depends on the specific equation. We will show that this "maximum value of y", which is called the "amplitude" of the sine wave, is numerically equal to the constant "A" in the specific equation.

11. All of the equations below are of the general form $\boxed{y = A \sin \theta}$, where A is a positive number.

$$y = 2 \sin \theta \qquad y = 5 \sin \theta \qquad y = \frac{1}{2} \sin \theta$$

The graph of each equation is a sine wave. All sine waves of this type are called "fundamental" sine waves.

In the equations which represent fundamental sine waves, the constant "A" must be a positive number. Which of the following equations represent fundamental sine waves? ________

(a) $y = 10 \sin \theta$ (b) $y = 1.7 \sin \theta$ (c) $y = -3 \sin \theta$

12. The equation $\boxed{y = \sin \theta}$ is called the "basic sine wave". The basic sine wave is a fundamental sine wave since its equation fits the form $y = A \sin \theta$. That is:

$y = \sin \theta$ can be written $y = 1 \sin \theta$.

Though the constant "A" is not explicitly written in $y = \sin \theta$, the constant "A" in that equation is really ______.

Answer to 11: Only (a) and (b)

In (c), -3 is a negative number.

13. The equation of the basic sine wave ($y = \sin \theta$) was graphed in an earlier chapter. In order to do so, we plugged various values of "θ" into $y = \sin \theta$, and found the corresponding values of "y". For example:

If $\theta = 0°$, $y = \sin 0° = 0.0000$
If $\theta = 60°$, $y = \sin 60° = 0.8660$
If $\theta = 90°$, $y = \sin 90° = 1.0000$

Using the same equation, complete each of the following and write the numerical value of "y":

(a) If $\theta = 150°$, y = ________. (d) If $\theta = 570°$, y = ________.

(b) If $\theta = 240°$, y = ________. (e) If $\theta = -30°$, y = ________.

(c) If $\theta = 360°$, y = ________. (f) If $\theta = -180°$, y = ________.

Answer to 12: 1

Answer to 13:

a) 0.5000 d) -0.5000

b) -0.8660 e) -0.5000

c) 0.0000 f) 0.0000

14. The graphs of the three equations below are fundamental sine waves:

$$y = 2 \sin \theta \qquad y = 5 \sin \theta \qquad y = \frac{1}{2} \sin \theta$$

To graph the equations, we must find pairs of corresponding values for "θ" and "y". In order to do so, the meaning of the right side of each equation must be clear. That is:

"2 sin θ" means: "(2) times (sin θ)" or "(2)(sin θ)"

In other words, "2 sin θ" means "multiply (sin θ) by (2)".

(a) 5 sin θ means "multiply (sin θ) by ______".

(b) $\frac{1}{2}$ sin θ means "multiply (sin θ) by ______".

---

15. Let's find pairs of values for: $y = 2 \sin \theta$

If θ = 0°, y = 2 sin 0° = 2(0.0000) = 0.0000
If θ = 30°, y = 2 sin 30° = 2(0.5000) = 1.0000
If θ = 90°, y = 2 sin 90° = 2(1.0000) = 2.0000

Using the pattern above, complete these:

(a) If θ = 75°, y = (2)(   ) = ______

(b) If θ = 210°, y = (2)(   ) = ______

(c) If θ = 400°, y = (2)(   ) = ______

(d) If θ = -50°, y = (2)(   ) = ______

Answers to 14:

a) 5

b) $\frac{1}{2}$

---

16. Let's find pairs of values for: $y = 3 \sin \theta$

If θ = 47°, y = 3 sin 47° = 3(0.7314) = 2.1942
If θ = 180°, y = 3 sin 180° = 3(0.0000) = 0.0000

Using the pattern above, complete these:

(a) If θ = 270°, y = 3(   ) = ______

(b) If θ = 360°, y = 3(   ) = ______

(c) If θ = 390°, y = 3(   ) = ______

(d) If θ = -120°, y = 3(   ) = ______

Answers to 15:

a) (2)(0.9659) = 1.9318

b) (2)(-0.5000) = -1.0000

c) (2)(0.6428) = 1.2856

d) (2)(-0.7660) = -1.5320

---

17. Let's find pairs of values for: $y = 5 \sin \theta$

If θ = 29°, y = 5 sin 29° = 5(0.4848) = 2.4240
If θ = 215°, y = 5 sin 215° = 5(-0.5736) = -2.8680

Using the pattern above, complete these:

(a) If θ = 0°, y = 5(   ) = ______

(b) If θ = 340°, y = 5(   ) = ______

(c) If θ = 450°, y = 5(   ) = ______

(d) If θ = -210°, y = 5(   ) = ______

Answers to 16:

a) 3(-1.0000) = -3.0000

b) 3(0.0000) = 0.0000

c) 3(0.5000) = 1.5000

d) 3(-0.8660) = -2.5980

18. Let's find pairs of values for: $y = 10 \sin \theta$

(a) If $\theta = 470°$, $y = 10($ $) =$ ________

(b) If $\theta = 720°$, $y = 10($ $) =$ ________

(c) If $\theta = -47°$, $y = 10($ $) =$ ________

(d) If $\theta = -310°$, $y = 10($ $) =$ ________

a) $5(0.0000) = 0.0000$
b) $5(-0.3420) = -1.7100$
c) $5(1.0000) = 5.0000$
d) $5(0.5000) = 2.5000$

---

19. Let's find pairs of values for: $y = \frac{1}{2} \sin \theta$

If $\theta = 30°$, $y = \frac{1}{2} \sin 30° = \frac{1}{2}(0.5000) = 0.2500$

If $\theta = 90°$, $y = \frac{1}{2} \sin 90° = \frac{1}{2}(1.0000) = 0.5000$

Using the pattern above, complete these:

(a) If $\theta = 37°$, $y =$ ________ (c) If $\theta = 570°$, $y =$ ________

(b) If $\theta = 180°$, $y =$ ________ (d) If $\theta = -300°$, $y =$ ________

a) $10(0.9397) = 9.3970$
b) $10(0.0000) = 0.0000$
c) $10(-0.7314) = -7.3140$
d) $10(0.7660) = 7.6600$

---

20. The pairs of values of θ and y which we found in the last few frames are necessary for graphing instances of the fundamental sine wave equation, $y = A \sin \theta$. Before doing such graphing, however, we will briefly examine the meaning of "basic cycle", "maximum value", and "amplitude". Such expressions will be used in discussing sine wave graphs.

The graph of any sine wave can be divided into "basic cycles". A "basic cycle" includes a positive loop on the left and a negative loop on the right. (See the diagram on the right.)

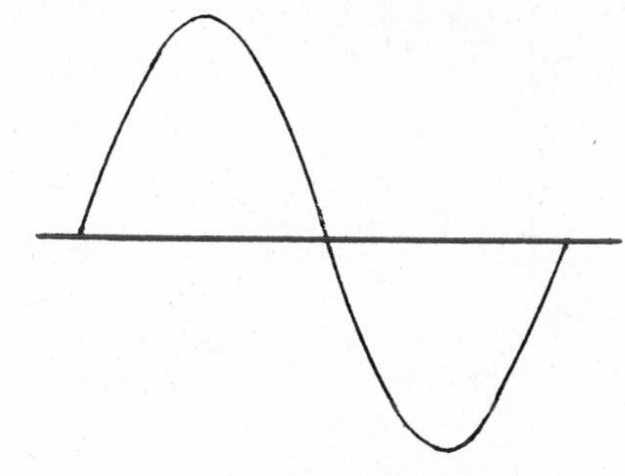

A BASIC CYCLE OF A SINE WAVE

(Continued on following page.)

a) $\frac{1}{2}(0.6018) = 0.3009$
b) $\frac{1}{2}(0.0000) = 0.0000$
c) $\frac{1}{2}(-0.5000) = -0.2500$
d) $\frac{1}{2}(0.8660) = 0.4330$

20. (Continued)

In the figure below, we have given the graph of the basic sine wave $y = \sin\theta$ from -360° to 720°.

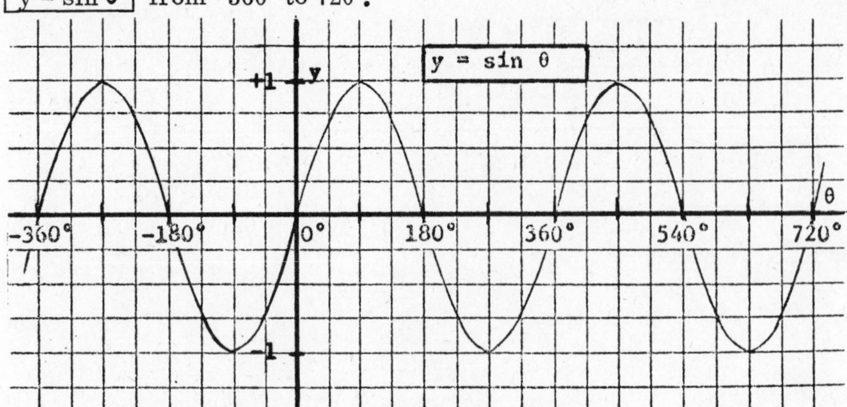

This part of the graph includes three basic cycles:

(a) The first basic cycle begins at -360° and ends at ________°.

(b) The second basic cycle begins at 0° and ends at ________°.

(c) The third basic cycle begins at 360° and ends at ________°.

---

21. In the figure at the right, we have graphed the basic cycle of the sine wave $y = \sin\theta$ from 0° to 360°.

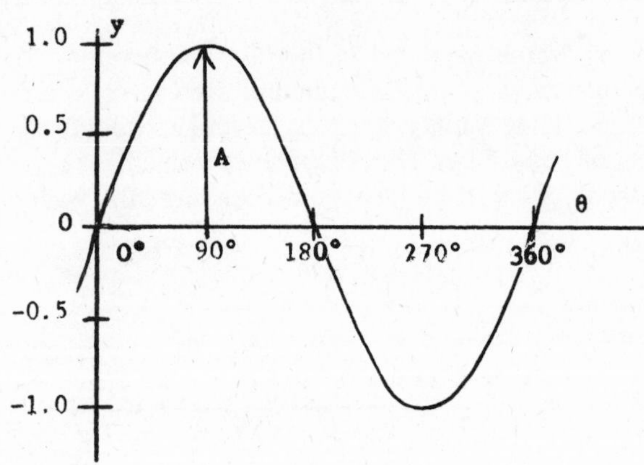

As you can see from the arrow in the positive loop (labeled "A"), the "maximum value of y" occurs when $\theta = 90°$. This "maximum value of y" is called the "amplitude" of the sine wave.

For $y = \sin\theta$: (a) The "maximum value of y" is ________.

(b) The "amplitude" of the sine wave is ________.

Answers to 20:

a) 0°

b) 360°

c) 720°

Answers to 21:

a) +1.0 or 1

b) +1.0 or 1

22. In the next few frames, we will examine the graphs of some fundamental sine waves. In order to simplify the discussion, only the parts of the sine waves from 0° to 360° will be shown.

The pairs of values in the table at the right satisfy this equation:

$y = 2 \sin \theta$

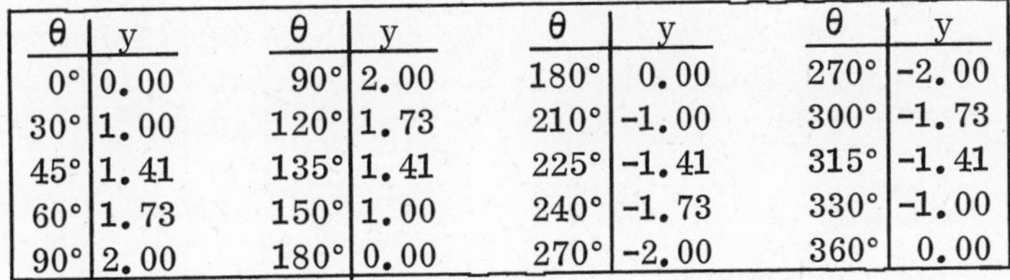

θ	y	θ	y	θ	y	θ	y
0°	0.00	90°	2.00	180°	0.00	270°	-2.00
30°	1.00	120°	1.73	210°	-1.00	300°	-1.73
45°	1.41	135°	1.41	225°	-1.41	315°	-1.41
60°	1.73	150°	1.00	240°	-1.73	330°	-1.00
90°	2.00	180°	0.00	270°	-2.00	360°	0.00

Using the pairs of values in the table, the graph at the right was constructed. You can see that there is a basic cycle from 0° to 360°.

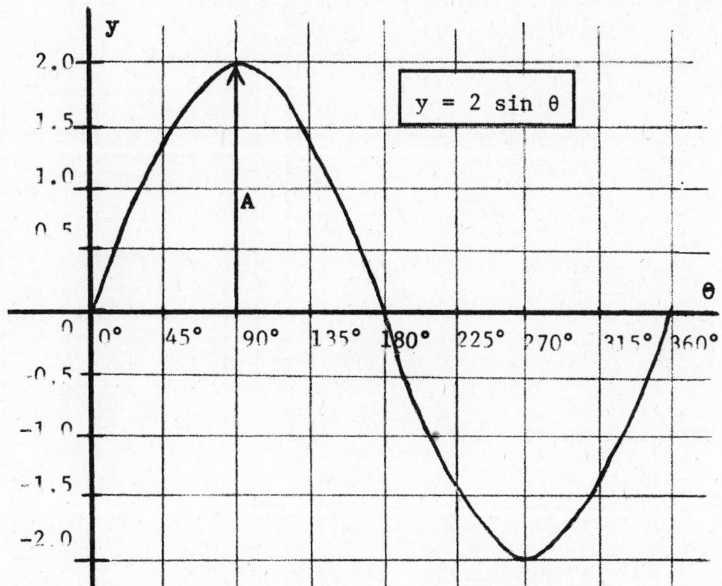

As you can see from the arrow labeled "A" in the positive loop, the "maximum value of y" or "amplitude" of the sine wave occurs at $\theta = 90°$.

For $y = 2 \sin \theta$: (a) The "maximum value of y" is __________.

(b) The "amplitude" of the sine wave is __________.

---

a) 2.0 or 2

b) 2.0 or 2

23. The pairs of values in the table at the right satisfy this equation:

$$y = \frac{1}{2} \sin \theta$$

θ	y	θ	y	θ	y	θ	y
0°	0.00	90°	0.50	180°	0.00	270°	-0.50
30°	0.25	120°	0.43	210°	-0.25	300°	-0.43
45°	0.35	135°	0.35	225°	-0.35	315°	-0.35
60°	0.43	150°	0.25	240°	-0.43	330°	-0.25
90°	0.50	180°	0.00	270°	-0.50	360°	0.00

Using the pairs of values in the table, the graph at the right was constructed.

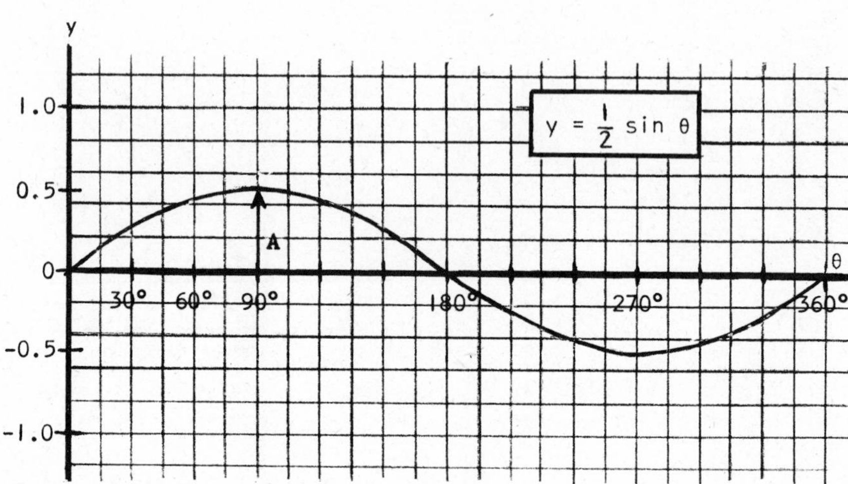

Examine the graph and answer the following:

(a) Is there a basic cycle between 0° and 360°? __________

(b) The "maximum value of y" occurs when θ = ________°.

(c) The amplitude of the sine wave is __________.

---

a) Yes

b) 90°

c) 0.5 or $\frac{1}{2}$

24. The following fundamental sine wave equations were graphed in the last three frames:

$$\boxed{y = \sin\theta} \quad \boxed{y = 2\sin\theta} \quad \boxed{y = \frac{1}{2}\sin\theta}$$

The graphs of these three equations are shown below on the same set of axes.

The three sine waves are <u>similar</u> because each has a "basic cycle" from 0° to 360°, with a positive peak at 90° and a negative peak at 270°.

The three sine waves are <u>different</u>, however, because each has a different amplitude.

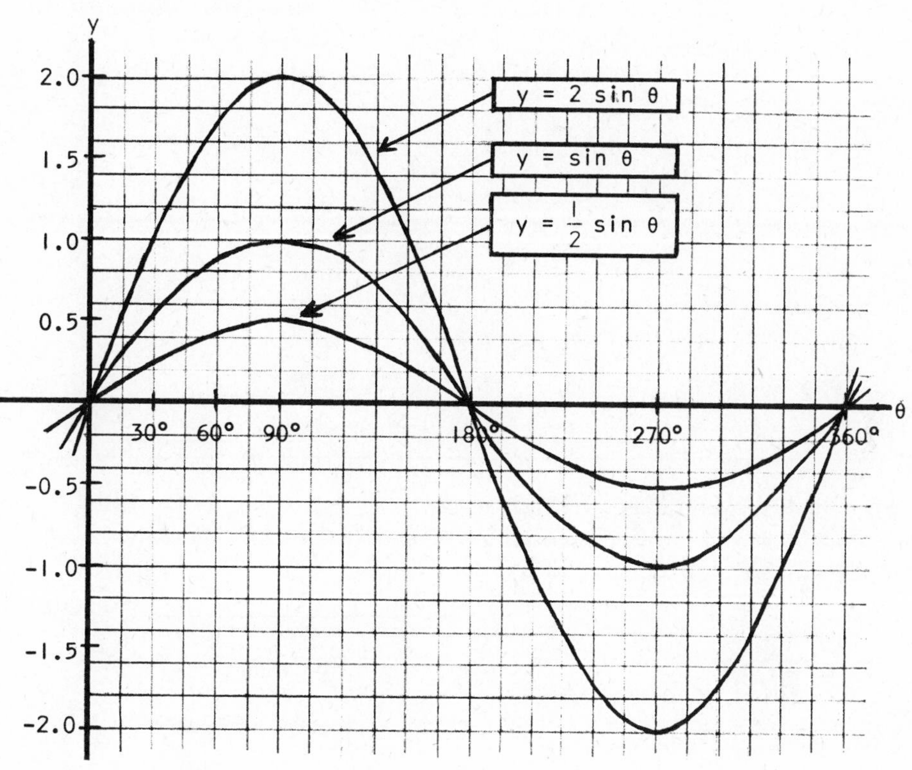

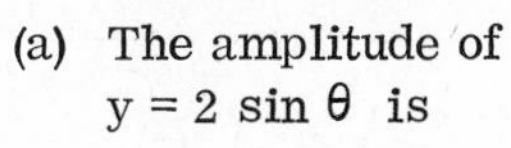

(a) The amplitude of $y = 2\sin\theta$ is __________.

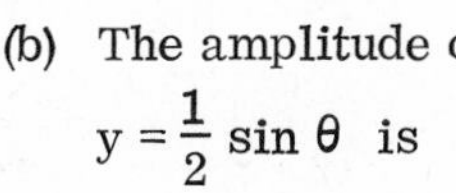

(b) The amplitude of $y = \frac{1}{2}\sin\theta$ is __________.

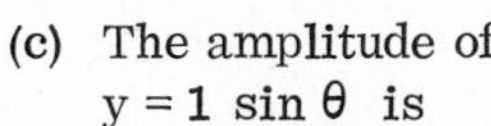

(c) The amplitude of $y = 1\sin\theta$ is __________.

---

a) 2.0 or 1

b) 0.5 or $\frac{1}{2}$

c) 1.0 or 1

25. A fundamental sine wave can be identified in either of two ways:

(1) Its equation fits the form $\boxed{y = A \sin \theta}$, where "A" is a positive number.

(2) Its graph has a basic cycle from 0° to 360°, with a positive peak at 90°.

In the general equation above, the constant "A" is numerically equal to the "amplitude" of the sine wave. This makes sense because the "maximum value of y" occurs when θ is 90° since $\sin 90° = 1$. Therefore, when $\theta = 90°$:

$$y = 10 \sin \theta = 10 \sin 90° = 10(1) = 10$$
$$y = \frac{1}{3} \sin \theta = \frac{1}{3} \sin 90° = \frac{1}{3}(1) = \frac{1}{3}$$
$$y = 4.7 \sin \theta = 4.7 \sin 90° = 4.7(1) = 4.7$$

If the following fundamental sine waves were graphed, what would the amplitude or maximum height of each be?

(a) $y = 3 \sin \theta$ ______ (c) $y = 200 \sin \theta$ ______

(b) $y = 5 \sin \theta$ ______ (d) $y = 1.92 \sin \theta$ ______

Answers: a) 3 b) 5 c) 200 d) 1.92

26. If you know the amplitude of a fundamental sine wave, you can easily write its equation. Write the equation of the fundamental sine wave:

(a) Whose amplitude is 50. ____________

(b) Whose amplitude is 0.36. ____________

Answers: a) $y = 50 \sin \theta$ b) $y = 0.36 \sin \theta$

27. Since the "maximum value of y" of a sine wave is the same as its amplitude, we can easily write the equation of a fundamental sine wave if we know the maximum value of "y".

Write the equation of the fundamental sine wave for which:

(a) The maximum value of "y" is 13. ____________

(b) The maximum value of "y" is $\frac{1}{5}$. ____________

Answers: a) $y = 13 \sin \theta$ b) $y = \frac{1}{5} \sin \theta$

28. Though the number of fundamental sine waves is unlimited, only one of them is called the "basic" sine wave. Write its equation.

____

Answer: $y = \sin \theta$ or $y = 1 \sin \theta$

SELF-TEST 2 (Frames 11-28)

If $\theta = 150°$, find the numerical value of "y" in each of the following:

1. $y = 4 \sin \theta$	2. $y = 100 \sin \theta$	3. $y = 0.50 \sin \theta$
y = ________	y = ________	y = ________

Given: $y = 50 \sin \theta$

4. What is the maximum value of "y"? ________
5. For what value of θ (between 0° and 360°) is "y" a maximum? ________

Write the equation of the fundamental sine wave whose amplitude is:

6. 170 ________
7. 0.025 ________

Refer to the graph at the right:

8. Write the equation of the sine wave which has the greater amplitude.

____

9. Write the equation of the sine wave which has the smaller amplitude.

____

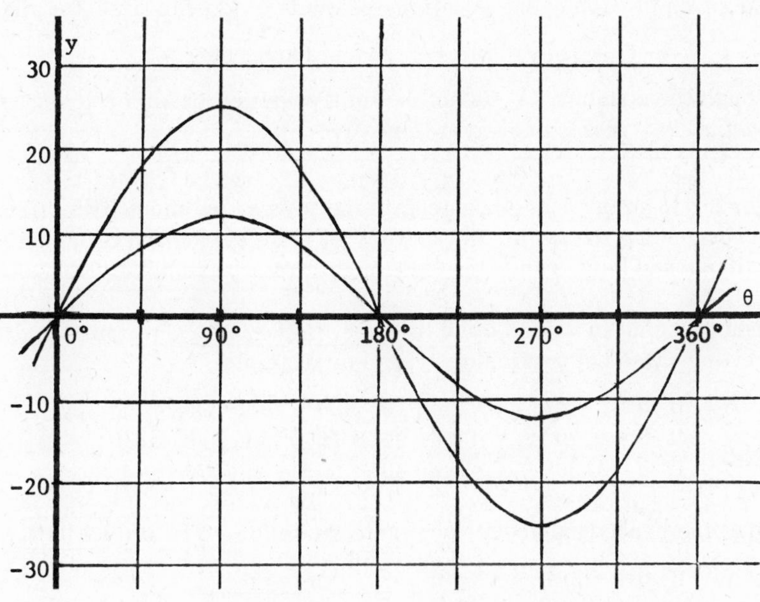

ANSWERS:

1. $y = 2$
2. $y = 50$
3. $y = 0.25$
4. 50
5. 90°
6. $y = 170 \sin \theta$
7. $y = 0.025 \sin \theta$
8. $y = 25 \sin \theta$
9. $y = 12 \sin \theta$

7-3 NON-FUNDAMENTAL SINE WAVES

The general form of the equations below is either $\boxed{y = A\sin(\theta + \varphi)}$ or $\boxed{y = A\sin(\theta - \varphi)}$.

$$y = \sin(\theta + 90°) \qquad y = 2\sin(\theta - 60°) \qquad y = 10\sin(\theta + 45°)$$

The graphs of all equations of this type are also sine waves. However, they are "non-fundamental" sine waves since they do not have a "basic cycle" between 0° and 360°. In this section, we will graph and discuss equations of this type. We will show that the amplitude of a non-fundamental sine wave is also numerically equal to the constant "A" in its equation.

---

29. The graph of any equation of the form $\boxed{y = A\sin(\theta + \varphi)}$ or $\boxed{y = A\sin(\theta - \varphi)}$ is a non-fundamental sine wave:

   (a) In $y = 2\sin(\theta + 45°)$, A is _______ and φ is _______.

   (b) In $y = 5\sin(\theta - 30°)$, A is _______ and φ is _______.

---

30. The graph of each of the equations below is a non-fundamental sine wave:

$$y = \sin(\theta + 90°) \qquad y = \sin(\theta - 60°)$$

Though the constant "A" is not explicitly written in either equation, the constant "A" in each equation is really _______.

Answers to 29:
a) 2 .... 45°
b) 5 .... 30°

---

31. In order to graph non-fundamental sine waves, we must find pairs of corresponding values for "θ" and "y". This evaluation procedure will be discussed before any non-fundamental sine waves are graphed.

When plugging in values for θ in $\boxed{y = \sin(\theta + 90°)}$, we must add 90° to every value before finding "y". For example:

If $\theta = 0°$, $y = \sin(0° + 90°) = \sin 90° = 1$
If $\theta = 90°$, $y = \sin(90° + 90°) = \sin 180° = 0$
If $\theta = 180°$, $y = \sin(180° + 90°) = \sin 270° = -1$

Using the procedure above, complete these for $\boxed{y = \sin(\theta + 90°)}$:

(a) If $\theta = 30°$, y = sin _______ ° = _________

(b) If $\theta = 210°$, y = sin _______ ° = _________

(c) If $\theta = 450°$, y = sin _______ ° = _________

(d) If $\theta = -60°$, y = sin _______ ° = _________

(e) If $\theta = -150°$, y = sin _______ ° = _________

Answer to 30:
1

Answers to 31:
a) sin 120° = 0.8660
b) sin 300° = -0.8660
c) sin 540° = 0
d) sin 30° = 0.5000
e) sin(-60°) = -0.8660

32. To find pairs of values for $\boxed{y = \sin(\theta - 30°)}$, we subtract 30° from each value plugged in for θ. For example:

If θ = 60°, y = sin(60° - 30°) = sin 30° = 0.5000

(a) If θ = 215°, y = sin(215° - 30°) = sin ______° = __________

(b) If θ = 425°, y = sin(425° - 30°) = sin ______° = __________

Answer to 32:

a) sin 185° = -0.0872

b) sin 395° = 0.5736

---

33. Here is the same equation: $\boxed{y = \sin(\theta - 30°)}$ If we plug in a negative angle or a positive angle less than 30° for θ, we must convert the subtraction within the parentheses to an addition in order to simplify the parentheses. For example:

If θ = -60°, y = sin(-60° - 30°) = sin(-90°),

since (-60°) - 30° = (-60°) + (-30°) = -90°.

(Note: We converted the subtraction to an addition by adding the opposite of 30°.)

Using the procedure above, complete these for $\boxed{y = \sin(\theta - 30°)}$:

(a) If θ = 0°, y = sin(0° - 30°) = sin ______° = __________

(b) If θ = -90°, y = sin(-90° - 30°) = sin ______° = __________

(c) If θ = -360°, y = sin(-360° - 30°) = sin ______° = __________

Answer to 33:

a) sin(-30°) = -0.5000

b) sin(-120°) = -0.8660

c) sin(-390°) = -0.5000

---

34. Here is another type: $\boxed{y = 2\sin(\theta + 60°)}$ To help understand what the right side means, we can write it in either of these ways:

y = (2) times [sin(θ + 60°)]

or

y = (2)[sin(θ + 60°)]

If θ = 10°, we get: y = 2 sin(10° + 60°)
= 2 sin 70° = (2)(0.9397) = 1.8794

If θ = 70°, we get: y = 2 sin(70° + 60°)
= 2 sin 130° = ( )( ) = __________

Answer to 34:

(2)(0.7660) = 1.5320

---

35. In this equation: $\boxed{y = 5\sin(\theta + 60°)}$

(a) If θ = 400°: y = 5 sin(400° + 60°)
= 5 sin 460° = ( )( ) = __________

(b) If θ = -200°: y = 5 sin(-200° + 60°)
= 5 sin(-140°) = ( )( ) = __________

Answer to 35:

a) (5)(0.9848) = 4.9240

b) (5)(-0.6428) = -3.2140

36. If $\theta = 150°$ in: $\boxed{y = 3\sin(\theta - 90°)}$

$$y = 3\sin(150° - 90°) = 3\sin 60° = 3(0.8660) = 2.5980$$

Find the value of "y" if:

(a) $\theta = 60°$ y = ___________ (b) $\theta = 600°$ y = ___________

Answer to Frame 36: a) $y = -1.500$ From: $y = 3\sin(-30°) = 3(-0.5000) = -1.5000$ b) $y = 1.5000$ From: $y = 3\sin 510° = 3(0.5000) = 1.5000$

---

37. The pairs of values in the table below on the left satisfy the equation:

$$\boxed{y = \sin(\theta + 90°)}$$

Using these points, the sine wave graph shown on the right was constructed.

θ	y	θ	y
-150°	-0.87	210°	-0.87
-120°	-0.50	240°	-0.50
-90°	0.00	270°	0.00
-60°	0.50	300°	0.50
-30°	0.87	330°	0.87
0°	1.00	360°	1.00
30°	0.87	390°	0.87
60°	0.50	420°	0.50
90°	0.00	450°	0.00
120°	-0.50	480°	-0.50
150°	-0.87	510°	-0.87
180°	-1.00	540°	-1.00

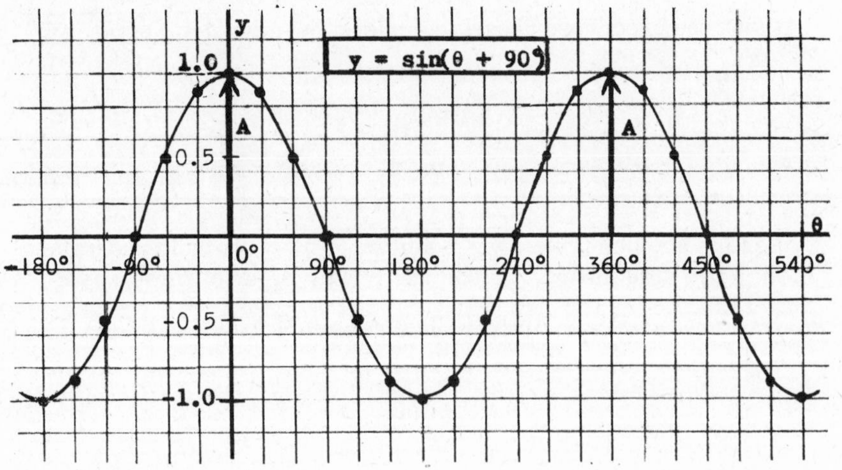

Notice these points about the sine wave:

(1) It is not a fundamental sine wave since it does not have a "basic cycle" from 0° to 360°.

(2) It peaks at 0° and 360°. The arrows (labeled "A") at these points represent the amplitude of the sine wave.

Complete: (a) The only "basic cycle" on the graph begins at −90° and ends at ________°.

(b) The amplitude of the sine wave is __________.

---

a) 270°

b) 1.0 or 1

38. The pairs of values in the table below on the left satisfy the equation:

$$y = 2\sin(\theta - 60°)$$

Using these points, the sine wave graph shown on the right was constructed.

θ	y	θ	y
-150°	1.00	210°	1.00
-120°	0.00	240°	0.00
-90°	-1.00	270°	-1.00
-60°	-1.73	300°	-1.73
-30°	-2.00	330°	-2.00
0°	-1.73	360°	-1.73
30°	-1.00	390°	-1.00
60°	0.00	420°	0.00
90°	1.00	450°	1.00
120°	1.73	480°	1.73
150°	2.00	510°	2.00
180°	1.73	540°	1.73

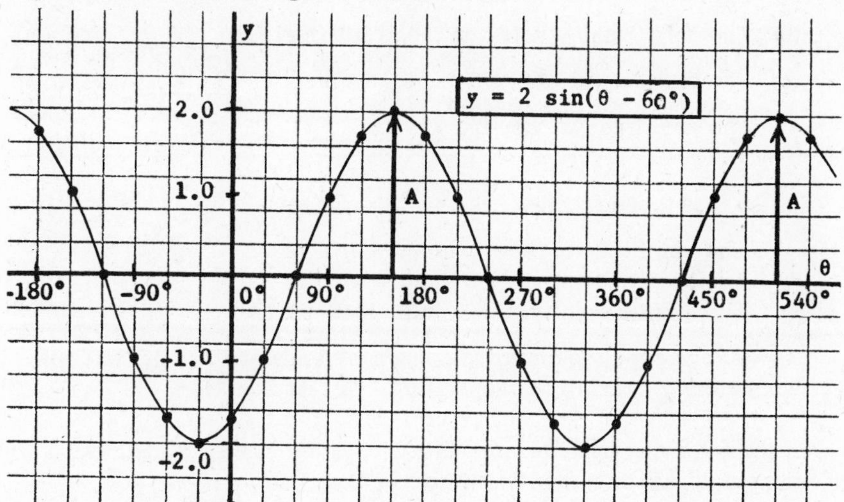

The sine wave is a non-fundamental sine wave because it does not have a "basic cycle" between 0° and 360°.

Complete: (a) The only basic cycle on the graph begins at ______° and ends at ______°.

(b) The sine wave has positive peaks when θ = ______° and when θ = ______°.

(c) The amplitude of the sine wave is ________.

Answer to Frame 38: a) 60° .... 420° b) 150° and 510° c) 2.0 or 2

---

39. The pairs of values in the table below on the left satisfy the equation:

$$y = 10\sin(\theta + 45°)$$

Using these points, the sine wave graph shown on the right was constructed:

θ	y	θ	y
-165°	-8.66	195°	-8.66
-135°	-10.00	225°	-10.00
-105°	-8.66	255°	-8.66
-75°	-5.00	285°	-5.00
-45°	0.00	315°	0.00
-15°	5.00	345°	5.00
15°	8.66	375°	8.66
45°	10.00	405°	10.00
75°	8.66	435°	8.66
105°	5.00	465°	5.00
135°	0.00	495°	0.00
165°	-5.00	525°	-5.00

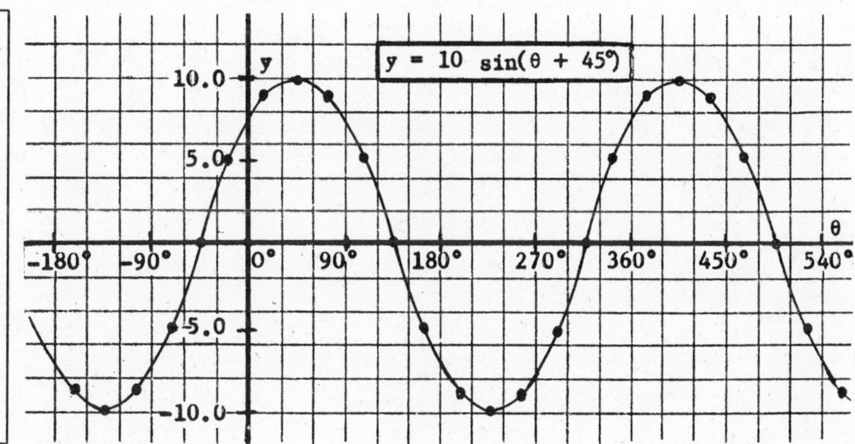

(Continued on following page.)

39. (Continued)

(a) How can you tell that the sine wave is a non-fundamental sine wave?

____

(b) The only "basic cycle" on the graph begins at _______° and ends at _______°.

(c) The sine wave has positive peaks when θ = _______° and when θ = _______°.

(d) The amplitude of the sine wave is __________.

Answers: a) Because it does not have a "basic cycle" from 0° to 360°. b) -45° .... 315° c) 45° and 405° d) 10.0 or 10

---

40. In the last three frames, we saw these facts:

The amplitude of $y = \sin(\theta + 90°)$ is 1.
The amplitude of $y = 2\sin(\theta - 60°)$ is 2.
The amplitude of $y = 10\sin(\theta + 45°)$ is 10.

As you can see, the amplitude of a non-fundamental sine wave is numerically equal to the constant "A" in its equation.

If the following non-fundamental sine waves were graphed, what would the amplitude or maximum height of each be?

(a) $y = 5\sin(\theta + 50°)$ ______ (c) $y = 2.68\sin(\theta + 75°)$ ______

(b) $y = 100\sin(\theta - 15°)$ ______ (d) $y = 67.5\sin(\theta - 29°)$ ______

Answers: a) 5 c) 2.68 b) 100 d) 67.5

---

41. How can you identify a non-fundamental sine wave from a graph?

____

Answer: By the fact that it does not have a "basic cycle" from 0° to 360°.

SELF-TEST 3 (Frames 29-41)

1. Complete the table at the right for the equation:

$y = 40 \sin(\theta + 30°)$

θ	y
-30°	
0°	
60°	
180°	

2. Complete the table at the right for the equation:

$y = 200 \sin(\theta - 90°)$

θ	y
-60°	
0°	
90°	
180°	

3. Which of the following equations have graphs which are non-fundamental sine waves? ________

(a) $y = 28 \sin \theta$ (c) $y = 6 \sin(\theta - 90°)$ (e) $y = \sin(\theta + 60°)$

(b) $y = \sin(\theta + 20°)$ (d) $y = \sin \theta$ (f) $y = 3.85 \sin \theta$

What is the amplitude of the graph of each of the following equations?

4. $y = \sin(\theta + 45°)$ ____ 5. $y = 70 \sin(\theta - 80°)$ ____ 6. $y = 7.3 \sin \theta$ ____

ANSWERS:

1.

θ	y
-30°	0
0°	20
60°	40
180°	-20

2.

θ	y
-60°	-100
0°	-200
90°	0
180°	200

3. (b), (c), (e)

4. 1
5. 70
6. 7.3

7-4 PHASE DIFFERENCES BETWEEN NON-FUNDAMENTAL AND FUNDAMENTAL SINE WAVES

Any non-fundamental sine wave is "out-of-phase" with a fundamental sine wave. The shift needed to put a non-fundamental "in-phase" with a fundamental is called the phase difference (or phase shift or phase angle) of the non-fundamental. We will discuss the concept of "phase difference" in this section. We will show that the phase angles of non-fundamental sine waves are related to the constant "φ" in the general equations of non-fundamental sine waves.

42. In each figure below, we have graphed two sine waves:

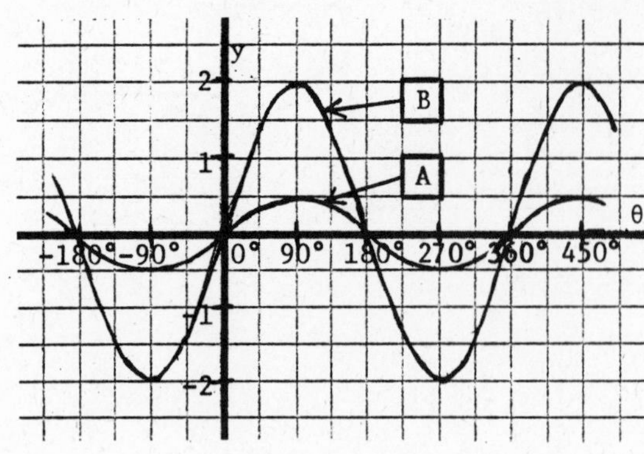

Figure 1

Figure 2

In Figure 1, both sine waves (A and B) are fundamental sine waves. That is, each has a "basic cycle" from 0° to 360° with a positive peak at 90°. Since both of these sine waves cross the horizontal axis at the same points and peak at the same points, we say that they are "in-phase".

In Figure 2, only one of the sine waves (C) is a fundamental sine wave. The other sine wave (D) is a non-fundamental sine wave since it does not have a "basic cycle" from 0° to 360° with a positive peak at 90°. Since these two sine waves do not cross the horizontal axis at the same points or peak at the same points, we say that they are "out-of-phase".

(a) If the two sine waves A and C were graphed on the same axes, would they be in-phase or out-of-phase? ____________

(b) If the two sine waves B and D were graphed on the same axes, would they be in-phase or out-of-phase? ____________

Answer to Frame 42:	a) In-phase, since both are fundamental sine waves.	b) Out-of-phase, since one is a fundamental and one is a non-fundamental sine wave.

43. The non-dashed line below is the graph of $y = \sin(\theta + 90°)$. For comparison, the graph of the fundamental sine wave $y = \sin\theta$ is shown in dashed lines. The two sine waves are out-of-phase.

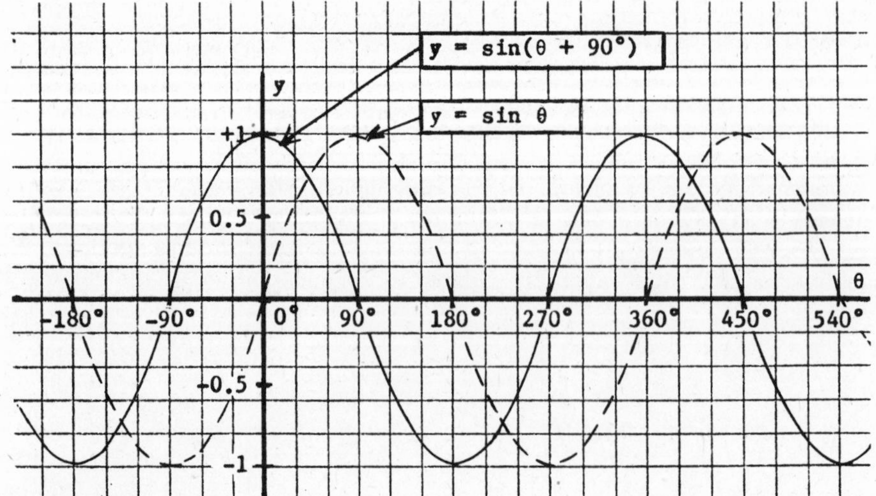

(Continued on following page.)

43. (Continued)

Note: (1) The two graphs are identical in shape.

(2) The maximum points and horizontal-axis crossings of $y = \sin(\theta + 90°)$ are to the left of the maximum points and horizontal-axis crossings of $y = \sin\theta$.

(3) The graph of $y = \sin(\theta + 90°)$ can be put in-phase with the graph of the fundamental sine wave by shifting the graph of $y = \sin(\theta + 90°)$ along the horizontal axis to the right.

Since the horizontal axis is calibrated in degrees, horizontal distances are stated in degrees.

How many degrees along the horizontal axis must the graph of $y = \sin(\theta + 90°)$ be shifted to the right to put it in-phase with the fundamental sine wave $y = \sin\theta$? ______ to the right

Answer to Frame 43: 90° to the right

44. The graph of $y = \sin(\theta + 90°)$ is given again below, together with the graph of $y = \sin\theta$.

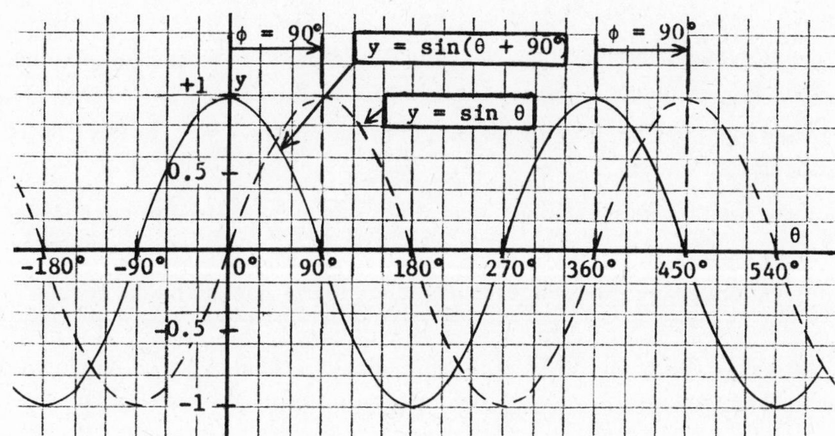

To diagram the horizontal shift needed to put $y = \sin(\theta + 90°)$ in-phase with $y = \sin\theta$, we pick any pair of adjacent peaks. We have diagramed the shift on the graph above in two places: (1) using the peaks at 0° and 90°, and (2) using the peaks at 360° and 450°. In each case, the vector φ (pronounced "fee") represents the shift needed to put $y = \sin(\theta + 90°)$ in-phase with $y = \sin\theta$.

The two sine waves above are out-of-phase. The amount they are out-of-phase is called their "phase difference". Their "phase difference" is the size of the shift along the horizontal axis needed to put the non-fundamental sine wave in-phase with the fundamental sine wave. Sometimes, the terms "phase angle" or "phase shift" are used instead of "phase difference".

In the example above, the vector φ represents the phase difference (or phase angle or phase shift). In this case, the phase difference is ______ ° to the right.

Answer to Frame 44: 90°

45. The graph of $y = \sin(\theta + 30°)$ is shown below. The graph of the fundamental sine wave $y = \sin\theta$ is shown in dashed lines for comparison. These two sine waves are also out-of-phase.

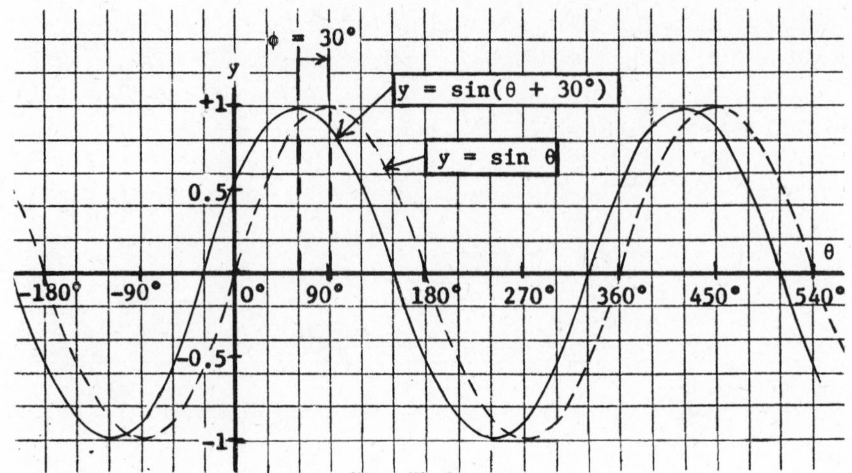

Using the two adjacent peaks at 60° and 90°, we have diagramed the shift needed to put $y = \sin(\theta + 30°)$ in-phase with $y = \sin\theta$.

(a) Are the peaks and horizontal-axis crossings of $y = \sin(\theta + 30°)$ to the left or right of those of the fundamental sine wave? ______________

(b) How many degrees to the right must $y = \sin(\theta + 30°)$ be shifted to put it in-phase with $y = \sin\theta$? ______°

(c) Therefore: φ, the phase difference or phase angle, is how many degrees? ______°

---

46. Two other technical terms are sometimes used instead of the term "phase difference". They are ______________________ and ______________________.

Answers to 45: a) To the left. b) 30° c) 30°

---

47. Up to this point, we have shown two phase shifts to the right. Since the shift is to the right, we say that the phase difference (or phase shift or phase angle) is positive. Therefore, the phase shift or phase angle can be represented by a positive angle. For example:

A phase shift of "90° to the right" is represented by +90°.

A phase shift of "30° to the right" is represented by ______°.

Answer to 46: phase angle and phase shift

Answer to 47: +30°

48. We can also have phase shifts to the left. As an example, the equation $y = \sin(\theta - 30°)$ is graphed below, together with the graph of $y = \sin\theta$.

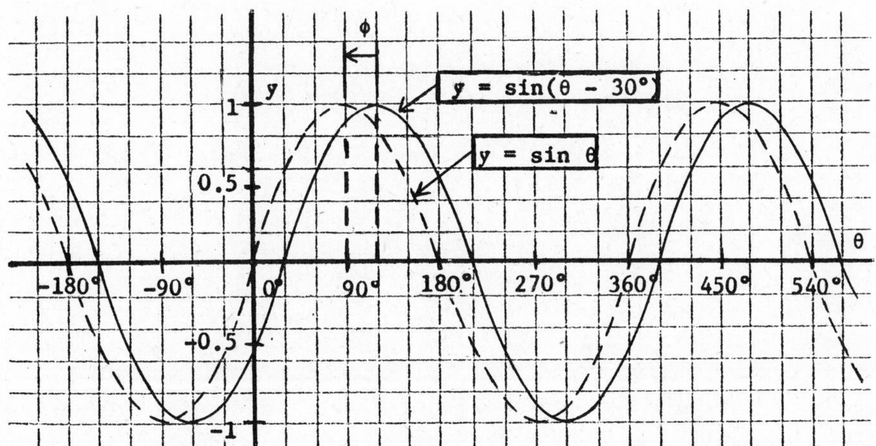

The peaks and horizontal-axis crossings of $y = \sin(\theta - 30°)$ lie to the right of those of $y = \sin\theta$. Using the peaks at 90° and 120°, we have diagramed the shift necessary to put $y = \sin(\theta - 30°)$ in-phase with $y = \sin\theta$.

(a) To put the two sine waves in-phase, $y = \sin(\theta - 30°)$ must be shifted ______° to the left.

(b) Since the vector representing this shift is to the left, the sign of the shift is ______________.

(c) Therefore: φ, the phase difference or phase angle, can be represented by what signed angle? ______°

Answer to Frame 48: a) 30° b) negative c) -30°

---

49. The graph of $y = \sin(\theta - 90°)$ is shown below, together with the graph of $y = \sin\theta$.

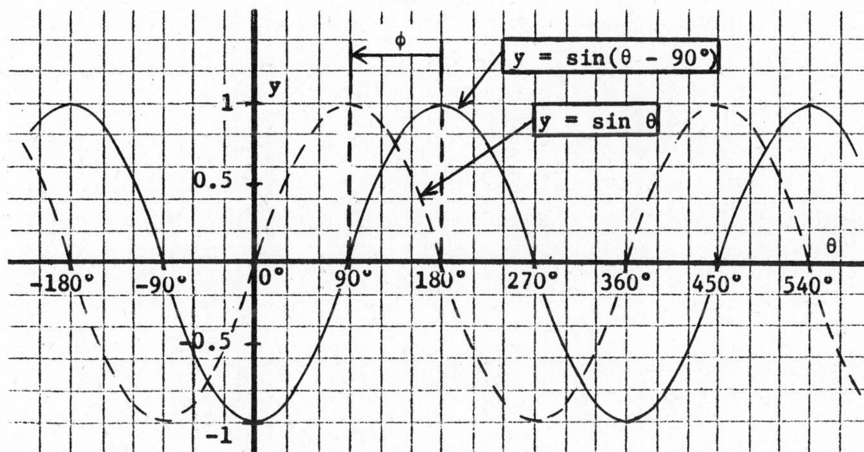

(a) Does the graph of $y = \sin(\theta - 90°)$ lie to the right or to the left of the fundamental sine wave? ______________

(b) Using the peaks at 90° and 180°, we can see that $\varphi = -90°$. What does a phase difference or phase angle of -90° mean? ______________________________

Answer to Frame 49: a) To the right. b) A phase angle of -90° means that $y = \sin(\theta - 90°)$ must be shifted 90° to the left to put it in-phase with $y = \sin\theta$.

50. There are four sine waves on the graph below. Though the three solid sine waves have different amplitudes, they are all in-phase with each other because their peaks occur at the same values of θ and because their horizontal axis crossings are at the same points.

The equations of the three solid sine waves are given on the graph. The dashed sine wave is the fundamental sine wave, $y = \sin\theta$.

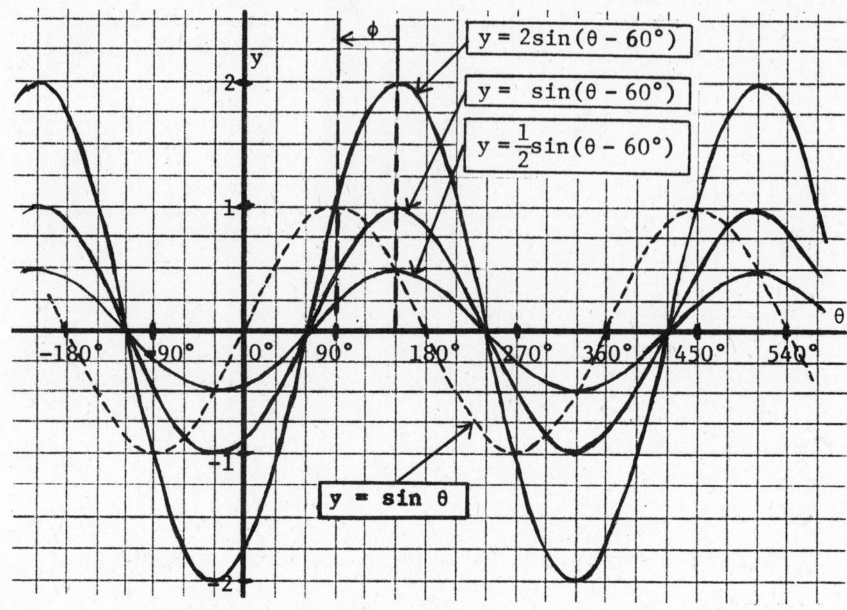

The graph also shows the shift needed to put each of the three solid sine waves in-phase with the fundamental sine wave. Note that the same shift is needed for each of the three solid sine waves.

Represent the phase difference (φ) for each of these three sine waves with a signed angle. φ = _______°

Answer to Frame 50: $\varphi = -60°$

---

51. The graph of $y = \sin(\theta + 90°)$ is given on the graph below, together with the graph of $y = \sin\theta$.

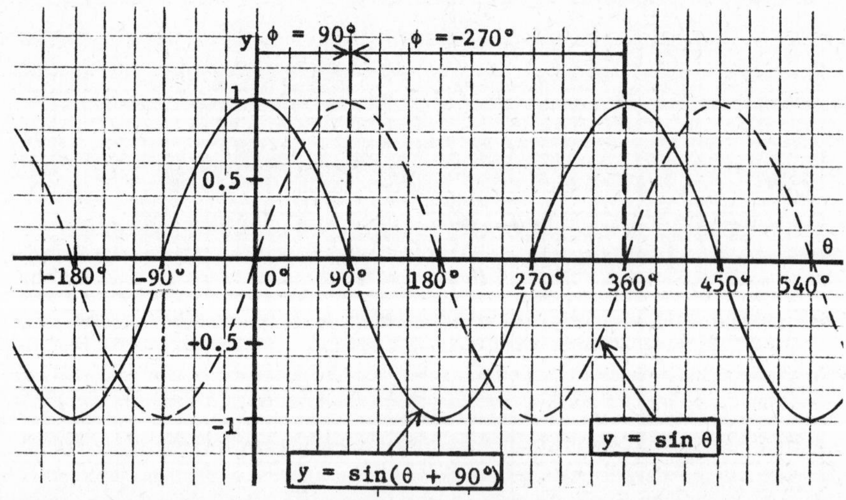

The non-fundamental sine wave can be put "in-phase" with the fundamental sine wave by a horizontal shift in either direction.

(1) Using the peaks at 0° and 90°, we have diagramed the shift to the right needed to put $y = \sin(\theta + 90°)$ "in-phase" with $y = \sin\theta$. For this shift, $\varphi = +90°$

(2) Using the peaks at 90° and 360°, we have diagramed the shift to the left needed to put $y = \sin(\theta + 90°)$ "in-phase" with $y = \sin\theta$. For this shift, $\varphi = -270°$.

Though two phase shifts are possible, we always use the one which involves the smallest number of degrees. Therefore, of the two possibilities in this case, we would report φ as _______°.

Answer to Frame 51: +90°

52. When shifting a non-fundamental sine wave to put it "in-phase" with a fundamental sine wave, two shifts are always possible. However, the phase shift which is used is always the one which involves the smallest number of degrees.

We have graphed below the non-fundamental sine wave $y = 2\sin(\theta - 150°)$ and the fundamental sine wave $y = 2\sin\theta$.

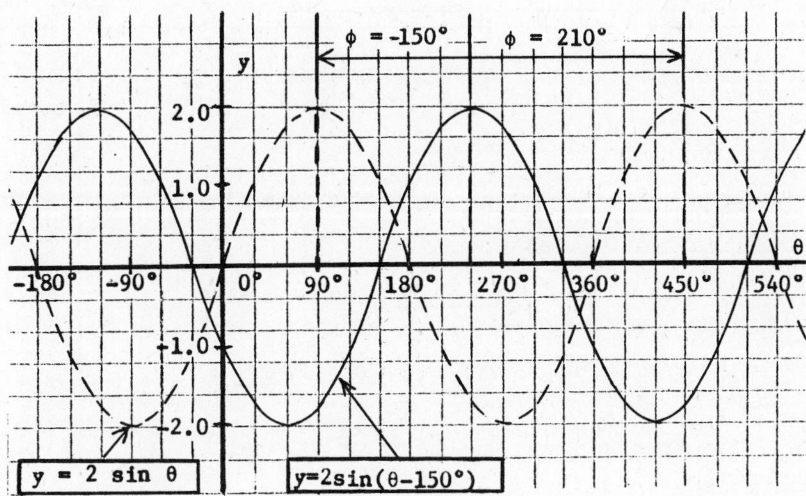

Two phase shifts are possible to put the non-fundamental sine wave "in-phase" with the fundamental sine wave:

$\varphi = -150°$ (See the peaks at 90° and 240°.)
$\varphi = +210°$ (See the peaks at 240° and 450°.)

Which one of these two phase shifts would be reported and used? ______________

Answer to Frame 52: $\varphi = -150°$, since it involves the smallest number of degrees.

53. On the graph below, the solid sine wave is a non-fundamental and the dashed sine wave is a fundamental.

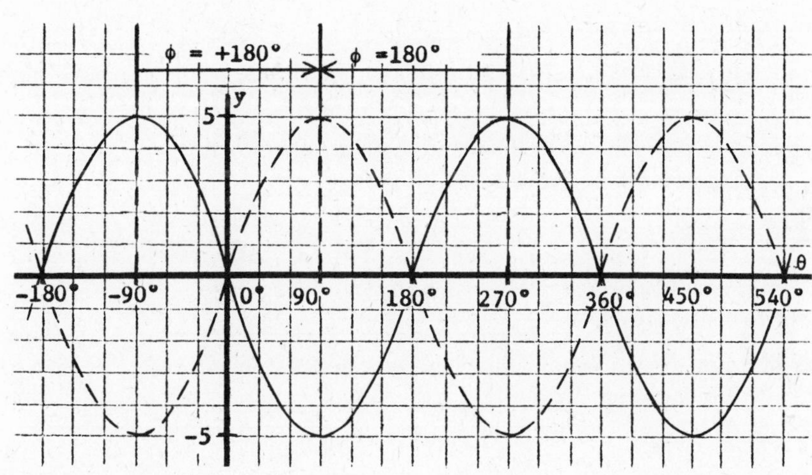

(Continued on following page.)

53. (Continued)

We have diagramed the two shifts possible for putting the non-fundamental "in-phase" with the fundamental. They are:

$\varphi = +180°$ (See the peaks at -90° and 90°.)
$\varphi = -180°$ (See the peaks at 90° and 270°.)

To maintain the principle of using the phase shift which involves the smallest number of degrees, which phase shift would we use? ______________________

---

54. The general equation of non-fundamental sine waves is either:

$$\boxed{y = A\sin(\theta + \varphi)} \quad \text{or} \quad \boxed{y = A\sin(\theta - \varphi)}$$

In any equation of this type:

(1) "A" is the amplitude of the sine wave.

(2) "φ" is the <u>phase difference</u> or <u>phase angle</u> of the sine wave compared to a fundamental sine wave.

A "+" in the equation means that the <u>phase difference</u> is <u>positive</u>.

A "-" in the equation means that the <u>phase difference</u> is <u>negative</u>.

Given this equation: $\boxed{y = \sin(\theta + 45°)}$

(a) The amplitude of the sine wave is _______.

(b) As a signed angle, its phase difference with a fundamental sine wave is _______°.

(c) To put the graph of $y = \sin(\theta + 45°)$ in-phase with a fundamental sine wave, the graph must be moved horizontally 45° to the ___________ (left/right).

Answer to 53: Either of them, since both involve 180°.

---

55. Given this equation: $\boxed{y = 0.67(\theta - 150°)}$

(a) The amplitude of the sine wave is __________.

(b) As a signed angle, its phase angle with a fundamental sine sine wave is _______°.

(c) To put the graph of this sine wave in-phase with a fundamental sine wave, the graph must be moved 150° to the ___________ (left/right).

Answers to 54:

a) 1

b) +45°

c) right

---

56. (a) If the amplitude of a non-fundamental sine wave is 5 and its phase angle is -45°, write its equation. ______________________

(b) If the amplitude of a non-fundamental sine wave is 1 and its phase angle is 135°, write its equation. ______________________

Answers to 55:

a) 0.67

b) -150°

c) left

Answers to 56:

a) $y = 5\sin(\theta - 45°)$

b) $y = 1\sin(\theta + 135°)$ or
$y = \sin(\theta + 135°)$

SELF-TEST 4 (Frames 42-56)

1. A non-fundamental sine wave of amplitude 3.85 has a phase angle $\varphi = +80°$ with respect to a fundamental sine wave. Write its equation. ______________

2. A non-fundamental sine wave of amplitude 170 has a phase angle $\varphi = -130°$ with respect to a fundamental sine wave. Write its equation. ______________

3. A non-fundamental sine wave of amplitude 58 lies 25° to the right of a fundamental sine wave. Write its equation.

____

4. A non-fundamental sine wave of amplitude 0.75 lies 152° to the left of a fundamental sine wave. Write its equation.

____

Write the equation of these sine waves:

5. Sine wave A:

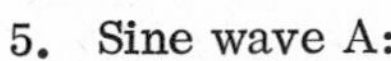

6. Sine wave B: ______________

7. Sine wave C: ______________

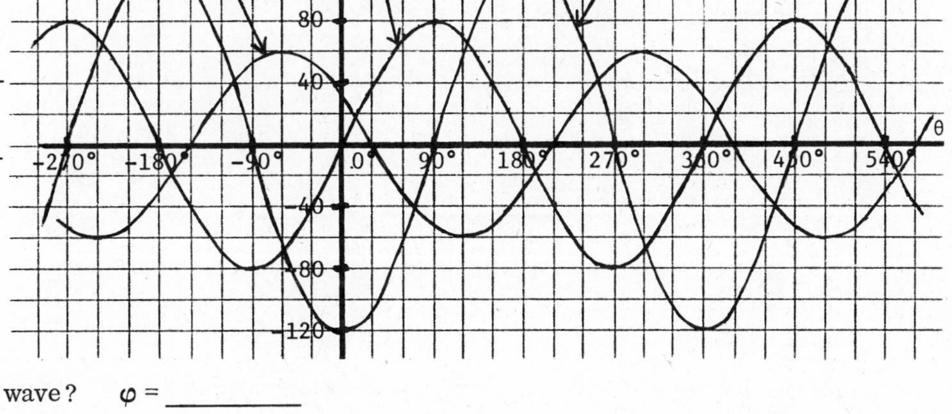

8. What is the phase angle of sine wave B with respect to a fundamental sine wave? $\varphi =$ ______________

9. What is the phase angle of sine wave C with respect to a fundamental sine wave? $\varphi =$ ______________

ANSWERS:

1. $y = 3.85 \sin(\theta + 80°)$
2. $y = 170 \sin(\theta - 130°)$
3. $y = 58 \sin(\theta - 25°)$
4. $y = 0.75 \sin(\theta + 152°)$
5. $y = 80 \sin \theta$
6. $y = 120 \sin(\theta - 90°)$
7. $y = 60 \sin(\theta + 150°)$
8. $\varphi = -90°$

   $\varphi = +150°$

7-5 THE CONCEPTS OF "LEAD" AND "LAG"

If a non-fundamental sine wave lies to the left of a fundamental, we say that it "leads" the fundamental. If a non-fundamental sine wave lies to the right of a fundamental, we say that it "lags" the fundamental. We will discuss the concepts of "lead" and "lag" in this section.

57. If the phase difference of a non-fundamental sine wave is positive:

(1) The non-fundamental sine wave must be shifted to the right to be put in-phase with a fundamental sine wave.

(2) Therefore, the non-fundamental sine wave lies to the left of a fundamental sine wave.

To put $y = 1.5 \sin(\theta + 90°)$ in-phase with a fundamental sine wave, it must be shifted 90° to the right.

Does $y = 1.5 \sin(\theta + 90°)$ lie to the right or left of a fundamental sine wave? ________________

---

58. If the phase difference of a non-fundamental sine wave is negative:

(1) The non-fundamental sine wave must be shifted to the left to be put in-phase with a fundamental sine wave.

(2) Therefore, the non-fundamental sine wave lies to the right of a fundamental sine wave.

To put $y = 52.6 \sin(\theta - 30°)$ in-phase with a fundamental sine wave, it must be shifted 30° to the left.

Does $y = 52.6 \sin(\theta - 30°)$ lie to the right or left of a fundamental sine wave? ________________

Answer to 57: To the left.

---

59. Which of the following sine waves lie to the left of a fundamental sine wave? ________________

(a) $y = \sin(\theta + 45°)$

(b) $y = 0.8 \sin(\theta - 45°)$

(c) $y = 15 \sin(\theta + 120°)$

(d) $y = 100 \sin(\theta - 140°)$

Answer to 58: To the right.

---

60. (a) If the phase difference of a non-fundamental sine wave with a fundamental sine wave is +100°, it lies 100° to the ____________ (right/left) of the fundamental sine wave.

(b) If the phase angle of a non-fundamental sine wave with a fundamental sine wave is −60°, it lies 60° to the ____________ (right/left) of the fundamental sine wave.

Answer to 59: Only (a) and (c).

Answer to 60:

a) left

b) right

61. IF A NON-FUNDAMENTAL SINE WAVE LIES TO THE LEFT OF A FUNDAMENTAL SINE WAVE, WE SAY THAT IT "LEADS" THE FUNDAMENTAL SINE WAVE.

IF A NON-FUNDAMENTAL SINE WAVE LIES TO THE RIGHT OF A FUNDAMENTAL SINE WAVE, WE SAY THAT IT "LAGS" THE FUNDAMENTAL SINE WAVE.

Sine wave A below lies 60° to the left of the fundamental sine wave. It leads the fundamental sine wave by 60°.

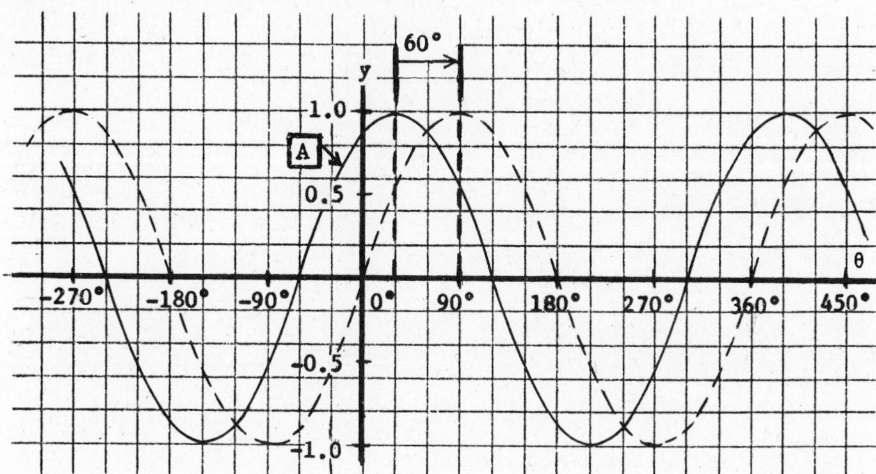

(a) What is the phase difference or phase angle of sine wave A with the fundamental sine wave? __________

(b) The amplitude of sine wave A is _______.

(c) Write the equation of sine wave A: ____________________

---

a) +60°

b) 1

c) $y = 1 \sin(\theta + 60°)$
or
$y = \sin(\theta + 60°)$

---

62. Sine wave B below lies 90° to the right of the fundamental sine wave. It lags the fundamental sine wave by 90°.

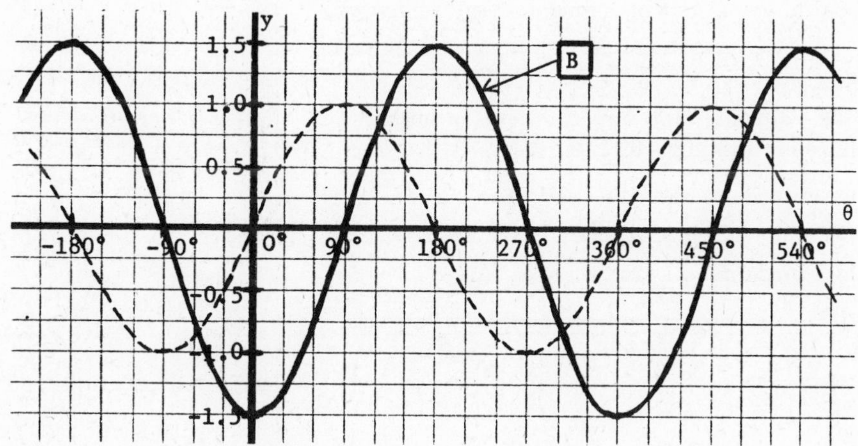

(Continued on following page.)

62. (Continued)

(a) What is the phase difference of sine wave B with the fundamental sine wave? __________

(b) The amplitude of sine wave B is _______.

(c) Write the equation of sine wave B: ____________________

---

63. (a) $y = \sin(\theta + 30°)$ lies 30° to the left of a fundamental sine wave. Therefore, we say that it __________ (leads/lags) the fundamental sine wave by 30°.

(b) $y = 1.9 \sin(\theta - 90°)$ lies 90° to the right of a fundamental sine wave. Therefore, we say that it __________ (leads/lags) the fundamental sine wave by 90°.

a) −90°
b) 1.5
c) $y = 1.5 \sin(\theta - 90°)$

---

64. Which of the following sine waves "lag" a fundamental sine wave?

____

(a) $y = 1.2(\theta + 70°)$ (c) $y = 37.1(\theta + 83°)$

(b) $y = 9(\theta - 27°)$ (d) $y = 200(\theta - 157°)$

a) leads
b) lags

---

65. Two sine waves are described below. Write the equation of each:

(a) Amplitude of 2.5; phase angle (with a fundamental sine wave) of +43°. ____________________

(b) Amplitude of 1; phase difference (with a fundamental sine wave) of −56°. ____________________

Only (b) and (d)

---

66. Write the equation of each of the following sine waves. Keep in mind whether the sine wave would have to be shifted to the right or left to put it in-phase with a fundamental sine wave.

(a) Amplitude of 5; lies 80° to the right of a fundamental sine wave.

____

(b) Amplitude of 0.7; lies 115° to the left of a fundamental sine wave.

____

a) $y = 2.5 \sin(\theta + 43°)$
b) $y = 1 \sin(\theta - 56°)$
or
$y = \sin(\theta - 56°)$

---

67. Write the equation of each of the following sine waves. Keep in mind whether it would have to be shifted to the right or left.

(a) Amplitude of 52.3; lags a fundamental sine wave by 60°.

____

(b) Amplitude of 10; leads a fundamental sine wave by 75°.

____

a) $y = 5 \sin(\theta - 80°)$
b) $y = 0.7 \sin(\theta + 115°)$

---

a) $y = 52.3 \sin(\theta - 60°)$
b) $y = 10 \sin(\theta + 75°)$

68. Referring to the graph at the right, write the equation of each of the following sine waves:

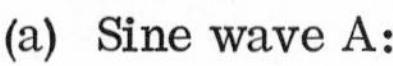

(a) Sine wave A:

(b) Sine wave B:

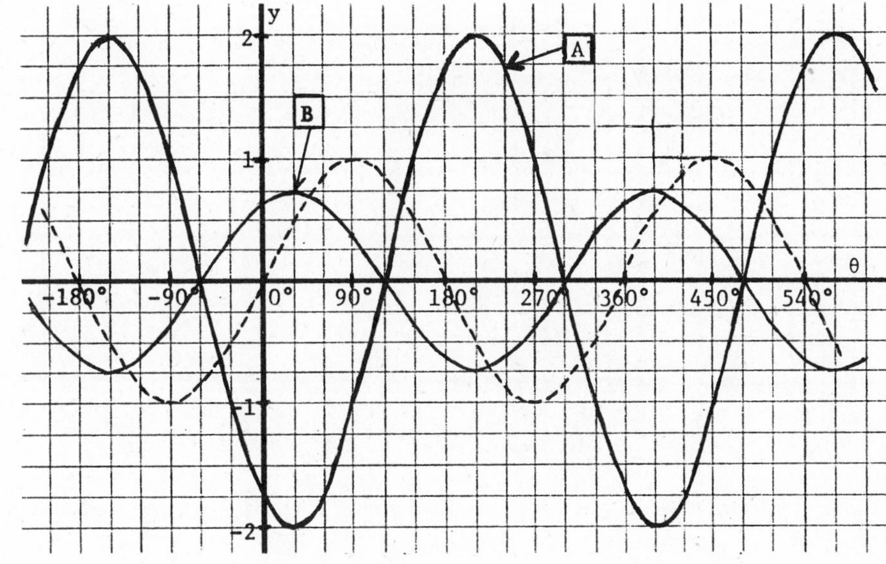

Answer to Frame 68: a) A: $y = 2 \sin(\theta - 120°)$ b) B: $y = 0.75 \sin(\theta + 60°)$

69. Referring to the graph at the right, write the equation of each of the following sine waves:

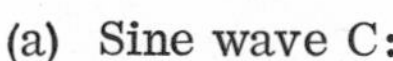

(a) Sine wave C:

(b) Sine wave D:

(c) Sine wave E:

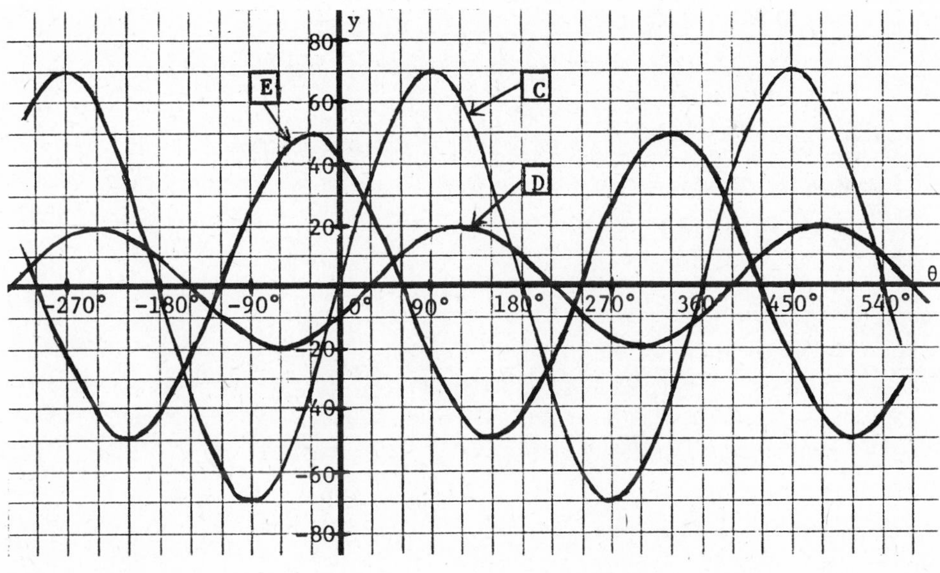

a) C: $y = 70 \sin \theta$

b) D: $y = 20 \sin(\theta - 30°)$

c) E: $y = 50 \sin(\theta + 120°)$

70.

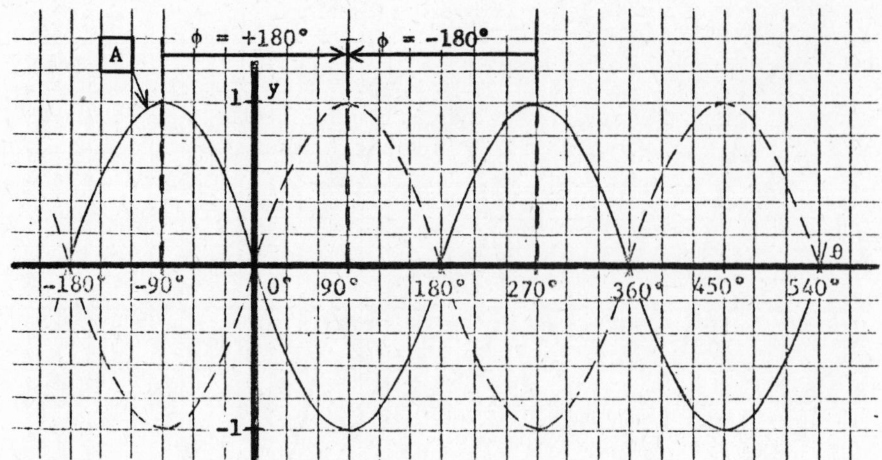

Examine the graph above. We can say either of two things about sine wave A:

(1) It <u>leads</u> the fundamental sine wave by 180°.

(2) It <u>lags</u> the fundamental sine wave by 180°.

Therefore, we can write either of two equations for sine wave A:

(a) If we consider it as "<u>leading</u>" a fundamental sine wave, its equation is: ______________________

(b) If we consider it as "<u>lagging</u>" a fundamental sine wave, its equation is: ______________________

---

a) $y = \sin(\theta + 180°)$

b) $y = \sin(\theta - 180°)$

SELF-TEST 5 (Frames 57-70)

1. A sine wave of amplitude 420 "lags" a fundamental sine wave by 135°. Write its equation. ______________

2. A sine wave of amplitude 28.7 "leads" a fundamental sine wave by 79°. Write its equation. ______________

Refer to the sine wave graphs below and answer the following:

3. Sine wave B __________ (leads/lags) fundamental sine wave A by ______°.

4. Sine wave C __________ (leads/lags) fundamental sine wave A by ______°.

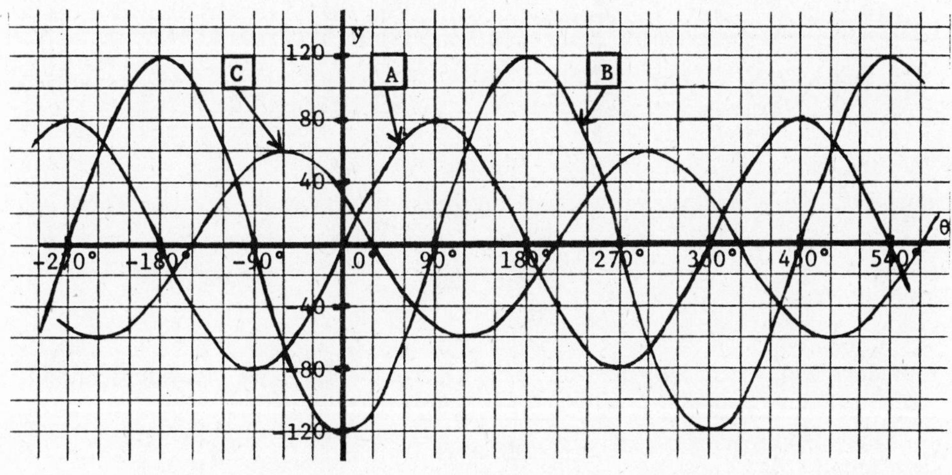

ANSWERS:

1. $y = 420 \sin(\theta - 135°)$
2. $y = 28.7 \sin(\theta + 79°)$
3. lags .... 90°
4. leads .... 150°

7-6 A BROADER INTERPRETATION OF PHASE DIFFERENCES AND THE CONCEPTS OF "LEAD" AND "LAG"

In the last two sections, we discussed phase differences and the concepts of "lead" and "lag". In those sections, we limited the discussion to phase shifts of non-fundamentals to fundamentals, since it is those particular shifts which appear as "φ" in the equations of non-fundamentals. This limitation was used in order to develop an understanding of the equations of non-fundamental sine waves.

In this section, we will show that there is a broader interpretation of phase differences and the concepts of "lead" and "lag". That is, we can speak of phase shifts of a fundamental to a non-fundamental, or phase shifts of one non-fundamental to another non-fundamental. When phase differences are used in this broader sense, we will see that the specific angles involved do not appear as "φ" in the equations of the sine waves.

71.

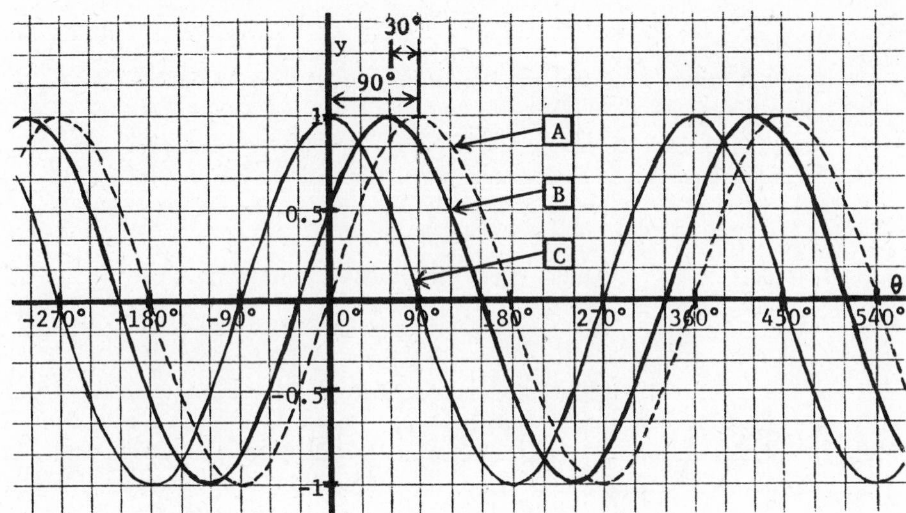

We have graphed one fundamental and two non-fundamental sine waves above. Their equations are:

A: $y = \sin \theta$
B: $y = \sin(\theta + 30°)$
C: $y = \sin(\theta + 90°)$

If we want to understand "φ" in the equations of non-fundamentals, we must determine the shift needed to put each non-fundamental in-phase with the fundamental.

For B, $\varphi = +30°$ since B leads A by 30°.
For C, $\varphi = +90°$ since C leads A by 90°.

However, the concept of phase shift can be used in a broader sense. That is, we can think of the shift needed to put the fundamental in-phase with each non-fundamental. In this broader sense:

The phase angle or phase difference between A and B is 30°.

It is +30° if we consider the shift needed to put B in-phase with A.
It is -30° if we consider the shift needed to put A in-phase with B.

Ordinarily, however, the use of a "signed" angle to represent a phase shift is limited to those cases in which a non-fundamental is shifted to a fundamental. When a fundamental is shifted to a non-fundamental, we use the words "lead" and "lag" instead of a "signed" angle. Therefore, we say:

A lags B by 30°

Similarly, we say: A lags C by ______°.

90°

72.

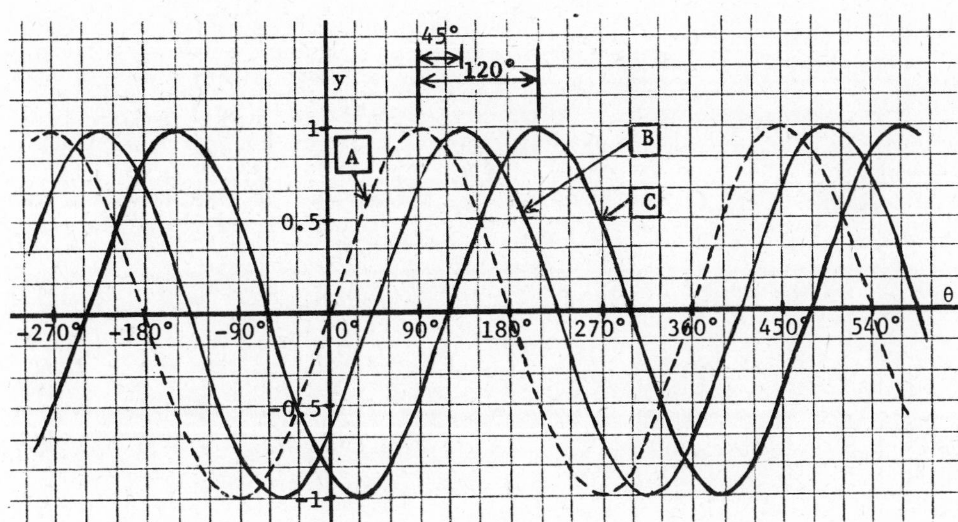

We have graphed one fundamental and two non-fundamental sine waves above. Their equations are:

A: $y = \sin\theta$
B: $y = \sin(\theta - 45°)$
C: $y = \sin(\theta - 120°)$

By examining the relationship of the fundamental and the two non-fundamentals, complete these:

(a) B ____________ (leads/lags) A by _______°.

(b) A ____________ (leads/lags) C by _______°.

(c) A ____________ (leads/lags) B by _______°.

(d) C ____________ (leads/lags) A by _______°.

---

a) lags .... 45°
b) leads .... 120°
c) leads .... 45°
d) lags .... 120°

73.

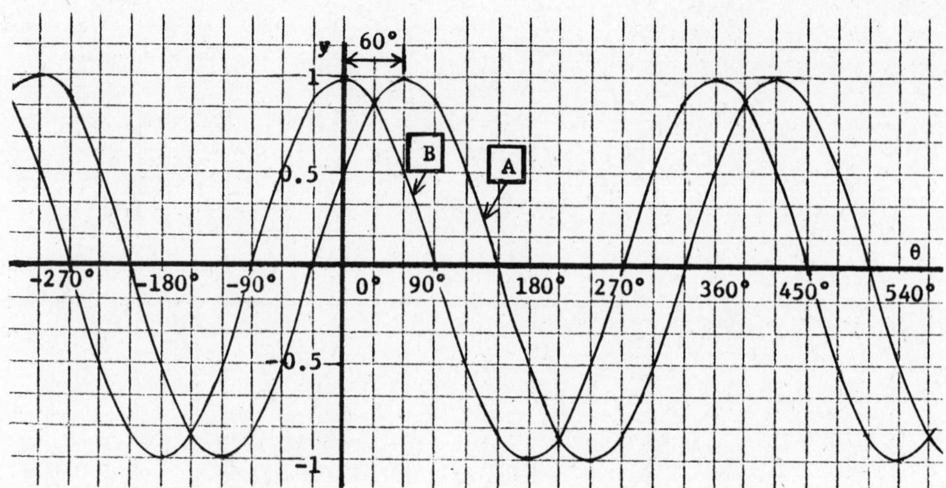

We have graphed two non-fundamental sine waves above. Their equations are:

A: $y = \sin(\theta + 30°)$
B: $y = \sin(\theta + 90°)$

Though the specific angle does not appear in either equation, we can speak of a phase angle or phase difference between these two non-fundamental sine waves.

The phase angle or phase difference between A and B is 60°.

It is +60° if we consider the shift needed to put B in-phase with A.

It is -60° if we consider the shift needed to put A in-phase with B.

Usually, however, we do not assign a "sign" to a phase angle between two non-fundamentals. We merely say that the phase angle is 60°. In order to indicate the direction of the shift, we use the words "lead" and "lag". Therefore, we say that:

(a) A ___________ (leads/lags) B by 60°.

(b) B ___________ (leads/lags) A by 60°.

---

a) lags

b) leads

74.

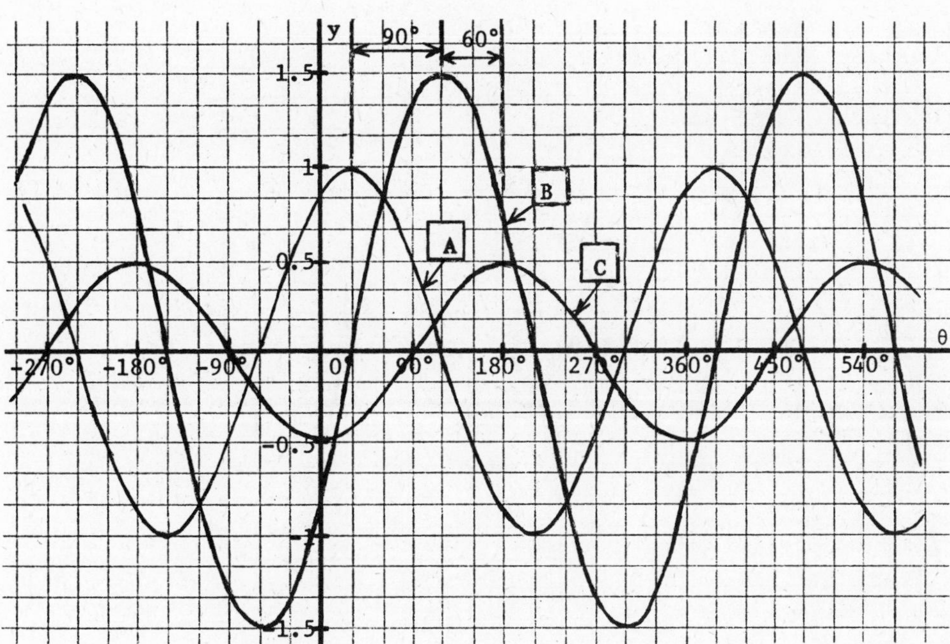

We have graphed the following three equations above:

A: $y = \sin(\theta + 60°)$
B: $y = 1.5 \sin(\theta - 30°)$
C: $y = 0.5 \sin(\theta - 90°)$

Since the phase angle between A and B is 90°, we say that:

A leads B by 90°.
B lags A by 90°.

(a) The phase angle between B and C is ________°.

(b) The phase angle between A and C is ________°.

(c) C ____________ (leads/lags) A by ________°.

(d) B ____________ (leads/lags) C by ________°.

---

a) 60°

b) 150°

c) lags .... 150°

d) leads .... 60°

75.

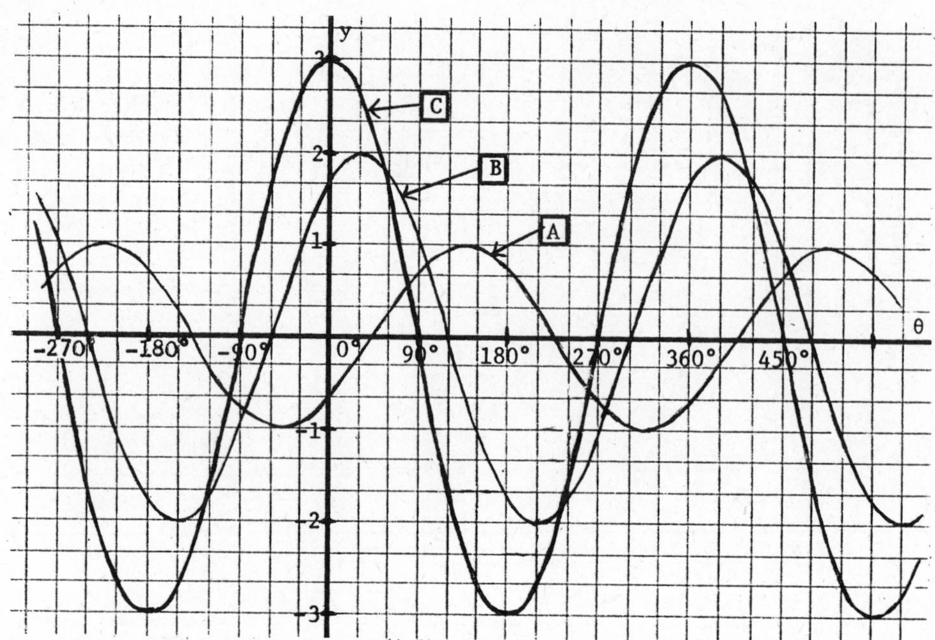

We have graphed the following three non-fundamental sine waves above:

A: $y = \sin(\theta - 45°)$
B: $y = 2\sin(\theta + 60°)$
C: $y = 3\sin(\theta + 90°)$

(a) The phase angle between C and B is ________°.

(b) The phase angle between B and A is ________°.

(c) C ____________ (leads/lags) A by ________°.

(d) A ____________ (leads/lags) B by ________°.

(e) B ____________ (leads/lags) C by ________°.

a) 30°

b) 105°

c) leads .... 135°

d) lags ..... 105°

e) lags ..... 30°

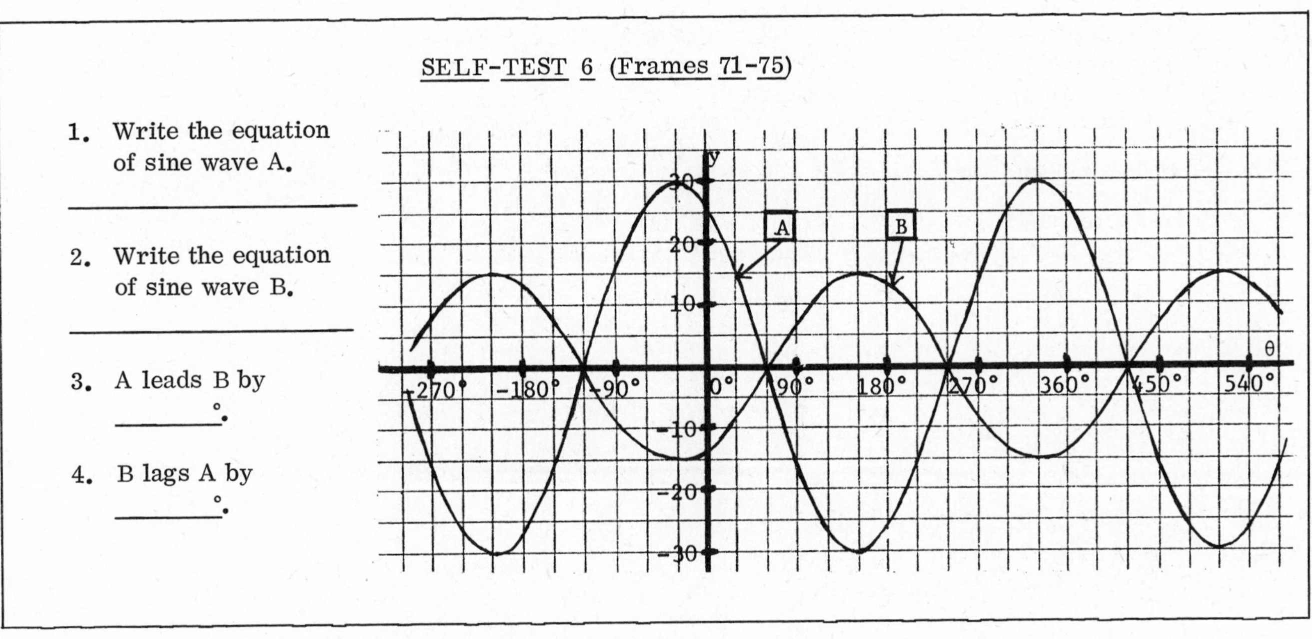

ANSWERS: 1. $y = 30 \sin(\theta + 120°)$ 2. $y = 15 \sin(\theta - 60°)$ 3. 180° 4. 180°

7-7 SINE WAVE HARMONICS

Up to this point, we have discussed fundamental and non-fundamental sine waves and their equations. In sine waves of both types, each "basic cycle" covers a 360° interval. There are other sine waves in which each "basic cycle" covers less than a 360° interval. In these sine waves, there are two or more "basic cycles" in each 360° interval. Sine waves of this type are called "harmonics".

In this section, we will discuss the "harmonics" of sine waves. Emphasis will be given to the harmonics of fundamental sine waves. The general form of their equations is: $\boxed{y = A \sin k\theta}$, where $k = 2, 3, 4,$ etc. A brief mention will be given to harmonics of non-fundamental sine waves. The general form of their equations is: $\boxed{y = A \sin(k\theta + \varphi)}$ or $\boxed{y = A \sin(k\theta - \varphi)}$. We will see that the new constant "k" in both types of equations is numerically equal to the number of basic cycles in each 360° interval in the "harmonic".

76. In order to graph sine-wave "harmonics", we have to prepare a table of pairs of values for "θ" and "y". Therefore, before graphing "harmonics", we will discuss the procedure for finding these pairs of values.

Here is the equation of a harmonic: $\boxed{y = \sin 2\theta}$

In order to find values of "y", we multiply by 2 each value plugged in for θ and then look up the sine of the new angle. For example:

If $\theta = 30°$, $y = \sin(2)(30°) = \sin 60° = 0.8660$

(a) If $\theta = 60°$, $y = \sin(2)(60°) = \sin$ _____ ° = __________

(b) If $\theta = 0°$, $y = \sin(2)(0°) = \sin$ _____ ° = __________

(c) If $\theta = 90°$, $y = \sin(2)(90°) = \sin$ _____ ° = __________

a) $\sin 120° = 0.8660$

b) $\sin 0° = 0$

c) $\sin 180° = 0$

---

77. Do not confuse $\boxed{y = 5 \sin \theta}$ and $\boxed{y = \sin 5\theta}$.

In $\boxed{y = 5 \sin \theta}$, we look up the sine of θ and then multiply the sine by 5. For example:

If $\theta = 45°$, $y = 5 \sin 45° = 5(0.7071) = 3.5355$

In $\boxed{y = \sin 5\theta}$, we multiply θ by 5 <u>and then</u> look up the sine of the new angle. For example:

If $\theta = 45°$, $y = \sin(5)(45°) = \sin 225° = -0.7071$

For each of the following equations, find "y" if $\theta = 30°$:

(a) $\boxed{y = 10 \sin \theta}$

y = __________

(b) $\boxed{y = \sin 10\theta}$

y = __________

a) $10 \sin 30° = 10(0.5) = 5$

b) $\sin 300° = -0.8660$

78. Here is another type: $\boxed{y = 2 \sin 3\theta}$

To find the value of "y":

(1) We multiply by 3 the value plugged in for θ.

If $\theta = 10°$, $y = 2 \sin(3)(10°) = 2 \sin 30°$

(2) We next look up the sine of the new angle <u>and then</u> multiply that sine by 2.

$y = 2 \sin 30° = 2(0.5) = 1.0$

Using the equation above, find the value of "y" if:

(a) $\theta = 45°$

y = ___________

(b) $\theta = 150°$

y = ___________

(c) $\theta = -50°$

y = ___________

---

a) $2 \sin 135° = 2(0.7071) = 1.4142$

b) $2 \sin 450° = 2(1.0) = 2$

c) $2 \sin(-150°) = 2(-0.5) = -1.0$

---

79. Here is another type: $\boxed{y = \sin(5\theta + 120°)}$

To find the value of "y":

(1) We multiply by 5 the value plugged in for θ.

(2) We then add 120° to this new angle before looking up the sine.

If $\theta = 20°$,

we get: $y = \sin[5(20°) + 120°] = \sin(100° + 120°) = \sin 220° = -0.6428$

Using the equation above, find the value of "y" if:

(a) $\theta = 0°$

y = ___________

(b) $\theta = 40°$

y = ___________

(c) $\theta = -120°$

y = ___________

80. Here is the most-complicated type: $\boxed{y = 10 \sin(2\theta - 60°)}$

To find the value of "y":

(1) We multiply by 2 the value plugged in for θ and then subtract 60°.
(2) We look up the sine of the new angle and then multiply that sine by 10.

If $\theta = 45°$, we get:

$$y = 10 \sin[2(45°) - 60°] = 10 \sin(90° - 60°) = 10 \sin 30° = 10(0.5) = 5$$

Using the equation above, find the value of "y" if:

(a) $\theta = 0°$

y = ___________

(b) $\theta = 100°$

y = ___________

Answers:

a) $\sin 120° = 0.8660$

b) $\sin 320° = -0.6428$

c) $\sin(-480°) = -0.8660$

---

81. If $\theta = 0°$, find the value of "y" in each of the following equations:

(a) $y = 7 \sin \theta$

y = ___________

(b) $y = \sin 7\theta$

y = ___________

(c) $y = \sin(7\theta + 120°)$

y = ___________

Answers:

a) $10 \sin(-60°) = 10(-0.8660) = -8.660$

b) $10 \sin 140° = 10(0.6428) = 6.428$

---

82. If $\theta = 10°$, find the value of "y" in each of the following equations:

(a) $y = 10 \sin \theta$

y = ___________

(b) $y = 10 \sin 3\theta$

y = ___________

(c) $y = 10 \sin(3\theta - 120°)$

y = ___________

Answers:

a) $7 \sin 0° = 0$

b) $\sin 0° = 0$

c) $\sin 120° = 0.8660$

83. If $\theta = 20°$, find the value of "y" in each of the following equations:

(a) $y = 5 \sin \theta$

y = ____________

(b) $y = \sin 5\theta$

y = ____________

(c) $y = 5 \sin(5\theta + 30°)$

y = ____________

a) $10 \sin 10° = 1.736$

b) $10 \sin 30° = 5$

c) $10 \sin(-90°) = -10$

Answer to Frame 83: a) $5 \sin 20° = 1.7100$ b) $\sin 100° = 0.9848$ c) $5 \sin 130° = 3.8300$

84. Before examining the graphs of "harmonics", let's briefly review the meaning of a "basic cycle" of a sine wave.

The graph of a "basic cycle" is given on the right. As you can see, there is a rise to a maximum peak on the left and a fall to a minimum peak on the right.

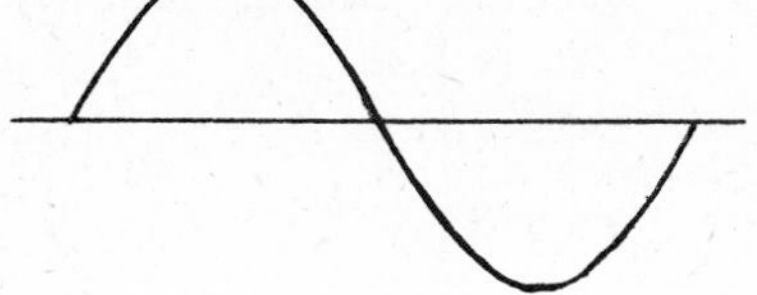

The sine wave on the right can be divided into two "basic cycles". We have done so with the vertical line.

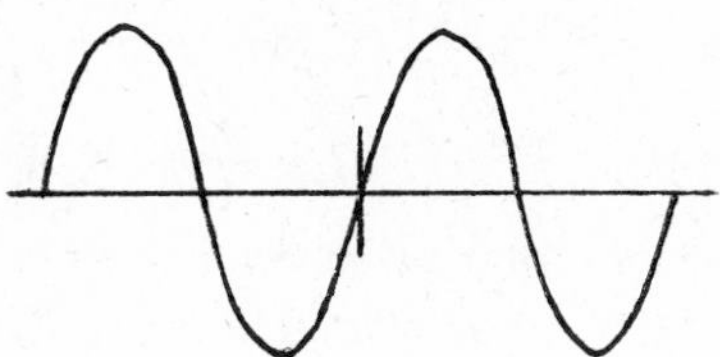

Into how many basic cycles can we divide each of the following sine waves?

____________ ____________

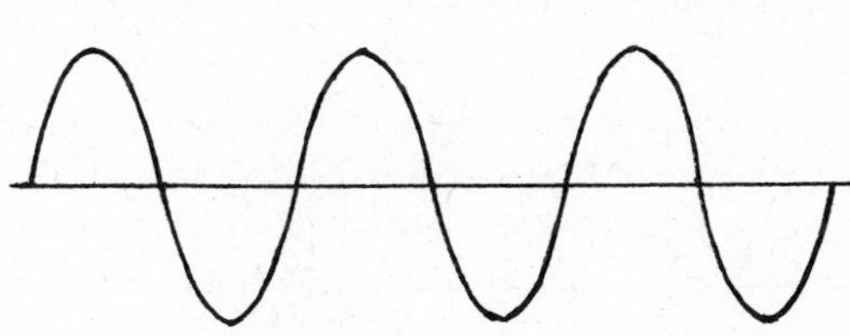

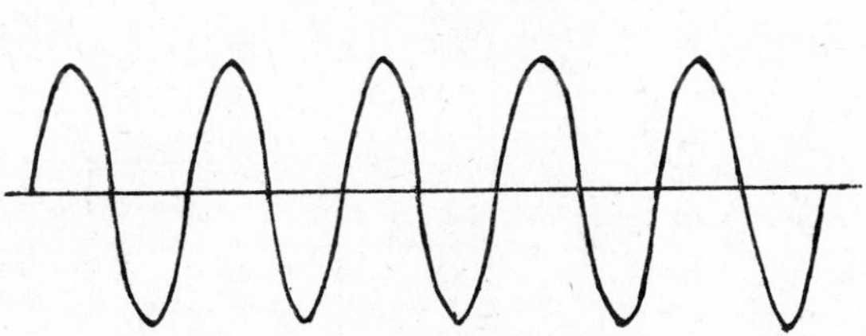

a) 3 basic cycles

b) 5 basic cycles

85. The following pairs of values satisfy this "harmonic" equation: $y = \sin 2\theta$

θ	y	θ	y	θ	y	θ	y
0°	0	90°	0	180°	0	270°	0
15°	0.50	105°	-0.50	195°	0.50	285°	-0.50
30°	0.87	120°	-0.87	210°	0.87	300°	-0.87
45°	1	135°	-1	225°	1	315°	-1
60°	0.87	150°	-0.87	240°	0.87	330°	-0.87
75°	0.50	165°	-0.50	255°	0.50	345°	-0.50
90°	0	180°	0	270°	0	360°	0

On the graph below, the first few points have been plotted. Plot the rest of the points. Then connect the points with a smooth curve to form the graph of $y = \sin 2\theta$.

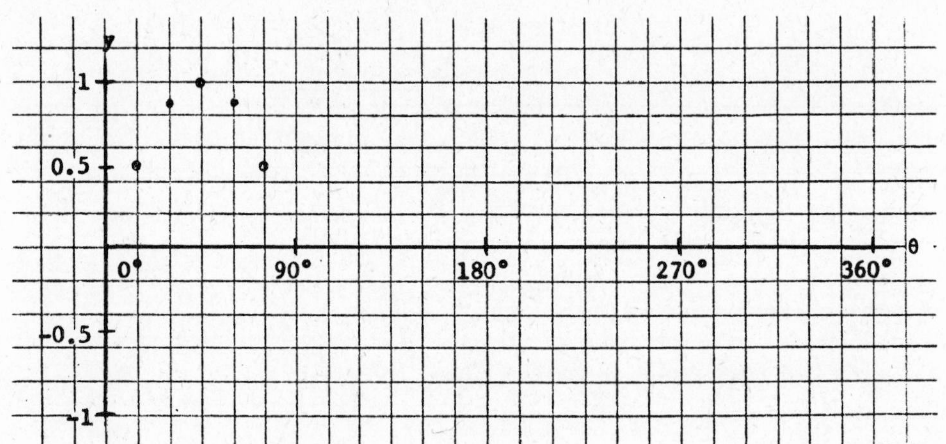

Answer to Frame 85: See the graph in the next frame.

86. Here is the graph of $y = \sin 2\theta$ from 0° to 360°. The graph of $y = \sin \theta$ is given as a comparison.

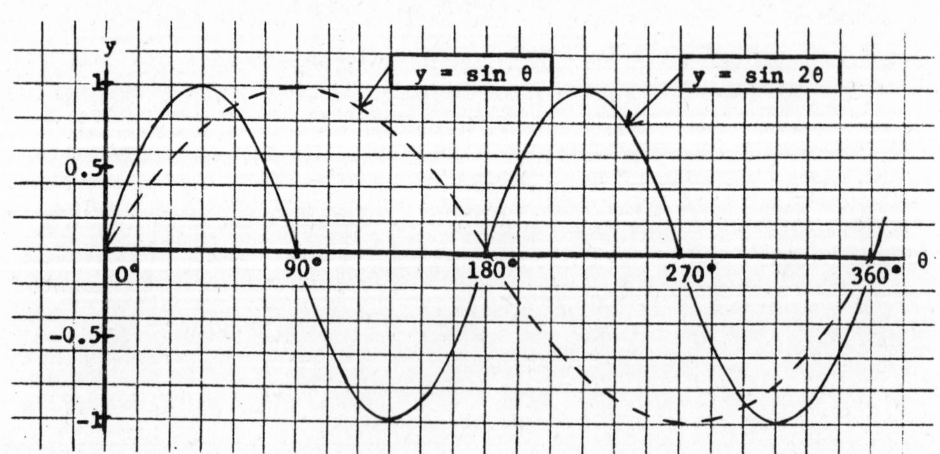

(Continued on following page.)

86. (Continued)

By examining the graph, answer the following:

(a) The amplitude or maximum value of $y = \sin 2\theta$ is _______.

(b) From $\theta = 0°$ to $\theta = 360°$, the graph of $y = \sin \theta$ (in dashed lines) has <u>one basic cycle.</u>

From $\theta = 0°$ to $\theta = 360°$, the graph of $y = \sin 2\theta$ has __________ basic cycles.

(c) Using $y = \sin 2\theta$, for what two angles between 0° and 360° does $y = 1$? _______ and _______

(d) Using $y = \sin 2\theta$, for what five angles from 0° to 360° does $y = 0$?

_______, _______, _______, _______, _______

<u>Answer to Frame 86:</u> a) 1 b) 2 c) 45° and 225° d) 0°, 90°, 180°, 270°, 360°

---

87. The "harmonic" equation $\boxed{y = \sin 3\theta}$ is graphed from 0° to 360° below. The graph of $y = \sin \theta$ is given as a comparison.

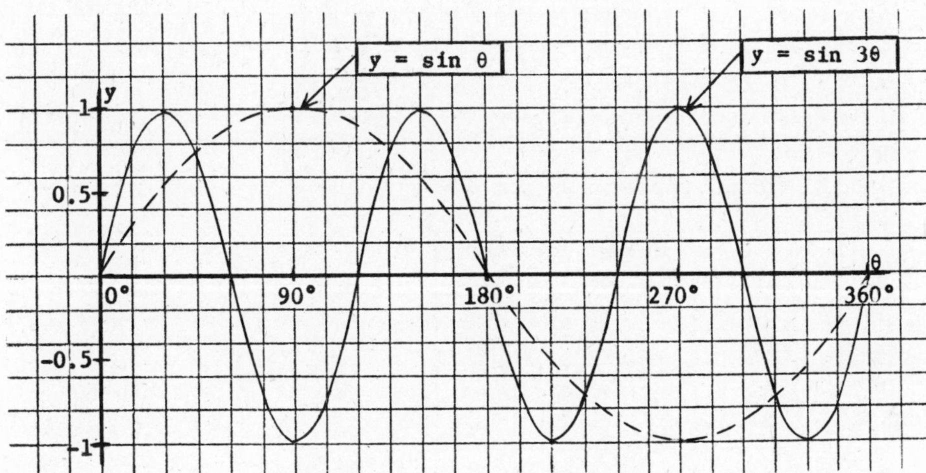

(a) What is the amplitude (maximum value) of the graph of $y = \sin 3\theta$? __________

(b) The graph of $y = \sin 3\theta$ has how many basic cycles from $\theta = 0°$ to $\theta = 360°$? __________

(c) At what three angles between 0° and 360° does the graph of $y = \sin 3\theta$ reach its maximum value?

_____, _____, _____

---

a) 1

b) Three cycles

c) 30°, 150°, 270°

88. The graph of $y = 2 \sin 4\theta$ is shown at the right.

(a) What is the amplitude or maximum value of the graph of $y = 2 \sin 4\theta$?

____

(b) The graph of $y = 2 \sin 4\theta$ has how many basic cycles from $\theta = 0°$ to $\theta = 360°$?

____

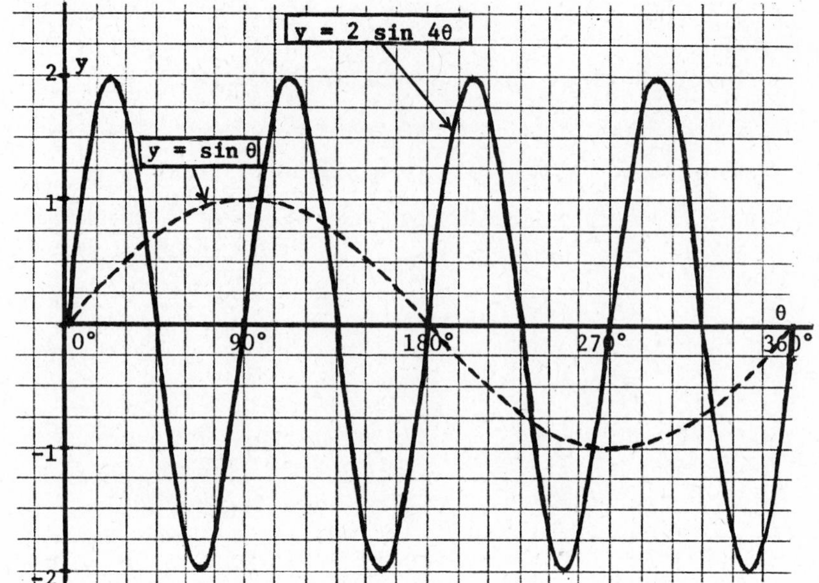

---

89. This table summarizes the graphs from the last few frames:

Equation	Amplitude	Number of Cycles Between 0° and 360°
$y = \sin \theta$	1	1
$y = \sin 2\theta$	1	2
$y = \sin 3\theta$	1	3
$y = 2 \sin 4\theta$	2	4

From the table, you can see this fact: If the general equation is written as:

$$y = A \sin k\theta$$

"A" is the amplitude.
"k" is the number of cycles between 0° and 360°.
(Note: $y = \sin \theta$ is the same as $y = \sin 1\theta$.)

(a) If $y = \sin 8\theta$ were graphed, there would be _______ basic cycles between 0° and 360°.

(b) If a sine wave has an amplitude of "1" with 7 basic cycles between 0° and 360°, its equation is _______________.

(c) If a sine wave has an amplitude of "3" with 10 basic cycles between 0° and 360°, its equation is ___________________.

Answers to frame 88:

a) 2

b) Four cycles

Answers to frame 89:

a) 8

b) $y = \sin 7\theta$

c) $y = 3 \sin 10\theta$

90. Any sine wave with two basic cycles between 0° and 360° is called a second harmonic. Here are some examples of the equations of second harmonics:

$$y = \sin 2\theta$$
$$y = 4 \sin 2\theta$$
$$y = 7 \sin 2\theta$$

Any sine wave with three basic cycles between 0° and 360° is called a third harmonic. Here are some examples of the equations of third harmonics:

$$y = \sin 3\theta$$
$$y = 5 \sin 3\theta$$
$$y = 10 \sin 3\theta$$

Similarly: (a) $y = \sin 4\theta$ is a _______________ harmonic.

(b) $y = 3 \sin 5\theta$ is a _______________ harmonic.

(c) $y = 0.7 \sin 10\theta$ is a _______________ harmonic.

---

91. Note: Though any fundamental sine wave is a first harmonic (one basic cycle between 0° and 360°), we do not use the term "first harmonic".

(a) The equation of a second harmonic of amplitude "50" is _______________.

(b) The equation of an eighth harmonic of amplitude "1" is _______________.

Answers to 90:

a) fourth

b) fifth

c) tenth

---

92. Write the equations of the following sine waves:

(a) A fifth harmonic with an amplitude of 0.67: _______________

(b) A seventh harmonic with an amplitude of 0.00125: _______________

Answers to 91:

a) $y = 50 \sin 2\theta$

b) $y = 1 \sin 8\theta$
or
$y = \sin 8\theta$

Answers to 92:

a) $y = 0.67 \sin 5\theta$

b) $y = 0.00125 \sin 7\theta$

93. The harmonics we have been discussing have the following general equation: $y = A \sin k\theta$. All harmonics of this type are harmonics of fundamental sine waves. Harmonics of non-fundamental sine waves are also possible. Harmonics of non-fundamental sine waves have either of the following general equations:

$$y = A \sin(k\theta + \varphi) \quad \text{or} \quad y = A \sin(k\theta - \varphi)$$

The graph of $y = \sin(2\theta + 60°)$ is shown below. It is a harmonic of a non-fundamental sine wave. The graph of the fundamental $y = \sin \theta$ is given as a comparison.

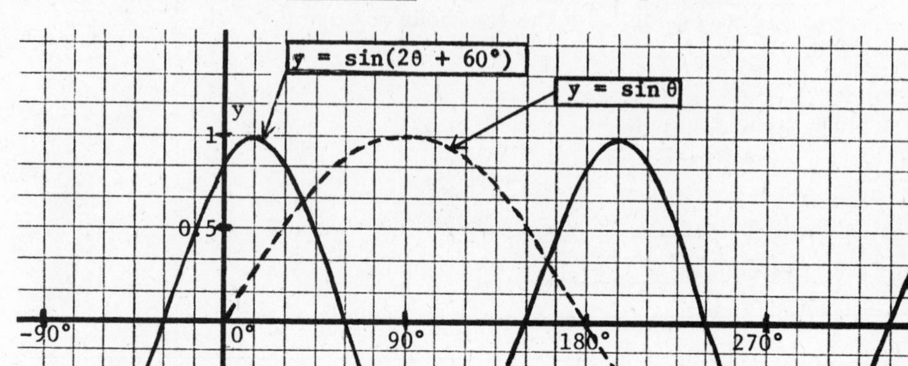

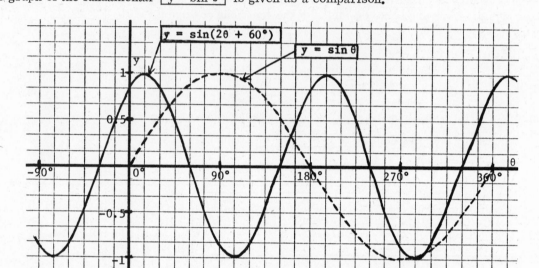

This sine wave also has two basic cycles within a 360° period. The 360° period, however, is between -30° and 330°. Its two basic cycles do not begin and end with the one basic cycle of the fundamental sine wave. That is, its two basic cycles do not begin and end at 0° and 360°.

A sine wave of this type is also called a harmonic. It is a second harmonic since it has two basic cycles within a 360° period. There are all sorts of harmonics whose basic cycles do not occur in a 360° period which begins and ends at 0° and 360°. Since harmonics of this type are not used in elementary sine wave analysis, we will not discuss them in this chapter.

SELF-TEST 7 (Frames 76-93)

Find the numerical value of "y" in each equation if $\theta = 60°$:

1. $y = 20 \sin 3\theta$ y = ________
2. $y = 8 \sin(2\theta + 90°)$ y = ________
3. $y = 300 \sin(5\theta - 30°)$ y = ________

4. Write the equation of a fifth harmonic whose amplitude is 0.50. ________

5. Write the equation of an eighth harmonic whose amplitude is 0.098. ________

6. The graph of $y = 6 \sin 3\theta$ has how many basic cycles between 0° and 360°? ________

7. A sine wave harmonic of amplitude 5 has two basic cycles between 0° and 360°. Write its equation. ________

8. A sine wave harmonic of amplitude 1.75 has one basic cycle between 0° and 90°. Write its equation. ________

ANSWERS:

1. $y = 0$
2. $y = -4$
3. $y = -300$
4. $y = 0.50 \sin 5\theta$
5. $y = 0.098 \sin 8\theta$
6. Three basic cycles
7. $y = 5 \sin 2\theta$
8. $y = 1.75 \sin 4\theta$

7-8 SKETCHING FUNDAMENTAL SINE WAVES

Graphing a sine wave exactly is tedious since a table of many pairs of values is needed in order to do so. Therefore, sine waves ordinarily are not graphed exactly. However, the ability to roughly sketch a sine wave is very useful. When sketching a sine wave, there are some key points which can be used as a guide. In this section, we will discuss the key points which can be used to sketch fundamental sine waves.

---

94. Here is the graph of the fundamental sine wave: $y = \sin \theta$ Though the graph only shows the cycle between 0° and 360°, other basic cycles could be added in either direction.

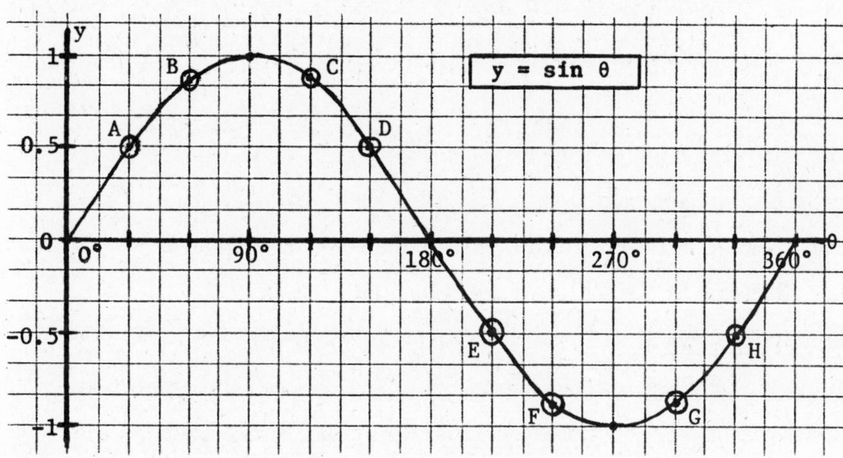

Notice these points about the graph of this one basic cycle:

(1) It crosses the horizontal axis at 0°, 180°, and 360°. Therefore, it crosses the horizontal axis every 180°.

(2) It reaches its positive peak (maximum positive value) at 90°. It reaches its negative peak (maximum negative value) at 270°.

(3) Since $\sin 30° = \sin 150° = 0.5$ or $\frac{1}{2}$, the graph is halfway up to its positive peak at 30° (point A) and 150° (point D).

(4) Since $\sin 60° = \sin 120° = 0.866$ or approximately $\frac{9}{10}$, the graph is approximately $\frac{9}{10}$ of the way up to its positive peak at 60° (point B) and 120° (point C).

(5) Since $\sin 210° = \sin 330° = -0.5$ or $-\frac{1}{2}$, the graph is halfway down to its negative peak at 210° (point E) and 330° (point H).

(6) Since $\sin 240° = \sin 300° = -0.866$ or approximately $-\frac{9}{10}$, the graph is $\frac{9}{10}$ of the way down to its negative peak at 240° (point F) and 300° (point G).

95. The graph of the fundamental sine wave whose amplitude is "2" is shown at the right. Its equation is:

$y = 2 \sin \theta$

Though the graph only shows the basic cycle between 0° and 360°, other basic cycles could be added in either direction.

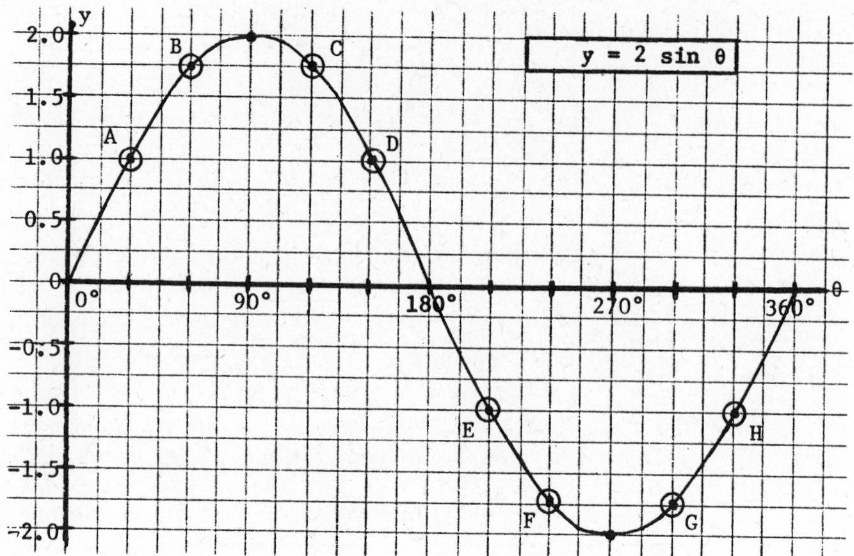

Notice these points about the basic cycle:

(1) It also crosses the horizontal axis every 180° (at 0°, 180°, 360°) and peaks at 90° and 270°.

(2) Halfway up or down in this case is +1 or −1 since the amplitude is "2".

It is halfway up at 30° (point A) and 150° (point D).

It is halfway down at 210° (point E) and 330° (point H).

(3) $\frac{9}{10}$ of the way up or down in this case is $\frac{9}{10}(2)$ or 1.8 since the amplitude is "2".

It is approximately $\frac{9}{10}$ of the way up at 60° (point B) and 120° (point C).

It is approximately $\frac{9}{10}$ of the way down at 240° (point F) and 300° (point G).

What is the only difference between this graph and the graph of $y = \sin \theta$? ______________________

Their amplitudes are different. In this case, the amplitude is "2".

96. Here is a diagram of the points you should plot when sketching a fundamental sine wave:

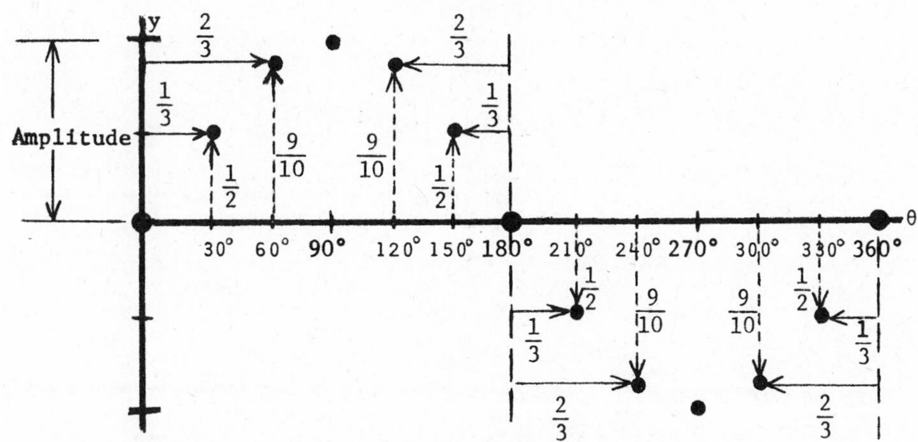

Besides the horizontal-axis crossings and the positive and negative peaks, you should plot:

(1) The points which are halfway up or halfway down.

Note: These "halfway" points are $\frac{1}{3}$ of the horizontal distance from each axis crossing to the peak points.

(2) The points which are approximately $\frac{9}{10}$ of the way up or $\frac{9}{10}$ of the way down.

Note: These "$\frac{9}{10}$ of the way" points are $\frac{2}{3}$ of the horizontal distance from the axis crossings to the peak points.

(a) The "halfway" up or down points are _______ of the horizontal distance from the axis crossings to the peaks.

(b) The "$\frac{9}{10}$ of the way" up or down points are _______ of the horizontal distance from the axis crossings to the peaks.

---

a) $\frac{1}{3}$

b) $\frac{2}{3}$

---

97. To determine the height of the "halfway" or "$\frac{9}{10}$ of the way" points, you must take the amplitude of the sine wave into account.

If the amplitude is "4":

The height of the halfway point is $\frac{1}{2}(4) = 2$.

The height of the "$\frac{9}{10}$" point is $\frac{9}{10}(4) = \frac{36}{10} = 3.6$.

If the amplitude is "$\frac{1}{2}$":

(a) The height of the halfway point is _______________.

(b) The height of the "$\frac{9}{10}$" point is _______________.

---

a) $\frac{1}{2}\left(\frac{1}{2}\right) = \frac{1}{4}$ or 0.25

b) $\frac{9}{10}\left(\frac{1}{2}\right) = \frac{9}{20}$ or 0.45

98. Let's sketch one basic cycle of the following fundamental sine wave:

$$\boxed{y = 3 \sin \theta}$$

(a) Plot the horizontal-axis crossings, peaks, "halfway" points and "$\frac{9}{10}$" points on the right.

(b) Then draw a smooth curve through the plotted points.

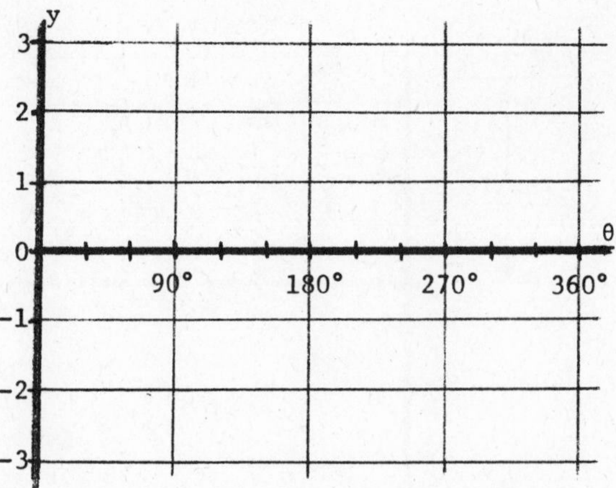

Answer to Frame 98: See next frame.

99. The sketch of one basic cycle of $\boxed{y = 3 \sin \theta}$ is given on the right:

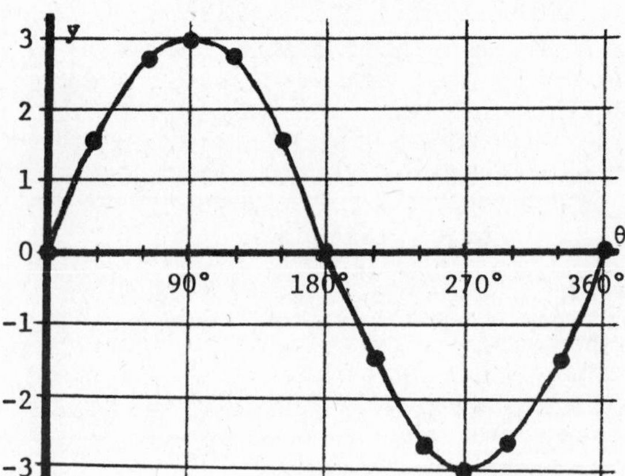

Notice the general shape of the curve of a sine wave:

(1) It is not a semi-circle like:

(2) It is not triangular like:

(3) It is not bent in on the sides like:

(4) It is not bent in on the sides with a cap on the top like:

100. Let's sketch the following fundamental sine wave on the axes below: $\boxed{y = 10 \sin \theta}$

After sketching the basic cycle between 0° and 360°, add a similar cycle between -360° and 0° and between 360° and 720°.

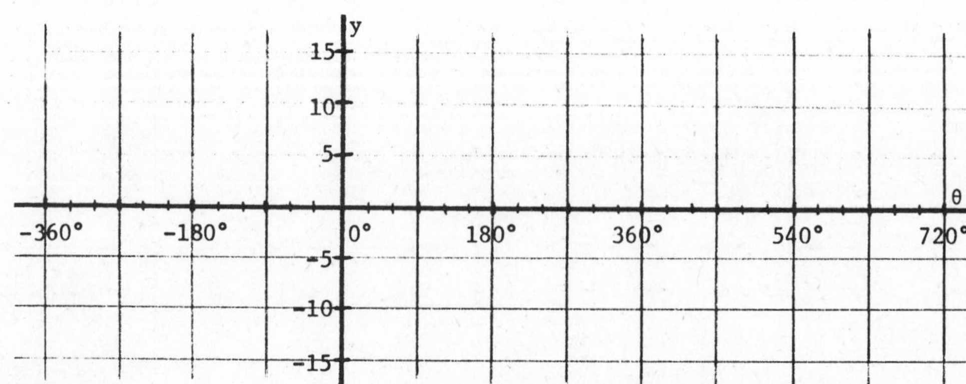

Note: No answer will be given for this frame.

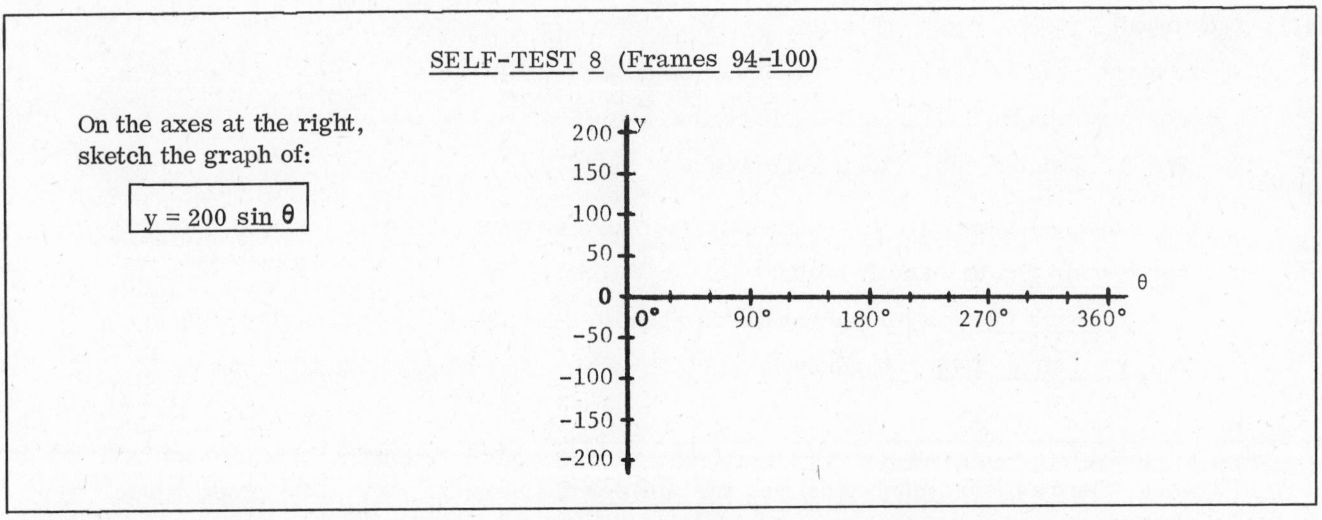

ANSWER: The graph is a fundamental sine wave of amplitude 200. It has one basic cycle between 0° and 360°. Check your sketch to make sure it passes through these points:

(30°,100) (60°,180) (120°,180) (150°,100) (210°,-100) (330°,-100)

7-9 SKETCHING NON-FUNDAMENTAL SINE WAVES

All sine waves have the same general shape. There is one major difference between the graphs of fundamental sine waves and non-fundamental sine waves. The graph of any fundamental sine wave has a basic cycle between 0° and 360°. The graphs of non-fundamentals do not have a basic cycle which begins at 0°. We will show the method for sketching non-fundamental sine waves in this section.

101. We have graphed the following two sine waves below:

$$y = \sin(\theta + 30°)$$
$$y = 2\sin(\theta - 90°)$$

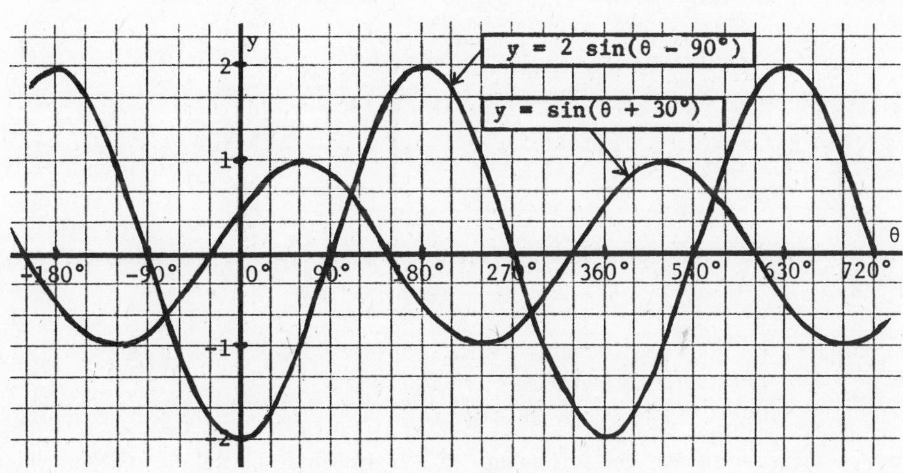

(Continued on following page.)

101. (Continued)

There is not a basic cycle between 0° and 360° on either graph.

For $y = \sin(\theta + 30°)$, a basic cycle begins at -30°.

For $y = 2\sin(\theta - 90°)$, a basic cycle begins at +90°.

A basic cycle begins when y is "0" at the start of a positive loop.

For $y = \sin\theta$, a basic cycle begins when $\theta = 0°$, since: $\sin 0° = 0$

For $y = \sin(\theta + 30°)$, a basic cycle begins when $\theta = -30°$, since: $\sin(-30° + 30°) = \sin 0° = 0$

For $y = 2\sin(\theta - 90°)$, a basic cycle begins when $\theta = +90°$, since: $2\sin(90° - 90°) = 2\sin 0° = 2(0) = 0$

---

102. To decide where a basic cycle begins, you must answer this question:

What value of θ makes the sum of the angles in parentheses equal 0°? For example:

A basic cycle begins for $y = \sin(\theta - 60°)$ at +60°, since $(\theta - 60°) = 0°$ when $\theta = +60°$.

A basic cycle begins for $y = 3\sin(\theta + 45°)$ at -45°, since $(\theta + 45°) = 0°$ when $\theta = -45°$.

At what value of θ would a basic cycle begin for each of the following?

(a) $y = \sin(\theta - 45°)$ Basic cycle begins at: $\theta =$ _______°

(b) $y = 10\sin(\theta + 90°)$ Basic cycle begins at: $\theta =$ _______°

(c) $y = 5\sin(\theta + 60°)$ Basic cycle begins at: $\theta =$ _______°

Go to next frame.

Answer to Frame 102: a) +45° b) -90° c) -60°

---

103. Once you know where a basic cycle begins, it is easy to determine the axis crossings and peaks. Here is a diagram of the major points of a basic cycle which begins at -60°:

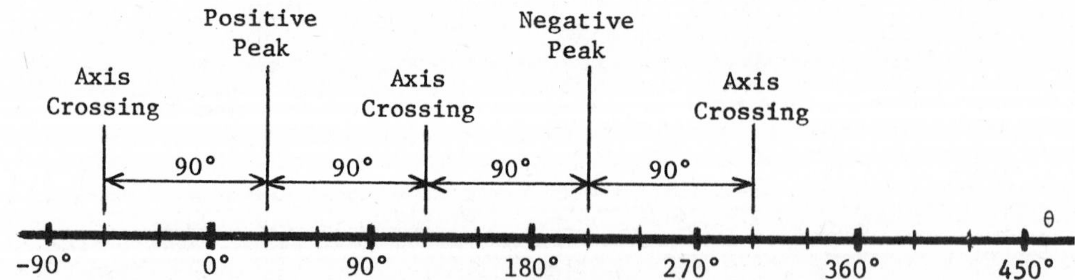

Notice these points:
(1) The basic cycle is still 360° along the horizontal axis.
(2) The axis crossings are 180° apart.
(3) The two peaks lie halfway between successive axis crossings.

Use the diagram above to answer the following. If a basic cycle begins at +60°, at what value of θ:

(a) Will the positive peak be? _______

(b) Will the negative peak be? _______

(c) Will the basic cycle end? _______

104. Below we have plotted the points which are necessary to draw a basic cycle of the following sine wave: $y = \sin(\theta + 90°)$

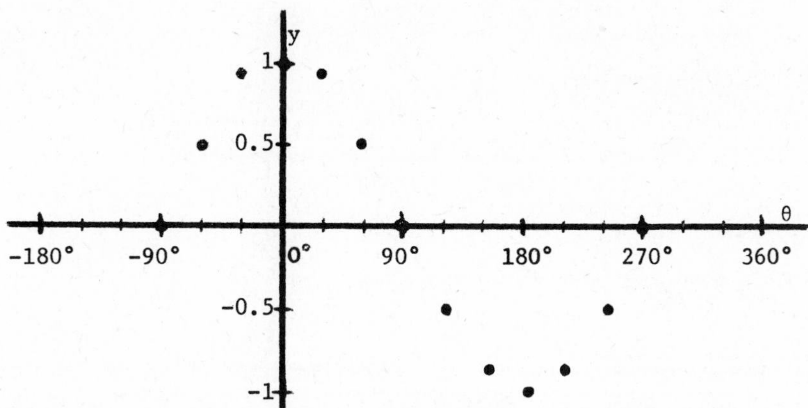

The basic cycle begins at -90° and ends at 270°. Besides the horizontal axis crossings and peaks, we have included:

(1) The halfway up and down points.

Note: These "halfway" points are $\frac{1}{3}$ of the horizontal distance from each axis crossing to a peak.

(2) The "$\frac{9}{10}$ of the way up and down" points.

Note: These "$\frac{9}{10}$ of the way" points are $\frac{2}{3}$ of the horizontal distance from each axis crossing to a peak.

a) 150° (60° + 90°)

b) 330° (60° + 270°)

c) 420° (60° + 360°)

---

105. Let's sketch a basic cycle of the following sine wave:

$y = \sin(\theta - 30°)$

(a) Plot the points needed to draw the cycle between 30° and 390°.

(b) Draw the curve through the plotted points.

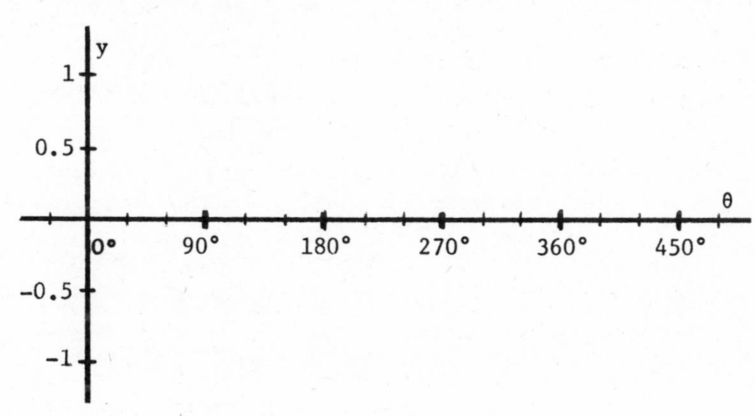

Answer to Frame 105:

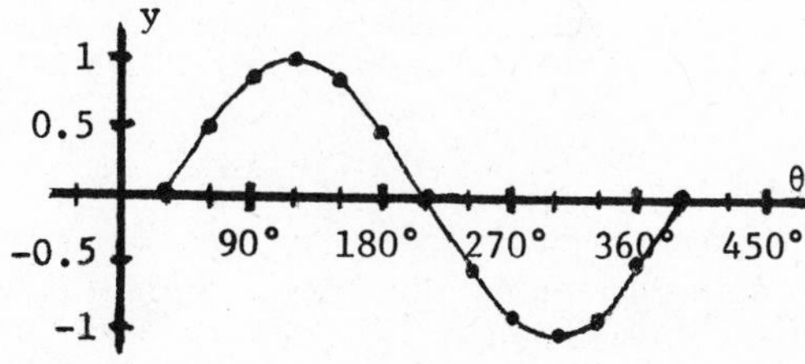

106. Let's sketch the following sine wave: $y = 5\sin(\theta + 60°)$

(a) The amplitude is _______.

(b) One basic cycle begins at _______° and ends at _______°.

(c) Plot the points needed to draw this one basic cycle.

(d) Draw the curve through the plotted points and then extend the curve in both directions to the ends of the horizontal axis.

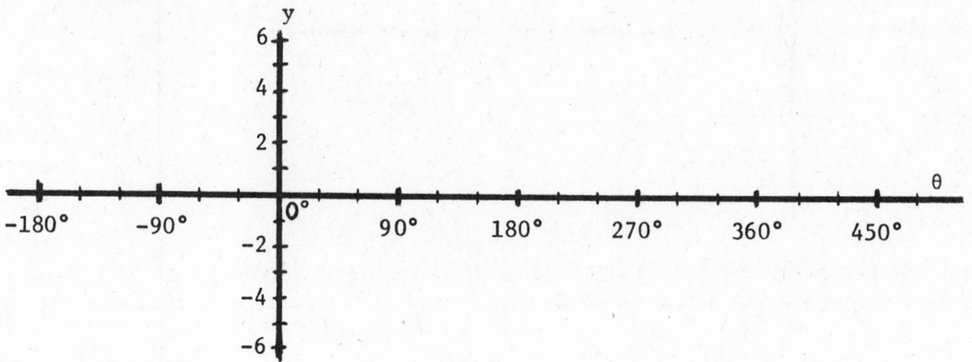

Answer to Frame 106:

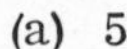

(a) 5

(b) -60° ... 300°

(c) See graph at right.

(d) See graph at right.

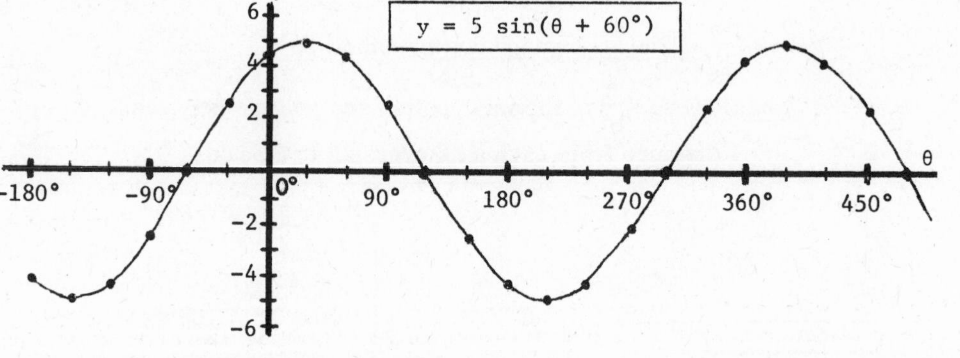

SELF-TEST 9 (Frames 101-106)

On the axes at the right, sketch the graph of:

$y = 8\sin(\theta + 30°)$

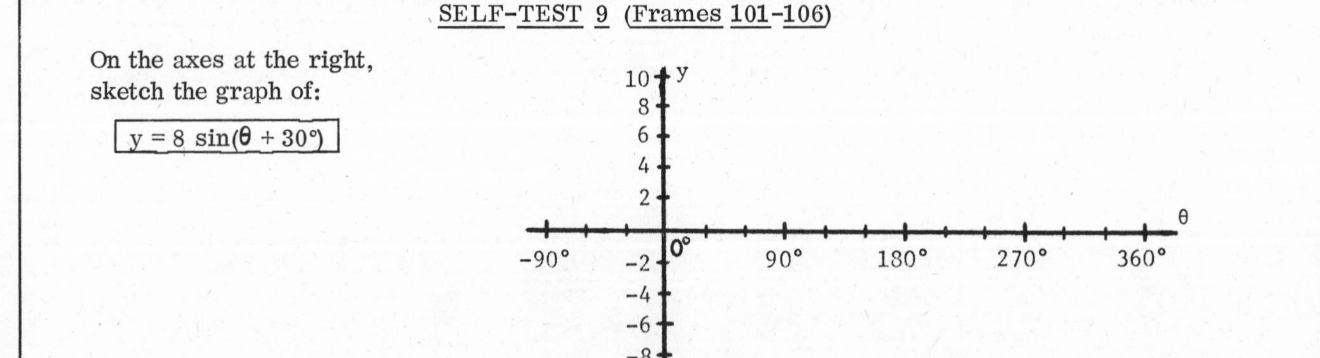

ANSWER: The graph is a non-fundamental sine wave of amplitude 8.
It has one basic cycle between -30° and 330°.
It has a peak positive value at 60°, and a peak negative value at 240°.
It crosses the horizontal axis at -30°, 150°, and 330°.

7-10 SKETCHING HARMONICS OF FUNDAMENTAL SINE WAVES

A fundamental sine wave has one basic cycle between 0° and 360°. A harmonic of a fundamental sine wave has more than one basic cycle between 0° and 360°. Except for the fact that more basic cycles appear within a 360° range, the method for sketching a harmonic is basically the same as the method for sketching a fundamental. We will show the method for sketching a harmonic in this section.

---

107. The basic equation for a harmonic of a fundamental sine wave is:

$\boxed{y = A \sin k\theta}$ where "A" is the amplitude
and "k" is the number of basic cycles between 0° and 360°.

If we graph the equation: $\boxed{y = \sin 3\theta}$

(1) There will be 3 basic cycles between 0° and 360°.

(2) Each one of the 3 cycles would cover 120° on the horizontal axis.

If we graph the equation: $\boxed{y = 4 \sin 2\theta}$

(a) There will be _______ basic cycles between 0° and 360°.

(b) Each one of the cycles will cover ________° on the horizontal axis.

---

108. How many degrees on the horizontal axis would one cycle cover if we graphed each of the following equations?

(a) $y = \sin 4\theta$ __________ (c) $y = 5 \sin 6\theta$ __________

(b) $y = 2 \sin 5\theta$ __________ (d) $y = 9 \sin 10\theta$ __________

Answers to 107:

a) 2

b) 180°

Answers to 108:

a) 90° (from $\frac{360°}{4}$)

b) 72° (from $\frac{360°}{5}$)

c) 60° (from $\frac{360°}{6}$)

d) 36° (from $\frac{360°}{10}$)

109. Below we discuss the method for sketching $\boxed{y = \sin 3\theta}$ between 0° and 360°.

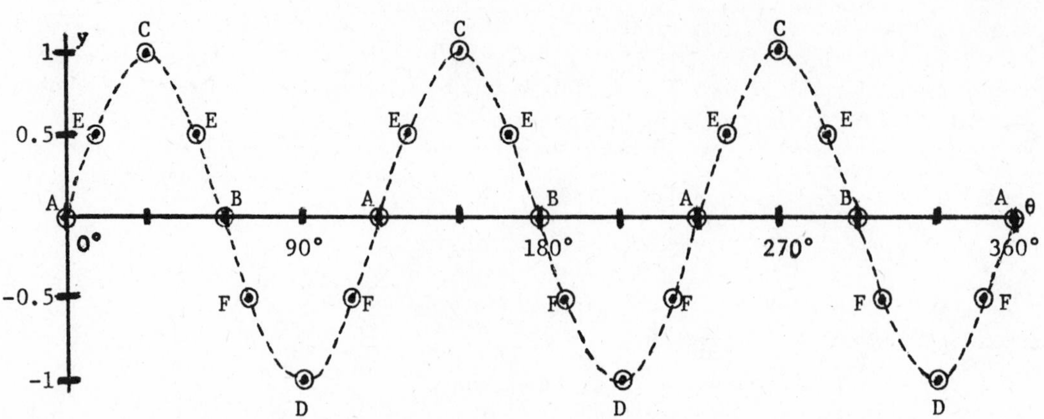

Step 1: Each basic cycle covers 120° on the horizontal axis. Therefore, we first plot points on the horizontal axis to separate the cycles. (See the 4 points labeled A above.)

Step 2: Since the curve also crosses the horizontal axis at the midpoint of each cycle, plot a point on the horizontal axis at the midpoint of each cycle. (See the 3 points labeled B above.)

Step 3: The peaks occur halfway between the end and midpoint of each cycle. Plot the positive and negative peaks. (See the 3 points labeled C and the three points labeled D above.)

Step 4: Plot the halfway up and halfway down points $\frac{1}{3}$ of the horizontal distance from each axis crossing to the peaks. (See the 6 points labeled E and the 6 points labeled F above.)

Note: Ordinarily, the graph is too crowded to plot the "$\frac{9}{10}$ of the way up and down" points. It may even be too crowded to plot the halfway up and down points for some graphs.

Step 5: Draw the curve through the plotted points. (See the dashed curve above.)

---

110. Let's sketch $\boxed{y = \sin 2\theta}$ between 0° and 360°.

(a) Plot the points needed to draw the sketch on the axes at the right.

(b) Draw the curve through the plotted points.

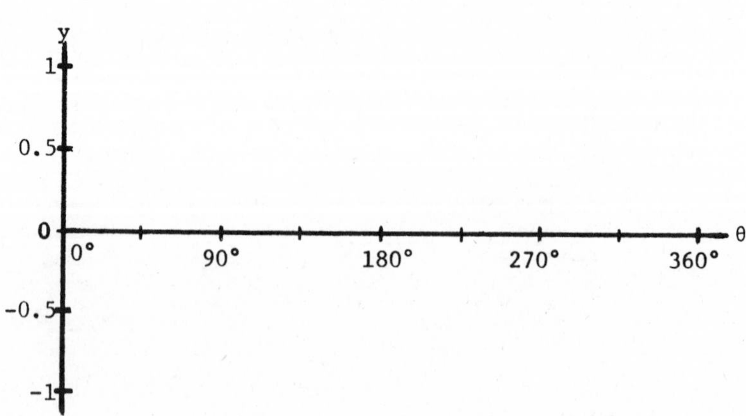

Answer to Frame 110:

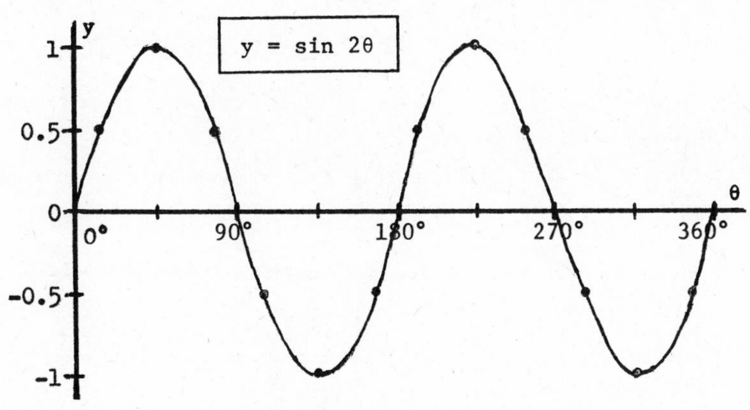

111. Let's sketch the following harmonic below: $y = 1.5 \sin 4\theta$

(a) Plot the points needed to draw the four basic cycles between 0° and 360°.

(b) Draw the curve through the plotted points.

(c) Extend the curve in both directions to the ends of the horizontal axis.

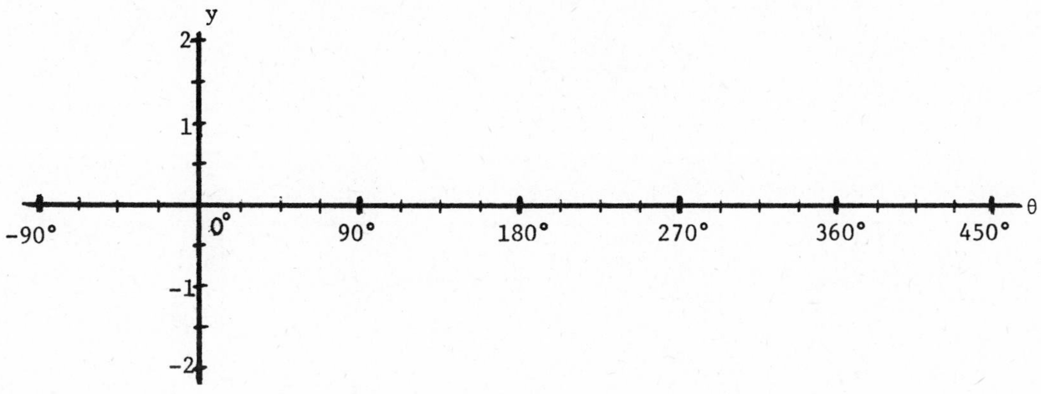

Answer to Frame 111:

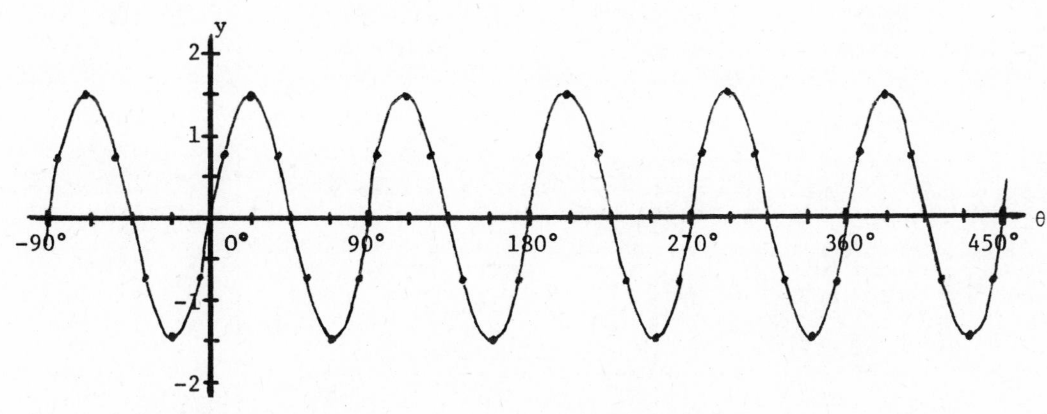

SELF-TEST 10 (Frames 107-111)

On the axes on the right, sketch the graph of:

$y = 4 \sin 3\theta$

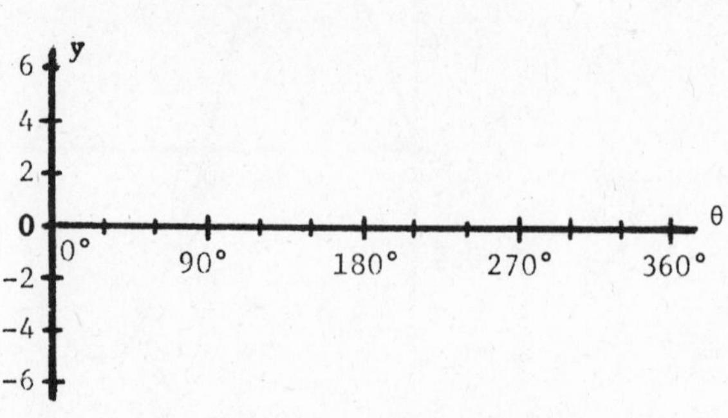

ANSWER: The graph is a sine wave harmonic, of amplitude 4, with three complete basic cycles between 0° and 360°. The first basic cycle begins at 0° and ends at 120°, with a positive peak at 30° and a negative peak at 90°. The second basic cycle begins at 120° and ends at 240°, with a positive peak at 150° and a negative peak at 210°. The third basic cycle begins at 240° and ends at 360°, with a positive peak at 270° and a negative peak at 330°.

7-11 SINE WAVES WITH NEGATIVE AMPLITUDES

In this section, we will discuss sine waves with negative amplitudes. We will show that any sine wave with a negative amplitude is the "opposite" of the sine wave with a corresponding positive amplitude.

112. $y = -\sin \theta$ is a sine wave equation with a negative amplitude. Though the numerical amplitude is not explicitly written, it is "-1". That is:

$y = -\sin \theta$ can be written: $y = -1 \sin \theta$

To graph this equation, we must find pairs of values for "y" and "θ" which satisfy the equation. For example:

(a) If $\theta = 0°$, $y = (-1)(\sin 0°) = (-1)(0) =$ __________

(b) If $\theta = 30°$, $y = (-1)(\sin 30°) = (-1)(0.5) =$ __________

(c) If $\theta = 90°$, $y = (-1)(\sin 90°) = (-1)(1.0) =$ __________

113. Here is another sine wave equation with a negative amplitude:
$y = -3 \sin \theta$

Let's find some corresponding pairs of values for "y" and "θ":

(a) If $\theta = 180°$, $y = (-3)(\sin 180°) = (-3)(\quad) =$ __________

(b) If $\theta = 210°$, $y = (-3)(\sin 210°) = (-3)(\quad) =$ __________

(c) If $\theta = 270°$, $y = (-3)(\sin 270°) = (-3)(\quad) =$ __________

Answers to 112:
a) 0
b) -0.5
c) -1.0

Answers to 113:
a) $(-3)(0) = 0$
b) $(-3)(-0.5) = +1.5$
c) $(-3)(-1.0) = +3.0$

114. The two tables below contain pairs of values for $y = \sin \theta$ and $y = -\sin \theta$. By comparing the tables, you can see that any value of "y" for $y = -\sin \theta$ has the <u>opposite sign</u> of the corresponding value of "y" for $y = \sin \theta$.

$y = \sin \theta$

θ	y	θ	y	θ	y	θ	y
0°	0.00	90°	1.00	180°	0.00	270°	-1.00
45°	0.71	135°	0.71	225°	-0.71	315°	-0.71
90°	1.00	180°	0.00	270°	-1.00	360°	0.00

$y = -\sin \theta$

θ	y	θ	y	θ	y	θ	y
0°	0.00	90°	-1.00	180°	0.00	270°	1.00
45°	-0.71	135°	-0.71	225°	0.71	315°	0.71
90°	-1.00	180°	0.00	270°	1.00	360°	0.00

Here is a graph of the two equations. When the values of "y" are <u>positive</u> for $y = \sin \theta$, they are <u>negative</u> for $y = -\sin \theta$ and vice versa.

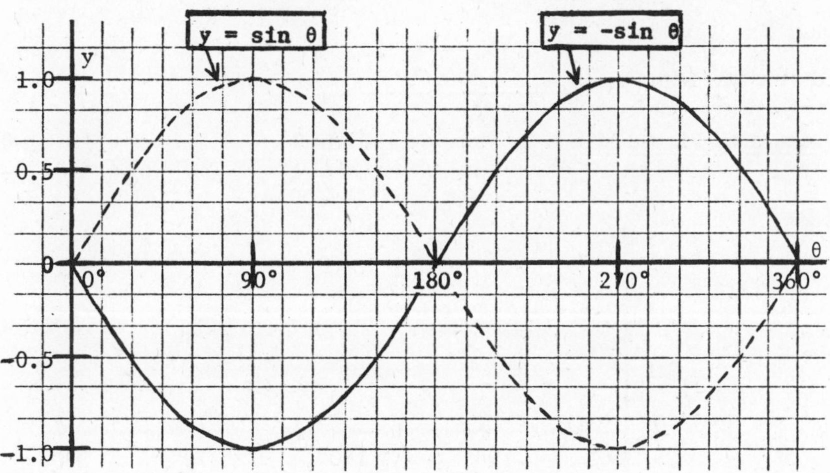

(a) At 90°: $y = \sin \theta$ has a <u>positive</u> peak and
$y = -\sin \theta$ has a ______________ peak.

(b) At 270°: $y = \sin \theta$ has a <u>negative</u> peak and
$y = -\sin \theta$ has a ______________ peak.

---

115. If a sine wave equation has a negative amplitude, its graph is the "opposite" of a sine wave equation with a corresponding positive amplitude. They are "opposites" in the sense that their positive and negative peaks are interchanged.

If $y = 2 \sin \theta$ is graphed: (1) It has a <u>positive</u> peak at 90°.
(2) It has a <u>negative</u> peak at 270°.

If $y = -2 \sin \theta$ is graphed: (a) It has a <u>positive</u> peak at ________°.
(b) It has a <u>negative</u> peak at ________°.

Answers (114):
a) negative
b) positive

Answers (115):
a) 270°
b) 90°

116. Examine the two sine waves graphed below:

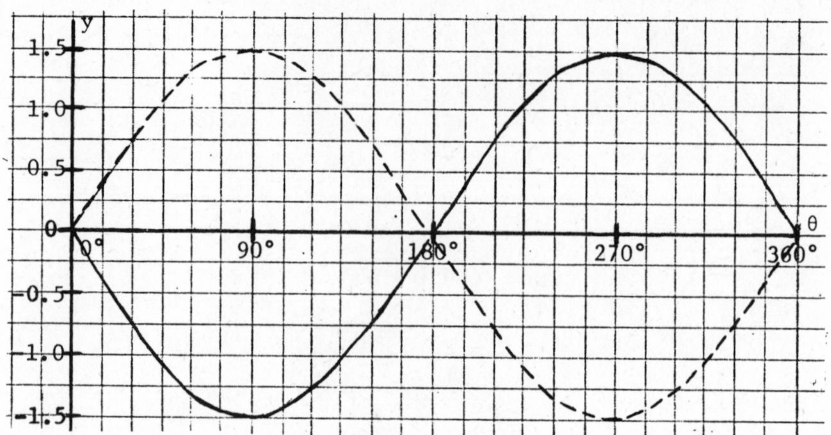

(a) The equation of the dashed sine wave is: ______________

(b) The equation of the solid sine wave is: ______________

Answer to Frame 116: a) $y = 1.5 \sin \theta$ b) $y = -1.5 \sin \theta$

117. At the right, sketch each of the following sine waves:

$y = -3 \sin \theta$

$y = -\frac{1}{2} \sin \theta$

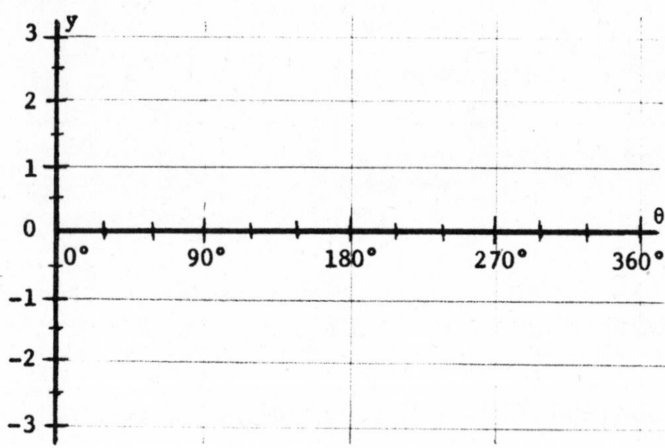

Answer to Frame 117:

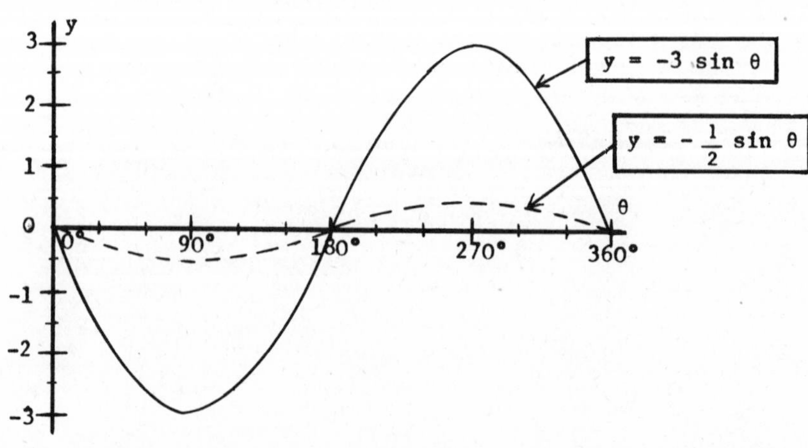

118. A harmonic with a negative amplitude is also the "opposite" of a harmonic with a corresponding positive amplitude. They are "opposites" in the sense that their positive and negative peaks are interchanged.

Below we have graphed the following two harmonics:

$y = \sin 2\theta$ (the dashed curve)
$y = -\sin 2\theta$ (the solid curve)

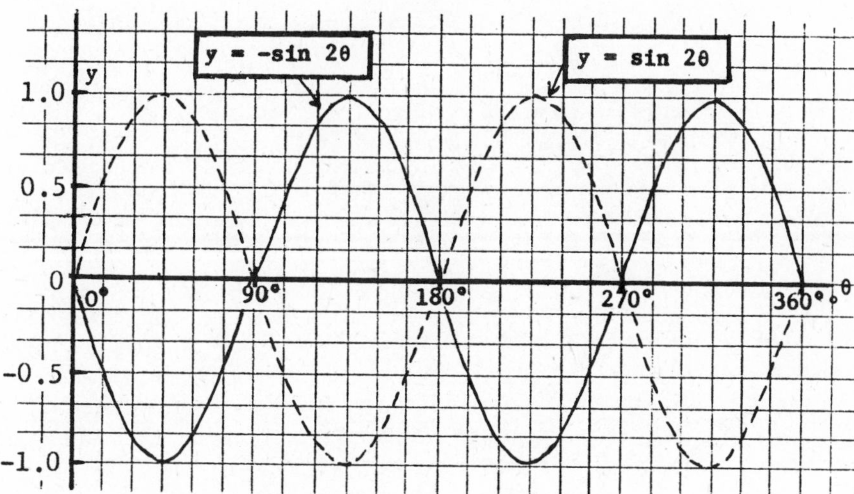

$y = \sin 2\theta$ has positive peaks at 45° and 225°.

$y = -\sin 2\theta$ has positive peaks at _____ and _____.

135° and 315°

119. Here are the graphs of two more harmonics:

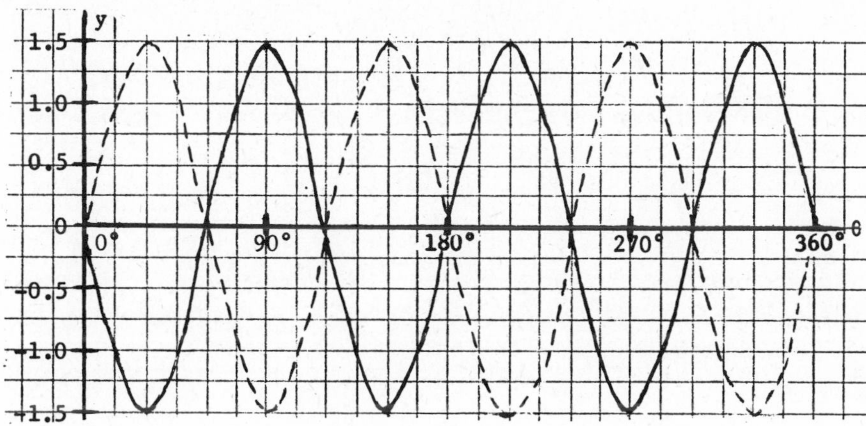

(a) The equation of the dashed harmonic is: ____________________

(b) The equation of the solid harmonic is: ____________________

a) $y = 1.5 \sin 3\theta$

b) $y = -1.5 \sin 3\theta$

120. When sketching harmonics with negative amplitudes, remember that their peaks are the "opposite" of the peaks of harmonics with positive amplitudes.

Sketch each of the following harmonics on the axes below:

$$y = -2 \sin 2\theta \quad \text{and} \quad y = -\frac{1}{2} \sin 4\theta$$

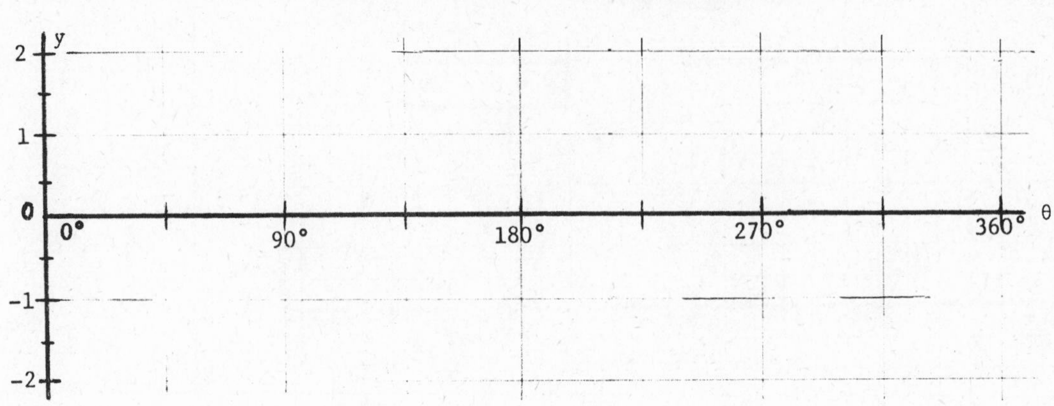

Answer to Frame 120:

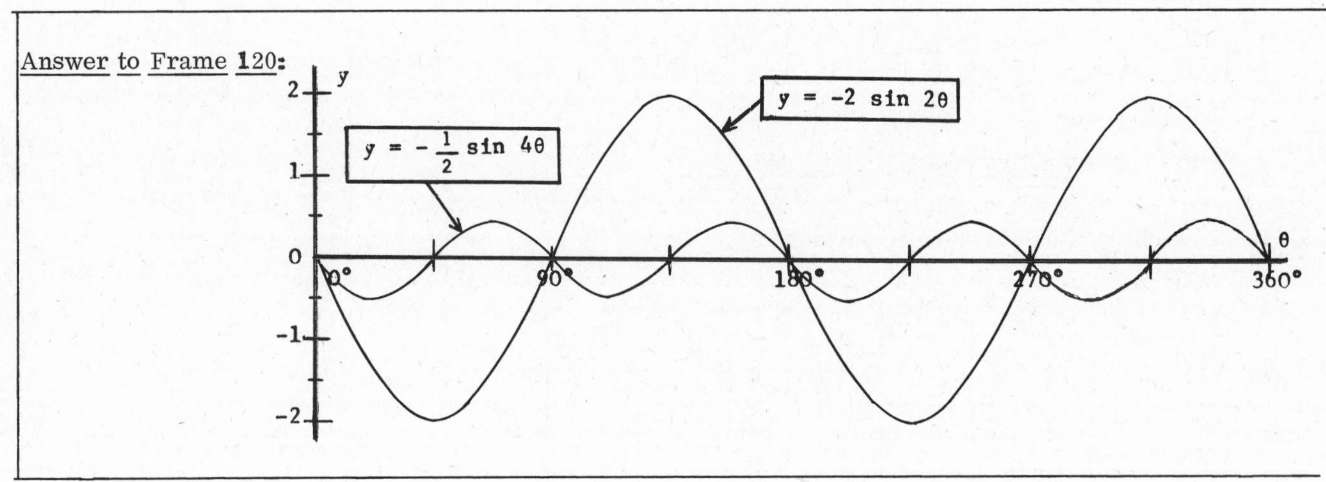

121. On the axes below, we have graphed the following two equations: $\boxed{y = \sin \theta}$ and $\boxed{y = -\sin \theta}$

You can see that $y = -\sin \theta$ is 180° out-of-phase with $y = \sin \theta$. We can look at it in either of two ways:

(1) $y = -\sin \theta$ "leads" $y = \sin \theta$ by 180° and $\varphi = +180°$.

(2) $y = -\sin \theta$ "lags" $y = \sin \theta$ by 180° and $\varphi = -180°$.

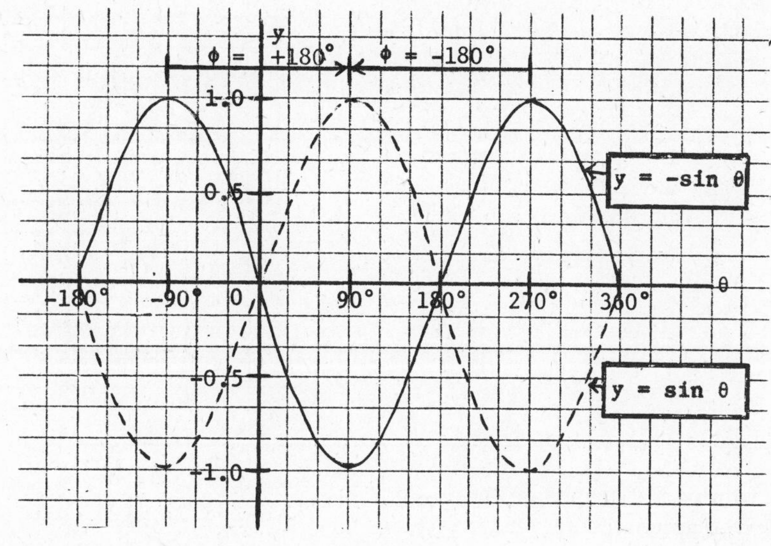

Since $y = -\sin \theta$ can be thought of as a sine wave which is 180° out-of-phase with a fundamental, we can write two equivalent equations with a positive amplitude instead of $y = -\sin \theta$.

If $\varphi = +180°$, we can write: $y = \sin(\theta + 180°)$

If $\varphi = -180°$, we can write: ____________________

---

122. Any sine wave equation which is the "opposite" of a fundamental can be thought of as a sine wave which is 180° out-of-phase with the fundamental. Therefore, we can always write two equivalent out-of-phase equations with positive amplitudes. That is:

For $\boxed{y = -A \sin \theta}$

the two equivalent equations are: $\boxed{\begin{array}{l} y = A \sin(\theta + 180°) \\ y = A \sin(\theta - 180°) \end{array}}$

For the same value of A, these equations have identical graphs.

Write $y = -5 \sin \theta$ in two equivalent forms:

____________________ ____________________

Answer to 121: $y = \sin(\theta - 180°)$

---

123. Write each of the following 180° out-of-phase equations as an equivalent equation with a negative amplitude:

(a) $y = \frac{1}{2} \sin(\theta + 180°)$ ____________________

(b) $y = 10 \sin(\theta - 180°)$ ____________________

Answer to 122: $y = 5 \sin(\theta + 180°)$

$y = 5 \sin(\theta - 180°)$

Note: There are equivalent harmonic equations with positive amplitudes for any harmonic equation with a negative amplitude. However, since there is little or no need for writing harmonics in these equivalent "positive" forms, we will not discuss the "positive" forms in this book.

Answer to 123:

a) $y = -\frac{1}{2} \sin \theta$

b) $y = -10 \sin \theta$

SELF-TEST 11 (Frames 112-123)

Examine the graphs shown at the right. Using a negative amplitude:

1. Write the equation of sine wave A: ______________

2. Write the equation of sine wave B: ______________

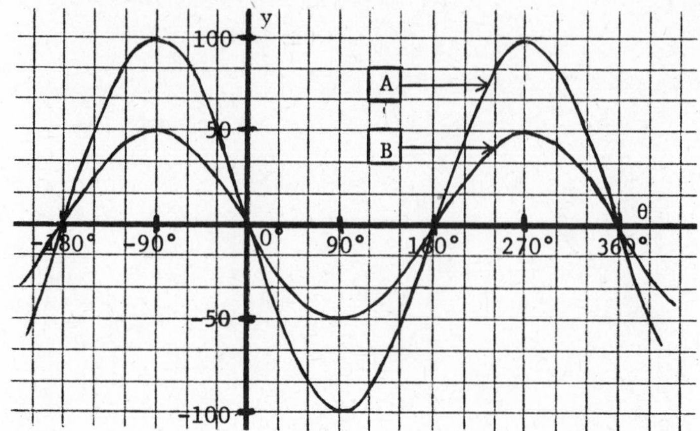

Examine the graphs of sine wave harmonics shown at the right. Write the equation of:

3. Graph C: ______________

4. Graph D: ______________

5. Graph E: ______________

6. Graph F: ______________

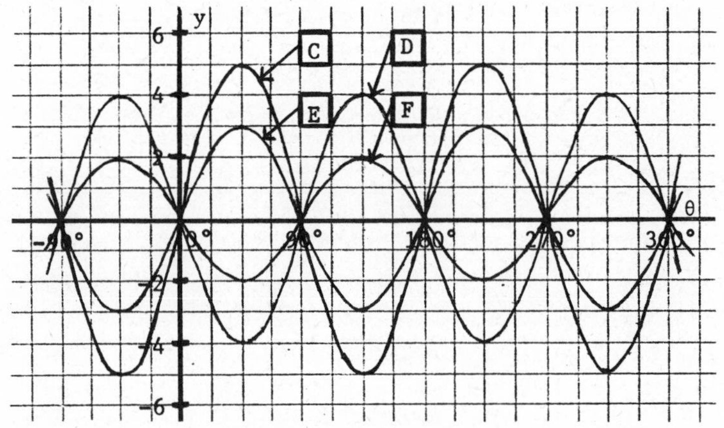

7. Which of the following have the same graph as $y = -20 \sin \theta$? ______________

(a) $y = 20 \sin(\theta + 180°)$ (b) $y = 20 \sin(\theta - 180°)$ (c) $y = -20 \sin(\theta + 90°)$

ANSWERS:

1. $y = -100 \sin \theta$
2. $y = -50 \sin \theta$
3. $y = 5 \sin 2\theta$
4. $y = -4 \sin 2\theta$
5. $y = 3 \sin 2\theta$
6. $y = -2 \sin 2\theta$
7. (a) and (b)

7-12 COSINE WAVES

In this section, we will briefly introduce cosine waves. We will see that a cosine wave has the same shape as a sine wave, and that it leads a corresponding sine wave by 90°.

124. The graph of the following equation is a cosine wave: $y = \cos\theta$

To graph this equation, we assign values to θ and determine the corresponding values of y. For example, if $\theta = 30°$, $y = \cos 30° = 0.87$. A table of pairs of values which satisfy the equation is given on the left below. Plot the points and draw the cosine wave on the axes below.

$y = \cos\theta$

θ	y	θ	y
0°	1	180°	-1
30°	0.87	210°	-0.87
60°	0.50	240°	-0.50
90°	0	270°	0
120°	-0.50	300°	0.50
150°	-0.87	330°	0.87
180°	-1	360°	1

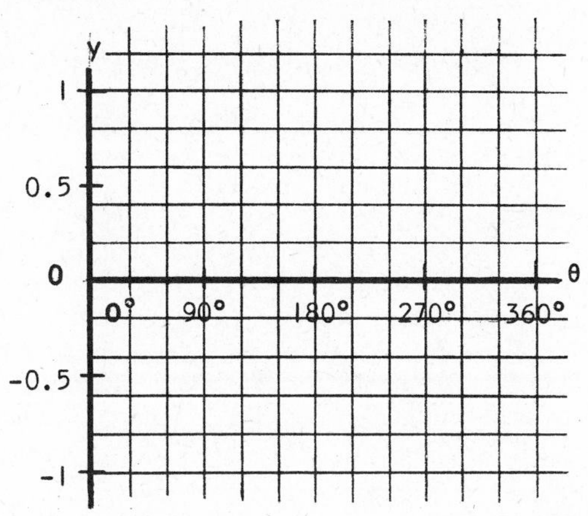

Answer to Frame 124: See the next frame.

125. Here is the graph of $y = \cos\theta$. It has been extended to the left of 0° and to the right of 360°. For comparison, the graph of $y = \sin\theta$ is also shown.

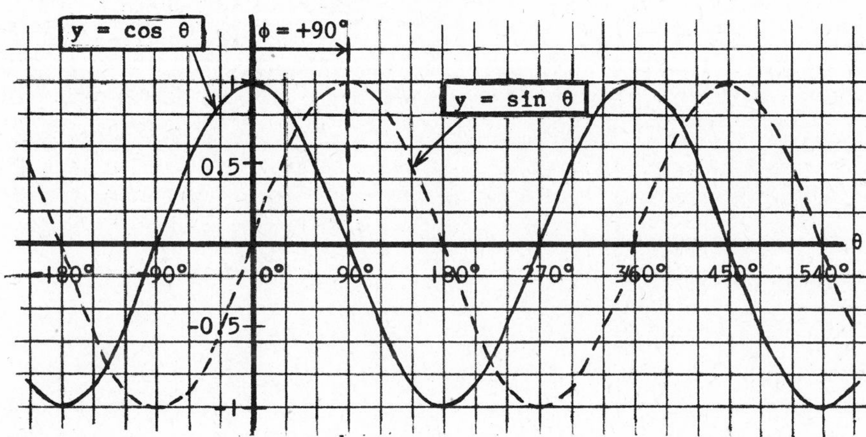

Note: (1) Both graphs have the same amplitude.

(2) The graph of $y = \cos\theta$ has the same shape as the graph of $y = \sin\theta$.

(3) The graph of $y = \cos\theta$ "leads" the graph of $y = \sin\theta$ by 90°. That is, the phase difference (φ) is +90°.

Since the cosine wave above is equivalent to a sine wave which "leads" the fundamental by 90°, there is a sine wave equation which is equivalent to $y = \cos\theta$. What is this sine wave equation?

$y = \sin(\theta + 90°)$

126. Any cosine wave "leads" the fundamental sine wave by ________.

---

127. The general equation of a cosine wave is:

$y = A \cos \theta$ where "A" is the amplitude.

Since any cosine wave leads a fundamental sine wave by 90°, the equivalent sine wave equation is:

$y = A \sin(\theta + 90°)$

(a) Write a sine wave equation which is equivalent to:

$y = 3 \cos \theta$ ____________________

(b) Write a cosine wave equation which is equivalent to:

$y = 1.5 \sin(\theta + 90°)$ ____________________

90°

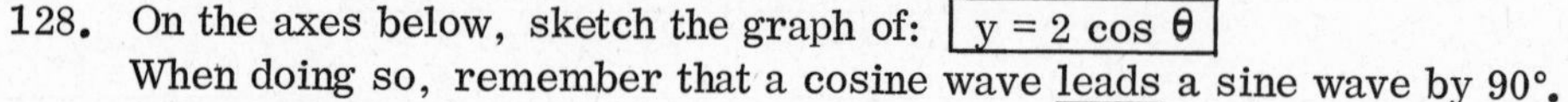

128. On the axes below, sketch the graph of: $y = 2 \cos \theta$
When doing so, remember that a cosine wave leads a sine wave by 90°.

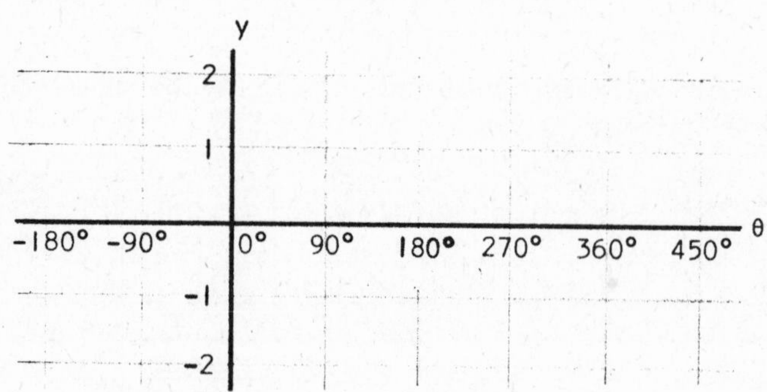

a) $y = 3 \sin(\theta + 90°)$

b) $y = 1.5 \cos \theta$

Answer to Frame 128:

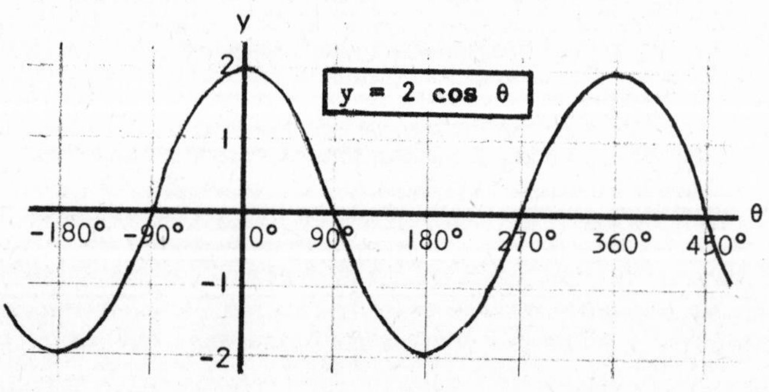

SELF-TEST 12 (Frames 124-128)

1. The graph of $y = \cos\theta$ ____________ (leads/lags) the graph of $y = \sin\theta$ by ________°.

2. Which of the following have the same graph as $y = 5\cos\theta$? ____________________

(a) $y = -5\cos\theta$

(b) $y = 5\sin(\theta + 90°)$

(c) $y = 5\sin(\theta - 90°)$

(d) $y = 5\sin(\theta + 180°)$

ANSWERS: 1. leads ... 90° 2. Only (b)

Chapter 8 COMPLEX NUMBERS

When vectors were discussed in earlier chapters, they were graphed on the "xy" coordinate system. In that system, vectors are represented by a pair of coordinates. Vectors can also be graphed on a coordinate system called the "complex plane". When graphed on the complex plane, vectors are represented by "complex numbers". When vectors are represented by complex numbers, the four basic operations with vectors are simplified.

In this chapter, we will discuss the use of the complex plane and complex numbers to graph and represent vectors. The procedures for adding, subtracting, multiplying, and dividing vectors by means of complex numbers will be shown. Since complex numbers can be written in either rectangular form or polar form, these four basic operations will be shown in both forms. The chapter is concluded with a brief section on formula evaluations which require combined operations with complex numbers.

8-1 REAL AND IMAGINARY NUMBERS

The word "COMPLEX" means that two or more parts are involved in one unit. A COMPLEX NUMBER is one unit with two parts: a real-number part and an imaginary-number part. Therefore, in order to understand the meaning of a complex number, the meaning of "real" and "imaginary" numbers must be understood. The meaning of "real" and "imaginary" numbers will be discussed in this section.

---

1. "Real" numbers are the ordinary numbers with which we are familiar. That is, any positive or negative whole number, decimal, or fraction is a "real" number.

   (a) 85, 6.4, 0.079, and $\frac{3}{11}$ are __________ numbers.

   (b) -85, -6.4, -0.079, and $-\frac{3}{11}$ are __________ numbers.

---

2. You probably are unfamiliar with "imaginary" numbers since they are not usually encountered in elementary mathematics work. Before discussing the meaning of an "imaginary" number, however, we must review some basic principles of squares and square roots.

   Whenever a number is squared, the "square" is always a positive number. For example:

   The "square" of both "+4" and "-4" is "+16".

   What number can we square to get "-16"? __________

   a) real

   b) real

There is none.

3. The square root of a number is some number whose square equals the original number. In finding the square root of any number, we are always looking for two equal factors. Any positive number has two square roots, a principal (positive) square root and a negative square root. We use the following symbols for each of these two square roots: $\sqrt{\quad}$ or $+\sqrt{\quad}$ for the principal square root, and $-\sqrt{\quad}$ for the negative square root. For example: $\sqrt{25} = 5$ $-\sqrt{25} = -5$ (a) The principal square root of 36 is _______. (b) The negative square root of 100 is _______.	
4. The two square roots of any positive number are real numbers. For example: $\sqrt{36} = +6$ and $-\sqrt{36} = -6$ Both "+6" and "-6" are called __________ numbers by mathematicians.	a) +6 b) -10
5. Let's examine the square root of a negative number like $\sqrt{-16}$. Here is one way of factoring $\sqrt{-16}$: $\sqrt{-16} = \sqrt{(+4)(-4)}$ (a) Are the two factors under the radical on the right equal? _______ (b) Is "+4" a square root of "-16"? _______ (c) Is "-4" a square root of "-16"? _______	real
6. What ordinary or real number is the square root of "-81"? __________	a) No. b) No. c) No.
7. Any positive number has two square roots, one positive and one negative. Both of these square roots are real numbers. Negative numbers do not have square roots in the same sense that positive numbers have square roots. That is, we cannot "take the square root of a negative number" and come up with some real-number root. However, mathematicians have developed a way of using the "square roots of negative numbers". To contrast them with real (ordinary) numbers, they have given them the name "imaginary" numbers. $\sqrt{-4}$, $\sqrt{-1}$, $\sqrt{-27}$ are called ______________ numbers by mathematicians.	There is none. $(+9)^2 = +81$ $(-9)^2 = +81$ If you square any real number (positive or negative), the result is always positive.
	imaginary

> The choice of the word "imaginary" for some numbers was probably not too good. Mathematicians are not happy with this choice because it suggests that the numbers do not exist or that they are useless. Both of these statements are false.
>
> Remember these two points:
>
> (1) The term "imaginary" is merely used as a contrast to "real". Imaginary numbers are different from the numbers which mathematicians call "real".
>
> (2) "Imaginary" numbers do exist and they are very useful, as we will see.

---

8. By using the principles for radicals, we can write any imaginary number as a real number times $\sqrt{-1}$. Let's use $\sqrt{-4}$ as an example:

Step 1: Factoring $\sqrt{-4}$ into two radicals, we get:

$$\sqrt{-4} = \sqrt{(4)(-1)} = \sqrt{4} \cdot \sqrt{-1}$$

Step 2: Then replacing the first radical by a real number, we get:

$$\sqrt{-4} = \sqrt{4} \cdot \sqrt{-1} = 2\sqrt{-1}$$

Complete this one:

$\sqrt{-16} = \sqrt{(16)(-1)} = \sqrt{16} \cdot \sqrt{-1} =$ ________

---

9. You will need your slide rule for this one:

$\sqrt{-55} = \sqrt{55} \cdot \sqrt{-1} =$ ____________

Answer to 8: $4\sqrt{-1}$

---

10. Write each of these as a real number times $\sqrt{-1}$:

(a) $\sqrt{-25} =$ ________ (b) $\sqrt{-100} =$ ________ (c) $\sqrt{-346} =$ ________

Answer to 9: $7.42\sqrt{-1}$

---

11. Instead of writing $\sqrt{-1}$, mathematicians substitute the letter "i". That is:

Instead of writing $7\sqrt{-1}$, they write 7i.

How would mathematicians write each of these?

(a) $15\sqrt{-1}$ ______ (b) $0.9\sqrt{-1}$ ______

Answers to 10: a) $5\sqrt{-1}$ b) $10\sqrt{-1}$ c) $18.6\sqrt{-1}$

---

12. Since "i" is substituted for $\sqrt{-1}$,

7i means $7\sqrt{-1}$ and 15i means ________

Answers to 11: a) 15i b) 0.9i

Answer to 12: $15\sqrt{-1}$

13. Since the letter "i" is used as the symbol for current in electrical work, engineers and technicians use "j" instead of "i" for $\sqrt{-1}$. They use "j" to avoid confusion.

Where a mathematician would write "3i", an engineer or technician would write "3j".

(a) Write $\sqrt{-49}$ in a mathematician's notation. _______

(b) Write $\sqrt{-49}$ in an engineer's or a technician's notation. _______

---

a) 7i

b) 7j

---

14. **In this chapter, we will use "j" instead of "i" for $\sqrt{-1}$.**

Here are two forms of the same imaginary number:

$$\underline{\sqrt{-9}} \text{ and } \underline{3j}$$

<u>We will always use the "non-radical" or "j" form because it is more efficient for mathematical operations.</u>

Just as we can add, subtract, multiply, divide and square real numbers, we can perform these same operations with imaginary numbers. The letter "j" is handled just as any other letter. Some examples are given below:

$2j + 3j = 5j$ $\quad$ $(4j)^2 = (4j)(4j) = 16j^2$

$7j - 4j = 3j$ $\quad$ $\dfrac{10j}{20j} = \left(\dfrac{10}{20}\right)\left(\dfrac{j}{j}\right) = \left(\dfrac{1}{2}\right)(1) = \dfrac{1}{2}$

$5(2j) = 10j$

Perform these operations involving imaginary numbers:

(a) $5j - 10j =$ __________ (c) $(10j)^2 =$ __________

(b) $(-7)(4j) =$ __________ (d) $\dfrac{40j}{5j} =$ __________

Answer to Frame 14: a) $-5j$ b) $-28j$ c) $100j^2$ d) 8

NOTE: As you have just learned, the imaginary number $\sqrt{-4}$ is written as 2j, where $j = \sqrt{-1}$. In many texts, however, 2j is written as j2, with the "j" first. The chief reason for using j2 is probably to indicate clearly that "j" does not represent a real number, as letters usually do in algebra.

In our work, however, we will always use the form 2j (or 5j, −20j, 7.83j, etc.). The form 2j is easier to write and to manipulate, since in algebra the numerical coefficient of a term is written first. Always remember, however, that "j" does not represent a real number, but rather is a symbol which represents $\sqrt{-1}$.

SELF-TEST 1 (Frames 1-14)

1. Which of the following numbers are imaginary numbers? ________

(a) 51.9 (b) -3.78 (c) $\sqrt{29}$ (d) $\sqrt{-29}$ (e) $-\frac{1}{2}$ (f) $\frac{3}{8}$ (g) -7

Write each of the following in "j" notation, where $j = \sqrt{-1}$:

2. $\sqrt{-100}$ = ________ 3. $-\sqrt{-64}$ = ________ 4. $\sqrt{-13.4}$ = ________

Perform these operations involving imaginary numbers:

5. 17j - 15j = ________ 6. (-7)(8j) = ________ 7. $\frac{5j}{8j}$ = ________

ANSWERS:

1. Only (d)
2. 10j
3. -8j
4. 3.66j
5. 2j
6. -56j
7. $\frac{5}{8}$

8-2 COMPLEX NUMBERS AND THEIR VECTOR REPRESENTATION

In this section, we will define what is meant by a "complex number". Then we will show how any complex number can be represented graphically by a vector on the complex plane.

15. Definition: A complex number is the sum of a real number and an imaginary number.

For example: "3 + 2j" is a complex number.

(a) The "real number" part is ________.

(b) The "imaginary number" part is ________.

16. Here is another example of a complex number: 5 + 4j

Just as we cannot simplify "5 + 4x" because we cannot add the two terms, we cannot simplify a complex number because we cannot add the two terms.

Though a complex number contains two terms, we think of it as one number. That is, "5 + 4j" should be thought of as ________ number.

Answer to Frame 15: a) 3 b) 2j

17. Here are two more complex numbers:

7 + (-3j) and (-8) + (-9j)

When the coefficient of "j" is negative, we ordinarily write the complex number in subtraction form. That is:

Instead of: 7 + (-3j), we write: 7 - 3j

Instead of: (-8) + (-9j), we write: ________

Answer to Frame 16: one

Answer to Frame 17: -8 - 9j

18. A complex number can be represented graphically by one vector. To do so, we use a set of axes which are different from the ordinary coordinate axes. This new set of axes defines what we call "the complex plane".

On the left below, we have drawn an ordinary set of coordinate axes. On the right, we have drawn the axes which make up the complex plane.

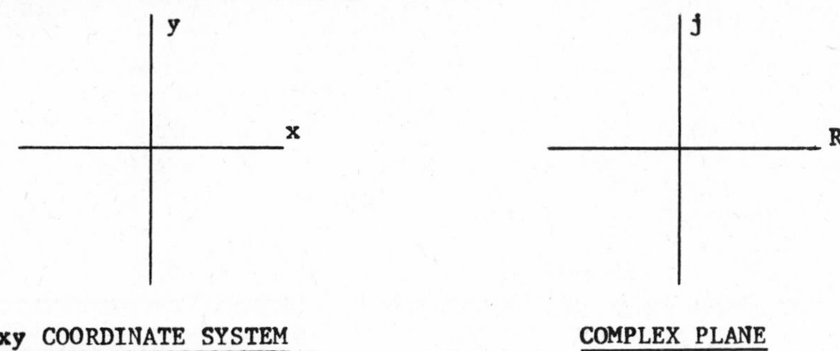

xy COORDINATE SYSTEM   COMPLEX PLANE

On the ordinary set of coordinate axes at the left:

(1) The horizontal axis is called the x-axis.
(2) The vertical axis is called the y-axis.
(3) Both "x" and "y" are real-number variables.

On the set of axes at the right which define the complex plane:

(1) The horizontal axis is called the R-axis (where "R" stands for "real" number).
(2) The vertical axis is called the j-axis (where "j" stands for "imaginary" number).
(3) Therefore, "R" is a real-number variable and "j" is an imaginary-number variable.

---

19. Using the set of axes which defines the complex plane, we can represent any complex number by a single vector.

On the graph at the right, we have drawn the vector which represents:

$3 + 2j$

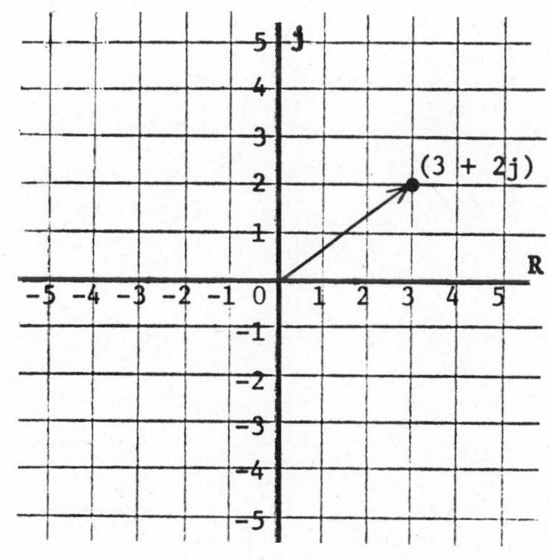

Notice these two points:

(1) The vector is drawn from the "origin" or intersection of the two axes.

(2) The real-number part ("3") gives us the horizontal component. "3" means 3-units to the right on the "R" or horizontal axis.

(3) The imaginary-number part ("2j") gives us the vertical component. "2j" means 2-units up on the "j" or vertical axis.

20. On the graph at the right, draw and label the vectors which represent the following complex numbers:

2 + 4j

-3 + 2j

5 - 4j

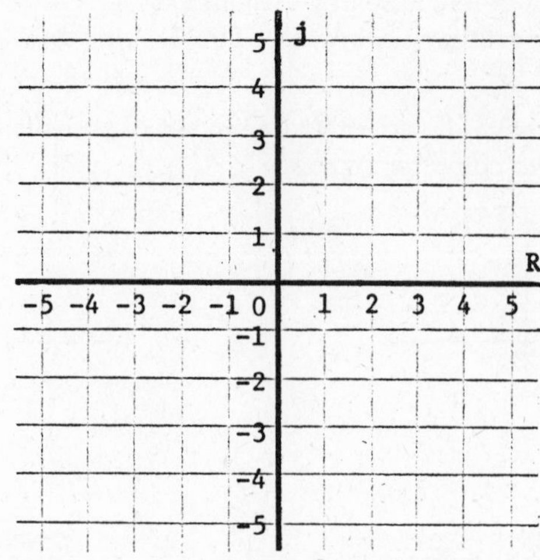

21.

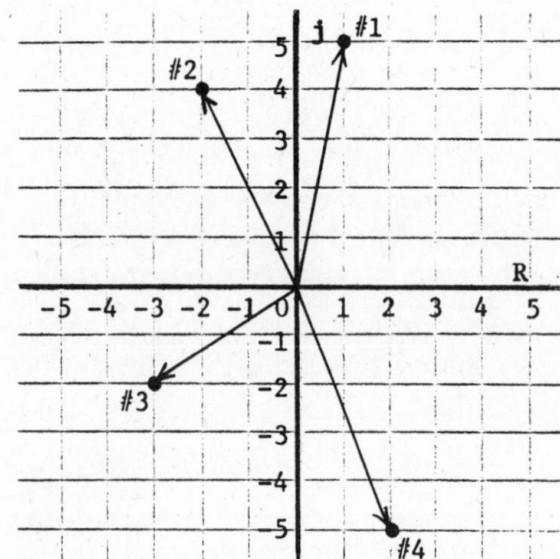

On the graph above, we have drawn and numbered four vectors. Write the complex number which each vector represents:

(a) #1: ____________

(b) #2: ____________

(c) #3: ____________

(d) #4: ____________

Answer to Frame 20:

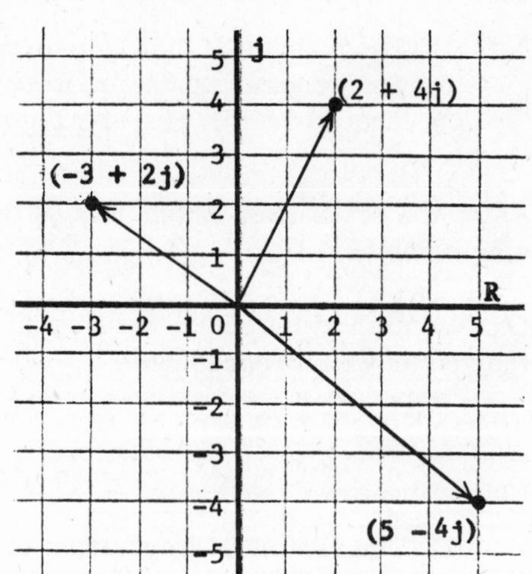

a) #1: 1 + 5j

b) #2: -2 + 4j

c) #3: -3 - 2j

d) #4: 2 - 5j

22.

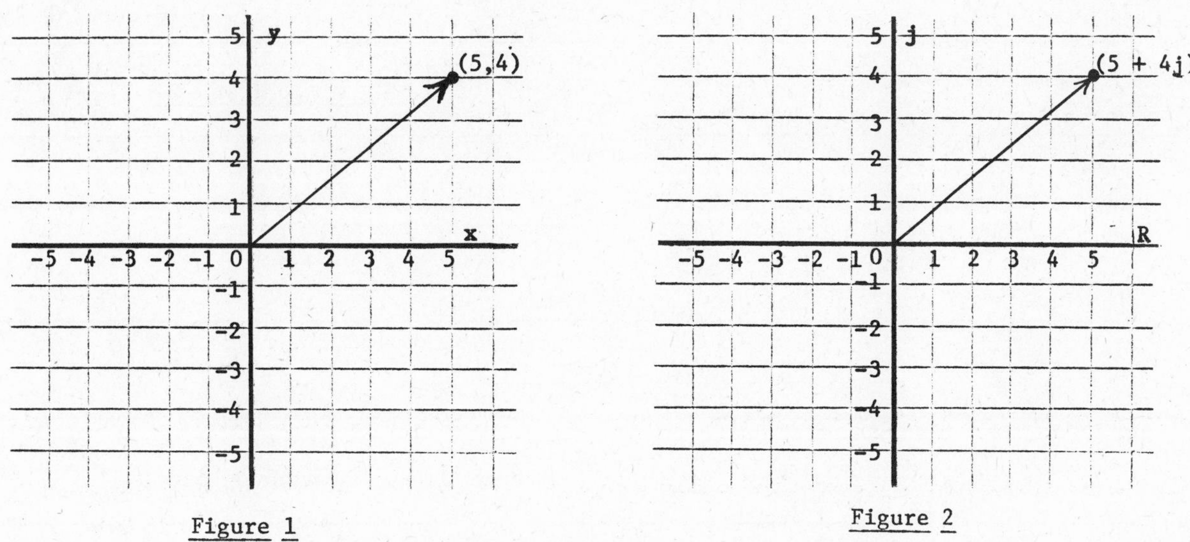

Figure 1 Figure 2

In Figure 1, we have drawn a vector on the regular (xy) coordinate system.

In Figure 2, we have drawn the same vector on the complex plane.

A vector can be drawn in either of the two ways above. However, there is this difference:

(1) When a vector is drawn on the regular coordinate system, it can only be represented by a pair of coordinates, like (5,4). It cannot be represented by a single number.

(2) When a vector is drawn on the complex plane, it can be represented by a single number, like 5 + 4j. However, the single number is a complex number.

> Operations with vectors are much easier if the vectors are represented by complex numbers. This fact will become obvious to you later in this chapter. Therefore, of the two alternatives, we will use complex numbers and the complex plane for vectors.

---

23. The quadrants of the complex plane are numbered in the same way as those of the "xy" system. Notice that we use Roman numerals to label them.

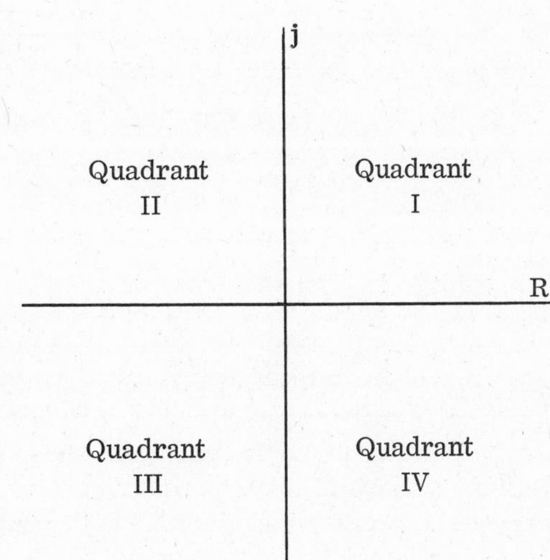

If you graphed each of the following complex numbers, in which quadrant would its vector lie?

(a) $-5 + 7j$ Quadrant _____

(b) $15 - 9j$ Quadrant _____

(c) $-20 - 17j$ Quadrant _____

(d) $10.8 - 1.9j$ Quadrant _____

24. The letter "j" is used in place of what radical? ________

a) II c) III
b) IV d) IV

Answer to Frame 24: $\sqrt{-1}$

SELF-TEST 2 (Frames 15-24)

For the complex number: -5 + 7j

1. The "real number" part is _____.

2. The "imaginary number" part is _____.

List the complex number which represents each vector shown on the complex plane at the right:

3. Vector #1: ____________
4. Vector #2: ____________
5. Vector #3: ____________
6. Vector #4: ____________

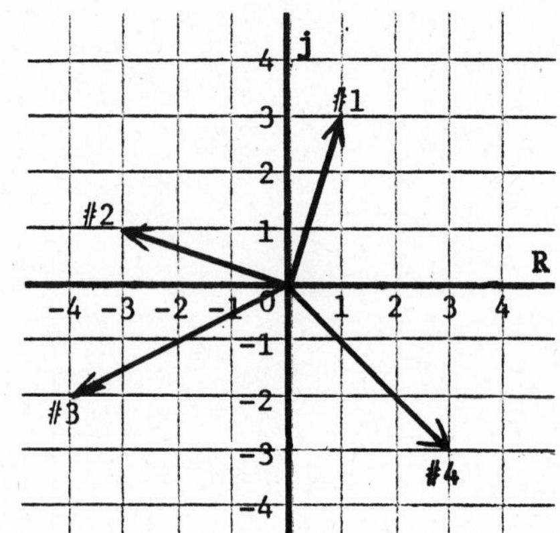

List the quadrant in which each vector lies:

7. 2 - 4j ______

8. -5 + 2j ______

9. -15.7 - 29.3j ______

ANSWERS:

1. -5	3. 1 + 3j	5. -4 - 2j	7. IV
2. 7j	4. -3 + 1j	6. 3 - 3j	8. II
			9. III

8-3 COMPLEX NUMBERS WHOSE VECTORS LIE ON AN AXIS

In the last section, we avoided complex numbers whose vectors lie on either axis of the complex plane. We will discuss complex numbers of that type in this section. We will show that vectors which lie on an axis of the complex plane can be represented by either a "complete" or an "incomplete" complex number.

25. On the complex plane at the right, we have drawn and numbered three vectors. In the blanks below, write the complex number which represents each vector. Notice that vector #3 lies on the horizontal (or <u>real</u>) axis.

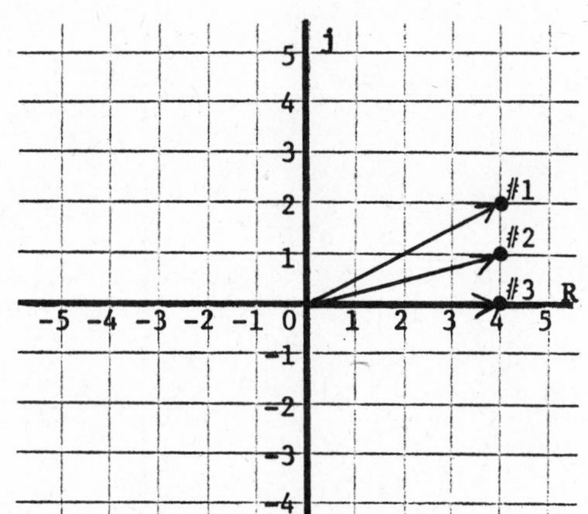

(a) #1: ____________

(b) #2: ____________

(c) #3: ____________

---

26. In the blanks below, write the complex number which represents each vector:

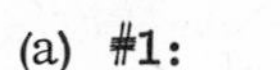

(a) #1: ____________

(b) #2: ____________

(c) #3: ____________

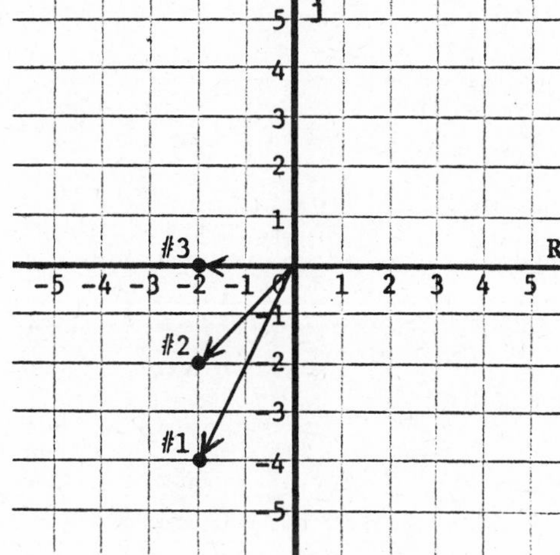

a) #1: 4 + 2j

b) #2: 4 + 1j

c) #3: 4 + 0j

---

27. When a vector lies on the <u>real</u> (horizontal) axis of the complex plane, the coefficient of the "j" term is "0". In the last two frames, you wrote these two complex numbers:

4 + 0j and -2 - 0j

Though "0j" = 0, we frequently include the "0j" term (instead of just "4" or "-2") <u>to signify that we have a complex number which can be represented as a vector on the complex plane.</u>

When writing a complex number for a vector that lies <u>on the horizontal axis</u>, the second term is "0j". Since "0j" = 0, it makes no difference whether we put an addition sign ("+") or a subtraction sign ("-") in front of it. That is:

-2 - 0j and -2 + 0j	represent the same vector, and they are the same complex number.

Which of the following represent the same vector on the complex plane? ____________

5 - 0j   -5 + 0j   5 + 0j

a) #1: -2 - 4j

b) #2: -2 - 2j

c) #3: -2 - 0j

28. Though "5 + 0j" and "5 - 0j" are two forms of the same complex number, we use the one with the addition sign, 5 + 0j.

Write the complex number which represents each of the horizontal vectors at the right.

(a) #1: ____________

(b) #2: ____________

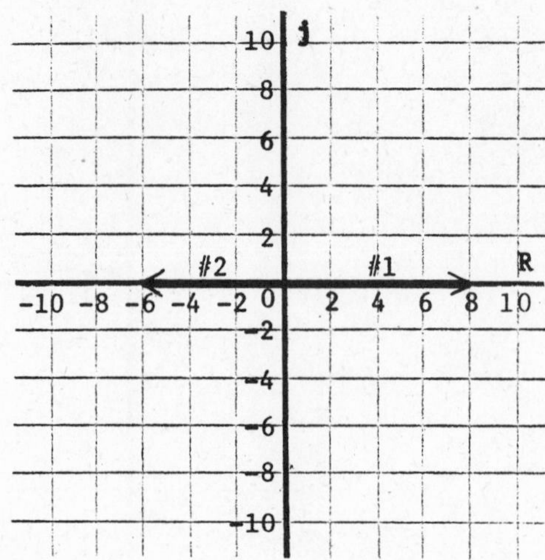

5 - 0j and 5 + 0j

---

29. On the complex plane at the right, we have drawn and numbered six vectors. Notice that two of them lie on the vertical (or "j") axis.

Write the complex number which represents each of the six vectors:

(a) #1: ____________

(b) #2: ____________

(c) #3: ____________

(d) #4: ____________

(e) #5: ____________

(f) #6: ____________

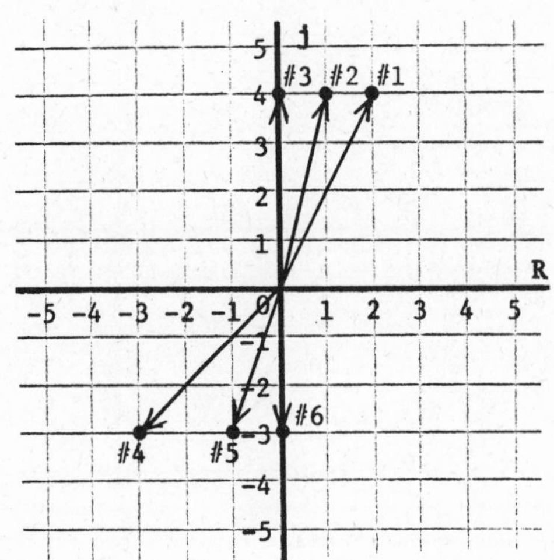

a) #1: 8 + 0j

b) #2: -6 + 0j

---

a) #1: 2 + 4j

b) #2: 1 + 4j

c) #3: 0 + 4j

d) #4: -3 - 3j

e) #5: -1 - 3j

f) #6: 0 - 3j

30. In the last frame, we represented the two vectors on the vertical axis with the following complex numbers:

$$0 + 4j \quad \text{and} \quad 0 - 3j$$

Even though the real-number term is "0", we frequently include the "0" (instead of just 4j or -3j) to signify that we are dealing with complex numbers which can be represented by a vector on the complex plane.

Write the complex number which represents each of the vectors on the complex plane at the right:

(a) #1: ____________

(b) #2: ____________

(c) #3: ____________

(d) #4: ____________

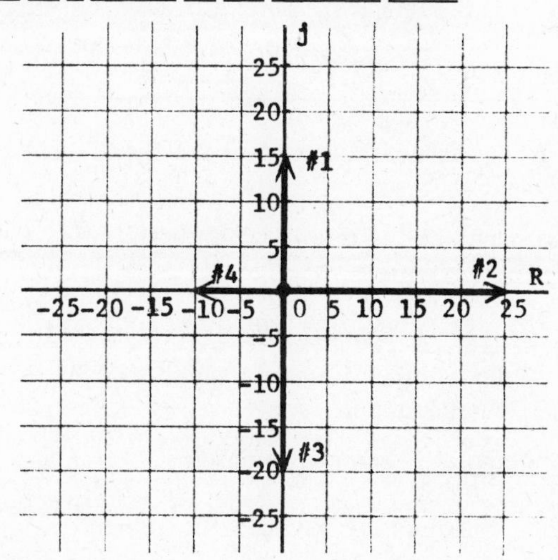

---

31. If the following three vectors were graphed on the complex plane, which one would lie on the horizontal axis? ____________

(a) 7 + 4j (b) -6 + 0j (c) 0 - 6j

a) #1: 0 + 15j
b) #2: 25 + 0j
c) #3: 0 - 20j
d) #4: -10 + 0j

---

32. If the following three vectors were graphed, which one would lie on the vertical axis? ____________

(a) 0 + 5j (b) 5 + 0j (c) 5 + 5j

b) -6 + 0j

---

33. The general form of a complex number is:

| a + bj | where both "a" and "b" are real numbers, and "b" is the coefficient of "j".

(a) If "a" is "0", we get complex numbers like:

0 + 3j 0 + 10j 0 - 5j

The vectors for all complex numbers of this type lie on the ________________ (horizontal/vertical) axis.

(b) If "b" is "0", we get complex numbers like:

4 + 0j 9 + 0j -6 + 0j

The vectors for all complex numbers of this type lie on the ________________ (horizontal/vertical) axis.

a) 0 + 5j

---

a) vertical

b) horizontal

34. When either "a" or "b" is "0" in a complex number, we frequently write it without the "0" term. For example:

Instead of (0 + 5j), we simply write "5j".
Instead of (8 + 0j), we simply write "8".

Write each of the following without the "0" term:

(a) 0 - 4j = _______ (c) -6 + 0j = _______

(b) 7 + 0j = _______ (d) 0 + j = _______

---

a) -4j c) -6

b) 7 d) j

35. When either "a" or "b" is "0" in a complex number, the "0" term may or may not be explicitly written.

A "complete" complex number is one in which both terms are explicitly written. For example:

0 + 4j and 5 + 0j are "complete" complex numbers.

An "incomplete" complex number is one in which the "0" term is not explicitly written. For example:

4j and 5 are "incomplete" complex numbers.

Which of the following are "incomplete" complex numbers? _________

(a) -7 + 0j (b) -9 (c) -3j (d) 0 - 5j

(b) and (c)

36. Write each of the following "incomplete" complex numbers as a "complete" complex number:

(a) 10j = ___________ (c) -7 = ___________

(b) 3 = ___________ (d) -j = ___________

---

a) 0 + 10j c) -7 + 0j

b) 3 + 0j d) 0 - j

37. (a) Since 15 = 15 + 0j, the vector for "15" lies on the ___________ (horizontal/vertical) axis of the complex plane.

(b) Since -4j = 0 - 4j, the vector for "-4j" lies on the ___________ (horizontal/vertical) axis of the complex plane.

---

a) horizontal

b) vertical

38. If the vectors of the following "incomplete" complex numbers were graphed on the complex plane, which ones would lie on the vertical axis? ___________

(a) -6j (b) 8 (c) -1 (d) j

Only (a) and (d)

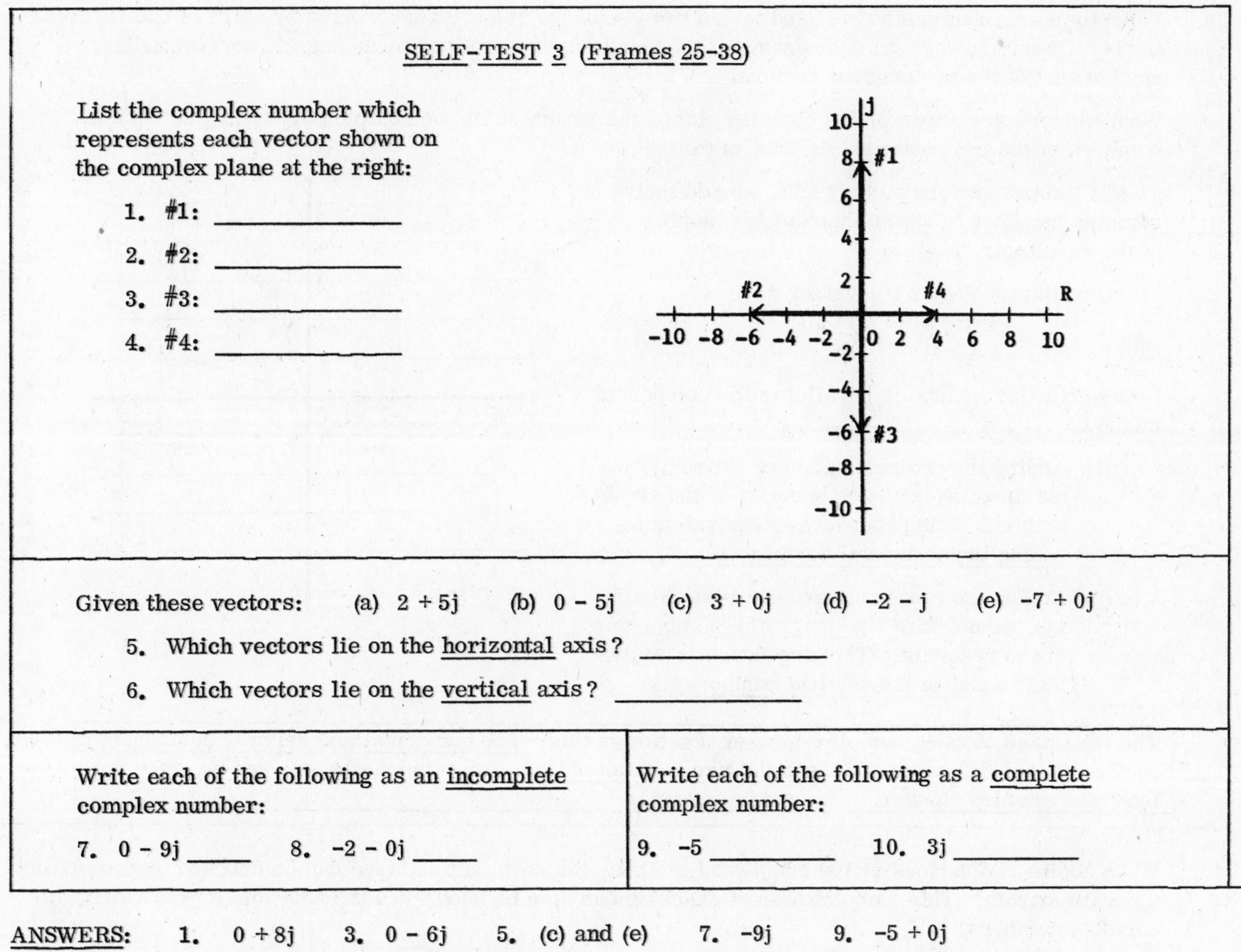

SELF-TEST 3 (Frames 25-38)

List the complex number which represents each vector shown on the complex plane at the right:

1. #1: ____________
2. #2: ____________
3. #3: ____________
4. #4: ____________

Given these vectors: (a) 2 + 5j (b) 0 − 5j (c) 3 + 0j (d) −2 − j (e) −7 + 0j

5. Which vectors lie on the horizontal axis? ____________
6. Which vectors lie on the vertical axis? ____________

Write each of the following as an incomplete complex number:

7. 0 − 9j ______ 8. −2 − 0j ______

Write each of the following as a complete complex number:

9. −5 ____________ 10. 3j ____________

ANSWERS:

1. 0 + 8j	3. 0 − 6j	5. (c) and (e)	7. −9j	9. −5 + 0j
2. −6 + 0j	4. 4 + 0j	6. (b)	8. −2	10. 0 + 3j

8-4 VECTOR-ADDITION BY MEANS OF COMPLEX NUMBERS

In an earlier chapter, we discussed vector-addition by both the parallelogram method and the component method. In that chapter, the vectors were represented by pairs of coordinates. In this section, we will discuss vector-addition when the vectors are represented by complex numbers. We will show the computational advantage of adding vectors by means of complex numbers.

39. When the component method is used to add two vectors which are represented by pairs of coordinates, the horizontal and vertical components of the resultant are obtained by adding the corresponding components of the two original vectors.

When vectors are drawn on the complex plane, the resultant can be obtained by adding the complex numbers of the two vectors. Here is an example:

To add the two vectors on the right, we add their complex numbers to obtain the complex number of the resultant. That is:

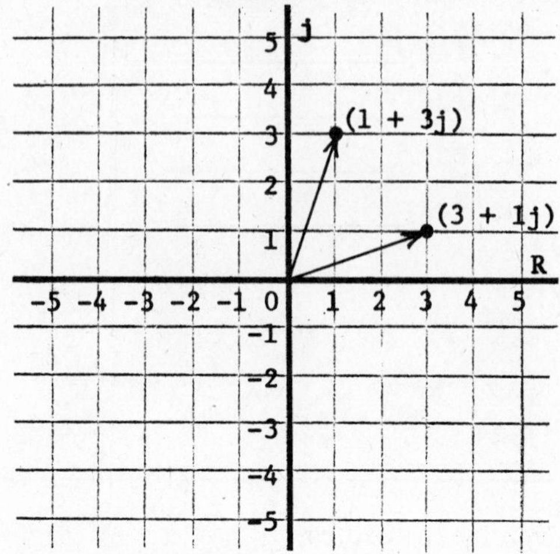

$$\begin{aligned}\text{Resultant} &= \text{Vector 1} + \text{Vector 2}\\ &= (1 + 3j) + (3 + 1j)\\ &= 4 + 4j\end{aligned}$$

Notice that this method is parallel to the component method.

(1) Adding the two real numbers (1 and 3), we obtained the real-number part of the resultant. This procedure is equivalent to adding the horizontal components.

(2) Adding the two imaginary numbers (3j and 1j), we obtained the imaginary-number part of the resultant. This procedure is equivalent to adding the vertical components.

The advantage of using complex-number notation is this: The horizontal and vertical components of each vector (including the resultant) are kept separate. The components are obvious when you look at a complex number.

---

40. When adding two vectors by the component method, the resultant obtained can be checked by completing a parallelogram. This same method of checking can also be used when the addition is performed with complex numbers.

In the last frame, we added the two vectors "1 + 3j" and "3 + 1j" and obtained the resultant "4 + 4j". On the graph at the right:

(1) Draw the three vectors.

(2) Then complete the parallelogram of the original two to show that the resultant is its diagonal.

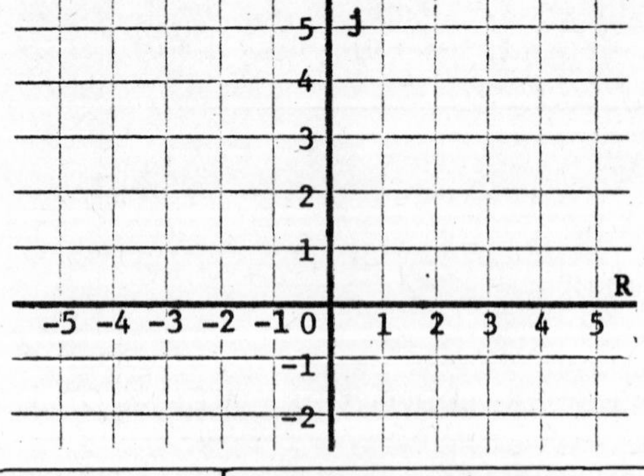

Your completed graph should look like this:

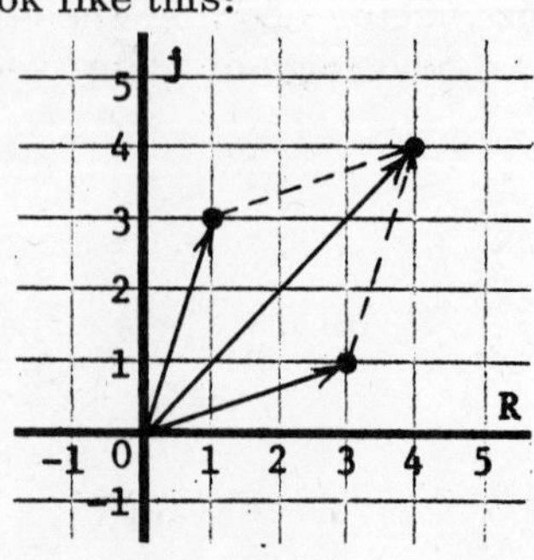

41. Two vectors are shown on the complex plane below.

Let's find their resultant by adding their complex numbers:

Resultant = Vector 1 + Vector 2
= (4 + 2j) + (2 - 6j)
= 4 + 2j + 2 + (-6j)
= 6 + (-4j)
= 6 - 4j

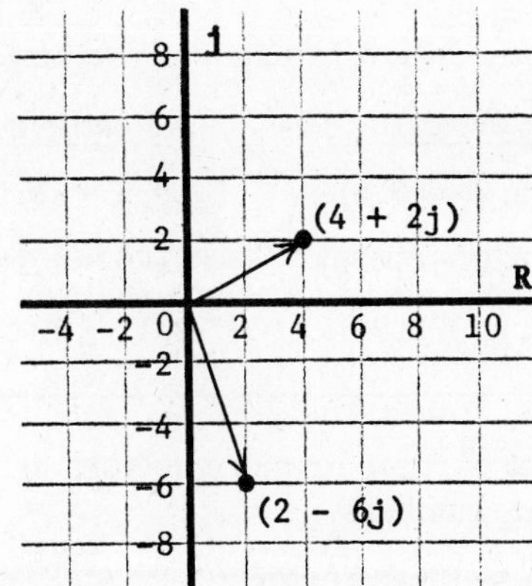

Draw the resultant on the complex plane, and then complete the parallelogram to justify that the obtained resultant is really its diagonal.

Your diagram should look like this:

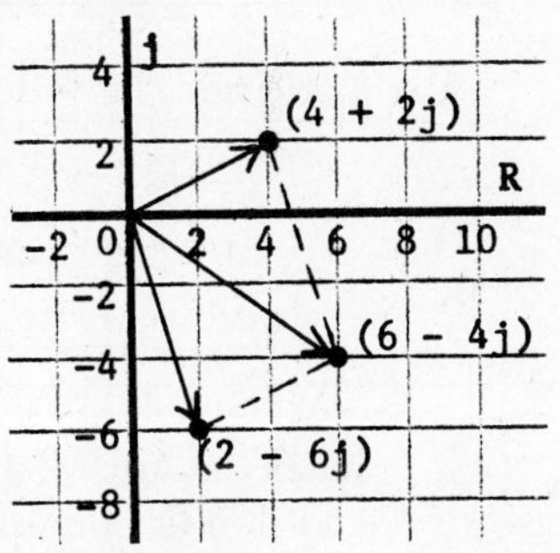

42. When vectors are given in complex number form, <u>the addition is ordinarily performed without using a graph.</u> The resultant is simply a complex number.

Let's add these two vectors: 3 + 4j and 7 + 2j

(a) The resultant of these two vectors is ____________.

(b) The resultant lies in what quadrant? ____________

---

a) 10 + 6j

b) Quadrant I

43. When one of the vectors contains a subtraction, you will avoid sign errors if you convert the subtraction to addition first. Here is an example:

The addition of: 5 + 2j and -10 - 4j

is equivalent to the addition of: 5 + 2j and (-10) + (-4j) .

(a) The resultant of the two vectors is ____________.

(b) In what quadrant does this resultant lie? ____________

---

a) -5 - 2j

b) Quadrant III

44. The coefficient "1" for the j-term is ordinarily not expressed. However:

5 + j means 5 + 1j
6 - j means 6 - 1j

Find the resultant of each pair of vectors and then state the quadrant in which the resultant lies:

	Vector 1	Vector 2	Resultant	Quadrant
(a)	-3 + j	-5 + 2j	____________	________
(b)	10 - 4j	-7 - j	____________	________
(c)	-9 - j	6 - 3j	____________	________

Answer to Frame 44: a) -8 + 3j (Quadrant II) b) 3 - 5j (Quadrant IV) c) -3 - 4j (Quadrant III)

45. In applied problems, the numbers involved in the complex numbers will frequently be large numbers or decimals. The procedure for adding them is still the same.

Add each of the following pairs of vectors and state the quadrant in which the resultant lies:

	Vector 1	Vector 2	Resultant	Quadrant
(a)	28.7 + 83.5j	10.5 - 89.7j	________	____
(b)	-4,320 - 2,650j	3,170 + 4,120j	________	____
(c)	-1.04 - 6.88j	-2.11 + 3.45j	________	____

---

46. The resultant of three or more vectors can easily be found by adding their complex numbers.

Let's add these three vectors: 9 + 6j, -6 + j, 5 - 8j

(a) Their resultant is ________.

(b) Their resultant lies in what quadrant? ________

Answers to 45:
a) 39.2 - 6.2j (Quadrant IV)
b) -1,150 + 1,470j (Quadrant II)
c) -3.15 - 3.43j (Quadrant III)

---

47. Let's add the following four vectors:

1.9 - 2.6j, 10.7 + 3.2j, -9.8 - 1.8j, -5.7 - 2.2j

(a) Their resultant is ________.

(b) Their resultant lies in what quadrant? ________

Answers to 46:
a) 8 - j
b) Quadrant IV

---

48. When "a" or "b" is "0" in one or both complex numbers, the vector addition is simplified. Here are some examples:

(5 + 4j) + (0 + 3j) = 5 + 7j
(7 + 0j) + (4 - 2j) = 11 - 2j
(0 - 6j) + (9 - 0j) = 9 - 6j

Write the resultant of each pair of vectors and state the quadrant in which the resultant lies:

	Vector 1	Vector 2	Resultant	Quadrant
(a)	(0 - 2j)	(-5 + j)	________	____
(b)	(-2 - j)	(7 + 0j)	________	____
(c)	(-5 + 0j)	(0 + 3j)	________	____

Answers to 47:
a) -2.9 - 3.4j
b) Quadrant III

Answers to 48:
a) -5 - j III
b) 5 - j IV
c) -5 + 3j II

49. When one or both vectors is represented by an "incomplete" complex number, the vector addition is also simplified. Here are some examples:

$$(7 + 2j) + (4j) = 7 + 6j$$
$$(-4) + (9 - 3j) = 5 - 3j$$
$$(-5j) + (7) = 7 - 5j$$

Write the resultant of each of the following pairs of vectors:

	Vector 1	Vector 2	Resultant
(a)	5j	3 - 4j	________
(b)	7j	8	________
(c)	-8	10 - 2j	________
(d)	4	-j	________

a) 3 + j

b) 8 + 7j

c) 2 - 2j

d) 4 - j

---

50. When both vectors are represented by incomplete complex numbers and both contain only real-number terms or only imaginary-number terms, the resultant is also an "incomplete" complex number. For example:

$7 + (-5) = 2$ (which is $2 + 0j$)

$-7j + 5j = -2j$ (which is $0 - 2j$)

(a) The resultant "2" lies on the ________ (horizontal/vertical) axis of the complex plane.

(b) The resultant "-2j" lies on the ________ (horizontal/vertical) axis of the complex plane.

a) horizontal

b) vertical

SELF-TEST 4 (Frames 39-50)

Refer to the two vectors shown on the diagram at the right:

1. By completing the parallelogram, sketch the vector sum or resultant.
2. Add the complex numbers of the two vectors shown. Their vector sum or resultant in complex number form is ________.
3. In what quadrant does the vector sum or resultant lie? ________

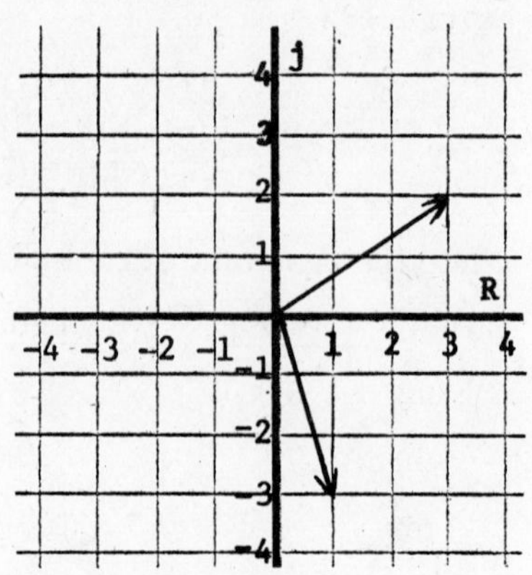

Find the vector sum or resultant of each of the following pairs of vectors, and list the quadrant in which each resultant lies:

	Vector 1	Vector 2	Resultant	Quadrant
4.	25 - 17j	-12 + 15j	________	________
5.	-6 + j	-8 - 2j	________	________
6.	-15.7 - 39.8j	10.9 + 54.6j	________	________

7. Two vectors in complex number form are 20 + 0j and 0 - 50j. Their resultant is ________.

8. Two vectors in complex number form are 0 + 18j and 0 - 15j. Their resultant is ________.

9. Three vectors in complex number form are 2 - j, -5 + 3j, and 6 - 2j. Their resultant is ________.

ANSWERS: 1.

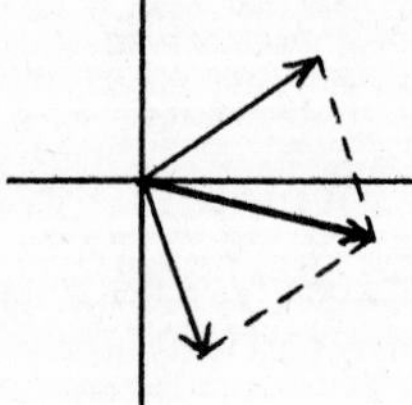

2. 4 - j
3. Quadrant IV
4. 13 - 2j Quadrant IV
5. -14 - j Quadrant III
6. -4.8 + 14.8j Quadrant II
7. 20 - 50j
8. 0 + 3j
9. 3 + 0j

8-5 VECTOR-SUBTRACTION BY MEANS OF COMPLEX NUMBERS

In a vector-addition, the two added vectors are called "vector-addends". Sometimes the resultant and one vector-addend are given and the second vector-addend must be found. In these cases, the second vector-addend is found by a subtraction. And when the vectors are represented by complex numbers, this subtraction is a subtraction of complex numbers. We will discuss the subtraction of complex numbers in this section.

51. When adding two vectors to find their resultant, the following formula is used.

Vector 1 + Vector 2 = Resultant

When the resultant and one vector-addend are given and the second vector-addend must be found, the formula above can be rearranged to this form:

Vector 2 = Resultant - Vector 1

Notice that the second formula requires a subtraction of the given vector from the resultant. This subtraction, of course, is a subtraction of one complex number from another. For example:

If the Resultant is $(7 + 5j)$ and Vector 1 is $(4 + 2j)$:
Vector 2 $= (7 + 5j) - (4 + 2j)$

If the Resultant is $(7 + 5j)$ and Vector 1 is $(3 - j)$:
Vector 2 $= (7 + 5j) - (3 - j)$

As you can see from the examples above, a subtraction of one complex number from another is equivalent to the subtraction of a grouping.

In the first example, the grouping is $(4 + 2j)$, which is an addition.

In the second example, the grouping is $(3 - j)$, which is a subtraction.

We will review the method of subtracting groupings in the following frames.

Go to the next frame.

---

52. The following subtraction of complex numbers is equivalent to the subtraction of a grouping which is an addition.

$$(9 + 5j) - (4 + 2j)$$

To perform this subtraction, we must convert the subtraction of the grouping to an equivalent addition. We do so by "adding the opposite of $(4 + 2j)$". We get:

$$(9 + 5j) + (\text{the opposite of } 4 + 2j)$$

Using the fact that the "opposite of an addition" is obtained by replacing each term with its opposite, we get:

$$(9 + 5j) + [(-4) + (-2j)]$$

Having converted the subtraction of a grouping to an addition, we can easily obtain the answer by combining like terms. Do so.

$$9 + 5j + (-4) + (-2j) = \underline{\hspace{3cm}}$$

5 + 3j

53. As we saw in the last frame, the opposite of a grouping which is an addition is obtained by replacing each term with its opposite. That is:

The opposite of $(a + b) = (-a) + (-b)$

Write the opposite of each of these additions:

(a) $5 + 4j$ ________ (c) $-5 + (-4j)$ ________

(b) $-5 + 3j$ ________ (d) $10 + (-9j)$ ________

---

54. Complete each of the following subtractions:

(a) $(7 + 4j) - (5 + 2j) = 7 + 4j + (-5) + (-2j) =$ ________

(b) $(9 + 2j) - (4 + 6j) = 9 + 2j + (-4) + (-6j) =$ ________

(c) $(-5 + 6j) - (3 + j) = (-5) + 6j + (\quad) + (\quad) =$ ________

(d) $(-3 - 4j) - (5 + j) = (-3) - 4j + (\quad) + (\quad) =$ ________

Answers to frame 53:
a) $(-5) + (-4j)$
b) $5 + (-3j)$
c) $5 + 4j$
d) $(-10) + 9j$

---

55. The following subtraction of complex numbers is equivalent to the subtraction of a grouping which is a subtraction.

$$(7 + 3j) - (5 - 2j)$$

To perform this subtraction, we must convert the subtraction of a grouping to an equivalent addition. The easiest way to make this conversion is to use the following two steps:

Step 1: Convert the grouping to an addition first.

$$(7 + 3j) - [5 + (-2j)]$$

Step 2: Then convert the subtraction of the addition-grouping to an equivalent addition in the usual way.

$$(7 + 3j) + (-5) + (+2j)$$

Complete the subtraction by combining like terms:

$$7 + 3j + (-5) + (+2j) = \text{________}$$

Answers to frame 54:
a) $= 2 + 2j$
b) $= 5 - 4j$
c) $(-3) + (-j) = -8 + 5j$
d) $(-5) + (-j) = -8 - 5j$

---

56. As we saw in the last frame, the easiest way to find the "opposite of a subtraction" is to convert the subtraction to an addition first. For example:

The opposite of $5 - 3j$ = the opposite of $5 + (-3j) = (-5) + (+3j)$

Each subtraction below has been converted to an addition. Write its opposite in the blank:

(a) $5 - 2j = 5 + (-2j)$ Opposite: ________

(b) $-9 - 4j = -9 + (-4j)$ Opposite: ________

(c) $-11 - j = -11 + (-j)$ Opposite: ________

Answer to frame 55: $2 + 5j$

Answers to frame 56:
a) $-5 + 2j$
b) $9 + 4j$
c) $11 + j$

57. Write the opposite of each complex number:

(a) 10 - 5j   Opposite: ________

(b) -2 - 7j   Opposite: ________

(c) -6 - j   Opposite: ________

Answers to 57:

a) -10 + 5j

b) 2 + 7j

c) 6 + j

58. In each case below, the subtraction within the second complex number has been converted to an addition. Complete the subtractions:

(a) (7 + j) - (5 - j) = (7 + j) - [5 + (-j)]

= (7 + j) + ( ) + ( ) = ________

(b) (10 - 4j) - (-3 - 2j) = (10 - 4j) - [(-3) + (-2j)]

= (10 - 4j) + ( ) + ( ) = ________

(c) (-7 - 5j) - (-9 - 8j) = (-7 - 5j) - [(-9) + (-8j)]

= (-7 - 5j) + ( ) + ( ) = ________

Answers to 58:

a) + (-5) + (j) = 2 + 2j

b) + (3) + (2j) = 13 - 2j

c) + (9) + (8j) = 2 + 3j

59. In each case below, convert the subtraction of the second complex number to addition, and find the answer:

(a) (7 + 4j) - (5 - 3j) = 7 + 4j + ( ) + ( ) = ________

(b) (9 - 2j) - (3 - j) = 9 - 2j + ( ) + ( ) = ________

(c) (-6 - j) - (-2 - 3j) = -6 - j + ( ) + ( ) = ________

Answers to 59:

a) + (-5) + (3j) = 2 + 7j

b) + (-3) + (j) = 6 - j

c) + (2) + (3j) = -4 + 2j

60. Complete these subtractions:

(a) (-6 + 4j) - (9 - j) = ________

(b) (-7 - 5j) - (-8 - 3j) = ________

Answers to 60:

a) -15 + 5j

b) 1 - 2j

61. Given the resultant and one vector, we subtract to find the second vector. For example:

If the Resultant is $(10 - 3j)$ and Vector 1 is $(5 + 5j)$:

$$\begin{aligned} \text{Vector 2} &= \text{Resultant} - \text{Vector 1} \\ &= (10 - 3j) - (5 + 5j) \\ &= 10 - 3j + (-5) + (-5j) \\ &= 5 - 8j \end{aligned}$$

Graph the two vectors and the resultant on the complex plane at the right. Then justify by the parallelogram method that the resultant is really the diagonal of the parallelogram whose sides are Vector 1 and Vector 2.

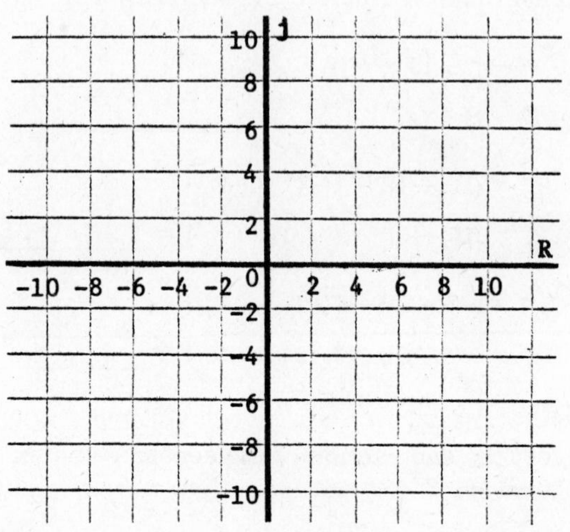

62. In each case below, the Resultant and Vector 1 are given. Find Vector 2 by performing a subtraction:

	Resultant	Vector 1	Vector 2
(a)	$10 - 7j$	$4 + 6j$	________
(b)	$15 + 10j$	$3 - 2j$	________
(c)	$-7 - 2j$	$-5 - j$	________

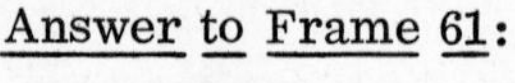

Answer to Frame 61:

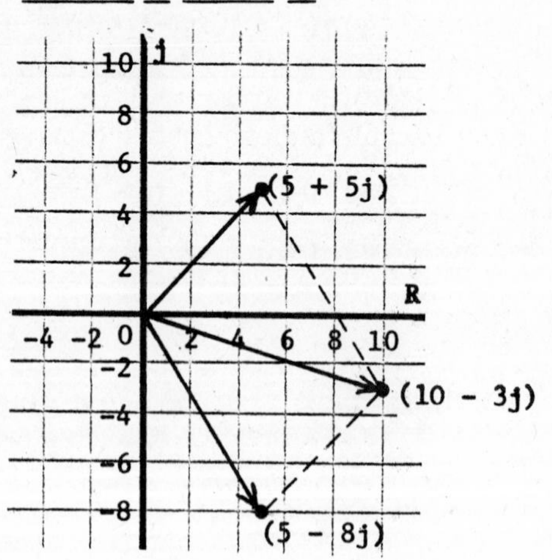

63. Find Vector 2 in each case below:

	Resultant	Vector 1	Vector 2
(a)	$10.8 + 3.2j$	$6.7 - 4.9j$	________
(b)	$2,420 - 980j$	$-1,560 - 2,140j$	________

Answers to Frame 62:

a) $6 - 13j$

b) $12 + 12j$

c) $-2 - j$

a) $4.1 + 8.1j$

b) $3{,}980 + 1{,}160j$

64. When the resultant lies on an axis, either "a" or "b" in its complex number is "0". For example:

> If a resultant is $5 + 0j$, it lies on the horizontal (real) axis to the right of the origin.
>
> If a resultant is $0 - 7j$, it lies on the vertical (j) axis extending downward from the origin.

In such cases, one of the terms in the resultant (the first complex number of the subtraction) is "0". Here are some examples:

$$(\overset{\downarrow}{0} + 5j) - (3 + j)$$

$$(7 + \underset{\uparrow}{0}j) - (5 - 2j)$$

Subtractions of this type are performed in the usual way. Therefore:

(a) $(0 + 5j) - (3 + j) =$ ________ (b) $(7 + 0j) - (5 - 2j) =$ ________

a) $-3 + 4j$

b) $2 + 2j$

65. When the resultant lies on an axis, we frequently write its complex number as an "incomplete" complex number. Here are some examples:

$$(-3j) - (7 - 5j)$$
$$(15) - (9 + 4j)$$

Subtractions of this type are also performed in the usual way by converting the subtraction of a grouping to addition first. Therefore:

(a) $(-3j) - (7 - 5j) =$ ________

(b) $(15) - (9 + 4j) =$ ________

a) $-7 + 2j$

b) $6 - 4j$

66. When the given vector-addend lies on an axis, either "a" or "b" in its complex number is "0". In such cases, one of the terms in the second complex number in the subtraction is "0". For example:

$$(5 + 4j) - (3 + \overset{\downarrow}{0}j)$$

$$(7 - 6j) - (\overset{\downarrow}{0} - 5j)$$

To perform subtractions of this type, it is easier to drop the "0" term and write the second complex number as an incomplete complex number. We get:

$$(5 + 4j) - (3)$$
$$(7 - 6j) - (-5j)$$

By doing so, we avoid the need to convert the subtraction <u>of a grouping</u> to addition. That is:

(a) $(5 + 4j) - (3) = 5 + 4j + (-3) =$ ________

(b) $(7 - 6j) - (-5j) = 7 - 6j + (+5j) =$ ________

a) $2 + 4j$

b) $7 - j$

67. When both the resultant and the given vector lie on an axis, each complex number contains one "0" term. For example:

$(5 + 0j) - (0 + 2j)$
$(0 - 7j) - (8 + 0j)$

In these cases, it is easier to write each complex number as an incomplete complex number. We get:

(a) $(5) - (2j) =$ ______________

(b) $(-7j) - (8) =$ ______________

---

a) $5 - 2j$
b) $-8 - 7j$

68. Find Vector 2 in each case below:

	Resultant	Vector 1	Vector 2
(a)	$0 + 5j$	$-3 + 4j$	______________
(b)	$7 + 3j$	$4 + 0j$	______________
(c)	$-6 + 0j$	$0 - 3j$	______________
(d)	$7j$	$8 + j$	______________
(e)	$4 + 5j$	-3	______________
(f)	$14j$	9	______________

---

a) $3 + j$
b) $3 + 3j$
c) $-6 + 3j$
d) $-8 + 6j$
e) $7 + 5j$
f) $-9 + 14j$

69. When subtracting to find the second vector-addend, sometimes we obtain an incomplete complex number. In these cases, the second vector-addend lies on an axis. Here are some examples. In each case, describe the position of the second vector-addend on the complex plane.

(a) $(5 + 4j) - (3 + 4j) = 2 + 0j$ (or 2)

____

(b) $(7 - 3j) - (7) = 0 - 3j$ (or $-3j$)

____

(c) $(9) - (9 - 5j) = 0 + 5j$ (or $5j$)

____

Answer to Frame 69:
a) $(2 + 0j)$ lies on the horizontal axis, to the right of the origin.
b) $(0 - 3j)$ lies on the vertical axis, below the origin.
c) $(0 + 5j)$ lies on the vertical axis, above the origin.

SELF-TEST 5 (Frames 51-69)

1. Resultant = Vector 1 + Vector 2 Therefore: Vector 1 = ____________________

Complete the following vector subtractions. Write each answer as a complex number:

2. (5 - j) - (1 + j) = ____________

3. (-60 + 30j) - (50 - 70j) = ____________

4. (9 + 0j) - (0 - 4j) = ____________

5. (12 - j) - (12 + j) = ____________

6. Vector A and Vector B are added, and their resultant is -7 + 3j. If Vector A is 5 - 8j, find Vector B.

Vector B = ____________

7. Two vectors are added, and their resultant is 200 - 100j. If one vector is 200 + 0j, find the second vector.

Second Vector = ____________

8. The vector sum of Vector P and Vector Q is -15 + 0j. If Vector Q is -20 + 8j, find Vector P.

Vector P = ____________

ANSWERS:

1. Vector 1 = Resultant - Vector 2
2. 4 - 2j
3. -110 + 100j
4. 9 + 4j
5. 0 - 2j
6. Vector B = -12 + 11j
7. Second Vector = 0 - 100j
8. Vector P = 5 - 8j

8-6 CONVERTING COMPLEX NUMBERS FROM RECTANGULAR FORM TO POLAR FORM

When a complex number is written in the form a + bj , we say that it is given in "rectangular form". A complex number can also be identified by the length and standard-position angle of its vector. When a complex number is represented by its length and angle, we say that it is given in "polar form".

In this section, we will show the method of converting complex numbers from "rectangular form" to "polar form". The conversions will be limited to complex numbers whose vectors do not lie on an axis.

70. The same vector is drawn on both graphs below. The graph on the left is the "xy" coordinate system; the coordinates of the vector are (3, 4). The graph on the right is the complex plane; the complex number of the vector is $3 + 4j$.

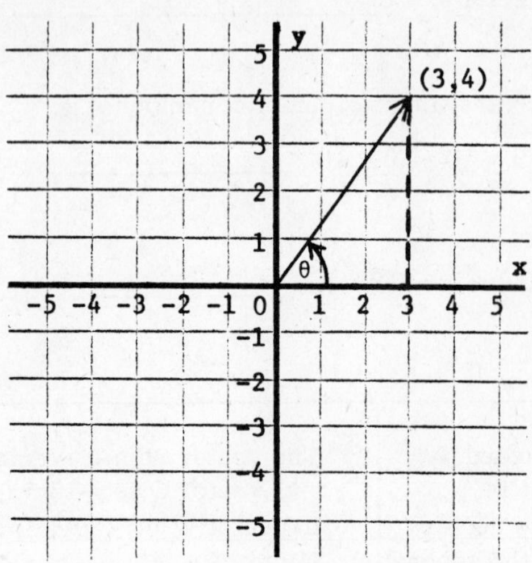

"xy" Coordinate System

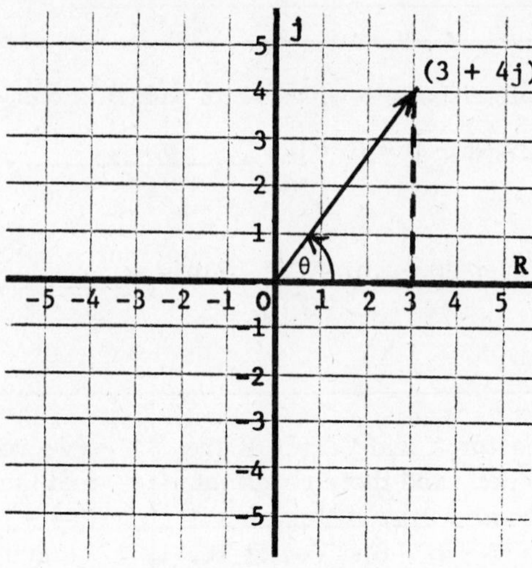

Complex Plane

In each case, the horizontal component is "+3" and the vertical component is "+4". Using these components, we can find both the length and the standard-position angle (θ) of the vector.

To find its length, we use the Pythagorean Theorem:

$$\text{Length} = \sqrt{3^2 + 4^2} = \sqrt{9 + 6} = \sqrt{25} = 5 \text{ units}$$

To find angle θ, we use the tangent ratio:

$$\text{Since } \tan\theta = \frac{4}{3} = 1.333,\ \theta = 53°.$$

---

71. The general form of a complex number is: $\boxed{a + bj}$ where both "a" and "b" are real numbers, and "b" is the coefficient of "j".

We have used these labels for the vector on the complex plane below:

Notice these points:

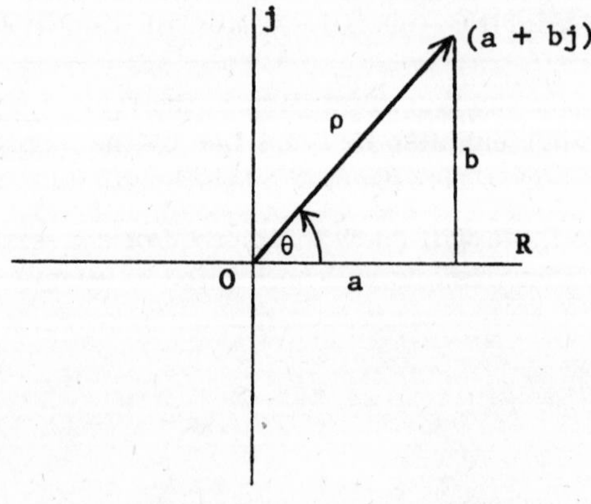

(1) The vector is "a + bj".

(2) "ρ" (pronounced "row") is the length of the vector.

(3) "θ" is the standard-position angle of the vector.

(4) "a" is the horizontal component of the vector.

(5) "b" is the vertical component of the vector.

Using the letters "a" and "b":

(a) Write a formula for "ρ", the length of the vector: = ____________________

(b) Write a formula for the tangent of angle θ: tan θ = ________

72. When the coordinates ("a" and "b") of a vector on the complex plane are given, we specify the vector in the form $\boxed{a + bj}$.

When the length (ρ) and standard-position angle (θ) of a vector on the complex plane are given, we specify the vector in the form $\boxed{\rho \angle \theta}$.

The length and standard-position angle of the vector on the right are:

$\rho = 10$ units
$\theta = 130°$

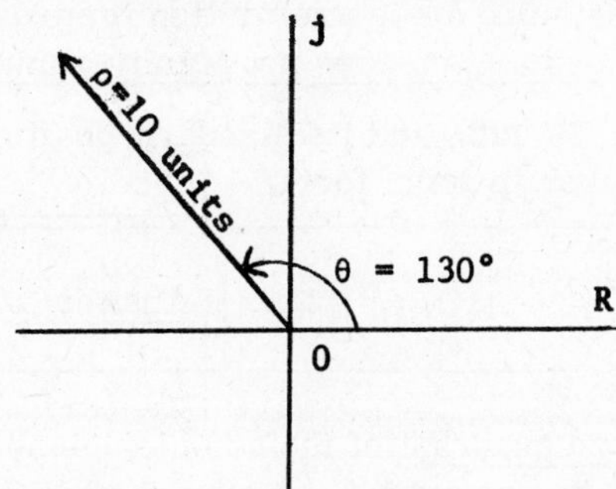

Using the form $\boxed{\rho \angle \theta}$, specify the vector.

____

a) $\rho = \sqrt{a^2 + b^2}$

b) $\tan \theta = \dfrac{b}{a}$

---

73. When a vector is shown on the complex plane, it can be represented in either of two ways:

(1) By a number in the form $\boxed{a + bj}$.

(2) By its length and angle in the form $\boxed{\rho \angle \theta}$.

These two different ways of representing the same vector are really <u>two different forms of the same complex number</u>.

$\boxed{a + bj}$ is called the "<u>rectangular form</u>" of the complex number.

$\boxed{\rho \angle \theta}$ is called the "<u>polar form</u>" of the complex number.

The vector on the right can be represented by a complex number in either of two forms:

(a) The "rectangular form" of the complex number is ____________.

(b) The "polar form" of the complex number is

____________.

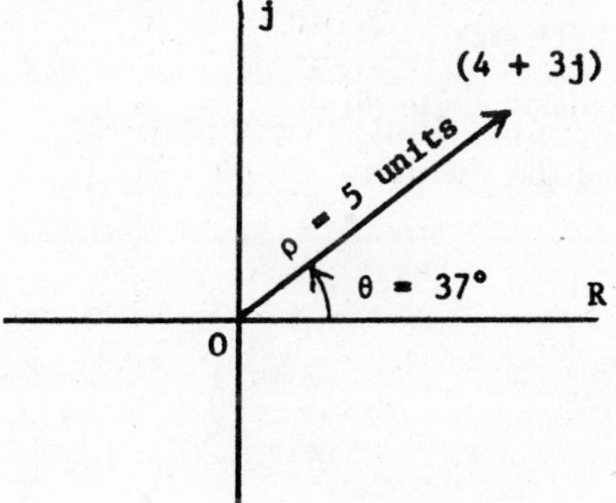

$10\angle 130°$

---

a) $4 + 3j$

b) $5\angle 37°$

74. Notice these points about the polar form of a complex number, like $10\underline{/130°}$.

(1) The symbol "$\underline{/\quad}$" means "angle".

(2) The "angle" symbol is extended underneath "130°".

(3) Even though the length and angle are written immediately next to each other, this symbolism does not signify a multiplication.

If the length of a vector is 1.78 units and its standard-position angle is 249°, write its complex number in polar form. ________

Answer to Frame 74: $1.78\underline{/249°}$

75. Here is a vector in the second quadrant of the complex plane. Its complex number is given in rectangular form. We want to convert its complex number from rectangular form to polar form.

We know these facts:

(1) Horizontal component (or "a") = -5

(2) Vertical component (or "b") = 4

(3) $\tan\alpha = \frac{b}{a} = \frac{4}{-5} = -0.8000$ ("α" is the reference angle.)

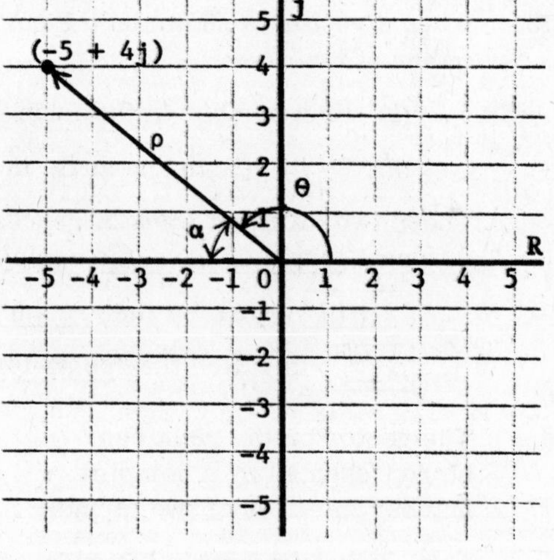

Calculate the following:

(a) The length (ρ) of the vector is ________ units.

(b) The reference angle (α) = ________.

(c) The standard-position angle (θ) = ________.

(d) The polar form of the vector is ________.

a) $\sqrt{41}$ = 6.40 units

b) 39°(Q2)

c) 141°

d) $6.40\underline{/141°}$

76. Here is a vector in the third quadrant of the complex plane:

(a) The length (ρ) of the vector is __________ units.

(b) The reference angle (α) of the vector is __________.

(c) The standard-position angle (θ) of the vector is __________.

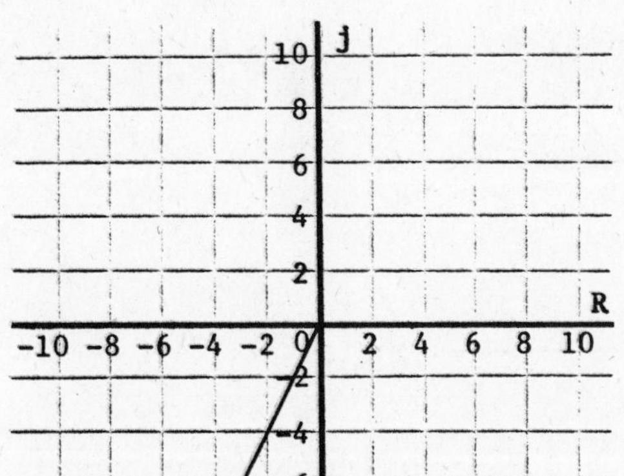

(d) The polar form of the vector is ______________.

---

77. Here is a vector in the fourth quadrant of the complex plane:

For this vector:

(a) ρ = __________

(b) α = __________

(c) θ = __________

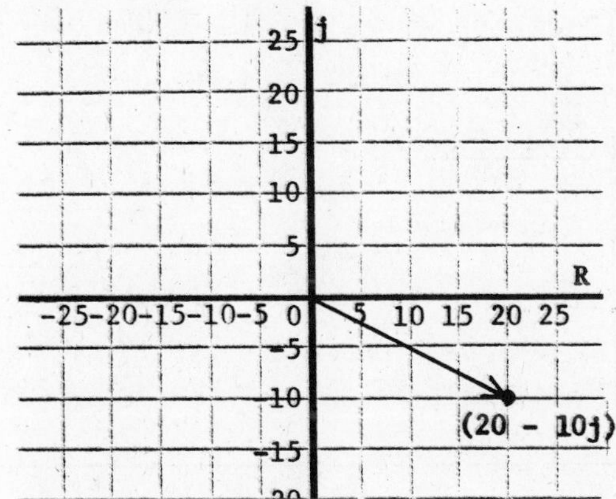

(d) The polar form is ______________.

Answers to 76:

a) $\sqrt{80} = 8.95$ units

b) 63° (Q3)

c) 243°

d) $8.95\underline{/243°}$

---

78. If we graphed the following complex numbers, in what quadrant would the vector for each lie?

(a) 10 - 7j _______

(b) -9 - 6j _______

(c) 17.8 - 5.9j _______

(d) -1,420 + 2,190j _______

Answers to 77:

a) $\sqrt{500} = 22.4$ units

b) 27° (Q4)

c) 333°

d) $22.4\underline{/333°}$

79. Find the polar form of each of the following complex numbers. Report their length to slide rule accuracy.

(a) 17 - 9j ______________

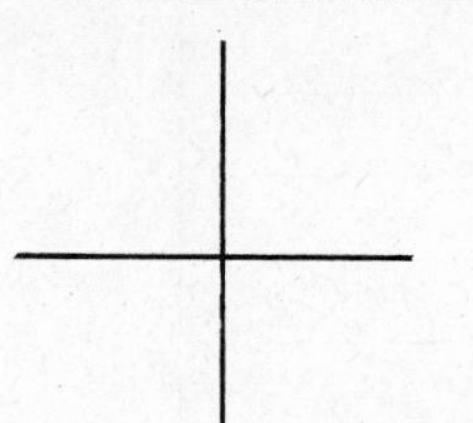

(b) -7 + 25j ______________

a) IV c) IV

b) III d) II

---

80. Find the polar form of the following complex numbers:

(a) 1.68 - 1.32j ______________

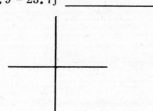

(b) -17.9 - 23.7j ______________

a) $19.2\angle 332^\circ$

b) $26.0\angle 106^\circ$

---

a) $2.14\angle 322^\circ$

b) $29.7\angle 233^\circ$

81. When writing a complex number in polar form, negative angles are sometimes used instead of standard-position angles. When negative angles are used, they are used for third and fourth quadrant angles, not for first and second quadrant angles.

In the figure on the right:

(a) "φ" is the negative angle corresponding to a positive angle of 225°. How large is angle φ ?

______

(b) "β" is the negative angle corresponding to a positive angle of 315°. How large is angle β ?

______

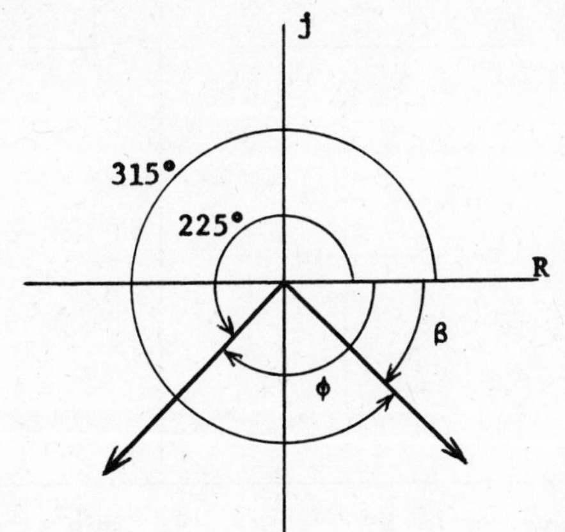

---

82. For each of the following, write the polar form with a negative angle instead of a positive angle:

(a) $15.7/\underline{330°}$ (b) $20.1/\underline{185°}$ (c) $789/\underline{297°}$

______ ______ ______

a) -135°

b) -45°

---

83. For each of the following, write the polar form using the corresponding positive standard-position angle:

(a) $1.77/\underline{-47°}$ (b) $16.9/\underline{-101°}$ (c) $76.7/\underline{-75°}$

______ ______ ______

a) $15.7/\underline{-30°}$

b) $20.1/\underline{-175°}$

c) $789/\underline{-63°}$

---

a) $1.77/\underline{313°}$

b) $16.9/\underline{259°}$

c) $76.7/\underline{285°}$

SELF-TEST 6 (Frames 70-83)

Write the polar form of each of the following vectors:

1. $4.00 - 3.00j$ ______________

2. $-2000 + 3000j$ ______________

3. The polar form of a vector is $57.3\underline{/320°}$. For this same vector, write the polar form using the corresponding negative angle. ______________

ANSWERS: 1. $5.00\underline{/323°}$ or $5.00\underline{/-37°}$ 2. $3610\underline{/124°}$ 3. $57.3\underline{/-40°}$

8-7 THE POLAR FORM OF VECTORS WHICH LIE ON AN AXIS

In the last section, we showed the method of converting complex numbers from rectangular form to polar form. In that section, we avoided the complex numbers of vectors which lie on an axis. In this section, we will discuss the polar form of vectors which lie on an axis. In doing so, we will show the method of converting incomplete complex numbers in rectangular form to polar form and vice versa.

84. On the complex plane at the right, we have drawn four vectors. Each vector lies on an axis. The rectangular form of the complex number of each vector is given.

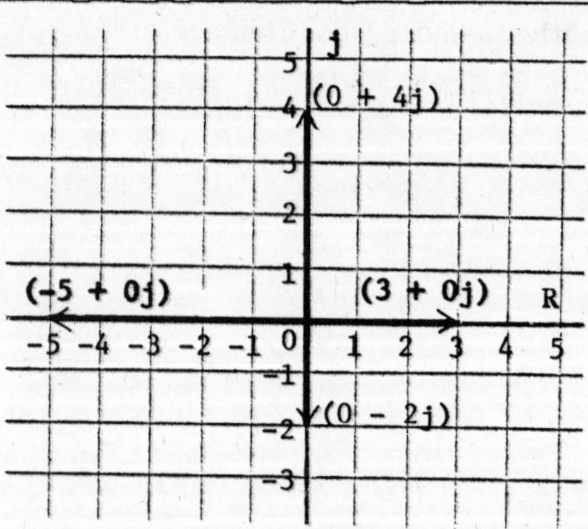

When a vector lies on an axis, it is easy to convert its complex number from rectangular form to polar form. For example:

The length of $(-5 + 0j)$ is 5 units.
The angle of $(-5 + 0j)$ is 180°.
Therefore: $(-5 + 0j) = 5\underline{/180°}$

Write the polar form for each of the following:

(a) $3 + 0j$ = ______________ (b) $0 + 4j$ = ______________ (c) $0 - 2j$ = ______________

a) $3\underline{/0°}$

b) $4\underline{/90°}$

c) $2\underline{/270°}$ or $2\underline{/-90°}$

85. When either "a" or "b" of a complex number in rectangular form (a + bj) is "0", the vector lies on an axis. In these cases, you should be able to convert to polar form immediately.

Convert these to polar form. First, make a sketch of each vector:

(a) 0 - 6j ________

(b) 17 + 0j ________

(c) -25 + 0j ________

(d) 0 + 15j ________

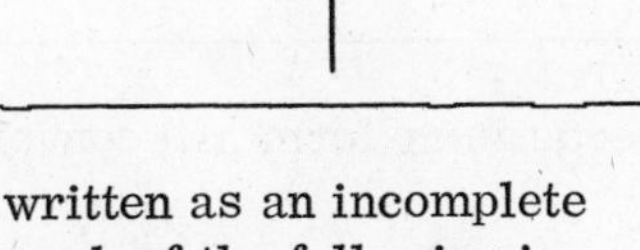

---

86. When a vector lies on an axis, it is sometimes written as an incomplete complex number in rectangular form. Convert each of the following incomplete complex numbers in rectangular form to polar form:

(a) 7 = ________

(b) -4 = ________

(c) 4j = ________

(d) -8j = ________

Answers to 85:

a) $6\angle 270°$ (or $6\angle -90°$)

b) $17\angle 0°$

c) $25\angle 180°$

d) $15\angle 90°$

---

87. Here are some vectors on the complex plane. The complex number of each is given in polar form.

Write the rectangular form of the complex number for each vector:

(a) $10\angle 0°$ = ________

(b) $10\angle 90°$ = ________

(c) $10\angle 180°$ = ________

(d) $10\angle 270°$ = ________

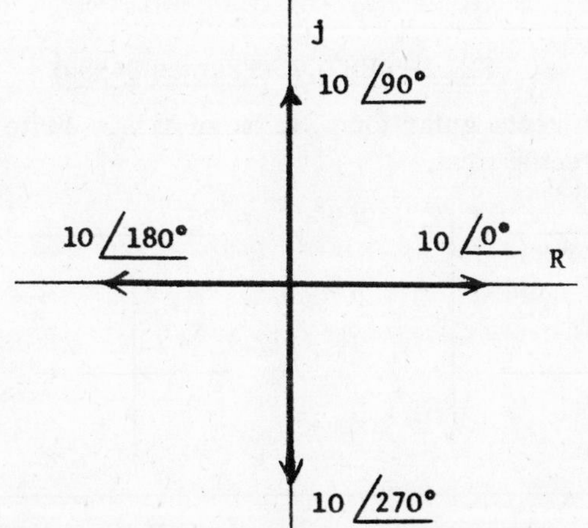

Answers to 86:

a) $7\angle 0°$

b) $4\angle 180°$

c) $4\angle 90°$

d) $8\angle 270°$ or $8\angle -90°$

---

88. If the angle of a vector in polar form is 0°, 90°, 180°, or 270° (or -90°), it lies on an axis. Write the rectangular form of the complex number which represents each of the following vectors. Make a sketch if you need one:

(a) $15\angle 180°$ ________

(b) $27\angle 270°$ ________

(c) $1.96\angle 0°$ ________

(d) $27.8\angle 90°$ ________

Answers to 87:

a) 10 + 0j (or 10)

b) 0 + 10j (or 10j)

c) -10 + 0j (or -10)

d) 0 - 10j (or -10j)

89. In the following problems, if a complex number is given in rectangular form, write its corresponding polar form. If the polar form is given, write its corresponding rectangular form:

(a) $0 + 17.7j$ __________

(b) $1.66\angle 0^\circ$ __________

(c) -15.9 __________

(d) $149\angle 270^\circ$ __________

a) $-15 + 0j$ (or -15)
b) $0 - 27j$ (or $-27j$)
c) $1.96 + 0j$ (or 1.96)
d) $0 + 27.8j$ (or $27.8j$)

---

90. Convert polar form to rectangular form and convert rectangular form to polar form:

(a) $16.5\angle -90^\circ$ __________

(b) $-289j$ __________

(c) $45.6 + 0j$ __________

(d) $8.66\angle 270^\circ$ __________

(e) $9.11\angle 90^\circ$ __________

a) $17.7\angle 90^\circ$
b) $1.66 + 0j$ (or 1.66)
c) $15.9\angle 180^\circ$
d) $0 - 149j$ (or $-149j$)

Answer to Frame 90: a) $0 - 16.5j$ (or $-16.5j$) b) $289\angle 270^\circ$ (or $289\angle -90^\circ$) c) $45.6\angle 0^\circ$ d) $0 - 8.66j$ (or $-8.66j$) e) $0 + 9.11j$ (or $9.11j$)

SELF-TEST 7 (Frames 84-90)

The following vectors in rectangular form lie on an axis. Write the polar form of each vector. If necessary, sketch the vector first.

1. $0 - 43.5j$ = __________ 2. $-200 + 0j$ = __________ 3. $7j$ = __________

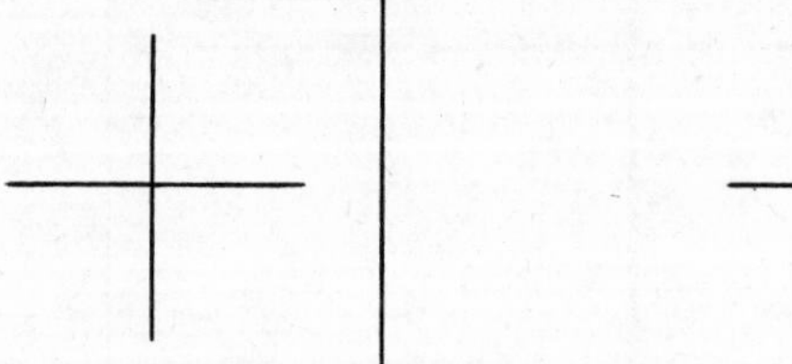

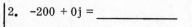

The following vectors in polar form lie on an axis. Write the rectangular form of each vector. If necessary, sketch the vector first.

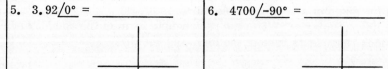

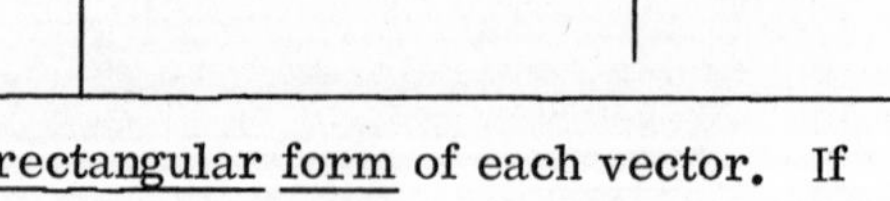

4. $78.6\angle 90^\circ$ = __________ 5. $3.92\angle 0^\circ$ = __________ 6. $4700\angle -90^\circ$ = __________

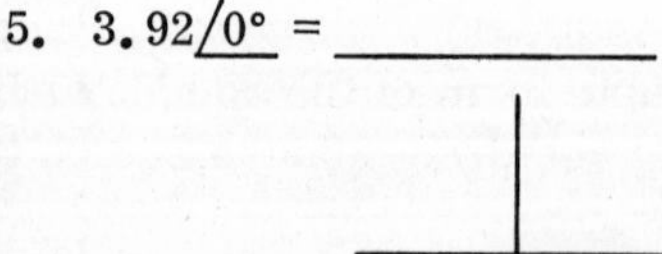

ANSWERS:

1. $43.5\angle 270^\circ$ (or $43.5\angle -90^\circ$)
2. $200\angle 180^\circ$
3. $7\angle 90^\circ$
4. $0 + 78.6j$ (or $78.6j$)
5. $3.92 + 0j$ (or 3.92)
6. $0 - 4700j$ (or $-4700j$)

8-8 CONVERTING COMPLEX NUMBERS FROM POLAR FORM TO RECTANGULAR FORM

In the last section, we saw that it is easy to convert complex numbers from polar form to rectangular form when they represent vectors on an axis. When a vector does not lie on an axis, however, trigonometric ratios must be used to convert from polar form to rectangular form. We will discuss this latter type of conversion in this section. A brief discussion of the method of adding or subtracting complex numbers in polar form is included.

91. A first-quadrant vector in polar form is shown at the right. The vector is: $10.0\underline{/50°}$

To convert the vector from polar form to rectangular form, we must find the components "a" and "b". These components can be found by using trigonometric ratios.

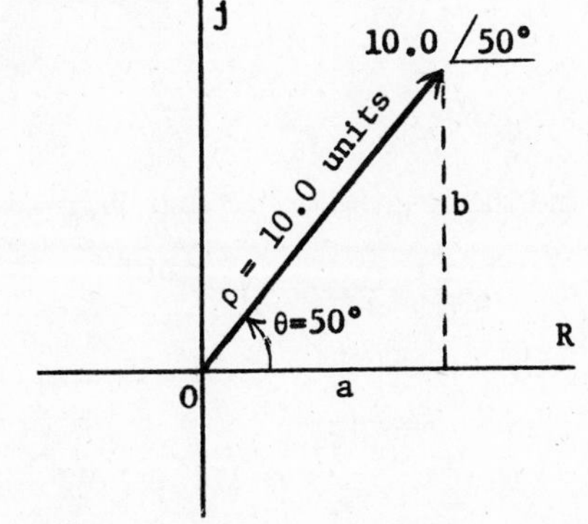

To find "a", we use cosine θ:

Since $\cos\theta = \frac{a}{\rho}$, $\cos 50° = \frac{a}{10.0}$

$$\begin{aligned} \text{and } a &= 10.0 \cos 50° \\ &= 10.0(0.6428) \\ &= 6.43 \text{ units} \end{aligned}$$

To find "b", we use sine θ:

Since $\sin\theta = \frac{b}{\rho}$, $\sin 50° = \frac{b}{10.0}$

$$\begin{aligned} \text{and } b &= 10.0(\sin 50°) \\ &= 10.0(0.7660) \\ &= 7.66 \text{ units} \end{aligned}$$

Since "a" = 6.43 and "b" = 7.66, the corresponding rectangular form of $10.0\underline{/50°}$ is ______________.

Answer to Frame 91: 6.43 + 7.66j

92. A second-quadrant vector in polar form is shown at the right. The vector is: $20.0\underline{/130°}$.

Its reference angle (α) is 50° (Q2).

$\cos\alpha = \cos 50°$ (Q2) $= -0.6428$
$\sin\alpha = \sin 50°$ (Q2) $= +0.7660$

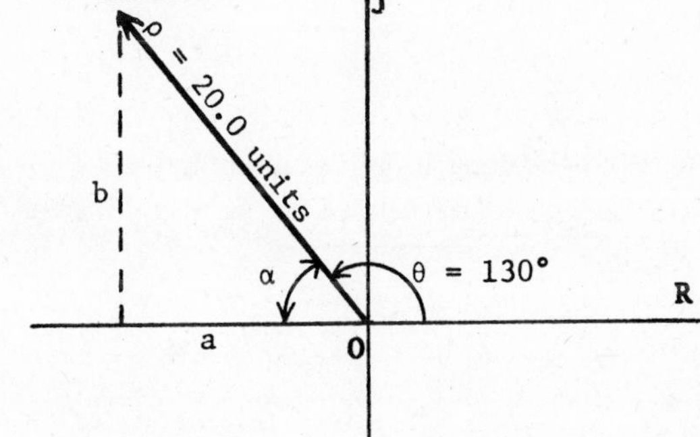

Find the lengths of "a" and "b" as follows:

(a) Since $\cos\alpha = \frac{a}{\rho}$, a = ______ units.

(b) Since $\sin\alpha = \frac{b}{\rho}$, b = ______ units.

(c) Therefore, the corresponding rectangular form of $20.0\underline{/130°}$ is ______________.

Answer to Frame 92: a) a = -12.9 units b) b = +15.3 units c) -12.9 + 15.3j

93. Here is a third-quadrant vector in polar form.
The vector is: $10.0\underline{/215°}$

Its reference angle (α) is 35° (Q3).

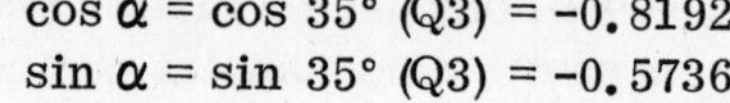

$\cos \alpha = \cos 35°$ (Q3) $= -0.8192$
$\sin \alpha = \sin 35°$ (Q3) $= -0.5736$

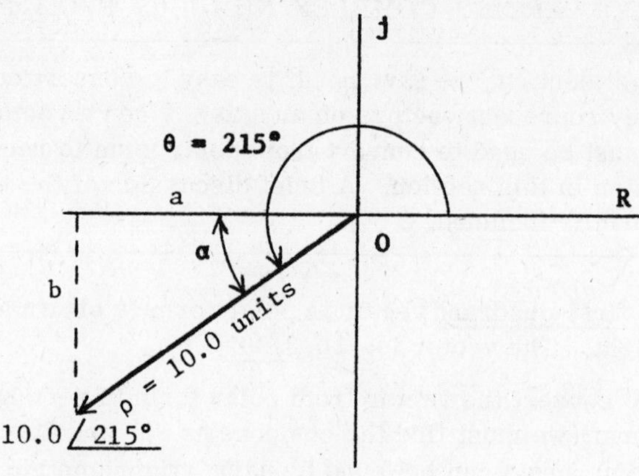

Find the lengths of "a" and "b":

(a) a = __________ units

(b) b = __________ units

(c) The corresponding rectangular form of $10.0\underline{/215°}$ is ______________.

Answer to Frame 93: a) a = -8.19 units b) b = -5.74 units c) -8.19 - 5.74j

94. Here is a fourth-quadrant vector in polar form.
The vector is: $20.0\underline{/300°}$

Its reference angle (α) is 60° (Q4).

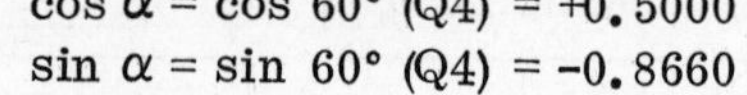

$\cos \alpha = \cos 60°$ (Q4) $= +0.5000$
$\sin \alpha = \sin 60°$ (Q4) $= -0.8660$

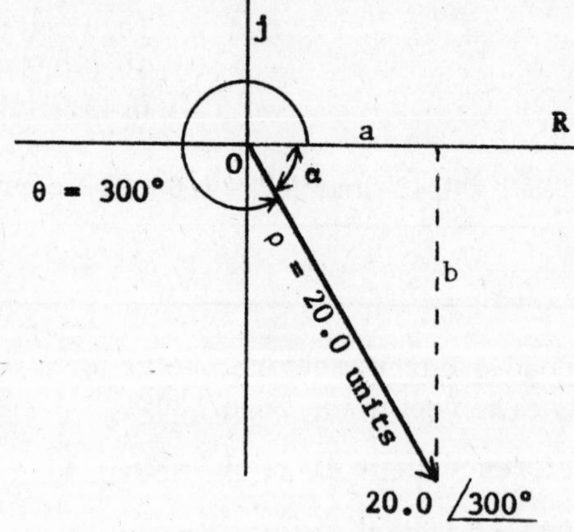

Find the lengths of "a" and "b":

(a) a = __________ units

(b) b = __________ units

(c) The corresponding rectangular form of $20.0\underline{/300°}$ is ______________.

a) a = +10.0 units

b) b = -17.3 units

c) 10.0 - 17.3j

95. When converting from polar form to rectangular form, the sine and cosine of the reference angle are used.

If the polar form contains a negative angle, be careful of the reference angle that you use. Below we have drawn a negative angle in the fourth quadrant (Figure 1) and in the third quadrant (Figure 2).

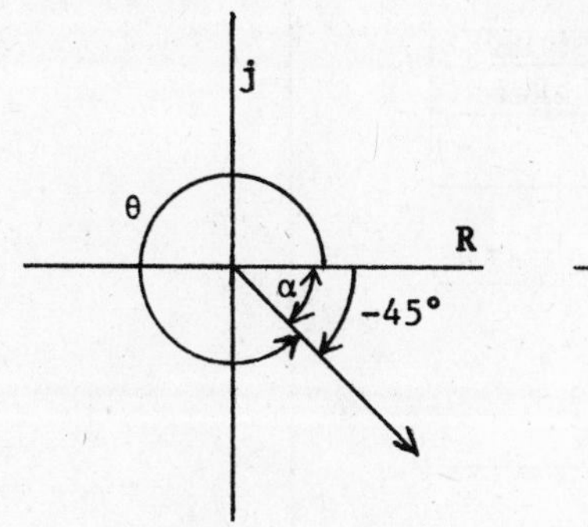

Figure 1

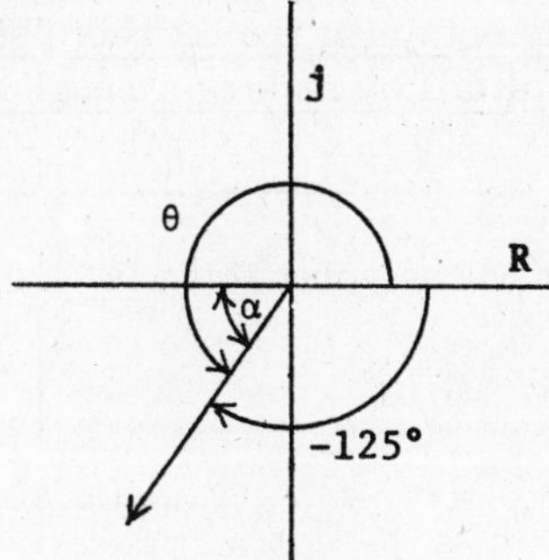

Figure 2

(a) How large is angle θ in Figure 1? __________

(b) The reference angle (α) for a -45° angle is __________.

(c) How large is angle θ in Figure 2? __________

(d) The reference angle for a -125° angle is __________.

---

a) 315°

b) 45° (Q4)

c) 235°

d) 55° (Q3)

96. For each of the following negative angles, write the corresponding standard-position angle (θ) and the reference angle (α):

(a) -79° θ = __________ α = __________

(b) -157° θ = __________ α = __________

---

a) θ = 281° α = 79° (Q4)

b) θ = 203° α = 23° (Q3)

97. Specify the reference angle (α) for each of the following negative angles:

(a) -175° __________ (b) -69° __________ (c) -99° __________

---

a) 5° (Q3)

b) 69° (Q4)

c) 81° (Q3)

98. When converting from polar form to rectangular form, the sine and cosine of the reference angle are used to find "a" and "b" of the rectangular form:

a + bj where "a" is the horizontal component and "b" is the vertical component.

The major pitfall in this conversion is obtaining the correct "signs" for "a" and "b". You will avoid this pitfall by making a rough sketch of the vector and its reference angle.

Find the rectangular form of a vector whose polar form is $6.00\angle 45°$. Sketch the vector first.

The rectangular form is ______________.

---

99. Find the rectangular form of a vector whose polar form is $30.0\angle 150°$.

4.24 + 4.24j

The rectangular form is ______________.

---

100. Convert the following vectors from polar form to rectangular form. Make a sketch of each vector first.

-26.0 + 15.0j

(a) $8.00\angle -60°$

The rectangular form is ______________.

(Continued on following page.)

100. (Continued)

(b) $20.0\underline{/190°}$

The rectangular form is ______________.

101. When vectors are given in rectangular form, they can be added or subtracted immediately. For example:

Given these two vectors: Vector 1 = 15.7 + 11.5j
Vector 2 = 25.9 - 32.6j

(a) Find the resultant of Vector 1 and Vector 2:

(15.7 + 11.5j) + (25.9 - 32.6j) = ______________

(b) Subtract Vector 2 from Vector 1:

(15.7 + 11.5j) - (25.9 - 32.6j) = ______________

a) 4.00 - 6.93j

b) -19.7 - 3.47j

102. When vectors are given in polar form, however, <u>they cannot be added directly</u>. To add them, you must:

(1) Convert each vector to rectangular form.
(2) Find the resultant in rectangular form.
(3) Convert the resultant back to polar form.

Here are two vectors in polar form: Vector 1 = $30.0\underline{/57°}$
Vector 2 = $20.0\underline{/300°}$

(a) Convert Vector 1 to rectangular form: ______________

(b) Convert Vector 2 to rectangular form: ______________

(c) Write the resultant of the two vectors in rectangular form:

____

(Continued on following page.)

a) 41.6 - 21.1j

b) -10.2 + 44.1j

102. (Continued)

(d) Write the polar form of the resultant: ______________

---

103. Similarly, vectors in polar form <u>cannot be subtracted directly</u>. To subtract them, you must also:

(1) Convert each vector to rectangular form.
(2) Perform the subtraction in rectangular form.
(3) Then convert the answer back to polar form.

Here are two vectors in polar form: Vector 1 = $10\underline{/0°}$
Vector 2 = $5\underline{/90°}$

(a) Convert Vector 1 to rectangular form: ______________

(b) Convert Vector 2 to rectangular form: ______________

(c) Subtract Vector 2 from Vector 1 in rectangular form: ______________

(d) Write the polar form of the answer: ______________

a) 16.3 + 25.2j

b) 10.0 - 17.3j

c) 26.3 + 7.9j

d) $27.5\underline{/17°}$

---

a) 10 + 0j or 10

b) 0 + 5j or 5j

c) 10 - 5j

d) $11.2\underline{/333°}$ or $11.2\underline{/-27°}$

SELF-TEST 8 (Frames 91-103)

Given these two vectors in polar form:

Vector A: $20.0\angle 37^\circ$
Vector B: $40.0\angle -75^\circ$

1. Convert Vector A to rectangular form: ______________

2. Convert Vector B to rectangular form: ______________

3. Find the resultant of Vectors A and B in rectangular form: ______________

4. Write the resultant in polar form: ______________

ANSWERS: 1. $16.0 + 12.0j$ 2. $10.4 - 38.6j$ 3. $26.4 - 26.6j$ 4. $37.5\angle 315^\circ$ or $37.5\angle -45^\circ$

8-9 VECTOR-MULTIPLICATION IN RECTANGULAR FORM

Just as vectors can be added or subtracted directly when they are given in rectangular form, they can be multiplied directly in rectangular form. The procedure for this type of vector-multiplication will be discussed in this section. We will see that the product of such a multiplication is also a complex number of the form $\boxed{a + bj}$. In this section, the multiplications will be limited to those in which neither "a" nor "b" is "0" in either the factors or the product.

104. A vector-multiplication in rectangular form is a multiplication in which both factors are complex numbers in rectangular form. Here are some examples:

$(3 + 5j)(9 + 4j)$ $(7 - j)(8 - 6j)$ $(-9 + 7j)(-8 - 5j)$

Since each factor is a complex number in rectangular form, each factor is equivalent to a grouping which is either an addition or a subtraction.

In order to multiply two groupings, we must use the distributive principle twice. The distributive principle is given in symbol form below:

$(\triangle + \bigcirc)(\square) = \triangle(\square) + \bigcirc(\square)$

When multiplying two groupings, the first use of the distributive principle breaks up the first grouping so that two simple instances of the distributive principle are obtained. For example:

$(3 + 5j)(9 + 4j) = 3(9 + 4j) + 5j(9 + 4j)$

Following the pattern above, complete these:

(a) $(3 + 4j)(5 + 6j) = 3(5 + 6j) + 4j(\quad)$

(b) $(2 + j)(4 - 3j) = 2(4 - 3j) + \square(4 - 3j)$

105. Three steps are needed to complete a multiplication of two groupings. Here is an example:

Step 1: $(2 + j)(3 + j) = 2(3 + j) + j(3 + j)$

Step 2: $= 6 + 2j + 3j + j^2$

Step 3: $= 6 + 5j + j^2$

Notice that the distributive principle is used in each of the first two steps.

Following the pattern above, complete this multiplication:

$(7 + j)(10 - j) = 7(10 - j) + j(10 - j)$

= ______________

= ______________

Answers to 104:

a) $(5 + 6j)$

b) $\boxed{j}$

106. Complete each of the following multiplications:

(a) $(4 + 3j)(2 + 5j) =$ ______________

(b) $(2 + j)(3 - 5j) =$ ______________

Answers to 105:

$= 70 - 7j + 10j - j^2$

$= 70 + 3j - j^2$

Answers to 106:

a) $8 + 26j + 15j^2$

b) $6 - 7j - 5j^2$

107. If the first grouping is a subtraction, an additional step is necessary in order to convert that subtraction to an addition. For example:

$$(7 - 2j)(5 + j) = [7 + (-2j)](5 + j)$$
$$= 7(5 + j) + (-2j)(5 + j)$$
$$= 35 + 7j + (-10j) + (-2j^2)$$
$$= 35 - 3j - 2j^2$$

Following the steps above, complete this multiplication:

$(10 - j)(5 - 3j) = [10 + (-j)](5 - 3j)$

= ______________________

= ______________________

= ______________

Answer: $= 10(5 - 3j) + (-j)(5 - 3j)$

$= 50 - 30j + (-5j) + 3j^2$

$= 50 - 35j + 3j^2$

---

108. Complete each of the following multiplications:

(a) $(2 - 3j)(2 + 4j) =$ ______________________

(b) $(1 - 2j)(2 - 3j) =$ ______________________

Answer: a) $4 + 2j - 12j^2$

b) $2 - 7j + 6j^2$

---

109. When two complex numbers in rectangular form are multiplied, the product contains three terms, including a "j^2" term. For example:

$$(2 + 3j)(1 + 4j) = 2 + 11j + 12j^2$$
$$(3 - 5j)(2 + j) = 6 - 7j - 5j^2$$

Since the product contains a "j^2" term, it does not fit the pattern of a complex number. However, each product can be reduced to a complex number. To do so, we must examine the meaning of "j^2".

What radical does the letter "j" stand for? ________

Answer: $\sqrt{-1}$

110. Since $j = \sqrt{-1}$, $j^2 = \sqrt{-1} \cdot \sqrt{-1}$

Following the pattern for multiplying two radicals which contain the same quantity, we can express "j^2" as a real (regular) number. That is:

Just as $\sqrt{3} \cdot \sqrt{3} = 3$,

$\sqrt{-1} \cdot \sqrt{-1} = -1$

Therefore, "j^2" equals what regular number? _______

---

111. Since "j^2" = -1, we can express "j^2" terms as real (regular) numbers by substituting "-1" for "j^2". For example: | -1

$$6j^2 = \quad 6(-1) = -6$$

$$-10j^2 = (-10)(-1) = +10$$

Express each of the following as a regular number:

(a) $14j^2$ = _______ (b) $-5j^2$ = _______ (c) $-j^2$ = _______

---

112. Even though the product of two complex numbers contains a "j^2" term, we can simplify the product to a complex number <u>by substituting</u> "-1" for "j^2". For example:

a) -14
b) +5
c) +1

$$(5 + 4j)(3 + 2j) = 15 + 22j + 8j^2$$
$$= 15 + 22j + 8(-1)$$
$$= 15 + 22j + (-8)$$
$$= 7 + 22j$$

$$(3 + 2j)(4 + 3j) = 12 + 17j + 6j^2$$
$$= 12 + 17j + 6(-1)$$

(a) $= 12 + 17j + \boxed{}$

(b) = _______________

---

113. When the product contains a <u>subtraction sign</u> in front of the "j^2" term, be careful with the signs. An example is given on the left. Complete the simplification on the right:

a) $\boxed{-6}$
b) 6 + 17j

$11 + 5j - 3j^2$		$7 - 6j - 9j^2$
$11 + 5j - 3(-1)$		$7 - 6j - 9(-1)$
$11 + 5j - (-3)$	(a)	$7 - 6j - (\quad)$
$11 + 5j + (+3)$	(b)	$7 - 6j + (\quad)$
$14 + 5j$	(c)	_______________

---

a) (-9)
b) (+9)
c) 16 - 6j

114. Simplify each of these products to a complex number:

(a) $10 + 7j + 6j^2 =$ ________ (c) $-9 + 5j + 4j^2 =$ ________

(b) $5 - 2j - 7j^2 =$ ________ (d) $-12 - 6j - 5j^2 =$ ________

---

115. In each case, multiply the two complex numbers and write the product as a complex number:

(a) $(5 + 3j)(6 - 2j) =$ ________

(b) $(2 - 6j)(1 - j) =$ ________

a) $4 + 7j$
b) $12 - 2j$
c) $-13 + 5j$
d) $-7 - 6j$

---

116. Write each product as a complex number:

(a) $(-2 + 3j)(4 - j) =$ ________

(b) $(-5 - 4j)(-6 - j) =$ ________

a) $36 + 8j$
b) $-4 - 8j$

---

a) $-5 + 14j$
b) $26 + 29j$

117. When a vector-addition is performed, there is an obvious geometric relationship between the vector-addends and the resultant. That is, if we use the two vector-addends to form a parallelogram, the sum or resultant is its diagonal.

When a vector-multiplication is performed, however, there is no immediately obvious geometric relationship between the factors and the product. Here is an example:

The factors and product of the following multiplication are graphed on the complex plane at the right.

$$\begin{aligned}(4 + 2j)(1 + 4j) &= 4 + 18j + 8j^2 \\ &= 4 + 18j + (-8) \\ &= -4 + 18j\end{aligned}$$

As you can see from the graph, there is no immediately obvious geometric relationship between the vectors of the factors and the vector of the product.

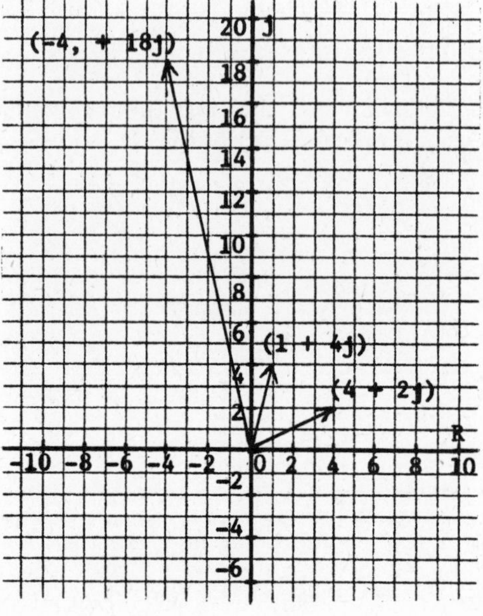

SELF-TEST 9 (Frames 104-117)

Write each of the following as a complex number:

1. $35 - 8j - 3j^2$ ________

2. $12 + 28j + 8j^2$ ________

3. $20 - 21j + 4j^2$ ________

Multiply. Write each product as a complex number:

4. $(1 - 3j)(-2 + j)$ ________

5. $(-40 - 20j)(30 - 10j)$ ________

ANSWERS: 1. $38 - 8j$ 2. $4 + 28j$ 3. $16 - 21j$ 4. $1 + 7j$ 5. $-1400 - 200j$

8-10 VECTOR-MULTIPLICATION INVOLVING INCOMPLETE COMPLEX NUMBERS

When either "a" or "b" is "0" in a complex number in rectangular form, we have seen that it is frequently written as an incomplete complex number. In this section, we will discuss vector-multiplications in which one or both of the factors is an incomplete complex number.

118. Since "b" is "0" in the first factor below, we can write the first factor as an incomplete complex number. We get:

$$(3 + 0j)(5 + 2j) = 3(5 + 2j)$$

And since this new form of the multiplication is a simple instance of the distributive principle, we can obtain the product by using the distributive principle only once. We get:

$$3(5 + 2j) = 15 + 6j$$

Is the product a complex number? __________

119. Perform each of the following multiplications:

(a) $5(3 + 2j) =$ __________ (b) $4(2 - j) =$ __________

Answer: Yes

120. Since "a" is "0" in the first factor below, we can write the first factor as an incomplete complex number. We get:

$$(0 + 2j)(1 + 3j) = 2j(1 + 3j)$$

And since this new form of the multiplication is a simple instance of the distributive principle, we can obtain the product by using the distributive principle only once. We get:

$$2j(1 + 3j) = 2j + 6j^2$$

The product as it stands is not a complex number. However, it can be simplified to a complex number by substituting "-1" for "j^2". We get:

$$2j(1 + 3j) = 2j + 6j^2 = 2j + (-6) = -6 + 2j$$

Notice that the real-number term is written first in the expression on the far right.

Using the steps above, complete each of these multiplications. Write each product as a complex number with the real-number term first:

(a) $3j(2 + 4j) =$ ______________________________

(b) $2j(5 - j) =$ ______________________________

Answers:
a) $15 + 10j$
b) $8 - 4j$

121. Write each product below as a complex number with the real-number term first:

(a) $-3j(1 + 2j) =$ ______________________________

(b) $-5j(2 - 3j) =$ ______________________________

Answers:
a) $-12 + 6j$ From: $6j + 12j^2$
b) $2 + 10j$ From: $10j - 2j^2$

122. Write each product below as a complex number:

(a) $5(2 - j) =$ ______ (c) $j(7 - j) =$ ______

(b) $-j(5 + j) =$ ______ (d) $-j(2 - j) =$ ______

a) $6 - 3j$

From: $-3j - 6j^2$

b) $-15 - 10j$

From: $-10j + 15j^2$

---

123. Since there is a "0" in one term of each factor below, we can write each term as an incomplete complex number. We get:

$$(3 + 0j)(0 + 2j) = 3(2j) = 6j \text{ or } 0 + 6j$$

Each of the factors below is an incomplete complex number. Write each product as an incomplete complex number and then as a complete complex number:

(a) $3(5j) =$ ______ $=$ ______

(b) $5(-2j) =$ ______ $=$ ______

(c) $2(4) =$ ______ $=$ ______

a) $10 - 5j$

b) $1 - 5j$

c) $1 + 7j$

d) $-1 - 2j$

---

124. Write each product below as an incomplete complex number and then as a complete complex number:

(a) $3(-5) =$ ______ $=$ ______

(b) $4j(-3j) =$ ______ $=$ ______

(c) $2j(5j) =$ ______ $=$ ______

a) $15j = 0 + 15j$

b) $-10j = 0 - 10j$

c) $8 = 8 + 0j$

---

125. Write each product as a complete complex number and then convert each to polar form:

	Complex Number		Polar Form
(a) $2(6j) =$	______	$=$	______
(b) $2(-4) =$	______	$=$	______
(c) $-2j(5j) =$	______	$=$	______
(d) $(-3j)(3) =$	______	$=$	______

a) $-15 = -15 + 0j$

b) $+12 = 12 + 0j$

c) $-10 = -10 + 0j$

---

a) $0 + 12j = 12\angle 90°$

b) $-8 + 0j = 8\angle 180°$

c) $10 + 0j = 10\angle 0°$

d) $0 - 9j = 9\angle 270°$

or $9\angle -90°$

SELF-TEST 10 (Frames 118-125)

Multiply. Write each product as a complete complex number:

1. (0 - j)(1 - j)	2. (-8 + 0j)(7 - 2j)	3. (5 + 0j)(0 + 4j)	4. (0 - 20j)(30 + 0j)
______	______	______	______

Multiply. Write each product in polar form:

5. (-3 + 0j)(4 - 0j)	6. (0 + 6j)(0 - 12j)	7. (7 + 0j)(0 - 4j)
______	______	______

ANSWERS:

1. -1 - j
2. -56 + 16j
3. 0 + 20j
4. 0 - 600j
5. $12\angle 180°$
6. $72\angle 0°$
7. $28\angle 270°$ or $28\angle -90°$

8-11 VECTOR-MULTIPLICATION OF CONJUGATES

In the last section, we saw some vector-multiplications in which the product was an incomplete complex number. In each case, the factors were also incomplete complex numbers. It is also possible to obtain a product which is an incomplete complex number when neither factor is an incomplete complex number. The major type of vector-multiplication in which this occurs is when the complex numbers are a pair of conjugates. We will briefly discuss this type of multiplication in this section.

126. Here is a multiplication in which the product simplifies to an incomplete complex number which does not contain a "j" term.

$$(7 + 2j)(7 - 2j) = 49 - 4j^2$$
$$= 49 - (-4)$$
$$= 49 + 4 = 53$$

In the multiplication above, the factors $(7 + 2j)$ and $(7 - 2j)$ are called a pair of conjugates. Notice these points about them:

(1) The real-number term is "7" in each.

(2) The imaginary-number term is "2j" in each.

(3) The only difference is this: One complex number is an addition.
The other is a subtraction.

"7 + 2j" is called the conjugate of "7 - 2j".
"7 - 2j" is called the conjugate of "7 + 2j".

(a) The conjugate of "5 + 7j" is ______________.

(b) The conjugate of "10 - j" is ______________.

---

127. Write the conjugate of each of the following:

(a) 3 + j __________ (b) 6 - 5j __________ (c) 12.8 - 13.6j __________

Answers to 126: a) 5 - 7j b) 10 + j

---

128. When the factors are a pair of conjugates, the product is an incomplete complex number for these reasons:

(1) A "j" term does not appear in the product:

$$(3 + 2j)(3 - 2j) = 3(3 - 2j) + 2j(3 - 2j)$$
$$= 9 - 6j + 6j - 4j^2$$
$$= 9 - 4j^2$$

(2) Since $j^2 = -1$, the product simplifies to a real number:

$$9 - 4j^2 = 9 - 4(-1) = 9 - (-4) = 9 + 4 = 13$$

Complete the following, simplifying each product to a real number:

(a) $(7 - 3j)(7 + 3j) =$ __________

(b) $(5 + j)(5 - j) =$ __________

(c) $(10.1 + 19.7j)(10.1 - 19.7j) =$ __________

Answers to 127: a) 3 - j b) 6 + 5j c) 12.8 + 13.6j

129. Write the conjugate of each of the following: (a) $-7 + 3j$ ________ (b) $-11 - j$ ________	a) 58 b) 26 c) 490
130. Write each product as a real number: (a) $(-4 - 2j)(-4 + 2j) =$ ________ (b) $(-6 + 5j)(-6 - 5j) =$ ________ (c) $(-90 + 70j)(-90 - 70j) =$ ________	a) $-7 - 3j$ b) $-11 + j$
131. Whenever a pair of conjugates is multiplied, the product is an incomplete complex number with the "j" term missing. In multiplications of this type, the product always simplifies to a <u>positive</u> real number. Since a <u>positive</u> real number represents a vector which lies on the <u>positive half of the horizontal axis</u>, the standard-position angle of each vector of this type is ______.	a) 20 b) 61 c) 13,000
132. For this multiplication: $(10 + 4j)(10 - 4j)$ (a) Write the product as a <u>real number</u>. ________ (b) Write the product as a <u>complete complex number</u>. ________ (c) Write the polar form of the <u>product</u>. ________	0°
133. For this multiplication: $(-9 - j)(-9 + j)$ (a) Write the product as a <u>real number</u>. ________ (b) Write the product as a <u>complete complex number</u>. ________ (c) Write the polar form of the <u>product</u>. ________	a) 116 b) $116 + 0j$ c) $116\underline{/0°}$
134. When multiplying two complete complex numbers which are not a pair of conjugates, we sometimes obtain an incomplete complex number as the product. Here is an example: $(3 + 4j)(8 + 6j) = 24 + 50j + 24j^2$ $= 24 + 50j + (-24)$ $= (0 + 50j)$ or $50j$ Write the polar form of the product. ________	a) 82 b) $82 + 0j$ c) $82\underline{/0°}$
	$50\underline{/90°}$

SELF-TEST 11 (Frames 126-134)

Write the conjugate of: 1. $-3 + 5j$ ________ 2. $1 - j$ ________ 3. $-6 - 9j$ ________

4. The product of $(1 - j)$ and its conjugate is ________.

Do the following multiplications:

5. $(5 - j)(5 + j)$ ________

6. $(-200 + 100j)(-200 - 100j)$ ________

7. $(1 - 2j)(2 - j)$ ________

Multiply $(1 - 3j)$ by its conjugate, and write the product as a complex number in:

8. Rectangular form: ________

9. Polar form: ________

ANSWERS:

1. $-3 - 5j$
2. $1 + j$
3. $-6 + 9j$
4. 2
5. 26
6. 50,000
7. $-5j$
8. $10 + 0j$
9. $10\underline{/0°}$

8-12 VECTOR-MULTIPLICATION IN POLAR FORM

When vectors are given in polar form, they cannot be added or subtracted directly. In order to perform the addition or subtraction, the polar form must first be converted to rectangular form. However, when vectors are given in polar form, they can be multiplied directly. The procedure for multiplying vectors in polar form will be discussed in this section. We will also briefly discuss the method for multiplying three or more vectors in both rectangular form and in polar form.

135. Let's examine the multiplication done in the last frame of the last section:

$$(3 + 4j)(8 + 6j) = 0 + 50j$$

Below we have written the polar form of each factor and the product:

$$3 + 4j = 5\angle 53°$$

$$8 + 6j = 10\angle 37°$$

$$0 + 50j = 50\angle 90°$$

Notice the relationship between the polar form of the two factors and the polar form of the product.

(a) We can obtain the length of the product-vector by ______________ (adding/multiplying) the length of the two factor-vectors.

(b) We can obtain the angle of the product-vector by ______________ (adding/multiplying) the angles of the two factor-vectors.

---

136. Here is another multiplication: $(5 + 4j)(2 + 3j) = -2 + 23j$

Converting each factor and the product to polar form, we get:

$$5 + 4j = 6.40\angle 39°$$

$$2 + 3j = 3.61\angle 56°$$

$$-2 + 23j = 23.1\angle 95°$$

(a) Can we obtain the length of the product-vector by multiplying the lengths of the two factor-vectors? __________

(b) Can we obtain the angle of the product-vector by adding the angles of the two factor-vectors? __________

Answers to 135:

a) multiplying
$5(10) = 50$

b) adding
$53° + 37° = 90°$

137.

TO MULTIPLY TWO VECTORS GIVEN IN POLAR FORM:

(1) MULTIPLY THEIR LENGTHS TO OBTAIN THE LENGTH OF THE PRODUCT.

(2) ADD THEIR ANGLES TO OBTAIN THE ANGLE OF THE PRODUCT.

Find the product of each of the following pairs of vectors:

(a) $2\angle 10°$ and $8\angle 60°$ ______________

(b) $5\angle 200°$ and $11\angle 150°$ ______________

Answers to 136:

a) Yes, since:
$(6.40)(3.61) = 23.1$

b) Yes, since:
$39° + 56° = 95°$

Answers to 137:

a) $16\angle 70°$

b) $55\angle 350°$

138. When multiplying vectors in polar form, we can obtain <u>an</u> <u>angle</u> <u>greater</u> <u>than</u> <u>360°</u> for the product. Here is an example:

The product of $7\underline{/305^\circ}$ and $12\underline{/80^\circ}$ is $84\underline{/385^\circ}$.

Ordinarily, we do not use angles greater than 360° when writing in polar form. We use angles between 0° and 360°. To obtain an angle between 0° and 360° which is equivalent to 385°, we simply subtract 360° and get 25°. Therefore:

Instead of $84\underline{/385^\circ}$, we write $84\underline{/25^\circ}$.

Write each of the following products in the preferred form:

(a) $17\underline{/405^\circ}$ ________ (b) $19.6\underline{/477^\circ}$ ________ (c) $2.11\underline{/367^\circ}$ ________

---

139. A multiplication in polar form is ordinarily written with parentheses around the polar form of each vector, as we have done below. Write the preferred form of the product for each multiplication:

(a) $(15\underline{/279^\circ})(5\underline{/146^\circ}) =$ ____________

(b) $(6.11\underline{/330^\circ})(7.89\underline{/106^\circ}) =$ ____________

a) $17\underline{/45^\circ}$
b) $19.6\underline{/117^\circ}$
c) $2.11\underline{/7^\circ}$

---

140. When a vector lies in the <u>third</u> or <u>fourth</u> quadrant, we sometimes write a <u>negative</u> <u>angle</u> in its polar form.

If a <u>negative</u> angle is given for one or both of the factors, you obtain the angle of the product by adding the "signed" angles. That is:

$70^\circ + (-30^\circ) = 40^\circ$ $60^\circ + (-80^\circ) = -20^\circ$ $(-30) + (-41^\circ) = -71^\circ$

Complete each multiplication below:

(a) $(10\underline{/30^\circ})(5\underline{/-19^\circ}) =$ ________ (b) $(4\underline{/-60^\circ})(3\underline{/-20^\circ}) =$ ________

a) $75\underline{/65^\circ}$
b) $48.2\underline{/76^\circ}$

---

141. For each multiplication below, write the product with a <u>negative</u> angle first and then write it with the corresponding positive angle between 0° and 360°:

(a) $(7\underline{/50^\circ})(6\underline{/-72^\circ}) =$ __________ = __________

(b) $(10\underline{/-17^\circ})(7\underline{/-67^\circ}) =$ __________ = __________

a) $50\underline{/11^\circ}$
b) $12\underline{/-80^\circ}$

---

142. Complete each of these multiplications:

(a) $(16.5\underline{/279^\circ})(32.6\underline{/116^\circ}) =$ ____________

(b) $(7.83\underline{/39^\circ})(5.42\underline{/-81^\circ}) =$ ____________

a) $42\underline{/-22^\circ} = 42\underline{/338^\circ}$
b) $70\underline{/-84^\circ} = 70\underline{/276^\circ}$

---

a) $538\underline{/35^\circ}$
b) $42.4\underline{/-42^\circ}$
or
$42.4\underline{/318^\circ}$

143. For this multiplication: $(6.02\underline{/57°})(7.99\underline{/-26°})$

(a) Write the product in polar form: __________

(b) Convert the product to rectangular form: __________

Answers to 143:

a) $48.1\underline{/31°}$

b) $41.2 + 24.8j$

144. For this multiplication: $(5.00\underline{/29°})(2.00\underline{/-64°})$

(a) Write the product in polar form: __________

(b) Write the product in rectangular form: __________

Answers to 144:

a) $10.0\underline{/-35°}$ or $10.0\underline{/325°}$

b) $8.19 - 5.74j$

145. Each factor below represents a vector which lies on an axis. Write each product in polar form and then convert the product to rectangular form.

		Polar Form		Rectangular Form
(a)	$(3\underline{/0°})(2\underline{/90°})$ =	______	=	______
(b)	$(5\underline{/270°})(2\underline{/-90°})$ =	______	=	______
(c)	$(4\underline{/180°})(2\underline{/270°})$ =	______	=	______

Answers to 145:

a) $6\underline{/90°} = 0 + 6j$ (or $6j$)

b) $10\underline{/180°} = -10 + 0j$ (or -10)

c) $8\underline{/90°} = 0 + 8j$ (or $8j$)

When two vectors are given <u>in rectangular form</u>, it is easier to multiply <u>by the rectangular method</u>, <u>even if you eventually want the product in polar form</u>.

When two vectors are given <u>in polar form</u>, it is easier to multiply <u>by the polar method</u>, <u>even if you eventually want the product in rectangular form</u>.

[In each case above, you would have to make two conversions (each factor) if you convert <u>before</u> multiplying. You have to make only <u>one</u> conversion (the product) if you convert <u>after</u> multiplying.]

146. Three vectors can be multiplied in either rectangular form or polar form. Here is an example in rectangular form:

$$(5 + 3j)(2 - 4j)(6 + j)$$

Step 1: We multiply the first two factors and simplify the product to a complex number:

$$(5 + 3j)(2 - 4j)(6 + j)$$

$$(10 - 14j - 12j^2)(6 + j)$$

$$(22 - 14j)(6 + j)$$

Step 2: We then multiply the simplified product of the first two factors and the third factor:

$$(22 - 14j)(6 + j) = 132 - 62j - 14j^2$$

$$= 146 - 62j \quad \text{(Answer)}$$

Do this multiplication: $(2 - j)(3 + j)(4 - j) =$ ____________

---

147. To multiply three or more vectors in polar form, we simply multiply their lengths and add their angles. Here is an example:

$$(10\angle 30°)(5\angle 27°)(3\angle 16°)$$

The length of the product is: $(10)(5)(3) = 150$
The angle of the product is: $30° + 27° + 16° = 73°$
The polar form of the product is: $150\angle 73°$

Find the product of each of the following in polar form:

(a) $(6\angle 30°)(5\angle -27°)(2\angle 17°) =$ ______________

(b) $(10\angle -80°)(2\angle 70°)(3\angle 30°)(5\angle -50°) =$ ______________

Answer to 146: $27 - 11j$

Answers to 147:

a) $60\angle 20°$

b) $300\angle -30°$ or $300\angle 330°$

SELF-TEST 12 (Frames 135-147)

Multiply. Write each product in polar form:

1. $(12\angle 72°)(4\angle 138°) =$ ______

2. $(105\angle 100°)(202\angle -43°) =$ ______

Multiply. Write each product in polar form, reporting each angle as a positive angle between 0° and 360°.

3. $(30\angle 250°)(20\angle 160°) =$ ______

4. $(2.50\angle -62°)(8.40\angle 37°) =$ ______

Multiply: $(17.5\angle 90°)(32.6\angle -60°)(28.4\angle 20°)$

5. Write the product in polar form: ______

6. Write the product in rectangular form: ______

Multiply: $(4 + 3j)(2 - j)(2 + j)$

7. Write the product in rectangular form: ______

8. Write the product in polar form: ______

ANSWERS:

1. $48\angle 210°$
2. $21{,}200\angle 57°$
3. $600\angle 50°$
4. $21.0\angle 335°$
5. $16{,}200\angle 50°$
6. $10{,}400 + 12{,}400j$
7. $20 + 15j$
8. $25\angle 37°$

8-13 VECTOR-DIVISION IN RECTANGULAR FORM

Just as vectors can be added, subtracted, or multiplied directly when they are given in rectangular form, they can be divided directly in rectangular form. We will discuss the procedure for this type of vector-division in this section. The divisions will be limited to those which do not involve incomplete complex numbers.

148. The division of vectors in rectangular form is based on various algebraic principles. One of these principles is "the principle of dividing a quantity by itself". This principle is stated symbolically below:

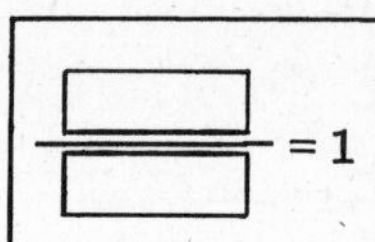

That is: $\frac{8}{8} = 1$, $\frac{2 + j}{2 + j} = 1$, $\frac{3 - j}{3 - j} = 1$

As you can see from the examples above, if we divide a complex number by itself, the quotient is _______.

---

149. Another principle used in dividing complex numbers is "the identity principle of multiplication". This principle is stated symbolically below: +1

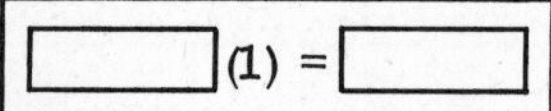

All of the examples below are instances of this principle, since any instance of $\frac{\square}{\square}$ equals 1:

$$\frac{2}{3}\left(\frac{5}{5}\right) = \frac{2}{3} \quad , \text{ since } \quad \frac{5}{5} = 1$$

$$\left(\frac{3 + 4j}{5 - 6j}\right)\left(\frac{5 + 6j}{5 + 6j}\right) = \frac{3 + 4j}{5 - 6j} , \text{ since } \frac{5 + 6j}{5 + 6j} = 1$$

$$\left(\frac{2 - 3j}{1 + j}\right)\left(\frac{1 - j}{1 - j}\right) = \frac{2 - 3j}{1 + j} , \text{ since } \frac{1 - j}{1 - j} = \_\_\_\_\_\_$$

---

150. In order to divide two complex numbers, we always write them as a fraction. For example: +1

Instead of $(6 + 5j) \div (3 - j)$, we write $\frac{6 + 5j}{3 - j}$.

Having written the division as a fraction, we must eliminate the "j" in the complex number in the denominator of the fraction. In order to do so, we multiply by the instance of $\frac{\square}{\square}$ which contains the conjugate of the denominator. For the fraction above, we multiply by $\frac{3 + j}{3 + j}$, since $3 + j$ is the conjugate of $3 - j$. We get:

$$\frac{6 + 5j}{3 - j} = \left(\frac{6 + 5j}{3 - j}\right)\left(\frac{3 + j}{3 + j}\right) = \frac{(6 + 5j)(3 + j)}{(3 - j)(3 + j)}$$

$$= \frac{18 + 21j + 5j^2}{9 - j^2}$$

$$= \frac{13 + 21j}{10}$$

(Continued on following page.)

150. (Continued)

Notice that the multiplication by $\frac{3 + j}{3 + j}$ produced a product in which the denominator is a real number.

Perform this multiplication:

$$\left(\frac{2 + 2j}{2 - j}\right)\left(\frac{2 + j}{2 + j}\right) = \underline{\hspace{3cm}}$$

Answer: $\frac{2 + 6j}{5}$

---

151. To eliminate the "j" in the complex number in the denominator of the fraction below, you must multiply by what fraction?

$$\frac{5 + 7}{13 - 2j} \qquad \underline{\hspace{3cm}}$$

Answer: $\frac{13 + 2j}{13 + 2j}$

(Since $13 + 2j$ is the conjugate of $13 - 2j$)

---

152. Having multiplied divisions by the instance of $\frac{\square}{\square}$ which contains the conjugate of their denominators, we obtain quotients in the following fractional form:

$$\frac{3 + 4j}{5} \qquad \frac{8 - 6j}{3} \qquad \frac{7 + 5j}{2}$$

Notice that each fraction above fits the pattern for the addition (or subtraction) of two fractions. These patterns are shown symbolically below:

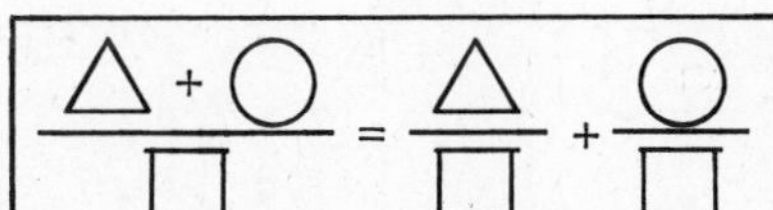

$$\frac{\triangle + \bigcirc}{\square} = \frac{\triangle}{\square} + \frac{\bigcirc}{\square} \qquad \frac{\triangle - \bigcirc}{\square} = \frac{\triangle}{\square} - \frac{\bigcirc}{\square}$$

That is: $\frac{3 + 4j}{5} = \frac{3}{5} + \frac{4j}{5}$

$\frac{8 - 6j}{3} = \frac{8}{3} - \frac{6j}{3}$

$\frac{7 + 5j}{2} = \underline{\hspace{3cm}}$

Answer: $\frac{7}{2} + \frac{5j}{2}$

153. In a division of complex numbers, we want to write a single complex number as the quotient. When we get the quotient in a fractional form with a real number as the denominator, we can convert it to a single complex number in two steps:

Step 1: Using the pattern for the addition (or subtraction) of two fractions, we separate the complicated fraction into two simpler fractions:

$$\frac{7+5j}{2} = \frac{7}{2} + \frac{5j}{2}$$

Step 2: Then we convert the numerical fractions to decimals:

$$\frac{7+5j}{2} = \frac{7}{2} + \frac{5j}{2} = 3.5 + 2.5j$$

Using the steps above, convert the following fraction to a complex number:

$$\frac{7+3j}{4} = \_\_\_\_\_\_\_\_ = \_\_\_\_\_\_\_\_$$

---

154. Use your slide rule to complete this one. Notice that the fraction on the left fits the pattern for the subtraction of two fractions.

$$\frac{17.3 - 25.9j}{14.6} = \_\_\_\_\_\_\_\_ = \_\_\_\_\_\_\_\_$$

Answer: $\frac{7}{4} + \frac{3j}{4} = 1.75 + 0.75j$

---

155. Write each of the following as a complex number with decimal coefficients. Do the necessary divisions on your slide rule:

(a) $\frac{13 - 23j}{17} = \_\_\_\_\_\_\_\_$ (b) $\frac{-5.23 - 8.47j}{1.69} = \_\_\_\_\_\_\_\_$

Answer: $\frac{17.3}{14.6} - \frac{25.9j}{14.6} = 1.18 - 1.77j$

Answer to Frame 155: a) $0.765 - 1.35j$ b) $-3.09 - 5.01j$

---

156. A complete division is shown below. Notice these two points:

(1) We multiply by $\frac{1+3j}{1+3j}$, where $1 + 3j$ is the conjugate of the denominator, in order to obtain a real number in the denominator.

(2) We convert the quotient to a complex number with decimal coefficients.

$$\frac{2+3j}{1-3j} = \left(\frac{2+3j}{1-3j}\right)\left(\frac{1+3j}{1+3j}\right) = \frac{2+9j+9j^2}{1-9j^2} = \frac{-7+9j}{10} = \frac{-7}{10} + \frac{9j}{10} = -0.7 + 0.9j$$

Following the steps above, complete this division:

$$\frac{3+6j}{4-8j} = \_\_\_\_\_\_\_\_$$

Answer: $-0.45 + 0.60j$

157. Do these divisions:

(a) $\frac{9 - 5j}{3 + 7j} =$ ______ (b) $\frac{-1.9 - 5.4j}{6.7 - 2.9j} =$ ______

Answer to Frame 157: a) $-0.138 - 1.34j$ From: $\frac{-8 - 78j}{58}$ b) $0.0563 - 0.782j$ From: $\frac{3.0 - 41.7j}{53.3}$

SELF-TEST 13 (Frames 148-157)

1. To do the division $\frac{7 + 3j}{2 - 5j}$,

we multiply by $\frac{\boxed{}}{\boxed{}}$.

2. Simplify the following to a complex number containing decimals:

$\frac{-15 + 11j}{20} =$ ______

Do these divisions, writing each answer as a complex number containing decimals:

3. $\frac{2 + j}{1 - j} =$ ______

4. $\frac{1 - 2j}{3 + j} =$ ______

5. $\frac{300 - 700j}{100 + 200j} =$ ______

ANSWERS:

1. $\frac{2 + 5j}{2 + 5j}$

2. $-0.75 + 0.55j$

3. $0.5 + 1.5j$ From: $\frac{1 + 3j}{2}$

4. $0.1 - 0.7j$ From: $\frac{1 - 7j}{10}$

5. $-2.2 - 2.6j$ From: $\frac{-110{,}000 - 130{,}000j}{50{,}000}$

8-14 VECTOR-DIVISION INVOLVING INCOMPLETE COMPLEX NUMBERS

In the last section, we avoided vector-divisions involving incomplete complex numbers. We will discuss divisions of that type in this section. We will see that divisions involving incomplete complex numbers do not always require as a first step a multiplication by an instance of $\frac{\square}{\square}$ which contains the conjugate of the denominator. A summary of the various "first-step" strategies for vector-divisions in rectangular form is included.

158. In an ordinary division of complex numbers in rectangular form, the first step is multiplying by an instance of $\frac{\square}{\square}$ which contains the conjugate of the denominator of the division. This first step is used to convert the denominator from a complete complex number to a real number.

When a vector-division contains an incomplete complex number only in its numerator, we must use this same first step in order to eliminate the complete complex number from the denominator. Then we proceed in the usual way. Here are two examples:

$$\frac{0 + j}{5 - 3j} = \frac{j}{5 - 3j}\left(\frac{5 + 3j}{5 + 3j}\right) = \frac{5j + 3j^2}{25 - 9j^2} = \frac{-3 + 5j}{34} = -0.0882 + 0.147j$$

$$\frac{5 + 0j}{3 - 2j} = \frac{5}{3 - 2j}\left(\frac{3 + 2j}{3 + 2j}\right) = \frac{15 + 10j}{9 - 4j^2} = \frac{15 + 10j}{13} = 1.15 + 0.769j$$

Following the steps above, complete each of these:

(a) $\frac{4 + 0j}{2 + j} =$ ________________ (b) $\frac{0 + 5j}{3 - j} =$ ________________

Answers to 158:

a) $1.6 - 0.8j$ From: $\frac{8 - 4j}{5}$

b) $-0.5 + 1.5j$ From: $\frac{-5 + 15j}{10}$

159. When a vector-division contains an incomplete complex number only in its denominator, however, we use a different first step since the denominator is not a complete complex number. That is, we do not multiply by the instance of $\frac{\square}{\square}$ which contains the conjugate of the denominator.

Here is a case in which the denominator is an incomplete complex number containing a "j" term. In the first step, we want to obtain a real number term in the denominator. The simplest way to do so is to multiply the division by "$\frac{j}{j}$". We get:

$$\frac{3 + 2j}{0 + 5j} = \frac{3 + 2j}{5j} = \left(\frac{3 + 2j}{5j}\right)\left(\frac{j}{j}\right) = \frac{3j + 2j^2}{5j^2} = \frac{-2 + 3j}{-5}$$

Multiply each of these divisions by "$\frac{j}{j}$" to obtain a real-number denominator. Be careful of the "signs".

(a) $\frac{5 - 2j}{0 + 2j} = \frac{5 - 2j}{2j}\left(\frac{j}{j}\right) = \frac{5j - 2j^2}{2j^2} =$ ________________

(b) $\frac{2 + 3j}{0 - 5j} = \frac{2 + 3j}{-5j}\left(\frac{j}{j}\right) = \frac{2j + 3j^2}{-5j^2} =$ ________________

160. Having eliminated the "j" term in the denominator, we can complete the problem in the usual way. Here is an example:

$$\frac{5+2j}{3j} = \frac{5+2j}{3j}\left(\frac{j}{j}\right) = \frac{5j+2j^2}{3j^2} = \frac{-2+5j}{-3} = \frac{-2}{-3} + \frac{5j}{-3} = 0.67 - 1.67j$$

Complete: $\frac{1-3j}{-2j} =$ ________________

a) $\frac{2+5j}{-2}$

b) $\frac{-3+2j}{5}$

---

161. Be especially careful of the "signs" in the first step when the numerator or denominator of the division contains a "j" or "-j". Complete the following divisions:

(a) $\frac{3-j}{0+j} = \frac{3-j}{j}\left(\frac{j}{j}\right) = \frac{3j-j^2}{j^2} =$ ________________

(b) $\frac{-5+j}{0-j} = \frac{-5+j}{-j}\left(\frac{j}{j}\right) = \frac{-5j+j^2}{-j^2} =$ ________________

1.5 + 0.5j

---

162. The first step of multiplying by an instance of $\frac{\square}{\square}$ is needed only to obtain a real number in the denominator. Therefore, if the denominator of the original division is an incomplete complex number which contains only a real-number term, we can skip the first step. For example:

$$\frac{5+2j}{4+0j} = \frac{5+2j}{4} = \frac{5}{4} + \frac{2j}{4} = 1.25 + 0.5j$$

Do each of these:

(a) $\frac{6-9j}{3+0j} =$ ________________ (b) $\frac{8+6j}{-4+0j} =$ ________________

a) $\frac{1+3j}{-1} = -1 - 3j$

b) $\frac{-1-5j}{+1} = -1 - 5j$

---

163. Complete: (a) $\frac{75-50j}{25+0j} =$ ________________

(b) $\frac{28+14j}{0-7j} =$ ________________

a) 2 - 3j

b) -2 - 1.5j

---

164. When both the numerator and denominator of a division are incomplete complex numbers, the division process is simplified.

<u>If the denominator contains only a real-number term</u>, we can simply reduce the numerical fraction to a whole number or decimal. For example:

$$\frac{0+8j}{2+0j} = \frac{8j}{2} = 4j \quad \text{(or } 0+4j\text{)}$$

$$\frac{-7+0j}{5+0j} = \frac{-7}{5} = -1.4 \quad \text{(or } -1.4+0j\text{)}$$

Complete: (a) $\frac{12+0j}{-4+0j} = \frac{12}{-4} =$ ________________

(b) $\frac{0-7j}{4+0j} = \frac{-7j}{4} =$ ________________

a) 3 - 2j

b) -2 + 4j

165. When both numerator and denominator are incomplete complex numbers <u>and the denominator is a j-term</u>, two different strategies are possible.

(1) <u>If the numerator is also a j-term</u>, we can simply reduce the division to a real number. For example:

$$\frac{0+8j}{0+4j} = \frac{8j}{4j} = \left(\frac{8}{4}\right)\left(\frac{j}{j}\right) = (2)(1) = 2 \quad (\text{or } 2+0j)$$

$$\frac{0-9j}{0+5j} = \frac{-9j}{5j} = \left(\frac{-9}{5}\right)\left(\frac{j}{j}\right) = (-1.8)(1) = -1.8 \quad (\text{or } -1.8+0j)$$

(2) <u>If the numerator is a real-number term</u>, we must multiply by $\left(\frac{j}{j}\right)$ to get a real-number denominator. For example:

$$\frac{10+0j}{0+5j} = \frac{10}{5j} = \left(\frac{10}{5j}\right)\left(\frac{j}{j}\right) = \frac{10j}{5j^2} = \frac{10j}{-5} = -2j \quad (\text{or } 0-2j)$$

$$\frac{-7+0j}{0+3j} = \frac{-7}{3j} = \left(\frac{-7}{3j}\right)\left(\frac{j}{j}\right) = \frac{-7j}{3j^2} = \frac{-7j}{-3} = 2.33j \quad (\text{or } 0+2.33j)$$

Complete these divisions:

(a) $\frac{0+20j}{0-4j} = \frac{20j}{-4j} =$ ________

(b) $\frac{18+0j}{0-6j} = \frac{18}{-6j} =$ ________

(c) $\frac{0-12j}{0-5j} = \frac{-12j}{-5j} =$ ________

(d) $\frac{14+0j}{0-3j} = \frac{14}{-3j} =$ ________

Answers (to preceding frame):
a) -3 (or -3 + 0j)
b) -1.75j (or 0 - 1.75j)

---

166. In each division below, both the numerator and denominator are incomplete complex numbers. Write each quotient as a <u>complete</u> complex number.

(a) $\frac{-4j}{8} =$ ________ (c) $\frac{-12}{8} =$ ________

(b) $\frac{15j}{6j} =$ ________ (d) $\frac{-8}{6j} =$ ________

Answers (to frame 165):
a) -5 (or -5 + 0j)
b) 3j (or 0 + 3j)
c) +2.4 (or 2.4 + 0j)
d) 4.67j (or 0 + 4.67j)

---

167. Write each quotient below in three forms: (1) as an incomplete complex number, (2) as a complete complex number, and (3) in polar form.

	Incomplete Complex Number	Complete Complex Number	Polar Form
(a) $\frac{24}{8}$ =	________ =	________ =	________
(b) $\frac{15}{5j}$ =	________ =	________ =	________
(c) $\frac{-20j}{10j}$ =	________ =	________ =	________
(d) $\frac{16j}{4}$ =	________ =	________ =	________

Answers (to frame 166):
a) 0 - 0.5j (from -0.5j)
b) 2.5 + 0j (from 2.5)
c) -1.5 + 0j (from -1.5)
d) 0 + 1.33j (from 1.33j)

168. Here is a summary of the "first-step" strategies needed to perform a vector-division in rectangular form:

(1) If the denominator is a complete complex number, we must multiply by the instance of $\frac{\square}{\square}$ which contains the conjugate of the denominator. This type of first step is needed for each of these divisions:

$$\frac{5+3j}{6-2j} \qquad \frac{7}{3+4j} \qquad \frac{8j}{9-2j}$$

(2) If the denominator is an incomplete complex number, multiplication by the conjugate of the denominator can be avoided. Two types are possible:

(a) In most cases, we can complete the division by a reduction to lowest terms, preceded by a "break-apart" if the numerator is a complete complex number. Here are some examples:

$$\frac{7-3j}{5} \qquad \frac{18j}{9j} \qquad \frac{-9}{6} \qquad \frac{-16j}{4}$$

(b) However, when the denominator is a j-term and the numerator is either a complete complex number or a real number, we must multiply by $\frac{"j"}{j}$. Here are some examples:

$$\frac{5+4j}{3j} \qquad \frac{7-2j}{j} \qquad \frac{7}{5j} \qquad \frac{-8}{-2j}$$

Answers:

a) $3 = 3 + 0j = 3\angle 0°$

b) $-3j = 0 - 3j = 3\angle 270°$ (or $3\angle -90°$)

c) $-2 = -2 + 0j = 2\angle 180°$

d) $4j = 0 + 4j = 4\angle 90°$

---

169. In each of the following:

(1) Find the quotient if multiplying by the conjugate of the denominator can be avoided.

(2) Write "Conjugate needed" if multiplying by the conjugate of the denominator cannot be avoided. (Do not find the quotient in those cases.)

(a) $\frac{12-8j}{4+0j} =$ ____________ (c) $\frac{0+18j}{3+6j} =$ ____________

(b) $\frac{15+10j}{0+5j} =$ ____________ (d) $\frac{24+0j}{0-6j} =$ ____________

Answer: Go to the next frame.

---

170. Write each quotient as a complete complex number:

(a) $\frac{0+9j}{6+0j} =$ ____________ (b) $\frac{18+0j}{9+0j} =$ ____________

Answers:

a) $3 - 2j$

b) $2 - 3j$

c) Conjugate needed.

d) $4j$ (or $0 + 4j$)

---

171. Write each quotient as a complete complex number:

(a) $\frac{-27j}{9j} =$ __________ (b) $\frac{14j}{7} =$ __________ (c) $\frac{28}{14j} =$ __________

Answers:

a) $0 + 1.5j$

b) $2 + 0j$

Answer to Frame 171: a) $-3 + 0j$ b) $0 + 2j$ c) $0 - 2j$

SELF-TEST 14 (Frames 158-171)

Write each quotient as a complete complex number:

1. $\frac{7 - 3j}{-2 + 0j} =$ ______ 2. $\frac{3 - 5j}{0 - j} =$ ______ 3. $\frac{-200j}{500j} =$ ______ 4. $\frac{0 + 4j}{1 + j} =$ ______

Write each quotient in polar form:

5. $\frac{-600}{200j} =$ ______ 6. $\frac{0 - 80j}{10 + 0j} =$ ______ 7. $\frac{0 + 30j}{0 - 20j} =$ ______

ANSWERS:

1. -3.5 + 1.5j 2. 5 + 3j 3. -0.4 + 0j 4. 2 + 2j 5. $3\angle 90°$ 6. $8\angle 270°$ 7. $1.5\angle 180°$

8-15 VECTOR-DIVISION IN POLAR FORM

Just as it is easy to perform a vector-multiplication in polar form, it is also easy to perform a vector-division in polar form. We will discuss the method in this section.

172. In each case below, the vector-division in rectangular form on the left has been converted to polar form on the right.

$$\frac{0 + 50j}{3 + 4j} = 8 + 6j \longrightarrow \frac{50\angle 90°}{5\angle 53°} = 10\angle 37°$$

$$\frac{8 + 6j}{0 + 2j} = 3 - 4j \longrightarrow \frac{10\angle 37°}{2\angle 90°} = 5\angle -53°$$

As you can see by examining the polar-form divisions on the right:

(a) The length of the quotient-vector is obtained by ______ (dividing/multiplying) the length of the numerator-vector by the length of the denominator-vector.

(b) The angle of the quotient-vector is obtained by ______ (adding/subtracting) the angle of the denominator-vector from the angle of the numerator-vector.

173.

> TO DIVIDE VECTORS IN POLAR FORM:
>
> (1) THE LENGTH OF THE QUOTIENT-VECTOR IS OBTAINED BY DIVIDING THE LENGTH OF THE NUMERATOR-VECTOR BY THE LENGTH OF THE DENOMINATOR-VECTOR.
>
> (2) THE ANGLE OF THE QUOTIENT-VECTOR IS OBTAINED BY SUBTRACTING THE ANGLE OF THE DENOMINATOR-VECTOR FROM THE ANGLE OF THE NUMERATOR-VECTOR.

Using the steps above, complete each of these divisions:

(a) $\dfrac{10\angle 50^\circ}{2\angle 10^\circ} =$ ________ (b) $\dfrac{45\angle 45^\circ}{5\angle 75^\circ} =$ ________

a) dividing $(\frac{50}{5} = 10)$
$(\frac{10}{2} = 5)$

b) subtracting
$(90^\circ - 53^\circ = 37^\circ)$
$(37^\circ - 90^\circ = -53^\circ)$

---

174. Do these: (a) $\dfrac{75\angle 90^\circ}{25\angle 0^\circ} =$ ________ (b) $\dfrac{64\angle 0^\circ}{8\angle 90^\circ} =$ ________

a) $5\angle 40^\circ$
b) $9\angle -30^\circ$ or $9\angle 330^\circ$

---

175. Do this one: $\dfrac{47.8\angle 37^\circ}{17.9\angle 52^\circ} =$ ________

a) $3\angle 90^\circ$
b) $8\angle -90^\circ$ or $8\angle 270^\circ$

---

176. Do these: (a) $\dfrac{36\angle -40^\circ}{9\angle 30^\circ} =$ ________ (b) $\dfrac{48\angle 70^\circ}{12\angle -30^\circ} =$ ________

$2.67\angle -15^\circ$ or $2.67\angle 345^\circ$

---

177. Quotients in polar form can sometimes be written in an equivalent form by replacing a negative angle with a positive angle or vice versa. For example, in the following division, the quotient $9\angle -270^\circ$ has been written as $9\angle 90^\circ$, since -270° and 90° have the same terminal side:

$$\frac{27\angle 70^\circ}{3\angle 340^\circ} = 9\angle -270^\circ = 9\angle 90^\circ$$

Though negative angles beyond -180° can occur in calculations with complex numbers in polar form, we generally do not use negative angles beyond -180° in final answers. For example:

$9\angle -270^\circ$ is written $9\angle 90^\circ$

Write each final answer in a preferred form:

(a) $\dfrac{35\angle 50^\circ}{7\angle 290^\circ} =$ ________ (b) $\dfrac{56\angle 335^\circ}{8\angle 15^\circ} =$ ________

a) $4\angle -70^\circ$ or $4\angle 290^\circ$
$[-40^\circ - 30^\circ = -70^\circ]$

b) $4\angle 100^\circ$
$[70^\circ - (-30^\circ) = 100^\circ]$

---

a) $5\angle 120^\circ$ (not $5\angle -240^\circ$)
b) $7\angle 320^\circ$ or $7\angle -40^\circ$

178. Write each quotient in polar form and then convert it to rectangular form:

		Polar Form	Rectangular Form
(a)	$\frac{15\angle 90°}{5\angle 0°}$ =	______ =	______
(b)	$\frac{28\angle 360°}{7\angle 180°}$ =	______ =	______
(c)	$\frac{36\angle 47°}{4\angle 47°}$ =	______ =	______
(d)	$\frac{25\angle 0°}{5\angle 90°}$ =	______ =	______

a) $3\angle 90° = 0 + 3j$
b) $4\angle 180° = -4 + 0j$
c) $9\angle 0° = 9 + 0j$
d) $5\angle -90° = 0 - 5j$

In general:

(1) If two vectors are given in complex-number form, it is easier to perform the division in complex-number form.

(2) If two vectors are given in polar form, it is easier to perform the division in polar form.

SELF-TEST 15 (Frames 172-178)

Write each quotient in polar form:

1. $\frac{60\angle 84°}{15\angle 23°}$ = ______

2. $\frac{338\angle 70°}{254\angle -90°}$ = ______

3. $\frac{29\angle 90°}{50\angle 330°}$ = ______

Write each quotient in polar form, and then convert it to rectangular form:

4. $\frac{8\angle -90°}{4\angle -270°}$ = ______ = ______

5. $\frac{6\angle 180°}{2\angle -90°}$ = ______ = ______

ANSWERS:

1. $4\angle 61°$
2. $1.33\angle 160°$
3. $0.58\angle 120°$ (not $0.58\angle -240°$)
4. $2\angle 180° = -2 + 0j$
5. $3\angle 270° = 0 - 3j$

8-16 FORMULA EVALUATIONS REQUIRING COMBINED OPERATIONS WITH VECTORS

There are various formulas related to electrical circuits in which the variables are vectors. To perform formula evaluations with formulas of this type, combined operations with vectors in either rectangular or polar form are necessary. We will examine formula evaluations of this type in this section.

179. Here is the formula for total impedance in a simple parallel alternating-current circuit:

$$Z_T = \frac{Z_1 Z_2}{Z_1 + Z_2}$$

Where Z_T, Z_1, and Z_2 are vectors in either rectangular or polar form.

If we know Z_1 and Z_2, we can find Z_T by a formula evaluation which requires various operations with vectors. For example:

If $Z_1 = 0 - 50j$ and $Z_2 = 10 + 25j$,

By substitution: $Z_T = \frac{(0 - 50j)(10 + 25j)}{(0 - 50j) + (10 + 25j)}$

Step 1: To simplify the numerator, we perform a vector-multiplication and get:

$$Z_T = \frac{1250 - 500j}{(0 - 50j) + (10 + 25j)}$$

Step 2: To simplify the denominator, we perform a vector-addition and get:

$$Z_T = \frac{1250 - 500j}{10 - 25j}$$

Step 3: Now to find Z_T as a single vector, we perform a vector-division and get:

$$Z_T = \frac{(1250 - 500j)(10 + 25j)}{(10 - 25j)(10 + 25j)}$$

$$= \frac{12,500 + 26,250j - 12,500j^2}{100 - 625j^2}$$

$$= \frac{25,000 + 26,250j}{725} = 34.5 + 36.2j$$

Convert the answer to polar form: $Z_T =$ ______________

$50.0\angle 46°$

180. Here is the same formula: $$Z_T = \frac{Z_1 Z_2}{Z_1 + Z_2}$$

We want to find "Z_T" when: $Z_1 = 25\angle 60°$ and $Z_2 = 50\angle 30°$

By substitution, we get: $$Z_T = \frac{(25\angle 60°)(50\angle 30°)}{25\angle 60° + 50\angle 30°}$$

Step 1: We can simplify the numerator by performing a vector-multiplication in polar form. We get:

$$Z_T = \frac{1250\angle 90°}{25\angle 60° + 50\angle 30°}$$

Step 2: We can simplify the denominator by a vector-addition. However, to perform this addition, the vectors must be converted to rectangular form first:

Since $25\angle 60° = 12.5 + 21.6j$

and $50\angle 30° = 43.3 + 25.0j$,

$$Z_T = \frac{1250\angle 90°}{55.8 + 46.6j}$$

Step 3: Now to perform the vector-division so that Z_T can be reported as a single vector, we convert the denominator from rectangular back to polar form. We get:

$$Z_T = \frac{1250\angle 90°}{72.7\angle 40°}$$

$Z_T =$ __________

---

181. The same formula is given on the left below. If Z_T and Z_2 are given and we are asked to find Z_1, we must rearrange the formula to solve it for Z_1 first. When we do so, we get the formula on the right. | $17.2\angle 50°$

$$Z_T = \frac{Z_1 Z_2}{Z_1 + Z_2} \qquad Z_1 = \frac{Z_T Z_2}{Z_2 - Z_T}$$

If $Z_T = 10 + 0j$ and $Z_2 = 0 + 5j$, find Z_1 by substituting into the formula on the right and performing various vector operations.

$Z_1 =$ __________

182. The formula on the right shows the vector-relationship among current, voltage, and two impedances in a particular electrical circuit.

$$I = \frac{E_A}{Z_1 + Z_2}$$

Find I if: $E_A = 5 + 10j$, $Z_1 = 3 + 4j$, $Z_2 = 6 + 6j$

I = ____________

$$Z_1 = \frac{(10)(5j)}{-10+5j}$$
$$= \frac{50j}{-10+5j}\left(\frac{-10-5j}{-10-5j}\right)$$
$$= \frac{-500j - 250j^2}{100 - 25j^2}$$
$$= \frac{250 - 500j}{125}$$
$$= 2 - 4j$$

---

183. The same formula is given on the left below. To find E_A when specific vectors for I, Z_1, and Z_2 are given, we should rearrange the formula and solve it for E_A first. When we do so, we get the formula on the right.

$$I = \frac{E_A}{Z_1 + Z_2} \qquad E_A = I(Z_1 + Z_2)$$

Using the formula on the right, find E_A in polar form when:

$I = 50\underline{/-40°}$, $Z_1 = 30\underline{/0°}$, $Z_2 = 40\underline{/90°}$.

E_A = ____________

$$I = \frac{5 + 10j}{9 + 10j}$$
$$= \frac{145 + 40j}{181}$$
$$= 0.801 + 0.221j$$

---

184. Here is a formula involving voltage drops in an alternating-current circuit:

$$E = I_1 Z_1 + I_2 Z_2$$

Find E if: $I_1 = 10\underline{/30°}$, $Z_1 = 5\underline{/60°}$, $I_2 = 10\underline{/50°}$ and $Z_2 = 5\underline{/-50°}$

(Note: Write your answer in rectangular form.)

E = ____________

$$E_A = (50\underline{/-40°})[(30 + 0j) + (0 + 40j)]$$
$$= (50\underline{/-40°})[30 + 40j]$$
$$= (50\underline{/-40°})(50\underline{/53°})$$
$$= 2500\underline{/13°}$$

$$E = 50\underline{/90°} + 50\underline{/0°}$$
$$= (0 + 50j) + (50 + 0j)$$
$$= 50 + 50j$$

185. Here is the same formula: $E = I_1Z_1 + I_2Z_2$

Find E if: $I_1 = 20 + 0j$, $Z_1 = 5 + 10j$, $I_2 = -25 + 5j$, $Z_2 = 4 + 0j$.

(Note: Write your answer in polar form.)

E = ________________

Answer to Frame 185: $E = (100 + 200j) + (-100 + 20j) = 0 + 220j = 220\angle 90°$

SELF-TEST 16 (Frames 179-185)

In the following formula, find Z_T if: $Z_1 = 30\angle 90°$ and $Z_2 = 60\angle -30°$

$$Z_T = \frac{Z_1Z_2}{Z_1 + Z_2}$$

Z_T = ________________

ANSWER: $Z_T = 34.6\angle 60°$ From: $Z_T = \frac{1800\angle 60°}{52.0\angle 0°} = 34.6\angle 60°$